The Little ICE AGE

A Little Ice Age winter scene (*An Inn by a Frozen River*, style of Isack von Ostade, National Gallery, London)

The Little ICE AGE

Jean M. Grove

METHUEN
LONDON AND NEW YORK

First published in 1988 by
Methuen & Co. Ltd
11 New Fetter Lane, London EC4P 4EE

Published in the USA by
Methuen & Co.
in association with Methuen, Inc.
29 West 35th Street, New York NY10001

Set by Hope Services, Abingdon
Printed in Great Britain
at the University Press, Cambridge

British Library Cataloguing in Publication Data
Grove, Jean
The little ice age.
1. Little Ice Age
I. Title
551.7'93 QC981.C5
ISBN 0 416 31540 2

Library of Congress Cataloging in Publication Data
Grove, Jean.
The Little Ice Age.
Bibliography: p.
Includes index.
1. Climatic changes. 2. Glaciers.
I. Title.
QC981.8.I23G76 1988 551.3'12'09031 87-14095
ISBN 0 416 31540 2

For Dick

Contents

Plates

Figures

Tables

Preface

Since I began to write this book interest in climatic change has greatly increased. The possible importance of climatic instability to mankind is now widely appreciated. Concern for the future has drawn attention not only to the possibility of forecasting but also to the need for a more adequate understanding of past fluctuations. The Little Ice Age of the present millennium is especially relevant here because there is more information to be gathered about it than about earlier periods of a similar length.

It is not possible in the compass of one book to cover all the available literature, and indeed it is increasingly difficult to keep up with it. Even since the text was handed to the publishers important new studies have appeared and others are imminent. It is particularly to be regretted that Friedrich Röthlisberger's major study of Holocene glacier fluctuations *10,000 Jahre Gletschergeschichte der Erde* (1986) was not available in time for detailed reference to be made to it in Chapter 10.

The account that follows is likely to be modified and extended as investigations continue. The present coverage is inevitably both uneven and incomplete; in particular the lack of detailed information from Russian, Chinese and Spanish sources must be acknowledged with regret.

This attempt to portray the characteristics, causes and consequences of the Little Ice Age could never have been made without the practical support of my husband or the forbearance of my children.

Spelling and transliteration of foreign names

It has not always been possible to achieve consistency between text and maps. The Icelandic letters thorn (þ) and eth (ð) are represented in the text by 'th' and 'd'. Both ø and ö are used in Norwegian maps. More than one system of transliteration is commonly used for Chinese and Russian.

Acknowledgements

Arthur Battagel played a crucial role and was wholly responsible for the translation of original documents from Norwegian. Rosemary Graham was responsible for translations from Icelandic and also translated some of the Norwegian printed material. Anita Dowsing assisted with the translation of some of the Norwegian printed material and nearly all the material in German.

Parts of the manuscript were read and helpful suggestions and criticisms provided by R. P. Ackert, R. S. Bradley, P. E. Calkin, A. E. Corte, A. F. Gellatly, J. L. Innes, W. Kick, G. D. Osborn, R. Randall, Ren Meier, M. Sharp, H. M. Spufford and P. Wardle. A. T. Grove and Rosemary Graham read and commented on the whole manuscript and A. Battagel on large parts of it.

My initial interest in the subject was aroused by Frank Debenham, Vaughan Lewis and Gordon Manley. Jean Mitchell was my early mentor in historical matters and I was encouraged and assisted by Alfred and Harriet Steers. Helpful information and advice were received from R. P. Ackert Jr, I. Allison, J. T. Andrews, J. R. Blyth, R. S. Bradley, N. Bhandari, C. Burn, C. J. Burrows, P. E. Calkin, T. Chinn, A. E. Corte, D. N. Collins, A. Dugmore, G. Farmer, F. Fridriksson, D. Fletcher, A. F. Gellatly, A. S. Goudie, H. Green, N. J. Griffey, R. H. Grove, A. Holmsen, J. M. R. Hughes, J. L. Innes, W. Karlén, W. Kick, H. H. Lamb, O. Liestøl, B. H. Luckman, P. McCormack, G. Manley, J. Mercer, G. H. Miller, J. A. Matthews, F. Müller, H. Nichols, D. Norton, A. E. J. Ogilvie, Y. Ono, G. Osborn, C. Pfister, S. C. Porter, F. Röthlisberger, H. Röthlisberger, J. Ryder, J. Sandnes, M. J. Sharp, Shi Yafeng, J. L. Sollid, A. Street-Perrott, E. Sulheim, S. Thórarinsson, A. M. Tvede, H. E. Waldrop, G. Wells, B. W. Whalley, M. Young and H. J. Zumbühl.

The photographs reproduced as Plates 7.2 and 7.3 were kindly given by Dr Wilhelm Kick; those reproduced as 6.2a, 6.2d, 6.5a, 6.5b, 6.5c and 6.5d by Dr H. Zumbühl; Plate 9.1 by Charles Harpum. The author and the publishers would like to thank the following organizations for allowing the use of plates: Nasjonalgalleriet, Oslo for 3.1, 3.5, 3.6, 3.7 and 6.2b; Bergen Billedgalleri for 3.2; Öffentliche Kunstsammlung Basel for 4.1; Landsmuseum Innsbruck for 5.1; E. T. H. Zürich and Dr H. Röthlisberger for 6.2c and 6.2d; Musée de la Marjorie, Sion for 6.4; Kunstmuseum, Basel for 6.5c; Denkschriften der Schweizerischen Naturforschenden Gesellschaft for 6.5a, 6.5b, 6.5c and 6.5d; Whyte Museum of the Canadian Rockies, Banff for 8.3, 8.4 and 8.5; NASA for 7.1, 7.4 and 11.1; Alexander Turnbull Library, Wellington for 9.2, 9.3, 9.4 and 9.6; Westland National Park, Franz Josef for 9.5 and 9.7. All plates not otherwise specified are from photographs taken by A. T. Grove.

The maps included as Figures 9.7, 9.8, 9.9 and 9.10 were drawn by A. S. Gellatly who had copied them from originals in New Zealand archives. The majority of the diagrams in the first five chapters, including the copies of old maps and sketches reproduced as Figures 2.10, 2.11, 2.12, 2.18, 2.19, 4.5, 5.3 and 5.4 were drawn by P. A. Lucas. Most of the diagrams in Chapters 6 to 12 were drawn by A. Shelley and the remainder by M. Young, S. Gutteridge and I. Gulley in the Geography Department, Cambridge University. I am very grateful for secretarial assistance from staff at Girton College.

The author and the publishers would like to thank the following copyright holders for permission to reproduce figures: Allen & Unwin for 3.4, 4.3 and 12.6; *Die Alpen* for 10.5; *American Scientist* for 12.3 and 12.4; Arctic & Alpine Research for 1.3, 1.5, 6.4, 7.6, 7.8, 8.2, 8.4, 9.11, 10.15, 10.16, 10.19 and 12.2; Balkema for 9.5 and 12.1; *Bollettino del Comitato Glaciologico Italiano* for 4.11 and 4.12; *Boreas* for 10.9; Cambridge University Press for 7.7 and 12.18; *Climatic Change* for 2.6, 12.7, 12.8, 12.11, 12.12, 12.16 and 12.17; Colorado Associated Press for 11.4; La Commission des Glaciers de la Société helvetique des Sciences Naturelles for 4.9 and 6.11; Fjell og Vidde for 3.16; Geografia Fisica e Dinamica Quaternaria for 4.8 and 4.13; *Geographical Bulletin* for 10.17; *Geographica Helvetica* for 6.6, 6.10 and 10.8; *Geographical Journal* for 3.8 and 9.7; *Geografiska Annaler* for 2.17, 3.8b, 3.15, 9.3b, 10.10, 10.11 and 10.12; Geographisches Institut der Universität Zürich for 4.4; *Geographischer Jahresbericht aus Österreich* for 5.5; *La Houille Blanche* for 4.10; International Association for Scientific Hydrology for 5.8, 6.8 and 8.9; *Jökull* for 2.3b, 2.4, 2.5, 2.7, 2.10, 2.14, 2.16, 12.9 and 12.10; *Journal of Glaciology* for 7.1, 8.3, 8.7 and 9.3a; *Journal of the Meteorological Society of Japan* for 12.19; *Middelelser om Grønland* for 2.2, 2.3a, 2.5 and 2.6; Methuen & Co. Ltd for 10.1; *Nature* for 1.2, 2.8, 11.2, 11.3, 11.6, 11.7, 11.8, 11.9 and 11.10; *Norsk Geografisk Tidsskrift* for 3.8c, 3.10, 3.11; Progress in Physical Geography for 10.3, 10.4, 10.6, 10.7, 10.13, 10.18, 10.20 and 10.23; *Quaternary Research* for 1.6, 10.2, 10.3a and 10.19; Reidel for 9.2; *Science* for 11.5; Scientia Sinica for 7.9; Snaelands útgáfan for 2.11, 2.12, 2.15, 2.18 and 2.19; *Transactions of the Institute of British Geographers* for 4.1, 4.7, 12.14 and 12.15; *Weather* for 1.4, 12.13; *Zeitschrift für Geomorphologie* for 4.6; *Zeitschrift für Gletscherkunde und Glazialgeologie* for 4.10, 5.7, 5.9, 10.14, 10.21 and 10.22.

The assistance of the archivists in Riksarkivet, Oslo and in Statsarkivet, Bergen has been unstinting. Other archival material has come from the Canton archives, Sion and the Alexander Turnbull Library, Wellington.

The librarians and staff of the Scott Polar Research Institute, the Geography Department of the University of Cambridge, Girton College and the Royal Geographical Society have given generously of their time.

The Sulheims at Spiterstulen in Norway provided support and encouragement in the initial stages of the work. A sabbatical term at the Geography Department of UCLA provided an opportunity to learn about the glaciers of the Sierra Nevada and Dr and

Mrs Jack Ives housed us in Boulder. Dr and Mrs David Norton afforded kind hospitality and a base during a visit to South Island, New Zealand and Camilla Nash afforded similar hospitality in Switzerland. Generations of undergraduates and research students have provided stimulus.

Finally I must acknowledge financial support from Girton College, Cambridge, the Smuts Commonwealth Fund and the University Travel Fund of Cambridge University and the British Council.

GIRTON COLLEGE, CAMBRIDGE
FEBRUARY 1987

Chapter 1

Introduction

Climatic changes come on several different timescales. Between short-term fluctuations lasting a few years and changes extending over thousands of years there are variations over a few centuries which may have profound effects on natural phenomena and human affairs. It is variations on this scale, stretching over several generations, with which we are concerned in studying the Little Ice Age. It was a period which may be seen as beginning in the thirteenth and fourteenth centuries (Porter 1986) and then, after an interval of more clement conditions, culminating between the mid-sixteenth and the mid-nineteenth century. It was also a period of lower temperature over most if not all of the globe, sufficiently marked to have had important consequences, especially in certain sensitive areas in high latitudes and at high altitudes where conditions for plant growth and agriculture are marginal.

For several hundred years climatic conditions in Europe had been kind; there were few poor harvests and famines were infrequent. The pack ice in the Arctic lay far to the north and long sea voyages could be made in the small craft then in use. Communications between Scandinavia, Iceland and Greenland were easier than they were to be again until the twentieth century. Icelanders made their first trip to Greenland about AD 982 and later they reached the Canadian Arctic and may even have penetrated the North West Passage. Grain was grown in Iceland and even in Greenland; the northern fisheries flourished and in mainland Europe vineyards were in production 500 km north of their present limits.

The beneficent times came to an end. Sea ice and stormier seas made the passages between Norway, Iceland and Greenland more difficult after AD 1200; the last report of a voyage to Vinland was made in 1347 (Gad 1970). Life in Greenland became harder; the people were cut off from Iceland and eventually disappeared from history towards the end of the fifteenth century. Grain would no longer ripen in Iceland, first in the north and later in the south and east. As the northern winters became colder, fish migrations took different tracks and life became tougher for fishermen as well as for farmers. In mainland

Europe, disastrous harvests were experienced in the latter part of the thirteenth and in the early fourteenth century, with famines in England in 1272, 1277, 1283, 1292 and 1311. The years between 1314 and 1319 saw harvests fail in nearly every part of Europe. Extremes of weather were greater, with severe winters and unusually hot or wet summers. In consequence the boundaries of cultivation contracted, though there were of course other forces as well as climate operating at the time.

In these late medieval times and in succeeding centuries, the impact of climatic fluctuations was felt most painfully and persistently in highland areas. Cultivated areas suffered especially in the uplands of oceanic regions, such as the Lammermuir hills of southeast Scotland. In the last few centuries glaciers have advanced in the mountains of Alpine Europe and Scandinavia, in the northlands, and indeed in most other moist and cold parts of the world. In the decades between the late sixteenth and late seventeenth centuries European glaciers swelled and their tongues advanced, destroying high farms and damaging mountain villages. Streams fed from glaciers flooded more frequently, sometimes catastrophically. In many areas this kind of hazard was compounded by landslides and avalanches triggered by increased precipitation and the greater glacial activity associated with it.

The relationship between glacial behaviour and meteorological control is delicate and complicated. The researches of H. W. Ahlmann (1949) and his pupils in Scandinavia and of the Innsbruck group led by H. H. Hoinkes (1970) laid firm foundations for understanding it. If the accumulation of snow and ice during a winter accumulation season exceeds the ice wastage in the following summer, glaciers increase in volume and are said to have positive mass balances. If the wastage is greater than the accumulation, glaciers shrink and their budgets are said to be negative. If a series of positive balances occurs, the volume of ice moving downslope increases and eventually the glacier front advances. The conditions most favourable for positive budgets or mass balances are given by plentiful precipitation in winter and short cool summers causing minimum wastage or ablation. There is a lag between the onset of a particular climatic change and the response of a glacier terminus, which depends on the topography of the glacier and its valley and the flow characteristics of the ice mass. Small temperate valley glaciers with a rapid turnover of ice will respond in a few years; valley glaciers of moderate size in Austria have a response time of about seven years. Large icecaps have much longer lag periods. Details may be found in Paterson's (1981) *Physics of Glaciers*.

Comparison of glacier behaviour and meteorological records over the instrumental period of the last two centuries provides a basis for extrapolating the climatic record into the past. Manley's (1974) long record of mean monthly temperatures for central England between 1654 and 1973, the result of careful integration of data from diverse records, is extremely valuable because it covers a good deal of the Little Ice Age and allows an impression to be gained of the regional temperature changes associated with the comparatively local glacial fluctuations in Europe. The behaviour of glaciers provides us with one of the most useful indicators of past climatic history for places and times lacking instrumental records, so long as the histories of sensitive glaciers with short response times are used.

Parry's (e.g. 1978) studies of the variations in the extent of cultivated land in upland Britain have provided a deeper understanding of the economic and social consequences of climatic fluctuations back into medieval times. Hubert Lamb and his successors in the Climatic Research Group at Norwich are rapidly extending knowledge of past climatic conditions (Lamb 1972, 1977, 1981). There is an increasing volume of evidence indicating the importance for agriculture and rural prosperity of long-term change in precipitation and temperature. This has necessitated a reconsideration of the influence of climatic change on the demographic and economic decline in the fourteenth century so long attributed to the Black Death (Lamb 1977, pp. 454–7).

Historians have long been inclined not only to overlook but deliberately to discount the influence of climatic change on human affairs (Russell 1948, van Bath 1963, Hoskins 1968). Even Le Roy Ladurie (1971), despite his substantial book on the history of glaciation in Europe, *Times of Feast, Times of Famine*, is undecided on whether a difference in secular mean temperature of 1°C has any substantial influence on agriculture and other human activities. These negative or agnostic attitudes may be explained in part by historians having made use of climatic chronologies that are now known to be incorrect (e.g. Britton 1937, Brooks 1949). Some early climatic historians were less rigorous in their use of sources than is now required and they often made extensive use of compilations and other secondary material, some of which was unreliable (Bell and Ogilvie 1978).

Studies of the Little Ice Age and climatic variations on a similar scale in the more distant past are also significant for the future. The early chapters which follow present a history of the Little Ice Age based on the records of glacier advance and retreat in northwest Europe. Information is then brought together about the behaviour of glaciers in this period in other mountainous parts of the world. The Little Ice Age is set in the longer time perspective of the Holocene, the period since the time of the Last Glaciation during which, in the words of a famous prehistorian, 'man has made himself' (Childe 1936). Finally an attempt is made to assess the significance of the results of this inquiry for an understanding of human affairs.

1.1 The term 'Little Ice Age'

The term 'Little Ice Age' is widely used to describe the period of a few centuries between the Middle Ages and the warm period of the first half of the twentieth century, during which glaciers in many parts of the world expanded and fluctuated about more advanced positions than those they occupied in the centuries before or after this generally cooler interval. A number of objections have been made to the employment of this term 'Little Ice Age' and these must be considered before the nature of the phenomena involved is explored. The main objections are that the term was originally applied to a quite different time period, that it is now used in several different ways by current authors, and that as far as the dominant usage is concerned it is a misnomer. It is also argued that it was not worldwide, that continental glaciation did not increase and that 'there was not any sustained low global temperature during the period' (Landsberg 1985), in short that it was not an Ice Age and was insignificant in scale (Landsberg 1984 and personal communication 1983).

It is certainly true that Matthes (1939), who introduced the term Little Ice Age into scientific literature, intended it to describe an 'epoch of renewed but moderate glaciation which followed the warmest part of the Holocene'. Matthes was a very close observer who was especially concerned with the glaciers of the Sierra Nevada of California, which he believed would not have survived the warmth of the climatic optimum of the Mid-Holocene. These little glaciers had at first been mistaken for snowfields and it was not until 1872 that John Muir was able to demonstrate that they were formed of moving ice. Matthes noted the fresh appearance of their frontal moraines:

> they are made up of many small terminal moraines, laid against and on top of each other, as is clearly shown in instances where individual moraines lie spread out in a series one behind another, with concentrically curving crests. They record for each glacier many repeated advances, all of approximately the same magnitude. How many centuries of glacier oscillation are represented by these moraine accumulations it is difficult to estimate.

Matthes had no means of dating the features he described but he concluded that the glaciers had reformed after the climatic optimum and estimated that they might have reappeared 'as recently as about 2000 BC', although he recognized that larger icestreams such as those on Mt Rainier had probably persisted since the Ice Age.

Matthes was well aware of the record of repeated glacial advances in Europe during the past 400 years and the extent of the recession after 1850, which he regarded as 'a turning point in the modern glacial history of central Europe; for since then the trend of climate, not only in Europe but throughout the world, has been distinctly milder and the glaciers have been in recession almost everywhere' (p. 155). Interestingly he commented that this general recession did not herald the end of his Little Ice Age but that it was much more likely that it represented merely one of the mild fluctuations which he surmised had occurred repeatedly during the last 4000 years. This surmise has recently received detailed verification as a result of investigations made in the Alps of Switzerland and Austria which will be discussed in Chapter 10.

If the concept of Mid-Holocene conditions so mild that at least in some regions small glaciers disappeared completely is correct, then there is room for a term to cover the subsequent period of several millennia during which glaciers have reappeared and fluctuated in extent. 'Little Ice Age' has still been used fairly recently in this way (e.g. Benedict 1968) but it has been generally overtaken by 'Neoglaciation' as used by Porter and Denton (1967) to describe the interval of rebirth or renewed growth of glaciers after a time of maximum hypsithermal shrinkage during a Holocene warm period lasting until about 2600 years BP. The use of this term 'Neoglaciation' is not without difficulty. Porter and Denton noted that 'rather complex low order changes of climate characterize the hypsithermal interval resulting in several early Neoglacial episodes of glacial expansion'.

The concept of Neoglaciation is not recognized as useful by all workers in the European Alps, where the sequence of Holocene glacial events has been worked out in most detail, and the term does not provide a separate label for the period of several centuries which saw the last expansion of glaciers in many parts of the world. Evidence of not one but a whole sequence of such events is, as Matthes surmised would be the case, accumulating rapidly and the advent of a whole range of dating techniques in the last few decades has made it possible to distinguish them one from another.

The term 'Little Ice Age' is now widely employed by geographers, geologists, glaciologists and, most significantly, climatologists, to describe the period of glacial advance of the last few centuries or 'the cold Little Ice Age climate of about 1550 to 1800' (Lamb 1977, p. 104). It is true that the dates assigned to it are not always identical, being influenced by the locational experience of individual workers and the volume and accuracy of the evidence available to them. Synchronous climate change over the globe cannot be assumed without proof, but the widespread indications of rapid twentieth-century glacial retreat and the striking similarity of much data coming from widely separated areas seem to indicate a coherence which justifies a single name. Alternatives to 'Little Ice Age' have been suggested. Ladurie (1971, p. 223) noted that German authors, including Kinzl and Mayr, use 'Fernau' but remarked that, as custom makes law, 'Little Ice Age' might well be accepted. More recently some German workers, for example Heine (1983), have begun to use a translation of the English term, '*Kleine Eiszeit*'.

Historical evidence of Little Ice Age events is much more plentiful in Europe than elsewhere but the documentation from other continents, though scantier, is supported by a great volume of field evidence (e.g. Hope *et al.* 1976, Hastenrath 1984) which is presented in Chapters 7, 8 and 9. It emerges that the Little Ice Age was a global phenomenon and it is shown in Chapter 10 that it was not unique in the Holocene. Involving fluctuations in temperature of 2°C or less, it was none the less sufficient to cause advance of the Greenland ice edge (Weidick 1967) and to be associated with measurable meteorological, geomorphological and vegetational changes (Chapter 12).

It is certainly true that lower temperatures were not sustained throughout the period. The Little Ice Age itself consisted of a series of frequent fluctuations, such as those exhibited in Manley's temperature curves for central England and worked out in great detail for Switzerland by Pfister (Chapter 6). Such fluctuations consist of individual years and clusters of years for which weather conditions depart strongly from longer-term means. Average conditions throughout the Little Ice Age were none the less such that mountain glaciers advanced to more forward positions than those they had occupied for several centuries, or in some areas even millennia, and fluctuated about those positions until the warming phase in the decades around the turn of the century brought them back to where they had been in earlier Holocene warm periods. It has been nicely demonstrated that certain Swiss glaciers such as the Ferpècle were of comparable extent in the 1980s and before about 3500 years BP (Röthlisberger *et al.* 1980 and Chapter 10). This finding provides excellent confirmation of Matthes's hunch that the recent recession represents merely 'one of the mild fluctuations that has occurred repeatedly in the last 4000 years'.

In the chapters that follow, the course of events in Europe is considered first, the emphasis being placed on historical evidence which is capable of providing the most complete and accurate record. The material has been deliberately selected so as to illustrate not only a variety of documentary source types but also to give some idea of the reactions of contemporaries. In particular, the somewhat anecdotal style of the earlier part of Chapter 5 has been adopted for this reason. Contemporary records are to be preferred to secondary sources when using historical evidence and great care has to be taken over both reliability and interpretation (Ingram *et al.* 1981). Paintings, sketches and lithographs, as well as written documents, can yield valuable evidence of glacial extent and character if their dates are known and if the artists were intent on accurate representation of nature (Zumbühl 1980).

European historical sources are more plentiful and go back further than those from other continents. Chronologies elsewhere have to depend heavily on the dating of moraines, which generally involve obtaining bracketing dates of greater or lesser accuracy, that is maximum and/or minimum estimates of the time when the debris forming a particular moraine was deposited (Figure 1.1).

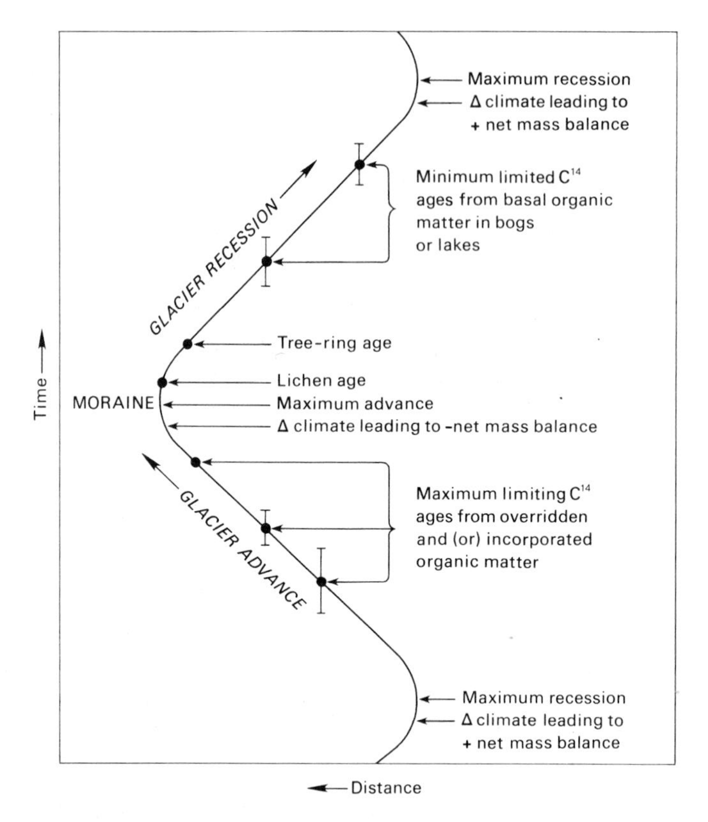

Figure 1.1 The relationship between changes in climate causing periods of positive and negative mass balance leading to glacial expansion and recession and possible methods of dating mentioned in the text. (From Porter 1979)

1.2 Dating methods

It is beyond the scope of this book to discuss critically and in full the variety of dating methods that are available (for a general survey see Porter 1981a and for a critique of many of the methods involved see Bradley 1985). It may however be useful to give a short introductory sketch of the principal methods involved, with their pitfalls and limitations.

Radiocarbon dating of organic material underlain or overlain by moraine nearly always suffers from the defect of time lags of unknown length between the period of moraine construction and the deposition or accumulation of the dated material. In many cases it is only possible to obtain a maximum or a minimum limiting date but not both. A major limitation of ^{14}C dating arises for the Little Ice Age period, quite apart from normal constraints such as sample contamination and variations in methodology and accuracy by both fieldworkers and laboratories. Calibration of ^{14}C values with calendar ages of rings from long-lived trees has revealed non-systematic relationships between them (Stuiver 1978, Klein *et al.* 1982). Within the last 500 years, it is not possible to obtain an unambiguous calendar age from a single radiocarbon date (Figure 1.2). It might be possible to eliminate the ambiguity by obtaining ^{14}C dates of several rings of the same log, but this has rarely if ever been done in this context. It must therefore be accepted that

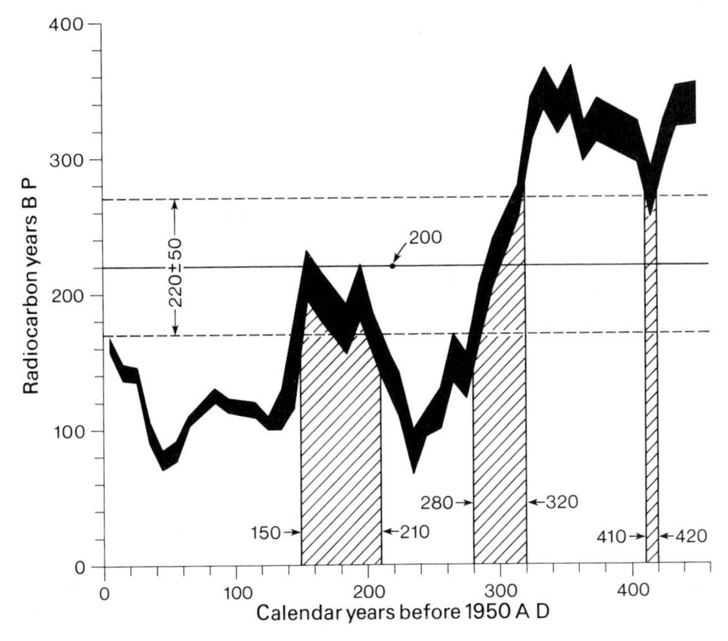

Figure 1.2 The relationship between ^{14}C years and calendar years calibrated from tree rings. The width of the curve, marked in black, is twice the standard deviation given by the laboratory. Any one radiocarbon year is equivalent to more than one calendar year. Thus a radiocarbon age of 220 ± 50 years is equivalent to all the calendar dates with the intervals AD 120–210, 280–320, and 410–420. (From Stuiver 1978)

while radiocarbon dating may be used to establish that glaciers in a particular region were affected by oscillations in extent or volume during the Little Ice Age, it cannot at present be used for identifying second-order advances and retreats within the last few centuries and more especially should not be used to prove or disprove synchroneity of such fluctuations from region to region.

Radiocarbon provides a most valuable tool for the dating of earlier Holocene periods of glacier advance (Röthlisberger *et al.* 1980) or retreat (Porter and Orombelli 1985) so long as it is used with care and its accuracy is not overrated. Maximum limiting dates can be obtained only from the ^{14}C ages of organic material beneath or within moraine. The most detailed and satisfactory reconstructions are provided if fossil wood in quantity is available (Furrer and Holzhauser 1984), though care must be taken over interpretation (Ryder and Thomson 1986). The clearest evidence comes from trees sheared off *in situ* by advancing ice, but this is exceptional in most regions (Holzhauser 1984).

Radiocarbon dating of buried soils presents complications because of the complex nature of soil organic matter (Matthews 1985). Pre-treatment of soil samples to separate soil organic fractions of different ages is employed in order to acquire the closest possible maximum or minimum age estimates. The ^{14}C age of the oldest uncontaminated organic fraction of the buried soil, which in some cases is known to consist of lichen (Geyh *et al.* 1985) provides the closest approximation to the time when soil formation began, while the age of the youngest uncontaminated fraction in the buried soil provides the best possible estimate of the time which has elapsed since burial. The two dates serve to bracket the interval between a time when the surface was exposed and a time when it was covered again by ice. Dates obtained from the top and the bottom of a 5 cm-thick A-horizon from beneath a moraine in southern Norway gave dates of 880 ± 35 and 3140 ± 55 (Matthews 1980); dating of several thin slices of a single horizon from beneath a moraine in southern Norway gave dates increasing from 485 ± 60 BP to 4020 ± 70 BP at the bottom (Matthews and Dresser 1983). The steep age gradient with depth revealed indicates both that a sample intended to provide a maximum age for the burial of a paleosol must be taken from as near the top as possible and the considerable error that may be involved should the surface layer be missing. The complications of ^{14}C dating of soils are least pressing when soils are buried at an early stage of their development. Investigation of immature soils within moraine sequences has provided satisfactory evidence on which to build complex Holocene chronologies in the Alps (Röthlisberger *et al.* 1980), the Himalaya and Karakoram (Röthlisberger and Geyh 1985) and New Zealand (Gellatly *et al.* 1985). However, reliability tests of over 300 ^{14}C ages of soils within moraines suggest a maximum resolution of ± 200 years, taking into account the advice of the International Study Group (1982) that the uncertainty should be taken as twice the laboratory value for the standard deviation (Geyh *et al.* 1985).

It is sometimes possible to date the culmination of a recent period of glacier expansion directly using dendrochronology at positions where advancing ice has tilted or damaged trees. Much more commonly, a minimum age for a moraine has been obtained by counting the number of rings in the oldest trees growing on it. Several possible sources of error are involved, including the assumption that the oldest tree has been found, and that this is a member of the first generation, the unknown length of time between the stabilization of debris and the establishment of seedlings, and the interval required for trees to grow to a height at which it is possible to obtain a core (Lawrence 1950, Sigafoos and Hendricks 1961, 1972). Dendrochronology, if used with care, regard being paid to variation in the length of time required for establishment of different species in a given area, can nevertheless be a valuable tool in the context of Little Ice Age investigations (e.g. Carrara and McGimsey 1981).

Lichenometry, a method of dating based on the assumption that the largest lichen growing on a given substrate is the oldest individual and that, if the growth rate for a given species is known, the maximum lichen size will provide a minimum age for the substrate, is a dating method of great potential (Beschel 1961). It is particularly valuable in treeless areas and those lacking material for ^{14}C dating. Successful use of lichenometry demands meticulous care. A growth curve should be set up individually for each region and the nature of the evidence available for this will determine its accuracy and the time depth in which it can be used within reasonable error limits. It has to be accepted that there is a delay between exposure of a surface and colonization and a further delay before thalli become visible to the naked eye. Lichen species grow at different rates and must be differentiated from each other in the field; this can be particularly difficult when thalli are very small (Calkin and Ellis 1984). Moreover, both rock surface characteristics and climatic factors affect growth rates (Figure 1.3).

Given favourable circumstances, lichenometry can provide a dating tool accurate to ± 5 years over the last 200 years (Porter 1981b). Dating error increases with increasing age, because lichen growth tends to diminish with time, perhaps at an exponentially decreasing rate (Porter 1981a) and also because of the extrapolation of growth curves to give time depth. Published lichen chronologies differ both in methodology, which is not always even stated, and in reliability. An extremely cautious approach must therefore be adopted as far as correlation from region to region is concerned.

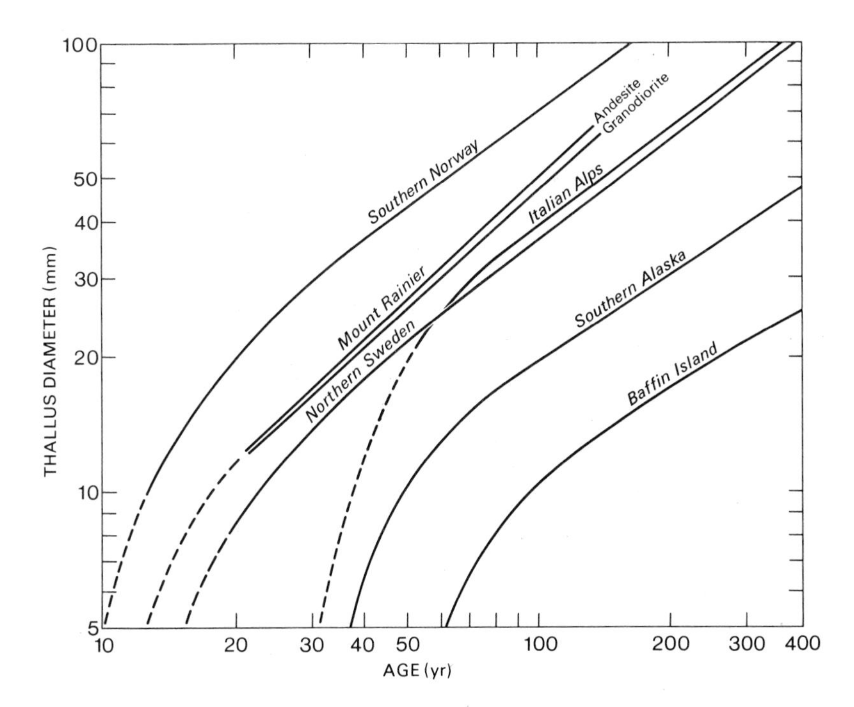

Figure 1.3 Lichenometric dating curves, relating thalli diameter to age, differ according to climate and also are affected by the rock type of the substrate. (Bradley 1985)

The sophistication of approach to lichenometry has increased markedly in the last decade. *Rhizocarpon geographicum* has been used in the majority of lichenometric studies but 29 species of *Rhizocarpon* have been recognized in Europe and 45 in the Arctic. Innes (1982, 1983a), working in Norway, has shown that *Rhizocarpon alpicola*, which colonizes deposits later than the *geographicum* group, has a faster growth rate, and that it is necessary to differentiate between the two to avoid important dating errors. This suggests that it may be necessary to re-examine the validity of some of the existing growth curves, including that of Denton and Karlén (1973). Techniques of sampling in the field and subsequent analysis are still undergoing detailed review (e.g. Innes 1983b, 1984, 1985).

The potential importance of lichenometry remains great. In the high Arctic, where other methods are liable to be least applicable, growth rates are low and the time range may conceivably exceed 5000 years. In maritime northern latitudes where growth rates are high, the time range is no more than a few hundred years.

While much of the earlier work on glacier chronology was based on a single method of dating, there has been an increasing realization of the value of a more broadly based approach (Burke and Birkeland 1979). In western North America, the additional evidence presented by tephra layers, themselves radiocarbon dated, has provided a useful framework for Holocene events (10.4.2). On Mt Rainier lichenometry has been used together with tree-ring dating to identify periods of glacier expansion during the Little Ice Age (Chapter 8.1 and Burbank 1981). In New Zealand a chronology has been built up using documentary records, lichenometry and radiocarbon dating of soils, supplemented by weathering-rind analysis and studies of soil and vegetation development (Chapter 10.7

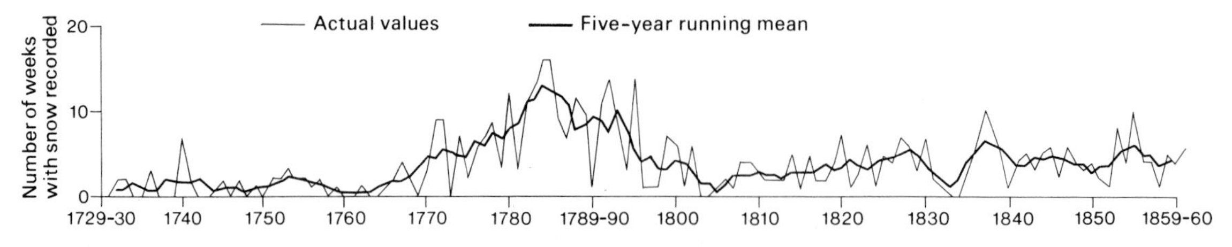

Figure 1.4 Winter snowfall in Scotland in the Little Ice Age. (From Pearson 1978)

and Birkeland 1981, Gellatly 1985a,b, Gellatly *et al.* 1985). The consistency achieved by using several techniques provides a more convincing result than could otherwise be obtained. There is a need to apply such multi-parameter approaches elsewhere to build up a satisfactory global picture of events in the Little Ice Age and earlier parts of the Holocene.

Little Ice Age chronologies from individual regions may be compared with evidence from other sources, such as documentary evidence of extent of winter snowfall (Figure 1.4) or time series of tree-ring widths which may yield estimates of summer temperature (Figure 1.5; Jacoby and Cook 1981). The most thorough comparison of this sort so far available comes from Pfister's reconstruction of Little Ice Age climate in Switzerland (Chapter 6.3), which provides an important line of approach to the problem of the extent of the impact of Little Ice Age climate on the population of mountain Europe. Comparison of the glacial evidence with that of other indicators can also be adopted for longer time periods (Figure 1.6) but an attempt to do this for the whole of the Holocene seems premature at the present time. The glacier chronology is not yet sufficiently securely based (Chapter 11.2) and complications arise because of the differing sensitivities of the various alternative indicators. The complexity of geomorphic responses inhibits simple correlation between glacial deposits of the Holocene and those of alpine and sub-alpine lake sediments (Harbor 1985). The conclusion of Davis and Botkin (1985), that fossil pollen deposits are not normally able to resolve climatic changes of the order of 100 to 200 years or to record very brief climatic events, underlines the potential value of obtaining a more soundly based glacier chronology for the Holocene than we have at present. All the indications are that the glacial record is likely to provide an unusually sensitive indicator of climatic changes over periods of the order of a few centuries.

1.3 The question of relevance

Investigations of the characteristics of the Little Ice Age and the incidence of other such intervals of cooler climate in the past are of practical as well as academic interest. Prediction of future climatic oscillations, dominated as they may be by increasing carbon dioxide (CO_2) in the atmosphere (Jones *et al.* 1986), demands greatly increased knowledge and understanding of the course of events in the past. The forcing factors, internal and external, which have operated in the past to prevent stability of the climatic system (Chapter 11.3), have not disappeared and will continue to influence the course of events. Mankind is by no means invulnerable to the effects of even minor variations (12.3). Semi-arid lands both north and south of the Himalayas depend crucially on

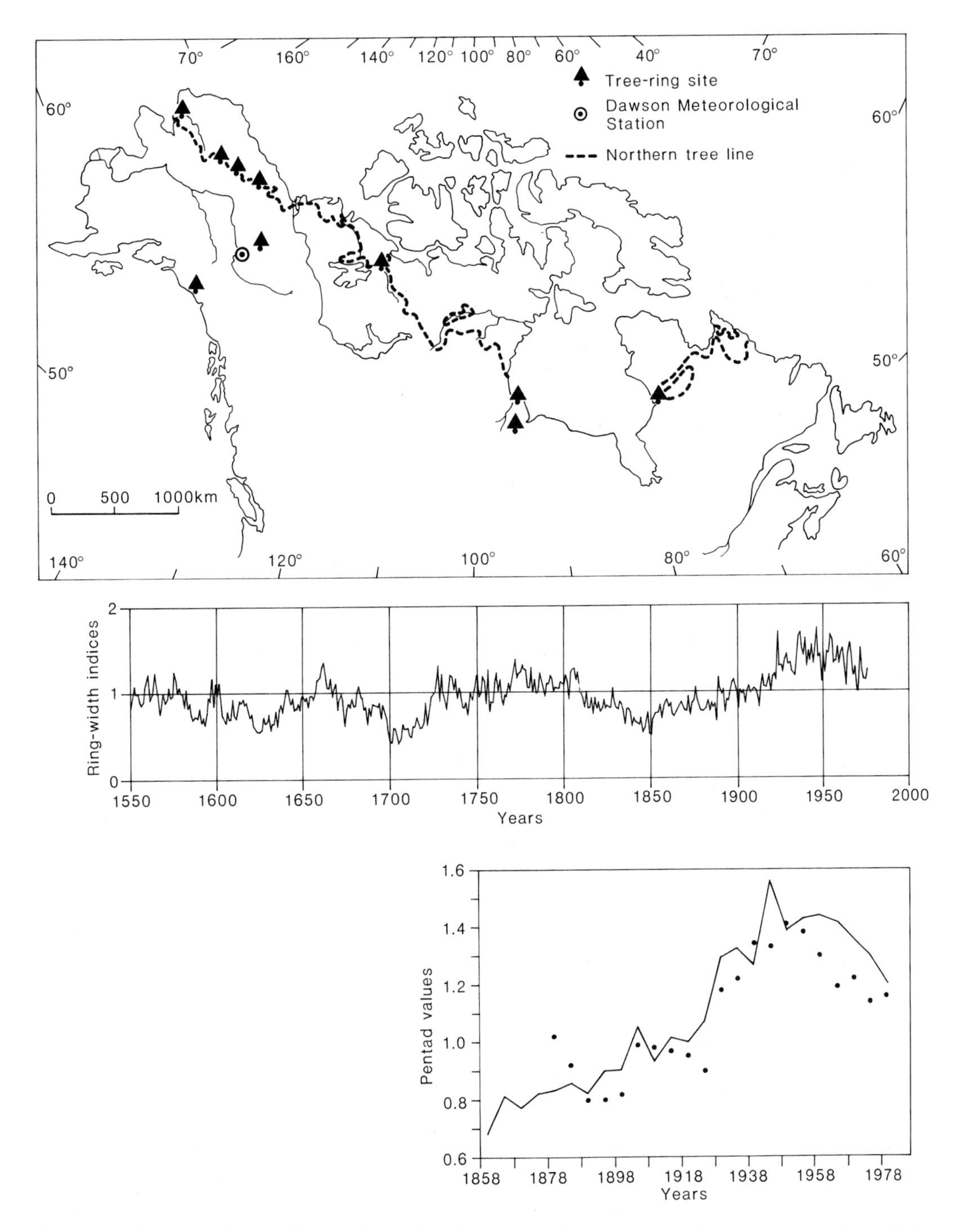

Figure 1.5 Tree-ring data collected from the sites shown have been used to construct ring-width indices for the period from the mid-sixteenth century to the late twentieth century. Pentad values for these indices are plotted against a temperature curve for the northern hemisphere derived independently from meteorological observations. (From Jacoby and Cook 1981)

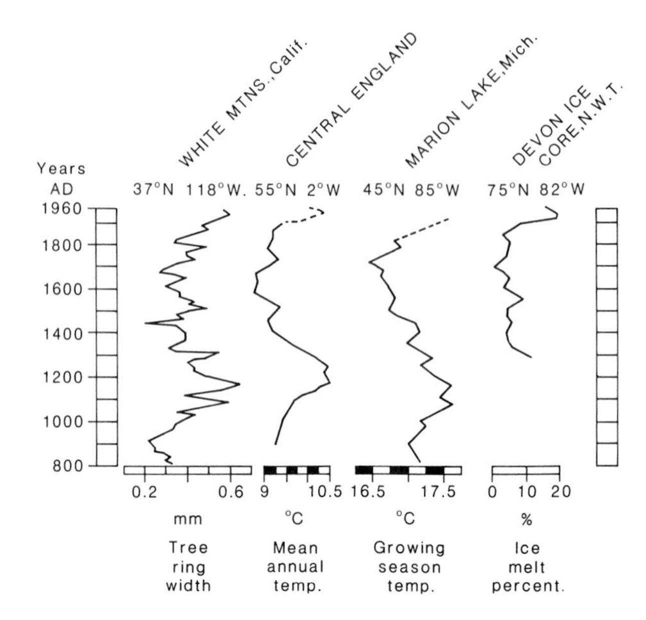

Figure 1.6 Comparison of independently derived indicators of environmental and climatic change covering both the medieval warm period and the Little Ice Age. (After Bernabo 1981)

glacier water, yet runoff from the ice is affected to a measurable extent even by such changes as have occurred in the last few decades (Collins 1984). Land ice and its changes provide a sensitive indicator of changes in the energy balance of the earth's surface and it is for this reason that the monitoring which is now regularly undertaken (Haeberli 1985) must be accompanied by continued research into past history.

Chapter 2

Icelandic glaciers and sea ice

Meteorological conditions around Iceland have an important bearing on the weather of northwest Europe and the fluctuations of Iceland's climate are symptomatic of those of a much wider area. The timing of glacial changes in this exceptionally sensitive climatic environment is therefore an essential ingredient of a study of the temporal behaviour of European glaciers.

Icecaps dominate the scene. Of the 11,800 km² which are ice-covered, over 11,600 km² are accounted for by the six largest icecaps: Vatnajökull, Langjökull, Hofsjökull, Mýrdalsjökull, Drangajökull and Eyjafjallajökull (see Figure 2.1), which are intermediate in form and size between Alpine icestreams and the ice sheets of Greenland and Pleistocene Europe.

Iceland was built by volcanic eruptions during the Cainozoic and is young both geologically and morphologically. Basalt areas in the east and west are separated by a central depression filled by palagonite, which consists of volcanic, glacial and aeolian deposits. This zone, bounded by local faults running north–south in the north and northeast–southwest in the south, is part of the enormous fault system extending far beyond the confines of the island as the mid-Atlantic Ridge, the boundary between the European and American plates running through central Iceland. During the Quaternary period large amounts of basalt were extruded and faulting gave vertical displacements of hundreds of metres. The lava plateaux in the east and west have been faulted into blocks tilting towards the central depression, with their greatest elevation towards the coast.

Iceland was first settled in the ninth century. Environmental conditions since have been so marginal that changes affecting their balance have always been of immediate concern to the inhabitants. A fall of 1°C in summer temperature in Iceland at the present day results in a 15 per cent reduction in crop yield (Fridriksson 1969). It is no wonder that long before the deterministic geographical writings of the early part of this century Icelanders were well aware of the interaction of climate and human activity in their own country, where the linkages are unusually direct.

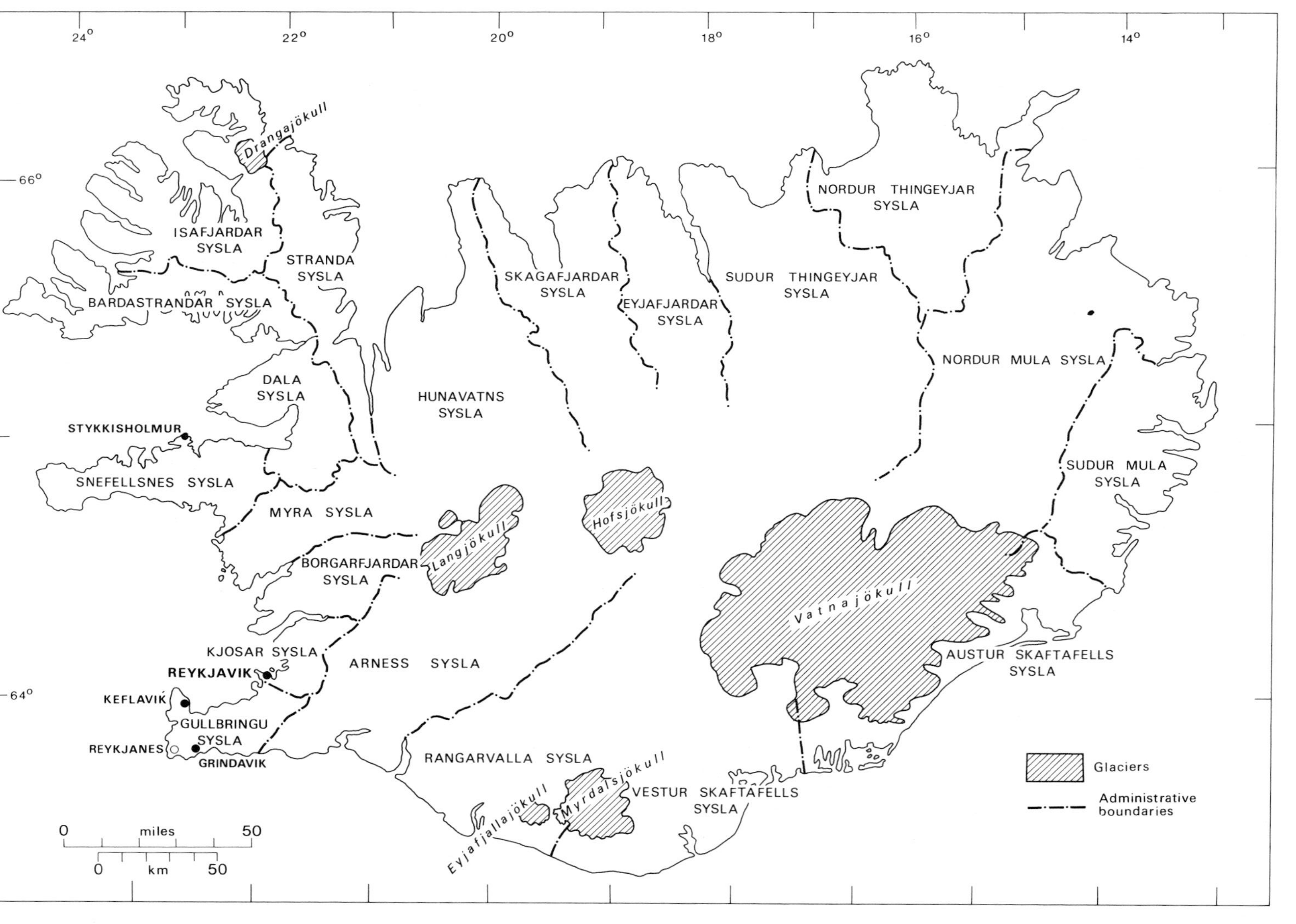

Figure 2.1 Iceland: major icecaps, administrative boundaries and some localities mentioned in the text

The sensitivity of the economy to climatic fluctuations, together with the extraordinary literary bent of the Icelanders and the emergence of a line of gifted field scientists (Thórarinsson 1960), mean that the Icelandic records relating to the Little Ice Age are of substantial length and richness. Nevertheless the record of Icelandic glacial fluctuations presents some special difficulties of interpretation associated with the geographical setting. In the first place, the climate of Iceland is greatly influenced by the distribution of sea ice and its variation through time. Sea ice is extensively developed in winter to the north of the country at present. During the Little Ice Age it often extended considerably further south. Secondly, being situated on the tectonically mobile mid-Atlantic Ridge, Iceland has several active volcanoes. The heat and tremors associated with these has at times affected the movements and extent of the ice. Thirdly, the icecaps themselves present special problems. Some of them are resting on rock surfaces far below the snowline of the present day and, if they were destroyed, would not re-form under present climatic conditions. Furthermore, some are so large that marginal lobes and tongues may respond at different times to the same climatic fluctuations.

After a consideration of sea ice and volcanism, attention is directed to the icecaps in the south and east of the country. The southern margins of Eyjafjallajökull, Mýrdalsjökull and the southern and eastern margins of Vatnajökull are bordered by lowlands that have long been settled. Both Mýrdalsjökull and Vatnajökull overlie volcanoes and several of the lobes of Vatnajökull are known to surge from time to time. Such surges are abnormally rapid advances not directly attributable to climatic events but rather to such features as instability related to the shape of the glacier bed and water at the base of the ice. It is therefore useful to check the findings from these southern icecaps against the record of glacier advance and retreat from Drangajökull, in the northwest, which is unaffected by volcanism and with outlet glaciers not known to surge.

2.1 Variations in the extent of sea ice around Iceland

Iceland is situated where warm water from the Atlantic and cold water from the Arctic oceans converge. A branch of the North Atlantic Drift, the Irminger current, is deflected westward by a submarine ridge to flow along the south and west coasts, before sinking beneath the East Greenland current (Figure 2.2). A branch of this cold current sweeps round the north and east coasts of Iceland and, at certain times, brings drift ice close to the land along the north and east and even the south coast. At other times, the sea ice lies further west, drifting down the coast of Greenland and through the Denmark Strait, the ice edge keeping well away from Iceland. The incidence of sea ice near the Icelandic coast has accordingly varied through the historic period. A 'normal' ice year has been defined as one when the ice edge is about 90 to 100 km away from Straumnes, in northwest Iceland, from January to April. In a 'mild' ice year the ice edge is about 200–240 km away, in a 'severe' ice year it extends along the northern coast and in an extremely severe year the ice is carried southward along the east coast by the East Greenland current and even reaches the south coast (Eythórsson and Sigtryggsson 1971).

An Irish monk, Dicuil, recorded that his brethren living in Iceland before AD 800 had found no ice along the south coast but had encountered it a day's voyage from the north coast. This has also been the position for most of this present century. Information for the medieval period is scanty and fragmentary with, for instance, laconic mentions of 'great

Plate 2.1 Landsat image of Vatnajökull taken on 22 September 1979

sea ice around Iceland in 1261' and 'sea ice surrounded nearly all Iceland in 1275' (Jóhannesson 1956). Thoroddsen published the first influential study of sea ice in 1874 and included in his book *The Climate of Iceland in a Thousand Years* (1916–17) a separate monograph in which he drew together a vast amount of information, ranging from the *Landnámabók* to Danish nautical and meteorological yearbooks. The recent literature was the most plentiful and the most accurate, the older more fragmentary and incidental in character. He surveyed the sagas and ancient annals but met a gap in the historical writings between the Nýi Annáll of 1430 and the Reformation of 1540–50. He had the few sixteenth- and seventeenth-century annals already printed at that time, that is *Gottskálksannáll*, *Biskupsannáll* and *Skardsárannáll*. For the seventeenth and eighteenth centuries he was able to use nearly all the important annals, but most of them were still in manuscript and he found them very disorganized and difficult to use. He also searched many other sources, both manuscript and printed, the nineteenth-century material including periodicals and newspapers as well as books, in an effort to gather together all the information he could find concerning climate and weather in Iceland; its analysis he left to others.

Koch (1945) reconstructed sea-ice variations since about AD 800 largely on a basis of Thoroddsen's compilation. He divided the coast into stretches, each 135 km long, and his ice index was the product of the number of weeks sea ice was noted and the number of stretches of coast from which it was seen (Figure 2.3a). He was moved to make this reconstruction by the publication of Thórarinsson's (1943) study of the oscillations of the

Icelandic glaciers. He wanted to investigate the dating of the climatic change which gave rise to the expanded glaciers of the eighteenth and nineteenth centuries. He was not entirely uncritical of Thoroddsen's methods, pointing out that Thoroddsen sometimes assumed the presence of sea ice on a basis of indirect evidence. Koch himself only made use of records which specifically mentioned sea ice. He concluded that there had been great changes in the climate of Iceland during historic times, although he recognized that cold periods were not necessarily all associated with the presence of sea ice.

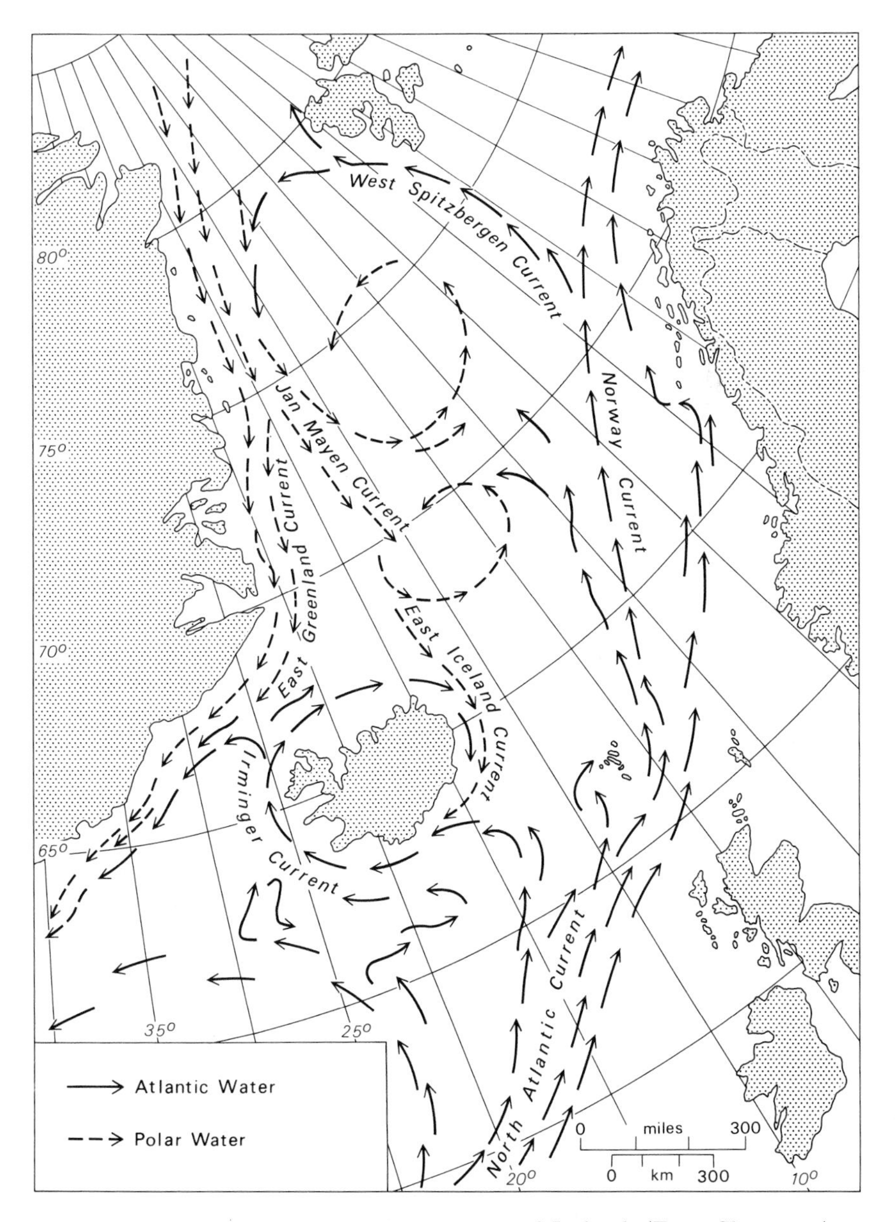

Figure 2.2 Surface currents in the seas around Iceland. (From Skov 1970)

Bergthórsson's (1969) graph of the severe ice years over the eleven centuries since the Norse settlement bears a considerable resemblance to Koch's (Figure 2.3b). Bergthórsson defined a severe sea-ice year as one in which sea ice is known to have reached the south coast or alternatively as one in which there were deaths from starvation. It could be questioned whether the association between severe sea-ice years and famine should be taken to be so close, and the continuity of the graph in periods where information is very sparse could be misleading. Bergthórsson found the lack of written history in the early fifteenth century excessive and left a gap in his graph for this period. Noting the good correlation between temperature and sea ice since the beginning of meteorological observations, he went on to infer past variations in temperature from the sea-ice data (Figure 2.4). A more detailed diagram showing year-to-year variability of sea ice was prepared by Sigtryggsson (1972) (Figure 2.5).

At the time when Thoroddsen was working, methods of dealing with historical sources

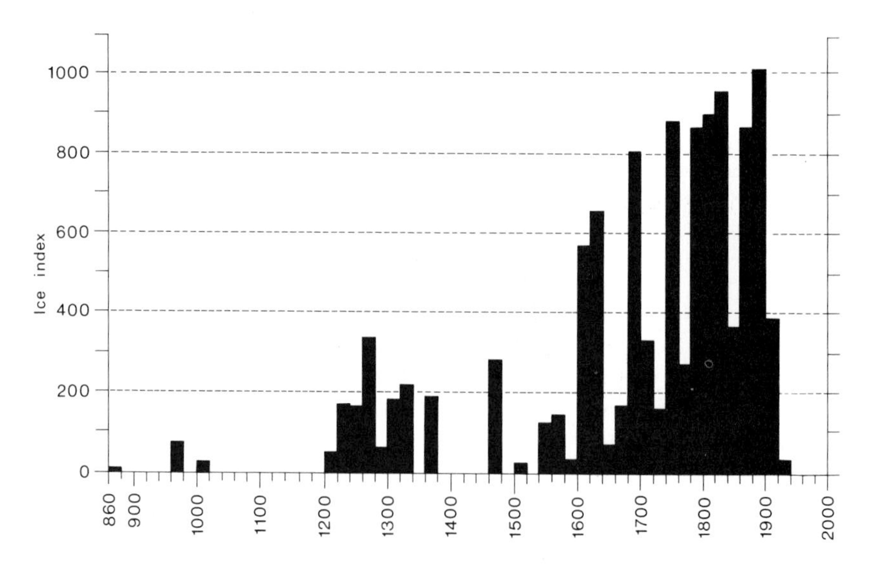

Figure 2.3a Koch's (1945) generalized diagram of the severity of sea ice incidence around Iceland from the ninth to the twentieth century. The index is the product of the number of stretches of coast (each about 135 km long) and the number of weeks when sea ice was noted along each of them

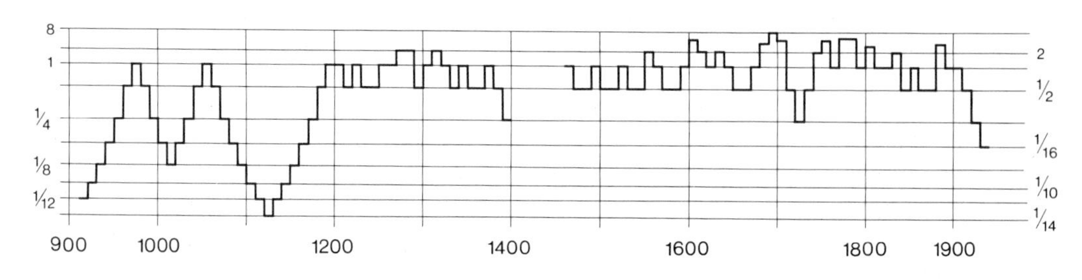

Figure 2.3b Bergthórsson's (1969) diagram of the decadal incidence of severe sea ice years around Iceland from the tenth to the twentieth century

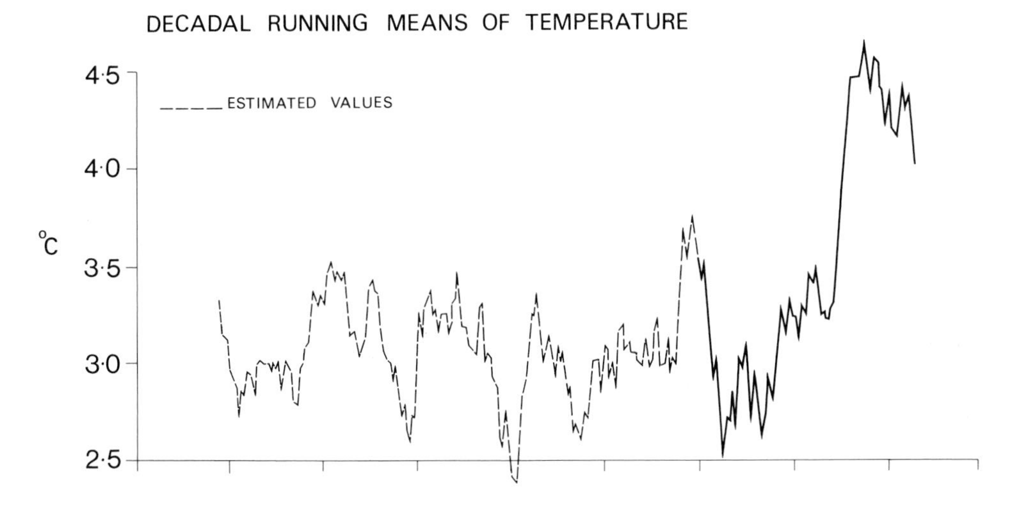

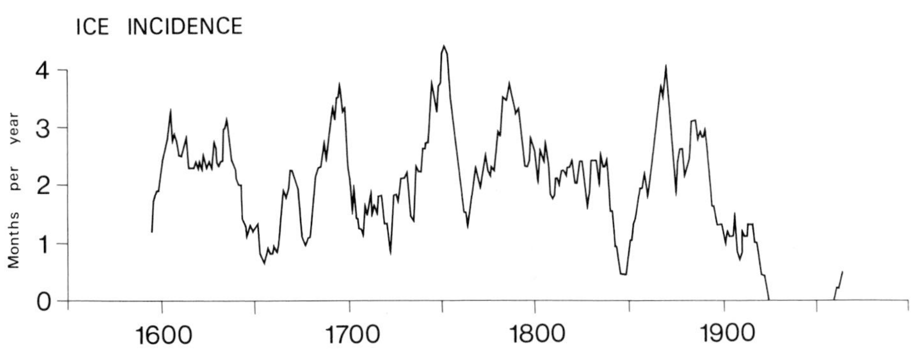

Figure 2.4 Decadal running means of annual mean temperature extended back from 1850 to 1600 on a basis of sea ice incidence. (From Bergthórsson 1969)

were less developed than they are today (Bell and Ogilvie 1978, Ingram *et al.* 1981). Vilmundarson (1972) investigated Thoroddsen's sources for some unusually cold years in the seventeenth century and found that while some were accurate others were misleading. It is possible, for example, that sea-ice conditions in 1639 were somewhat less severe than Thoroddsen supposed. He used the account from the *Skardsárannáll*, 'winter hard from Christmas onward. Sea ice drifted round the country the whole winter. It came along the east coast, then along past Sudernes' (the part of the southern peninsula in the vicinity of Keflavík; see Figure 2.1). 'It was accompanied by severe weather.' The *Skardsárannáll* was a contemporary description of events but, Vilmundarson pointed out, it is not necessarily adequate to determine which annal contained a contemporary account and then use that to determine conditions in a given year. The *Skardsárannáll* was composed by a man who lived far away on the north coast of Iceland and he was writing in general terms rather than with precision. Another account written at Grindavík, south of Keflavík, in 1639 was

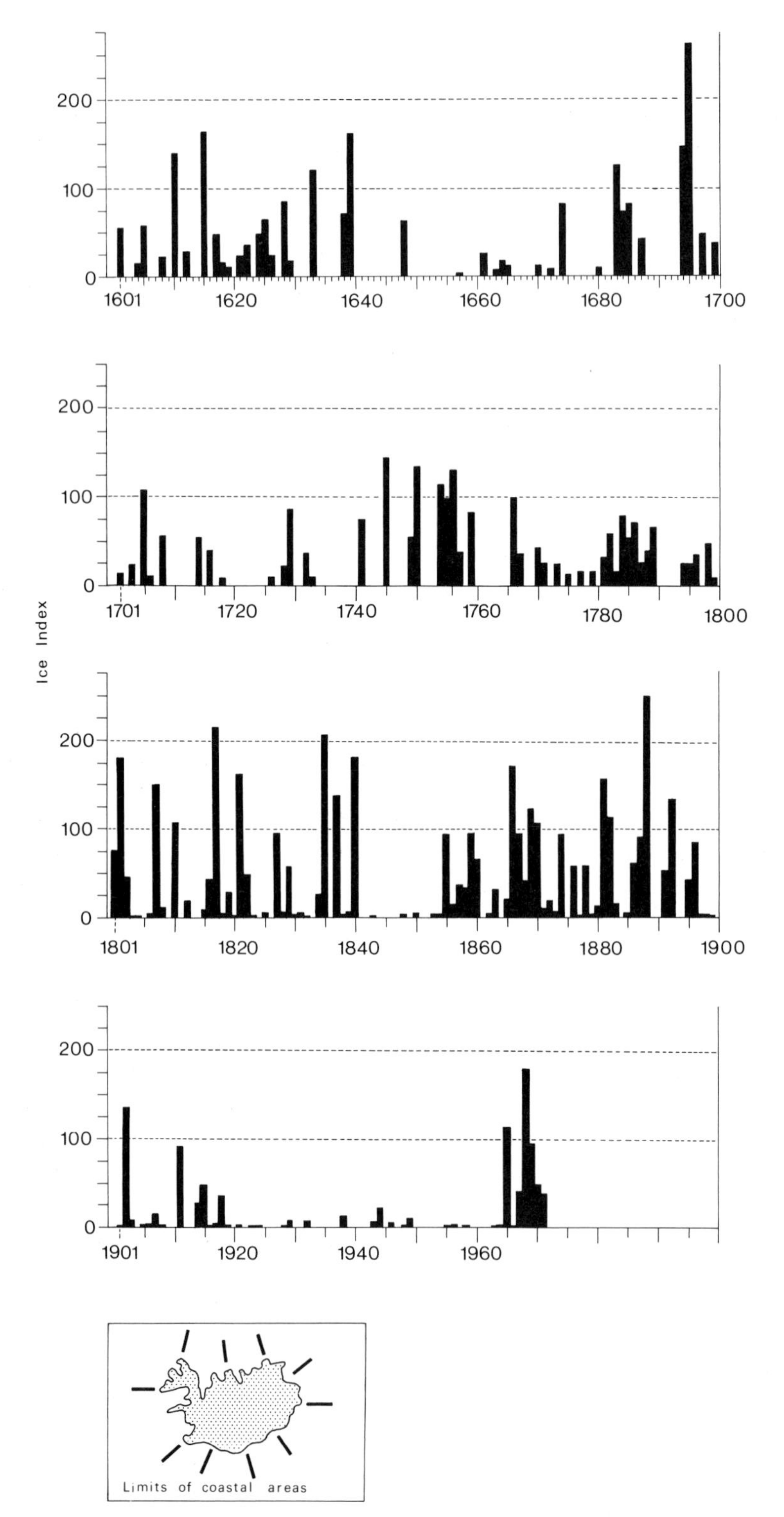

Figure 2.5 Incidence of sea ice around Iceland from the seventeenth to the twentieth century. (From Sigtryggsson 1972)

consulted by Thorlákur Magnússon when he wrote the *Sjávarborgarannáll* between 1727 and 1730, from which it appears that in 1639 the sea ice drifted as far as Reykjanes but no further (Figure 2.1).

> Terrible sea ice arrived three weeks after Easter and moving round the country from the east drifted ashore along the south coast all the way to Reykjanes. It drifted back and forth, east or west along the coast according to wind and current, past the Flitting days [late in May]. No use could be made of the ice either by seal hunting or by driftwood gathering. It cut away all the seaweed on the beaches at Grindavik, not a strand of any kind remaining.

Clearly conditions were severe but Thoroddsen's account, taken up by Koch, gives a somewhat exaggerated idea of the extent of ice round the southern peninsula towards Keflavík.

The section of the *Eyrarannáll* dealing with the period from 1673 to 1703 was considered by Vilmundarson to have been written concurrently with events and to be reliable, at any rate for the earlier years, though the later section must be used cautiously as the author, Magnús Magnússon of Eyri, in Seydisfjördur in northwest Iceland (1630–1704), was by that time suffering from the effects of old age. The picture which the *Eyrarannáll* presents of 1695 is of considerable importance, for it came in a period when glaciers were advancing in Norway as well as Iceland:

> The winter was fairly good with periods of little snow and favourable weather on land, still there were periods of extremely severe frost in between, so that all the fjords were frozen and fishing was hampered . . . sea ice came also to the north and west coasts, into the fjords after the New Year and stayed until summer . . . the drift ice froze together so that one could travel on horseback from one promontory to another across all the fjords also out beyond Flatey in Breidifjördur. No one could remember such ice cover, nor had anyone heard of such ice from older people . . . they also told of such a girth of ice in the sea round this country that ships could hardly reach the shore except in a small area in the south. The same frosts and severe conditions came to most parts of this country; in most places sheep and horses perished in large numbers, and most people had to slaughter half their stock of cattle and sheep, both in order to save hay and for food since fishing could not be conducted because of the extensive ice cover. . . . In Stranda Sýsla people fished through the ice 1.5 to 2 Danish miles offshore and shark, flounder and other fish were transported on horseback to the shore.

Vilmundarson not only looked critically at Thoroddsen's use of sources but also pointed out some he had overlooked, especially the regular reports on the climatic and other conditions, made by the sheriffs of the various areas to the Danish Governor of Iceland, which became customary in the eighteenth century.

Ogilvie (1984) made a new reconstruction of the sea ice and climatic record of Iceland from medieval times to 1780, based on critical examination of all the available documentary sources and careful selection of reliable data (Figure 2.6). She not only rejected many of Thoroddsen's sources because of dating errors but she discriminated between alternative versions of the same event, eliminated unreliable and spurious items and incorporated a great deal of new material, much of it from manuscript sources.

Evidence for the early medieval period Ogilvie found to be scanty and fragmentary.

Indications that the climate was mild at the time of the Norse settlement and for some time afterwards, though essentially circumstantial, she accepted as pointing to a mild climatic period. The first indications of increasing rigour of the climate appear in the 1180s; references appear to cold years in the next three decades but not from 1212 to 1223. Cold seasons came sporadically over the next thirty years and then the 1280s and 1290s saw more continuously harsh conditions. Ogilvie cites a statement of 1287 that 'at this time many severe winters came at once and following them people died of hunger'. Much more information is available for the fourteenth century, which seems to have been very variable climatically. The first two decades were rather mild; then the early 1320s were harsh, with sea ice close at hand and severe weather in 1320, 1321 and 1323. In the next two decades only two winters were severe, but from the 1350s temperatures were lower, with hard winters in 1350, 1351, 1355 and 1362. In 1365 there began the most severe period of the fourteenth century, with particularly hard winters recorded in 1370, 1371, 1374, 1376 and 1379. Though sea ice is mentioned only for 1374, this cannot be taken as proof of its absence. Arngrímur Brandsson had written around 1350 that 'on the sea are great quantities of drift ice that fill up the northern harbours'. By the 1360s it seems likely that it had become recognized that the ice from the north was disrupting the old sailing routes from Norway. But Ogilvie's findings do not agree with Bergthórsson's graph, which shows the second half of the fourteenth century as relatively mild. He had based his temperature curve on the occurrence of sea ice, mentions of which, for this period, are rare.

An independent study cited by Ogilvie is that of Teitsson (1975), which is based on the number of polar-bear skins mentioned in church inventories. They are held to indicate a period of great cold from 1350 to 1380, when the priests were particularly glad to have a warm fur rug on which to stand in church and sea ice was near enough to the coast for bears from Greenland to come ashore.

Comparatively mild conditions seem to have returned from about 1380 to 1430, though winters were long and severe in 1424 and 1426. Contemporary sources are rare for the period 1430 to 1560 and Bergthórsson's graph indicating mild conditions has little factual basis. However, trade with England at this time, involving cloth rather than grain, suggests that food was not particularly short. The latter part of the sixteenth century in contrast was undoubtedly cold. Oddur Einarsson wrote in the late 1580s that

> the Icelanders who have settled on the northern coasts are never safe from this most terrible visitor. The ice is always to be found between Iceland and Greenland although sometimes it is absent from the shores of Iceland for many years at a time . . . sometimes it is scarcely to be seen for a whole decade or longer . . . sometimes it occurs almost every year.

Ogilvie used the wealth of documentary information for 1601 to 1780 to construct a decadal sea-ice index (Figure 2.6). The main features of its variations correspond to those distinguished by Koch: a remarkably mild period from 1640 to 1660 was followed by one of increasing severity, with the decade from 1690 to 1700 being outstanding in this respect and the 1740s being similarly frigid.

Ogilvie finds that the variations in the decadal ice index generally agree with those in a winter/spring thermal index she constructed for northern Iceland. Her documentary sources were voluminous enough to provide five to ten indications of conditions on land for each season. The 1630s and 1730s, it might be noted, though cold on land, saw little

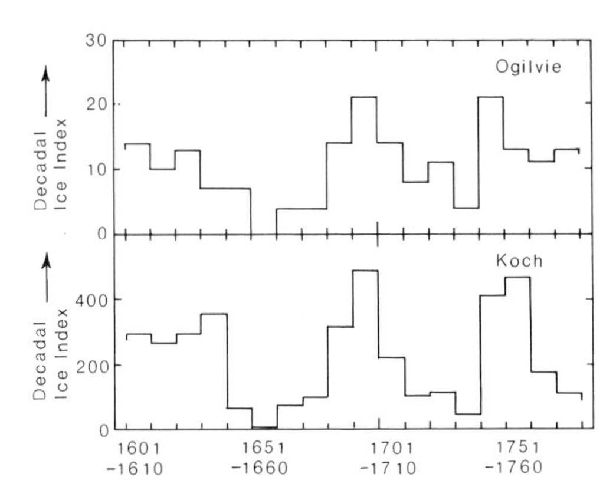

Figure 2.6 Decadal sea ice index for Iceland from 1601 to 1760. The upper values are from Ogilvie (1984) and are derived from the number of seasons of sea ice occurrence (winter, spring and/or summer) weighted by the number of regions reporting ice (i.e. 1 to 4). The lower values are an equivalent index based on Koch (1945).

sea ice; in these decades the response of the sea ice to climatic fluctuation seems to have lagged behind that of conditions on land. She was also able to show that fluctuations in the severity of winters were not simultaneous throughout the country; although there was a good deal of correspondence overall, the 1630s, for example, were milder in the west than in the north and south; the reverse was true in the 1640s. The cooling trend in the latter part of the seventeenth century, so obvious in the north, is less so in the south; that of the early eighteenth century is more evident in the south than the north.

It is apparent from Sigtryggsson's diagram (Figure 2.5) that the sea ice continued to be extensive around Iceland from 1780 until the early twentieth century, except for a spell of fourteen years between 1840 and 1854. The last year in which sea ice grounded along the north coast of Iceland was 1877. It happens that this was also the year in which the Danish Meteorological Institute began collecting reports on sea ice from around Iceland and in the Greenland Sea. These observations were later collected for the whole of the Arctic Ocean from ships, sealing vessels and coastal stations. The work was continued by the Icelandic Meteorological Office and during the Second World War by the British Meteorological Office. Since 1887 there has been no gap in the records, though those for 1945 to 1951 are less full than some others, partly because seal catching off the east Greenland coast did not revert to its pre-war level. It should also be added that interest in the whole subject of sea ice around Iceland was then at a low ebb because for several decades it had scarcely affected the country.

Jón Eythórsson was one of those who considered this situation unsatisfactory and took a lead in organizing regular drift ice observations, publishing them in the journal *Jökull* from 1953 to 1966. Late results were published in the same journal from 1967 to 1971 by Sigtryggsson and Sigurdsson, and after that in an official publication, *Hafís vid strendur Islands*, devoted to the detailed presentation of sea-ice data. Sources now include daily satellite pictures of the main ice areas with infra-red imagery for the dark winter months. Icelandic interest in the subject reawakened as the ice returned to nearby sea areas (Jónsson 1969).

The arrival of the ice came rather abruptly in 1965 and after that it invaded Icelandic waters every spring. The ice conditions in 1968 were worse than in any year since 1888; Figure 2.7 allows a comparison to be made between the ice years 1965 to 1972 and the last

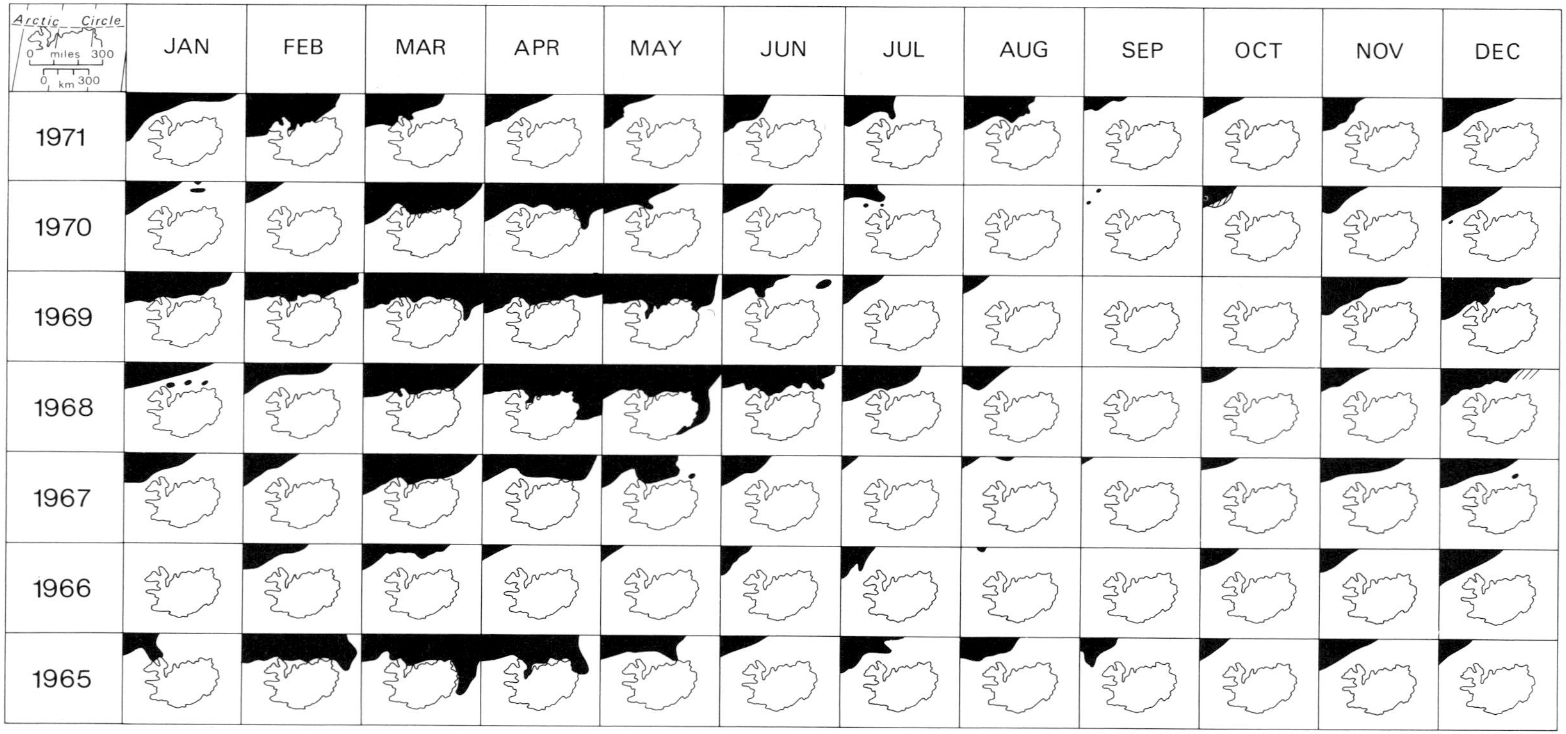

Arctic Circle 0 miles 300 0 km 300	JAN	FEB	MAR	APR	MAY	JUN	JUL	AUG	SEP	OCT	NOV	DEC
1971												
1970												
1969												
1968												
1967												
1966												
1965												

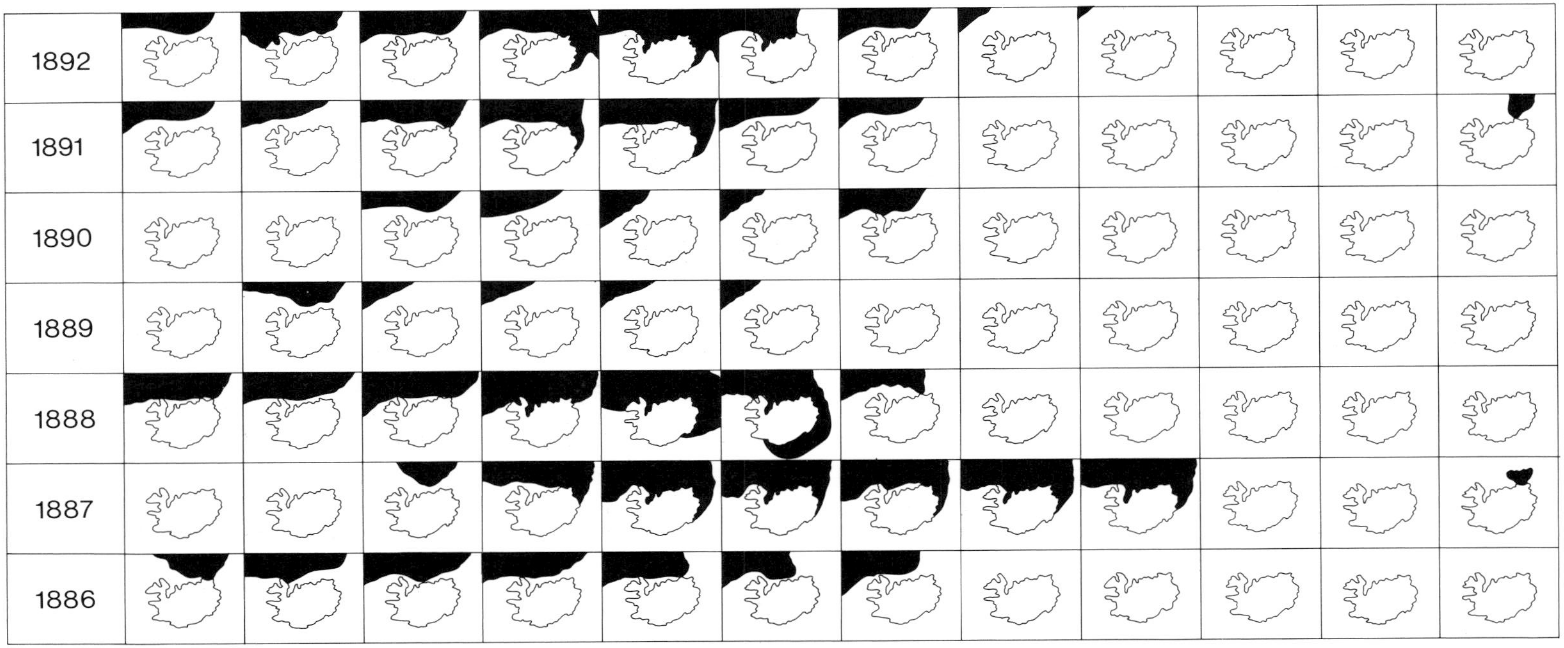

1892
1891
1890
1889
1888
1887
1886

Figure 2.7 Monthly sea ice positions around Iceland during the periods 1886–92 and 1965–71. Icing was more severe in the October–December quarter in the later period. (After Sigtryggsson 1972)

set of years of comparable severity, from 1886 to 1892. The south coast was not as severely affected as it was in June 1888 and in no year did the drift ice ground on the north coast as it did in 1887. But the greater proximity of the ice to the northern coast in November and December, as compared to the nineteenth century, is evident.

The change in ice incidence in the 1960s was associated with the development in the 1950s of an anomalous ridge of high pressure over Greenland, which strengthened in the early 1960s to reach a peak intensity of 12 mbars above the 1900–39 'normal'. It was coupled with a slight decrease in pressure over the Norwegian and Barents seas, causing a strong pressure gradient promoting northerly airstreams, and a steep decline of mean winter air temperatures over the Norwegian–Greenland seas. The mean annual temperature at Stykkishólmur in western Iceland dropped in 1965 and succeeding years until it was well below the average value, since observations began in 1846, for the first time since 1925 (excepting only 1943 and 1949) (Sigfúsdóttir 1969). Mean surface temperatures also dropped over the North Atlantic. As the northerly winds increased in strength, the East Icelandic Current changed from being an arctic current, as it was in 1948 to 1958, and became a polar current from 1964 to 1971 (Malmberg 1969 and Figure 2.8). An arctic current does not promote the preservation of sea ice, because the existence of a slight vertical stratification gradient in the surface layer prevents the water from cooling to freezing point before vertical convection starts, but a polar current with a salinity of less than 34.7 per cent can cool to the freezing point of sea water, 1.8°C, before

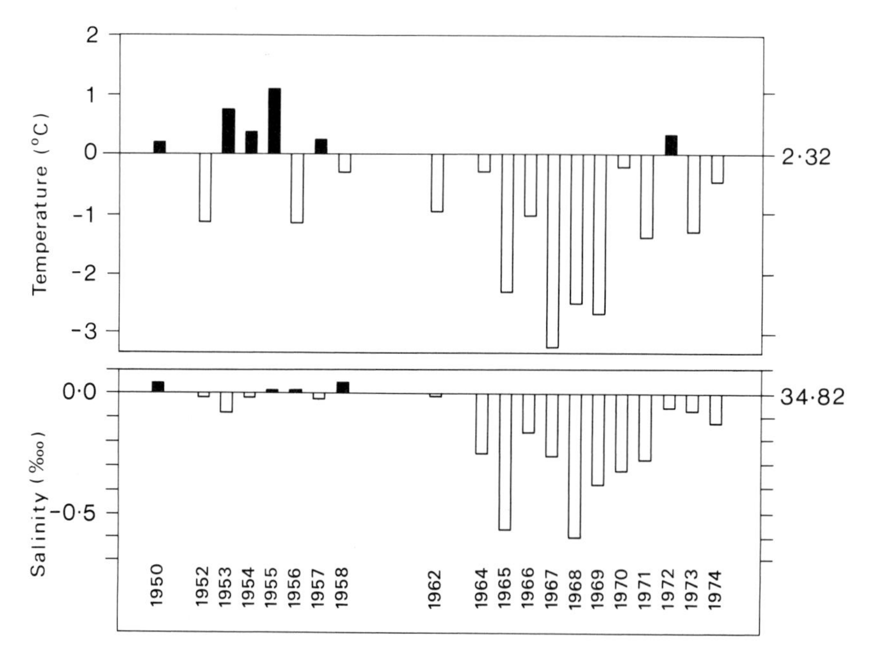

Figure 2.8 Temperature and sea ice in the Icelandic sea area, 1950–75. Arctic water has a slight vertical stratification gradient in the surface layer which prevents surface water cooling to the freezing point before vertical convection starts. Polar water, with a salinity of less than 34.7 per cent cools to the freezing point of sea water at −1.8 °C before convection starts and so drift ice forms. (From Dickson *et al.* 1975)

convection starts, and so drift ice forms more readily and remains so long as the current retains its surface character.

After 1970 the high pressure anomaly over Greenland collapsed dramatically, the mean pressure at sea-level in winter falling by 9.6 mbars between 1966–70 and 1970–4, and the polar influence off northern Iceland weakened. Polar water of low salinity failed to reach as far south and Atlantic waters returned to the sea areas north of Iceland.

The economic significance of the variations in the extent of sea ice round the country is considered at a later stage in conjunction with the associated history of glacier fluctuations in Iceland (see Chapter 12).

2.2 The glacial history of Mýrdalsjökull and Eyjafjallajökull

Mýrdalsjökull and Eyjafjallajökull together formed a single icecap in the Little Ice Age which separated into a larger eastern portion and smaller western one only in the middle of this century (Figure 2.9). Mýrdalsjökull, with an area of 700 km², overlies Katla, the second most active volcano in Iceland. Katla's eruptions have been accompanied by catastrophic floods from beneath the ice, known as *jökulhlaups*. Other jökulhlaups have been caused by the emptying of ice marginal lakes.

The area to the south and east of Mýrdalsjökull was the first part of Iceland to be fully settled. The history of the coastal settlements has been one of a struggle to survive volcanic eruptions, floods, ice advances and avalanches. Many farms have had to be abandoned and the rest have become concentrated in the safer localities, giving a clustered pattern of settlement unusual in Iceland.

The Mýrdalsjökull ice lens is 200 to 370 m thick (Rist 1967a) and rises to over 1400 m. Vík, the nearest meteorological station, records the second highest mean annual precipitation in Iceland (2256 mm, 1931–60) and on Mýrdalsjökull accumulation and ablation both reach very high values. The northern and eastern margins of the ice are lobate and relatively gently sloping. On the south, valley-type outlet glaciers descend over rock steps and display ogives in their lower sections. Smaller, steeper tongues of ice drain towards Thórsmörk in the west. The margins of Eyjafjallajökull are also rather precipitous. The highest point, Eyjafjall, rises to over 1600 m, a height exceeded in Iceland only by Öræfi, the great volcano south of Vatnajökull.

Katla is situated on a line of fissure eruptions that can be traced running southwest from Vatnajökull and disappearing under Mýrdalsjökull. Its crater, Kötlugjá, is normally beneath the ice but after eruptions it can readily be discerned. Such eruptions have caused a great deal of damage, with lava flows, ash falls and jökulhlaups (Kötluhlaups, that is jökulhlaups coming from Kötlugjá) destroying farms and ruining pastures. Earlier jökulhlaups had carried great loads of sediments from beneath the ice and deposited them to form a gently sloping apron, a sandur, running down into the sea. It was upon these sandur that the first settlements were established. Most of the early settlements were destroyed by hlaups from Kötlujökull between the ninth and eleventh centuries. Eruptions became even more frequent after the fourteenth century (Table 2.1) and various accounts of these give a vivid impression of their consequences.

The Kötluhlaup of November 1660 carried away all the houses and the church of Höfdabrekka, so that hardly a stone was left on the original site and so much material was carried down to the shore that a dry beach appeared where previously fishing boats had

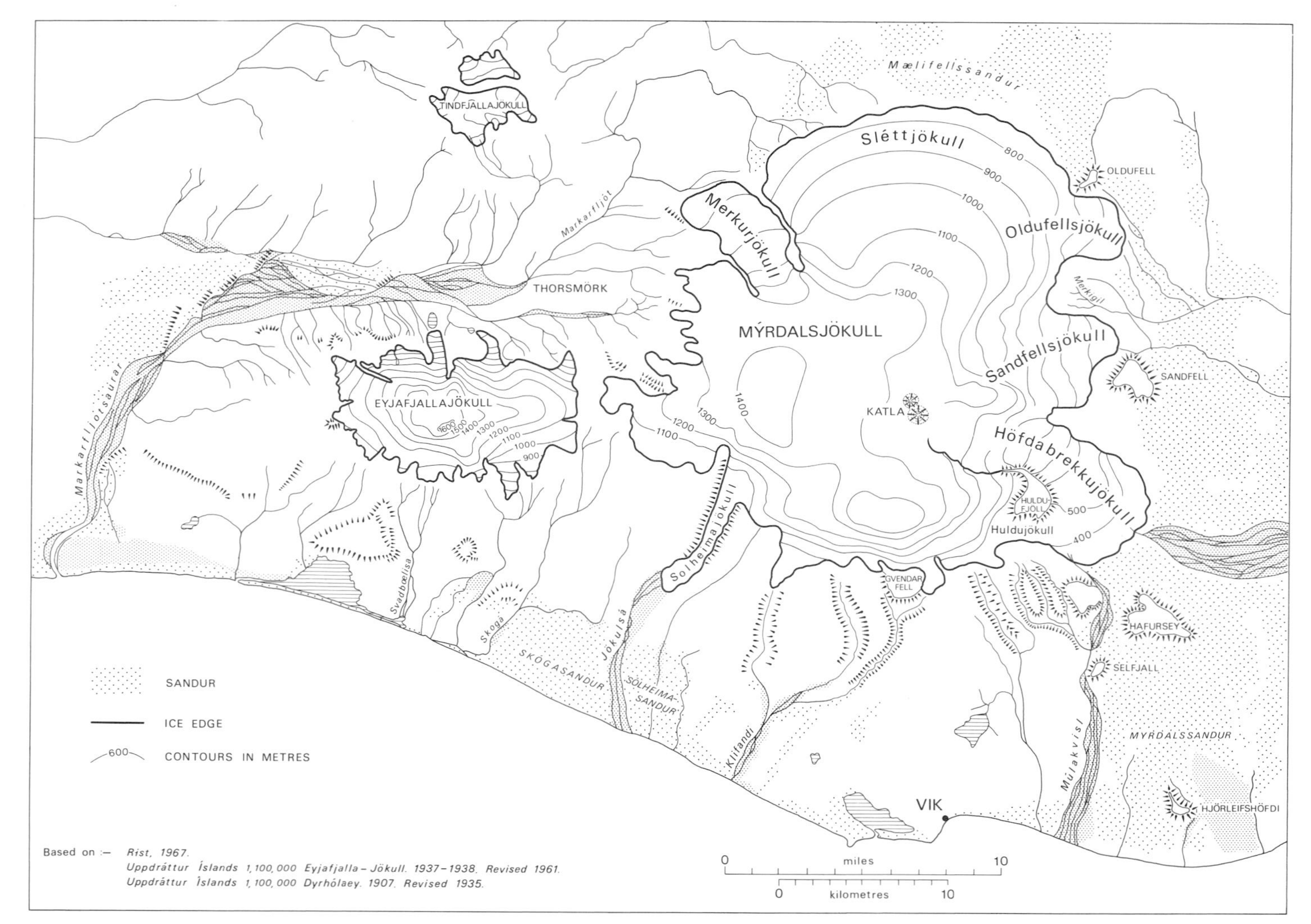

Figure 2.9 Mýrdalsjökull and Eyjafjallajökull. (Based on Rist 1967b)

operated in waters 20 fathoms deep. In 1721 masses of ice from the front of the icecap were floated out to sea and formed a floating barrier 3 nautical miles offshore. Ice, clay and solid rock were hurled into the sea for a distance of more than 15 miles in 1755: 'in some places where it was formerly forty fathoms deep, the tops of the newly deposited rocks were now seen towering above the water' (Henderson 1819, p. 248). But the principal damage, so far as the survivors were concerned, was the destruction of pasture.

Thórarinsson (1957) has estimated that, even discounting their ice content, the discharge of Kötluhlaups may reach maximum values for a short time of at least 100,000 m^3 per second, comparable with the mean discharge rate of the Amazon. A Kötluhlaup in 1955, not a large one, was preceded by an earth tremor five hours before the flood reached the ice margin. It is suspected that both the tremor and the hlaup may

Table 2.1 Dates of eruptions of Katla since AD 844

Dates at which eruptions may have occurred		*Eruptions for which there are eyewitness accounts*
894		1625
900		1660
934		1721
1000		1755
1179		1823
1245	hlaups from	1860
1262	Sólheimajökull	1918
1311	'Sturluhlaup' – legendary	1955 miniature eruption
1332		
1416	'Höfdahlaup' – legendary	
1485	tephra evidence	
1580		

Sources: Thoroddsen (1905–6), Rist (1955, 1967) and Thórarinsson (1959)
Note: There is not complete agreement between available authorities on the dates of the eruptions, especially the earlier ones

have been caused by a subglacial eruption melting the adjacent ice. The water is estimated to have flowed under the ice at a velocity of about 10 m s^{-1} on a gradient of about 4°.

The most destructive jökulhlaups in Iceland have emerged from lobes of ice east of Katla, Höfdabrekkujökull and Huldujökull, which have advanced and then collapsed on such a large scale in consequence that they are of little value for tracing events of climatic significance. The northern lobes are too remote from settlements for details of their movements to have been observed. The most helpful information comes from the southern outlets of Mýrdalsjökull and Eyjafjallajökull and the ice-dammed lakes and floods associated with them, though they have not been altogether free of jökulhlaups and the effects of volcanism.

Most information is available for the long outlet glacier in the south west of Mýrdalsjökull, Sólheimajökull, which is drained by one of the most dangerous rivers in Iceland, known as the Jökulsá á Sólheimasandi. It has another name, Fúlilækur, which means stinking river, attributable to the smell of hydrogen sulphide emanating from it. Bárdarson (1934, p. 41) mentions jökulhlaups from Sólheimajökull in 930 and 1262 and attributes them to volcanic eruptions, but he gives no references to his sources and his explanation is not necessarily the only one available.

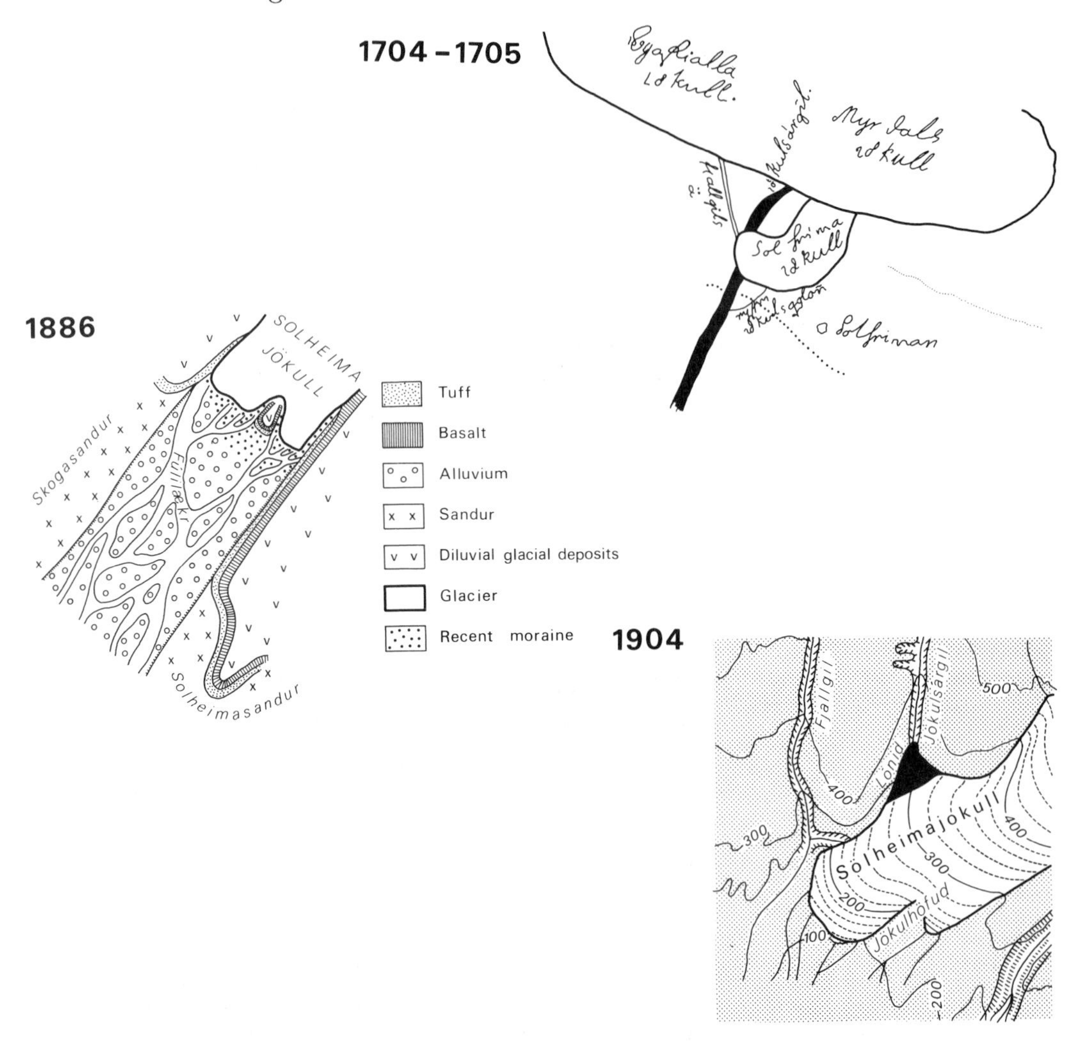

Figure 2.10 The tongue of Sólheimajökull mapped in 1704–5 after Thórarinsson (1960) from Árni Magnússon; in 1886 after Keilhack (1886); and in 1904 after Thórarinsson (1960) from the Danish General Survey map of 1904

According to Árni Magnússon's description of Sólheimajökull in 1704–5, it was 'a flat, low outrunner of Mýrdalsjökull, extending in a bend southwards from that glacier and then to the west'.[1] The description is accompanied by a sketch (Figure 2.10) which is the earliest known map of an Icelandic glacier. Magnússon explains that the glacier had grown towards the west and in doing so had blocked a canyon carrying meltwater from Mýrdalsjökull. The water escaped by a tunnel under the tongue of Sólheimajökull.

When the tunnel is blocked up, the water which would otherwise have run out of the

[1] Árni Magnússon (1663–1730), the founder of the Arnamagnaean collection, travelled through Iceland between 1702 and 1712 and recorded many geographical and topographical details in his 'Chorographica islandica'. The Sólheimajökull material is probably in the handwriting of his secretary and dates from his visits to Vestur-Skaftafellssýsla in 1704 and 1705.

canyon mentioned above, forms an enormous deep lake. When the ice blocking up the tunnel can no longer withstand the pressure of the water, the tunnel is opened *cum impeta* and everything inside it is crushed. Thus arise the jökulhlaups in Jökulsá. . . . The jökulhlaups usually occur once a year and are smaller the shorter the intervals between them. . . . Sólheimajökull both lifts and slides over the surrounding land, so that the difference can be seen from one year to the next. People say of the main glacier [Mýrdalsjökull] that it does not grow visibly except in places where it descends into ravines, which are gradually filled up by small glaciers.

From this passage and from the sketch it is clear that Sólheimajökull had blocked the Jökulsá canyon a good many years before 1704. It continued to advance, swelling and thickening very greatly after Katla's 1755 eruption according to Ólafsson and Pálsson (1772, p. 763).[2] Then it retreated, and Thoroddsen (1905–6, p. 183) recorded that by 1783 it had withdrawn and the Jökulsá was flowing freely between the glacier and Skógafjall, which means that Sólheimajökull was considerably smaller than in 1705.

By 1794 the glacier had advanced again, reaching Skógafjall, blocking the Jökulsá and damming back a lake which according to Sveinn Pálsson (1945)[3] flooded especially often

[2] Eggert Ólafsson, a naturalist and poet, travelled through the country every summer from 1750 to 1757 with Bjarni Pálsson, a physicist, and eventually published a very comprehensive, if not especially original, account of the observations they made. He was too much affected by the abstruse scientific theories then current, but fortunately he recorded the views of the country people, which were often based on detailed practical knowledge, although himself rejecting some of the soundest of them. Thus he gave the local view on the alimentation of glaciers which 'reached high up in the air, where it is much colder than on the flat low-lying land. On them, rain will change to snow and ice, and as they always attract rain, clouds and fog, they will maintain their size and grow unless the sun can every year melt as much as is added to them.' While Ólafsson's informants had grasped the basic notion of glacier mass balance, he succeeded in convincing himself that glaciers were probably nourished by penetration of sea water through subterranean passages.

[3] Sveinn Pálsson (1762–1840) was the most notable glaciologist of eighteenth-century Iceland. He was the son of a farmer from Skagafjördur in the north. He studied natural history and medicine in Copenhagen from 1787 to 1791 and then, in the succeeding four summers, he travelled widely through Iceland making geographical, geological and botanical observations. In 1795 he married Thórunn, the daughter of Bjarni Pálsson, who had travelled with Eggert Ólafsson. In 1799 he was appointed doctor to the south of Iceland, serving an area stretching from Hellisheidi to Skeidarársandur, as well as the Westmann Islands. He held this appointment till 1824 and for most of the time lived and also farmed at Sydri Vík in Mýrdalur. Despite this double occupation in two outstandingly onerous fields, he made careful weather records from 1791 to the end of his life and undertook a great deal of research and writing. His most important glaciological work was the treatise which he sent in 1795 to the Natural History Society in Copenhagen, which financed his travels. He wrote this in Danish. Parts of it were published in Den Norske Turistforening's Year Books for 1881, 1882 and 1884, but it was not published in full until it was edited and brought out in Icelandic by Jón Eythórsson in 1945. As a result it was entirely overlooked elsewhere in Europe and played no part in the mainstream of glaciological thought. Had it done so, it must surely have ranked as the most important and illuminating work on the subject written during the eighteenth century.

Pálsson had read Strøm's *Physisk og oeconomisk Beskrivelse over Fogderiet Søndmør* published in 1762, which contained Wiingaard's description of the fluctuations of some of the Jostedal tongues (see Chapter 3), and also the very much less outstanding discussion of glacier motion included in Fleischer's *Forsøg til en Natur-Historie*. Through this, but not directly, he learnt of the work of Walcher, Gruner, de Saussure and other Alpine glaciologists. He made such use as he could of existing knowledge gained in this way but treated it critically. Unlike his predecessors he thought it worthwhile to give detailed accounts of individual glaciers. He also drew many maps, that of Vatnajökull being especially impressive (see Fig. 2.19).

Pálsson was the first glaciologist to perceive that convection rather than radiation could dominate the ablation process on temperate glaciers, writing that 'I also venture to affirm that the rays of the sun do not play as much part in melting the firn of the plateau ices, not even in cold weather when they fall directly on them, as misty weather without precipitation does' (quoted by Thórarinsson 1960, p. 14). This view was eventually to be confirmed by the work of the Swedish-Icelandic investigations of Vatnajökull in the 1930s.

It is not surprising that Pálsson was the first to differentiate clearly beween the oscillations of the Vatnajökull margins caused by climatic changes and those caused by volcanic activity. His work is a major source for any study of the Icelandic glaciers in the eighteenth century. Thórarinsson (1960) summed up his

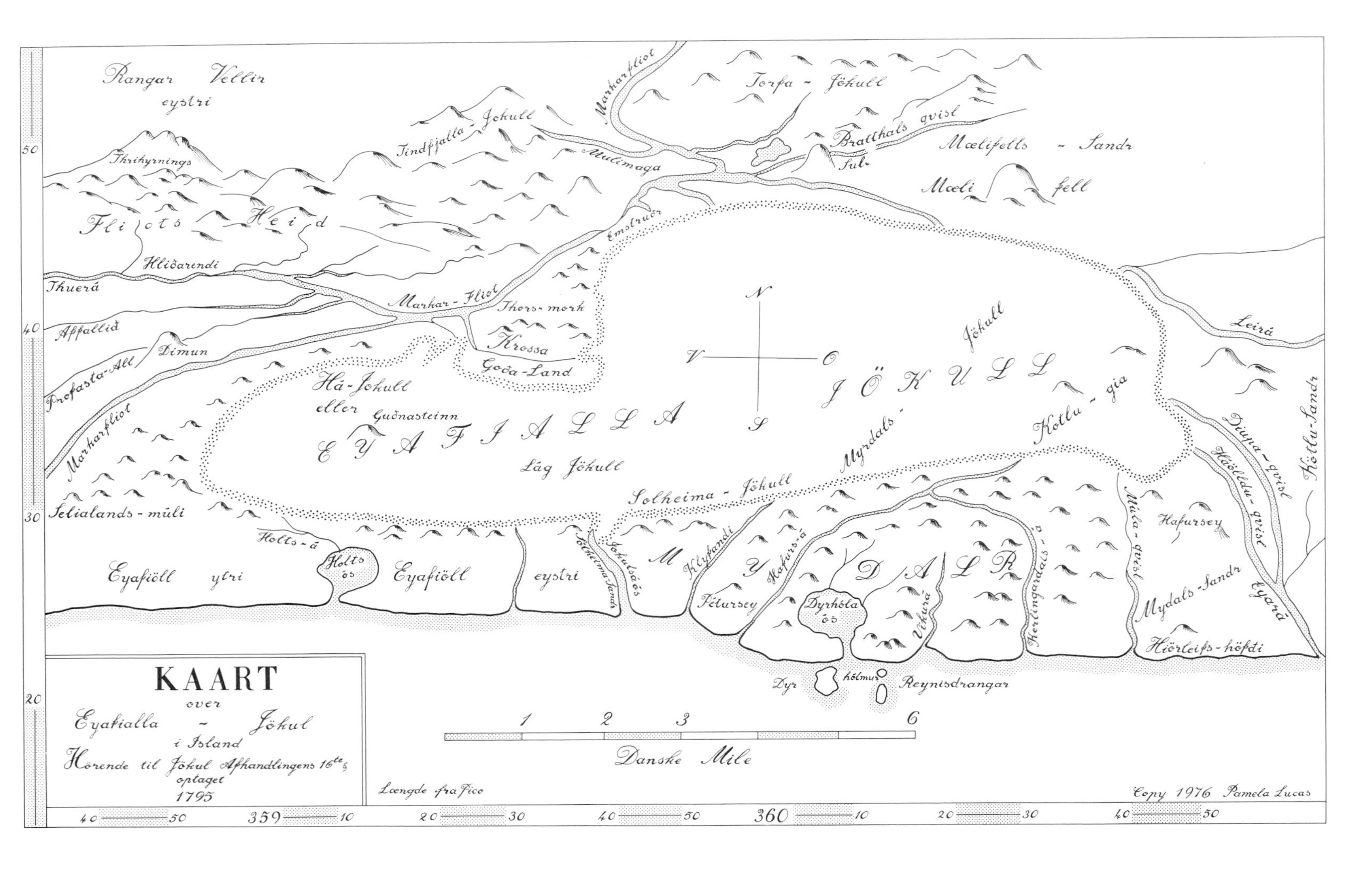

Figure 2.11 Eyjafjallajökull–Mýrdalsjökull in 1795 as mapped by Pálsson (1945 edn)

in 1794. Neither his map nor his panoramic view (Figures 2.11 and 2.12) provides further detail but it would seem likely that Sólheimajökull's tongue was then in a similar position to that of 1704. Eythórsson (1931) recorded that a reliable local man, eighty years old, told him that the ice covered the rock called Jökulhöfud by 1820; according to Thoroddsen (1905–6, p. 183) it was still covered in 1860. The glacier may then have begun to recede because local people told Eythórsson in 1930 that the ice had not completely covered Jökulhöfud for sixty or seventy years. Certainly by 1883 the greater part of Jökulhöfud was free of ice, for a map drawn by Keilhack (Figure 2.10) shows the glacier dividing into a wider western and a narrower eastern part either side of Jökulhöfud, with moraine labelled recent, and possibly dating from 1860, extending about 100 m in front of the ice. The situation seems to have been much the same in 1893 (Thoroddsen 1905–6, p. 183).

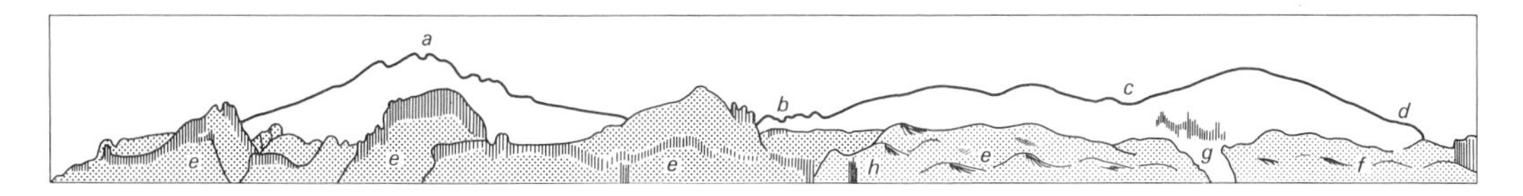

a.	Håjökull or Gudnasteinn	*e e e e.*	Eyjafjöllin
b.	Lågjökull	*f.*	Mýdalsfjöllin
c.	Sólheimajökull	*g.*	Jökull, which Fúlilækur flows under
d.	Mýdalsjökull	*h.*	Skógafoss

Figure 2.12 Cross-section view of Eyjafjallajökull–Mýrdalsjökull in 1795 by Pálsson (1945 edn)

A Danish map of 1904 (Figure 2.10) shows that the front of Sólheimajökull had retreated to a height of about 100 m above sea-level, as compared with 50 m when Keilhack had mapped it. Jökulhöfud was free of ice on its northern side and the eastern terminus of the glacier has retired about 200 m since 1883, though the western one had remained more or less stationary. The retreat continued with a mean rate of recession of 30 to 40 m per year between 1930 and 1937, and thinning of the ice becoming very rapid. After 1930 the positions of the front of Sólheimajökull and other glaciers were measured every year and, as Figure 2.13 shows, the retreat decelerated until the early 1960s, since when there have been slight advances.

Summing up the oscillations of Sólheimajökull in 1939, Thórarinsson decided that in 1705 it was of

> about the same extent as in 1904. In 1783 it was considerably smaller than in 1904. In 1794 it had resumed its advance; at about 1820 it had reached its 1705 position and was advancing or stagnant until about 1860. Since then it has, on the whole, been in recession. . . . Its maximum extent in the first half of the 18th century –

contribution most effectively: 'His treatise on glaciers constitutes a last phase and a culmination of a glaciology that may be called Icelandic in the sense that it was principally based on knowledge of glaciers in Iceland.' That knowledge was to a large extent common to the country people who lived along the southern margin of Vatnajökull, in close contact with its advancing outlet glaciers and its glacier rivers and sandur. It was a knowledge which had gradually accumulated during nine centuries, because this people was in large measure endowed with 'Man's nature to wish to see and experience the things that they had heard about and thus to learn whether the facts are as told or not' (Larson 1917). What is more remarkable is that Pálsson should have been able to distil, interpret and build upon this knowledge whilst living for forty years in extreme poverty, working as a doctor, fisherman and farmer, caring for his people, his land and his large family.

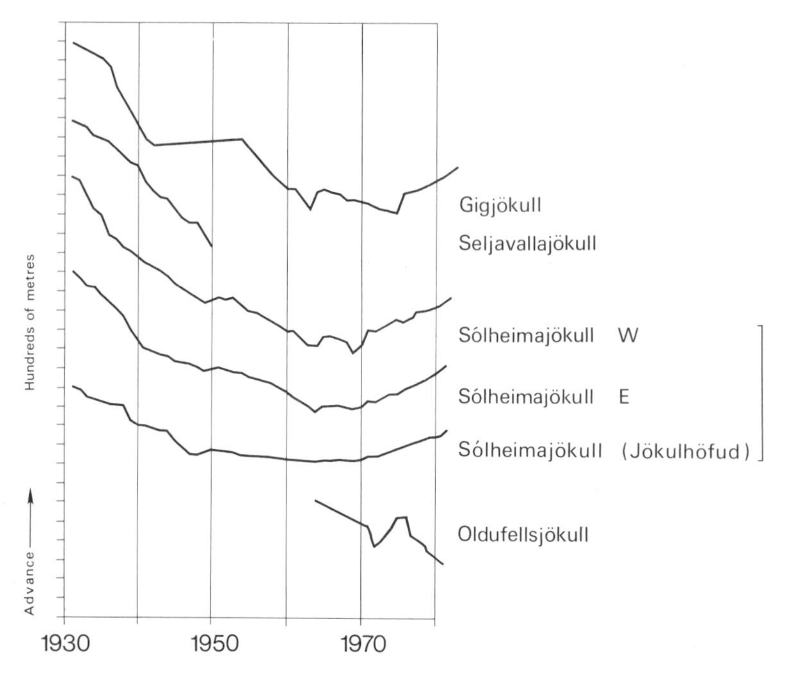

Figure 2.13 Frontal fluctuations of Eyjafjallajökull–Mýrdalsjökull glaciers, 1932–82. (From figures published in *Jökull* by Eythórsson and Rist)

> probably approximately the same as in the middle of the 19th century – was the largest in historical times. (1939, p. 236)

No major jökulhlaups occurred over the period of the Sólheimajökull record, though its expansion about 1755 was certainly associated with volcanic activity. The timing of the advances and retreats are believed to be of climatic significance and may now be compared with those of neighbouring glaciers.

Gígjökull, the glacier east of Jökultungur, and another near it, came down onto the plain in 1756 and were described by Ólafsson (Bárdarson 1934, pp. 44–5) as alternating and advancing every year. But he also commented that Eyjafjallajökull was smaller than it had been earlier and that two hills which had not been seen for a long time were sticking out of the ice. He thought that this thinning was due to the eruption of Katla in 1755, but Katla is a considerable distance away. Gígjökull and its neighbour still extended down to the plain when Sveinn Pálsson saw them in 1794. For the nineteenth century we have no information. When it was mapped in 1907 the front of Gígjökull was 200 m from the outermost moraines of historic times (Bárdarson 1934, p. 44). Local people reported that it had been advancing somewhat for a few years before 1930, when a permanent marker was erected. Subsequently, retreat was rapid in the 1930s and late 1950s, and small oscillations followed in the 1960s.

Einar Einarsson (1966), a farmer of Skammadalshóll in Mýradalur, was told by his father and grandfather that the glaciers of southern Mýrdalsjökull advanced between 1870 and 1890 but after that retreat was general until 1966. Local people living on the south side of Eyjafjallajökull reported to Eythórsson (1931) that Seljavallajökull retreated between 1907 and 1930 and that the glaciers had become much thinner. All of this information corresponds to the data for Sólheimajökull.[4]

2.3 Vatnajökull

Vatnajökull, which was also known as Klofajökull, is the largest icecap in Europe, and is also the world's largest temperate icecap (Figure 2.14). It exhibits a closer juxtaposition

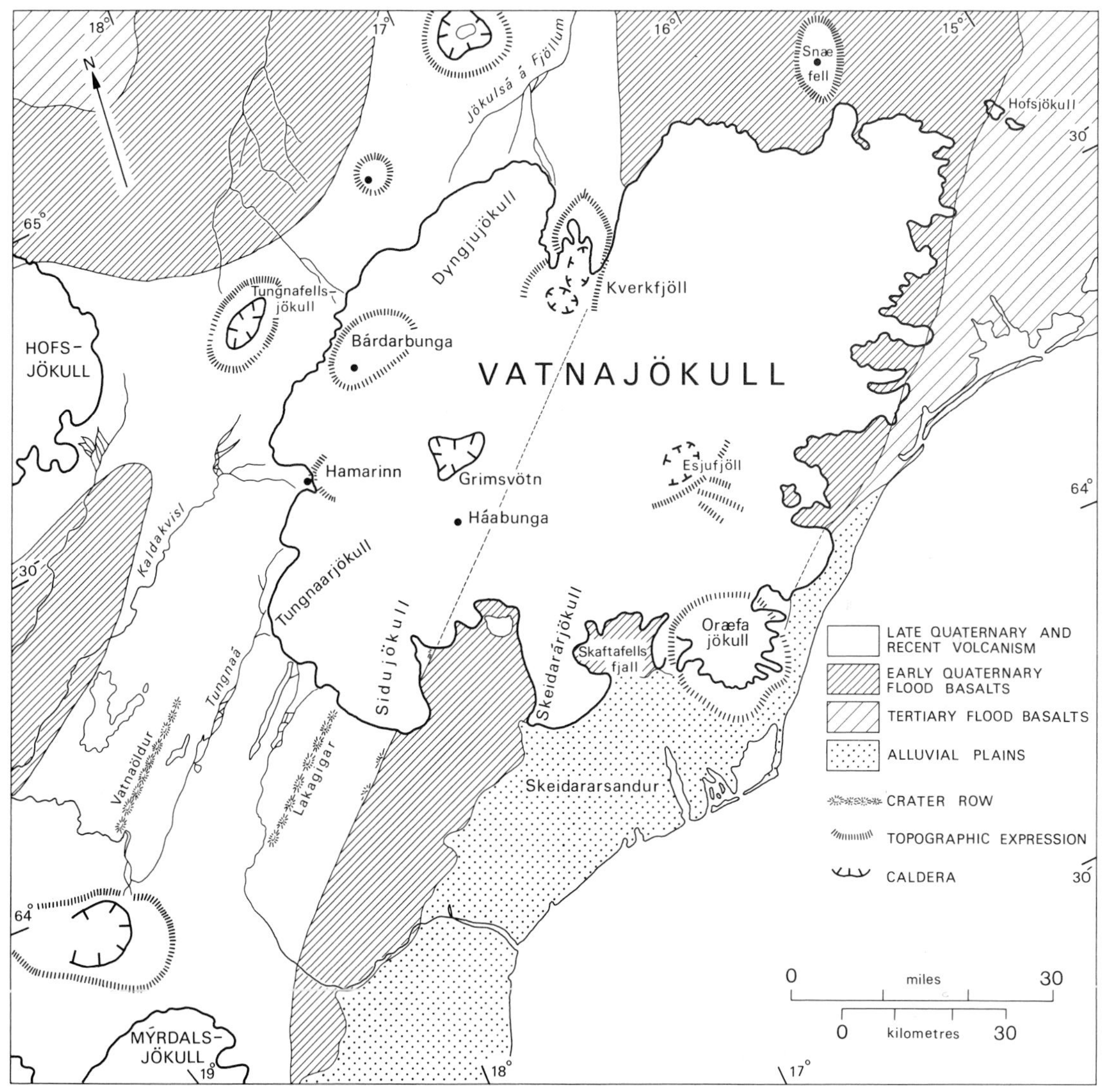

Figure 2.14 Vatnajökull in relation to geological boundaries and principal volcanic features as revealed by satellite imagery. (From Thórarinsson *et al.* 1973)

[4] The only exception seems to have been the 'glacier east of Jökultungur which advanced in the period 1907–30' (Eythórsson 1931). Unfortunately no details are available.

and a more intimate interplay between vulcanism and glaciation than is to be found anywhere else on such a large scale. It rests on a group of active volcanoes centred on Grímsvötn. To the east of the Grímsvötn caldera is an extinct caldera in the little-known Esjufjöll area; to the northwest a volcano underlies Bárdarbunga. Jökulhlaups in the Jökulsá á Fjöllum led Thórarinsson (1957) to suggest that there might be an eruption centre in Dyngjujökull in the northwest. Satellite imagery shows two elliptical features, presumed to be calderas, in the Kverkfjöll area a little further east, and also a volcano–tectonic line running from these caldera SSW across Vatnajökull, through Grímsvötn, to Sídujökull. To the west of this line the volcanoes appear to be active; to the east they are extinct, with the very important exception of Öræfajökull.

In 1362 an eruption of Öræfajökull[5] was the largest explosion in Europe since that of Vesuvius in AD 79; it erupted again in 1727. The most disastrous outburst of all was that of the Laki craters immediately west of Vatnajökull, in the fault zone southwest of Sídujökull, which took place in 1783. It lasted seven months and emitted 15 km^3 of lava that covered 450 km^2 between Vatnajökull and the sea. More destructive to life and property was the ash which was deposited over most of Iceland, poisoning the pasture lands and killing half the cattle and three-quarters of the ponies and sheep. More than 9000 people died of famine, one-fifth of Iceland's population, and the total abandonment of Iceland was seriously considered at this time.

The first map to represent correctly the whole periphery of Vatnajökull (except Skaftárjökull) was compiled by Ahlmann and Thórarinsson in 1937 (Ahlmann and Thórarinsson 1937, p. 200, and Figure 2.15). The ice was estimated at this time to cover 8800 km^2, about 8 per cent of the country; more recent measurements give a value of 8538 km^2 (Thórarinsson 1958). The surface of the icecap is made up of several flattish domes rising to about 1600 m with a broad shallow depression, running NNE to SSW from Brúarjökull to Breidamerkurjökull, about 20 km east of the volcano–tectonic line. The margins of the ice consist of great lobes except in the southeast, where, from Breidamerkurjökull to Eyjabakkajökull, a series of valley glaciers flow down towards the coast. The glaciers descending from Öræfajökull, 2119 m, form a largely independent system in the southeast.

Several of the outlet glaciers of Vatnajökull, especially those on the north and west sides, are surging glaciers (they are numbered 1 to 6 on Figure 2.15). The anomalous behaviour of Brúarjökull and Eyjabakkajökull attracted the attention of Nielsen (1937) and Thórarinsson (1938). At first glaciologists were inclined to explain their sudden advances in terms of subglacial volcanism or seismic activity (Bárdarson 1934, Thórarinsson 1943). By 1960 enough information had been collected to discount these explanations and to show that none of the Vatnajökull surges corresponded in time to volcanic activity and earthquakes, except for those of Dyngjujökull and Sídujökull in 1934, when there was an eruption in the Grímsvötn caldera (Thórarinsson 1964, 1969).

The Vatnajökull glaciers that have surged are shown in Table 2.2. Brúarjökull figures prominently; early surges of this glacier are pinpointed by records left by Ólafsson (1772, p. 792) and Thoroddsen (1914b, p. 297). Others have followed at intervals of between 73 and 95 years. In 1964, when rates of movement of the Brúarjökull front were measured for the first time, it was remarked that the noise made by the surging glacier could be heard 2 or 3 km away, but during the winter of 1889/90 the rumbling advance had been audible

[5] Before 1362 Öræfajökull was called Knappafell or Knappafellsjökull.

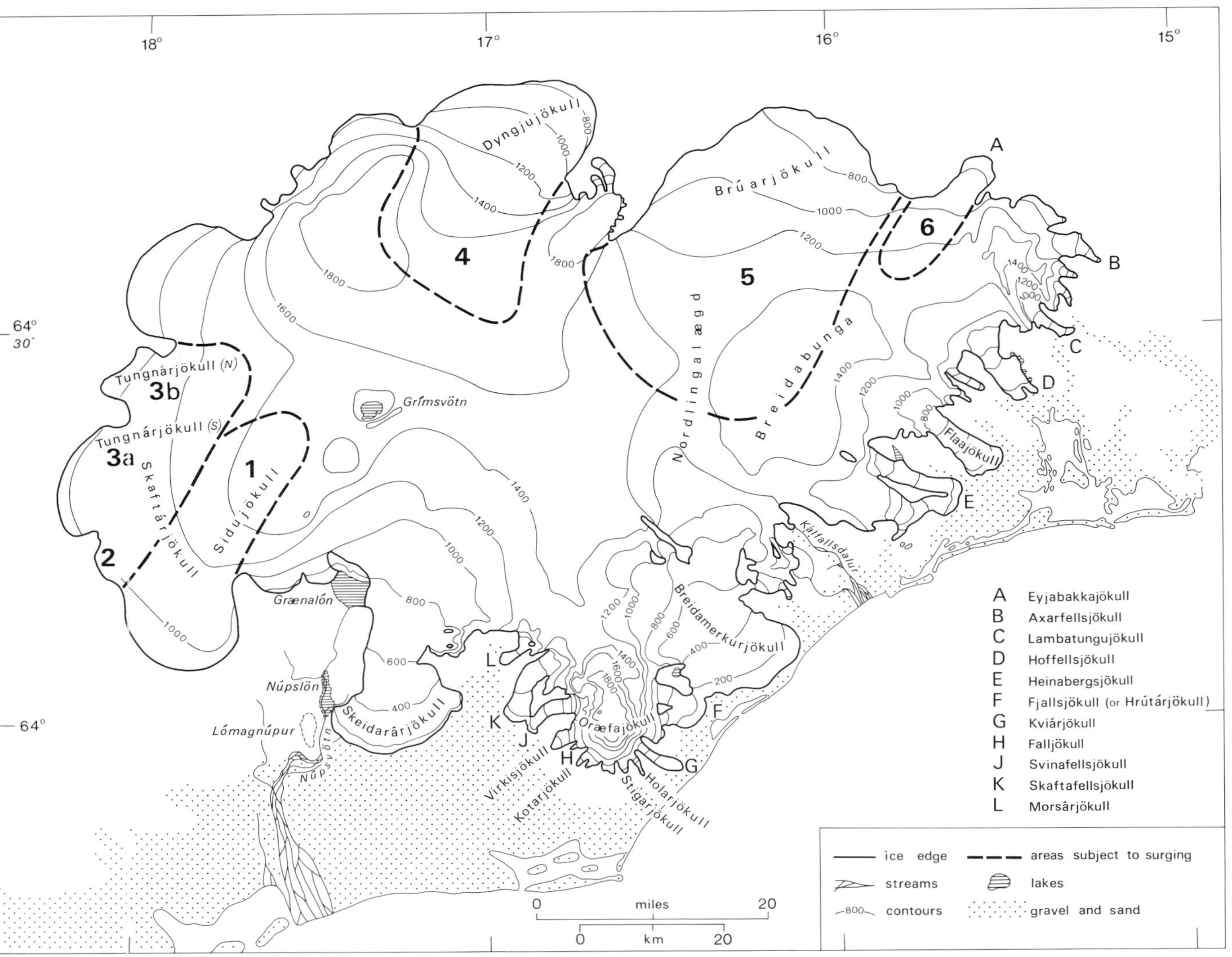

Figure 2.15 The most important surging lobes of Vatnajökull, numbered 1–6. (From Thórarinsson 1969)

Table 2.2 Dates when Vatnajökull's outlet glaciers surged

Year	*Glacier*	*Area affected* (km^2)	*Max. advance (km)*
1978–9	Dyngjujökull	?	?
1963–3	Sídujökull	480	about 0.5
1963–3	Brúarjökull	1,400	about 8.0
1951	Dyngjujökull	(700)*	about 0.3
1945	Skaftárjökull	450	about 0.6
1945	Tungnaárjökull	?	about 1.0
1934	Sídujökull	480	about 0.6
1934	Dyngjujökull	(700)*	about 0.3
1890	Brúarjökull	1,400	about 10.0
1890	Eyjabakkajökull	110	about 0.6
1810	Brúarjökull	1,400?	?
late 1720s	Brúarjökull	?	?
1625	Brúarjökull	?	?

Source: Thórarinsson (1969)
Note: * Values in brackets are rough estimates

at a distance of 50 to 60 km. The scale of the phenomenon is very great, with as much as 700 km^3 of ice being involved, spreading over an area of 1400 km^2 (Thórarinsson 1969).

Brúarjökull is only the extreme example of surging glaciers emanating from Vatnajökull. Most of the other glaciers around the periphery, except those in the southeast, are known to have surged during the last hundred years, which has been predominantly a period of glacial recession. In periods of expansion surging may well have been more frequent. Its history is not easily established because of the remoteness of the inland margins of Vatnajökull and also because there is not much evidence left on the ground. Moraines formed by surging glaciers here are generally small (Thórarinsson 1964), though according to Todtmann (1960) those of 1890 are big, and it is only with the availability of remote sensing data that it has been possible to begin to assemble a general picture of the activity in recent years.

Some of the surges of Vatnajökull could have been climatically induced; the low gradients of the surfaces of the glaciers make them sensitive to changes in temperature and precipitation, and Thórarinsson (1964) has argued that their morphology may result in stress accumulating until a critical point is reached in the balance between accumulation and ablation, when flow suddenly accelerates. By confining our attention mainly to the glaciers of the southern and eastern margins, for which it happens that most historical information exists, we can avoid some of the uncertainties attributable to surging, and concentrate on climatically controlled glacier oscillation.

Bárdarson (1934) made the first attempt to survey in a critical manner the recent history of Icelandic glaciers. He used data first brought together by Thoroddsen, plus the results of field studies made by Eythórsson and Eiríksson, and various cartographic data. He did not compare the oscillations of the various glaciers, nor did he attempt to relate the fluctuations to changes in climate. Thórarinsson's (1943) synthesis made use of material resulting from his own and Ahlmann's investigations of Vatnajökull in the 1930s, plus information from such early sources as Ísleifur Einarsson's Land Registers and descriptions of Austur-Skaftafellssýsla from 1708–9 and 1712[6] (Figure 2.1). Since 1950, the journal of the Icelandic Glaciological Society, *Jökull*, has published the results of current glaciological research.

Although detailed information about the extent of the glaciers before 1700 is sparse, it is clear that at the time of the settlement and for long afterwards the southern and eastern tongues of Vatnajökull were smaller than in the eighteenth and nineteenth centuries. Farms built between Vatnajökull and the coast by the first generations of settlers were damaged by later ice advances.

Skeidarárjökull is further west than the other glaciers that are to be discussed and its movements are affected by the eruptions of Grímsvötn. Nevertheless it is worth noting some features of its history, which is relatively well known because it was commonly observed and remarked upon by travellers on their way to the districts of Mýrar and Lón and farms near Öræfajökull, such as Skaftafell (Figure 2.14).

Skeidarárjökull holds back a lake, Grænalón, at 635 metres on its west side as it emerges from Vatnajökull (Figure 2.15). When the ice front is in an advanced position it touches the ridge of Lómagnúpur and dams up the Súla to form another lake known as Núpslón. Grænalón is the largest glacier lake in Iceland, with a volume of 1500–2000 million m^3, and causes jökulhlaups from time to time, as it did in 1898, 1935 and 1939. Núpslón has never been known to have an area greater than 5 km^2, which, with a depth of about 20 m, gives a maximum water volume of about 100 million m^3. Both Thoroddsen (e.g. 1892, p. 111) and Sveinn Pálsson (1882, p. 53) thought that a flood in 1201, described in an MS edition of *Biskupasögur* written between 1212 and 1220, and in *Sturlungasaga*, was due to a hlaup from Núpslón. Thoroddsen considered this one of the most important proofs that Icelandic glaciers in saga days were as large as in his own. The argument is rejected by Thórarinsson (1939) on the grounds that there are no moraines or shorelines round Skeidarárjökull, indicating that the glacier was larger in the thirteenth century than at its greatest extension in the nineteenth century. Moreover a lake of Núpslón's size will fill in four to eight days and probably run off early in the summer. It would not re-form in the same year because the large run-off from Skeidarárjökull would keep the tunnel open. But the description in the sagas indicated that the 1201 hlaup lasted some days and that it must have occurred in mid-August, for the bishop, whose journey was impeded by it, reached Lón on St Bartholomew's Day, that is 24 August. Thórarinsson took the view that it is far more likely that this hlaup, the earliest for which there are records, was due to the emptying of Grænalón rather than Núpslón. He argues that a thinner Skeidarárjökull would have allowed Grænalón to empty subglacially, even without being affected by a volcanogenic hlaup. There is strong evidence that volcanism was dormant in the Grímsvötn area until the fourteenth century. There are no records of eruptions of Öræfi before 1362 or from the Skeidarárjökull area before 1389 (Thoroddsen 1925, p. 83). The whole history of settlement in Skaftafellssýsla indicates volcanic inactivity in saga times. In particular the Eyrarhorn farm (see Figure 2.16), one of considerable size situated on land now covered by the Skeidará lagoons, was being worked up until the late fifteenth century, and in Thórarinsson's view, as this area has been completely flooded by all the known volcanogenic jökulhlaups from Skeidarárjökull, no farm could possibly have existed there if Grímsvötn had been active. This being the case, he concluded that the 1201 hlaup provides strong evidence that 'the purely

[6] Unfortunately the section of Magnússon's and Vídalín's *Jardabók* (1702–12) for Skaftafellssýsla and Múlasýsla are lost, probably burnt in the great Copenhagen fire of 1728.

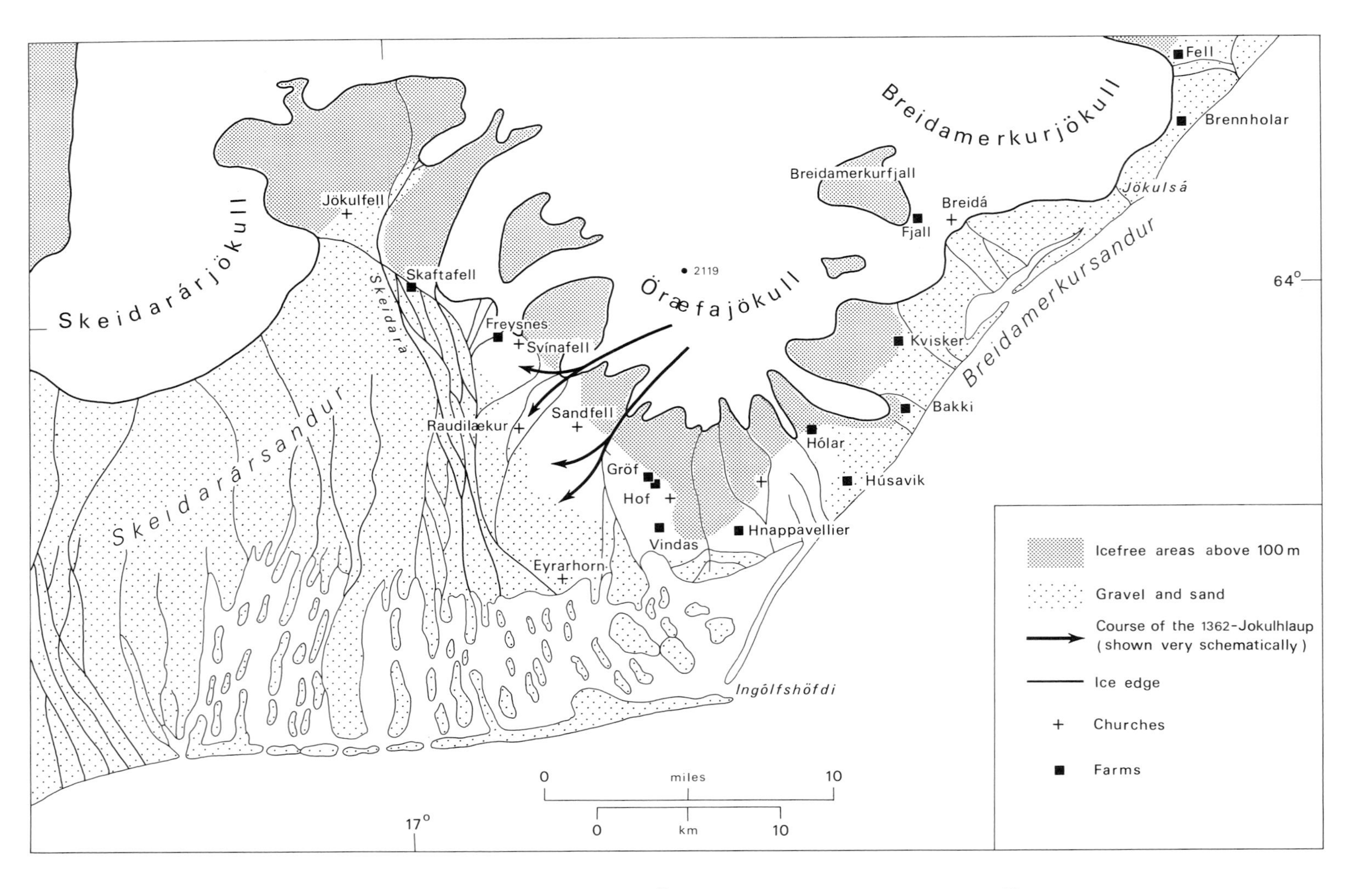

Figure 2.16 Skeidarárjökull, Breidamerkurjökull and the glaciers of Öræfajökull. (From Thórarinsson 1958)

climatologically conditioned outlet glaciers of Vatnajökull must have been considerably less extensive in the saga period than they are now' (1939, p. 231).

Taking this lack of activity in the earlier centuries into account, Thórarinsson advanced an interesting supportive argument for climatic deterioration since the end of the fourteenth century. He noted that twenty-eight jökulhlaups were recorded from Skeidarársandur between the end of the fourteenth century and 1938.

A hlaup which occurred in 1934, although not reckoned amongst the largest, was estimated to have involved some 15 km^3 of water. Thórarinsson took 10 km^3 of water as a safe value for the average total discharge of a hlaup, considering that this was probably a very substantial underestimate in fact. This gives an aggregate water loss for twenty-eight hlaups of 280 km^3 of water or about 300 km^3 of ice. Now this is of the order of twice the mid-twentieth-century volume of Skeidarárjökull below the firnline. Thórarinsson concluded that, as Skeidarárjökull did not recede despite these enormous losses, the climate must have been more favourable to glacier growth after volcanic activity began in the fourteenth century than it had been in saga times. The general drift of these arguments is substantiated by positive evidence from the pre-Little Ice Age period from Breidamerkurjökull.

Land adjacent to and now covered by the ice of Breidamerkurjökull was settled in Landnám times (AD 870–930) and is said to have been occupied by many farms and to have been well covered by grass until 1100 or later (Ólafsson and Pálsson 1772, pp. 786–8). There was certainly a large area of scrub woodland, 'Breidamörk', on the western slopes of Breidamerkurjökull, and two farms, Fjall and Breidá, in the area which was eventually to be first devastated and then covered by ice from Breidamerkurjökull, and its neighbour, Hrútárjökull[7] (see Figure 2.16). Fjall, at the southern end of Breidamerkurfjall, was first cultivated in about 900 by Thórdur Illugi, who had taken in the land according to the *Landnámabók*. In 1179, when the pastures were rather good, the church of Raudilækur, in Hérad,[8] had grazing for 160 wethers on Fjall lands (*Diplomatarium Islandicum*, I, p. 148, cited Thórarinsson 1943). Fjall was mentioned again in 1387 and was still being cultivated (*Diplomatarium Islandicum*, III, p. 401). Indeed, agricultural activity probably had another two centuries to run before trouble came. As late as 1660 Brennhólar to the east of Breidamerkurjökull is mentioned as being leasehold of Fjall (Bárdarson 1938, p. 27). Breidá seems to have been a large prosperous farm in saga times, the home of chieftains of noble birth (Eythórsson 1952). It was adopted by the author of *Burnt Njál's Saga* as the home of Kári, one of the heroes of that sad tale. Whilst the saga is to be considered as fictional, the choice of Breidá as Kári's home would scarcely have been made if it had not been recognized as an important and prosperous place. However that may be, there is no doubt of the existence of Breidá and that it was situated at the southern end of Breidamerkurfjall. Breidá was still prosperous in 1343 when, according to documentary evidence, the church at Breidá owned all the farmlands, including a wood and two other farms at Hólar and Hellir Eystri as well as other property (Bárdarson 1934, p. 28).

The great eruption of Öræfi in 1362 (see Figure 2.16) had consequences that were both widespread and dramatic. According to an approximately contemporaneous account of the year 1362,

[7] This glacier is now called Fjallsjökull.
[8] This district was known as Hérad or Hérad milli Sanda as long as Raudilækur existed, but later came to be known as Litlahérad (Thórarinsson 1958).

> Volcanic eruption in three places in the South and kept burning from flitting days [i.e. end of May] until the autumn with such monstrous fury as to lay waste the whole of Litlahérad as well as a great deal of Hornafjördur and Lónshverfi districts, causing desolation for a distance of some 100 miles. At the same time there was a glacier burst from Knappafellsjökull [i.e. Öræfajökull] into the sea carrying such quantities of rocks, gravel and mud as to form a sandur plain where there had previously been thirty fathoms of water. Two parishes, those of Hof and Raudilækur, were entirely wiped out. On even ground one sank in the sand up to the middle of the leg, and winds swept it into such drifts that buildings were almost obliterated. Ash was carried over the northern country to such a degree that foot-prints became visible on it. As an accompaniment to this pumice might be seen floating off the west coast in such masses that ships would hardly make their way through. (*Diplomatarium Islandicum*, p. 226, cited Thórarinsson 1958, p. 26)

Thórarinsson (1958) made a very detailed study of the 1362 eruption which was based not only on careful assessment of documentary evidence but especially upon measurements of the very extensive tephra layer associated with it. Volcanogenic ice advances and hlaups from Fallsjökull and Róturfjallsjökull destroyed some farms in Hérad, but despite the long-established tradition that they were the chief cause of devastation, Thórarinsson showed that the hlaups probably swept west, not directly south to the sea, and that the major cause of destruction was the tephra fall. He estimated that the volume of tephra was at least 10 km^3, corresponding to about 2 km^3 of solid rhyolitic rock. Hérad was ruined permanently and the farms of the districts of Sudursveit, Mýrar and Nes, east of Öræfajökull, were abandoned for some years.

We have no direct evidence as to whether Breidá and Fjall were abandoned after the eruptions, but in 1387 Breidá had no longer any cattle and the church had lost its ornaments (Eythórsson 1952). This certainly reflects the results of the eruption, but equally clearly that Breidá survived it. Documents dated 1525 indicate that the farm was still in cultivation then, although no mention of a church is to be found (*Diplomatarium Islandicum*, IX, pp. 158, 247). Farming is again mentioned in documents of 1587 and 1697 (*Blanda*, I, p. 49) but 1697 may have been the last year during which anyone managed to cultivate this land.

The evidence of the existence of farms near Breidamerkurjökull and Hrútárjökull is sufficiently concrete and reliable to substantiate the general proposition that the Vatnajökull tongues were substantially smaller at the time of the settlement of Iceland and for several centuries thereafter than they have been for the last three centuries or so. Nothing very much is to be gleaned, however, about the time at which the swelling leading to the Little Ice Age maxima may have begun. The occurrence of events such as the 1362 tephra fall, causing widespread environmental deterioration, would in any case make the identification of the onset of the climatic change involved more difficult, except perhaps as far as changes in the distribution of sea ice were concerned.

2.3.1 EARLY AND MID-EIGHTEENTH-CENTURY ADVANCES OF VATNAJÖKULL

By the end of the seventeenth century the Vatnajökull glaciers were already much enlarged. The Fjall farm seems to have been abandoned by about 1695 at the latest. In the land register of 1708–9 we read that '14 years ago tún [meadows near the house] and ruined buildings were still to be seen, but all that is now in the ice'. Again in 1712 this farm appears in the register of abandoned farms in Öræfi. 'Fjall was the name of a farm west of Breidamörk. It is now surrounded by ice. Twelve years ago ruins could still be seen.' The date of final abandonment is not known; there was still some activity at Breidá in 1697, but it was completely deserted by 1698/9 (Bárdarson 1934, p. 29).

Breidamerkurjökull and Hrútárjökull had united in front of Breidamerkurfjall at the end of the seventeenth century, according to a document of 1700–9 (Jónsson 1914, p. 44, cited Bárdarson 1934, p. 29). In 1709 there was another reference to Breidá being deserted but it was said to have had the use of wood at Breidamerkurmúli, the part of Breidamerkurfjall farthest to the north, which was surrounded that year by the glacier (Ísleifur Einarsson 1709, cited Bárdarson 1934, p. 29). Einarsson's cadastral register of 1712 mentions that 'the ruins of the farm buildings can still be seen', though the ice was evidently then close to them. There was said to have been a stone slab in the ruins, supposedly on the tomb of Kári Sölmundarson, the original settler, who had gone there about 1017 (*Blanda* I, pp. 49–50), but this had disappeared under the ice by 1712.

An early map of Vatnajökull, made by Knopf who surveyed Skaftafellssýsla in 1732, shows the ice margins about 6 km from the shore at Breidá. A verbal account of Iceland by Knopf, written down by a Swede in Norway and dated 14 August 1741, contains a description of Breidamerkurjökull:

> On the south side of the country, in Skattefialls Syzel, a quarter of a mile from the sea, there is a glacier called Breide Märker Giökel or Giakel, which moves as much as 30 ells forward once or more times every year, and then as much back again after a short time. The movement is so much the more obvious as this glacier, as it were, rests on the sandur, and in advancing pushes the sandur in front of it like high walls or breastworks. (Kålund 1916, cited Thórarinsson 1943, p. 25).

The question now arises as to whether the behaviour of Breidamerkurjökull and Hrútárjökull during the seventeenth and early eighteenth centuries was typical of the Vatnajökull outlets as a group.

A grassy area of Hafrafell between Skaftafellsjökull and Svínafellsjökull, on the west of Öræfi, traditionally used for grazing goats, became more and more difficult to reach as the two glaciers advanced towards each other (Figure 2.15). By the early eighteenth century, according to Ísleifur Einarsson's Land Register of 1708/9, 'The Farm has the right of summer grazing on the part of Freysnes Farm called Hafrafell, but this right cannot now be executed as everything is covered by ice.' Árni Magnússon's 'Chorographica' is rather more explicit. 'Between Svínafell and Skaftafell lies Hafrafell, a large grassy hill, formerly accessible by paths, and with grazing for sheep in summertime; now this hill is so surrounded by glacier tongues that it can only be reached on foot and with great difficulty' (quoted Thórarinsson 1943, p. 32). The position was described in similar terms in 1746, 'Further east lies Hafrafell, surrounded by glaciers' (Stefánsson 1746, cited Thórarinsson 1943, p. 32). The ice had clearly not retreated to any marked extent at this time.

When Ólafsson and Pálsson visited the Stígárjökull in 1756, it descended through a gorge to the pastures near the tún of Knappavellir, and the people at the farm told the travellers that the ice had been advancing in recent decades. Thórarinsson (1956, 1943) concludes that Stígárjökull, like the other Öræfi tongues, was larger in the mid-eighteenth century than in the twentieth.

Heinabergsjökull, to the northeast of Breidamerkurjökull, was also in an advanced position in 1756, for Ólafsson reported that 'Heinabergsjökull has destroyed the central part of Hornafjördur, above which it is situated, one arm having reached right down on the plain to the farm Heinaberg, which nevertheless is still inhabited' (Ólafsson and Pálsson 1772, p. 790) (see Figure 2.17). A manuscript description of Austur-Skaftafellssýsla dated 1746 by Sigurdur Stefánsson (quoted by Thórarinsson 1943, p. 22) mentions that 'In Heinabergsjökull there is a hill called Hafrafell, with some grass and great precipices'. This account confirms that the snout had come right down to the plain, and states that Heinabergsvötn (or Dalvatn), the ice-dammed lake formed when this tongue is in a forward position, had 'spoilt many adjacent farms, devastated many of them completely, washing away grassland, leaving only shingle, clay and sand'. This shows that the ice had already dammed the lake during the first half of the eighteenth century and was near its maximum extension in the middle of the century (Figure 2.17).

Svínafellsjökull in Lón was also enlarged in the mid-eighteenth century, for according to Sigurdur Stefánsson, who lived in the neighbourhood, 'A glacier comes down on both

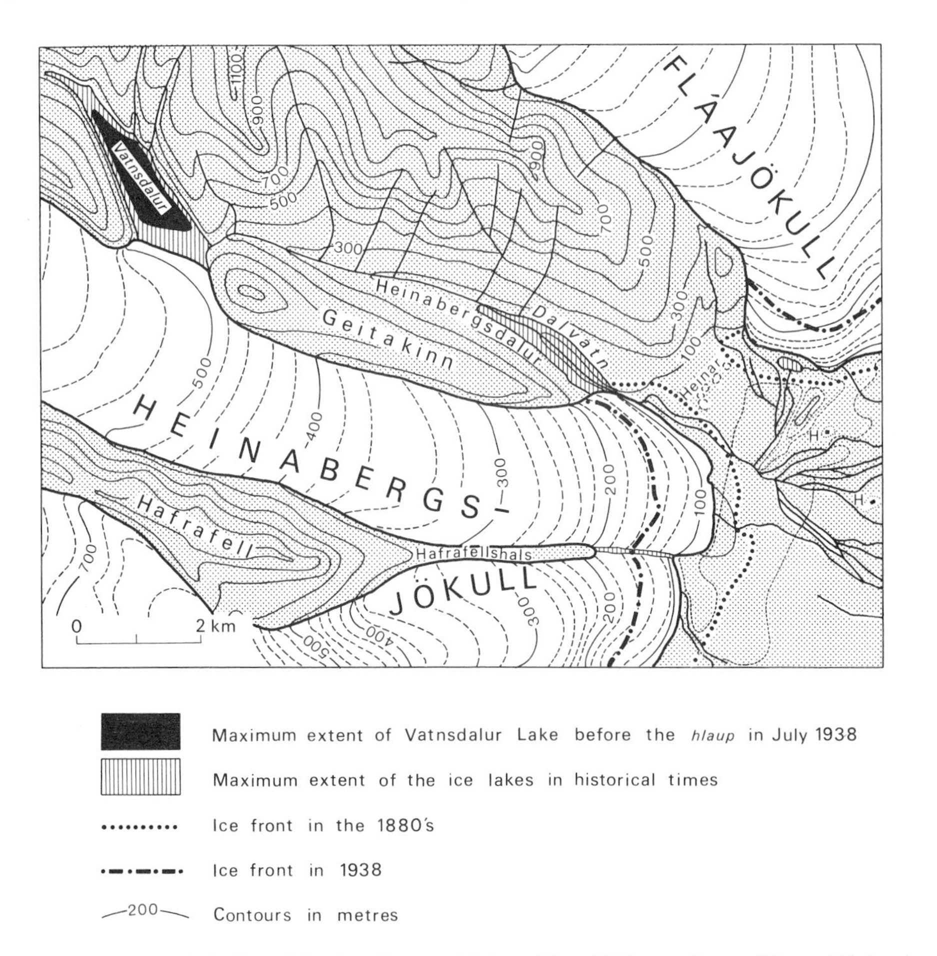

Figure 2.17 Heinabergsjökull and its ice-dammed lakes. H = Hólar, a farm. (From Thórarinsson 1939)

sides of Svínafell, which is a low but precipitous hill, with some grass, ling and forest vegetation' (Thórarinsson 1943, p. 19).

Skeidarárjökull, like the other Vatnajökull glaciers, advanced during the first half of the eighteenth century. When Knopf mapped the southern margin of Vatnajökull in 1732, he wrote that Skeidarárjökull 'slid out 600 years ago. The same glacier moves back and forth'. The first statement is interesting but unsubstantiated; the second is certainly to be taken seriously. Öræfi had erupted in 1727 (Thorláksson 1740, cited Henderson 1819, pp. 208–12).

> In 1727, when both the *Öræfi* and *Northern Skeiderá* volcanic Yökuls were in activity, this low Yökul began to rock, to the great danger and consternation of some people who happened to be travelling on the sand before it. According to the account they afterwards gave, it moved backwards and forwards, undulating at the same time like the waves of the sea, and spouting from its foundations innumerable rivers, which appeared and vanished again almost instantaneously, in proportion to the agitation of the Yökul. As the progress it made was inconsiderable, the spectators saved themselves on a sand-bank, but the suddenness and unexpectedness with which the rivers continued to rush forth, rendered it impossible to travel any more that way the whole summer. (Ólafsson and Pálsson 1772, p. 780, cited Henderson 1819, p. 215)

By 1756 things had settled down again, with the glacier in a forward position, for when Ólafsson and Pálsson crossed the Skeidarársandur they found that 'the mountain slope north of the Núper is beautiful and fertile and lush grass and birches are growing there, which is all the more astonishing as the glacier is so close to it and the cold Núpsvötn glacier river, rushing out of the glacier crevasses, floods the bottom of the said slope' (Ólafsson and Pálsson 1772, p. 777). The obvious interpretation of this passage is that Skeidarárjökull was sufficiently enlarged to have reached Lómagnúpur and so dammed up Núpslón. This most probably reflects climatic conditions and not only volcanic activity.

It may be concluded that the Vatnajökull glaciers advanced rapidly between 1690 and 1710 and were more extensive than at any time earlier in the historic period. For the next few decades their fronts were stationary or fluctuated within fairly narrow limits. Towards 1750 to 1760 marked advances or readvances took place. Thórarinsson (1943, p. 47) considered that most of them reached their maximum Little Ice Age positions at that time, though it seems difficult to prove on the evidence so far available that the mid-eighteenth-century advances were much greater than those around 1710.

2.3.2 THE MID-EIGHTEENTH- TO LATE NINETEENTH-CENTURY PERIOD OF ENLARGED GLACIERS

In the second half of the nineteenth century, the southern margins of Vatnajökull were visited by several foreign travellers and scientists. In 1857 the Swedish geologist Otto Torell visited the southeast coast, studying the glaciers and fluvio-glacial deposits. He was followed by C. W. Paijkull, whose account (1866) of his observations in 1865 aroused interest in jökulhlaups outside Iceland. Although there were no systematic measurements of any of the Icelandic glaciers, useful measurements at the margins of a number of Vatnajökull outlets were made by Helland in 1881.

The Skeidarárjökull lobe was bulging markedly in 1782 and the following year (Figure 2.18):

> I was told that before the last hlaup occurred in 1784, Skeidarárjökull was so high that the foremost point of Lómagnúpur could be seen only as a small cliff above the glacier from the highish mountain near Skaftafell sæter, since the glacier reached right up to the above-mentioned moraines lying below its front. Now, [i.e. ten years later] on the contrary, almost the whole of the upper half of the same [Lómagnúpur] can be seen above the glacier from the farm itself. (Pálsson 1882, p. 51)

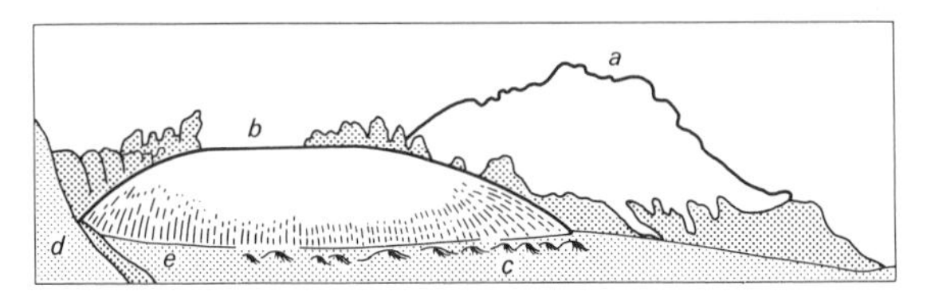

a. Öræfajökull

b. Skeidarárjökull has advanced between Færiness to the east and Súlutinda to the west

c. Skeidarársandur and some heaps of pebbles at the terminus of the glacier

d. The lower part of Lómagnúpur

e. Núpsvötn

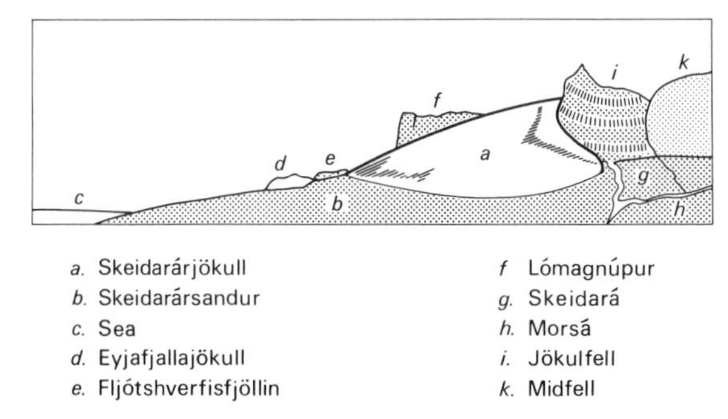

Figure 2.18 Skeidarárjökull and its environs in 1794, showing the convex shape of the tongue and the existence of the lake Núpsvötn or Núpslón at that time. (From Pálsson's (1945 edn) sketches)

The flood in 1783 probably caused thinning, which would account for the retreat of the ice observed by Pálsson when he was mapping Vatnajökull in 1793–4 (Figure 2.19). Pálsson (1882, p. 47) found that

> up at the glacier, large oblong bands of gravel follow the glacier margins at a distance of a couple of hundred fathoms almost all the way. These have obviously been tumbled or pushed up by an advancing glacier and indicate how far it extended before it began to recede.

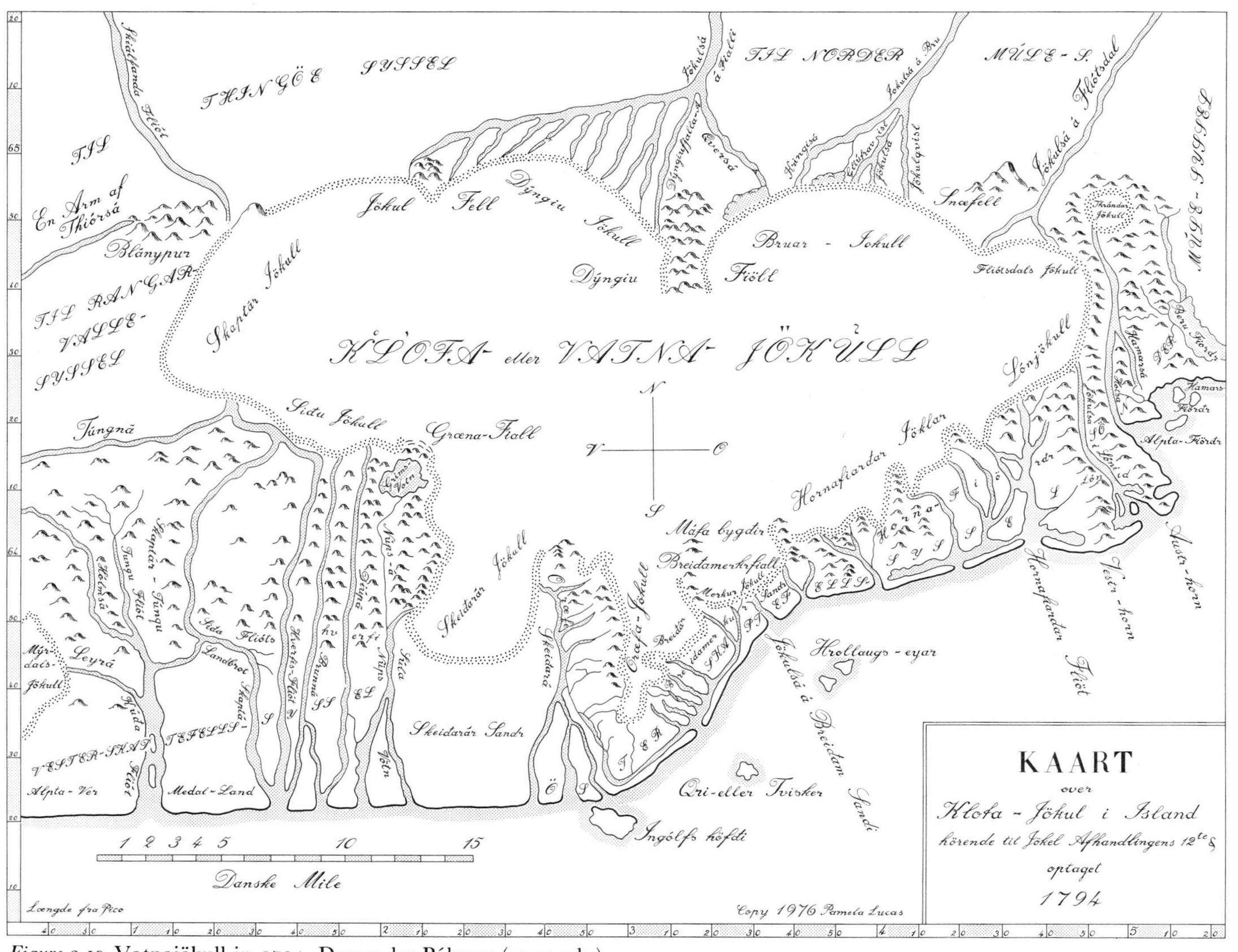

Figure 2.19 Vatnajökull in 1794. Drawn by Pálsson (1945 edn)

Although by 1794 Núpslón had disappeared, because of the recession, Sveinn Pálsson made the first specific mention of the lake: 'the southwest corner of Skeidarárjökull, from which the Súla comes, sometimes glides forwards till it collides with Lómagnúpur in front: Súla and Núpsá are then completely blocked and dammed up into a lake which finally breaks out, causing great floods' (1882, p. 53). He showed part of the lake in his diagram (Figure 2.19).

Thórarinsson (1939, p. 228) thought Núpslón probably did not come into existence before 1700, and judged from the indistinct shorelines and moraines formed by the mid-eighteenth-century advance that it did not last for very long. These moraines, which appear on the General Staff map of 1904, were in part swept away by later hlaups, and although some, such as the ridge known as Sangígar on the western sandur, remain, reconstructing past events is now difficult.

Henderson, travelling across Skeidarársandur in 1815, had to traverse several acres of dead ice 'the surface of which has the same appearance as the rest of the sand'. He described the ice as

> the remains of the projection which took place in 1787. . . . This region may be about three quarters of a mile from the present margin of the Yökul; and near the middle of the intervening space are a number of inferior heights which have been left on the regress of the Yökul in 1812, the last time it was observed to be in motion. (Henderson 1819, pp. 215–16)

So it would appear that there was an advance in 1787 before the retreat in the 1790s observed by Pálsson, and that the subsequent recession was followed or interrupted by a stationary period from 1812 to 1815.

The Revd P. M. Thorarensen's description of the parishes of Öræfi, dated 6 July 1839, indicates that by 1830 Skeidarárjökull was advancing again. Helland gives the distance between the southwest corner of the glacier and the mountainside west of Núpsvötn as 113 m in 1857, although it is not clear from where he got this information (Helland 1882b, p. 208). Advance was not long sustained after that, however, and the formation of the lake in the mid-nineteenth century was probably temporary, if, as is probable, it occurred. A description of Kálfafell parish, dated 12 July 1859 (cited Thórarinsson 1939, p. 228) by Jón Sigurdsson of Kálfafell, includes a statement that

> Many years ago this glacier advanced rapidly westwards towards the mountains on the inner side of the so-called Lambhagi cave, E. of and inside Lómagnúpur, but it gradually receded again. A few years ago it again advanced westwards as far as Núpsvötn, but has now almost receded, and since last year is still moving in the same direction.

The Háalda moraine,[9] rising 40 m above the sandur surface and running half a kilometre from (and within) the eighteenth-century moraine, was probably formed by the mid-nineteenth-century advance, which presumably ceased about 1850. Thórarinsson (1943, p. 35) reckoned that in 1850 the glacier front was about 800 m forward of the position shown in the 1904 map.

According to Helland, Skeidarárjökull retreated 450 m between 1857 and 1880, and 188 m in the year 1880/1. He estimated that the ice had thinned some 62 m along the line

[9] The Háalda moraine, shown on sheet 87, Öræfi SW, of the 1904 General Staff Map, was swept away by a hlaup in 1922.

between Sandfell in Öræfi and Súlutindar, so that there was now a clear view across the glacier in that direction, whilst in the summer of 1880 only the highest peaks of Súlutindar could be seen. The distance between the southwest corner of the glacier and the mountainside west of Núpsvötn was now 750 m, so the ice had retreated 637 m between 1857 and 1881; the edge of Skeidarárjökull was now 60 m above sea-level at the lowest.

Between 1892 and 1894, when Thoroddsen visited the area, the tongue thinned again. There was a major jökulhlaup in 1892, and the thinning of the tongue in 1893 may well have been due to this (Thoroddsen 1914b, p. 243).

It has now to be seen whether the general picture of mid-eighteenth-century advances and retreat at the end of the century, followed by mid-nineteenth-century advances to less forward positions again followed by retreat, was reflected by other glaciers in the region and so is to be considered as climatic in origin. The observations of Skeidarárjökull were made by particularly reliable observers, but it is essential that any ideas of the sequence of glacier fluctuations gained from its behaviour should be substantiated by ample evidence from elsewhere before being accepted as significant climatically.

Information about Svínafellsjökull and Skaftafellsjökull in the eighteenth and nineteenth centuries is much sparser than for Skeidarárjökull, but in general they appear to have behaved in the same way. Sveinn Pálsson found that

> East of Svínafell farm there are two large glaciers which are said to be always growing. Coming down two separate clefts, they unit, and descend to a fair distance out on to the flat land without the slightest melting in the summer . . . . The southern arm comes close to Svínafell farm . . . but the northern close to Skaftafell, the most northerly farm in Öræfa parish.' (1882, p. 41)

Thus these glaciers were still enlarging in the last decade of the eighteenth century and, according to Thórarinsson's estimate, probably about the same size as in 1904.

According to the parish descriptions of the 1830s by Thorarensen, these Öræfi outlet glaciers, like their larger neighbour, were advancing in around 1830 and by 1865 the ice was only five minutes' walk from Svínafell farm; the position was much the same in 1904 (Paijkull, cited Thórarinsson 1943, p. 33). In 1881 the margin of Svínafellsjökull, according to Helland, was 98 m above sea-level.

Unfortunately we have little or no eighteenth-century evidence, but the Öræfi outlets were evidently generally affected by advances in the 1830s. P. M. Thorarensen was probably referring in fact to Kótárjökull or Virkisjökull when, in 1839, he wrote that Sandfellsjökull was advancing to the WSW, though not as rapidly as Skeidarárjökull, and that the Öræfi glaciers in general were advancing strongly. Helland (1882b, p. 206) found a 137-foot-high moraine in front of Fallsjökull–Virkisjökull in 1881, the snout of which was 110 m above sea-level (Figure 2.15). The ice seems to have been thinning immediately before 1881 and stagnant from 1881 to 1884. By 1904 the ice front was about 600 m behind the moraine, but that ice was dead; the moving ice was 1000 m from the moraine. Helland evidently saw the first stages of real recession here. These glaciers were not affected by advances in the late nineteenth century.

When the Hrútárjökull (Fallsjökull) was investigated by Sveinn Pálsson in 1794, he recorded that it covered the entire Fjall estate and a few years earlier had joined Breidamerkurjökull. The two glaciers had been united earlier in the century (p. 30), so there must have been a recession sometime between 1709 and 1793/4. Hrútárjökull was still confluent with its neighbour in 1894, when Thoroddsen saw it and found it to be advancing, unlike the other Öræfi glaciers.

Eggert Ólafsson wrote of his visit to the Breidá area in 1756 that 'As we were riding back, the guide took us to the western edge of the glacier to the place where the church had been, but we saw no other traces of the farm but a couple of green cultivated meadows and a heap of stones pushed together' (Ólafsson and Pálsson 1772, pp. 784–8, 840–4). It is clear from his description that Breidamerkurjökull must have been advancing in 1756.

Sveinn Pálsson travelled along the front of Breidamerkurjökull in both 1793 and 1794. In 1793 he found that the

> Jökulsá divides it not externally but internally, into two parts. The part lying to the west of the river is steeper and has a lower, narrower and more uneven margin. It is more crevassed and is covered with a quantity of grit and has some considerable mounds of grit in front of it. The part to the east, on the other hand, is quite compact and without crevasses. It has a steep margin, 16 to 20 fathoms in height, no grit on its surfaces except such loose sand as has been deposited on it by strong winds, and this gives this part a greyish-white colour. There are no, or very few, grit mounds in front of it which shows that the western part has been and still is in motion, though no withdrawal has so far been observed, as with Skeidarárjökull, but rather a gentle advance. The eastern part, however, so far, shows no sign of movement. . . . Such was this glacier in the summer of 1793 when I visited it for the first time, but next summer, or in 1794, things had changed. The part to the east of the river was clearly seen to have advanced over 200 fathoms from its position of the previous year. The smooth, unbroken margin described by Ólafsson and Pálsson 40 years ago was now quite unrecognisable. Here were large crevasses and sharp-pointed pyramids; there it was excavated like some filigree work, great pieces of ice thrust out, and where the margin still seemed more or less smooth and unchanged, it had greatly increased in height and had bulged out in its centre like a wall of greensward at the point of bursting, or slit down its length on account of the water which had accumulated behind it. There was too a constant rumbling from the entrails of the glacier. Here and there were small streams jetted from the fissures, accompanied by an exceedingly nasty, damp mist from cavities in the glacier, and a very penetrating cold. The eastern part of the glacier at Vedurá had advanced furthest . . . but now appears to have started to withdraw, as mounds of grit were clearly to be seen in front of its margin. The advance is said to have begun suddenly, so to speak, and to a large extent without any appreciable discharge of water, at Whitsuntide in the said year 1794. Since that time people in the neighbourhood of Hornafjördur have complained of the continual mist, cold and sleet which is said to have come from this uneasy glacier. (Pálsson 1882, pp. 33–5)[10]

It was very fortunate that Pálsson visited Breidamerkurjökull at exactly the right time to observe this late-eighteenth-century advance, which evidently took place later here than in other Vatnajökull outlets. From his description it must be suspected that when he saw it in 1793, the western part of the lobe had already begun to retreat from its maximum, although it was still more advanced than the eastern part which was to advance so decisively the following year.

[10] Henderson's (1819, pp. 198–9) translation of Pálsson is somewhat inaccurate.

The 1794 withdrawal of the eastern part of Breidamerkurjökull was only temporary. Henderson travelled across the sandur in 1815 and reported renewed evidence of advance:

> Of its progress towards the sea, I was furnished with the amplest proof on passing along the margin. About the distance of a quarter of a mile from the south-east corner of the Yökul, I was surprised to find it traversing the track made in the sand by those who had travelled this way the preceding year; and, before reaching the point, I again discovered a track, which had been made only eight days previous to my arrival, lost and swallowed up in the ice. The same fact is confirmed by a comparison of the present length of the river, with what it was about fifty years ago. Olafsen and Povelsen, describing it as the shortest river in Iceland, state it to have been scarcely a Danish mile, or about five British miles, from its egress to its junction with the sea at the time they passed it; whereas it does not now appear to exceed a British mile in length. (Henderson 1819, p. 196)

Henderson had some odd ideas on the origin of Breidamerkurjökull which are without scientific merit, but his statement that 'it is only in summer it advances, after a strong thaw on the snow-mountains; at which time, also, the river which it discharges, is poured forth, now at one place, and now at another' (1819, p. 197) rings true. He expected the advance to continue:

> if this field of ice be not entirely carried away by some awful convulsion in the mountains behind it, the progress it is making will soon bring it to the sea; and, in the course of a few years, all communication between the southern and eastern districts by this route will be cut off.

This very rapid advance was still continuing in 1820, when, according to Thienemann (cited Bárdarson 1934, p. 25), central Breidamerkurjökull advanced about 1000 m and its movement was so fast that on some days it advanced 4 to 8 m. It is not completely clear whether this was a surge or perhaps the result of a hlaup. Thórarinsson (1969) thought it was probably a surge. When Thienemann crossed the sandur in 1821 the glacier was retreating, leaving frontal moraines 10 to 13 m high, and the distance from the margin to the sea was about ¼ Meile, or 2 km (Thienemann 1824, pp. 311–13, cited Bárdarson 1934). From all this it is clear that the Breidamerkurjökull fluctuations in the 1790s and early nineteenth century were dominated by rapid advances, interspersed with recessions which were probably slower, and it is possible that at least one surge occurred.

Gunnlaugsson visited Öræfi in 1835 and surveyed Austur-Skaftafellssýsla in 1839. On his map he showed Breidamerkurjökull 2200 m from the sea at Jökulsá, and the distance rather less to the east of the river. In the west the margin is shown 800 m in front of Breidamerkurfjall. Thus the ice was probably further from the sea at Jökulsá in 1835 than it had been in 1794.[11]

In 1836 a Frenchman, Gaimard, travelling in Iceland, reported that the shortest distance from Breidamerkurjökull to the sea was about 400 m (Gaimard 1838, II, p. 237). When Thorarensen wrote his description of Hof and Sandfell parish in 1839, the glacier was enlarged, for he wrote, 'nobody crosses these glaciers except for driving sheep to Hafrafell and Breidamerkurfjall, both of which are surrounded by glaciers'.

[11] Gunlaugsson probably had access to the data of Scheele and Frisak, who had surveyed there in 1813, and that of Aschlund from 1817 (Thórarinsson 1943, p. 27).

Alternate advance and retreat near the maximum extension seem to have continued for the rest of the nineteenth century. Paijkull (1866, pp. 122–5, quoted Thórarinsson 1943) saw the tongue in 1865 after the glacier had advanced in 1861, forming a moraine, and then receding slightly once more. In 1869 the margin expanded rapidly so that in June and July it was encroaching on the beach and the moraines on its eastern edge. Great floods which devastated the farm, Fjall, were associated with the advance (Figure 2.16). Kålund found the ice barely 1000 ells, or 600 metres, from the sea (Kålund 1872–82, II, p. 278).

According to Helland (1882b, pp. 208–25), Breidamerkurjökull retreated after 1875, and was still retreating in 1881, by which time the ice edge was about 20 m above sea-level at Jökulsá and about 11 to 12 m above sea-level further east. Thoroddsen, a decade later, found the centre of the lobe 256 m from the sea, although the ice was 1000 m from the sea at Jökulsá. Breidamerkurjökull was then at, or very close to, its greatest extension during the 1890s, and although its margins had fluctuated a number of times during the preceding century, there is no evidence of really substantial retreat between the rather delayed advances of the 1790s and those of the 1890s. Despite the unusually rapid advances of this glacier, especially around 1820, and the possibility that not all the intermediate fluctuations between 1790 and 1890 have been identified, the main outlines of the oscillations of Breidamerkurjökull may be identified with reasonable certainty.[12]

Most of the rest of the information about Vatnajökull's eastern outlets is concerned with the fluctuations of Heinabergsjökull, Fláajökull and Hoffellsjökull. Heinabergsjökull (see Figure 2.17) was apparently enlarged throughout the period between the mid-eighteenth century and the late 1880s, when it reached its maximum extension in historic times (Thórarinsson 1943, p. 23). The lakes dammed up by this glacier in both Vatnsdalur and Heinabergsdalur have been notorious for their hlaups. Thórarinsson reckoned that settlement in the western part of Mýrar (Figure 2.15) would not have been possible if Heinabergsjökull had been large enough to cause hlaups. He could find no evidence of interruption of settlement in Mýrar before 1700; but the glacier evidently became sufficiently enlarged to dam up lakes sometime between 1708 and 1783. It is probably significant that a list of abandoned farms in Austur-Skaftafellssýsla made by Jón Helgason of Hoffell in 1783 lists three farms in Mýrar which had been occupied in 1708, according to Ísleifur Einarsson's Land Register (Thórarinsson 1939, pp. 223–4).

In 1839 both Fláajökull and Heinabergsjökull were advancing and had been doing so for some years. The Revd Jón Bjarnason wrote in his description of the parish of Einholt, dated 24 December 1839 (quoted Thórarinsson 1943, p. 21):

> that these glaciers are growing and gradually sliding out over the low-lying land is evident, great is their unrest, and loud crashes can be heard far away, when the lakes are drained off underneath them, rushing forth with terrible speed and doing great damage to the soil on the plains, meadows and grazing grounds. I presume they are impassable almost everywhere, especially near the settlements . . . when the rivers are dammed, which often happens, especially in Heinabergsdalur, the water collects in a kind of lake.

Gaimard, travelling through Skaftafellssýsla three years earlier, had described the two

[12] Many of the figures quoted for the distance between Breidamerkurjökull and the sea at various dates are merely estimates. Moreover this sandur coastline is not static and so statements of distance of the ice front from the water must be treated with caution.

glaciers and their junctions: 'Nous côtoyâmes encore, pendant deux à trois heures, deux grands glaciers qui, après être descendus des Jökulls Skálafels et Heinabergs, se réunissent en un seul. Semblables aux autres par leurs aiguilles, ils offrent cependant cela de remarquable, que leur réunion où leur point de contact est indiquée par une ligne ou crête de moraine à la surface de la glace' (Gaimard 1838, II, p. 240).

Fláajökull almost reached Heinabergsjökull in 1857 when Torell visited the area. According to local tradition the two glaciers were nearly united in the 1860s. This fits in with Paijkull's description (1866, p. 63). During his visit in 1865 he was told of a place 'where the jökulhlaup is due to two glaciers coming from different valleys blocking up a third valley with no glacier but running water. The water collecting there will eventually break through the glacier and the jökulhlaup will begin.' The lake mentioned in the 'third valley' was probably Dalvatn. Thórarinsson (1939, pp. 217–24) collected oral evidence about the lakes from the local farmers and accepted their account that in the 1860s and 1870s Heinabergsjökull was large and thick and its front close to the outwash moraines on the sandur. The southeast corner of the tongue nearly met Fláajökull. Vatnsdalur was full of water up to the col at 464 m, and it had been full as long as the oldest inhabitants could remember. Its outlet was over the col to Heinabergsdalur, which in its turn was dammed by Heinabergsjökull, forming the Dalvatn lake. This emptied almost every year, usually in the early summer, the flood rarely lasting more than one day.

Heinabergsjökull was nearly stationary in the 1870s and 1880s, tending to advance rather than recede, and attained its maximum late in the 1880s. Thórarinsson found that the moraines from which the ice began to recede in the 1880s were everywhere the outermost on the sandur. The ice finally began to recede in 1887. As the tongue thinned at the end of the nineteenth century, so Dalvatn began to shrink. By 1897 the Vatnsdalur lake was too low to drain into Heinabergsdalur. By 1889 Heinabergsjökull was thin enough for the lake to drain out under it. Between 1898 and 1938, when Thórarinsson visited the area, the Vatnsdalur hlaup was an annual event, occurring earlier in the summer as the glacier tongue thinned.

We have no information about pre-nineteenth-century Fláajökull, but it is known to have oscillated between 1882 and 1894, advancing and retreating three times in those twelve years. Thoroddsen (1914b, p. 221) found it extended to its moraines in 1894, but it is uncertain which moraines he meant. It is clear that advance until well into the 1890s, with a strong tendency to fluctuate near the position of maximum extension, was not a peculiarity of the Breidamerkurjökull lobe.

There is little data to be had about Hoffellsjökull (Figure 2.15) before the eighteenth century. This glacier is divided into two by a hill, and so is double-tongued. In the description of Austur-Skaftafellssýsla by Sigurdur Stefánsson of Holt in Hornafjördur, dated July 1746 (quoted Thórarinsson 1943, p. 19), he wrote that 'then follows the small hill, Svínafell, which incidentally was formerly called Gölltur [The Boar]. A glacier comes down to the plain on both sides of Svínafell, which is a low but precipitous hill with some grass, ling, and forest vegetation.' Thórarinsson concluded that Hoffellsjökull was probably about the same size in 1746, or slightly larger, than it was in 1903 when it was first mapped. There are no moraines that might suggest that it was larger in the eighteenth century than it was round about 1890.

A parish account by the Revd Th. Erlendsson (quoted Thórarinsson 1943, p. 20) mentions that hlaups from ice-dammed lakes on the east side of Hoffelsjökull–Svínafells-jökull made the river impassable when they drained all the way down the main road from

Bjarnanes, where Erlendsson lived, to Holt. This indication of thick ice is supported by the statement that 'it comes close to the north side of Svínafell and even extends up its slope'. Erlendsson recorded that the outlet glaciers from the eastern part of Vatnajökull 'go slightly forward and back, but more forwards'. This further substantiates the impression that the glaciers of Vatnajökull were tending to advance in the decades before 1840 and were thick in the mid-nineteenth century. Svínafellsjökull was still damming up lakes and causing hlaups in 1873, when one of its branches reached the plains east of Svínafell (parish account by Bergur Jónsson, 1873, cited Thórarinsson 1943, p. 20) but was said to have 'looked plane', so it may have been thinning or receding a little then. According to local residents, it advanced slightly in the 1880s and reached its maximum extension in 1890 or soon after that date.

Summing up the evidence from the mid-eighteenth- to late-nineteenth-century period, it is fair to say the information available from the outlets of southern and eastern Vatnajökull gives a consistent picture. All the glaciers were in an enlarged state in the 1750s and were subject to minor fluctuations throughout the next 150 years, although they remained generally enlarged. The mid-eighteenth-century advances of Breidamerkurjökull were perhaps a little delayed, while Skeidarárjökull was already retreating in the latter part of the nineteenth century, having reached its maximum extent in the mid-eighteenth and not in the late nineteenth century like many of the other tongues. This consistency of the pattern which emerges is pleasing and a little surprising in view of the possible complications.

2.3.3 THE TWENTIETH-CENTURY RETREAT

Reliable mapping of the Icelandic glaciers began in 1902, and the Danish General Staff map of 1904 provides a very useful basis for comparisons of the extensions of Vatnajökull ice at different times. Regular measurement of the frontal positions of glaciers in Iceland was initiated by Eythórsson[13] in 1932 and continued under his guidance till 1967, when responsibility passed to Sigurjón Rist. The important work of the Swedish–Icelandic expeditions to Vatnajökull of 1936–8, reported in *Geografiska Annaler*, was the forerunner of much scientific investigation of this icecap which has taken place since, and been reported in large part in *Jökull*. Air photography, which began in 1937/8, made possible the extension of mapping to the whole of Vatnajökull, and in the 1970s satellite images added the possibility of monitoring the ice front at regular intervals.

In general, Vatnajökull outlet glaciers melted back rapidly in the 1890s. Skeidarárjökull's margin rose about 15 m in altitude between Helland's measurements of 1881 and the 1904 map. Of the Öræfi glaciers, the Virkisjökull front rose 48–58 m between 1881 and 1904, involving a horizontal recession of at least 600 m, and Svínafellsjökull's margin rose 22–30 m. Breidamerkurjökull receded 250 m east of Jökulsá, about 500 m at Jökulsá and

[13] Jón Eythórsson (1895–1968) first studied natural history in Copenhagen and then meteorology in Oslo in the Bjerknes school. He worked with Ahlmann in Jotunheimen, helping to set up the first high-altitude meteorological station in the northlands. Returning to Iceland he became a forecaster with the meteorological office, an association which lasted practically all his life. He was not only a full-time meteorologist but also wrote and translated a large number of books, and was responsible for the publication of such important works as Pálsson's treatise (1945). He made his greatest impact as a glaciologist. He initiated systematic observations of the margins of the major Icelandic glaciers in 1932 and, with the help of local volunteers, carried these on for nearly forty years. He published regular reports of the drift ice situation from 1953 to 1966, founded the Icelandic Glaciological Society, and was editor of its journal *Jökull*.

about 300 m in the southwestern part of the lobe. Hrútárjökull was exceptional in that it appears in much the same position on the 1904 map as it was when Thoroddsen saw it in 1894 and found it advancing. It is possible that a late-nineteenth-century advance here, followed by retreat, may have brought the margin back to roughly the 1894 position in the early twentieth century. In about the period 1890–1903 Svínafellsjökull retreated 420 m and Hoffellsjökull also withdrew. Fláajökull retreated 300 m between its maximum positions in 1894 and 1903, and Heinabergsjökull 420 m between 1897 and 1903. (Thórarinsson 1943). Figure 2.20 shows the course of recession in the present century. Clearly the last eighty years have seen catastrophic recession and thinning of Vatnajökull, and this has brought the fronts of its outlet glaciers back towards positions comparable with those they occupied at the time of the Norse settlement.

Twentieth-century recession is best documented for Breidamerkurjökull, though this glacier may not be entirely representative because it spreads out into a broad terminal lobe. Maps on a scale of 1:100,000 by the Danish Geodetic Survey are available for 1904 and on a scale of 1:50,000 by the US Army Map Service for 1945/6. Durham University Exploration Society made a plane-table map of the tongue on a scale of 1:25,000 in 1951, the Department of Geography, Glasgow University, produced a 1:30,000 map in 1965 from 1:15,000 air photographs, Iceland's Survey Department, Landmælingar Íslands, photographed the area in 1980/1. Using the maps and supplementing the information they provided with aneroid barometer measurements of the heights of trimlines marking former glacier margins, Sigbjarnarson (1970) made estimates of the changes in Breidamerkurjökull's surface area and volume from 1894 to 1968 as shown in Table 2.3. Price (1982) re-examined the evidence and, making use of the latest air photograph cover, found rates of retreat somewhat faster than those given by Sigbjarnarson persisting to 1980. Precision in these matters where the glacier front is lengthy and uneven cannot be expected, but detailed and frequent surveys at west Breidá by groups from the University of East Anglia tend to support Price's figures.

Breidamerkurjökull constitutes about 14 per cent of the total area of Vatnajökull (Thórarinsson 1958) and seems to have diminished in volume by about 49 km^3 between 1894 and 1968. On this basis Sigbjarnarson calculated that the shrinkage of Vatnajökull has been somewhere between 268 and 350 km^3, that is some 8 to 10 per cent of its entire mass. Should this continue it would take only 600 years for Vatnajökull to waste away entirely. Whether melting at a similar rate over a period as long as six centuries would

Table 2.3 The recession of Breidamerkurjökull since the late nineteenth century according to (a) Sigbjarnarson (1970) and (b) Price (1982)

Period	*Frontal retreat (m)*	*m/yr*	*Loss in area (km^2)*	*km^2/yr*	*Loss in volume (km^3)*	*10^6 m^3/yr*
(a)						
1894–1904	120	12	3.0	0.3	2.8	280
1904–1945	1130	28	25.5	0.6	21.2	520
1945–1968	1046	45	25.3	1.1	25.0	1119
1894–1968	2296	31	53.8	0.7	49.0	660
(b)						
1903–1948		30–40				
1945–1965		53–62				
1965–1980		48–70 (greater on west than east side)				

ever be likely to occur is, of course, quite another matter. Even the precipitate retreat of Breidamerkurjökull since detailed measurements began in the 1930s has not been unbroken, though the tendency for standstill or even slight advance, which has occurred in the period since 1940, has been less marked than with some other Vatnajökull tongues. The slight advances which occurred (see Figure 2.20) in the 1960s were not by any manner of means restricted to Vatnajökull, and are therefore of all the greater significance (Chapter 11).

2.4 Drangajökull

Drangajökull is a relatively small icecap in the far northwest of Iceland. While it has attracted rather little attention compared with Vatnajökull, the behaviour of its outlet glaciers is of particular interest because no volcanic eruptions have occurred either under it or near it in historic times. Eythórsson (1935) took the view that it was particularly worthy of investigation because 'its variations must therefore chiefly be due to climatic changes'. This remains, in general, probably true, though Thórarinsson (1969) has pointed to at least two advances in the 1930s which are difficult to explain climatologically.

Drangajökull (see Figure 2.21) has an area of about 166 km^2, the long axis runs northwest to southeast and its outlet glaciers radiate towards the sea, which is only a short distance away, except to the south. They enter valleys which were settled, until recently, Kaldalón to the southwest, Leirufjördur to the northwest and Reykjafjördur and Tharalátursfjördur to the northeast. Data about these valleys were assembled by Bárdarson (1934, pp. 51–8). Eythórsson made some detailed investigations in 1931, fixing markers and instituting regular measurements of the main outlet glaciers, which have since continued.

Drangajökull was visited by many of the earlier Icelandic glaciologists. Ólafsson and Pálsson travelled in Vestfirdir in 1753, 1754 and 1757, and Olavius in 1775. Sveinn Pálsson unfortunately never went there and depended upon information from Ólafsson and Pálsson; it was probably reliable because Eggert Ólafsson, like Olavius, was a Vestfirdir man. Gunnlaugsson published a map showing Drangajökull in 1844, but he was obliged to base that part of it dealing with the interior, including Drangajökull, on some earlier nineteenth-century geodetic observations made by others, on older maps and on local information. He did not himself visit Drangajökull and so his map cannot be used for comparative purposes or to elucidate changes in the area of the icecap (Thórarinsson 1943, p. 13). Thoroddsen investigated Drangajökull in the summer of 1886 and 1887, and it was mapped by the Danish General Staff in 1913/14.

2.4.1 THE MAIN SEVENTEENTH- AND EIGHTEENTH-CENTURY ADVANCE PHASE

The narrow valleys of Kaldalón, Leirufjördur, Reykjafjördur and Tharalátursfjördur all contain some lowland which was farmed for centuries, and in each case there is evidence of early Little Ice Age devastation of cultivated farms and farmhouses. The outermost moraines in the Kaldalón valley are certainly pre-Little Ice Age (Eythórsson 1935, John and Sugden 1962). Two farms, Lónhóll and Trimbilsstadir, inside these moraines, were ruined before 1710 according to Árni Magnússon. John and Sugden recorded that 'local people believe that Trimbilsstadir was destroyed by the glacier around 1600'. This would

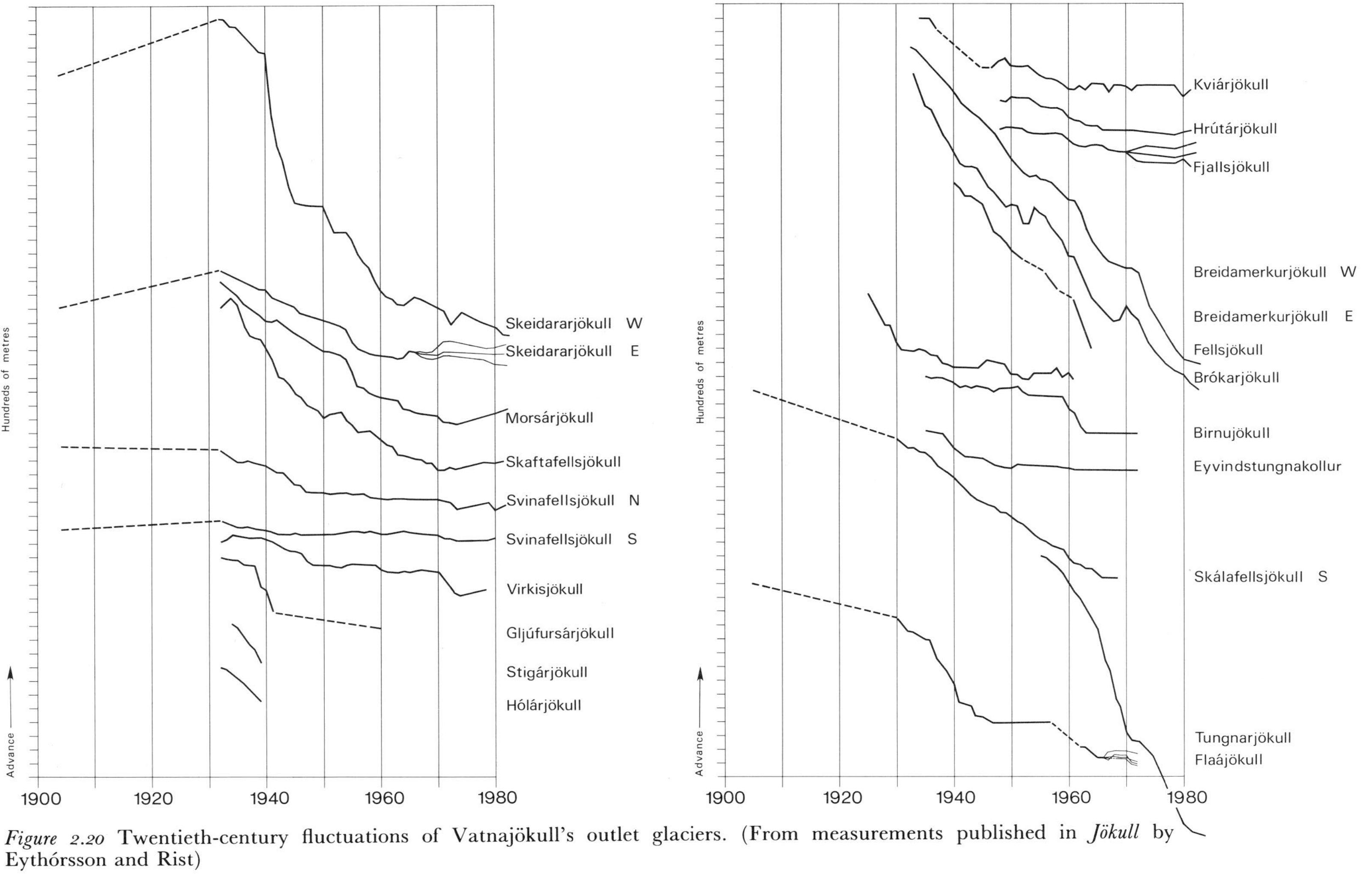

Figure 2.20 Twentieth-century fluctuations of Vatnajökull's outlet glaciers. (From measurements published in *Jökull* by Eythórsson and Rist)

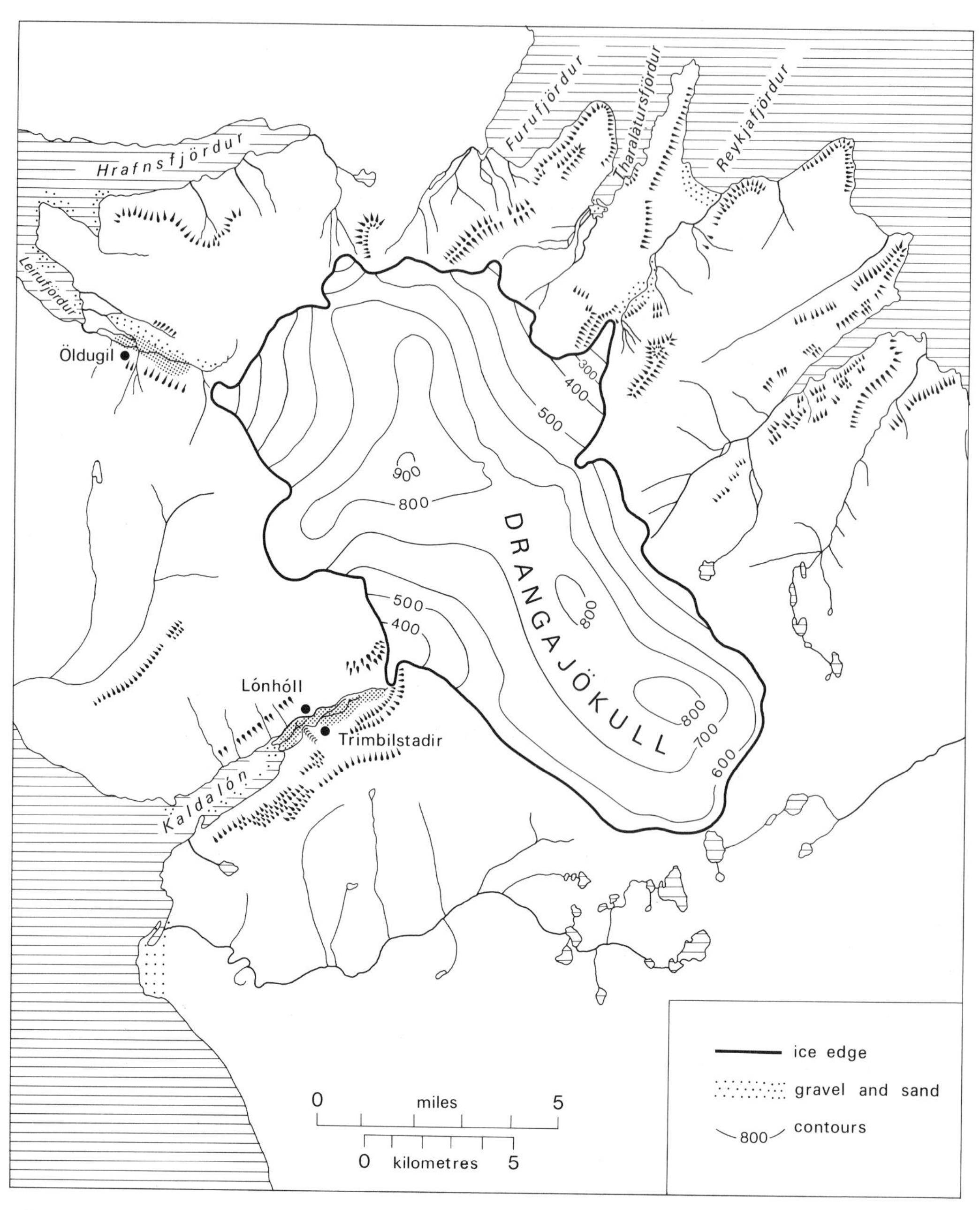

Figure 2.21 Drangajökull. (From Eythórsson 1931)

suggest that Drangajökull's expansion was much swifter than that of Vatnajökull, but it is much too vague to be accepted as real evidence. John and Sugden (1962, p. 352) query whether the farms were actually destroyed, as 'even if they were inhabited in 1700–56, a glacier snout less than half a mile away and inevitable flooding of the valley floor periodically, would have rendered all agriculture impossible and would surely have forced the inhabitants to move'. They may indeed have moved early in the Little Ice Age, for although Olavius records that Lónhóll was devastated by a great 'glacier-burst' in 1741, it may have been untenanted at the time.

Ólafsson and Pálsson visited the valley in 1754 and 'all the local people reported that the ice now covers areas where there was green grassy ground twenty years ago. . . . The inhabitants also assured me that the ice had retreated from time to time' (1772, p. 516). So the front must have been more advanced in 1754 than in the 1730s, and probably more than in 1710, although expansion was not continuous in the first half of the eighteenth century.

In Leirufjördur there are said to be two main terminal moraines, with remains of smaller ridges on the flat below them (Eythórsson 1935, p. 54).[14] The outermost of these crosses the valley opposite Öldugil, where a small tributary stream enters the valley from the south. There was once a farm near this stream, also called Öldugil. In Magnússon and Vídalín's *Jardabók* of 1710, Öldugil is said to have been deserted between 1400 and 1500, 'and glacier-bursts and floods destroyed this dwelling according to common opinion, so that what is to be seen of the ruins is now [1710] situated quite close to the margin of the glacier; according to statements of people still alive, the glacier has overwhelmed the whole area of the former farm' (Magnússon, cited Eythórsson 1935, p. 128). Olavius mentioned several other deserted farms near Öldugil, of which, one, Svidningsstadir, was specifically said to have been destroyed by the glacier (Olavius, cited Eythórsson 1935, p. 129).

It is uncertain from all this whether Öldugil was ruined by floods or by ice, though it is more likely that it was flooded. Very substantial advance of the ice had certainly taken place by 1710.

Deserted farm sites are also to be found in the Reykjafjördur valley, where again a series of old frontal moraines cross the floor. The outermost, not more than 300 m long, on the north side of the river, was found by Eythórsson just below the narrowing of the valley at Stórahorn. This moraine formed the limits of the meadowland when Eythórsson saw it; upstream were only gravel flats with scanty vegetation. There was one farm, Kirkjuból, in the grassy valley below this moraine. It was said to have been deserted finally about 1740, although recorded as untenanted in 1525. (Eythórsson 1935, p. 130). Árni Magnússon describes an old untenanted farm, Knittilstadir, close to the glacier, but Eythórsson (1935, p. 130) was informed locally that Knittilstadir was in the outer part of the valley. Olavius mentions two other farms, Nedra-Horn and Fremra-Horn, leaseholds of Kirkjuból, which were devastated in the seventeenth century or earlier. At Fremra-Horn the damage was said definitely to have been caused by the glacier (Olavius 1780, p. 63).

Clearly, the rather wide valley of Reykjafjördur was once well occupied and farmed despite its remoteness. The desertions which took place in the sixteenth century may not have had anything to do with deterioration of climate or glacial advance; there is no evidence either way. But farms in the upper part of this valley, and specifically Fremra-Horn, were apparently destroyed by glacial advances and perhaps associated floods in the seventeenth century or possibly earlier.

The glacier in Tharalátursfjördur is described by Árni Magnússon as having destroyed the inhabited land in the valley before living memory (Magnússon and Vídalín 1710)

[14] According to Bárdarson there are four moraines here, according to Thoroddsen three. Presumably the less developed moraines mentioned by Eythórsson account for the discrepancy.

while Olavius (1964–5, p. 63) says that Drangajökull reached to the head of the fjord. Evidently the glacier was in a very advanced position in the mid-eighteenth century, but there is no evidence as to when the advance began. Eythórsson estimated that the front was at the order of 2000 m in advance of its 1913 position in 1770.

The situation in the mid-eighteenth century was nicely summed up by Ólafsson:

> In view of the size of this ice mountain which is twelve [Danish] miles long by six miles broad, and of its situation so close to the settlements and the sea on all sides, it is not astonishing for it to cause snow and fog, wind, cold and unsettled weather. The waxing and waning of the glacier are also remarkable. All the residents agree in saying that the glacier covers ground which was green and fertile twenty years ago. The constant winds, which for some years at a time may be either easterly or northeasterly from the ice mountain, or westerly or southwesterly from the sea, are presumably the prime cause of this. (Ólafsson and Pálsson 1772, pp. 516–17)

There is no doubt that the Drangajökull outlets had advanced over previously farmed land by the end of the seventeenth century, and by the mid-eighteenth century were more enlarged than they had been since the surrounding valleys were settled.

2.4.2 THE MID-EIGHTEENTH- AND LATE NINETEENTH-CENTURY CONDITION OF DRANGAJÖKULL

Unfortunately there is very little information about Drangajökull between the visit of Olavius in 1775 and Thoroddsen's in 1887. Thoroddsen thought that the Kaldalón ice had retreated 500 or 600 m in total between the mid-eighteenth century and 1887 (Thoroddsen 1905–6, p. 175). He wrote that

> From the head of the fjord to the glacier there is a distance of about half a Danish mile; the ground is perfectly smooth except for three moraines which stretch in half circles between the hillsides. The moraine nearest to the sea is grass-grown and was formerly covered with brushwood; the second moraine is bare of vegetation and such is naturally the case with the innermost moraine. Twenty to thirty years ago the glacier extended as far as the innermost moraine, but it has since retreated so that now there are 2–3,000 fathoms between the moraine and the glacier tongue. (Thoroddsen 1911, p. 19, cited Eythórsson 1935, p. 125)

The glacier front was swollen and 130–70 m high when Thoroddsen saw it, so it was probably advancing again at that time.

Local people told Thoroddsen that the Leirufjördur glacier formed the outermost moraine in the valley in the decade 1837–47, that is, the ice reached nearly as far as it had in 1710 (Thoroddsen 1905–6, p. 36). He cut a cross in a basalt block, then 90 m downstream from the front, in the hope of providing the opportunity to measure future oscillations.

When he visited Reykjafjördur in 1886 Thoroddsen was told that the glacier there had advanced very noticeably in about 1840, and had reached its greatest extension in 1846, pushing a moraine ahead of it (Thoroddsen 1914, II, p. 74). It probably did not begin to retreat till about 1850, but by 1886 had retreated about 1500 m from its maximum. Thoroddsen collected rather less information about the Tharalátursfjördur glacier, but he did note that it retreated several hundred fathoms between 1860 and 1880, and that its horseshoe-shaped moraines might date from 1840 or 1860–70.

The evidence available is certainly insufficient to permit identification of all the fluctuations occurring between the mid-eighteenth and late nineteenth centuries. It seems likely, however, that the marked advance which occurred around 1840 may have been preceded by a slight recession, while there is no doubt that the mid-nineteenth century advance was succeeded by a retreat which was both rapid and substantial.

2.4.3 THE GREAT RECESSION OF THE LATER NINETEENTH AND TWENTIETH CENTURIES

When Eythórsson worked on the Drangajökull in 1931 he took the opportunity to compare the positions and characteristics of the outlet glaciers in that year with those recorded by Thoroddsen, at the same time as initiating measurement of frontal oscillations.

By 1931 the Kaldalón glacier had retreated about 340 m since 1887 and its thickness had diminished even more drastically. Eythórsson found the tongue thin and not very steep, whilst Thoroddsen had described it as precipitous. The Leirufjördur and Reykjafjördur glaciers had retreated much more in the same period. The Leirufjördur tongue had become 'short and insignificant' and it hardly extended more than 200 m beyond the main margin of Drangajökull when Eythórsson saw it. He reckoned that the front had retired about 3000 m between 1837–47 and 1931, with the most rapid retreat of about 85 m per year during the period 1898–1913. Retreat was accelerated at this time, as several hundred metres of the lower glacier became isolated. The Reykjafjardarjökull reached almost to the valley floor in 1931. Eythórsson judged from the 1913/14 map that this tongue must have retreated about 100 m between 1886 and 1913. Moreover, Thoroddsen gave the distance from the 1840 moraines to the 1886 position as 1500 m, but the distance between these positions on the 1914 map is also shown as 1500 m. Thoroddsen's figure in this case was probably guesswork.

The Tharalátursfjördur glacier was thinning rapidly in 1931 but, judging from the map, had retreated only a little since 1913/14. Its tongue still reached to 201 m above sea-level.

Thoroddsen and Eythórsson thus witnessed two stages in the main retreat from the Little Ice Age extensions of the Drangajökull outlets. This retreat has continued since 1931, although not without interruptions; the results of the measurements made since then are shown on Figure 2.22. No evidence has been forthcoming from Drangajökull of any substantial advances in the late nineteenth century comparable with those of Vatnajökull glaciers, but between 1939 and 1942 the Leirufjördur ice suddenly advanced 1 km of which 540 m took place in the winter of 1938/9, while Reykjafjardarjökull advanced 750 m during 1934–6. It is difficult to explain these advances on meteorological grounds, or except as possible surges. Finally it has to be noticed that, after slackening their retreat in the decade around 1970, the glaciers here, like those in Vatnajökull, resumed their retreat.

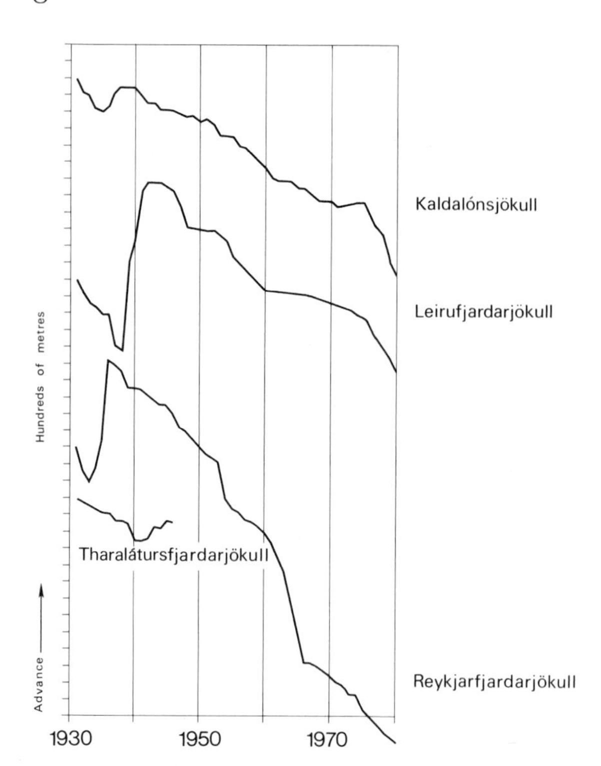

Figure 2.22 Fluctuations of Drangajökull's tongues, 1932–80, showing advances in the 1930s and 1940s. (From measurements published in *Jökull* by Eythórsson and Rist)

2.5 Variations in glaciers and sea ice

Although data on sea ice and glacial conditions in Iceland in early medieval times is very fragmentary, one gains a general impression from it that in settlement times, from AD 870 to 930, and saga times, from AD 930 to 1030, the climate was milder than it has been since. This impression is strengthened when one turns one's attention to Greenland. Farming settlements established there when the settlement of Iceland was proceeding, throve for a time and acted as bases for the Vinland voyages (Gad 1970) (see Chapter 8.7). In Iceland the population expanded for a century or two and a new society emerged with its own culture and institutions, notably the Althing, the 'grandmother of parliaments'. Every line of evidence points to the environmental circumstances in which the Norse settlement of Iceland took place and prospered as having been favourable. The sea ice lay far from sight, glaciers were withdrawn, grass generally grew well and sea voyages in open boats in the North Atlantic and adjacent Arctic seas were not too difficult.

Evidence of climatic deterioration about the thirteenth century is forthcoming from a number of different sources. The incidence of sea ice was increasing. By the end of the twelfth century cereals were no longer being cultivated in northern and eastern Iceland. Thórarinsson (1944; 1956, p. 16) quotes from a document of 1350 to the effect that 'grain, but only barley, is grown in a few places in the south country' and showed, using both documentary evidence and pollen analysis, that grain-growing persisted in a few places along the southern shore of Faxaflói until the end of the sixteenth century. Meanwhile, Greenland lost contact with the other Norse lands. There is evidence that the ground was frozen more severely, bodies buried in the graveyard at Herjolfsnes between 1350 and 1500 being better preserved than those buried earlier (Gad 1970, p. 154). The Norse settlements in Greenland were finally extinguished in mysterious circumstances towards the end of the fifteenth century (see Chapter 12).

Confirmation of the authenticity of the thirteenth-century climatic deterioration and some indication of its cause are provided by oxygen isotope analysis of the upper parts of a deep ice core from the Greenland ice sheet (Dansgaard *et al.* 1969). Oxygen isotope analysis of such cores provides a record of mean annual surface air temperature (Robin 1976, Bradley 1985, Chapter 5, p. 159) and in the core from Camp Century a strong cooling around 1200 followed by a further cooling around 1300 is indicated.

Very little information is forthcoming about glaciers or sea ice in the fourteenth and fifteenth centuries. It seems that harsh conditions were experienced in the late sixteenth century. After a remarkably mild interval from 1640 to 1660 the cold greatly intensified and reached a maximum in the last decades of the century. The glaciers Vatnajökull and Drangajökull were already much enlarged and they readvanced about 1750 to 1760 at a time when the sea ice index was also high. A fluctuating retreat of the ice was followed by advances about 1840 to 1850, which happens to have been the only decade in the nineteenth century when the sea ice index was low. At the end of the century, when sea ice was more in evidence than ever before on record, some of the Vatnajökull glaciers reached their historic maxima whilst those of Drangajökull continued to retreat. In the 1930s some of the Drangajökull glaciers staged a dramatic advance while those of Vatnajökull continued their retreat.

By the 1960s nearly all the Iceland glaciers had shrunk to occupy smaller areas than at any time since the seventeenth century, or even earlier, and the sea ice was far removed from the north coast. There was a sudden return of sea ice in the years around 1970; some of the glaciers ceased to retreat and a few even readvanced. But by the early 1980s, general glacier retreat had been resumed.

Chapter 3
Scandinavia

The Scandinavian glaciers are scattered over several mountainous areas between 60° and 71° north, mainly in Norway and all within 180 km of the west coast (Figure 3.1). They are nourished by snow brought by cyclonic depression from September to April, the weather conditions of the autumn and spring months being critical for the glacier budgets. A large proportion of the total area under ice, which exceeds that of the Alps, is made up of Jostedalsbreen, 486 km^2, the largest icecap of mainland Europe (Wold 1982). Many of the glaciers are in the far north, in remote, sparsely populated areas, and it is fortunate for our purpose that Jostedalsbreen is well to the south, where the adjoining valleys have been cultivated for centuries and records detail critical features of their history.

3.1 Jostedalsbreen

3.1.1 THE QUESTION OF ICE ADVANCES IN THE FOURTEENTH CENTURY

Jostedalsbreen outlet glaciers overlook the cultivated fields of Oldendalen, a valley in the north tributary to inner Nordfjord, and also finger down into Jostedalsbreen itself in the east, and into the heads of Veitestrandsvatnet and Fjærlandsfjorden (Figure 3.2). All these glaciers respond to changes in the mass balance of the icecap feeding them, some more sensitively than others.

The farming economy of the valleys was based, as it still is, on pastoralism, with reliance being placed on summer pastures (*saeter*) on the higher ground, plus hay and cereals near the farmsteads at lower levels. Settlements were established in the coastal areas of Nordfjord at an early stage and tracks to them crossing over Jostedalsbreen are known to have been in use in the twelfth century. Most of the farms with which we shall be concerned in Oldendalen and Lodalen were in existence in the early fourteenth

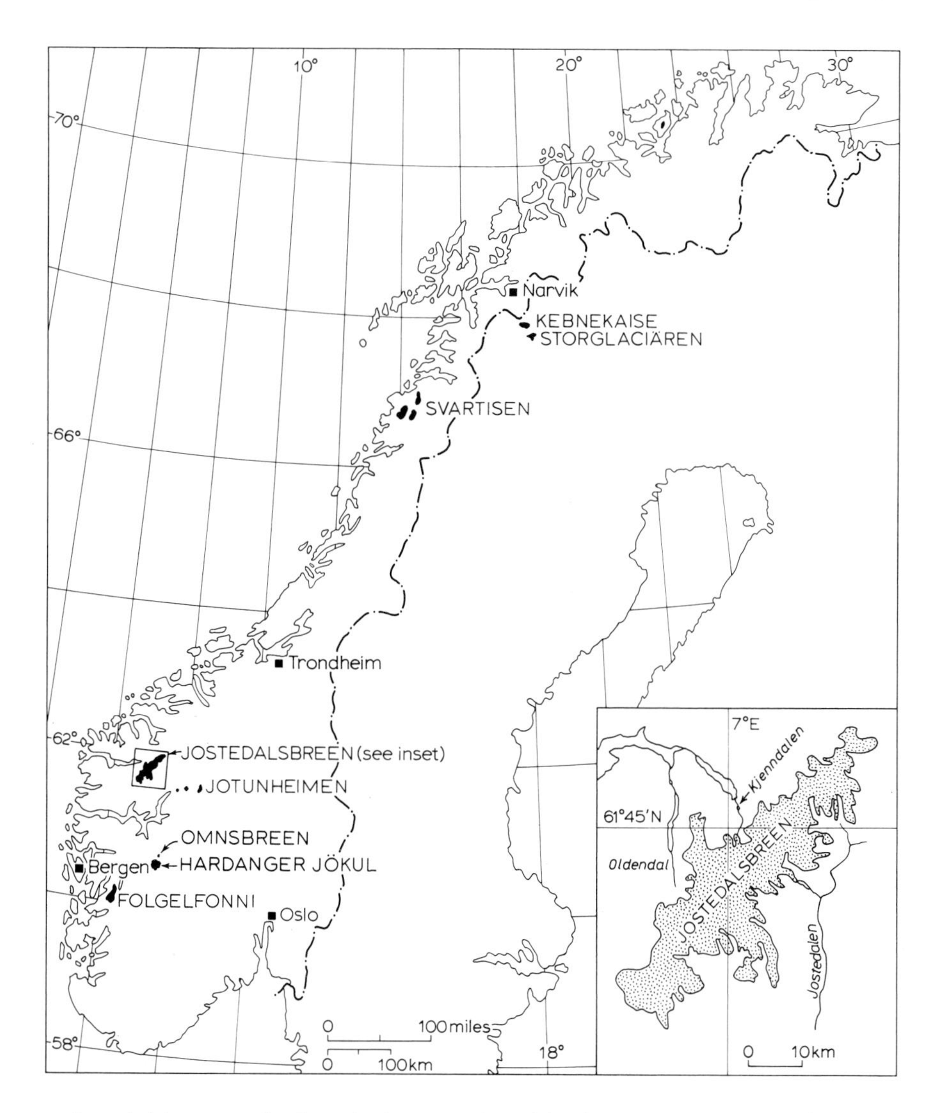

Figure 3.1 Jostedalsbreen and other glaciers mentioned in the text

century and a church had been built in Jostedal by 1332. It has been generally supposed that the climate at this time was relatively benign (Øyen 1907, Hoel and Werenskiold 1962). Nevertheless, life for most of the people must have been hard with natural hazard never far away.

Some kind of disaster seems to have afflicted the area in the first half of the fourteenth century, judging from a letter of 1340 in the Norwegian archives (*Diplomatarium Norvegicum*, IX, no. 120) which lists farms that had been deserted and tax reductions that had been ordered. The farms deserted or in part abandoned, listed as *audnir*, that is wasteland, may have been out of cultivation for many years as a result of long-term

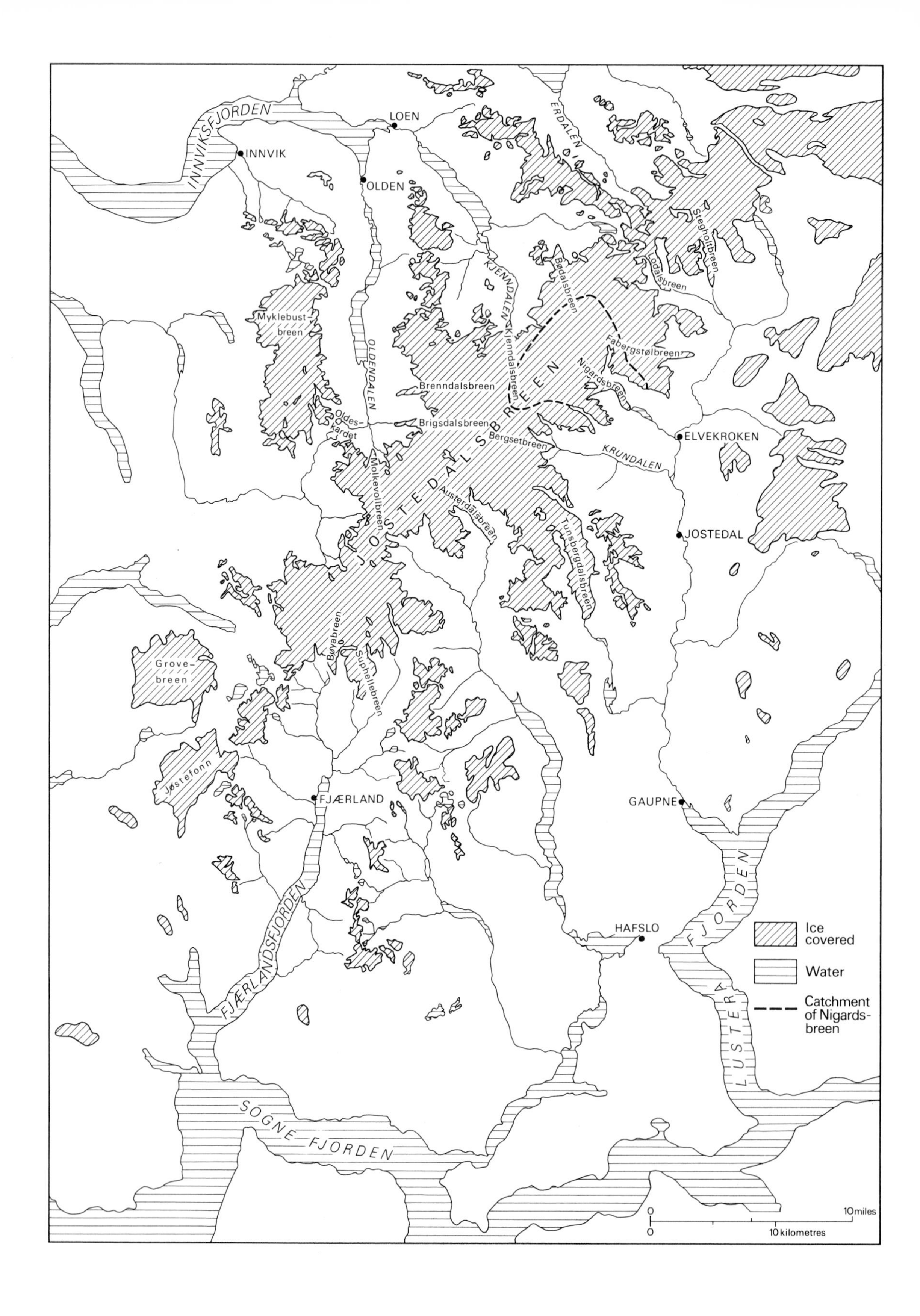
INNVIKSFJORDEN
LOEN
INNVIK
OLDEN
ERDALEN
KJENNDALEN
Bødalsbreen
Steghoibreen
Lodalsbreen
Myklebust-
breen
OLDENDALEN
Kjenndalsbreen
Fabergstølbreen
Nigardsbreen
Brenndalsbreen
Oldes-
kardet
Brigsdalsbreen
Bergsetbreen
JOSTEDALSBREEN
ELVEKROKEN
KRUNDALEN
Melkevollbreen
Austerdalsbreen
Tunsbergdalsbreen
JOSTEDAL
Bøyabreen
Supheilebreen
Grove-
breen
Jøstefonn
FJÆRLAND
GAUPNE
FJÆRLANDSFJORDEN
HAFSLO
LUSTERAFJORDEN
Ice
covered
Water
Catchment
of Nigards-
breen
SOGNE FJORDEN
0
10 miles
0
10 kilometres

difficulties of some kind (Figure 3.3).[1] The tax reductions would only have been made in response to some sudden event in the years immediately preceding 1340 causing serious environmental disruption. Exactly what happened is unknown but it is significant that the farms that suffered most, namely Bødal and the two Nesdals, were situated at the head of Lovatnet; Tyva and Seta were at the lower end of the lake and others somewhat less seriously affected were along the lake's northeast shore. This suggests the possibility that an enormous rockfall at the precipitous southern end may have created a great wave that swept down the lake from one end to the other. Damage on a lesser scale in Oldendalen and on Innviksfjorden could conceivably have occurred at the same time. A rockfall into the head of Lovatnet in 1905 caused a wave which destroyed Bødal farm, which was then rebuilt on a safer site. It was finally destroyed by an even larger rockfall in 1936.

The Black Death, which caused a catastrophic decline in population throughout Norway, is commonly seen as having been responsible for many farms being abandoned in marginal areas in the fourteenth century. However, the abandonment of farmland in the Nordfjord area and the damage mentioned above took place before the Black Death. These events could be explained by climatic deterioration and associated mass movements of the kind which, it will be shown, affected the same sites in the Little Ice Age of the late seventeenth and eighteenth centuries.

There is some stratigraphical evidence suggestive of fourteenth-century ice advance. The thinning of the central areas of a small glacier called Omnsbreen, north of Hardangervidda (Figure 3.1), revealed in the 1970s an extensive area of fresh till (glacial ground moraine) overlying humus and plant remains which were dated to 550 ± 110 (T-1485), 440 ± 100 (T-1578) and 430 ± 100 BP (T-1479) (Elven 1978). The undisturbed nature of the plant remains points to rapid climatic deterioration, causing the plants to become embedded in increasingly permanent snowbeds. The species involved grow at somewhat lower altitudes today and Elven (1978) concluded that the change in climate had been too swift for vegetation to respond before being overwhelmed by the ice. Having calibrated the ^{14}C ages of the plant fragments, according to the procedure of Ralph *et al.* (1973), Elven concluded that the cooling had taken place in the fourteenth century.

There are other local studies that indicate the possibility that the glaciers began to advance in the late fifteenth and sixteenth centuries. For example, Andersen and Sollid (1971) described peat protruding from beneath morainic material of the eighteenth century in front of Tverrbreen (now called Tuftebreen), one of Jostedalsbreen's outlet glaciers, and obtained a date from it of 410 ± 60 BP (T-779). Calibrating it according to Damon *et al.* (1974), they took it to indicate a date of about AD 1540 for the peat and went on to assume that this was also the date of initial moraine formation, even though they noted this was not compatible with historical records of ice advance. Moss from beneath moraine fronting Storbreen in Jotunheimen, radiocarbon dated to 664 ± 45 (SRR-1083)

[1] *Diplomatarium Norvegicum*, IX, no. 120 gives in full a letter of 19 January 1340 concerned with proceedings held at the church in Olden on the day before the feast of Fabian and Sebastian in the twenty-first year of the reign of King Magnus Eriksson (1319–50). Deserted lands were listed and so were farms for which tax reductions were ordered. The reductions totalled in all 81¾ *månadsmatsleigar*, one månadsmatsleiga equalling one *laup* of butter or its equivalent (see page 75). The exact date and nature of the event or events causing reductions were not specified. Lands designated as *audnir* were recognized as having gone permanently out of cultivation.

Figure 3.2 Jostedalsbreen. (From Norges Geografiske Oppmåling 1:50,000 maps 1418.III, 1418.IV, 1318.II and 1318.I)

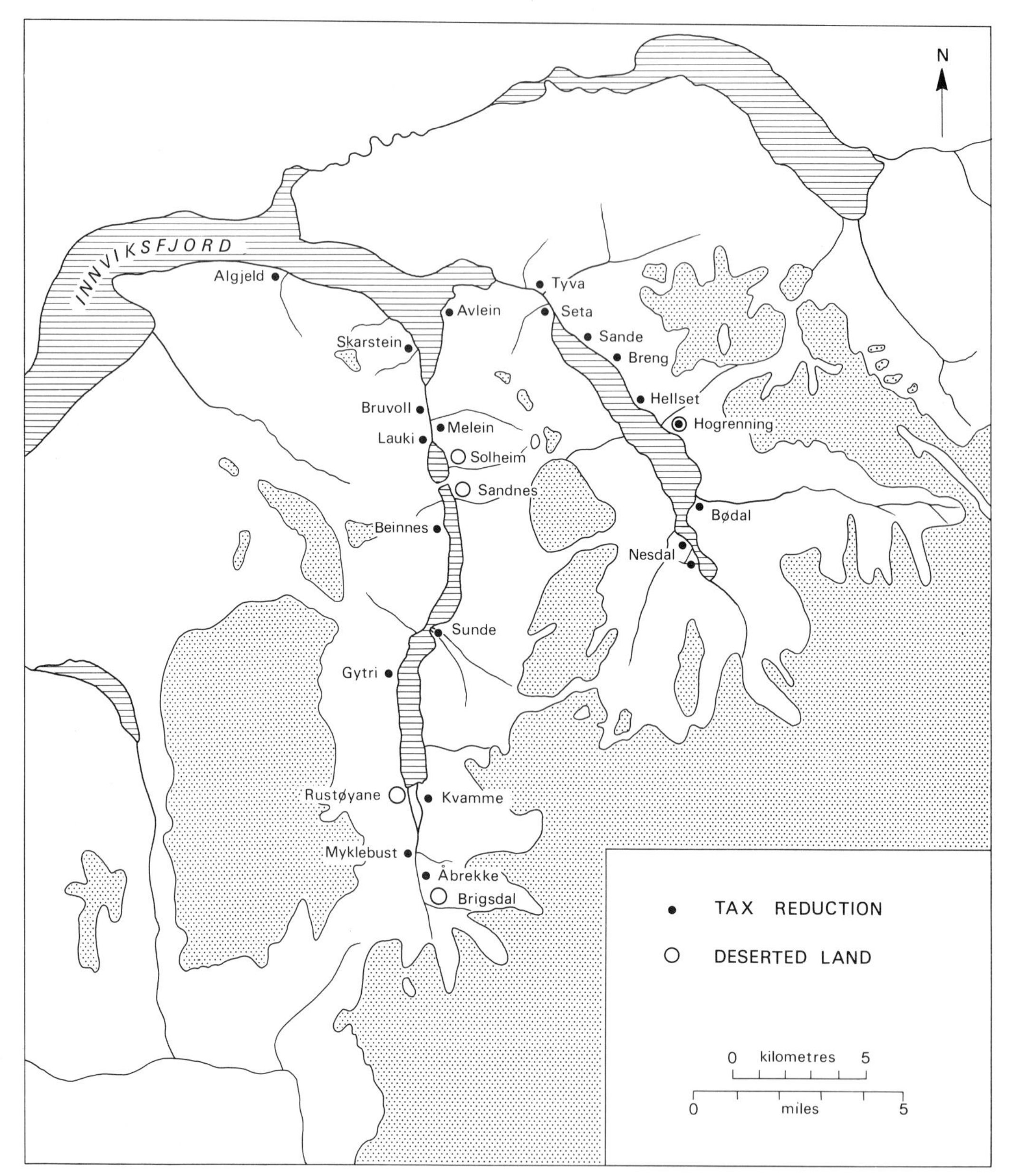

Figure 3.3 Farms in the parishes north of Jostedalsbreen where there were tax reductions in 1340 or where the land was deserted. (Data from *Diplomatarium Norvegicum*, IX, no. 120)

and 532 ± 40 (SRR-1084), again calibrated according to Damon *et al.* (1974), was also originally taken to show glacial expansion in southern Norway in the second half of the fifteenth or the first half of the sixteenth century (Griffey and Matthews 1978). However, the upper of the two moss layers sampled has the older date and the authors concluded that 'no importance can be attached to the difference between the two dates'.

In all three cases it must be recognized that the radiocarbon dates refer to peats and other organic material beneath the glacial sediments and the time interval between the two is quite unknown. Furthermore, radiocarbon dates for the last few centuries are ambiguous in terms of sidereal or calendar years and have errors which may well be as much as three times the laboratory errors stated (Stuiver 1978, International Study Group 1982) (see Figure 1.2). Thus field studies depending on radiocarbon dating cannot at present reveal unambiguously the century in which Little Ice Age advances began in southern Norway (Matthews 1982, Grove 1985) and we are left dependent on historical sources.

3.1.2 THE EXPANSION OF JOSTEDALSBREEN 1680–1750

According to Schøning (1761) there were crop failures in Norway in 1600, 1601 and 1602 and again in 1632 and 1634.[2] Gerhard Schøning's compilation has not yet received full critical analysis but, according to Sandnes (1971, p. 336), he used contemporary seventeenth-century records. There were certainly crop failures in the mid-seventeenth century: a petition made in June 1644 by people in the Jølster area, immediately west of Jostedalsbreen, explained 'there is such great misery here that many in this little skipreide are starving and will soon die of hunger, some having to eat bark mixed with chaff instead of bread' (*Samlinger til det norske Folks Sprog og Historie*, V, p.493). In 1648, King Christian IV addressed an open latter to the farmers of Norway, in which he granted 'the same relief granted them in previous years'. This was because of 'crop failure, poor fisheries and cattle pestilence' (*Norske Rigsregistranter*, VIII, p. 602). A letter in almost identical terms was sent the following year (*Norske Rigsregistranter*, IX, p. 278). By this time it seems likely that Jostedalsbreen was enlarging, for in the 1680s ice began to creep down into the surrounding valleys.

The earliest reliable evidence of direct damage to farmland by advancing ice in Scandinavia comes from Jostedal in a brief account, dated 1684, of the arraignment before the local court or *ting* of two farmers, Knut Grov and (illegible) Berset, for non-payment of *landskyld* by the proprietor of their farms, a widow called Brigitte Munthe (Tingbok for Indre Sogns Fogderi, no. 14, 1684). They pleaded that they could not pay because their high pastures had been covered by advancing ice. The exact position of these *saeter* grasslands and the huts which stood on them is unknown but Grov and Berset farms were, as they still are today, in Krundalen, a right-hand tributary of Jostedalen. There is not a great deal of space in upper Krundalen and the implication is that if the land here was suitable as pasture it must have been ice-free for a considerable period. It is significant that Brigitte Munthe submitted that the two farmers could use pastures at Kriken and Espe, also her properties, as there was sufficient grass there. Kriken and Espe lay on the eastern side of Jostedalen, the side of the valley away from Jostedalsbreen.

The advance of the ice after 1684 was swift. According to the records of an inquiry held in August 1742, Knut Grov's son Ole, born about 1678, and a neighbour Ole Bierch, born about 1672, 'men of good repute',

> explained that they remembered that in their youth, the said glacier (Tverrbreen) had been only high up on the mountain in the narrowest neck of Tufteskar, but it had forced it way through the gap, and down onto the flat fields towards the river, and,

[2] Schøning (1761) lists a number of cold years from 1294 to about 1340 and several years of acute harvest failure in Norway from 1315 onwards, but this part of his compilation has yet to receive critical analysis.

> according to Rasmus Cronen's explanation it had advanced 100 fathoms [c. 200 m] in only 10 years.

They made no mention of temporary retreats or halts between 1684 and 1742, or of earlier advances of the glacier. This is hardly surprising in view of the history of settlement in the valley. However, the assessors did look to the future and measured the distance from the glacier to a cairn erected near the river 'so that it may be noted how much the glacier may advance in future years' (Kongelige Resolusjoner, Bergens Stift, 1740–3).

In Norway the effects of the Black Death, though severe, were uneven (Hovstad 1971). Many upland areas were left with little or no population for prolonged periods. After the plague of 1349–50 Jostedalen was deserted. There was still one farm occupied in Krundalen in 1374, but eventually this too lay empty. Survivors commonly migrated to richer lowland areas or to coastal situations. It was not until the population had risen substantially in the sixteenth century that remote valleys like Jostedalen were reoccupied. The resettlement of Jostedalen can be traced by examination of the tax rolls (Laberg 1944). When a farm was taken in from the wild a few years' grace was allowed before it was assessed for land rent taxes or tithes. It may be assumed that farms were reoccupied during the decades preceding the dates at which they first appear in the tax rolls. The results of Laberg's investigations are shown on Figure 3.4. Jostedalen was not farmed again after its desertion until the years immediately preceding 1585, when several immigrants came in and settled down. The Krundalen farms, Grov and Berset, were taken in just at the time, as will appear later, when the glaciers around Mont Blanc are known to have been expanding to cover or damage farms and high villages in the Arve valley (Grove 1966, Ladurie 1971, and Chapter 4). The grasslands at the head of Krundalen must not only have been clear of ice in the late sixteenth century but also for a long period before that, as they were available for summer grazing. Conditions in Jostedalen were sufficiently attractive for further newcomers to arrive in the early part of the seventeenth century; several farms, including Kriken and Espe, appear in the tax rolls in 1611.

When Jostedalsbreen eventually expanded in the late seventeenth century, the damage it caused to farmland led to on-the-spot investigations (*avtaksforretninger*) and courts of inquiry and hence to an accumulation of documents. From these it is clear that the advance of Tverrbreen was by no means exceptional in its extent or timing. Foss, vicar of Jostedal, gave an account of his parish in 1744 in which he noted that ice had covered six fields (*jorder*) near the head of Mjølverdalen and explained how the ice there and in Krundalen, as it pushed forward, had widened as the valleys opened out.

> Its colour is sky blue and it is as hard as the hardest stone ever could be with big crevasses and deep hollows and gaps all over and right down to the bottom. Nobody can tell its depth although they have tried to measure it. When at times it pushes forward a great sound is heard, like that of an organ and it pushes in front of it unmeasurable masses of soil, grit and rocks bigger than any house could be, which it then crushes small like sand. In summer there is an awful cold wind blowing off it. The snow which falls on it in winter vanishes in summer but the ice glacier grows bigger and bigger.

The summer cold damaged the crops and so chilled the people working in the fields that they had to go clad for winter even in the summer weather. 'The volumes of water coming

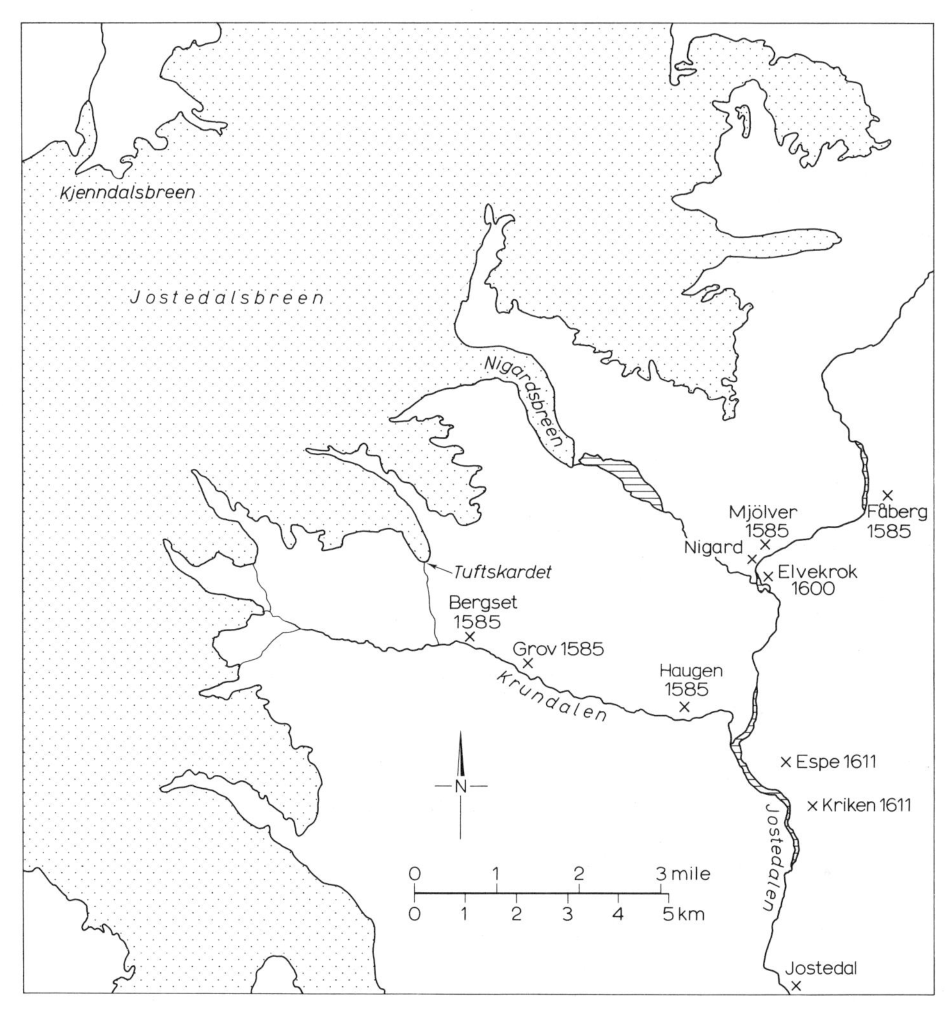

Figure 3.4 Farms in upper Jostedalsbreen resettled in the late sixteenth and early seventeenth centuries, together with the dates at which they reappeared in the tax rolls. (Data from Laberg 1944)

from the two valleys make an enormous river that not only damages what poor roads there are, but also undermines the soil itself, damaging meadowland and removing large trees' (Foss in Laberg 1948, p. 207).

An inquiry into the tax assessment of Mjølver farm on 16 November 1735, shortly before the farm buildings of the highest *bruk*[3] disappeared beneath Nigardsbreen, provides more detail.

> Guttorm Johanssen Mielvær of Jostedal Skipreide appeared before the court and, with the consent of the court, requested the common people to give a true account of his farm in Jostedal, of its situation and of the extraordinary damage which the well-

[3] These farms were compound; each family occupied separate buildings in the farm cluster and farmed a separate piece of land known as a *bruk*.

known Jostedal snow or ice glacier inflicts each year on Mielvær farm on which he lived. To which the people replied that they could in all God's truth, state and testify that the well-known Jostedal glacier, which extends over seven parishes, has grown so much, year after year, that it has carried away not only the greater part of the farmed meadowland but also that the cold given off by the glacier prevents any corn growing or ripening so that the poor man, Guttorm Johanssen who lives there, and his predecessors have each year had to beg for fodder and seedcorn. Nor for some years has he been able to pay his landlord his dues which have been in arrears for many years and he has nothing whatever with which to pay them. His landlord, Hr Christopher Munthe, has had nothing from this farm, except relief of taxes which this poor man does his best to pay, mostly by begging. They also explained that providing the above-mentioned glacier continued to grow as it had been doing for some years and was now within a stone's throw of the house, the farm would be completely carried away within a few years and would never again be habitable. These were the true facts according to the Jostedal people present. (Eide 1955, p. 18)

Thus although Nigardsbreen had taken longer to reach the grasslands in Mjølverdalen, between 1710 and 1735 the front had in fact moved forward some 2800 m.

The forecast made in 1735 proved to be accurate. By 22/23 August 1742, when Mjølver farm was again examined for tax relief purposes, two of its three *bruk* were derelict. Only the one worked by Ole Ottesen was still being farmed, probably because 'this part lies behind a large hill which in some measure shields it from the strong cold winds which the glacier exhales'. On Anna Aasen's *bruk*, corn could no longer be ripened 'because of the cold given out by the glacier'. The officials found that

the glacier has advanced to within 20 alen [12.15 m] of the cabins or derelict farm buildings onto Rev. Chr. Munthe's part which was abandoned by their people four years ago. The said glacier comes down from the great ice-mountain and covers the whole valley from one mountainside to the other and is some thousand alen wide at its margin where, immediately in front of the farm, it is over 140 alen [88 m] in height measured by instruments and reckoning. . . . This glacier has, as we saw and had explained to us, covered all the farm's hayfields, grazing lands and summer pastures over the distance of a quarter of a mile [equivalent to 2.7 km] in which the grazing land, as well as the upper part of the Mielvær Farm has been carried away and the little remaining of the derelict pasture and arable land belonging to the Rev. Chr. Munthe has also become derelict by reason of the cold that rises from the glacier and is so ruined that it could not provide forage for one beast. (Kongelige Resolusjoner, Bergens Stift, 1740–3)

Nigardsbreen continued to advance and Foss tells us that by 1743

the glacier had carried away buildings; pushing them over and tumbling them in front of it with a great mass of soil, grit and great rocks from the bed and had crushed the buildings to very small pieces which are still to be seen, and the man who lived there has had to leave his farm in haste with his people and possessions and seek shelter where he could.

Foss went on to note that the glacier 'had approached another farm called Bjerkehougen and carried away its fields and meadows, only the buildings remaining, and it is thus uninhabitable' (Foss 1750, pp. 17, 18).

In Krundalen, at about the same time, Bersetbreen

> filled the whole of the valley from side to side and from the top down to the stream. . . . Further, the glacier called Vetledalsbreen has, from the south, forced itself between the mountains into the valley in front of the mountains called Høye Nipen, and it is by these glaciers (which are all joined to the great glacier or mountain of ice covering all the mountains, which we were told, were said to be one and a half [Norwegian] miles [16 or 17 km] wide. . .), that part of Berset farm's hayfields and grazing lands such as their cattle enclosures and summer pastures, has been almost completely taken away.

The farm's arable land and meadows were next examined:

> as regards the arable land, all was still quite green, part in ear and part just come into ear, so that there was little or nothing in them, as it advances more than formerly down the valley. . . . The freeholder, Anders Berset, then stated that during these years of crop failure he had got no seedcorn but the meagre crop he had won had been of a kind of light corn or chaff which, with straw and pine bark, he had ground to flour and mixed in order to keep alive. (Kongelige Resolusjoner, Bergens Stift, 1740–3)

Farms in the main valley of Jostedalen were affected by summer cold, avalanches and floods in the eighteenth century. Ormberg, wedged into a narrow flat on the floor of the valley, was badly damaged by flooding in 1741. The farmer told a court of inquiry of many years with poor harvests; in some he had got back what he had sown, in others a little more, so that like his neighbours he had to mix pine and elm bark with his corn. Much of the arable land which lay along the river was eroded away in 1741 so that a strip 1334 alen long by 210 alen wide lost 84 alen of its breadth; further up the river a strip 336 alen long by 63 wide was also lost (1 alen equalled 0.63 m) and much of the remaining land was covered with sand (Laberg 1944).

Elvekrok was damaged by the river in 1742 and the court of inquiry reported that

> it was apparent to us that it was the nearness of the glacier which is the cause of crop failure on this farm, for in the fields where the crops were now in ear, the ears on the side towards the glacier from the west were quite brown, and on the other side green, though some ears were not even in bud because the cold and the strong cold winds which the glacier exhales freezes it away in one night's still weather. The damage to this farm by flooding by the stream which comes from the glacier could not be measured, as we could not cross it, as it runs in several branches, but as far as it was possible to see and carefully observe, more than the half part of the farm's pastures had been washed away and removed, so that it can never recover and be improved, nor, because of the cold, can there be any corn harvested for food let alone seed.

Taxes here were reduced by half (Kongelige Resolusjoner, Bergens Stift, 1740–3). In sum, evidence from the farm histories of Jostedalen dates the descent of the ice from the high tributary valleys to the 1680s and the subsequent encroachment onto permanent farm sites to the period between the 1680s and 1745, a period of increased winter snowfall and short cold summers.

Witnesses of the glaciers' behaviour, it must be emphasized, were not lacking between 1585 and 1684 and they would have left records of any sizeable expansion affecting farm and grazing land. We have records of other natural events from the seventeenth century. For instance it is known that Kruna or Kronen, one of the largest farms in Krundalen, passed from one Sjur to his son Klaus in 1653 after the son's farm had been destroyed by flooding (Laberg 1944). But there is no hint of damage caused by glaciers in documents written before the late seventeenth century. The grasslands overwhelmed by Tverrbreen and Nigardsbreen in the late seventeenth and early eighteenth centuries, it may be concluded, had been free of ice since the resettlement of Jostedalen.

The likelihood that Jostedalsbreen advanced earlier than the late seventeenth century diminishes further in the light of historical evidence from Oldendalen, on the northern edge of the icecap. This valley was not completely deserted after the Black Death, a number of farms remaining occupied without interruption. People began to move back earlier than into other valleys around Jostedalsbreen, especially between 1530 and 1620 (Aaland 1973). By 1563 there were twenty-three farms within Olden parish and a further sixteen had been added by 1667. Not only were old farms resettled but new farms were taken in and the valley was fully occupied. Farming settlements or *gårder*, consisting of a group of farmhouses and their associated buildings, more like hamlets than unit farms, were increasing in size. Each of the families in a *gård* occupied a *bruk*, a portion of land; Kvamme *gård*, for example, was composed of three *bruk* in 1602, five in 1606, and six in the 1640s.

The most significant histories from the point of view of the present inquiry are those of two farms, Åbrekke and Tungøen (Tungøyane), which both held grazing land up in Brenndalen, a hanging valley tributary to Oldendalen.[4] These farms, established before 1563, huddled beneath the rock step that terminates Brenndalen at its western end on either side of the stream from Brenndalsbreen, which joins the river from the Brigsdal and Melkevoll glaciers in upper Oldendalen (Figure 3.5). The neat farm buildings of Åbrekke can still be seen today at the foot of the step where the stream descends from the hanging valley of Brenndalen. According to tradition, after the settlement of the farms the ice was just visible on the skyline and did not reach the floor of Brenndalen. Both Åbrekke and Tungøen certainly used Brenndalen for pasturing their cattle. Documents from the period 1563 to 1670 do not reflect any anxiety about the future of the farms and no damage to their lands is recorded. In the tax register (*matrikkel*) of 1667[5] Tungøen is listed as a farm of three *bruk* and with the taxable value in butter, as was commonly the case, of 2 *laup* and 1 *pund*: 24 *merker* of butter were equivalent to 1 *pund*, and there were 3 *pund* to 1 *laup*. A *laup* was about equal to 18 kg, valued for tax purposes at 2 *riksdaler* in coin, although the market value of 18 kg of butter must, by the seventeenth century, have been higher.[6] So

[4] Rekstad (1901) published long quotations from documents concerned with the history of Tungøen and Åbrekke, but was not conversant with the complete histories of the two farms. Aaland (1932, p. 34 ff.) gives a more complete but summary account without references. Eide's (1955) version was the fullest but unfortunately he also gave no references.

[5] The assumption of absolute monarchy by the Dano-Norwegian Crown in 1660 brought about a measure of centralization in government and some uniformity to a complex of ancient taxes and dues. The whole of Norway was reassessed for landrent (*landskyld*) between 1165 and 1670. The reassessment of Nordfjord in 1667 provides a very useful index against which to measure tax reliefs granted to many farms during the subsequent period of climatic deterioration.

[6] In medieval times taxes and rents were paid in kind. Most of the goods were sent to Bergen, where they were weighed and measured at Bergenhus. Some of the goods were used to provision the fortress, some may have been sent to Copenhagen, but the main bulk was sold to private merchants in Bergen. The transition to payment in cash appears to have taken place in the seventeenth century. The *foged* or sheriff now had to forward cash to Copenhagen. Because many of the peasants could not pay taxes in cash, the sheriff had to trade the payments he received in kind for cash. The course of the transition from payment in kind to payment in cash has not been

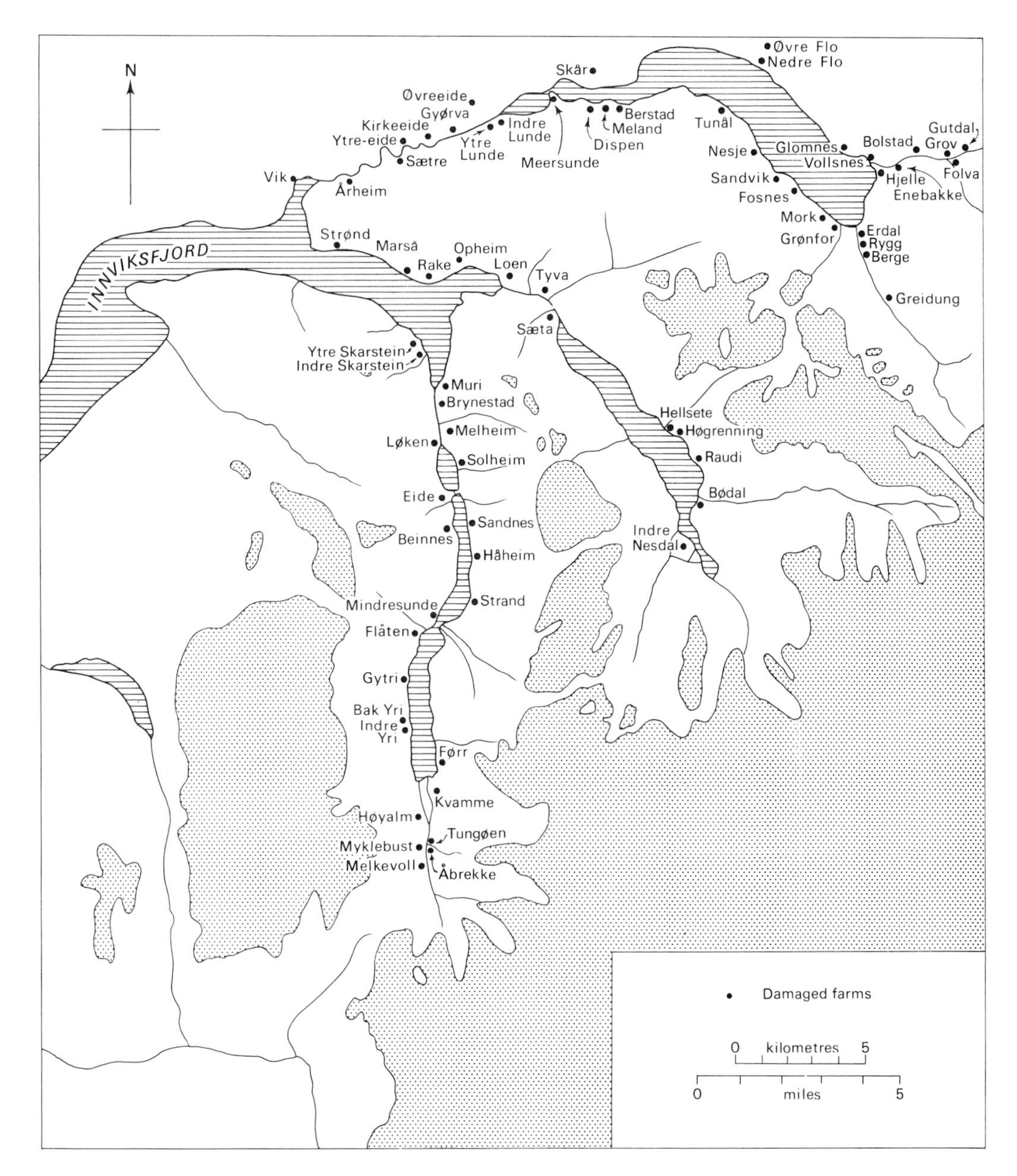

Figure 3.5 Farms in Stryn and Olden Skipreider which suffered serious physical damage during the Little Ice Age as recorded in *avtak* records

investigated in detail for the Nordfjord area. (Private communication from Dr Håkon Hovstad.)

The *mark*, the basic unit for butter, was fixed at the equivalent of 249.875 g and the *bismerpund* at the equivalent 5.997 kg in May 1683. By a decree of 4 March 1684, tenants were given the chance of paying in cash or kind, but if they chose cash then the amount paid was to be an official evaluation of the *landskyld* in kind. A decree of 5 February 1684 urged payment in cash and this, incorporated in a law of Christian V, remained in force until 1965 (*Norsk historisk Leksikon* 1974, p. 193). Equivalents were generally recognized between units of butter and other goods. Thus in the Nordfjord area, in the eighteenth century, one *laup* of butter equalled two *vog* of fish, equalled two hides, equalled one *skippund* of flour. Because of fluctuations of the *landskyld* over time, a reduction of *landskyld* or tax on one *laup* in the seventeenth and eighteenth centuries indicates more damage than a reduction of one *månadsmatsleiga* in 1340. (Private communication from Professor A. Holmsen.)

the taxable value of Tungøen was about 42 kg of butter. The farm was described as consisting of cultivated fields and meadows. It supported thirty-eight cattle and three horses and produced 29 *tønner* of corn from 9 *tønner* of seed.[7] Farms like this were almost entirely self-supporting. The cattle were raised for their milk and sold for slaughter. Any surplus corn and dairy produce was also sold. Farming families with sheep clothed themselves from the wool. They built their own houses and outbuildings. But the standard of living must have been low at the best of times and near penury when crops failed or disaster struck, as happened at Tungøen in the late seventeenth century.

In the 1680s the ice was spilling over into Brenndalen and accumulating below the rockstep at its head. The glacier stream was enlarged and by 1685 Tungøen had suffered great damage from flooding (Aaland 1932, p. 433). A court of inquiry held in 1693 found serious damage to both arable land and pastures. Rockfalls continued and, though the farm buildings had been moved, it was recorded that 'great rocks continue to fall towards the buildings, so the occupants are in danger of their lives'. The court recommended a reduction of the taxable value by 2 *pund* and 8 *merker*, leaving an assessment of 4 *pund* 16 *merker*. It had much other business to do in the Olden and Stryn area that year (Rentekammeret, Ordningsavdeling, Affældnings Forretninger, Bergens Stiftamt, Pakke 3, 1702–84); next door at Åbrekke more than half the arable land and pasture had been carried away by

> two terrifying rivers which come from two glaciers . . . in addition, great damage is being done by rocks and grit which frequently fall from the mountain above their arable and pasture and because of this the farm's tenants, one by one, have become impoverished and have had to leave the farm.

The damage to Bødal in upper Lodalen in 1693 was explicitly attributed to glacier advance:

> this farm, Boedahl, has suffered much damage this year to its hayfields outside the fenced lands by a landslide, called the glacier landslide, and last year . . . another landslide called Espe. . . . Immediately inside the fenced land the river has broken out and made a new course across the farm's best meadows and fields, damaging up to five mæler of seed ground and the closer the river approaches the great freshwater lake, the more it spreads out on each side, causing great damage to their meadows and ploughland.
>
> A valley called Tiørdal, half of which they formerly owned with the Indre Næsdal farmers, has now been wholly lost and completely ruined by the destructive glacier and by the great river running from it, so that where they formerly had five enclosures and six barns they and their contents have been completely carried away so that they have neither the benefit of their hay and grass nor can they henceforward obtain any because of the great mass of rocks and boulders that can never be cleared away. (Rentekammerarkivet, Fogedregnskaper, Sunnfjord–Nordfjord, 1702)

'Tiørdal' was Kjenndalen and the damage was caused by Kjenndalsbreen (Figure 3.6). The six barns and other property would have been included in the 1667 assessment as

[7] In Nordfjord a *tønne* was a measure of about 162 litres in the 1660s. A *tønne* varied in volume from district to district and from one time to another.

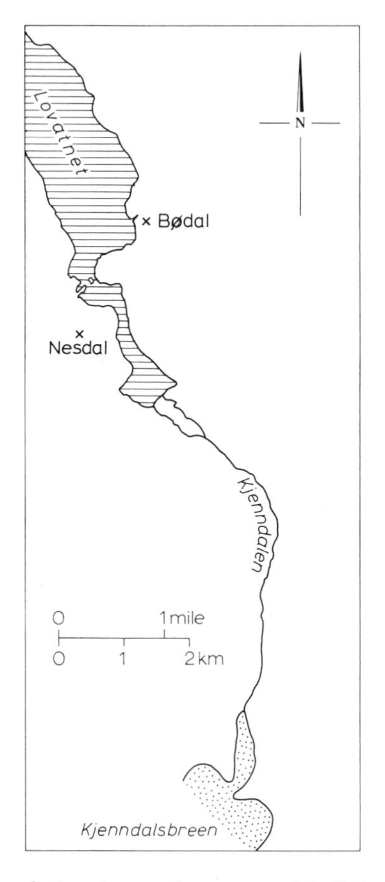

Figure 3.6 Bødal and Nesdal farms in relation to Kjenndalsbreen

being in no danger and it follows that the advance of the glacier had taken place rapidly between 1667 and 1692/3.

The documentary evidence shows that the advance of the glaciers was associated with increased damage from landslides, rockfalls, avalanches and flooding (Grove 1972, and see Chapter 12). Not only was farming made more difficult but fisheries were also damaged. The 1693 assessors examining Muri near the entrance to the fjord recorded that

> this farm in times past had again and again had increases made to its *landskyld* by His Majesty's sheriffs, because it had been blessed by God with fisheries, for their land reached them, and they and many others living in the valley were able to supply themselves, but many years have now passed since the fish came to the land, and they now have only the meagre growth of the soil and the produce of their cattle by which to meet the royal taxes and dues and with which to support their wives and children.

After 1700 the ice advanced rapidly down Brenndalen towards the rock step under which Tungøen and Åbrekke stood. Rekstad (1901, p. 26) records the tradition that on one occasion pieces of ice came down through the smokehole at Åbrekke. But the tax and *landskyd* reductions recommended by the court of 1693 were not approved by the higher authorities. As conditions worsened the farmers became desperate and Hugleik Tungøen, a man of unusual initiative and energy, journeyed to Copenhagen twice to seek the King's authority for a further inquiry to be held into the damaging of thirty-six farms in Olden and Stryn. He was successful and delivered the writ authorizing the inquiry to the Provincial Governor in Bergen in July 1702. Then, tragically, he lost his life, freezing to death in a storm on his way home (Eide 1955). Memories are long in Oldendalen and a commemorative pillar was erected in the valley during the present century.

The reports of the 1702 commissioners and of the further courts of inquiry held over the next half century show that conditions at Tungøen, Åbrekke and Bødal were by no means unusual and that the climatic deterioration and advance of the ice after 1667 had been dramatic and swift. Events at Yri (Figure 3.5), which had been prosperous in the early seventeenth century, with its *landskyld* being increased as late as 1674, showed that Myklebustbreen, an outlying fragment of Jostedalsbreen overlooking the farm, expanded in the last decades of the century. Yri's dues were reduced on account of landslides and glacier encroachment in 1687 (Aaland 1932, pp. 473–80). The 1693 commissioners found that

> the farm Indre-Yri has time and again been caused even more damage, always by the same means, and in particular, last winter when a glacier (Blaabræde) lying above the farm and its adjacent lands (which is of snow as hard as ice and which has lain there summer and winter as long as any man can remember) gradually comes further and further down and finally falls and causes damage, which has now happened here, bringing with it a great mass of rocks, grit and sand from the mountain which has taken away almost all the arable land here and partly covering and ruining it, so it seems impossible ever to be able to improve it or ever bring it into any use. Down by the farm buildings some of the said snow or glacier was still lying, not having been thawed during the summer by sun or rain, and the occupants sorrowfully deplore and confirm the same on oath.

Near by at Bak-Yri 'the said glacier and landslides have not only covered and ruined this farm's arable land with grit and rocks, but also carried away two mill houses with their gear and broken the mills to pieces, item a smithy and two boathouses down by the lake' (Rentekammeret, Ordningsavdeling, Affældnings Forretninger, Bergens Stiftamt, Pakke 3, 1702–84). Avalanches compounded the trials of the local people, who eventually had to leave their homes and take shelter under nearby cliffs. Sunde in lower Oldendalen, a large farm at the narrowing of Oldevatnet, was destroyed by a fragment toppling from the edge of the icecap overlooking it (Eide 1955, p. 27). On the coast at Loen, where the 1667 assessment had taken the value of the fisheries into account, their yield diminished and by 1702 it was recorded that 'a new river course had carried with it a great quantity of grit and earth, thereby spoiling and ruining the best herring-ground'. Expansion of the ice continued into the early decades of the eighteenth century. An inquiry at Stryn in 1726, to see whether the tax reliefs granted earlier should be rescinded, reported that conditions had not improved anywhere and that most of the farms had suffered further damage. In addition,

> there is another farm in the Skipreide, called Erdal, which for some years now has suffered irreparable damage to pasture and arable from two large rivers and a glacier which is breaking down onto it, so the 5 poor occupants can never remain there permanently unless a most gracious tax relief can be granted. (Tingbok for Nordfjord Sorenskriveri, no. 8, 1726–8)

The ice in Brenndalen had continued to advance. In July 1722 three men were sent in accordance with Treasury instructions to inspect farms granted tax relief in 1702. They reported absolutely no improvement in the condition of any of them, noting in particular that Tungøen 'has clearly suffered great damage from the rivers, rock falls and the glacier to both *Inn* and *Utmark*. Further, this farm and its buildings are always in danger from the advancing great glacier.' Åbrekke, it was recorded, 'has suffered greater damage than the other farms from the encroaching river and the glacier. Some of the buildings have been moved and the occupants are now running around the parish begging their bread, so that no taxes have been paid for the greater part of the farm for the past two years'. (Tingbok for Nordfjord Sorenskriveri, no. 6, 1721–3). The following year a general tax commission examined all the farms and in their report more detail is given:

> at Åbrekke they have been keeping themselves alive for several years largely by begging and on the parish. Three years ago the farmed arable land and pastures were completely ruined and carried away by an immoderate burst of water which burst out from beneath the advancing glacier, causing a great trench across the land, a good 30 to 40 alen in length so that only a very small part of their land now remains. The two poor occupants had to move their buildings hurriedly but one of them is still in the greatest danger of his life, remaining there although his house was quite ruined but because of his great poverty he has not been able to leave it and so remains there in an extremely dangerous situation. (Matrikkel over Nordfjord Fogderi, 1723, vol. 146, fos. 126–39)

After Hugleik Tungøen's death, Rasmus Andersen came to Tungøen, together with his neighbour Rasmus Olsen. He petitioned the Provincial Governor for a further investigation and reduction of his assessment, the substantial relief recommended by the 1723 commissioners not having been implemented, and Andersen explained how the farm had

> in our forefathers' time and in our own suffered great damage both to arable land and meadows by reason of a dangerous and destructive glacier above it and below by the cruel river running through the length of Oldendal. Thus death daily stares us in the face, most particularly from the great and horrid glacier, so that henceforth we are not able, without great risk to our lives, to venture to take up our abode there, and far less are we able to support ourselves and our poor women and children by the miserable remnants of land left to us. (Eide 1955, pp. 10–11)

Before the inquiry was held on 12 October 1728, 'These poor and needy people, for fear of the danger and disaster hanging over them in the form of the terrible glacier mentioned, now in 1728 have had to move their buildings and dwellings to a lower place which, with God's support and protection they think to be safer' (Rekstad 1901). The tax was reduced from 2 *laup* to 1 *laup*.

Six years later there was need for yet another inquiry. In the previous year, at midsummer, Tungøen's fields and meadows had been flooded and the glacier now overhung the farm and had destroyed most of its remaining arable land. Only two *mæler* each of barley and oats[8] could now be sown, instead of two *tønner*, and only six cows kept instead of eighteen.

[8] A *mæle* was a measure of corn which varied in size over time and also differed from one area to another.

> The glacier has time after time advanced so far that it is to be feared that in a few years it will come right down to the mainstream and if that should happen (which God in his mercy will surely avert) they can expect nothing but the total ruin and destruction of the whole neighbourhood. In the meantime, the mainstream has also broken out into another river course so that in high summer it goes over the whole field and right up to their poor little buildings which were moved to the place where they now stand in 1728. The other river, which previously came out under the glacier is now completely blocked at that place, but has dug itself another course right down over the arable land and meadows which were intact until 1733. But now that both these rivers, together with numerous lumps of ice from the glacier, mixed with gravel and large stones have joined and broken out over most of the best part of the farm's arable fields and meadows, these poor people with their wives and children both in the years 1733 and in the following year have had to beg for food in order to stay alive and, therefore, have to an even lesser extent been in a position to pay or to guarantee payment of their taxes for that same exceptional year. There was nothing but wretchedness and misery to be seen here, for where there were previously great level fields, both arable and meadow, there is now nothing to be seen but terrible large, wide and deep pits and trenches, which no one can cross but has to take another long way round, either high up on the mountains where there is continual fear of the same glacier, or down on the stones and grit which have been discharged by the river. . . . Things have finally reached such a pass that total ruin and destruction seems imminent. (Rekstad 1901)

Complete exemption from taxes for 1733 and 1734 was recommended and the assessment for the future reduced to half a pund.

The fears of the court were justified. The years around 1740 were cold, with heavy winter snows and short cold summers. The glacier in Brenndalen, up to now fed by ice falling from the edge of Jostedalsbreen, now thickened so that the ice flowed directly from the icecap 4 kilometres down Brenndalen. On 1 December 1743 a great mass of water, ice, stones and gravel swept down and obliterated the new houses of Tungøen, killing all but two of the occupants, a soldier and a small boy. We learn from the military inquiry that followed, on account of the soldier losing his musket and sword, that

> the soldier, Anders Pederssen Møchleoen, was then working on the farm and at the same time had his musket and sword with him for cleaning and no one on the farm was saved but for the soldier mentioned and a little boy, but all the other people, with everything that was in the buildings, were destroyed and covered by stones and grit so that there are no means of recovering any of it. (Eide 1955, p. 13)

All that was left to the heirs were two injured cows, two waistcoats, a pair of socks, a blanket and an old skirt (Rekstad 1901). Tungøen was removed from the tax list and the remnants of its fields were taken into the neighbouring farm of Aaberg. Its site is now marked only by a large mound of moraine lying under the Brenndalen step.

In December 1743 there were eleven days of very heavy rain which left a train of devastation up the west coast of Norway from around Stavanger to north of Nordfjord and inland as far as Jotunheimen. Over a hundred successful requests for tax reductions followed (Øvrebø 1970, Battagel 1981).

The tongue of Brenndalsbreen was not the only one to collapse around this time. There is also a record of Vetlefjordbreen, a tongue of Jostefonni at the head of the Vetlefjorden branch of Fjærlandsfjorden. Because of the great damage caused by a 'terrifying flow of water from the celebrated glacier' on 14 August 1741, a detailed inspection of six farms in the valley above Vetlefjorden was carried out starting in October 1741 and, because of bad weather ('the rains had so swollen the river that we could not properly measure the damage suffered'), continued in September 1742. The assessors

> all went out into the field and went northwards, right up to the well-known and celebrated glacier. From the front of its descending tongue a piece had broken off down between the two great rock-walls, being – as well as we could estimate – 40 fathoms [78 m] wide and 100 fathoms [190 m] long and of enormous thickness.

The collapse of the snout may have taken place at the time of the water burst, but when the assessors went up the valley in October 1741 they made no report of a detached piece, so it may well have occurred between then and September 1742 (Kongelige Resolusjoner, Bergens Stiftskontor, 1740–5).

Landslides and other gravitational movements, as well as floods, continued to destroy farmland in the areas surrounding Jostedalsbreen and consequent tax reductions were registered (see also 12.1.3). In 1755 seven farms in Olden Skipreide[9] made a joint petition for relief, explaining that 'the unusually heavy snowfall of last winter caused heavy damage and devastation to our lands'. Helleseter [Helect] on Lovatnet was destroyed. At a time

> when the freeholder was out fishing in the fjord, a great avalanche broke out from the mountain and not only swept all the farm buildings and cattle into the lake but also his wife, children and servants, eleven people in all and all of them being killed. The avalanche had also devastated and torn away all the land. . . . Nevertheless the freeholder has started to build again on the outer edge of the farm where he thinks the buildings will be somewhat safer. (Rentekammeret, Affældnings Forretninger, Nordfjord Fogderi, Pakke 5, 1734–61)

All the farms in Olden and Stryn Skipreider which are known to have suffered serious physical damage and were accordingly allowed relief on either *landskyld* or taxes, or both, are shown on Figure 3.5. The similarity of the distribution to that shown on Figure 3.3 of farms damaged in the fourteenth century is evident.

There are no tax documents mentioning glaciers advancing after 1743, the year of Tungøen's final destruction.

3.1.3 THE PERIOD OF WITHDRAWAL, 1750 TO THE PRESENT

Foss, writing in 1750, dates the beginning of ice retreat in Jostedalen to 1748: 'one has experienced how, from 1748 it has retired, only very slowly but still noticeably' (Laberg 1948, p. 209). He was probably basing his account on the behaviour of the large ice tongues near Jostedal vicarage, especially Bersetbreen and Nigardsbreen. A decade later,

[9] A *skipreide* was an administrative district approximately equal to an English hundred.

Wiingaard (1762) wrote of the ice retreating from the mounds of stones left behind by the glacial advances in Krundalen. So the glaciers of Jostedalsbreen seem to have reached their most advanced positions in the Little Ice Age between about 1743 and 1750. From that time until the present the glaciers of Jostedalsbreen and the surrounding areas have retreated, with minor interruptions.

The precision of the record of glacier retreat increases towards the present day. Between 1750 and 1890, Jostedalsbreen was visited by many travellers, climbers and field scientists, whose observations, including for example those of Forbes, are both vivid and reliable. Artists ventured into this part of Norway and their paintings can be compared with later photographs. Regular surveys of the positions of the glacier fronts began at the end of the nineteenth century, organized by Øyen and Rekstad. Since the early 1960s, mass balance studies have been made, with annual volumetric changes being measured for several glaciers, including a number of the Jostedalsbreen tongues. The retreat from the moraines formed in about 1740–50 was rather slow and seems to have occasioned little comment at the time. Conditions in Jostedalsbreen and Scandinavia generally seem to have been cool in the 1770s and 1780s. Slåstad's (1957) index of tree growth in Gudbrandsdalen, converted into a temperature index by Matthews (1976), shows cool summers in the 1770s. (It has to be noted that Slåstad's data is of unknown reliability; the methods used in its collection and analysis were less meticulous than would be employed now.) Sommerfeldt (1972) quotes a contemporary account of the Lensmann (sheriff) of Vågå to the effect that large masses of snow and ice accumulated in the mountains and the glaciers grew markedly, but we have no precise record of any glaciers readvancing as a result. Winters were on the whole dry, and spring came late and lasted for a long time. The records of the duration of ice in the Danish Sound, which are reasonably complete for the 1770s and 1780s, show that the means for these decades of the numbers of days with ice have never been exceeded.

The winter of 1778/9 was particularly severe, the hardest in Norway for many years, with intense cold before Christmas continuing into April and freezing the soil to a great depth. In the Sound there were 134 days with ice, a value only once exceeded since then (Lamb 1977, p. 589). Sea ice persisted near the Icelandic coast from May to early June and in the open waters between the Faroes and southeast Iceland surface water temperatures in the summer were still low, 1 to 1.6 °C below the average for 1921–38. However, to the southwest of Iceland and also at the entrance to the Baltic, temperatures rose above normal and a warm summer in southern Norway culminated in 'almost tropical heat of July' (Jarrman, 25 July 1789, quoted by Sommerfeldt 1972). Then came a great flood in Gudbrandsdalen as a result of unusually heavy rains over a very large area and probably of snowmelt in the mountains. It left a trail of devastation from landslides, rockfalls and flooding over the eastern and central parts of the country (Blyth 1982). Successful *avtak* (tax or rent relief) were registered for 953 farms damaged on 22 July 1789 (Øvrebø 1970). No ice was recorded in the Danish Sound in the following winter, nor in the one after that, and the decadal average for the 1790s was only half that of the preceding decade (Lamb 1977, p. 589).

It seems likely that after this eventful year glacial retreat was resumed at a greater pace. A young botanist called Christian Smith, two years later to become Professor of Botany at Christiania (Oslo), who visited Jostedalsbreen in 1812, described the glaciers as being in retreat (1817). Bohr wrote of Nigardsbreen in 1818 or 1819 that

> The mighty accumulation of moraine which this very glacier at Nigard had formerly pushed before it is now about 1726 [Norwegian] feet [541 m] below its margin whilst the bare sides of the mountain show its depth, now more than 200 feet less than it had once been. . . . The crops at Elvekroken this year were very good, while nothing but the moraine stood between the glacier and the ripe corn. (Bohr 1820, p. 257)

At about the same time he went on to visit Lodalsbreen and Stegholtbreen, the other two glaciers at the head of Jostedal, and recorded that 'the moraine showed clearly that these glaciers too had formerly descended about 1700 feet [538 m] further down; while the dark naked sides of the mountains, as if the surface had been shorn off, showed that they had been almost 200 feet deeper' (Bohr 1820, p. 259). By 1821, according to Naumann (1824, p. 201), Lodalsbreen was 1853 feet (586 m) from the 1740s moraine. The following year he tells us that in Mundal, near the southern extremity of Jostedalsbreen, the two tongues of Suphellebreen, which he calls Veslebreen and Storebreen, had retreated respectively 1500 and 750 paces in the course of the preceding 100 years. Close by, Bøyabreen had retreated 900 paces over the same period, which Rekstad (1904, p. 10) interprets as a retreat of 500 m. Kraft (1830, p. 808) recorded that Mundalsbreen and Vetlefjordbreen had withdrawn considerably during the previous century and that it was evident from the moraines in front of Vetlefjordbreen that it had once reached 2000 paces further down the valley. Vetlefjordbreen had again caused a great deal of damage in 1820 when a large volume of water burst from it, devastating farms downvalley. This water may have been dammed up by the glacier in Svartvassdalen until the ice withdrew sufficiently to release it. (Vetlefjordbreen has since completely disappeared.)

The descriptions of the glacier fronts of Bøyabreen and Nigardsbreen given by Bohr and Neumann[10] suggest the possibility that the glaciers advanced a little in the 1820s. The tongue of Nigardsbreen was 'cut off at the lower edge at a height of 20–30 feet' and this edge was 'almost black from soil, dust and gravel', while the edges of Bøyabreen formed an 'ice wall of about 50 alen in perpendicular height' (Neumann 1923, pp. 540–2). The painting of the terminus of Tverrbreen by Johannes Flintøe, made between 1822 and 1835, shows a smooth but swollen tongue which could be advancing (Plate 3.1). Rekstad (1902, p. 13, and 1904, p. 37) states that in Krundalen a moraine was deposited in 1830, 50 m inside the 1750 moraine of Bersetbreen (Plate 3.2). However, the first clear account of renewed advance since the mid-eighteenth century is given by a botanist, Lindblom, who visited Nigardsbreen in 1839. He saw the glacier only vaguely through the rain, but recorded that the people living close by had noticed that in recent years it had begun to advance slowly, although it was at a standstill in 1839 (Plates 3.5 and 3.6). The advance can only have been a modest affair because in 1845 Durocher found the distance from the snout of the glacier to the outermost moraine had increased from 541 m in about 1819 to 700 m (Durocher 1847, p. 104). Durocher also visited Krundalen and drew a picture of its glaciers (Plate 3.3). While Bersetbreen at the head of the valley had now retreated more than 600 m, Tverrbreen entering from the north, had retired only 350–400 m. Lodalsbreen, according to his measurements, had continued to retreat since Bohr had seen it and was, by this time, 600–700 m from its outer moraine (Plate 3.4).

[10] Bishop Neumann travelled in the Sogn area in 1823 and recorded the state of this part of his diocese in considerable detail, but his figures for time and distance are so rounded that they have to be discounted.

Plate 3.1 (on facing page—above) Tverrbreen, Jostedalen, by Johannes Flintøe, 1822–33. The smooth but swollen state of the tongue suggests that Tverrbreen may have been advancing when Flintøe saw it; however, the abandoned left lateral moraine shows that it had retreated from its maximum extent. (Nasjonalgalleriet, Oslo, Inv. no. 1085, cat. 1968, no. 649)

Plate 3.2 (on facing page—below) Bersetbreen in Krundalen, by J. C. Dahl, 1844. The front is steep and possibly advancing but a moraine on the left of the picture has been abandoned. (Rasmus Meyers Collection. Cat. no. 92, Bergen Billedgalleri)

Plate 3.3 Schematic sketches of glaciers at the head of Krundalen, from Durocher (1847, Figure 5). AA = Tverrbreen; BB = Bergsetbreen; CC = Vetlebreen. The portrayal of Tverrbreen suggests that the tongue was substantially withdrawn from its most extended position

Plate 3.4 Lodalsbreen to the left and Stegholtsbreen (T), by Durocher (1847). Durocher found Lodalsbreen 600–700 m from its outer moraine. K = Lodalskåpa; N = skyline of Jostedalsbreen; M = medial moraine; L = well-marked lateral moraines. The whole of the medial moraine shown here is now free of ice.

Nigarsbreen : 1839

Plate 3.5 (on facing page—above) Nigardsbreen seen from Elvekroken in 1839, by J. C. Dahl. The original sketch has the artist's notes written on it. (Nasjonalgalleriet, Oslo)

Plate 3.6 (on facing page—below) Nigardsbreen in 1839 by J. C. Dahl, dated 1847 but based on the sketch of 1839 shown as Plate 3.5. The *sæter* huts must have been erected after the ice retreated as Elvekroken was damaged by the ice. Dahl visited and sketched Nigardsbreen in 1839, the same year as it was visited by Lindblom. The front of Nigardsbreen in that year was said to be at a standstill following a slow advance (p. 83). (Nasjonalgalleriet, Oslo, Inv. no. 1477, Cat. 1968, no. 367)

Plate 3.7 Nigardsbreen by Joachim Frich, from an illustration in *Norge fremstillet i Tegninger*, Christiania, 1855. It shows two terminal moraines, the outer one with small trees growing on it. The tongue appears to be in retreat

In 1851 Forbes,[11] following his Alpine expeditions (see Chapter 4), made an excursion to Norway, where the ogives of the outlet glaciers of Jostedalsbreen particularly attracted his attention. He measured the intervals between the ogives on Bersetbreen and found their breadth to be somewhat unequal but an average of 167.7 feet, which 'represents, I have no doubt, very nearly the average annual movement of the glacier'. He counted twenty ogives beneath the icefall, which makes the length of the tongue of the glacier on the flattish valley floor about 100 m in 1851 (see Figure 3.7). The terminus was, Forbes found, 900 yards from the moraine, 'a great moraine evidently modern. Its limits may be at once traced all round for no birch woods grow within them. Beyond question, it is of the same date with the great extent of the Nygaard glacier . . . of which the date is unknown. The Tvaer Brae has a corresponding moraine.' He was also impressed by the evidence of the recent retreat and downwasting of Nigardsbreen, of which he made and published a drawing (Forbes 1853, p. 169). He knew of the advances currently taking place in the western Alps but makes no mention of any such occurrence in Jostedalen. Prominent moraines below some of the largest glaciers flowing from Jostedalsbreen have been dated to 1848–50 by lichenometry (see Figure 3.8). However, the accuracy of the methods involved was probably not great enough to allow their precise dating.[12]

The glaciers retreated through the 1860s as Doughty (1865, p. 143) reported for Nigardsbreen, Blytt (1869, p. 37) for Nigardsbreen and Bøyabreen, and de Seue (1870, p. 15) for Bersetbreen. However, de Seue found Bøyabreen and Lodalsbreen advancing in

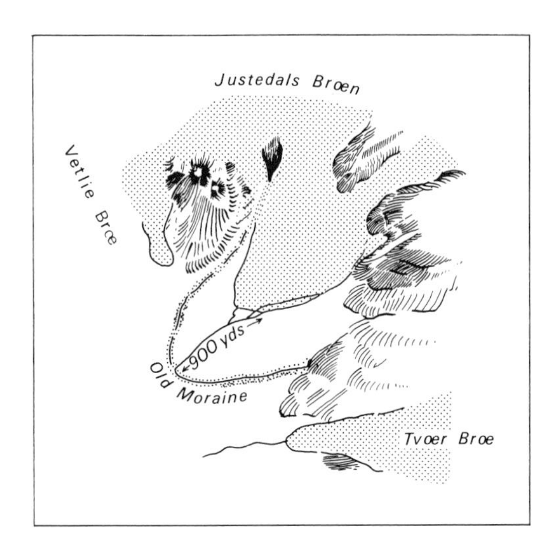

Figure 3.7 Bersetbreen in 1851, showing the tongue in retreat. Sketched by Forbes (1853, p. 163)

[11] This was Forbes's last scientific expedition; the illness which began in the autumn of 1851 put an end to his active fieldwork (Shairp *et al.* 1873). Forbes had made pioneer studies of the alternate light and dark bands formed beneath the icefall of the Mer de Glace in France and the relationship between such bands or ogives and the speed of flow of the ice. When he visited Norway he had already observed the expansion of Alpine glaciers towards 1850, e.g. 'I found the Rhône glacier much enlarged since I last visited it' he wrote in his journal for 1846 (Shairp *et al.* 1873, p. 329).

[12] Dating methods were not the same in the three cases. Quite apart from the techniques involved we may notice that the outer moraine of Nigardsbreen here was dated to 1850, although Foss (Laberg 1948) recorded that the tongue started to retreat in 1748. The only pre-1900 dated surface used by Mottishead and White (1972) at Tunsbergdalsbreen was for the outermost moraine 'dated from historical evidence by Rekstad (1901) presumably by analogy with Åbrekkebreen and Nigardsbreen. Contemporary descriptions exist telling of the advance of the glaciers up to 1743. . . . This terminal moraine is thus dated to 1743.'

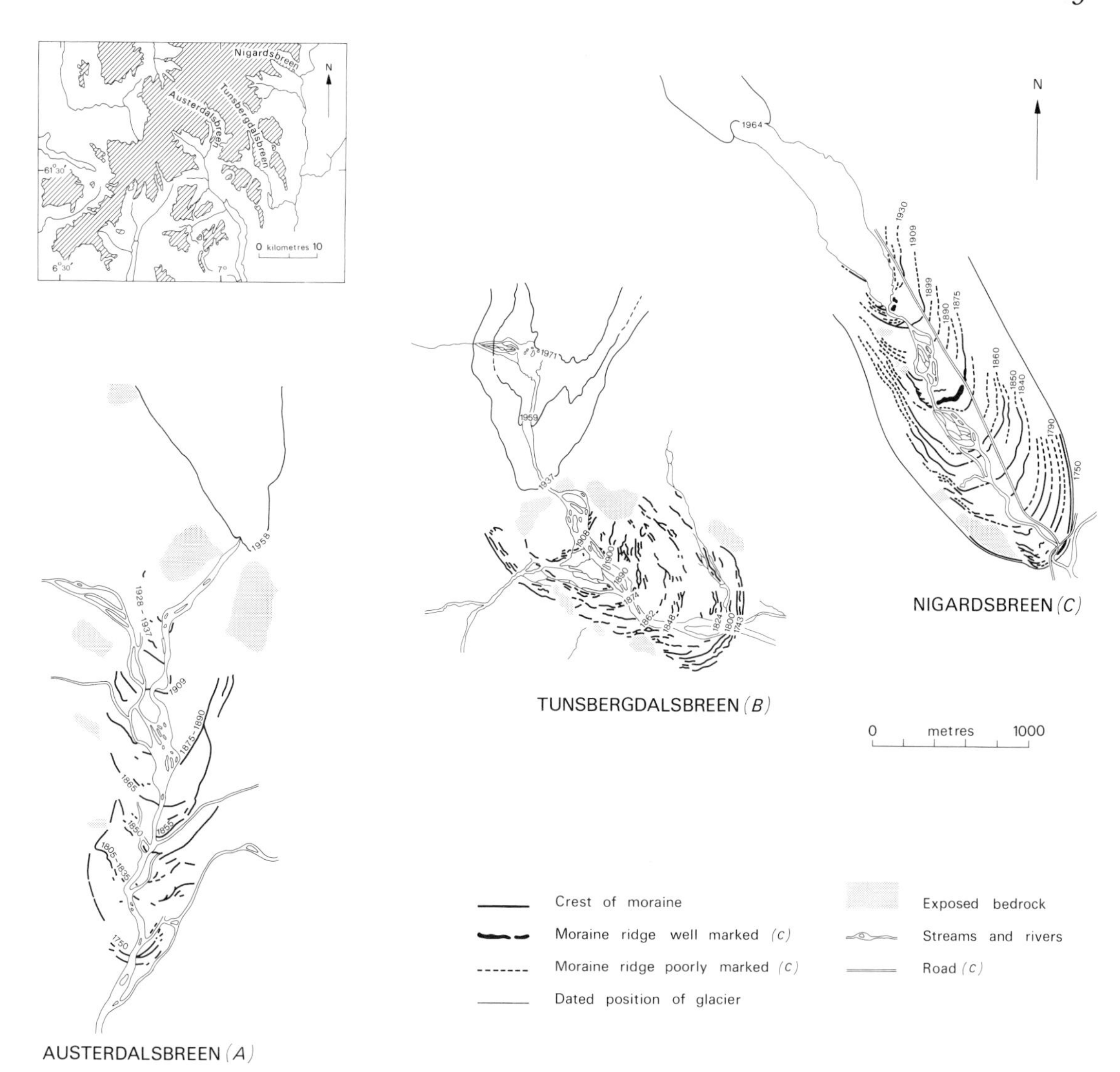

Figure 3.8 The forefields of (a) Austerdalsbreen (after King 1959), (b) Tunsbergdalsbreen (after Mottishead and White 1972) and (c) Nigardsbreen (after Andersen and Sollid 1971), all three providing evidence of rapid retreat after 1930

1869 and his wording suggests that Nigardsbreen may also have been advancing. A large moraine had formed in front of Nigardsbreen by 1873 (Rekstad 1902, p. 15) when Larson said the ice was advancing (Larson 1875, p. 11). Rekstad noted (1901, p. 13) that Brigsdalsbreen was also advancing between 1869 and 1872, though it failed to produce a moraine; he recognized that variations between the characteristics of different glaciers would result in a lack of uniformity in the morainic record. This lack of uniformity is exemplified by Bøyabreen, which advanced from 1880 to 1890 (Rekstad 1904), at a time when there are no reports of other Jostedalsbreen glaciers advancing. Thus the glaciers were almost continuously retreating for a century and a half, without substantial readvances of the kind that have been recorded from the Alps (see Chapters 4, 5 and 6). Small icecaps and isolated glaciers bordering Jostedalsbreen were also shrinking. Bing (1899) wrote that Rauddalsbreen, which eighty years previously 'went out over the valley to the other side', had now retreated and thinned so that it scarcely reached the valley bottom.

Access to the Jostedalsbreen icecap at the present day is difficult for cattle but throughout much of the latter part of the eighteenth century and until well into the nineteenth century ice projected down into the valley heads and provided convenient routeways from Jostedal to the lowlands around Nordfjord. The people, wrote Foss (in Laberg 1948, p. 209),

> used to go year after year, arranging beforehand so that many could go together. They take with them wooden things made in the winter, such as troughs, baskets, bowls, ladles, wooden spoons and so on, which they sell very easily for a good price. They do not bring much back with them, apart from some oats, herrings and corn. . . . Also drovers go there to buy cattle which they bring back with them and also horses except for those who live in Valders in Christiania Stift who, when they go this way over the glacier and trade in Nordfjord with big droves of cattle and horses do not usually go back this way but by a more convenient route.

Foss also described several ways over the icecap including one from

> Berset to Aalden [Olden], where the first farm you reach is called Qvame. . . . Here the glacier is so big that the clouds blow against the top of it and the highest mountains lie far below as in an abyss so that you can hardly see them, even less arrange your route on them, so, in order not to get lost, wooden posts have been put into the glacier to mark the route.

Norvik (1962) enumerates no less then ten main routes over the ice, pointing out that the alternative to a 30 km trek to the farms in Oldendalen was a 340 km journey down to the fjord, round to Vadheim and then overland.

Slingsby (1904, p. 249) noted that at the beginning of the nineteenth century many passes were used by traders crossing between Sogn, Jølster and Nordfjord, and suggested that their disuse was due to the shrinkage of the glaciers, though by then other routes were easier and new means of transport were available.

The systematic measurement of the ice fronts of Jostedalsbreen was begun by P. A. Øyen and J. Rekstad. From 1900 to 1940 annual reports appeared in *Naturen*, published by the Bergen Museum. Rekstad handed over responsibility to K. Fægri, of the Bergen Botanical Museum, in 1932. He tabulated the consolidated results from 1900 to 1947 (1948, pp. 301–2). Since then data have appeared in the yearbooks of the *Norsk Polarinstitutt* and the work has been in the hands of O. Liestøl. A complete set of Rekstad's glacier photographs is held by the Geology Museum of the University of Bergen.

The main features of the fluctuations after 1890 were two readvances of the glaciers of Jostedalsbreen, which reached their most forward positions in 1910 and 1930 (Figure 3.9). The retreat that had continued without major interruptions since 1750 was halted or greatly slowed and then, after 1930, was resumed at a greater rate than ever before.

It has long been recognized in Norway that glaciers respond to climatic events at different rates (Rekstad 1902, p. 14). In the early 1920s the shorter, steeper glaciers of Jostedalsbreen stated to advance about three years earlier than the longer ones, and generally speaking they are more sensitive to short-term changes in the climate. The shifts

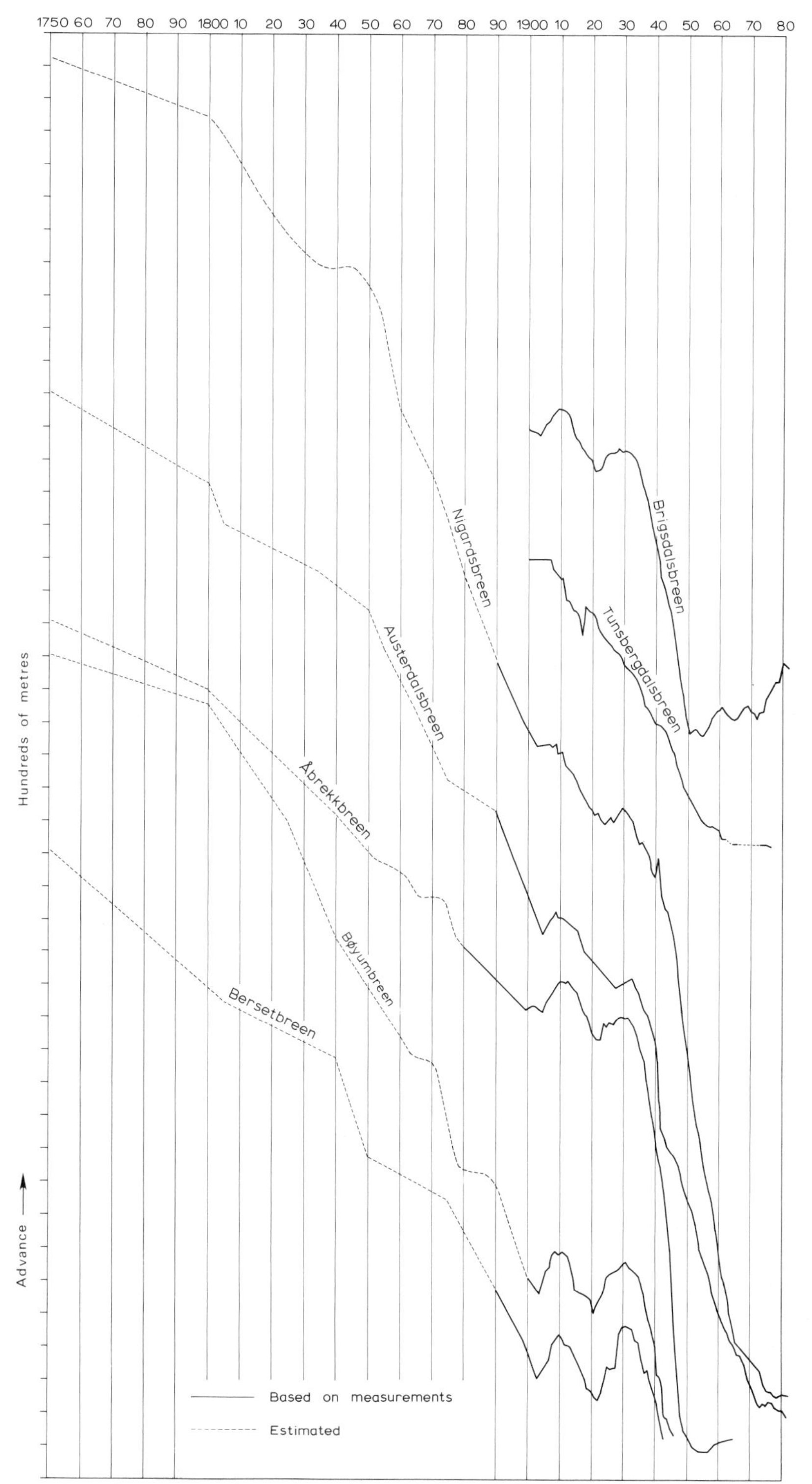

Figure 3.9 Retreat of Jostedalsbreen tongues from 1750 to 1980, based on Fægri (1933), Liestøl (1963), Kasser (1967 and 1973), Müller (1977) and Liestøl (1976, 1977a, b, 1978, 1979, 1980, 1982a, b)

in the positions of the termini reflect changes in the total volume of the ice, caused by variations in the relationship between accumulation and ablation, and between snowfall and snowmelt.

Olaf Rogstad (1941), of the Norsk Vassdrags og Elektrisitetsvesen (Norwegian Water Power and Electricity Board), made a study in 1940 in which he compared the discharge of the glacial rivers emerging from Oldendalen and Kjenndalen with the runoff from the nearby valley of Hornindal, which is free of ice. By this means he obtained some indication of the mass of water being stored in glacier ice or released from the ice reservoir as water in individual years between 1900 and 1940. His results showed that the advances and retreats of the ice in the two valleys lagged four years behind changes in their volumes. The response time is brief because the glaciers involved, including Brigsdalsbreen and Melkevollbreen, are relatively short and steep. Rogstad went on to show that advance of the tongues was associated with the ice masses increasing as a result of low summer temperatures and higher winter precipitation. He also attempted to obtain a figure for the enhanced thickness of the glaciers at the time of the Little Ice Age 1740–50 maximum; the value he computed was 60 m.

Between 1886, when regular meteorological observations in Norway began, and 1940, mean annual temperatures rose markedly, especially in the north of the country. The rise in temperature was 0.4 °C in Bergen, 1 °C in Finnmark, and 2 °C in Spitsbergen (Hesselberg and Birkeland 1944, 1956). It was enough to cause mean annual isotherms to be elevated 110 m, or to shift 300 km northwards. The warming, which was most marked in winter, was associated with a reduction in atmospheric pressure in northern Europe and an increased frequency of westerly winds. While this goes a long way towards explaining why the glaciers retreated, the understanding of the details of their variations involves more than generalized meteorological information and recording of tongue positions. Fortunately, regular mass balance observations on the Norwegian glaciers, begun in the 1940s, have provided much of the data required.

Glaciers act as reservoirs by storing winter snow and releasing meltwater in the summer when it is needed for agriculture. Glacier water has been used for irrigating valleys around Jostedalsbreen for several centuries and the notion of the icecap as a reservoir appears in the literature as early as 1758: 'One could consider it as one of nature's peculiar reservoirs which the valleys profit from whilst they are short of rains' (Jostedalsbreden, *etc.* 1758). Now water is required for generating electricity and, with a view to more thorough exploitation of the energy resources provided by the ice, studies of glacier mass balance have been made in various parts of the country since 1962. The methods used are described in Pytte (1969) and Østrem and Stanley (1969). It was at just about the time that these mass balance observations began on a selection of glaciers in both maritime and continental environments that a quite important fluctuation in climate was taking place. Records from stations examined by Hesselberg and Birkeland (1956) show that the mean annual temperature in the period 1941–50 was as much as 0.4 °C lower than in the previous decade, and the cooler conditions, accompanied by northerly winds and an increase of annual precipitation near the west coast of about 15 per cent, were to continue for several years.

One of the glaciers that has received most attention is Nigardsbreen, which constitutes almost 10 per cent of the total area of Jostedalsbreen. Østrem, Liestøl and Wold (1976) reconstituted the longitudinal profile of the glacier as it was in 1748 by mapping the trimlines on the valley slopes, which had attracted the attention of Forbes and Bohr,

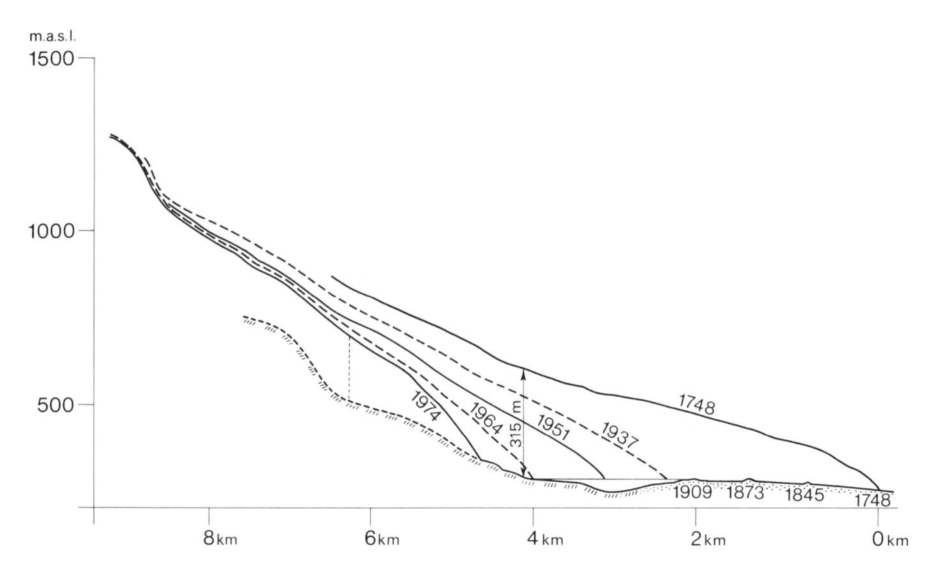

Figure 3.10 Thinning of Nigardsbreen since the mid-eighteenth century. (From Østrem *et al.* 1976)

together with the terminal moraine and other moraines on nunataks. Figure 3.10 makes it plain that the lower glacier thinned more between 1937 and 1974 than between 1748 and 1937. In the 1960s, however, shrinkage of the glacier diminished and between 1962 and 1975 mass balance studies showed that the total volume of Nigardsbreen increased by a mean annual thickness of 0.4 m of water equivalent, with positive mass balances amounting to about 2 m for both 1961/2 and 1966/7 (Figure 3.11). Up to 1975 the front of the glacier continued to retreat at a rate of about 40 m annually. It is a long glacier with a long response time. Between 1976 and 1981 the total volume of Nigardsbreen decreased by a mean annual thickness of 0.16 m of water equivalent, with a negative balance of 1.22 m in the single year 1979–80. But just at the time when the curve for cumulative mass balance for the glacier was tending to diminish, the front of the glacier ceased to retreat and even advanced slightly. Steeper tongues like Brigsdalsbreen and Åbrekkebreen had ceased to retreat in the early 1950s, as had Tunsbergdalsbreen by 1960.

3.2 Comparison of the Little Ice Age record from Jostedalsbreen and glacier fluctuations elsewhere in Scandinavia

3.2.1 JOTUNHEIMEN

Documents relating to the fluctuations of other ice masses in Scandinavia are scanty. Many of the valley and small cirque glaciers in Jotunheimen discharge into the Bøvra river, which is known to have caused severe flood damage in 1708 and 1743 (Rekstad 1900, Kleiven 1915). Floods at these times could well have been associated with glacial advances of the kind that were taking place around Jostedalsbreen in that period. However, there is need for caution in ascribing flooding to glacial advances unless there is specific evidence to that effect. The Bøvra also flooded severely in 1760 and 1763. The greatest flood of all, Store Offsen in 1789, which caused widespread devastation over much of southeastern Norway, was primarily caused by heavy rains that began in July

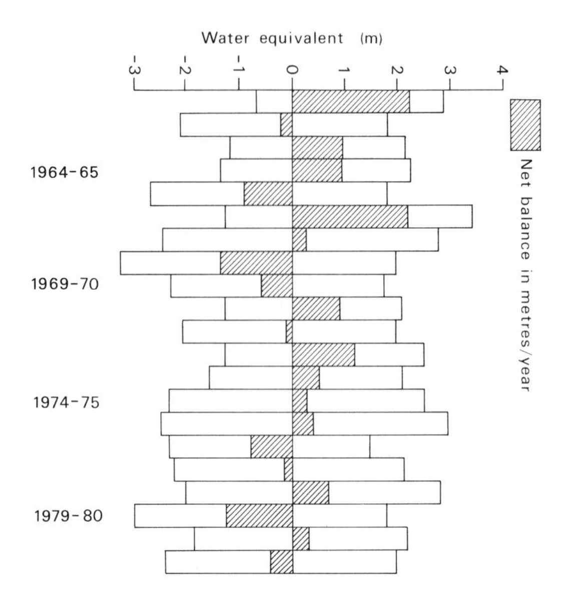

Figure 3.11 Mass balance of Nigardsbreen, 1962–82 (From Roland and Haakensen 1985)

and lasted for a fortnight, though, as we have seen, it can be argued that sluicing of snow from the glaciers may have augmented the flow of rivers draining glacial valleys. The Lensmann of Vågå may have been familiar with Jotunheimen when he stated that the glaciers grew markedly in the years before 1789 (Sommerfeldt 1972).

The few accounts of the Jotunheim glaciers available after 1820 were summarized by Øyen (1894). He himself worked in Jotunheimen between 1892 and 1909, but the measurement of frontal positions of glaciers which he initiated then lapsed until Werenskiold took it up again from 1933 until 1948. Such measurements as are available suggest a general accordance of behaviour of the glaciers of Jotunheimen and Jostedalsbreen. This view is supported by the dates obtained lichenometrically for the moraine sequences of Storbreen by Liestøl (1967) and Matthews (1974) (Figure 3.12) and for those of Svellnosbreen by Green (1981).

Mass balance studies of Storbreen in Jotunheimen, starting in 1948, provide one of the longest such records available for any glacier (Liestøl 1967, 1973 and 1978b) (Figure 3.13). A comparison of the records of Storbreen and Nigardsbreen shows that for every one of the years 1962 to 1974 the signs of the mass balances of these glaciers were the same, being either both negative or both positive. The cumulative mass balances for each of the two glaciers is very different, however, with Nigardsbreen thickening by 6 m over the period 1960/61 to 1974/5 while Storbreen thinned by 2.37 m. Between 1948/9 and 1974/5 the surface of Storbreen was lowered by 7.37 m and the front retreated every year between 1959 and 1975 (Kasser 1967 and 1973, Müller 1977). The differing cumulative mass balances of Nigardsbreen and Storbreen between 1960 and 1975 typify the contrasting behaviour and economies of the maritime glaciers of Norway on the one hand and the more continental ice masses on the other (Figure 3.14) (see Haakensen 1982, 1984).

3.2.2 FOLGEFONNI

Somewhat more data comes from Folgefonni, the third largest sheet of ice in Norway, on

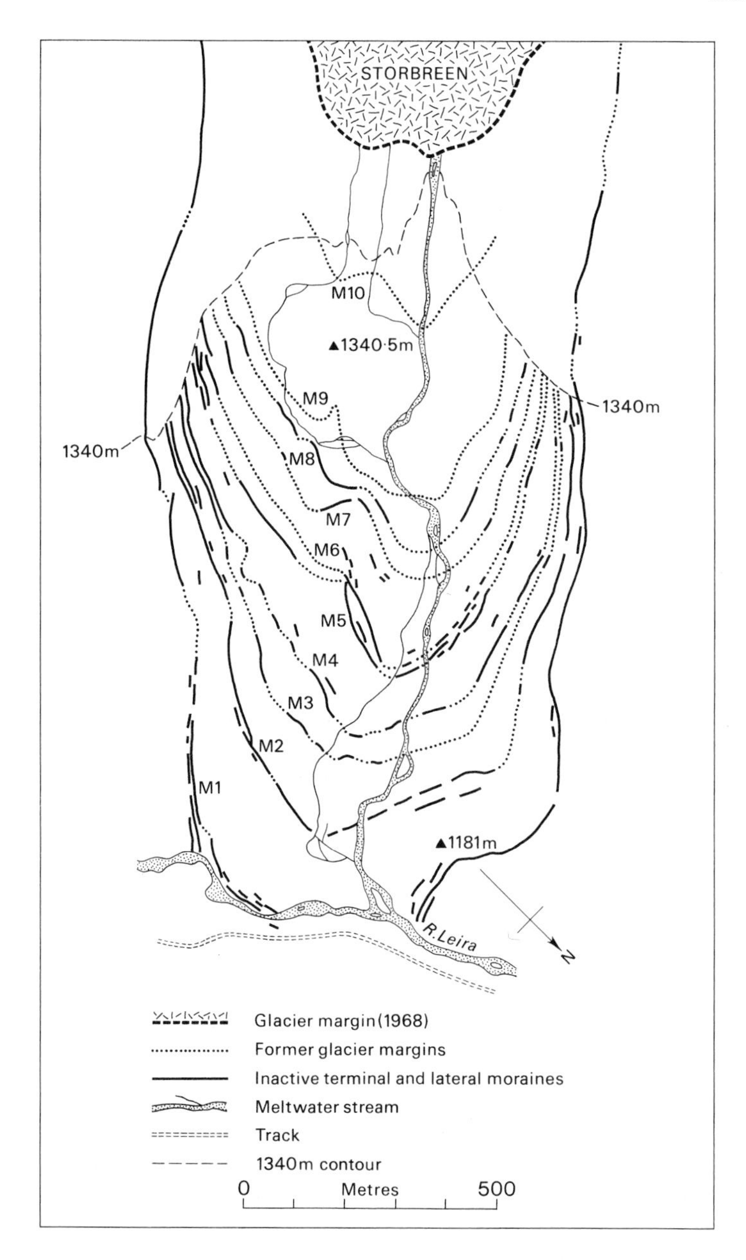

Figure 3.12 The forefield of Storbreen, Jotunheimen. Lichenometric dates of moraines: M1-1750, M2-1807/12, Ms-1833, M4-1858, M5-1871. (After Matthews 1974)

the peninsula between Åkrafjorden, Kvinnheradsfjorden and Sørfjorden (Figure 3.1). Two glaciers, Buarbreen in the east and Bondhusbreen in the west, flow from it down towards the coast and populated valleys. Most information therefore relates to these two tongues.

Hoel and Werenskiold (1962), in the acknowledgement section of their book, refer to a

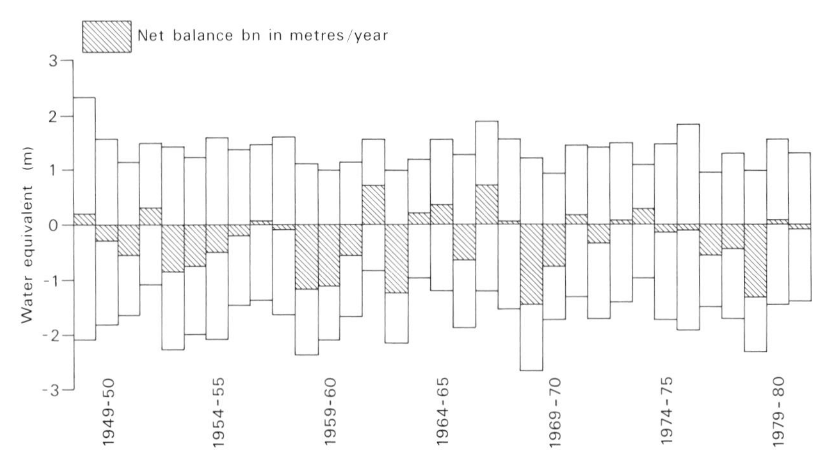

Figure 3.13 Mass balance of Storbreen, 1949–81. (After Liestøl's figures in *Norsk Polarinstitutt, Årbøker*)

1677 document dealing with an advance of Buarbreen 'in the period immediately preceding that year', but it has not proved possible to locate such a manuscript in Statsarkivet in Bergen. A court of inquiry held in 1677 found serious damage to Buar farm: 'a cruel mountain had fallen on the pasture . . . and the best part of it taken away, besides which the river had broken out of its course and carried away all the arable' (Tingbok for Hardanger Fogderi, 1677, fos. 14b and 15). The report is reconstituted but the missing parts are small and the original entry evidently quite brief. There is no reference to Buarbreen, nor is there any other reference to Buar in 1677 to be found in the Tingbok (A. Battagel, personal communication).

All the *avtak* reports from the period 1665 to 1815 for the area surrounding Folgefonni have been examined (Rentekammeret, Affældnings Forretninger, Hardanger og Sunnhords Fogderier, 1702–84). Many of the farms in the surrounding zone were damaged in the late seventeenth and the first half of the eighteenth century, as were those around Jostedalsbreen (see Figure 3.15). It is surprising to find that there is no record of Buarbreen having caused direct damage to Buar farm, although a court was held at Buar in 1744 because fields had been destroyed there by the great storms of December in the preceding year.

Øyen (1899) collected together many references to Folgefonni in the literature. He cited Hertzberg, a famous dean in Hardanger, to the effect that

> an old man in Strandebarm parish whose farm is so situated that Folgefonni can be seen from it, rising above a lower mountain has told me that his father said and he himself had noticed, that the glacier had increased in height for in his youth only the upper part of the glacier could be seen from the farm's buildings but now the same part of the glacier could be seen from a good way below the buildings. (Hertzberg 1817–18, p. 720, quoted Øyen 1899, p. 167)

Figure 3.14 Accumulated mass balances of Scandinavian glaciers. The glaciers in the interior, Storglaciären in northern Sweden, and Gråsubreen, Hellstugubreen and Storbreen in Jotunheimen, have been diminishing in volume, while the more maritime glaciers, Hardangerjøkulen, Nigardsbreen and Ålfotbreen in southern Norway and Engabreen in northern Norway have been increasing in volume. (From Kasser 1967, Kasser 1973, Müller 1977, Haeberli 1985)

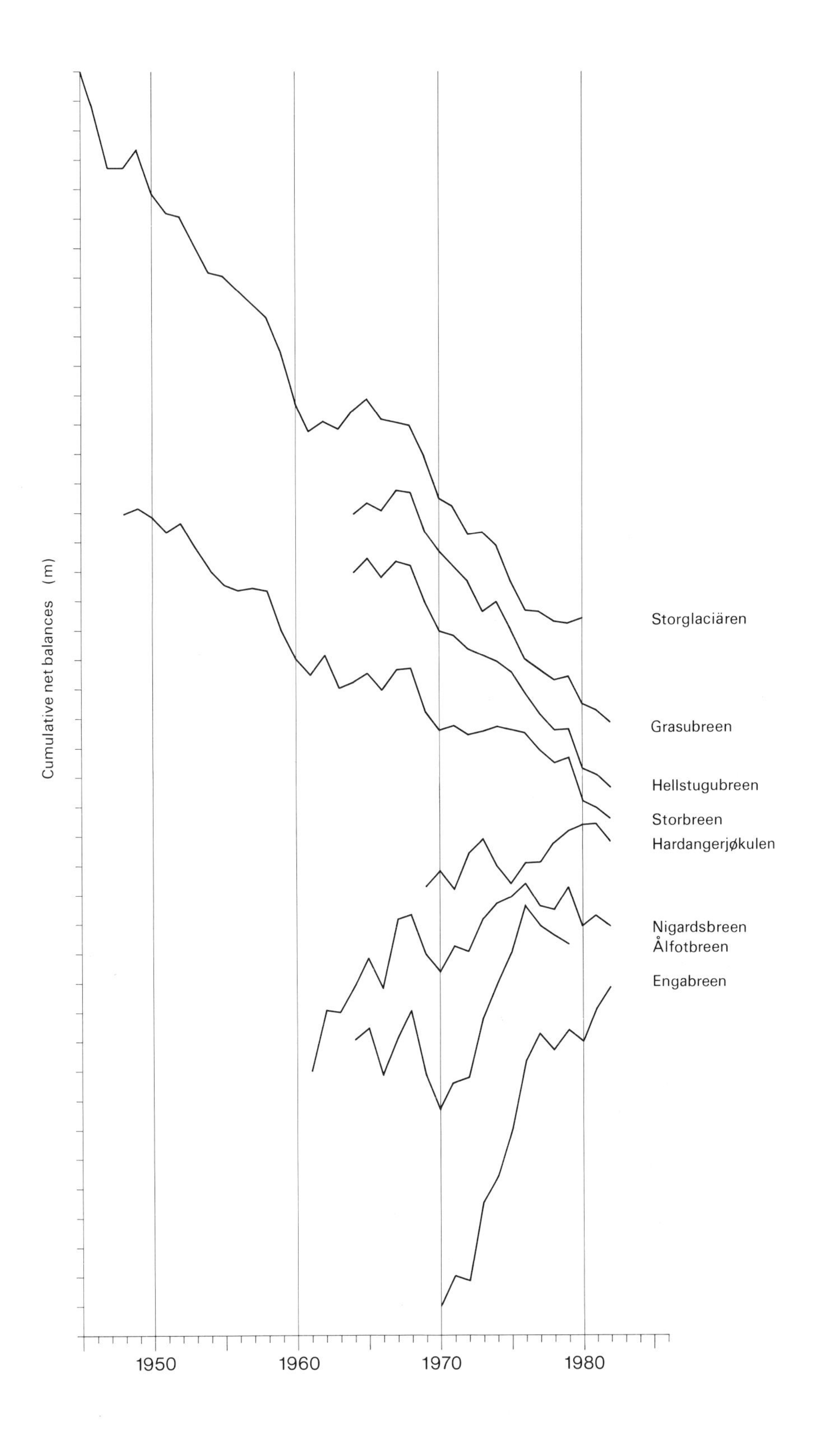
Cumulative net balances (m)
Storglaciären
Grasubreen
Hellstugubreen
Storbreen
Hardangerjøkulen
Nigardsbreen
Ålfotbreen
Engabreen
1950
1960
1970
1980

The reference here was probably to Bondhusbreen and the farm Bondhus farm (see Figure 3.14). It is unclear whether 'in his youth' refers to the old man or to his father, but the period of expansion in question would not seem to have been later than 1750 to 1760.

Bondhusbreen was advancing again in the early nineteenth century, for Hertzberg

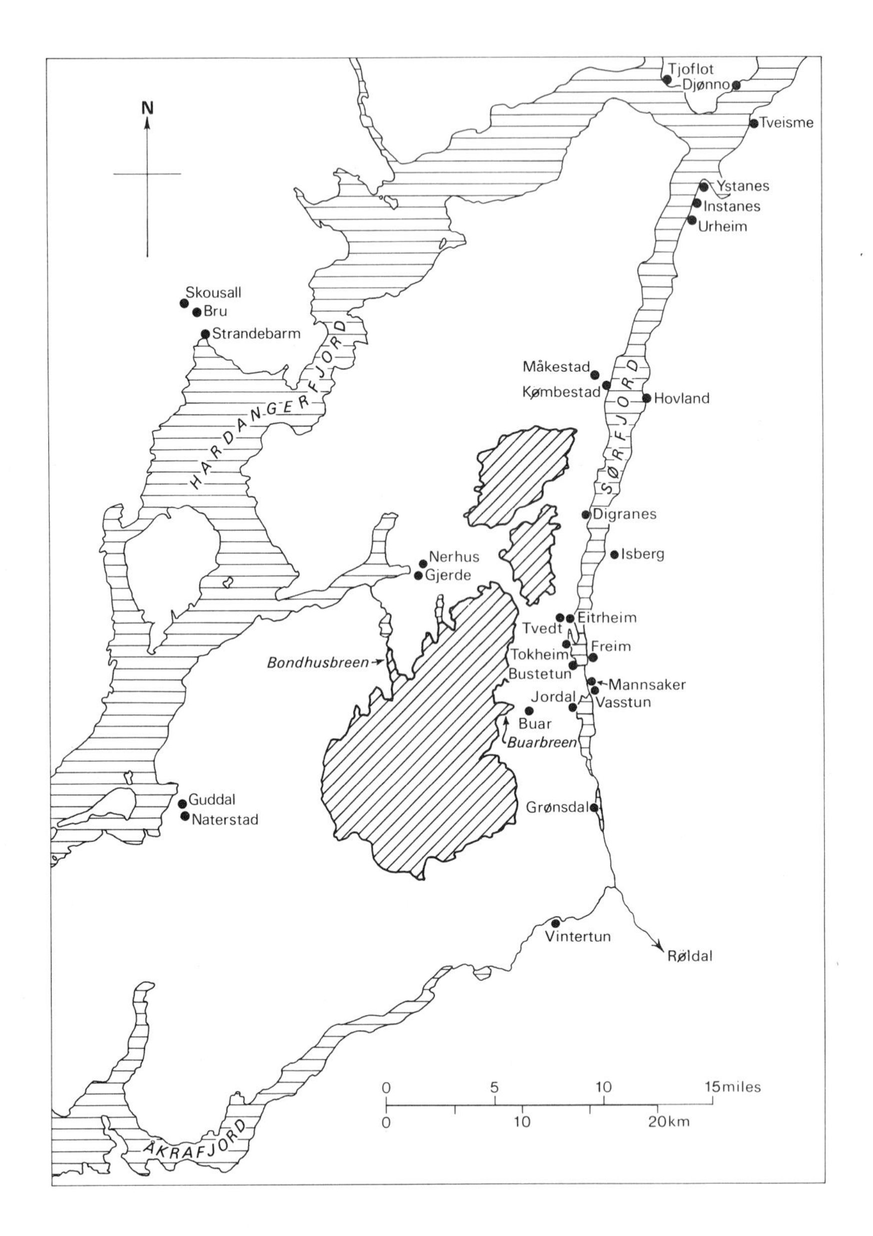

Figure 3.15 Folgefonni: farms in the vicinity seriously damaged by landslides, avalanching or flooding between 1665 and 1815. (Drawn from Norges Geografiske Oppmåling 1315,III and 1314,IV. Farm data from *avtakforretninger* reports)

reported that 'when I was up there in 1807 and took a look at this glacier, I saw clearly that a mass of ice had advanced, for a bank or collection of large and small rocks lay below the glacier's lowest edge, it being driven forward by the ice' (Hertzberg 1817–18, p. 720, quoted Øyen 1899, p. 170). Like Nigardsbreen, Bondhusbreen was retreating before the middle of the nineteenth century, for in 1845 Konow noted a 'significant retreat' (Øyen 1899, p. 176) and Bing reported that

> Folgefonni has thawed significantly during the last 25 to 30 years. . . . This applies particularly to the southern part above Mauranger and Odda and also to the central part, where rock now appears which formerly was always covered by the snowfield . . . whether the glacier tongues have retreated or advanced during the same period, the people here cannot say definitely. (Bing 1895, quoted Øyen 1899, p. 183)

The Folgefonni tongues advanced in the later part of the last century. According to Yngvar Nielsen, Buarbreen 'advanced more than 80 m in 1870, and 4 m in one week in 1871, but has now retreated 30 to 40 m'. In 1872 it advanced '70 yards' (Wilson 1872), while in 1874 it 'lay above the greensward and cows were grazing close beneath it' (Nielsen, quoted Øyen 1899, p. 188). By 1877, Sexe reported that 'since 1832 Buarbreen has advanced significantly some 2000 paces [*c.* 1700 m] and has covered a not inconsiderable stretch of cattle and sheep pastures' (Sexe, quoted Øyen 1899, p. 189). The front was still advancing in 1878 when Holmstrøm 'measured the distance from the front to Buar farm and found it to be about 947 m'. (The 1974 NGO 1:50,000 map, sheet no. 1315/III shows this distance to be 1350 m.) The following year, 1879, Nielsen visited the glacier again for the fourth time, twenty years after his first visit and five years after his third.

> I thus had a good opportunity to judge its advance, which had never been so apparent as now. The glacier had ploughed up the fresh turf and had rolled it in front of it up onto a small moraine which it continues to push forward. Turf, rock and grit and uprooted trees formed a moraine wall which was certainly remarkable while also being most unpleasant in a way, showing as it did the destructive elements of natural forces. Never had I seen Buarbreen like this. In 1874 it lay above greensward and cows were grazing close beneath it. It did not then appear to be moving, and it was not then easy clearly to imagine its rapid advance. Now, in 1879, the first glance was enough to show that powerful forces were at work here, and the first question I had to ask myself was, how long it would take before the ice tumbled into Buar farm's Innmark, the buildings lying only ten minutes' walk from the present terminal moraine. The glacier would soon reach a projecting rock-spur, which would presumably offer some resistance. It would either have to surmount it or take a course down through the narrow river bed before spreading out again over Buar's hayfields. Assuming it maintains its present rate of advance, this might reasonably take three years, a period which will be very interesting. (Quoted by Øyen 1899, p. 90)

Bondhusbreen was also advancing in 1879, though Nielsen wrote that 'this advance appears to be correspondingly slight as compared with the rate of Buarbreen'. He also got the impression that the icecap itself was simultaneously thinning, because new areas of exposed rock were appearing, and in 1897 he was told that the route across Folgefonni from Sørfjorden down to Gjerde in Mauranger was as good as over unbroken rock, as it still is in the 1980s.

Of Mysevatn, Yngvar Nielsen reported in 1879, 'the glacier is said once to have gone down into the lake, but has now retreated further from its bank', while Bing in 1896 recorded that Mysevatnbreen was by then 100 m from the lake, while fifty or sixty years earlier it had stretched so far into the water that it was not evident where the end of the lake was. In view of this it is not entirely surprising that Bondhusbreen was retreating in 1892 and had been in overall retreat since 1882. Buarbreen was still advancing in 1892–3 but retreated steadily in 1894, 'perhaps most in the last two years; it has retreated *c.*50 m and its height is much less. The same in Mauranger, Gjerde and Bondhusdalen' (Øyen 1899, p. 199). Richter visited the area in 1895 and formed the view that 'there cannot have been a large-scale advance for centuries as there is old vegetation to be found immediately behind the new moraine'. When he saw it, Bondhusbreen 'had a new terminal moraine 5 or 6 m in height not more than 50 m from the front'. In 1897 the glacier was still not as small as some of the older people remembered it to be. Bondhusbreen continued to retreat slowly until 1899 (Øyen 1899, pp. 217–18).

The late-nineteenth-century advances of the Folgefonni tongues appear to have brought them close to the position of maximum Little Ice Age extension or even to have equalled it. Those of the 1870s were apparently in phase with those of the Jostedalsbreen tongues, and measurements of the positions of Buarbreen and Bondhusbreen in the present century (e.g. Fægri 1948, p. 303, and Liestøl 1963, p. 188) show fluctuations paralleling those of the Jostedalsbreen tongues, with small advances in 1905 and 1910 and more marked advances in the 1920s, followed by much swifter retreat. This slowed down and was replaced by small advances at times until 1957.

The fluctuations of Blomsterskardbreen, a remote outlet of the southern part of Folgefonni, merit particular attention, although nothing is known about this glacier before the time when it was visited and photographed by Rekstad in August 1904. A photograph taken from the same vantage point in August 1971 showed it 200 to 250 m forward of its 1904 position (Tvede and Liestøl 1977). Examination of available maps, sketches and photographs reveals that the advance took place between 1920 and 1940, and probably mostly within the 1930s. Air photographs taken in 1959 and 1976 show only insignificant changes, with retreat of less than 50 m.

The southern part of Folgefonni receives, according to Tvede and Liestøl (1977), some of the heaviest precipitation in Norway. They calculate the mean annual precipitation to be between 5000 and 5500 mm on the higher parts of Blomsterskardbreen. The precipitation is estimated to be between 3500 and 4000 mm on the northern part of the icecap feeding Buarbreen and Bondhusbreen. Not only is the precipitation greater on the southern névéfield but Blomsterskardbreen is longer and less steep than the northern outlets. Tvede and Liestøl argue that it is to be expected that Blomsterskardbreen would have a longer response time. It is therefore reasonable to categorize the advanced position reached in 1940 as a response to the strong positive net balances around 1920 which, as we have seen, caused noticeable advances of some of the Jostedalsbreen tongues. However, Blomsterskardbreen is, according to Tvede and Liestøl (1977, p. 321), 'the only glacier in Scandinavia where a net advance has been documented within the last 70 years'.

3.2.3 NORTHERN SCANDINAVIA

Glaciers and icecaps in northern Norway and Sweden are generally remote and far from farming settlements, and consequently very few records are to be found concerning their positions at times earlier than the late nineteenth century. Evidence for earlier glacial events consequently depends on radiocarbon and lichenometric dating (Karlén 1982).

Svartisen (Figure 3.1), the second largest ice sheet in the country, lies on the Arctic Circle and is divided into two sections by Vesterdalen, running north and south. From the top 2 cm of a peat deposit beneath the outermost moraine of Fingerbreen, an eastern tongue of Svartisen (Figure 3.16), Karlén (1979) extracted a sample consisting of 70 per cent wood fragments. This gave a date of 695 ± 75 (I-10364) as a maximum for the deposition of the moraine. Another date of 600 ± 100 (St-6757) was obtained from the upper 2 cm of a peat layer under the forest bed of a delta formed in a lake dammed by the expansion of one of the tongues of the western section of Svartisen into Glomdalen. This sample was considered to predate the glacier expansion closely. However, Karlén (1979) pointed out that peats may be deposited over a long period of time; he also illustrated the considerable disparities that may occur between dates within the top few centimetres of a single peat layer and noted that the uppermost section of a peat layer could be missing. In view of these points, maximum dates obtained from peat layers cannot be expected to indicate directly the age of the ice advance succeeding them. These two ^{14}C dates from Svartisen could conceivably be compatible with thirteenth- or fourteenth-century advances, but they cannot be used as proof of such advances.

Numerous sets of moraines in the Sarek and Kebnekaise mountains of Sweden between 67° and 68°N, as well as in the Svartisen area of Norway, were mapped by Karlén (1973, 1979) and Karlén and Denton (1976) and dated by radiocarbon analysis of soils and by lichenometry. Measurements made on both *Rhizocarpon geographicum* and *Rhizocarpon alpicola* thalli were consistent, with moraines known to be older on a basis of geomorphological mapping bearing larger thalli. Maximum lichen diameters on control surfaces of known age in the Sarek area agreed closely with those of similar age in Kebnekaise. Thus in both regions surfaces dating from 1900 to 1916 bore lichens 21 to 27 mm in diameter, while lichens on surfaces from the seventeenth century gave maximum values of 66 to 85 mm. Only a few surfaces of known age were available in Sarek National Park and so, in view of the close agreement found with Kebnekaise, a lichen growth curve was constructed using twenty-one control points drawn from both Kebnekaise and Sarek. Twelve of these datable surfaces were twentieth century, mostly exposed at times determined from travellers' accounts and photographs. The older control points were provided by copper and silver mine tips from the seventeenth and eighteenth centuries. Most of them were datable only to within a certain span of years and two of them were ambiguous. One of the copper tips was formed between 1884 and 1902 and another either between 1745 and 1751 or between 1699 and 1702.

The growth curve used to date the moraines around Svartisen and the little glaciers of the Okstindan and Saltfjellet areas was based on the ages of fourteen surfaces, the great majority late-nineteenth-century or early-twentieth-century mining tips (Figure 3.17). Only one pre-nineteenth-century control point could be found: a silver mine tip at Nasa, in operation in the seventeenth century. Karlén (1979) gathered lichenometric data from about 125 moraines and concluded that the glaciers in northern Norway had reached advanced positions in the early 1300s and had subsequently retreated. He identified

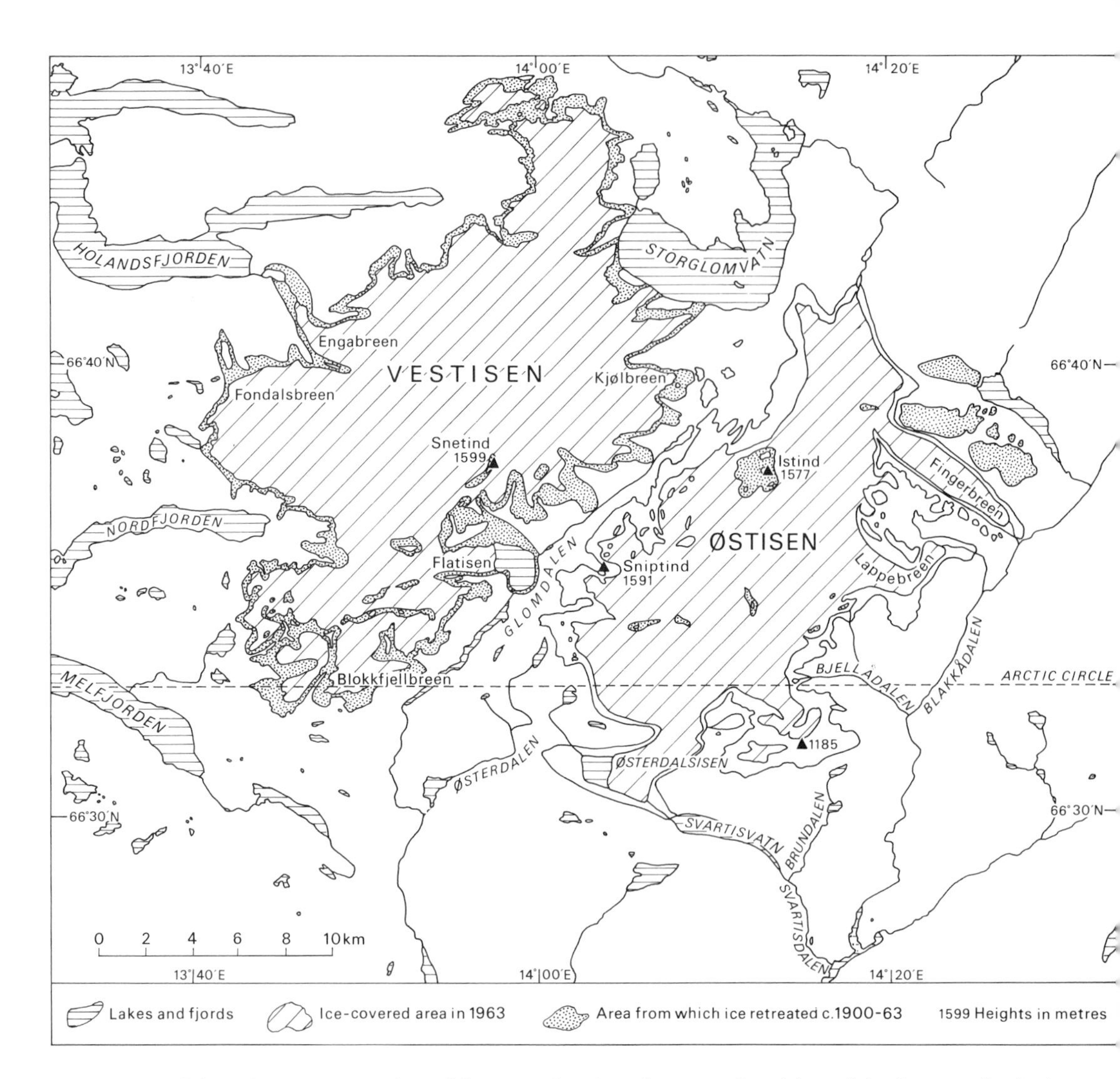

Figure 3.16 Svartisen area, northern Norway, showing changes of position of the ice margin during the twentieth century. (After Theakstone 1965)

further periods of glacier expansion towards the end of the sixteenth and in the mid- and late seventeenth centuries. 'These results also permit relatively good dating of several of the youngest general periods of glacial maxima. These occurred at AD 1780, *c.* 1800, 1810–1820, 1860, *c.* 1880, 1900–1910, and then *c.* 1930.' These results compare closely with those from Kebnekaise and Sarek, where individual Little Ice Age advances culminated about 1590–1620, 1650, 1680, 1700–20, 1780, 1800–10, 1850–60, 1880–90 and 1916–20 (Denton and Karlén 1976).

Karlén's field investigations certainly demonstrated the complex nature of the Little Ice Age in northern Scandinavia. The dates of many of the minor episodes of glacial expansion he identified correspond to those in the Alps (e.g. Ladurie 1971, Messerli *et al.* 1978, Bray 1982). However, lichenometric dating is insufficiently precise to differentiate minor episodes in the last few centuries unless the lichen growth curves employed have very narrow error limits, and it does not seem that the error limits for the curves for northern

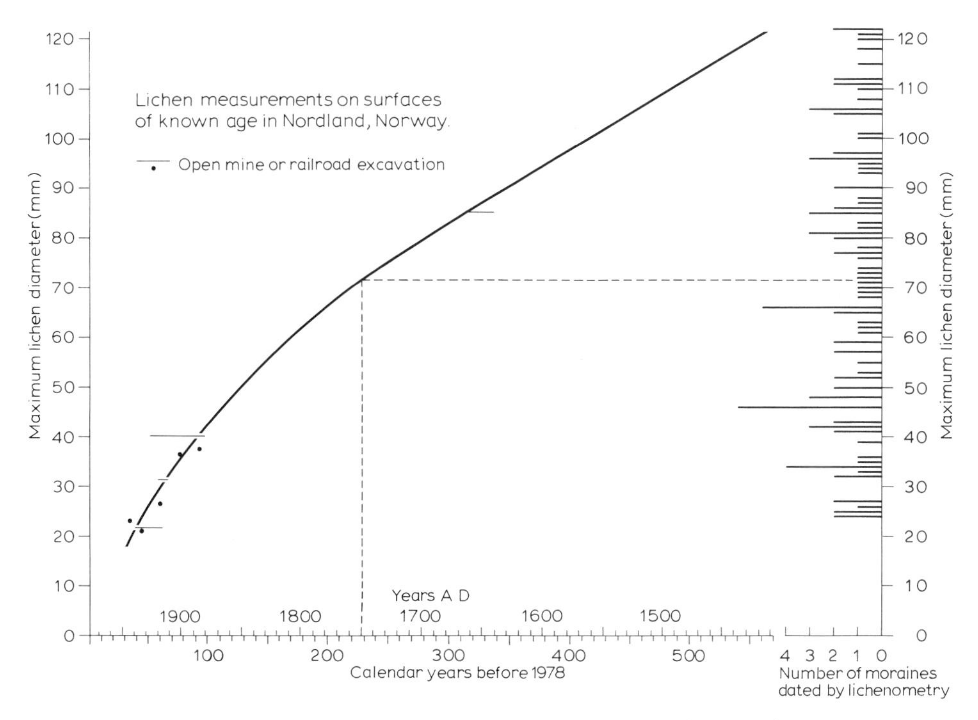

Figure 3.17 Lichen growth curve for the Svartisen area. (From Karlén 1979)

Sweden and Norway are sufficiently narrow. Innes (1982, 1983) has pointed to differences in the growth rates of *Rhizocarpon alpicola* and species within the *geographicum* group which result in dating errors if these groups are not treated separately. This work indicated the need for verification of the curves for northern Scandinavia. Innes (1985) subsequently derived an independent growth curve for the area immediately south of the Svartisen icecap and applied it to some of the moraines of Blockfjellbreen (a southern outlet of the western section of Svartisen) which had also been dated by Karlén. This curve was primarily based on measurements of *Rhizocarpon* section *Rhizocarpon* lichens on acidic igneous and basic igneous gravestones in two cemeteries in Dunderdalen (between 45 and 68 m above sea-level) and upon a railway embankment built in 1934 at 130 m above sea-level. It indicates that the lichens grew faster than Karlén had supposed; this is confirmed by photographs taken in 1910 which show Blockfjellbreen extending into an area which, according to Karlén, had been occupied by ice as long ago as 1780–1830. Innes attributes the discrepancies between the curves to differences in sampling methods and in particular Karlén's reliance on mining tips and ^{14}C dated surfaces. The tips are often the products of intermittent mining activity and therefore difficult to date precisely. Furthermore, as Innes shows, lichen growth rates differ from mica schist to granite–gneiss and may be affected by the metal in ores. The activity of carbon in soils is known to vary with depth (Matthews 1980, 1981, 1984), uppermost horizons are liable to be destroyed, and laboratory error ranges are quite wide. Consequently it seems that Karlén's dating of moraine sequences, though a considerable advance on what has been done before in northern Scandinavia, is not sufficiently accurate to allow them to be used as a basis for comparing Little Ice Age sequences there with other regions.

Of the glaciers in northern Norway, most information comes from Svartisen. It is clear from the 1743 Matrikkel that Engabreen was by then already far advanced. According to Petter Dass (1647–1707), the glacier reached the shore fifty years earlier (cited Liestøl 1979). The 1723 draft Matrikkel lists three small farms in the Svartisen area, Funøren, Fundahlen, and Storsteenøren, two of which had their taxes reduced. Storsteenøren (Figure 3.18) was described as being damaged day by day by the river and the glacier and had been deserted, the farm buildings having been carried away by the ice. Fundahlen, with frostbound soil, 'is damaged year by year by the glacier'. According to Øyen (1899),

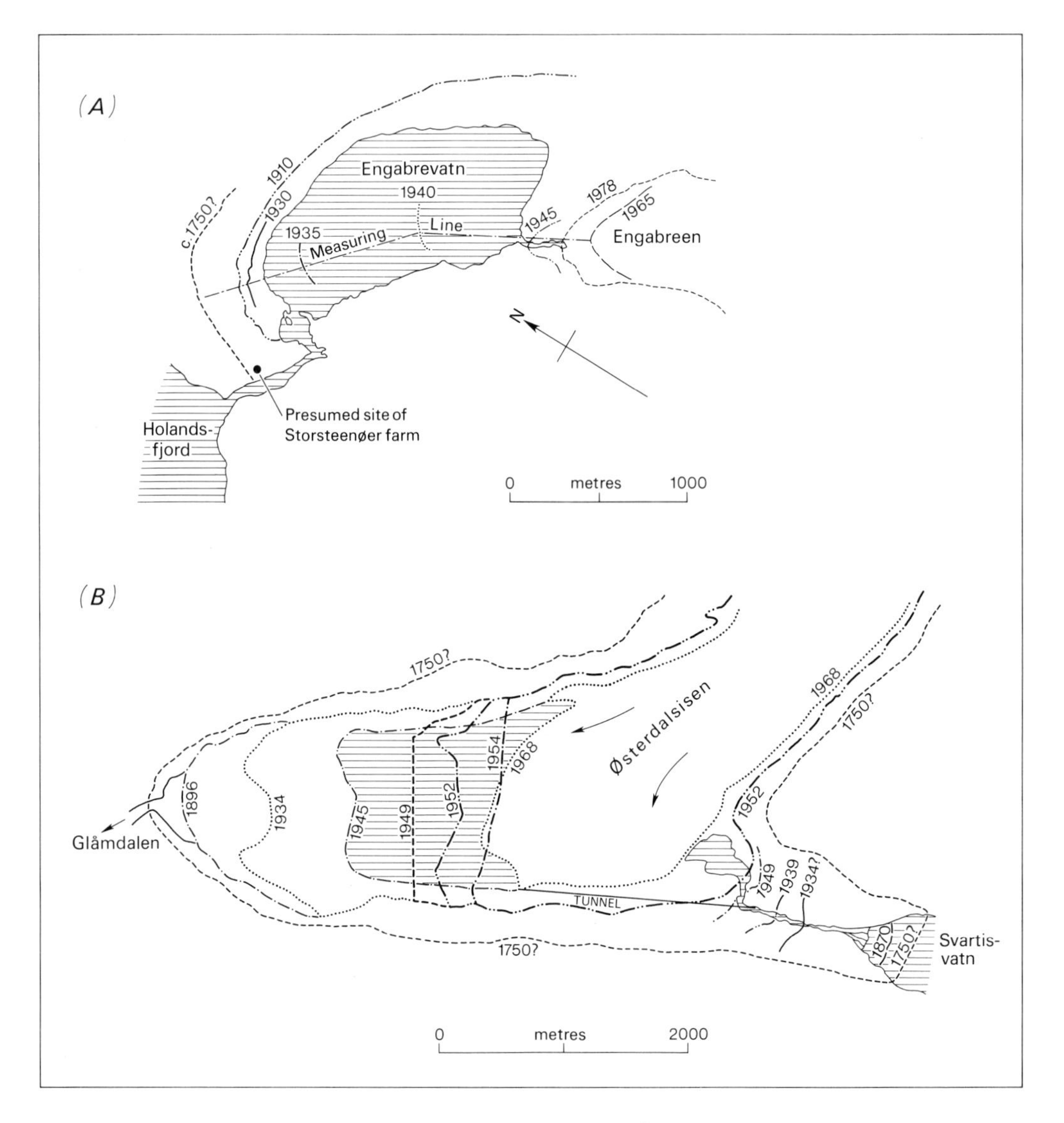

Figure 3.18 The frontal positions of (a) Engabreen and (b) Österdalsisen since the mid-eighteenth century. (From Liestøl 1979)

there was still a tale amongst the people at Rødøya and Meløya that 'when the corn was about to be cut the ice came so close to the buildings that the inhabitants had to leave them and shortly afterwards the glacier buried everything'. The *landskyld* which had amounted to 31 *laup* 1 *pund* for seventeen farms in the district in 1667 was reduced by 6 *laup*, 2 *pund* and 16 *merker* in total on the 1723 draft Matrikkel, presumably reflecting the deterioration of climate that had taken place. The rolls of 1740 and 1745 listed Storsteenøren as still derelict.

Theakstone (1965) summarized the data available on the glaciers of Svartisen and their fluctuations. He found little evidence for the period 1723 to 1800. In 1800 Engabreen was so close to the sea that it was reached by the water at flood tides (Rekstad 1893). By 1800 slow retreat was under way and the ice was a hundred feet from the outermost moraine. In 1865 Geikie found it ending in a small lake with a plain of shingle and alluvium in front of it. Østerdalsisen, the southernmost of the Svartisen outlets, was visited by de Seue in 1873 and found to have advanced since 1870, but by 1881 Rabot found it had retreated to its 1870 position again, while Engabreen was now 1 km from the fjord. In 1891 Østerdalsisen still reached the edge of Svartisvatnet, as it had in 1873, but no longer carved into it as it had then, and the glacier surface was 15 m lower than it had been twenty years earlier. Fingerbreen, on the east of Østisen, was also in slow retreat. Rekstad examined much of Svartisen in 1890 and 1891 and found widespread evidence of recent retreat. This diminution continued through the next decade. Kaiser Wilhelm II measured a retreat of Engabreen of 60–80 m in 1898–9 and in 1898 Rabot was told by a local farmer that the glacier had retreated continuously for the last fifteen years.

Small advances occurred in the early twentieth century. Engabreen began to advance in 1903. In 1909 Rabot found that it had advanced 100 m and Fonndalsbreen by about 60 m between 1907 and 1909. This expansion was not long-lived. Marstrander made extensive observations of the Svartisen glaciers in 1910. Østerdalsisen was thinning rapidly and local people told him that it had retreated 300 to 400 m in the previous twenty years. Flatisen had also retreated considerably since Rekstad's visit in 1890 and Fingerbreen's tongue was only half as wide as it had been when noted by Rabot in 1882. Marstrander reported signs of a new advance in a number of small, steeply sloping glaciers but found all the major trunk glaciers unaffected.

Annual measurements of the fluctuations of Engabreen and Fondalsbreen were made from 1909 to 1943 and published in the *Bergens Museums, Årbok* (see Figure 3.19). Although minor advances occurred in the period up to 1930, retreat predominated and after 1930 accelerated as it did in the Jostedalsbreen tongues. By 1950 the lower part of Fonndalsbreen was detached and Engabreen had retreated 400 m since it was last measured in 1943 and 2 km since 1909. Mean annual surface lowering of Østerdalsisen was about 7 m from 1960 to 1963 and this glacier, of all the Svartisen tongues, was suffering the most serious diminution. By 1965 Vestisen and Østisen were surrounded by a number of stagnating ice masses which had separated off from these two icecaps into which Svartisen is divided. The period of rapid retreat was broken in the 1960s in maritime northern Norway, as it was further south near the coast. By the mid-1960s Engabreen was advancing once more. Positive mass balances were recorded in five of the six years 1970/71 to 1975/6 but must have started some years earlier, before observations began. The contrast between the increases in volume registered for Engabreen and the contemporary decreases registered for Storglaciären further east in Sweden echoes that between maritime and continental glaciers in southern Norway in recent years (Figure

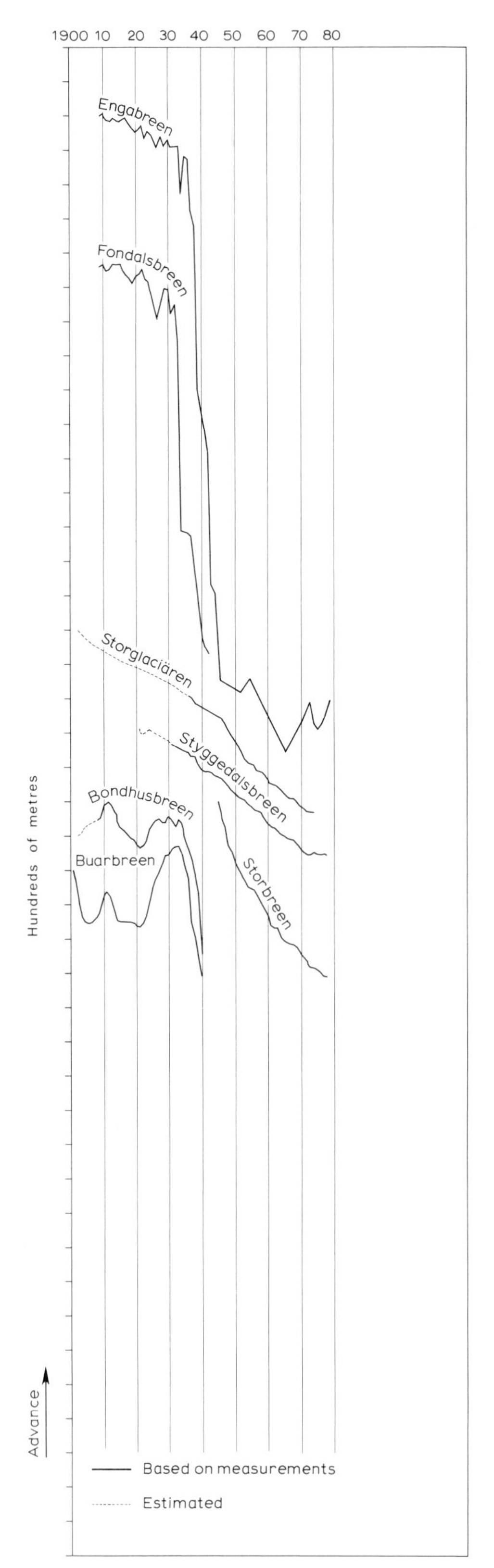

Figure 3.19 The frontal positions of selected tongues of Folgefonni (Buarbreen and Bondhusbreen) and Svartisen (Engabreen and Fonndalsbreen), together with valley glaciers in Jotunheimen (Styggedalsbreen and Storbreen) and in northern Sweden (Storglaciären). (Data from Fægri 1933, Liestøl 1963, Kasser 1967 and 1973, Müller 1977 and Liestøl 1976–82)

3.14). Summer temperature has been the main factor affecting the continental glaciers such as Storglaciären in the north and Gråsubreen and Hellstugubreen in the south, whereas in the more maritime areas, although variations in temperature have been almost the same as in the interior, increased accumulation has played the decisive role.

3.3 Summary

No convincing evidence of expansion of Scandinavian glaciers before the seventeenth century has been found. The enlarged state of glaciers and ice sheets in the eighteenth century was common to both north and south, as was the great recession of the twentieth century. The great majority of the Scandinavian glaciers about which we have information reached their maximum size in the mid-eighteenth century. Waning after this was at first slow and broken by halts or minor readvances. Rekstad recorded an advance of Brigsdalsbreen between 1869 and 1872, while de Seue found Bøyabreen and Lodalsbreen advancing in 1869. When he visited northern Norway in 1873 he found Østerdalsisen larger than it had been in 1870. The small advance and subsequent stationary state of Engabreen in the first three decades of the twentieth century coincided with the time when many of the Jostedalsbreen outlets advanced or halted. Superimposed on the generally coherent pattern are variations caused by differing degrees of exposure to oceanic influences. Figure 3.14 shows the cumulative mass balances of a selection of glaciers and demonstrates very clearly the contrast between the decreasing volumes of the more continental ice masses in both Norway and Sweden and the increasing volumes of those in more oceanic situations near the Norwegian coast. Mass balance studies carried out in recent decades have shown that the extent of winter accumulation has dominated the total change in volume of ice masses such as Folgefonni, while summer temperature has been the dominant control for those situated further inland (Haakensen and Wold 1981). It is significant that the most deviant behaviour of any of the glaciers has been that of Folgefonni, one of the most oceanic of the Norwegian glaciers, with Buarbreen and Bondhusbreen advancing almost if not as far in the late nineteenth as in the mid-eighteenth century. Blomsterskardbreen is apparently the only known glacier in either Iceland or Norway to have reached its maximum extent in the twentieth century.

The accordance of the major features of the history of Scandinavian and Icelandic glaciers is apparent. It is particularly noteworthy that the eastern tongues of Vatnajökull in Iceland not affected by surging reached advanced positions, even maxima, in the late nineteenth century at the same time as the glaciers of Folgefonni in Norway: both depend on very heavy precipitation. Blomsterskardbreen in 1980 is something of an anomaly, being near its maximum position; its expansion in the 1930s, it might be noted, overlapped with an anomalous, extraordinarily rapid advance of the Leirufjördur tongue of Drangajökull in northwest Iceland.

Chapter 4

The massif of Mont Blanc

The snowcap of Mont Blanc is the highest summit of a mountain massif stretching 50 km from near Martigny in the northeast to St Gervais in the southwest. The group imposes its 20-km wide bulk between France and Italy, and protrudes into the Swiss canton of Valais in the north (Figure 4.1).

The massif is a part of the Hercynian system, in which sedimentary rocks were metamorphosed by granitic magmas. The centre of the massif is granite and it is this rock which gives the near vertical slopes of such famous peaks as the Dru, the Grandes Jorasses and the Aiguille du Géant. At very high levels the granites are covered by crystalline schists and other rocks, which are exposed in the Grands Mulets and near the summit of Mont Blanc. The massif is delimited by valleys closely associated with geological structures. On the French side the deep trough valley of Chamonix is drained by the upper Arve and the more gently pastoral valley of its tributary the Bon Nant. The main watershed lies on the eastern side of the massif, where the Italian face towers above the valleys of Allée Blanche – Val Veni and Val Ferret in a series of precipices on a Himalayan rather than an Alpine scale. The Italian valleys are drained by the Doire streams, joining at Entrèves to form the Dora Baltea, which flows down the Val d'Aosta to the Po. Between the curving arms of the eastward and westward valleys, intricate chains of splendid peaks and aiguilles separate the valley glaciers fingering their way down from the domes and basins of the high snowfields. The mountains are festooned and flanked by numbers of smaller glaciers, some tributary to the larger ice streams, others now isolated.

An unusual variety and volume of evidence is available about the fluctuations of the glaciers of the Mont Blanc massif. Although cartographic coverage was poor until the nineteenth century and, as late as 1863, Adams-Reilly (1864) found it 'strange that the chain of Mont Blanc should be the most visited and at the same time the worst mapped

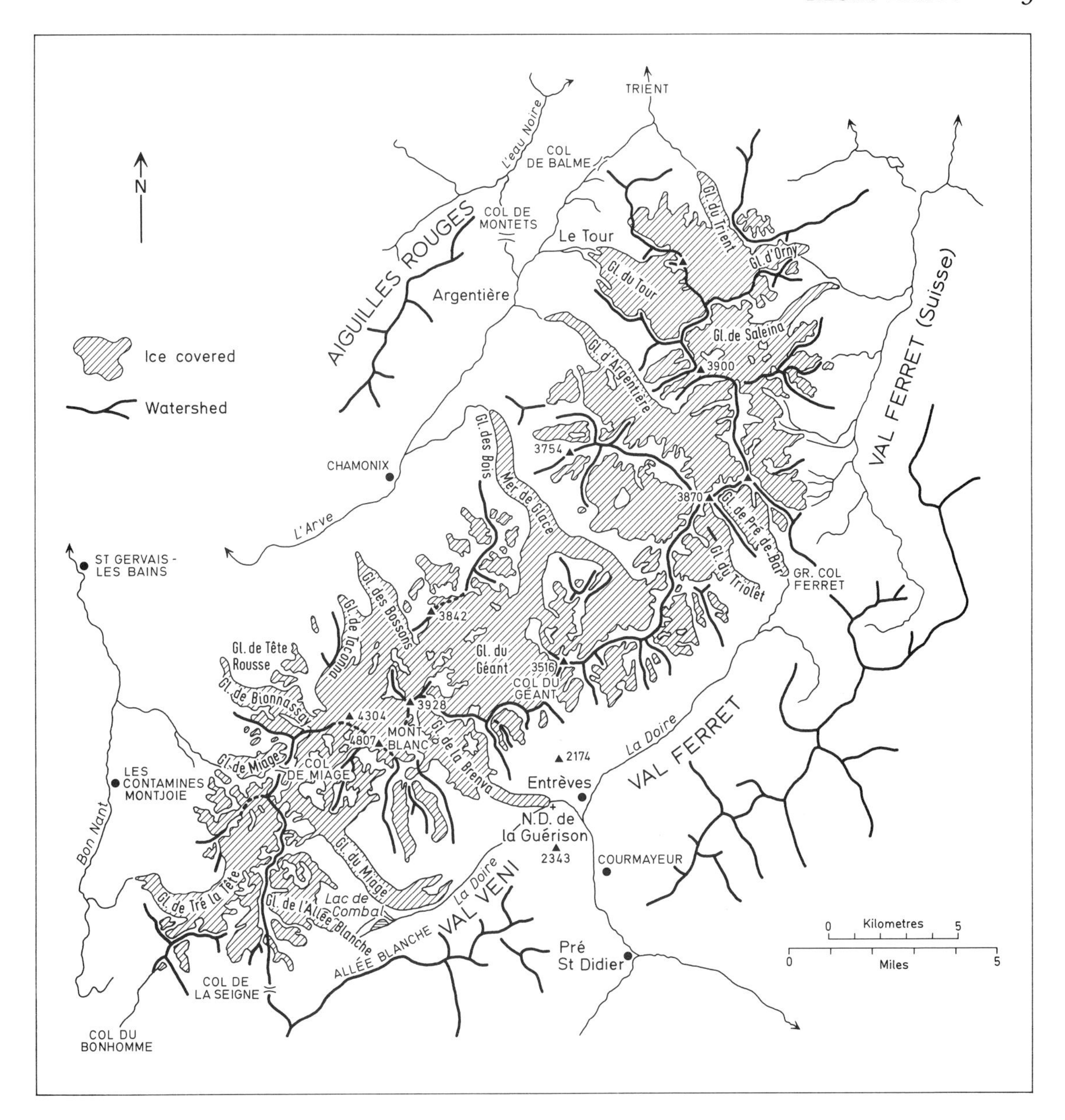

Figure 4.1 Glaciers of the Mont Blanc massif in the mid-twentieth century

portion of the Alps', a wealth of good maps has appeared since.[1] The glaciers have been the subject of well over a hundred papers as well as several books (Bordier 1773, Bourrit 1773, 1776a and b, Viollet-le-Duc 1876) and monographs (Marengo 1881, Mougin and

[1] Martel (1744) gave a sketch of the Arve area on which the glaciers of Chamonix are at least represented. That the map is rudimentary is hardly surprising as it was the product of three hours spent after dinner at Montanvers 'which time I was employed in making a plan of the glaciers'. Favre (1867, vol. 1) lists seventy-one maps earlier than that of Martel showing the Chamonix valley if not necessarily its glaciers. Sacco (1918) provided a useful sketch of the Italian cartography of the area, while Forbes presented a more detailed and

Bernard 1922, Vallot 1900). Some of these give valuable systematic accounts of the oscillations of particular glaciers for various periods (Rabot 1902, Mougin 1925, Bouverot 1958), or even of all the glaciers of a particular national sector for the whole of the Little Ice Age (Sacco 1918). Ladurie (1971) gives by far the most complete account available of the behaviour of the Mont Blanc glaciers as portrayed in original archival sources, while Vivian (1975) presents a valuable résumé of modern knowledge about glaciers of the French sector, with an extremely detailed discussion of changes of volume, area, length and flow during the period of scientific measurement, since the late nineteenth century. Documentary evidence for the Italian versant of Mont Blanc is scantier than for the French side. However, the fluctuations of the Brenva have been reconstructed in more detail than for the other Italian glaciers, using maps, photographs and more particularly paintings and other illustrations, as well as documents and instrumental surveys (Orombelli and Porter 1982).

4.1 The glaciers of Mont Blanc in the sixteenth century and earlier

Some notion of the state of the glaciers during the sixteenth century and earlier may be gathered from local tradition and myth. The story goes that the Brenva did not always occupy the bottom of Val Veni, as it does even today beneath its covering of moraine; in its stead were cultivated fields and meadows. But on 15 July of some year unknown, when the hay was dry and the weather fine, the villagers of St Jean de Pertuis failed to observe the Feast of St Margherita. Next day the glacier came down, engulfing the village and all its inhabitants (Forbes 1843, p. 207) (Figure 4.2) A tale mentioned by Virgilio (1883) gives another version of the catastrophe, whereby it was some old chalets 'de Pertus', on the slopes of Mont Noir de Péteret, which were carried away by the glacier.

Apparently there was once a village named after the martyr St Jean, who was killed by the Gauls in 'Pertu' during the reign of Emperor Maximian, at the end of the third or beginning of the fourth century (Viollet-le-Duc 1915, cited Orombelli and Porter 1982, p. 17). Pertu was probably near the site of modern Pertud. Dollfus-Ausset (1867) records the existence of a manuscript dating from 1300 which documents the existence of St Jean de Pertuis on the south side of Mont Blanc in the valley bottom, below the modern chapel of Notre Dame de la Guérison, but no such village is to be found in any tax list of the late Middle Ages from the Val d'Aosta (Ladurie 1971, pp. 221 and 327). According to Dollfus-Ausset, the village was destroyed by a landslide or rockfall. Orombelli and Porter concluded that his interpretation is the most probable and indicate that the Brenva may have advanced sometime after 1300. Glacial advances in the main period of the Little Ice Age have, they found, been associated with large rockfalls (Porter and Orombelli 1980, 1981, see Chapter 12). However, there seems little justification for ascribing the destruction of the village to the sixteenth century (Sacco 1918) or to around 1600 (Matthes 1942).

Radiocarbon dating of wood from the uppermost parts of the massive lateral moraines of the Brenva does not clarify matters very much. Orombelli and Porter (1982) pointed

critical account of that of the whole massif (1900 edition). In 1922 Vallot published a paper on the evolution of cartography of Savoie and Mont Blanc. A good coverage of the Mont Blanc area is provided by three 1:50,000 French sheets (Carte de France en 50,000, file XXXVI, 30, 31, 32) and the Institut Géographique National has produced a superb set of maps of the whole chain, based on air photography, on the 1:10,000 scale.

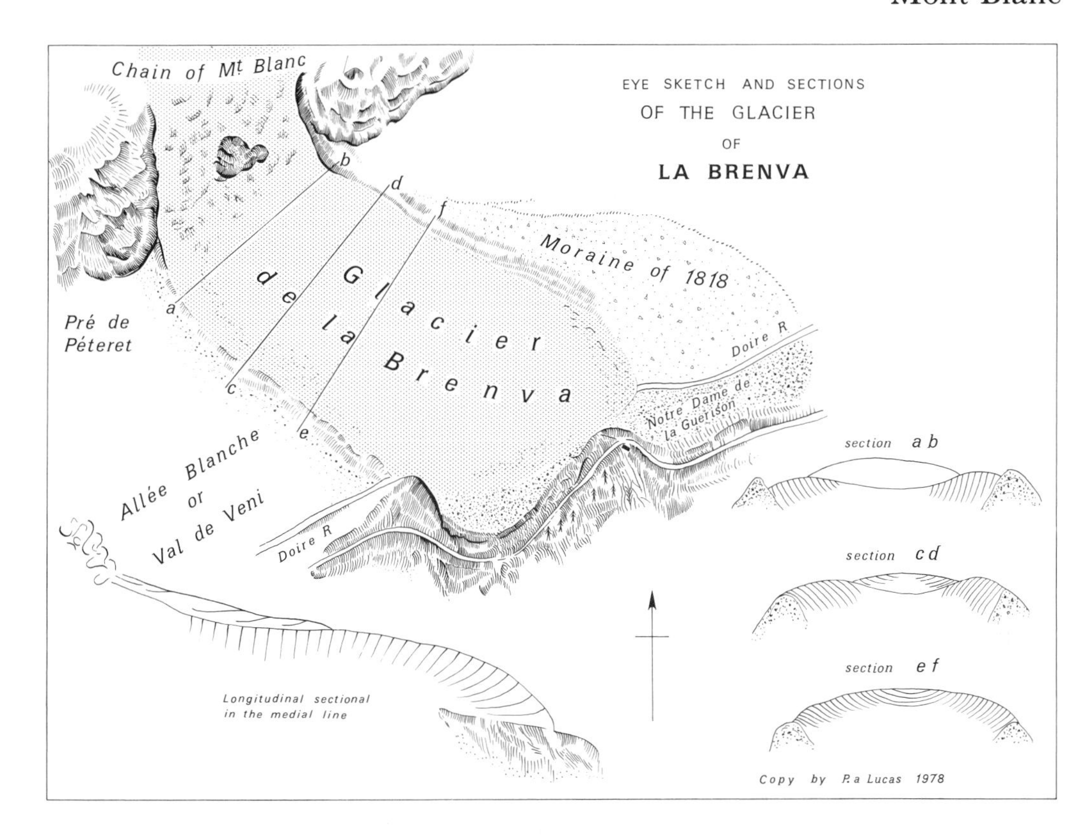

Figure 4.2 Sketch of the Brenva in 1742. The village of St Jean de Pertuis was presumably on the floor of Val Veni, somewhere within the area covered by the lower part of the tongue near the chapel of Notre Dame de la Guérison. Copied from Forbes (1900 edn)

out that a sample with an age of 285 ± 60 (UW-464), when corrected for atmospheric variation in ^{14}C could indicate an advance before 1490 or as late as AD 1800. It is worth stressing that Orombelli and Porter also found a log only 6 m below the crest of the Brenva lateral moraine with a radiocarbon age of 1170 ± 55 (UW-465) which, when corrected according to Stuiver (1982) at one standard deviation, gives a calendar age of AD 760 to 980. Thus the complex moraine system of the Brenva, which protrudes right across Val Veni at right angles, cannot be attributed merely to the oscillation of the terminus over the last few centuries but represents a sequence of accretions of drift, some of which may predate the Holocene. The moraines of the Miage glacier extend across Val Veni upstream of the Brenva and from time to time a lake, the Lac de Combal, formed behind them. Rabot (1902) suggested that the waters of Combal, overflowing at the same time as those of Rutour and uniting with them at Pré-St-Didier, caused the floods which destroyed the fortified houses of Rubbily and Rovary at Margex sometime before 1340. These events suggest the possibility of late medieval advances of the Miage and Brenva. On the other hand, no evidence has been found of ice advances at this time on the other side of Mont Blanc, in the valley of the Arve.

Ladurie (1971, p. 155) provides indirect evidence that the glaciers of the Arve valley began to swell in the second half of the sixteenth century. In the Chamonix area, tithes of villages and hamlets well away from the glaciers, or protected from them, rose during the sixteenth and early seventeenth centuries. But the nominal value of the tithes of those settlements that were later destroyed or damaged by glacial advances remained generally stationary in the sixteenth century and especially in the latter part of it. This stagnation during a period of inflation suggests a decrease in real productivity. For example, at La Rosière, which first appeared in the accounts in 1390 and was eventually destroyed by the Argentière glacier, the tithes fell from 50 florins in 1577–1600 to 32 florins in 1622, while at Vallorcine at a much lower altitude, the tithes were 92 florins in 1520, 160 in 1578–90, 300 in 1600 and 400 florins in 1622–5. The population in the Chamonix area decreased by half or more between 1458 and 1680, although population in the seventeenth century in southern France was generally higher than in the fourteenth and fifteenth centuries. At Le Châtelard, the number of taxpayers had already fallen from forty in 1458 to eighteen by 1559. Environmental decline is also reflected in reductions in dues paid in respect of mountain pasture. The pasture of Blaitière, under the glacier of Blaitière, for instance, was paying 5 to 7 lbs of cheese to the priory of Chamonix between 1540 and 1580 but only 3 lbs between 1580 and 1602. The deliveries of cheese to the priory fell from 52 lbs in 1540, to 45–47 lbs in 1550–70 and to 39 lbs in 1622.

Not only were the Brenva and other large glaciers safely tucked away in their valleys until the sixteenth century, but the passes were said to have been much easier. There was even a local tradition that the Col du Géant was once passable, or at least much more practicable than in 1788 when de Saussure made his famous observations there (1779–96, vol. IV). Windham, who with his companions was the first to describe a journey to the Chamonix valley in English (1744), wrote that

> Our guides assured us in 1741 that in the time of their fathers the Glacier [Mer de Glace] was but small, and that there was even a Passage through these Valleys, by which they could go into the Val d'Aosta in six hours, but that the Passage was then quite stopped up, and that it went on increasing every year.

This tradition was first recorded by a tax official called Arnod (1691), who tried vainly to force a way over to Chamonix in 1689, accompanied by three hunters 'avec des grappins aux pieds, des hachons et des crocs de fer à la main'. But Arnod made his attempt at least a hundred years too late, a circumstance which was not fully appreciated by Montagnier in his scholarly examination of the Col Major legend (1920–1, 1921–2).[2]

These stories certainly cannot be accepted at face value, especially as in the course of time they lost nothing in the telling; it was even sometimes said that the Chamoniards went over to Courmayeur to attend mass there! – though why they should have done this when there was a priory in Chamonix is not clear. But the tales are symptomatic, and

[2] Montagnier's exploration of the cartography and documentary sources of the Col Major legend led him to conclude that the tradition rested partly upon the identification of the 'Col Major' with the Col du Géant. Montagnier traced the cartographic vicissitudes of the Col Major in great detail and established that despite the contrary opinion of the great Alpine historian Coolidge (1908, p. 202) it could not be regarded as an exact geographical term synonymous with the Col du Géant. Montagnier concluded that the Col du Géant could not have been crossed within 75 or 100 years of the date of Arnod's attempt to reopen it. He thought that above the snowline much the same conditions must have been offered to travellers in the Middle Ages as in the twentieth century, and saw little reason for assuming that the glaciers had undergone any great change within the last thousand or more years at these higher levels. Montagnier took too little account of the effects of glacier thinning and thickening and of the fluctuations of the snowline associated with retreat and advance. The Little Ice Age in the Alps was well advanced by the time of Arnod and conditions certainly very different then from those a century or more earlier.

their essential basis is confirmed by an examination of seventeenth-century documents. A good start has been made (e.g. Röthlisberger 1974) but much research remains to be done on the early history of Alpine passes and cols, and the results would be of glaciological and climatic, as well as historical, interest.

The earliest useful account of the Mont Blanc glaciers occurs in a description of the valley of Chamonix by Bernard Combet, Archdeacon of Tarantais, who visited it in 1580 as arbiter in a tax dispute. He had, remarkably, sufficient interest in the local topography and scenery to record what he saw with some exactness:

> To the right, as you approach from the south, these mountains are white with lofty glaciers, which even spread through rifts in the mountains themselves, and descend almost to the said plain [of the Arve] in at least three places. One thing is clear: those rifts which people call moraines have sometimes caused unavoidable floods, both in the regions through which the waters descend and in the middle of the valley, where they swell the stream said to rise in the Alpages du Tour and which then forms into quite a large river [the Arve]. (Translated from the original by Ladurie 1971, p. 135)

Despite its apparent simplicity, there has been some disagreement over the interpretation of this text. Rabot (1915) identified Combet's three glaciers as Bossons, Argentière and Tour, and concluded that the advances of the Little Ice Age had not yet begun in 1580, if these were the only glacier tongues which could be seen from the plain. Blanchard (1913) thought the Tour much too far from the other two glaciers to have been intended, and also dismissed the Tacconaz because its tongue is always higher than that of the Bossons, its alimentation being more restricted. He therefore identified the third tongue as that of the Glacier des Bois, the terminal part of the Mer de Glace. But the Glacier des Bois is now concealed from view in a deep ravine, and if Combet saw it reaching 'almost to the plain' in 1580, this would imply that the advance was already under way. Ladurie (1971) pointed out that Combet specified his exact route, and that nowhere on it could he have seen the Tour glacier. Combet must therefore have been referring to the Argentière, Bossons and Mer de Glace. So the advance of the ice had already begun by 1580 and the Mer de Glace was then more extensive than at any time in the twentieth century.

4.2 The onset of the Little Ice Age 1580–1645

Between Combet's visit and the turn of the century there was a regular glacial invasion. Important tracts of cultivated land and forest were covered by ice and further areas ruined by glacier torrents. Flooding seems to have increased in direct proportion to the advance of the ice, the plentiful meltstreams from the enlarged tongues being reinforced from time to time by bursting water pockets and outflows from glacial dams. Crops in the valleys failed and yields became meagre with the deterioration of climate. The economy of the local people was severely affected and a series of supplications for tax relief were made. Conditions were officially investigated by both ecclesiastical and civil authorities. Surprisingly precise documentary evidence remains in the archives of Chamonix and Haute Savoie, from which the main pulsations of the advance can be dated, at least for the French side of the mountains. Both the extent and destructive character of the invasion, and the nature of the evidence upon which our knowledge of it is based, may be illustrated from documents concerning the Mer de Glace.

In 1605 the 'Chambre des Comptes de Savoye' made an inquiry into the justice of various requests for tax reduction, and found that by the time of the tillage reform, that is by 1600, the glaciers had spoilt 195 'jornaulx' of land in various parts of the parish of Chamonix, of which 90 belonged to the village of Le Châtelard (Figure 4.3) where twelve houses had been destroyed (Blanchard 1913). Le Châtelard, situated midway between Les Tines and Les Bois, was assessed for tithes, commonly paid in wheat between 1384 and 1640. In 1564–5 Chamoniards of the valley had bought houses there, and a widow called Perrette purchased some land there for her children (Letonnelier 1913). So Le Châtelard had very recently been a flourishing place and local people had had no doubts for the future. The document of 1605 described the glaciers as still advancing ('the said glaciers, whose ravages continue and progress from one day to the next') and mentioned that Les Bois had had to be abandoned because of danger from the ice, although the greatest advances seem already to have taken place in 1599/1600. In 1610 one Nicolas de Crans (Letonnelier 1913) reported again to the ecclesiastic authorities. By now the Glacier des Bois, 'which is terrible and frightening to look at', had almost completed the destruction of Le Châtelard and a good part of its land. It had also damaged another little settlement called Bonnenuict,[3] but trouble was not limited to that

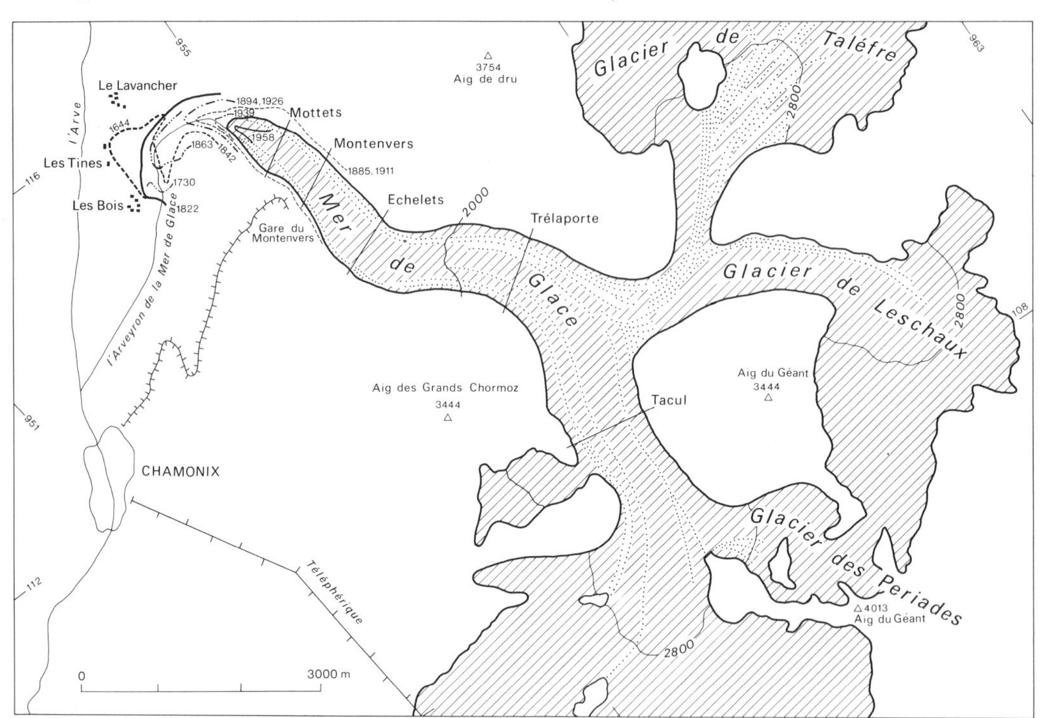

Figure 4.3 Tongue of the Mer de Glace (after Reynaud 1977). The rises and falls of the ice surface between 1890 and 1981 at the cross-sections shown on this map are illustrated on Figure 4.10. Details of the frontal positions from 1644 onwards are after Lliboutry (1965 p. 726) and Ladurie (1971, p. 145)

[3] Ladurie (1971), following and extending the work of Rabot (1920), investigated the positions of the villages destroyed by the Mer de Glace and concluded that Le Châtelard was midway between the present villages of Les Tines and Les Bois, while Bonnenuict or Bonanay was immediately north of the slope of Le Piget (Figure 4.3). The ruins of Le Châtelard were still to be seen in 1920.

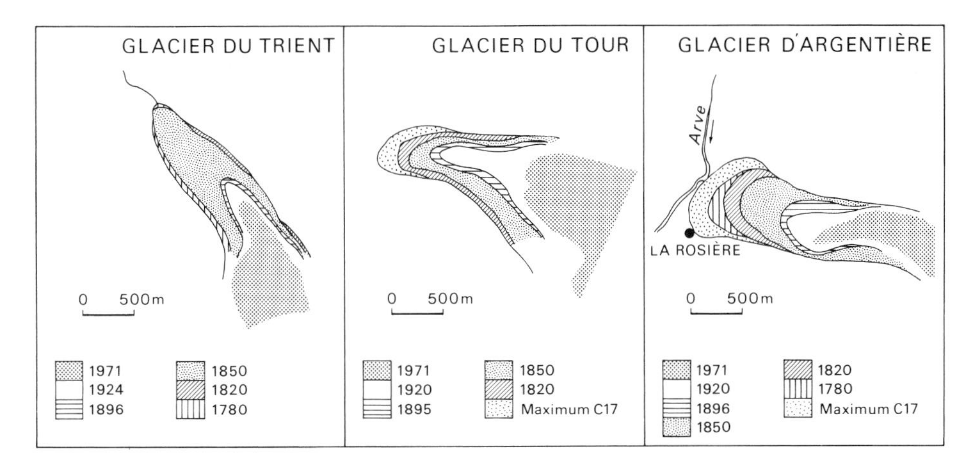

Figure 4.4 Tongues of the Trient, Tour and Argentière glaciers at various times since their maximum Little Ice Age extents. Note the greater extent of the ice in 1780 than 1820 and in 1820 than 1850. (After Bless 1984)

caused directly by the ice. Meltwater streams carried away whole houses and barns and ruined valuable land out on the plain. Ladurie (1971, p. 144) cites another document from 1610 which quantifies the types of damage. 'The streams have spoiled seventeen journaux of land since the tillage reform (1600). . . . The Arve has spoiled eight . . . and the glaciers two hundred and four and a half.'

There is no evidence from the documents of any recession between 1605 and 1610. The ice fronts may well have remained more or less stationary during the interval. Undoubtedly there was another advance in 1610. The Argentière glacier had already damaged the hamlet of La Rosière as well as the village of Argentière, covering seven houses by 1600 according to the 1605 inquiry (Figure 4.4). But 'it happened that on the twenty second of June 1610 by the overflowing of the glacier of La Rosière, eight houses and forty-five journaux of land were completely destroyed' (cited Ladurie 1971, p. 148).[4] In 1616 Nicolas de Crans

> went to the village of Châtelard where there are still about six houses, all uninhabited save two, in which live some wretched women and children, although the houses belong to others. Above and adjoining the village, there is a great and horrible glacier of great and incalculable volume which can promise nothing but the destruction of the houses and lands which still remain.

Crans visited the Argentière tongue and found

> The great glacier of La Rosière every now and then goes bounding and thrashing or descending; for the last five or six years . . . it has been impossible to get any crops from the places it has covered. . . . Behind the village of Les Rousier, by the impetuosity of a great horrible glacier which is above and just adjoining the few houses that remain, there have been destroyed forty three journaux [of land] with nothing but stones and little woods of small value, and also eight houses, seven barns,

[4] The extent of a 'journal' at this time is not known.

> and five little granges have been entirely ruined and destroyed. (Cited Ladurie 1971, pp. 148–9)

So in 1616 both the Mer de Glace and the Argentière had moved forward again and their steep termini were about a kilometre outside their 1980 positions.

These two glaciers were in phase with their neighbours. Ladurie (1971, p. 151) cites texts to show that not only was the Bossons glacier similarly enlarged but so also were the glaciers on the Italian side of Mont Blanc. A notary living in Aosta was visited by one Jacques Cochet of Les Bois on 6 April 1600. Cochet inquired whether it was true that the parishioners of Courmayeur had sent to Rome to request the Pope to pray that the glaciers might withdraw, and whether it was true that the Italian glaciers, and particularly the Brenva, had actually retreated. The people of Courmayeur had not appealed to Rome and, Cochet was told, the glaciers had not retreated and were as threatening as ever.

Thirty years later, in 1641–3, came another advance almost as great as before. The Tour, Argentière and Bossons glaciers were all menacing established settlements in the valley and the Mer de Glace had advanced in the previous two years, 'contre la territoire du village des Bois, des Prés et du Chastellard ença d'une mousquetade' (Blanchard 1913) (150 to 200 m). But the tongue only threatened and did not reach Les Bois, although it had been so nearly reached by ice in 1605 that its inhabitants had to abandon it. It follows that the glacier must have receded somewhat in the intervening years, perhaps between 1610 and 1628. In 1628 there was a great flood of the Arve 'caused by the glaciers' and a text of 1640, cited by Ladurie (1971, p. 340) refers to the 'third of good and cultivable land lost in about the last ten years through avalanches, falls of snow and glaciers'. It is improbable that we have evidence of all the individual incidents which occurred during the great seventeenth-century advances of the ice. Indeed Jean Duffong, administrator of the priory of Chamonix, described in 1643 how the glaciers had advanced 'again and again' (Blanchard 1913). Duffong related that he had led various processions to exorcise the local glaciers, the most recent having taken place in May 1643. The situation of the Chamoniards was certainly one of considerable distress; another passage of the same date refers to 'les personnes y sont si mal nourries quils sont noirs et affreux, et ne semblent que languissants' (Blanchard 1913). The burden of refugees from the more directly exposed settlements must have been heavy on their neighbours in the valley, struggling with diminished crop yields and recurring floods. In 1645 the Glacier des Bois was so much extended that it was feared that the Arve itself would be dammed, and Charles August de Sales, Bishop of Geneva (nephew of St Francis de Sales), decided to exorcize the glacier again. We learn from a document of 1663 that this was successful, for the glaciers retired little by little (Blanchard 1913).

The Courmayeur archives have unfortunately twice been burnt down, so we have little documentary information about the behaviour of the Italian glaciers between 1580 and 1645. Most is known about the Lac de Combal, which was formed by the Miage glacier whenever it sprawled out across Val Veni and blocked the Doire (Figures 4.1 and 4.5). Outbursts from Combal are recorded from 1594, 1595, 1629/30, 1640 and 1646; these dates fit in well with the glacier evidence from the Arve valley (Baretti 1880, de Tillier 1968). Apart from these Combal records, there is a seventeenth-century tradition of a route through Val Ferret being cut by the Toula glacier (Vaccarone 1884).

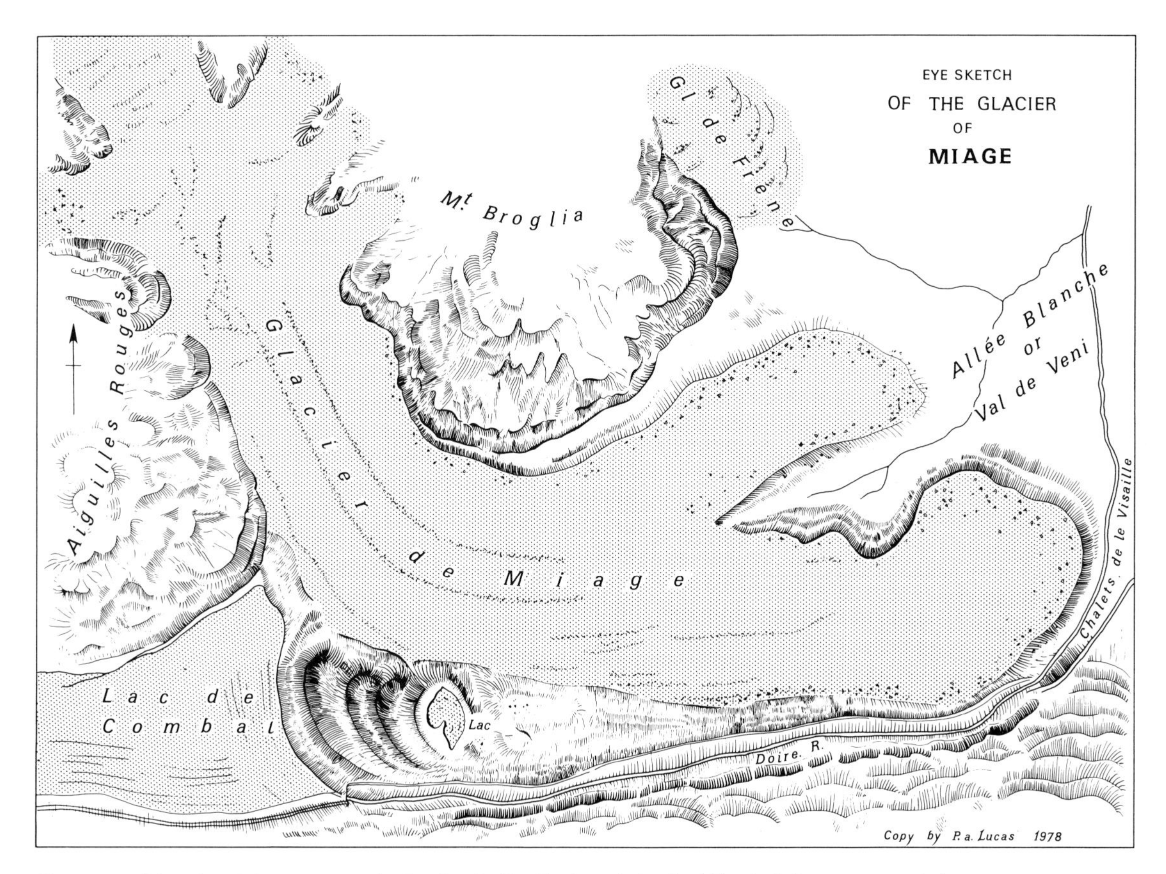

Figure 4.5 Sketch of the Miage glacier by J. D. Forbes (1853). (Copied from 1900 edn)

4.3 The enlarged glaciers of the mid-seventeenth to the mid-eighteenth century

The initial period of advance was followed by one of rather greater quiescence, during which the glaciers remained much larger than before 1600 but were by no means stable. Slow withdrawals, such as that which set in after the visit of the Bishop of Geneva in 1644, were interrupted from time to time by fresh advances, which were less damaging than those which had come earlier, as the ice was now moving over ground already spoilt. The glacier fronts remained uncomfortably close to villages such as Les Bois and Argentière.

Between 1678 and 1680 there were further outbreaks from the Lac de Combal.[5] Ten years later, the people of Chamonix were in a state of such apprehension that they begged the Bishop of Geneva, then Jean d'Arenthon, to come to see them again, their faith in his powers of exorcism being sufficient for them to offer to pay his expenses. This was clearly not his first visit; according to Ladurie he had already exorcized the glaciers in 1664. Perhaps this was the occasion after which 'the glaciers had retired more than eighty

[5] Corbel and Ladurie (1963) obtained a radiocarbon age from wood within the Tacconaz moraines but in view of the ambiguity of radiocarbon dates of the last few centuries their result cannot be taken as evidence of an advance around 1680.

paces'. Although we learn that after his final visit in 1690 'the glaciers have withdrawn an eighth of a league (about 500 m) from where they were before and they have ceased to cause the havoc that they used to do' (Le Masson 1697, p. 147). The Brenva was far advanced across Val Veni in 1691 when it was seen by Arnod, a judge from Val d'Aosta, for there was then only a 'very narrow pass' between it and the opposing hillside, near the site of the modern Chapel of Notre Dame de la Guérison.[6] But the balance of evidence is that the Mont Blanc glaciers were retreating slowly during the last decade of the seventeenth century.

The Mont Blanc glaciers do not appear to have enlarged in any very pronounced way in the early eighteenth century. The Chamoniards, now subjects of the King of Sardinia, had not ceased to dispute about their taxes and supported their demands for reduction with vivid details of the meteorological and other handicaps under which they were labouring. The syndics of Chamonix made such a supplication in 1716, mentioning that 'doten que leur paroisse devins toujour plus inculte à cause des glacier qui advancent seur leur terre, et qui inonde partie en faisent des grand débordement d'eau, en vuyden leur lac, et même il i a pleusieur village qui sont en grand danger de périr' (Letonnelier 1913). The very general terms of this document contrast with the more precise evidence embodied in the seventeenth-century sources. The syndics were surely sufficiently wise to quote details had any been available. But there is no impression of any sudden change in the situation.

Records of an ice and rock fall from the Triolet glacier are dramatic and more detailed:

> the highest mountain of the aforesaid Trioly, along with rocks and ice, suddenly collapsed in the night of the 12th September 1717. Boulders, water and ice, all mixed together, rushed with great force over the aforesaid mountains or alps, so that there were covered in the depths all moveable chattels, one hundred and twenty oxen or cows, cheeses and men to the number of seven who perished instantly. (Translation from Latin citation by Sacco 1918)

The slopes overlooking Val Veni are so steep that even minor advances are very likely to cause rock or ice falls, so this occurrence could well have marked a small forward pulse. There is no clear evidence on this point. De Tillier (1968, cited Porter and Orombelli 1980) recorded that the debris which originated from the fall of a high ice-covered rock onto the Triolet rushed violently downslope, rose against the valley side and then travelled on for a league. When de Saussure saw the glacier in 1781 it was still covered with granitic rock debris. A massive deposit of large angular boulders, extending some 2 km downvalley from moraines dated lichenometrically by Porter and Orombelli to the eighteenth century, was considered by Sacco (1918) to be moraine from the sixteenth to nineteenth centuries. This deposit was the subject of careful investigation by Orombelli and Porter. They concluded that it was formed early in the eighteenth century and has characteristics which permit it to be distinguished satisfactorily from moraine. They inferred that it resulted from the rockfall which wiped out the settlement of Ameiron and Triolet in 1717 (Figure 4.6).[7]

[6] According to Drygalski and Machatschek (1942, p. 214) the Brenva covered the floor of Val Veni from 1691 to 1694. But these two dates merely relate to the manuscript of Arnod, which was dated April 1691 and had notes added in 1694. This manuscript, in the archives in Turin (and which was published in 1968) was not available to Drygalski and Machatschek, who were quoting it via Vaccarone (1881).

[7] De Saussure (1779–96, vol. 4, p. 18) thought that the catastrophe could have been caused by an earth tremor. Forbes (1843, p. 245) mistakenly recorded the event as having taken place in 1828, and as having been caused by an avalanche or sudden descent of the whole glacier.

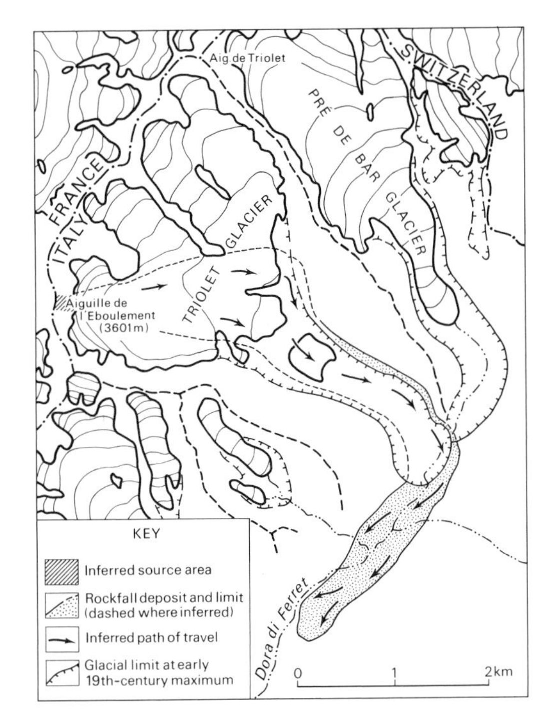

Figure 4.6 Rockfall from Triolet. The debris of the giant rockfall from the Aiguille de l'Eboulement which swept the Triolet glacier in 1717 is estimated to have had a volume of 16–20 million cubic metres and to have descended 1860 m over a distance of 7 km in a few minutes. The resulting deposit was thought by Sacco (1918) to be an eighteenth-century moraine and by Mayr (1969) to be an early sub-Atlantic moraine. Porter and Orombelli's (1980) rejection of its status as a moraine depends not only on lichenometric dating but also on characteristics such as lithology, surface gradient, thickness and anomalous extent. The Pré de Bar and probably the Triolet glacier extended slightly further downvalley in the mid-eighteenth century than in the early nineteenth century. (From Porter and Orombelli 1980)

The frontal positions of all the main glaciers in the Arve valley are marked on a detailed plan of land-holdings in the Chamonix area made between 1728 and 1732. Although the general state of cartography still left much to be desired, this cadastral survey was exceptional, and it is of great value as a primary source (Guichonnet 1955). All tongues were still much enlarged but had retreated so far that they were no longer immediately threatening such villages as Le Tour and Argentière. The Mer de Glace was still less than 400 m from the last houses of Les Bois. As the positions of glacier fronts and recent moraines were marked, it is possible to calculate from the survey that the Tour glacier had retreated 414 m from its last advanced position by 1730 and was more than 700 m forward of its 1911 position. The Argentière was 257 m and the Glacier des Bois 250 m back from the last pre-1730 moraines, but still 675 m and 1330 m forward of their 1911 positions. So the evidence of the survey of 1728–32 is that while the glaciers were then much enlarged compared with their state in the twentieth century, they were somewhat smaller than they had been.

A document dated by Letonnelier as written after 1730 gives no indication of further advances but makes it clear that conditions had not otherwise much improved. In this further plea for tax relief to the King of Sardinia much is made of the exposure of the land to rock, ice and snow falls, and the danger of floods and corrosion by the Arve and its tributaries, as well as the strong winds and 'cet air glacial cause une certaine aridité et sterilité . . . et malgré tous les soins d'un vigilant labeur, le terrain ne produira que d'avoine et fort peu' (Letonnelier 1913). The villagers, whose fathers had paid their tithes in wheat, were now reduced to growing oats, and were harassed by low yields even so.

So from the mid-seventeenth to the mid-eighteenth centuries, the glaciers of Mont

Blanc were subject to several fluctuations, some probably undated, during which they moved backwards and forwards over the outer fringes of the zone which had been occupied during the great advances of the late fifteenth and early sixteenth centuries. For the local people the time of great disasters had been succeeded by another of cold, hard climate, poor crops and continued floods.

4.4 The advances of the mid-eighteenth to the mid-nineteenth centuries

When Windham (1744) visited Savoy in 1741, his guides told him that the Mer de Glace 'went on increasing every year', although his description of his descent to the ice over a morainic slope which was 'exceedingly steep, and all of a dry crumbling earth, mixed with gravel and little loose stones, which afforded us no firm footing' makes it clear that the glacier had not regained its maximum thickness, and soon afterwards Martel (1744) reckoned that the glaciers 'must have been eighty feet higher than they are now', though the Mer de Glace still spread out below the bar of Mottets and at its tongue was the famous grotto of Arveyron.

Windham's visit was definitely in the pre-tourist age. His party was 'assured on all hands that we shall scarcely find any of the Necessities of Life in those Parts', and so took horses, loaded with provisions, and a tent 'which was of some use to us'. They found the terrible description of the country which they had been given much exaggerated, although they noticed some flood damage; the good stone bridge near Bonneville had 'suffered in the late inundations of the Arve'. During the next hundred years Windham and Martel were followed by a gradually increasing number of tourists and by some of the most distinguished of the early glaciologists and field scientists, such as H.-B. de Saussure and J. D. Forbes, both of whom had a passion for accurate measurement and a particular interest in the Mont Blanc area. Writings about the Mont Blanc glaciers during the period accordingly proliferated.

The last three large advances of the Little Ice Age culminated between 1770 and 1780, around 1818 and 1820 and about 1850, the maxima for different glaciers ranging between 1835 and 1855 in this final expansion. We have some clues as to when the glaciers began to swell again, although the advance of the Mer de Glace which had begun in 1741 may have been merely a minor oscillation. De Saussure visited the Tacconaz in 1760 and again in 1778, and found that it had augmented greatly in the interval (1779–96, vol. 4, pp. 432, 463–4), while the Glacier de l'Allée Blanche blocked the Doire river by advancing onto the plains of Combal, and the new lake which it formed on part of the bed of the old Lac de Combal was already known as the Lac de l'Allée Blanche by 1765. This was a matter of importance locally, as it meant that shepherds had to cross the ice to reach their pastures on the slopes to the east of the glacier tongue (Virgilio 1883, Sacco 1918).

By 1780 the Chamonix glaciers were again quite close to the Arve. Engravings by Charles Hackert, of this date, show the Argentière tongue reaching practically to the river, and the source of the Arveyron at the snout of the Glacier des Bois, in a position comparable with that of the much better-known advance of 1820 (Forel 1901). The representation of other detail on these pictures is so accurate that full dependence may be placed upon them as a source of evidence. The Italian glaciers were also much enlarged. The Triolet was still increasing in 1781 and the shepherd at Pré de Bar told de Saussure that it had been advancing for the previous eight years (de Saussure 1779–96, vol. 2,

p. 293). The Brenva was much swollen and overflowing its moraines by 1767 and reached close to the fields of Entrèves (de Saussure 1779–96, vol. 2, p. 293). In 1776 it probably reached the south side of Val Veni; the Doire flowed under the glacier and emerged from a 'beautiful arch of ice' (Bourrit 1776b). But the glacier probably did not reach the Doire in the last decades of the eighteenth century. A drawing of the Brenva about 1795 by Jean-Antoine Linck, an artist from Geneva, shows the glacier still extended over the main valley floor and apparently terminating close to its position in 1980, but only two small outcrops of bedrock appear in the middle of the icefall, suggesting that the Brenva was much thicker than now and probably advancing (Orombelli and Porter 1982, Figure 7).

The Mer de Glace is one of the best-studied glaciers in the Alps. It was advancing in 1778 and this movement culminated about 1780. A minor retreat followed. In 1784 de Saussure noted that the front had withdrawn since 1777 and was 300 m as the crow flies behind the moraine of 1600. It was about 1000 m beyond its 1958 position (Ladurie 1971, p. 343). The retreat about 1780 was swiftly followed by one of the main Little Ice Age maxima around 1820. Already in 1816 the Glacier des Bois was described as 'every day increasing a foot, closing up the valley' (Shelley).

Detailed observations of structure and flow were made by J. D. Forbes, who wrote that 'the hameau des Bois . . . is almost in contact with the glacier, and, indeed, in 1820 it attained a distance of only sixty yards from the house of John Marie Tournier, the nearest in the village, where its further progress was providentially stayed' (Forbes 1843, p. 61) (Plate 4.1). But Forbes's interest in dating glacier variations was only secondary, despite the distinction of his work in other respects; he incorrectly took 1820 to be the date of maximum extension of all the glaciers in the area. According to Tournier, a local guide, the maximum of the Mer de Glace was in 1825 (Forel 1889, pp. 462–4), while Venance Payot, who began systematic observation of the fluctuations at its frontal position, dated its greatest extension to 1826. Other French glaciers reached their culminations earlier, in about 1818. The Tour glacier reached the village fields in 1818, closely threatening the village itself (Mougin 1910, p. 6) and was stationary between 1823 and 1826 (Figure 4.7).

In 1842 Forbes surveyed the tongue of the Mer de Glace. He describes how

> when we approach the foot of the glacier . . . we are at no loss to perceive that the ice has retreated. The blocks of the moraine of 1820 . . . lie scattered almost at the doors of the houses and have raised a formidable bulwark at less than a pistol shot of distance where all cultivation and all verdure suddenly cease (Forbes 1843, p. 62)

Forbes's map of the Mer de Glace, which shows the terminus 1300 m outside its position in 1960 but 370 m less extended than it had been in 1826, was not only so well executed and so detailed as to provide a useful basis for future work but also gave valuable impetus to French topographical mapping, as it was thought disgraceful that he should have had to do the work himself (Adams-Reilly 1864).

By 1850 the Glacier des Bois, the tongue of the Mer de Glace, had expanded again. No detailed measurements were made, but Vallot (1908) summarized the condition described to him by Couttet of Chamonix, who was 'capable de donner des renseignements précis', a sufficient testimony from Vallot, the quality of whose observations was far ahead of his time. The tongue was again only 50 m from Les Bois; in its extension it had destroyed a larch wood on the valley bottom and the frontal moraines were so completely filled up that boulders fell down the outer side of them. Above the snout the ice rose up as high as the steps of the Mauvais Pas, and it was impossible to see across the glacier because of the

Plate 4.1 The Mer de Glace reached out on to the floor of the Arve valley in 1823 when it was painted by Samuel Birmann. (*Au village des Prats*, Öffentliche Kunstsammlung Basel, Kupferstichkabinett, Inv. Bi. 30. 125)

convexity of its surface. Favre (1867) gives 1855 as the date of maximum extension, but by then the upper part of the snout was already thinning and it was necessary to go down about 15 m over newly exposed moraine to get onto the ice near the Cabane de Burret.

The neighbouring Bossons glacier waxed and waned in much the same way. In 1812 the tongue was advancing again. In that year a series of cold summers began (Mougin 1910, p. 10) and by 1817 or 1818 the ice had covered 4 or 5 hectares of cultivated land and a wood, and was menacing the village of Monquart. The population, as had their forebears in 1643, processed to the snout and placed a cross on the frontal moraine. This proved a most useful marker for later observers. There was a second advance in 1845 but this was not so well developed as that of the Mer de Glace, and it only interrupted temporarily the recession which had begun in 1820. Mougin was able to construct a detailed table of the fluctuations of the tongue of the Bossons from 1818 to 1908, based mainly on the observations which Payot made so faithfully until his death (Mougin 1908–9).

The history of the Brenva (Figures 4.2 and 4.8) can be traced more adequately than the other Italian glaciers of Mont Blanc (Orombelli and Porter 1982). Not much is known about its behaviour in the latter part of the eighteenth century, although de Saussure

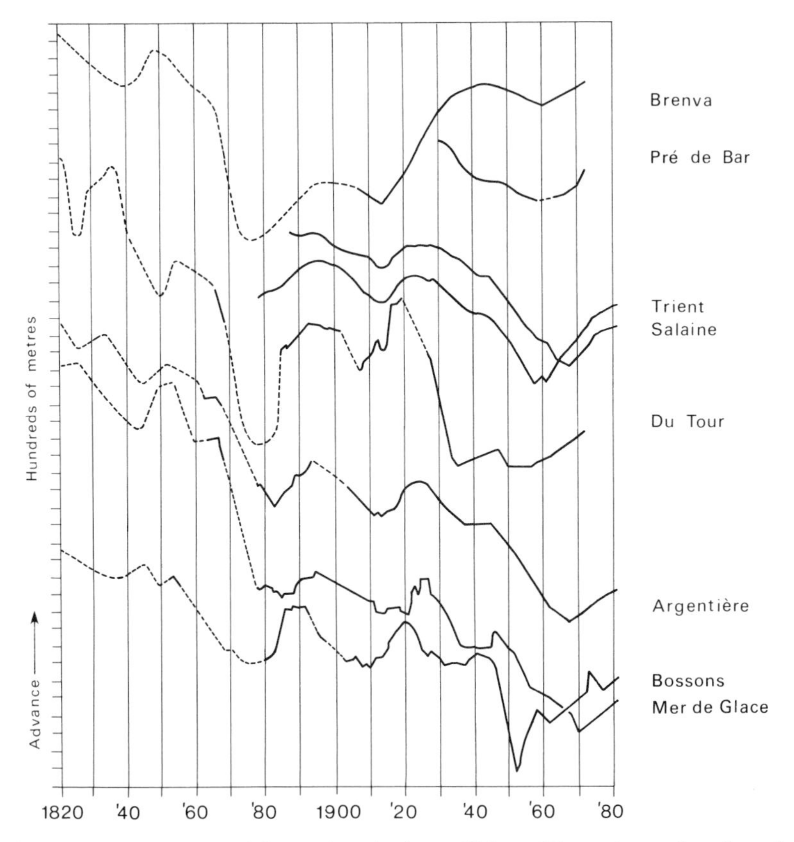

Figure 4.7 Advances and retreats of the major glaciers of Mont Blanc since 1820, based on Mougin (1908/9 and 1910), *Les Variations des glaciers Suisses* (Nos 101–2), Guex (1929), Capello (1941), Bouverot (1957), Vanni (1942–1971), Reynaud (1984)

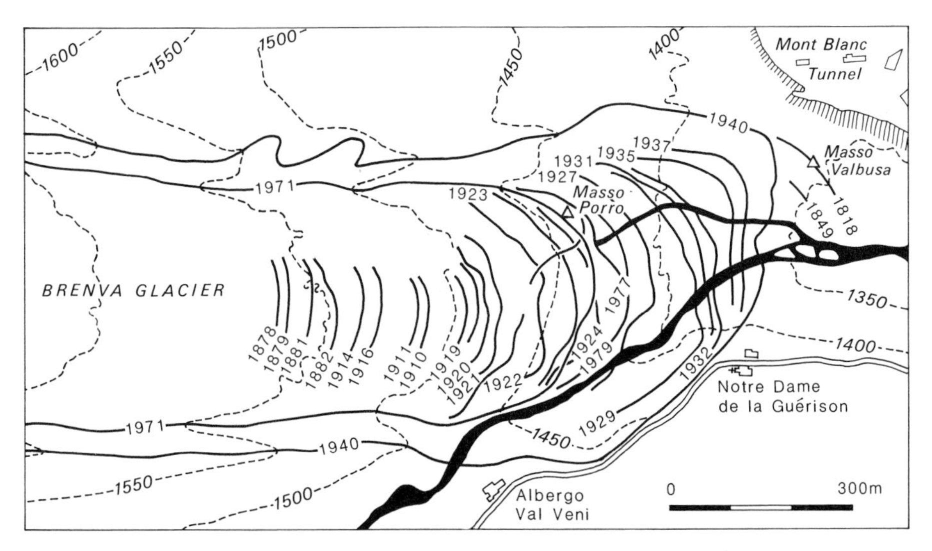

Figure 4.8 Frontal positions of the Brenva, 1818–1979. (From Orombelli and Porter 1982)

visited it in 1765, 1774 and 1781, but at the beginning of the nineteenth century it began to swell rapidly and there is more information to be gleaned about this advance than about those which were contemporaneously affecting the French glaciers. D'Aubuisson (1811) reported that in 1811 the ice front was at 1440 m above sea-level and was about 1000 metres from Entrèves, which it had menaced much more immediately some years earlier, presumably around 1780. Canon Carrel of Aosta was reported by Favre (1867, p. 74) to have said that the glacier was 2000 metres from the village in 1810–12 and did not reach to within 1000 metres till 1818, but it seems safest to accept the contemporary account of d'Aubuisson. From this it appears that the glacier had already reached almost as close to the village in 1811 as it was to go in 1818. In the course of this advance the Brenva felled two trees, one 200, the other 220 years old (Venetz 1833, p. 18), so this expansion was more extensive than any of those in the preceding centuries. The Brenva tongue enters Val Veni at right angles and by 1818 it had expanded sufficiently not only to reach the opposite southern slopes but also to rise some way up them, so that it was above the level of Notre Dame de la Guérison (Figure 4.8). This chapel, still a notable place of pilgrimage today, had been reconstructed as lately as 1781, so we may conclude that the 1780 advance had not affected the Brenva to the same extent as the French glaciers. Forbes (1843, p. 206) was told that in 1818 'the hermitage connected with the chapel was supplied with water from a conduit which descended from the ice of the glacier which was then at a higher level'. The pressure of the ice on its foundations caused such serious cracks that the building had to be demolished in 1820. It was rebuilt a short distance away. By the time that Forbes visited it in 1842 (Figure 4.2), the glacier had thinned considerably but still covered the whole floor of Val Veni, with the Doire flowing beneath the ice. Forbes concluded that it was still much more extensive than it had been in 1767 when de Saussure (1779–96, vol. 2, Plate III) sketched it. After 1842 the glacier advanced rapidly, the enlarged tongue approaching the 1818 moraines again. This advance culminated in 1849 (King 1858, p. 42). The hermit of Notre Dame de la Guérison described how, in 1850, the surface of the ice was only a few metres below the floor of the new chapel and the top of the ice higher than the crest of the right-hand lateral moraine (Sacco 1918, p. 55). After this the ice receded year by year, although only 10 km away the Allée Blanche continued to expand for another decade.

The pattern of advances with culminations around 1780, 1818 and 1850 was common to the Mont Blanc glaciers. But while the mid-nineteenth-century advances of some of the Italian glaciers brought them close to their frontal positions of 1818, some of the French glaciers reached lower altitudes in 1780 than in 1820 and in some later nineteenth-century advances scarcely interrupted the general retreat which was now setting in.

4.5 The great recession of the glaciers – from the mid-nineteenth century to the present

All over the Alps mountaineering huts stand high above the ice and must be approached up steep moraines, fixed ladders or even ropes. They were not built deliberately to ensure an awkward scramble at the end of the day; their isolation is due to the wasting of the ice during the last century. The glaciers, large and small, sprawling around the mountains of Mont Blanc, have withdrawn high into their valleys, leaving their moraines and debris spread behind them. The gleaming tongues of Argentière and Mer de Glace (Plate 4.3)

Plate 4.2 Miage moraines with the bed of the Lac de Combal, August 1958

have disappeared from the view of travellers almost as abruptly as they appeared at the beginning of the Little Ice Age. The swelling tongue of the Brenva has melted away from the chapel of Notre Dame de la Guérison.

Regular measurement of glacier fluctuations began during the deglaciation. Payot led the way with his observations of the Tour, Argentière, Mer de Glace and Bossons, published in the *Revue Savoisienne* and *Revue Alpine*. His efforts were augmented by the photographic record made by Tairraz of Chamonix. In 1892 widespread damage, including the destruction of part of the town of St Gervais, was caused by the bursting of a pocket of water in the little Tête-Rousse glacier. The incident was investigated by the Administration des Eaux et Forêts (Mougin and Bernard 1922) and the outcome was that this organization undertook the regular observation of the larger Chamonix glaciers, excepting at first the Mer de Glace, where Vallot was already working (Vallot 1900, 1908). Charles Rabot succeeded in putting the programme on a more permanent basis in 1907. When Mougin was in charge he published the very useful *Etudes glaciologiques en Savoie*. After 1930 the results appeared in a series of reports on the variations of European glaciers, collected by Mercanton and published in the snow and ice volumes of the *Hydrological Section of the International Union of Geology and Geophysics*. Unfortunately around 1960 observation on many of the glaciers monitored by the Administration des Eaux et Forêts ceased, but in 1977 it was announced that regular aerial surveys of twenty French glaciers were to be made at three-year intervals. They were to include the glaciers of

Plate 4.3 The shrunken Mer de Glace seen from Montenvers, August 1958

Tour, Argentière, Mer de Glace, Bossons, Taconnaz, Bionassay and Tre-la-Tête (Valla 1977).

In Switzerland Forel started the excellent series *Les Variations des glaciers suisses* in 1880. These annual reports were published in the *Annuaire de Club Alpine Suisse* till 1924 and afterwards in *Les Alpes*, and have included figures for the frontal position of Trient, Orny and Saleina, the largest glaciers in the Swiss sector of the Mont Blanc massif. The Guexes, father and son, interested themselves in the Trient (Figure 4.6), and Jules Guex was able to build up a complete account of the fluctuations of its tongue for the period 1878 to 1928 (Guex 1929) (Figure 4.9).

The Italian glaciers also attracted greatly increased attention towards the end of the last century, but observations were unfortunately less systematic. The Comitato Glaciologico Italiano was set up in Turin in 1890; Porro's work (1898, 1902, 1914) on the Miage was some of the most important that followed this impetus. Revelli (1911, 1912) and Valbusa (1921, 1931) all made notable contributions during the next few decades, but regular observations of the fluctuations of the Italian glaciers did not begin till 1927 (Vanni 1958). Work in the Mont Blanc area was influenced after that especially by Capello (1936, 1941, 1958, 1966, 1971). Publication of reports on the variation of the Italian glaciers was undertaken by Vanni for several decades (e.g. Vanni 1942, 1970, 1971).

When the data from the larger Mont Blanc glaciers are plotted from these various sources, a surprisingly consistent pattern emerges (Figure 4.7). Retreat was general between 1850 and 1880. The fronts advanced again from about 1880 to 1895, though not

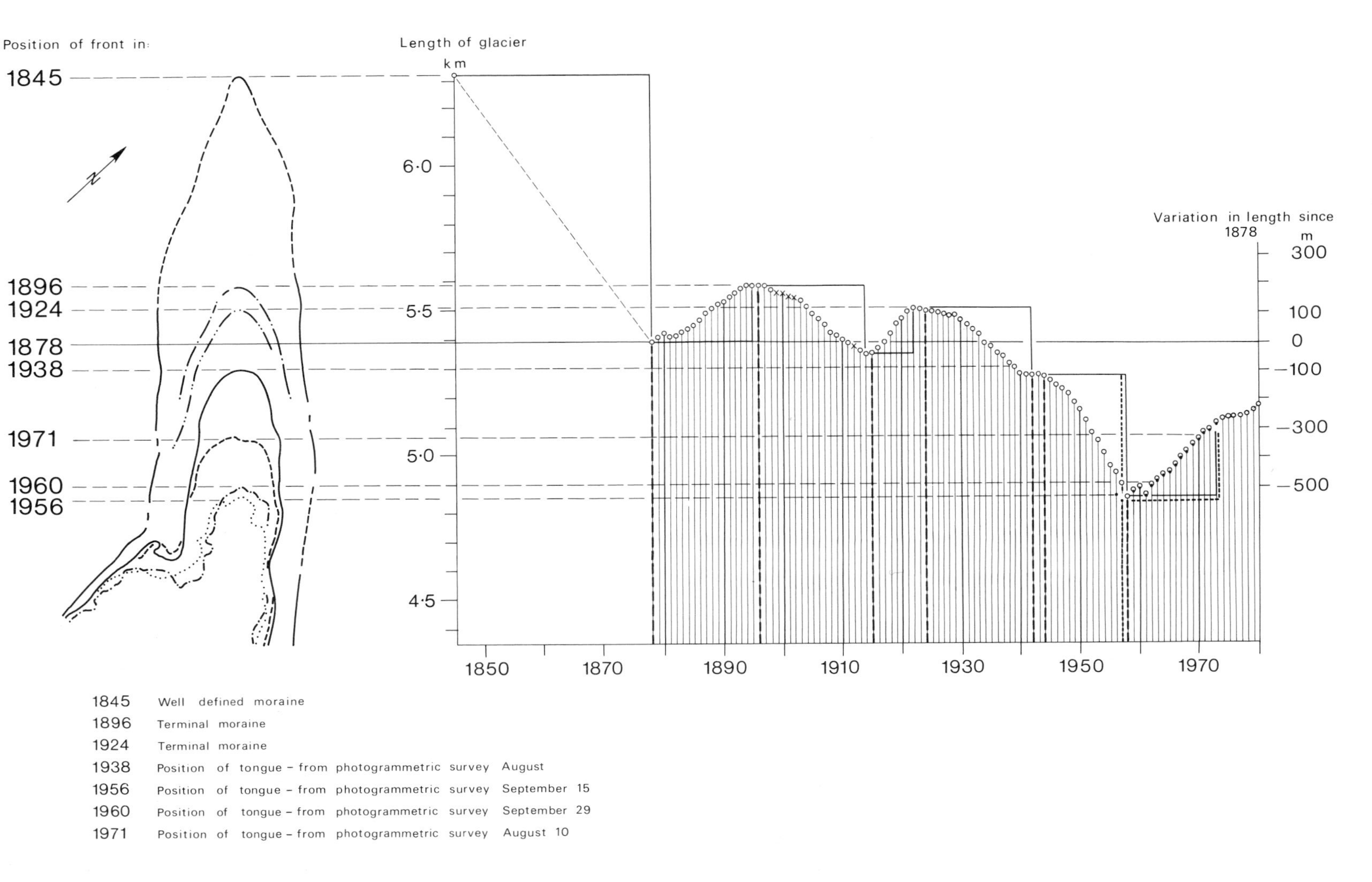

Figure 4.9 Glacier de Trient, 1845–1980. Yearly observations made by the Swiss Glacier Commission make it possible to trace the last three oscillations of the tongue of the Trient glacier. (After *Les Variations des glaciers Suisses*, nos 98–102)

so much as they had previously retreated. From 1895 to 1915 recession continued but was not so rapid as it had been from 1850 to 1880. The advances of 1915 to 1925 were generally insufficient to bring the snouts back to their 1895 positions, and after that recession continued more rapidly in many cases, with a stationary period and even slight advances about 1945.

The most obvious anomaly which emerges is the very well-authenticated advance of the Brenva from 1925 to about 1940, when its neighbours were all in retreat (Capello 1941). The Italian side of the Mont Blanc chain is especially prone to rock and ice falls (Capello 1959, Porter and Orombelli 1981). No one who has stood at Entrèves and looked up at the majestic rock wall beneath the Tour Ronde, Aiguille du Géant or Grandes Jorasses can fail to understand this. On 14 November 1920 a mass of rock and ice, including the whole of a small hanging glacier, slid from the side of Mont Blanc de Courmayeur, fell 2800 m onto the upper Brenva, and shot another 5000 m down to the tongue (Valbusa 1921, 1931). Five days later another fall swept practically the whole glacier even more dramatically. This slide, of which a much higher proportion was rock, was about fifteen times greater in volume than its predecessor. Valbusa (1921) estimated that something of the order of 5 million cubic metres of rock were deposited on the lower tongue alone. Much of the old forest of Pertud was buried beneath masses of ice and rock, which had overshot the high lateral moraine. The Doire was dammed for a time, forming a temporary lake which practically lapped the walls of the chalets of Pertud. The Brenva was already advancing in 1920, in common with the other glaciers of the area, but was alone in continuing to swell after 1925. By 1931 its ice was in contact with the cliff below Notre Dame de la Guérison, and the tongue went on enlarging till 1940. By 1943 waning had begun, and this continued until the 1960s. As the Brenva was in phase with the other Mont Blanc glaciers until 1925, there is no doubt that it was the protective effect of debris from the avalanches and rockfalls of 1920 which caused its anomalous movement from 1925 to 1940.

The Miage too has a protective debris cover. Névé from the great ice mass of the Dôme de Goûter is canalized into the deep trough valley of the Miage, which cuts three-quarters of the way through the Mont Blanc massif. It is not perhaps surprising that, even now, the resultant icestream shoots out across Val Veni and then, swinging round through 90 degrees, penetrates far down the main valley. The enormous scale of the Miage moraines (Plate 4.2), even in comparison with those of Brenva, is particularly evident from the *téléferique* station at the top of Mont Frêty, below the Col du Géant. The tongue is debris-covered for more than 5 km above the forked snout. Sacco thought that the Miage was, because of this, particularly insensitive to climatic fluctuations. Revelli found that by 1911 it had retreated only a few tens of metres from the 1879 position mapped by Marengo (Revelli 1911, 1912). There was a slight advance in the early twentieth century; Sacco found the glacier swelling and overriding some of its moraines in 1917. After that the Miage slowly retreated until the late 1960s. The debris cover may well have damped down frontal oscillations during the last century or so; and its presence has made observation of such changes as have occurred more difficult.

The Tour glacier has been characterized by extraordinarily large seasonal fluctuations in length, both during advance and retreat, because for much of the period since 1818 its tongue has hung on the great south-facing rock step which dominates Le Tour village. Sudden advances, like that of 267 m in July 1916, have been counterbalanced by equally drastic retreats, for example of 228 m in three months of 1915 (Mougin 1925, pp. 2–9).

During the 1920 advance the glacier was prolonged from time to time by the incorporation of a reconstituted ice mass at the foot of the step. Instability continued as the ice withdrew up the step during the subsequent retreat phase, which was therefore unusually swift. On 14 August 1949 a crack formed across the hanging tongue, and half a million tons of ice, equivalent to about forty years' accumulation, moved forward, broke into blocks, and poured down the precipitous face to the moraine below, killing a party of picnickers as it did so (Glaister 1951). The snout had probably retreated out of a gully, where it had been firmly entrenched, and the crack formed as it hung unsupported on the slope above (Messines de Sourbier 1950). The instability, so characteristic of a glacier of this type during the periods when the front is hanging, clearly limits the value of individual observations of advance and retreat, but it is striking how well the general pattern of fluctuations of the Tour fits in with those of others in the massif.

The Glacier des Bossons also advances and retreats very swiftly. A withdrawal of nearly 600 m between 1943 and 1951 was succeeded by an expansion of 350 to 400 m between 1952 and 1958, which was measured photogrammetrically. Finsterwalder (1959) suggested that this advance was probably due to increased accumulation in the firnfield during the period 1953–6, causing a kinematic wave to reach the tongue in only three years, travelling three times as fast as the glacier itself. Lliboutry (1958) was able to use Vallot's observations (1900, 1908) to demonstrate the development of such a pressure wave on the Mer de Glace between 1891 and 1895 (Figure 4.10). Movement of such pressure waves down the Bossons is especially rapid as the bed is so steep. It is clear from Figure 4.7 that each general advance of the Mont Blanc glaciers has been preceded by a forward surge of the Bossons, coming about ten years earlier (Grove 1966, Vivian 1971).

The Bossons glacier began to advance, together with its consort, the Tacconaz, in 1953, following a period of rapid retreat. Advance was halted briefly around 1960 but by 1976 the terminus had moved forward by over 700 m (Veyret 1974). The other glaciers of the massif duly followed, each in its own time. The swelling of the upper parts of the glaciers (Vivian 1971, 1975) which caused these advances has been particularly closely observed in the cases of the Mer de Glace and Argentière, which were amongst the last to advance in 1970 and 1971 (see Figure 4.7 and Reynaud 1977). Vallot surveyed not only the front but also five cross-profiles of the Mer de Glace. A sixth profile was added by Mougin in 1923, and a seventh by the Administration des Eaux et Forêts in 1970 (Figures 4.3 and 4.10), but thinning and withdrawal of the front in the mid-twentieth century was so great that the two lowest profiles at Le Chapeau and Mauvais Pas have been lost. Figure 4.10 shows three periods during which the surface of the ice rose and a wave passed down the glacier, eventually causing an advance at the front. The first occurred between 1890 and 1900, the second between 1910 and 1930, and the third started in the 1960s and still continues (Lliboutry and Reynaud 1981, Reynaud 1977, 1984). Seismic observations and surveys made on several of the Italian glaciers have allowed us to follow the passage of waves and the thickening preceding advances on them as well (e.g. Lesca and Armando 1972, Lesca 1972).

The first glaciers to react on the Italian side were the cirque glaciers with steep slopes, such as those of the Grandes Jorasses and Planpincieux, which in 1963 had already covered markers installed 20 m in front of them in 1961 (Vivian 1975). Particularly spectacular evidence of renewed activity was provided by the Brenva, which by 1968 had destroyed a road built across the frontal moraines in 1960. The front of the Glacier de l'Allée Blanche was already 1090 m in advance of the 1954 position by 1973 (Figure 4.11),

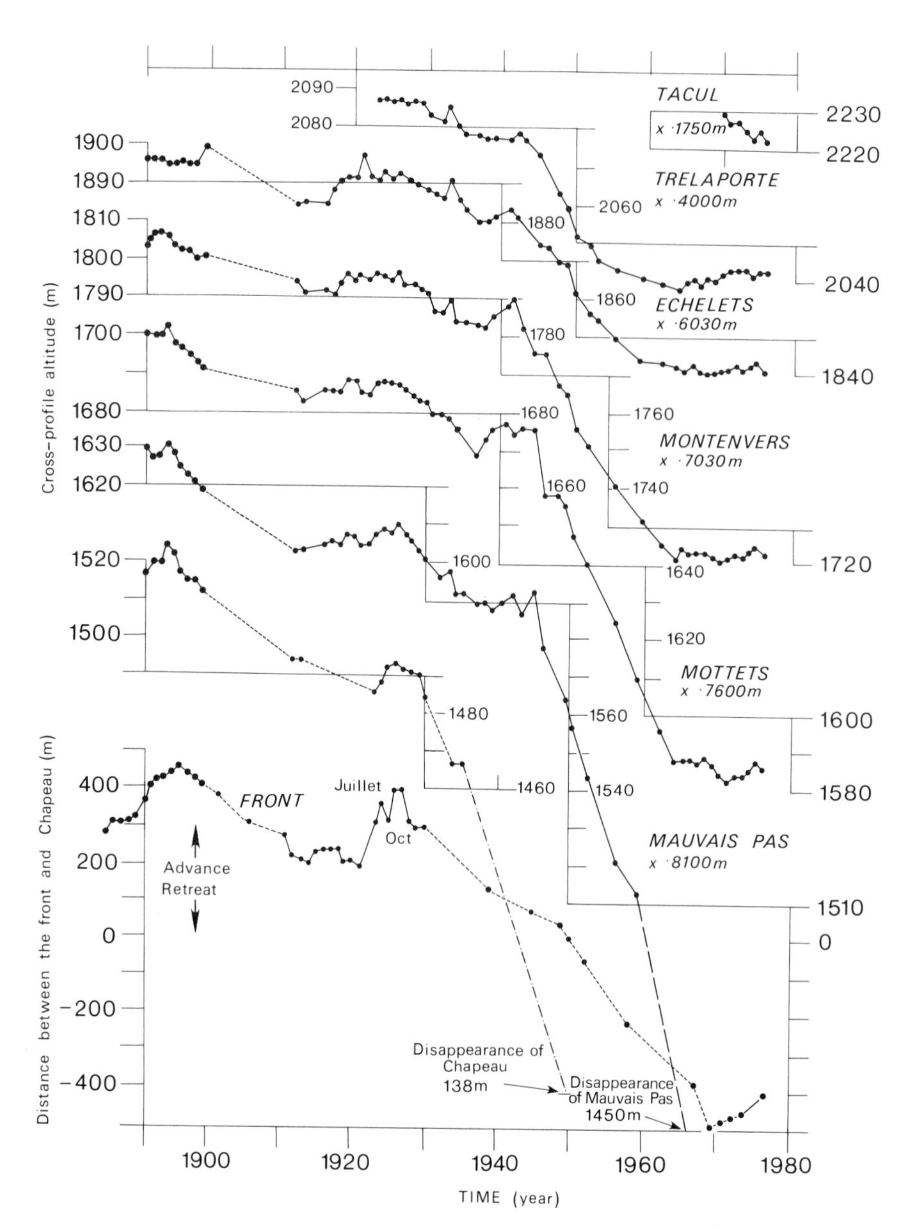

Figure 4.10 Changing height of cross-profiles at specified positions indicates the passage of waves down the Mer de Glace since 1900, which accounts for the episodes of advance shown on Figure 4.7. Vallot's observations were far more precise than those of contemporaries; no other such long series of observations are available elsewhere to demonstrate the relationship between variations in mass balance, glacier flow and frontal advance. (After Reynaud 1977 and 1984)

although the Pré de Bar had still not managed to regain its 1952 position in 1975 (Figure 4.12). The difference in the speed of reaction of these two glaciers is in no doubt associated with the steepness of the slope on which the tongue of the Allée Blanche lies compared with that of the Pré de Bar. The Trient glacier also advanced rapidly and by 1976 its terminus had regained over a third of the ground lost since the maximum of 1896 (Figure 4.9).

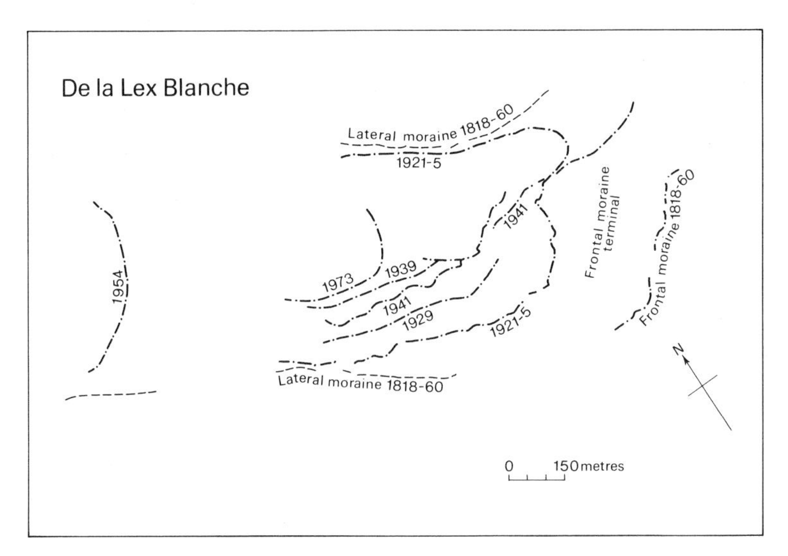

Figure 4.11 Frontal positions of the L'Allée Blanche (Lex Blanche). Note the great retreat of the mid-twentieth century and the advance between 1954 and 1973. (After Cerruti 1977)

The swing from retreat to advance has not only caused the glaciers to modify their forefields during their reoccupation of ground abandoned several tens of years earlier but also to change very markedly in appearance (Veyret 1971, 1974, 1981). Retreating tongues are characterized by gentle profiles, caused by the excess of melting over delivery of ice to the front. Advancing tongues are bulging or even cliffed. By the autumn of 1970 the general state of advance was immediately obvious to the informed observer.

It is generally accepted that the advances of the 1960s were associated with lower temperatures, especially in the summer ablation season (Vanni 1970), and also heavy snowfall. Corbel (1963) pointed out that since most precipitation falls in summer in the

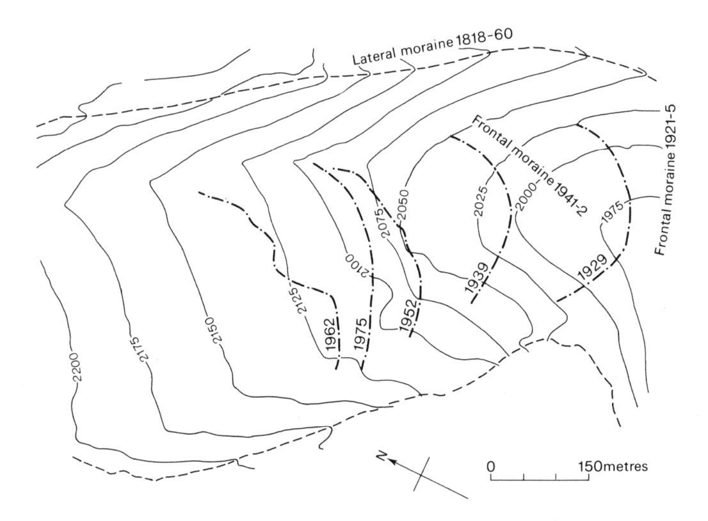

Figure 4.12 Frontal positions of the Pré de Bar in the twentieth century. (After Cerruti 1977)

Chamonix–Mont Blanc area, a lowering of temperature in the May to September period leads to an important increase of snowfall at high altitudes. Between 1940 and 1950 the rate of glacial retreat was already lessening and some fronts were stationary. Glacial expansion was promoted by the conditions of the 1960s. In 1962, for instance, May temperatures were the lowest for a century. The average temperature for June was 2.1 °C below the normal for the previous thirty years and minimum temperatures the lowest on record. July remained cold and snowy, and then, after a brief fine, dry period in August and September, temperatures fell again in October. There were exceptionally heavy snowfalls in December. These low temperatures and heavy snowfalls in summer, typical of a period of glacier expansion, were reinforced by voluminous winter snowfall in 1962–3. The repetition of such sequences led to the situation in 1971 when, despite heavy ablation in summer, none of the glaciers on the Italian side of Mont Blanc were retreating (Vanni 1971). The general advances which followed were maintained into the 1980s (Sessiano 1982).

It seems that earlier intervals of expansion of the Mont Blanc glaciers also took place when temperatures were below average. Cerutti (1971) considered the relationship between their fluctuations and climatic conditions, using meteorological data from the Gt St Bernard Pass assembled by Janin (1970). She found that glacial advances had followed colder periods with more precipitation, and retreat followed warmer periods. The dominant influence of temperature is well exemplified by the advance period of the 1880s and the retreat period of the 1920s and 1930s. Annual average precipitation on the pass

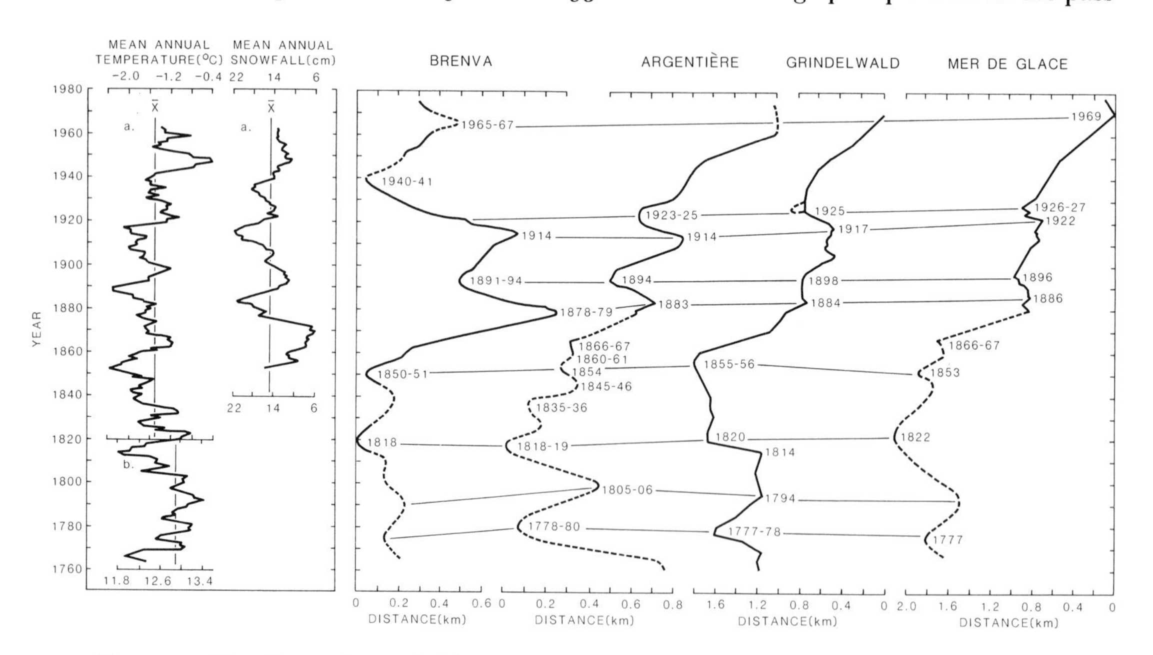

Figure 4.13 The fluctuations of Mont Blanc glaciers are closely paralleled by those of the Grindelwaldgletscher (see Chapter 6). Note the relationship between periods of advance and retreat and the incidence of temperatures above and below the mean (measured at Great St Bernard between 1820 and 1980 and at Milan between 1763 and 1820) and decrease and increase of snowfall (measured at Great St Bernard between 1850 and 1960). (From Orombelli and Porter 1982)

was 1903 mm in 1876–90, while it was 2183 mm in 1920–40, but the increased precipitation was insufficient to counteract the rise in mean annual temperature of 0.4 °C. The general relationship between glacial behaviour and mean temperature is shown in Figure 4.13. Orombelli and Porter (1982) noted that the temperature record, kept by the Osservatorio Astronomico di Brera in Milan, closely parallels that from the Gt St Bernard over the last 160 years. If glacial variations are compared with temperature variations as recorded in Milan for the period 1763 to 1820, then it emerges that the advances culminating in the 1770s and around 1820 coincided with periods of low mean annual temperature, while in the intervening retreat period temperatures in Milan were as much as 1.5 °C higher.

The history of climate during the Little Ice Age has been traced back further and in more detail for Switzerland, but before discussing this in Chapter 6, the behaviour of the Mont Blanc glaciers is compared with that of certain glaciers in the eastern Alps.

Chapter 5

The Little Ice Age in the Ötztal, eastern Alps

The largest group of glaciers in the eastern Alps lies at the head of Ötztal, where mountains on the frontier with Italy rise to over 3600 m (Figure 5.1). The main névéfields extend from the Weisskugel, northeast towards the Mittelbergferner, and east towards the Hochwild. Many of the glaciers with which we are concerned flow southeast into the Rofen section of Ventertal, and most of the rest northwards into the Rofen and Spiegler section of Ventertal, or into the head of Gurglertal (Figure 5.2). Both Ventertal and Gurglertal are steeply sloping with small cultivated basins separated by deep gorges.

When the Vernagtferner advances across the floor of Rofental it obstructs the flow of streams draining the glaciers further up. The Gurgler expanding down its valley blocks the mouth of Langtal. In both cases lakes form and, given sufficient time, are liable to spill violently. The consequent floods have caused much damage in Ötztal and inquiries made by government authorities, the earliest at the end of the sixteenth century, are the main sources of information about the early part of the Little Ice Age in the eastern Alps.[1] These documents give a very clear impression of the problems and dangers faced by people living in the area and their reactions. The literature will therefore be discussed in some detail, so that the human implications of the more extreme events involved in Little Ice Age glacier expansion may be revealed more explicitly than was possible in earlier chapters dealing with northern Europe or the western Alps.

Several glaciers in Ötztal were selected for detailed observation towards the end of the

[1] The history of the Rofen and Gurgl lakes was first placed in the general context of Alpine glacier fluctuations by Richter (1891). He later provided full information on the documentary sources in the Gubernium at Innsbruck and in the Ferdinandeum (1892), and in the same paper corrected many details mentioned in his book of 1888. The sources consist mainly of reports from the Wardens of Petersberg and Castellbell, letters from local clergy, reports from special emissaries, and submissions from Innsbruck to the Imperial Court. Richter underestimated his own contributions, writing 'As these documents had been seen by other authors, I did not expect any new information from my investigations. Nor did I find any.' In fact Stotter, writing in 1846, had not seen all the documents, whilst Sonklar (1960) made some serious errors in his account (Richter 1891, p. 3).

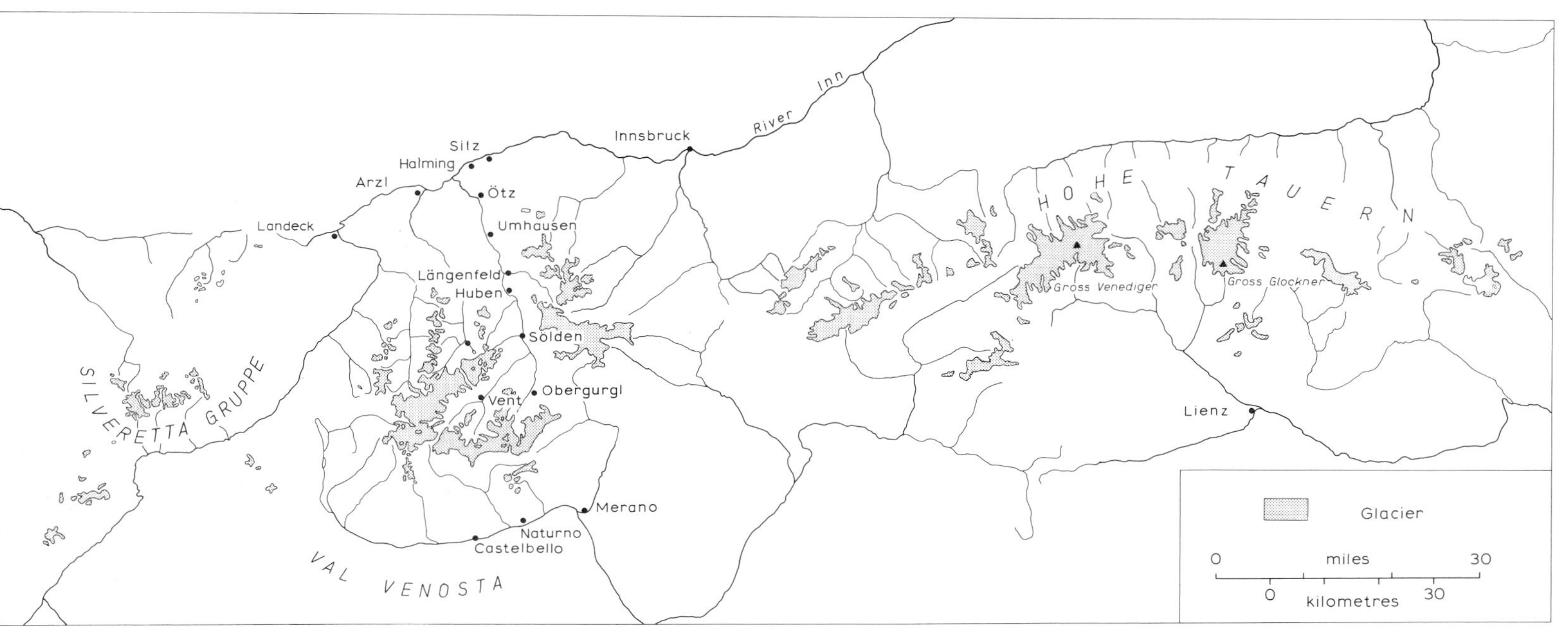

Figure 5.1 The principal glacier concentrations in the Austrian Alps

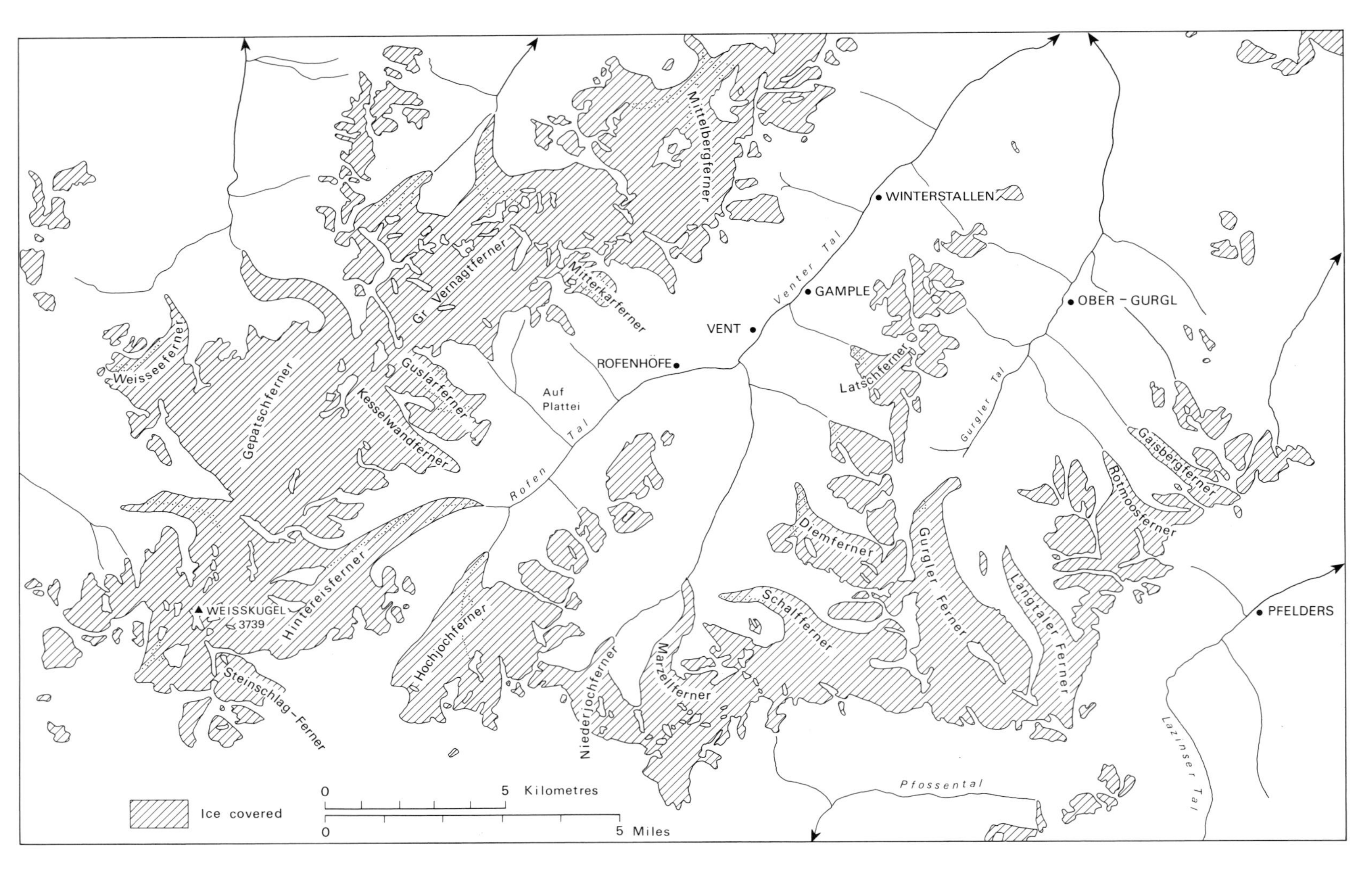

Figure 5.2 The glaciers of the Ötztaler Alpen from 1:25,000 maps 'Gurgl' and 'Weisskugel-Wildspitze' (published by Österreichischer Alpenverein 1951, based on photogrammetrical survey of 1942/3, revised 1950). Note the very short tongue of the Vernagtferner in the mid-twentieth century

nineteenth century by Finsterwalder, Blümcke and Hess (Finsterwalder 1897a and b, Blümcke and Hess 1895, 1897, 1899). More recently mass balance studies have been made of some of the key glaciers, and the investigation of the relationship between glacier fluctuations and meteorological controls made by Hoinkes (1970) and his associates are particularly noteworthy. The voluminous literature was listed by Rudolph (1963); the Ötztal clearly provides a concentration of material for the eastern Alps fully comparable in richness with that available for the Mont Blanc massif in the west.

5.1 The Vernagtferner 1599–1601: the onset of the Little Ice Age

For this early period there is detailed information for only one glacier, the Vernagtferner. This, it must be admitted, provides only a very slender basis for a glacial chronology, all the more as up to the mid-nineteenth century this glacier surged a number of times, the tongue thickening to an abnormal extent and advancing with exceptional speed for a European glacier. The probable reasons for these characteristics will be discussed later in this chapter. Meanwhile, it will be of interest not only to follow local reactions to the consequent hazard but also to examine the timing of its surges in relation to the major advance phases of the glaciers in the western Alps discussed in Chapter 4.

The Vernagtferner flows southeast and unites with the adjacent Guslarferner and, entering the Ventertal at right angles, dams back the rivers coming from the Hintereisferner, Kesselwandferner and Hochjochferner, to form a lake in Rofental.

An account of the first outburst of the Rofensee is given in the chronicle of Benedikt Kuen.[2] According to this,

> in 1600, so our ancestors tell us, the big glacier behind Rofen after it had come into the valley according to its habit, broke out on the feast of St James [25 July], did great damage to the fields in the Ezthal [Ötztal], spoilt the roads and streets and carried away all the bridges. In the parish of Langenfeld [Figure 5.1] the water flooded the ground from Rethlstain to Lener Kohlstatt.

Kuen's wording suggests the possibility that this advance of the Vernagt was not unprecedented, but as he was writing after several later advances it may well have been these that he had in mind. He made no mention of the situation in 1601 about which Richter (1892) found a great deal of evidence surviving.

The ice was still in an advanced position in 1601. The Rofensee threatened to bring even greater destruction not only to Ötztal but also to the Inn valley. A report was made by the government of Tirol and Vorarlberg at Innsbruck, dated 30 July 1601, and sent to Emperor Rudolph II.[3]

> In the morning of the 11th inst. the said Clerk of the Works [Jäger] reported to us verbally how he and each of the others . . . had found the lake and the glacier, and had found the situation so alarming that the said lake might break out on the 13th of

[2] The 'Chronicle of Benedikt Kuen', written by Johann Kuen in 1683 and his son Benedikt in 1715, was the only known account of the 1601 advance until Richter unearthed a mass of contemporary documents. There is an earlier account of the first outburst of the Rofensee in the 'Fugger Korrespondenzen' (Klein 1968). The Fugger family owned land in Inntal.

[3] The report, together with various supplements, was discovered by Richter in the Innsbruck State Archives, AVIII 19.

> the said month, and that no human means, help or advice could avert the danger, for all would be in vain, and the destruction that would follow would necessitate a great sum of money being required for the damage. Bearing in mind that the glacier which had advanced and broken out before this one, also in Ötztal, had caused damage of over 20,000 gulden, it is easy to imagine what this glacier, which is at least six times as big as the above-mentioned one, would result in, as it would doubtless run down through Ötztal to the river Inn and tear away some 200 houses, much fine land as well as bridges, levees, roads and paths. (Richter 1892, p. 359)

It is clear that the glacier that had broken out before this one was not the Vernagt. 'Also in Ötztal' strongly suggests that two different glaciers and lakes were involved. It is reasonable to assume that the earlier disaster was caused by an outbreak from the Gurglersee a few years earlier; the wording suggests a recent date but not the previous year. As the Gurgler does not surge it can be taken that valley glaciers in the Ötztal area were already enlarged by 1600.

The report of 30 July 1601 on the Vernagt included supplements by Jäger, who noted that the glacier

> fills the valley like a big round pate . . . it is not like others, smooth ice, but is full of cracks and crevasses with strange colours, so that one cannot marvel enough. Furthermore, this glacier has swelled and dammed a lake where there used to be nothing but a beautiful alp and grazing land. It is 625 fathoms [1312 m] long; its greatest width from one mountain to the other is 175 fathoms [332 m] and its depth according to Hans Rothen is 60 fathoms [114 m] and it is rising a man's height in 24 hours. (Richter 1892, p. 361)

The Rofensee, he adds, 'has come in the course of two years, 1599 and 1600, and still grows higher, longer and broader with each day'. A number of streams were flowing into the lake but no water was escaping from the ice dam and therefore 'a great danger is to be feared'. Orders were given on 12 July to 'clear the banks of the river and remove dangerous heaps of wood and put out guards'. The authorities of the upper and lower Inn valley were directed 'in so far as they were exposed to any danger, to take measures to have processions with the Cross arranged everywhere and further to do all that is humanly possible to prevent too much damage caused by a possible eruption of the lake' (Richter 1892, pp. 362–3). A procession was arranged to the abbey at Wilten, near Innsbruck, where a special sermon was given by a preacher who, it may be noted incidentally, had been provided beforehand with a copy of Jäger's report.

The Burgomaster and town council of Innsbruck wanted to have more information about the situation and they selected two experienced men, Peter Puppel, a butcher of Innsbruck, and Martin Griessetter, chief miner of Hetting. They visited the Vernagt on 14 July 1601, finding it 'easy to get up to the ice' even though the valley downstream of the glacier was 'no wider than one fathom . . . quite a deep valley, running between hard mountains and rocks, so that one cannot get up anywhere but has to go through the defile'. They recorded that the previous week, 'according to the Rofen farmer and others living near the glacier, a piece of the glacier 200 fathoms [380 m] wide had fallen into the lake. Therefore . . . the water was pressed out.' On 12 July the lake had

> run over a bit, then between morning and midnight the lake found an outlet. A stream . . . ran out between the mountain and the glacier, visible for about 30 paces

> ... then ... down through the glacier so that it can no longer be seen but only heard. Then much further on it comes out again under an avalanche and runs on, quite muddy, through the Öz valley .... Almost in the middle of the lake, on the left bank, looking upward there is a post, so ... people can see with their own eyes that in one hour, while they have gone round it, the lake has fallen a handsbreadth.... It seems that the outlet will become deeper so that the lake will fall little by little. As the glacier is so crevassed and cracked, it is expected that it will melt away soon. (Richter 1892, pp. 363–5)

They estimated that almost twice as much water was running out of the lake as the tributaries were bringing in.

Most of the Ötztal above Sölden came under the jurisdiction of the Court of Petersberg at Haiming in the seventeenth century. The Warden of Petersberg was ordered by the government to make investigations of the dangerous glacier and lake but excused himself and commented that 'all human help is impossible'. He asked the local Rofen farmer to report if anything should happen and petitioned the Margrave of Burgau asking to be relieved of the commission as he did not feel experienced or skilled enough to avert the danger. Furthermore he claimed that there was no one in Petersberg whose advice could be asked in this matter (Richter 1892, pp. 365–6).

The Margrave wrote to the government of Upper Austria explaining the Warden's unwillingness to approach the glacier and the government ordered Count Daniel Felix of Spaur and Valor to bring pressure to bear on the Margrave. Count Daniel Felix encountered the Margrave on the bridge at Aichwaldel on 23 July 1601, as he was returning from a fishing expedition, and showed him Jäger's sketch and account of 11 July. The Margrave laughed at Jäger's assertion that it was dangerous to approach the glacier; he had visited it himself. He pointed out that the glacier was in the jurisdiction of Castellbell, not Petersberg. His need to do this is significant, for if the previous outbreak in Ötztal had been from Vernagtferner, the Petersberg people would have known it was not within their jurisdiction. The Margrave recommended that people with long sticks and sharp blades should be sent to widen the outlet of the lake and to crush the bits of ice blocking the course of the water. Advice should be taken, he emphasized, from suitably qualified people, those conversant with mining, building of sluices and dealing with water generally, for 'they could much sooner and much more easily find a suitable remedy'. 'We do not mind our Warden helping you', he concluded, 'but he must not be troubled by large costs.'

The Warden of Castellbell was Maximillian Hendl. He went into action as soon as he received orders to do so, taking local people, but he 'could not in his hurry find any good miners, either for working on the ice or the rocks or sluices'. With a party of six, including an ecclesiastical judge, Christian Mayr, and his steward, Adam Rainer, he arrived at Rofen on 11 August 'in a terrible storm, with snow and rain'. The judge and the steward had been to Rofen only three weeks earlier to inspect their cattle, oxen and sheep, which had been grazing above and below the lake. They found that since then the water had fallen by '13 good meat fathoms' leaving behind 'blocks of ice as big as houses and also smaller ones, which had been floating on the water and are now melting little by little'. He estimated that a third more water was flowing out of the lake than was entering it. The outflow had excavated a hole in the ice forming an arch one and a half times the height of a foot soldier's pike (a *Landsknecht*'s pike at this time was about 4 m long). 'The course and

Plate 5.1a The advancing Vernagtferner, drawn by a Capuchin friar, Georg Respichler, portraying the glacier as it was on 16 May 1678, from the manuscript 'Über den Ausbruch des Vernagtferner in Vent im Jahre 1678', dated 28 July 1678 (Fb 3631) in the Landsmuseum, Innsbruck

Plate 5.1b (on facing page) The original key to Respichler's drawing

outlet of the glacier seem to be getting easier and more comfortable all the time.' The construction of sluices would not be any great help and, as he pointed out, the cost would have been considerable. Nevertheless the peril might not have passed entirely and 'we must pray to God Almighty and try to placate him with pious processions' (Richter 1892, pp. 366–7).

Despite more rain and snow Abraham Jäger returned to Vernagt on 9 September 1601. He found the glacier no longer the same shape as it had been. The lake was 'half emptied and since the 5th of this month it has become 12 mining feet lower. . . . From this it is to be supposed that this lake will disappear completely in a short time.' Georg Wurmbser, carpenter of the salt works, accompanied him and also Michael Grasel and Melchior Wanner, 'who know about sluices'. Jäger concluded that it was not possible to place sluices in a useful position and 'it would be much better to leave everything to God'.

The order to put out guards had been revoked as early as 30 July in view of the reported emptying of the lake. For the present the threat was over. The next severe flood did not come until 1678, and by then the witnesses of the 1600 disaster were dead.

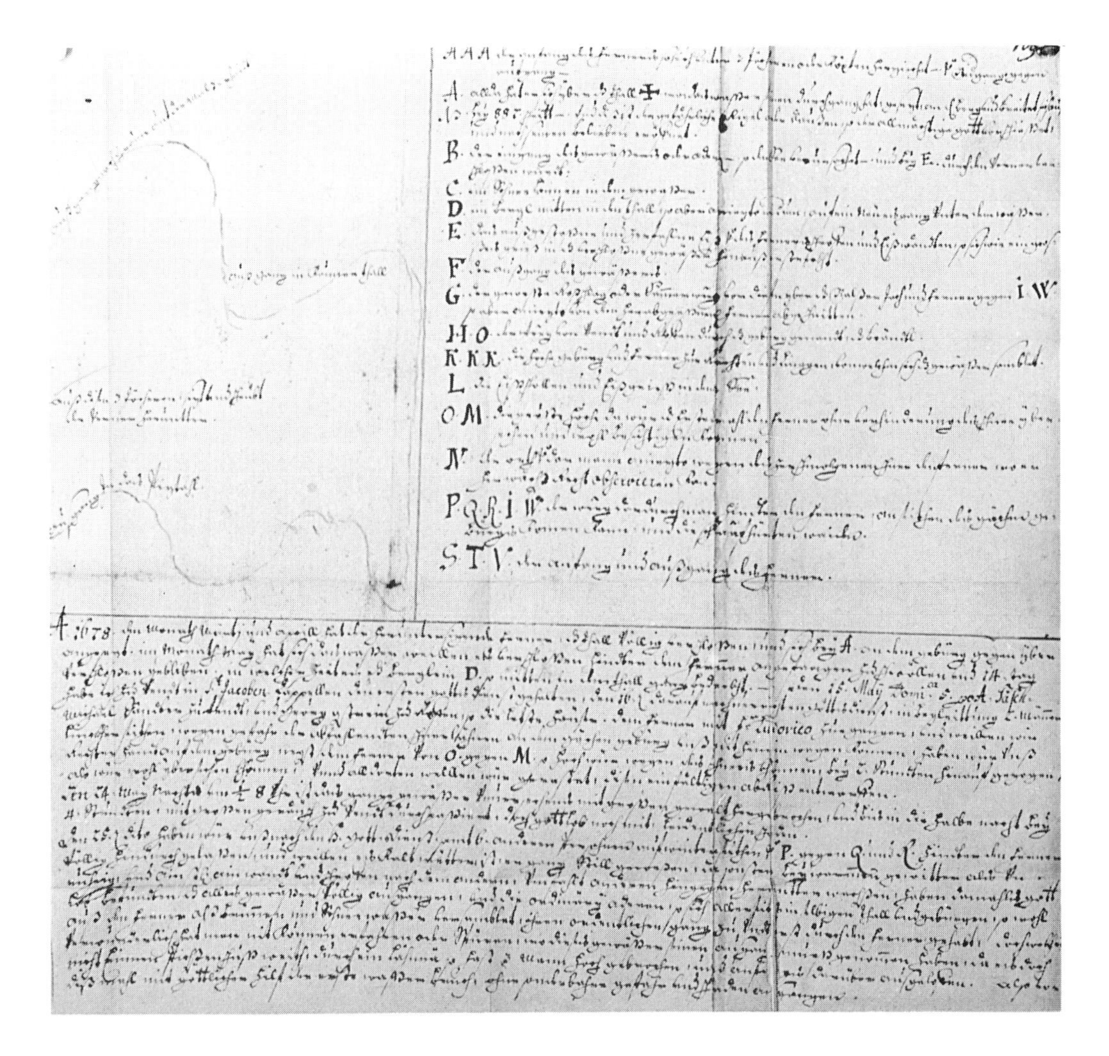

5.2 The Ötztal glaciers 1678–1725

Disquiet in Ötztal may have existed for a year or two before the 1678 flood. Johann Kuen of Langenfeld, who was prominent locally and was often consulted by visiting emissaries and officials, wrote a short account in 1683 according to which he had heard in 1676 that the Vernagt was advancing; by the autumn of 1677 'it had reached the mountain opposite . . . so that it dammed the water'. However, a Capuchin friar, dispatched to Vent by the Bishop of Brixen in 1678 to calm the population and hold services, had written to this same Kuen of Langenfeld on 1 July 1678, immediately after he had visited the Rofensee with his congregation, and he dated the formation of the ice dam to March or April 1678 and the beginning of the lake to May (Plate 5.1). The contemporary account is likely to be more reliable than Kuen's, written five years later; the congregation had recently repaired the path which was in common use by travellers and herdsmen and had watched the formation of the lake 'with sad hearts'. The friar had taken a knotted measuring rope with him and found the lake to be 744 ellen (588 m) long and 250 paces broad alongside the ice.

> At a point 120 ellen [95 m] from the end of the lake which abuts on the glacier, we measured the depth of the lake and found it to be 10 ellen [18 m]. From this it can be deduced how deep it must be in the middle or at the end near the glacier. We have also observed that during the two hours we spent there, the water rose more than one span. So it is easy to see how dangerous the situation is especially as we fear that the glacier has not yet stopped advancing. (Richter 1892, pp. 375–6)

Within a week, on 6 July 1678, the lake was visited again by a Father Sabinus with a small party including an official from Petersberg. They found the lake much enlarged. According to a letter written by the Capuchin friar to his superior on 18 July (Richter 1892, pp. 376–9) the ice dam had broadened to 840 ellen (654 m) and the lake was 1380 ellen (1074 m) long, and men who knew the place said it was as much as 100 fathoms (190 m) deep. The water level rose half an ellen and three breadths of a finger (44 cm) in the three hours the party was at Rofen and 'the glacier grows on all sides except the side where the water touches it'; here the water had undercut the ice. Nevertheless the dam was believed to be '78 ellen [60 m] lower than it had been years ago'. Presumably the traces of the level reached by the 1600–1 lake were still recognizable.

A regular series of visits was arranged to give warning of imminent danger, Kuen recording that 'every week a suitable person is sent to the glacier to bring back a report to the authorities'. On 14 July the friar walked up beside the glacier for three hours and was much impressed by its size: 'No messenger would be able to run round its circumference in one day, for it goes incredibly far and only the Good God knows how far it extends up the mountains.' So, as the weekly reports came in, 'the hay was mown somewhat earlier and brought in, the cattle sent into the alps and other preparations made'. Then, after several days of rain, the ice dam gave way.

'On 17th July the glacier split and the water broke through, preceded by a stinking fog and a terrible roar'.[4] Before dawn the flood crest had reached Huben (Figure 5.1) and it passed Langenfeld early in the morning. In a letter of 28 July 1678 Georg Respichler wrote that all the bridges except for the high one at Rofen 'have carried away so that nothing looks as it did before'. Some villages such as Hopfgarten, Östen and Tumpen were almost wiped out; others suffered only slightly less damage. Winkl and Platten lost eight or nine houses, together with barns and byres as well as goats and pigs. The damage to the land itself was so great that Respichler doubted whether the place would ever again be habitable. 'The roads and paths from Hochlrain to Vendt . . . have been spoilt completely in many places so that one can neither drive, nor ride, nor walk on them, and in some places it will be very difficult to mend them.'

Despite this trail of devastation there were few casualties, though in one group of villages, where the people were surprised by the flood before dawn and isolated from help for four days, two children were lost. Providence had been kind. There had been processions with the cross in various places and, as Kuen observed in his account of the flood,

> We can observe the beneficial results of these devotions in that, despite so much danger to life, what with the overflow of water during the night, and with the many dangerous jobs at the bridges, paths and levées, we have all preserved our lives, and although some people have fallen into the water, they have got out again, which is certainly a miracle and a wonder.

[4] In Richter, 1892, pp. 381–91, from J. P. Hurn (1770).

The cost of the 1678 floods was high; 45,000 florins for Umhausen, 115,000 florins for Langenfeld, and 22,000 for Sölden. Benedikt Kuen, son of Johann, writing forty-four years later, thought the cost had been unnecessarily high; he considered that more should have been done in earlier times to control the streams and rivers. But there had seemed to be little need for strenuous measures for very many years. The advanced position of the glaciers from 1678 and for nearly half a century afterwards presented a new challenge for him and his generation which they faced with energy and determination.

While work was beginning on repairing the damage of 1678 there was another flood in 1679 which did not seriously interrupt the work. Next year, however, the water broke out from behind the ice and did 'terrible damage to houses, barns, lands and bridges all over the Ez valley and beyond the mountains. . . if all the water had broken through at once, not only the lower Ez valley but also the lower Yn valley might have been completely ruined' (Richter 1892, p. 383).

According to the Kuen chronicle in 1681 'the glacier came very low, it was hard and blue'. Another outburst was expected,

> Therefore on 8th July twelve men were sent out from Langenfeld to cut a ditch through the ice, which, thank God, went well, so that the water ran out little by little through this ditch. So this time we could live without danger or care, repair the damaged things and lands in comfort and pleasure though with hard work.

Kuen's account was written two years after the cutting of the ditch, which was evidently a complete success. Such measures have not always been so; a similar attempt to deal with a nineteenth-century hazard at Mauvoisin in Valais ended in disaster (Mariétan 1959 and Chapter 6).

The 1681 Rofen flood, as on previous occasions, was followed by an official inquiry. A judge, Jeremias Rumblmayr, the court architect and master of the Board of Works, Martin Gump, and several others made a survey of the glaciers in June. They reported that the lake had been dammed back in the course of the winter, between October 1680 and July 1681, so that the water rose '50 to 60 fathoms and spread a good hour's walk up the valley, here 1000 paces wide' (Richter 1892, p. 401). The spillway through the ice had been cut by thirteen men in three days under the direction of Johannes Kain, Steward of Langenfeld. The water ran away between the ice and the mountain in two or three days; Rumblmayr was inclined to think that the same thing would have happened if no action had been taken. The remaining ice was calving rapidly into what was left of the lake and a medium-sized stream was discharging underneath it. The situation could still become dangerous but at any rate there did not seem to be any threat to the Inn valley, which was Rumblmayr's main concern. He reckoned that a flood reaching the Inn would take ten hours to traverse the Ötztal and therefore lose its power.

Rumblmayr noted that the damage of the preceding years had been accentuated by tributary streams flooding and causing landslides. Whether their flooding was due to rain or snowmelt is not clear. At Langenfeld and Ästan, where fields had been cut away in 1678 and 1679 old flood protection works had been revealed and the river 'had resumed a course it had taken a hundred years or more earlier'. The countryside had suffered severely. 'All the bridges have been pulled down for fear of the overflowing of the lake and the ordinary roads are impassable and the people have partly left their houses on the flat land and retired with their cattle to some miserable cabins at the foot of both mountain slopes.'

The Commission was anxious to find a permanent solution to the flood hazard and Martin Gump, in one section of the report (Richter 1892, pp. 402–7), considered the possibility of cutting a long gallery or tunnel through the mountain side. He saw the difficulties there would be in providing light and air for the excavators and the high cost of upkeep. The worst problem, he thought, would be prevention of blocking by ice from water trickling through the rock and freezing, or the stream itself freezing in the tunnel and not melting in summer 'because it is always colder in such places'. He concluded that a gallery would not be advisable.

Another member of the commission of inquiry, a joiner called Paul Heuber, suggested that a dam might prevent flooding. It could be built on one of three places; at Lenersegg below Zwieselstein where the Ötztal narrows, at Gämpel (Figure 5.2) where the stream flows through a narrow defile, or behind the Rofen fields. These proposals were put to local farmers, who found objections to all of them but eventually came down in favour of the dam at Gämpel. The cost of this was too high and some members of the Commission were not convinced that it would hold back the floodwater. Cutting through the ice was judged to be impractical. In fact no remedy commended itself.

By this time 'the glacier seemed to be retreating all the time' and the ice dam had shrunk to half its maximum height. With some relief the Commission reminded the communities of Ötztal of their promises and those of their forefathers to make devotions for the remission of floods: 'For just as God Almighty has had such a monster come into existence during one winter so his divine Goodness could be moved by earnest prayers so that he let it retreat in the same or a shorter time.' It recommended that the money that might have been spent on one or other of the suggested remedies should be devoted to relief of those who had suffered from the floods; 2000 florins already granted for projects were immediately available. It was also proposed that the province of Tirol should remit taxes for a certain period: 'the Abbess of Kiemsee, and especially the convent of Stams which gets most tithes, could convert the tithes into corresponding aid for a period of some years; and the government might give the sufferers new land instead of the spoilt land and houses' (Richter 1892, p. 398).

The Vernagt tongue remained extended into the Rofental for some thirty years, until 1712 when the remains of the ice dam finally melted away. Very much earlier, by 1683, we learn, 'the people can now live without danger or worry' (Richter 1892, p. 394).

Scarcely had the last remnants of the Rofen ice dam melted away than in 1717 the Gurglerferner, advancing down its valley, blocked the Langen tributary (Figure 5.2) and a lake formed in Langtal. This seems to have been the culmination of an advance that had been under way for many years. J. Cyriak Lachemyr, who visited the glacier on 2 July 1717, mentioned that an old hunter 'who had worked the area for more than forty years affirmed that in the course of that time the Ferner had driven further down into the valley' (Richter 1892, p. 414). Again there was concern about the danger of flooding in Ötztal. The Warden of Petersberg reported to Innsbruck that the 'Gurglerferner had formed a lake quite unexpectedly and without anyone knowing anything about it until three days ago'. However, Sonklar (1860), who based his account on Walcher (1773),[5] states that the Gurglerferner had extended past the junction with the Langtal for the first time in 1716.

[5] Josef Walcher, a Jesuit, Professor of Mechanics at the University of Vienna, visited Ötztal in 1772 and afterwards published an interesting little book, *Nachrichten von den Eisbergen in Tirol*. Many of the documents unearthed by Richter were unknown to him and he based his account of the events of the earlier part of the eighteenth century on Kuen.

At any rate, by 1717 the ice was holding back a lake, 1600 paces long, 500 paces wide and about 70 fathoms deep. Avalanches of snow made it difficult to tell whether or not the water was escaping between the ice and rock, so guards were posted and prayers, processions and masses were arranged.

The Petersberg Warden was told by his messenger, a native of Sölden, that there was a story that the Gurglerferner had erupted three centuries earlier, causing damage. Kuen had also heard the story but was unable to find any support for it. Richter was not inclined to believe it. Nevertheless, it is quite possible that there is some substance in the story. Walcher (1773, p. 4) heard it said that glaciers appeared in the Tirol for the first time in the thirteenth century, when several cold winters followed each other and so much snow and ice collected in the high mountains that the sun was unable to melt it. Stoltz (1928, p. 19) cites a text of 1315 according to which farms in the Ötztal were granted respite from their taxes in that year as their lands had been 'ravaged ex alluvionibus et inundationibus'. But there is no direct evidence that such damage was connected with a glacial advance.

As a consequence of the governor's report, a Commission from Innsbruck arrived on 2 July 1717, only to find that the lake had emptied the previous day. The damage was not very great, though a number of bridges had been wrecked. Lachemayr, a member of the Commission, pointed out that the Gurgler, not the Vernagtferner, was to blame: 'your Excellency might graciously note that this one is not the glacier of which there is a model at Innsprugg'.[6] The Rofental, 'according to reliable information . . . is now entirely free of water'. Lachemayr visited the lake held back by the Gurglerferner and reported that it was '1600 paces long and 500 paces wide'. He went past the lake and up to the foot of the glacier in Langtal. He hoped that the two glaciers might soon be confluent, thereby obliterating the lake.

The Commission recommended that a relief channel should be cut to drain the lake, that guards should be posted, and the river channels of the Ötztal and Inn valleys cleared. In the event it proved too costly to dig the relief channel, but observers were certainly posted and orders were given to expedite protective works, the managers of the salt works at Hall being told to give free wood to the people between Zwieselstein and Gurgl so they could build fortifications and bridges. The lake had emptied by 3 August, but on 16 September the Warden of Petersberg wrote that he was expecting another outburst within a year.

Guards visited the lake trapped by the Gurgler through the winter of 1717–18, receiving 30 florins on each occasion. On 1 April the Warden of Silz reported that, according to Erhard Prugger, 'the glacier has grown both longer and broader, not so much or so suddenly as last autumn, but it might advance if the weather grows warmer or there is a long period of rain'. Prugger's experience had evidently led him to a good understanding of glacier movement.

By May 1718 the lake was 1100 paces long and only 540 paces from the Langtal tongue. On 2 July, Langenfeld and Umhausen sent representatives to join a procession from Sölden to the lake. Jacob Kopp, vicar of Sölden, said mass and the representatives met to discuss possible remedies. On 6 July the vicar wrote to Lachemayr,

> I have already said mass there on the ice for the third time and done everything spiritual for the prevention of this evil. But I have not been able to find any means,

[6] As far as is known this early model has not survived, nor is there any record of its maker.

> nor have other competent people of whom many have been up there, with which to improve the situation or remove the imminent danger by human power.

However, he went on to suggest that it might be possible to widen the narrow gorge above the village of Gurgl using gunpowder.

By 14 July 1718 the lake was rising 2 inches every two hours and was expected to overflow in six day's time. But two days later water began to escape under the ice on the right-hand side, and on 1 August Kopp was able to tell Franz Lachemayr, a government secretary, that 'at the lake where I say mass every week 17 fathoms, or half the water, has run off without causing any trouble. And we have good hopes that the rest will run away between the ice and the rock without causing any damage at all.'

The sequence of the lake filling and emptying seems to have become a recurrent event, at least until 1724. After that there is no information about the Gurgler and its lake for fifty years, except that in 1740 when 'Christian Gstrein measured the Ferner at Gurgl . . . the breadth of the lake was 145 fathoms, the depth 66 fathoms and the length 500 fathoms' (Richter 1892, p. 388).

5.3 The advance episode 1769–74

The Ötztal glaciers were said by Stotter (1846, p. 26) to have retreated for a while around 1750, but whether or not this was in fact the case by the 1770s both the Vernagt and Gurgler glaciers, which had been out of phase earlier in the century, were in advanced positions. There had been several floods in the preceding decade in the Ötztal and other parts of the Tirol. In 1763 the Härlachbach fed by the Grasstallersee, the Stralkogler and the Grieskoglerferner, extensively damaged Umhausen, carrying away or covering with debris sixty-two houses, 'so that the next morning their owners could no longer recognize the place where the houses had been' (Walcher 1773, p. 62). In 1769 the Farstrinnerbach, which was usually very small, covered the streets of Oesten, near Umhausen, with debris, so that one had to descend to the door of the church instead of climbing six steps up to it.

Walcher (1773) gives a detailed description of the condition of Ötztal in the third quarter of the eighteenth century. Maize, wheat and other crops were grown around Ötz and Umhausen; Langenfeld specialized in flax. Further up the valley, barley and oats were grown but ripened completely only in warmer years. Grass and herbs could be cut only every second year towards the head of the valley, though the pastures near the ice were particularly rich 'so that it is difficult to find richer milk or better butter than at Fender [Vent] or Rofen'. Rofen was the last inhabited place.[7]

By 1769 the Gurglerferner's advance was causing disquiet. It held back a lake that threatened to overflow and, although it was only 500 fathoms long in early September 1770, when Johann Peter Hürn, a district inspector of highways, visited it, he noted that it had been three times that length not long before (Richter 1892, p. 422). Walcher wrote that the glaciers in the area were increasing all the time. When he visited Ventertal he found that even the lesser glaciers such as Latschferner were swollen, much crevassed and with steep tongues reaching far down into the main Ventertal (Plate 5.2). The tongue of the Vernagt advanced more than 100 fathoms between the spring and autumn of 1770 and a further 25 fathoms in a few weeks in 1771. Not only the Vernagt but also the neighbouring Guslarferner were advancing and the local people were well aware of the serious implications of this conjunction of events (Walcher 1773, pp. 35–7). An

Plate 5.2 Latschferner in August 1772. This illustration, reproduced from Walcher (1773), shows the enlarged tongue reaching far down towards Ventertal. A and B are mountain peaks, between them seracs or 'ice pyramids' C and D are shown. Walcher noted that the space between D, the source of the Latschbach and the ice pyramids was partially covered by old snow and firn. The tongue was thus protected by snow cover very late into the ablation season

Plate 5.3 The Gurglersee in 1772 (A). The Langthalerferner on the left of the picture is marked B. The mountain in the centre is the Schwarzerberg (C). A number of travellers are making their way over the Gurglerferner (D) on one of the recognized long-distance routes (E). (From Walcher 1773)

inspection made on 12 June 1771 showed that the gap through which water could drain down the main valley was only a hundred fathoms wide and a crevasse 800 fathoms long separated the 'Fernaggerferner' from 'the mother ice' (Richter 1892, pp. 424–5).

In Innsbruck on 21 July 1771, the Governor of Upper Austria conferred with various experts including a professor of mathematics, Weinhart, the judge of Petersberg, Kirchmayr, and a clerk of the works called Nenuer. It was decided to prevent at all costs the threatened devastation. A new commission with extensive powers was set up and provided with a copy of a new map of the Tirol.[7] Included on the commission were administrators, technicians and people with local knowledge: the manager of the mint and salt works, a Jesuit father, the clerk to the board of works, the judge of Petersberg, district highway inspector Hürn, a carpenter and a wood merchant. They reckoned the glaciers were advancing as a result of wedges of ice breaking away from the crevassed masses on the upper slopes and it was not considered that much could be done to prevent this.

On 29 July 1771 the commissioners went to inspect the lake held back by the Gurglerferner and found it to be 1600 paces long, 500 wide and 50 fathoms deep (Plate 5.3). The water was running away more slowly than usual and there were fears the outlet might become blocked altogether and freeze up. Cairns were erected to mark the position of the ice front and a post was set in place for observing the changing level of the water. If it were to rise a few fathoms more, than it was agreed that twenty or thirty workmen should dig a trench through the ice 280 fathoms long and 6 fathoms deep to let the water out. Reports made in July, August and September have been lost. It is known that Hürn was given the task of hiring men to prevent blocks of ice obstructing the stream flowing out of the lake. They carried on with this work until the end of October, though by then Hürn had died.

The main threat was from the Vernagtferner, which had continued to grow through the winter of 1771–2 and by April was 60 to 70 fathoms high and 400 to 460 fathoms broad. It had obstructed the Rofen stream since November 1771 and by April the lake behind was 900 fathoms long and 30 to 40 fathoms deep. Prantl, Steward of Sölden, mentions that the glacier was making a great cracking and rushing sound that could be heard at times in the houses of Rofen (Richter 1892, p. 429). By 8 June the lake was rising fast and a crack had formed 10 to 15 fathoms from the side of the valley. It was found impossible to dig an outlet channel because the ice was constantly moving. In a report to the government, Menz proposed using artillery to batter a way through the ice. But guns could not be brought to such a remote place, so bridges were torn down and a path was built down the valley above the level of any possible flood. By 3 July 1772, a flood was expected within a few days and Menz regretted that people had not been persuaded to work on digging a canal. On 20 July the ice barrier was as high as it had been in 1678 and the lake was 130 fathoms wide. Kirchmayr reported on 20 July that 'in all three valleys the ice is more crevassed right up to the passes than it was last year' (Richter 1892, pp. 430–1). By mid-August, when Walcher visited Rofen, he was astonished by the size of the mass of ice damming back the lake (Plates 5.4 and 5.5).

[7] Walcher noted that not only the larger villages, but 'also most of the houses, streams, bridges and roads' were marked on 'a new map of Tirol' then shortly to appear, which had been made with great diligence and exactitude by Ignos Weinhard, SJ, official lecturer in Mathematics and Mechanics at the University of Innsbruck, and Peter Annich and George Heuber, two Tirolese peasants (Walcher 1773, p. 2)

Plate 5.4 The Rofnersee in 1772 as seen from a position on the Zwerchwand. The seracs of the Vernagtferner in the right foreground block the Rofental and hold up the lake (A). Note that the Vernagt tongue is not only very steep but fills its valley to a high level. The Plateikogl is marked H and I. The Kesselwandferner is marked N and the Hochjochferner D. F is the higher part of the Vernagtferner, M is the Guslarberg. (From Walcher 1773)

Plate 5.5 The Vernagt ice (E) in 1772 seen from a distance. The Rofental gorge below the dam (S) is in the centre foreground. The gentler slopes of the Platei (H) were used to collect hay, which was stored locally (T). Stone cairns (P) mark the routeways. I is the Plateykogl. (From Walcher 1773)

> The whole of the Vernagt valley is filled with especially big blocks right up to the path, and more still are coming. Standing on the Plattei summit . . . an incredible amount of ice stretches as far as the eye can see and it will certainly come down as soon as the other ice clears away. (Walcher 1773, p. 26)

Walcher attributed the many crevasses to the very hot summer. The ice continued to advance into the Rofental and by 19 September it reached 17 fathoms beyond the marker posts that had been erected. Then to everyone's surprise and relief the water found a way out between the ice dam and the side of the valley and over a period of eight or nine days drained quietly away.

In the summer of 1773 the lake rose 5 or 6 fathoms higher than in the previous year, according to Prantl, and then emptied suddenly, falling 13 fathoms between 11 and 12 July, and 30 fathoms in five hours on 23 July. But the other streams happened to be low and, surprisingly, 'the flood did no damage worth mentioning' (Richter 1892, p. 422).

In June 1774, the Vernagt was well forward, 'reaching the old marks in many places' and even exceeding them. By 22 June it was only 10 paces below its highest recorded level of 1681. Two days later, heavy rain caused all the rivers and streams to rise and on 26 June the water found an outlet. Until 4 July it drained away slowly and then at six in the evening Prantl at Sölden saw the river rise suddenly and continue to flow strongly until next morning. The lake fell 31 fathoms in 12 hours but no damage was done. The danger was over for that year, but 'what will happen next year God only knows'.

An anticlimax; in 1779 the bookkeepers of the Imperial Government recorded that 1920 florins had been spent on the dismantling of twenty-two bridges in the early 1770s in Ötztal and for rebuilding them in 1779. The Vernagtferner had evidently melted back and the danger of flooding had disappeared.

5.4 Glacial advances in Ötztal 1845–50

For over thirty years little information is forthcoming about the Ötztal glaciers. Sonklar (1860, cited Richter 1891, p. 31) tells us that the years from 1812 to 1816 were cold and damp, and 1816 and 1817 were years of famine (Richter 1891, p. 31). The Niederjoch and Gurglerferner advanced (Rohrhofer 1953–4, p. 58). The Hintereisferner reached a maximum in 1818 (Richter 1891, p. 28). The Vernagt advanced with the others but not nearly as far as in 1770–4. Ensign Hauslab, later to become a prominent cartographer, was sent to survey the Vernagt area in 1817 and produced a fine map (Plate 5.6) which shows the Vernagt tongue enlarged but still 1.4 km away from the Rofental. Its advance continued until 1820–2 but the terminus did not reach into the main valley (Stotter 1846).

The advances of the Ötztal and other Tirolese glaciers in the early nineteenth century were less important than those that took place in the Mont Blanc massif about the same time. However, the mid-century advances were as important as those of the Brenva and others on the Italian side of Mont Blanc and the expanded condition of the glaciers in 1850 is shown on a sketch map published by the Schlagintweit brothers (Plate 5.7 and Figure 5.3). The Marzellferner, Mutmalferner and Schalfferner came together at the head of Niedertal; the Hintereis joined the Kesselwand and the Vernagt came down across the Rofental.[8] The Niederjoch reached down beyond the 1770 limit in 1845, retreated a little

[8] Some confusion can arise because the names used for glacier in Ötztal vary. Schlagintweit called the Schalff-Marzell the 'Stock-Marcell', the Mutmal the 'in den-Schwarse' and the Gurgler the 'Grosser Ötzthaler', whilst Sonklar called the Marzell the 'Marzoll'.

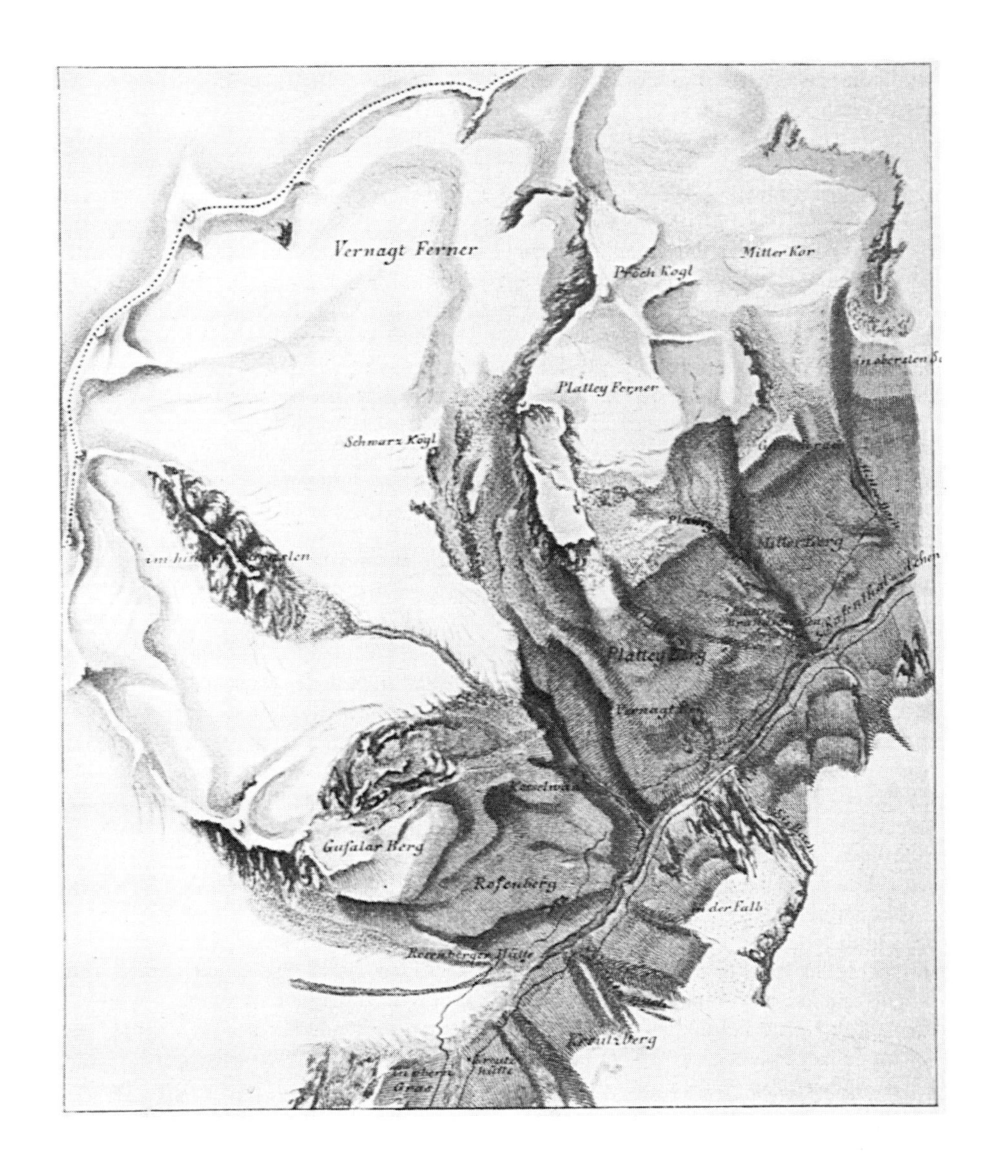

Plate 5.6 Hauslab's map of Vernagtferner in 1817. (From Finsterwalder 1897)

Plate 5.7 The end of the Vernagtferner in 1847. (From Schlagintweit and Schlagintweit 1850, Figure 55)

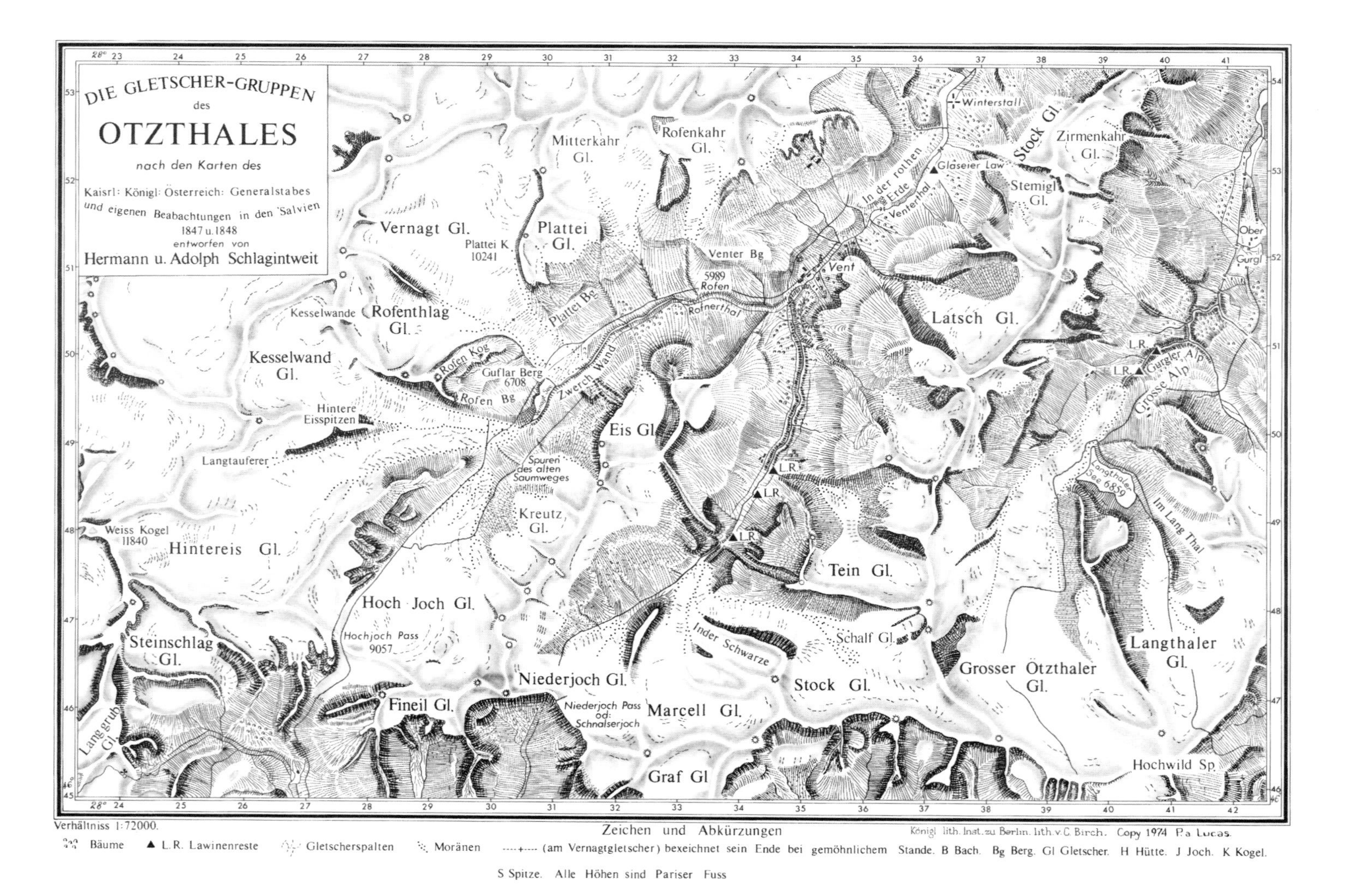

Figure 5.3 The Schlagintweits' map of the mid-nineteenth century (Schlagintweit and Schlagintweit 1850). Note the extended state of the Vernagtferner

and then, in 1850, returned to its 1845 position, remaining there for six years (Schlagintweit and Schlagintweit 1850, Rohrhofer 1953–4, p. 59). The Gurgler pushed well past Langtal and the emptying of the Langtalersee for a time became an annual event (Schlagintweit and Schlagintweit 1850, p. 145). Both the Schalfferner and the Hintereis resembled the Niederjoch in reaching a first maximum position about 1845–8 and a second about 1855–6 (Schlagintweit and Schlagintweit 1850, Richter 1891, p. 34).

The behaviour of the Vernagtferner as usual demands special consideration. Between 1822 and 1840 it had retreated a good deal 'more than one hour's walk' writes Stotter (1846, p. 28). Then it began to advance and for the first time measurements were made of the rate at which this was happening. A farmer of Rofen called Nikodemus Klotz, who was also a chamois hunter, had noticed in 1840 that the névé of the Gurglerferner had thickened and that there were enormous quantities of snow on the upper Vernagt. Encouraged by the parish priest of Vent, he kept records which show that the terminus of the Vernagt advanced 3 m between 2 and 9 April 1843, and 200 m between 18 June and 21 August, its tongue filling the old lateral moraines (Frignet 1846, p. 88, and Figure 5.4 here). It continued to advance through the following winter, pushing a wad of snow in front of it and gradually accelerating (Table 5.1).

On 1 June 1845 the schoolmaster at Vent reported that the Vernagt had blocked the Rofental. A technical commission was set up 'to examine the situation at the Vernagtferner scientifically and practically and to make proposals for preventing or at least controlling the threatened danger'. The party of officials and experts, assembled near Vent on 13 June 1845, was joined by representatives from Sölden and other villages. They set off for Rofen but 'did not dare venture into the crevasses and seracs at the Zwerchwand, 'so it seemed that they would only be able to make a cursory survey. However, a small party of local mountaineers came to their assistance, and, crossing the Rofen ravine on a snow bridge, 'skilfully and surely climbed the often almost vertical rock. We shuddered to watch them as they went. Hr Bergratt Zottl had instructed them in the use of instruments and had indicated the lines which were to be measured' (Stotter 1846, p. 49).

The party had completed its measurements of the lake and was crossing the Rofen meadows to reach the bridge when the stream suddenly changed colour and became dark brown, bringing with it lumps of ice. The cry 'the lake is breaking out' went from mouth to mouth. They hurried to the bridge and found the water rising gradually at first but soon much more rapidly. 'The lake with all its power had suddenly broken through the ice dam' (Stotter 1846, p. 51). Next day Klotz went up to the ice dam and returned to Vent by the time of Sunday mass to tell them that the water had not cut through the ice

Table 5.1 Mean daily rates of advance of the Vernagtferner from 18 June 1844 to 1 June 1845

Period	*Mean daily rate of advance (m)*
18 June–18 October 1844	1.0
18 October 1844–3 January 1845	2.1
3 January–19 May 1845	3.3
19 May–1 June 1845	12.5

Source: Hoinkes (1969)

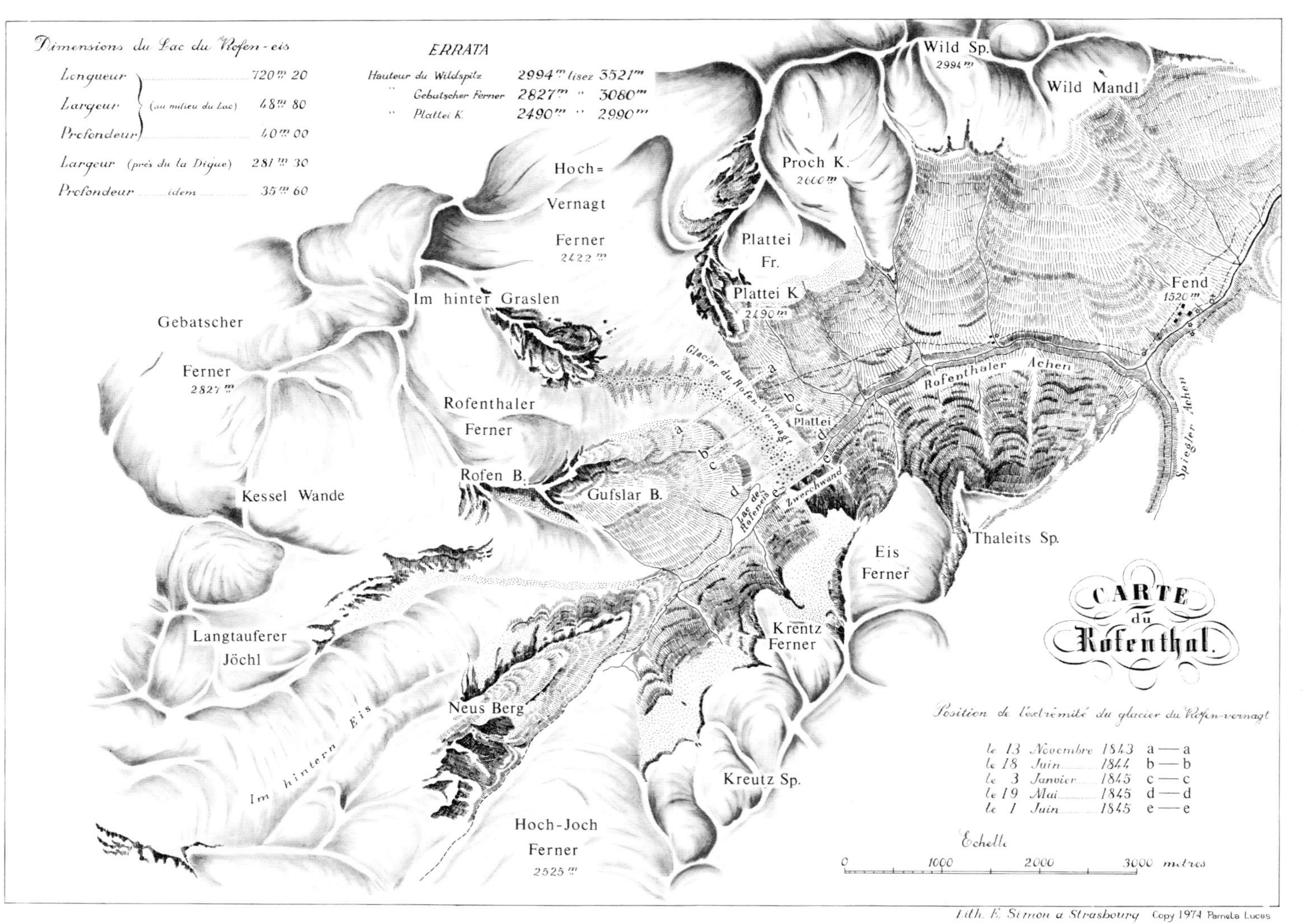

Figure 5.4 The 1843–5 advance of the Vernagtferner. This map was published by both Frignet (1846) and Stotter (1846)

barrier but had found a way underneath it; the opening was already blocked and the lake beginning to refill.

As they returned down the valley they saw the damage that had been done already. 'From Vent to Umhausen scarcely one tenth of the track along which we had ascended was passable.' Out of twenty-one bridges between Rofen and Umhausen only three remained undamaged. In the rocky gorge between Zwieselstein and Sölden debris hung in the trees many fathoms above the stream bed. The floodwater had destroyed the dykes protecting the meadows and fields in the Sölden basin, carrying away everything in its path. Though the buildings were all on the edge of the valley or on slopes rising from its floor, they were quite badly damaged. Farmland had been flooded as far downstream as Langenfeld. The flood wave had reached the Rofen bridge shortly after five in the evening, Sölden at seven; it was between one and two the following morning that it reached Innsbruck; 'thus', wrote Stotter, 'the water traversed the 22 hour way from Vent to Innsbruck in about eight hours'.

The lake formed again in 1846 and drained away slowly. There was another sudden emptying on 25 May 1847 and again on 13 June 1848. Hoinkes (1969) has estimated that the volume of the 1847 lake was comparable with that of 1678, about 10×10^6 m^3; that of 1848 contained about 3×10^6 m^3 of water (Hess 1918).

The velocity of the Vernagt had greatly diminished. When the Schlagintweits measured the surface movement between 28 August and 20 September 1847, at two cross-profiles 483 m and 840 m up the glacier, they found it to be 0.13 m and 0.09 m per day respectively, a great deal less than the 2.3 m per day in the period July 1844 to June 1845. The surface of the glacier had bulged up and had been deeply crevassed (Sonklar 1860, p. 37). After 1847 the surface got lower and when he saw it in 1852 Sonklar was able to appreciate the change that had taken place:

> The bulge in the middle had disappeared and by looking at the sides of the valley it was possible to see that the surface of the ice had collapsed no less than 250 to 350 feet or 80 to 100 metres since the time of its maximum. . . . The strongest impression was produced by the enormous extent of the two lateral moraines . . . the glacier appeared like the bottom of a valley between these two moraines.

Since 1848 there have been no more floods. The catastrophic emptyings of the Vernagt lake took place between about 1600 and 1848 and in addition there were six occasions when there was reasonable cause for alarm in Ötztal but the water drained out slowly (Table 5.2).

Table 5.2 Dates of filling and emptying of the Vernagt lake between 1599 and 1846

Lake first formed	*Slowly emptied*	*Rapidly emptied*
1599	1601	25 July 1600
1678	1679	16 July 1678
	1681	14 June 1680
1771	1772	23 July 1773
	1774	
1845	1846	14 June 1845
		28 May 1847
		13 June 1848

Source: Hoinkes (1969)

5.5 The retreat of the glaciers after 1850

The glaciers of the Ötztaler Alpen continued to diminish in volume from 1850 until about 1964, though there were periods between 1890 and 1900 and again around 1920 when many advanced.

Although there is little precise information for the years between 1850 and 1890, all available accounts agree that the glaciers were retreating (Kerschensteiner and Hess 1892). Towards the end of the 1880s it became known that the Swiss glaciers were beginning to advance again and this resulted in a widespread effort to record the positions of the Tirolean ice tongues.

In the summer of 1883, Richter started a programme of mapping in the Ötztal. He intended to base his work on the 1:25,000 military map of 1870 but found that it was not of much use except for the Vernagtferner. He therefore concentrated on making simple tape measurements of the positions of the fronts of nine glaciers in addition to the Vernagt. They showed substantial differences. The Hintereisferner, which terminated in a narrow gorge, had retreated only 150 m since the mid-century maximum, whereas the retreat of the Mittelberg was quite obvious from a distance, for the terminus had receded 880 m since Sonklar saw it in 1856 (Sonklar 1860, p. 134) (Table 5.3). The Taschach had gone back 490 m between 1856 and 1883 whereas the Marzell–Schalfferner tongue, ending in a ravine, had retreated only 72 m though it had thinned by about 100 m. The surface of the Gurgler was 20 to 25 m lower than it had been, though the terminus had not retreated very much. Richter was intrigued by these variations and realized that more knowledge was required about the relationships between the frontal positions of the ice, the height of the névéline and the form of the subglacial surface. Many of the glaciers seemed to have got thinner until about 1870 and then their fronts had begun to retreat rapidly. The Vernagt (Plate 5.8) was a law unto itself. In the 1880s its tongue was remarkably small in relation to its overall area. It occupied over 16 km^2 above the 2800 m contour and only 103 ha below that height.

Finsterwalder, Schunck and Blümcke from Munich selected the Gepatschferner for particular attention in 1886 and 1887 (Finsterwalder and Schunck 1888). Collaborating with Kerschensteiner of Nuremberg, they mapped the tongue of the Vernagt (Finsterwalder

Table 5.3 Nineteenth-century retreat of the Vernagt, Mittelberg, Taschach and Gepatsch glaciers, Austria

		Retreat (m)	*m/yr*
Vernagt (Richter 1885)	1847–83	2092.5	58.1
Mittelberg (Richter 1885)	1856–70	162.5	11.6
	1870–83	717.5	55.2
Taschach (Richter 1885)	1856–78	353	16.0
	1878–83	137	27.4
Gepatsch (Finsterwalder 1928)	1856–88	460	14.3
	1886–99	153	51.0
	1891–96	127	25.4

Sources: Richter (1885) and Finsterwalder (1928)

1897b), and they also attempted, rather unsuccessfully, to make velocity measurements. The Hochjochferner was mapped in 1890 and the Hintereisferner in 1893–4 (Richter 1893–4).

As early as 1874 Richter advocated regular observations of the frontal positions of a large number of glaciers in the eastern Alps. His scheme was eventually adopted by the Deutscher und Österreichischer Alpenverein in the 1890s. By this time the recession that had begun in the 1850s was slowing down or had halted. The research committee of the Alpenverein requested its members and sections to collaborate in making regular observations and to continue with those already being made (Finsterwalder 1891).

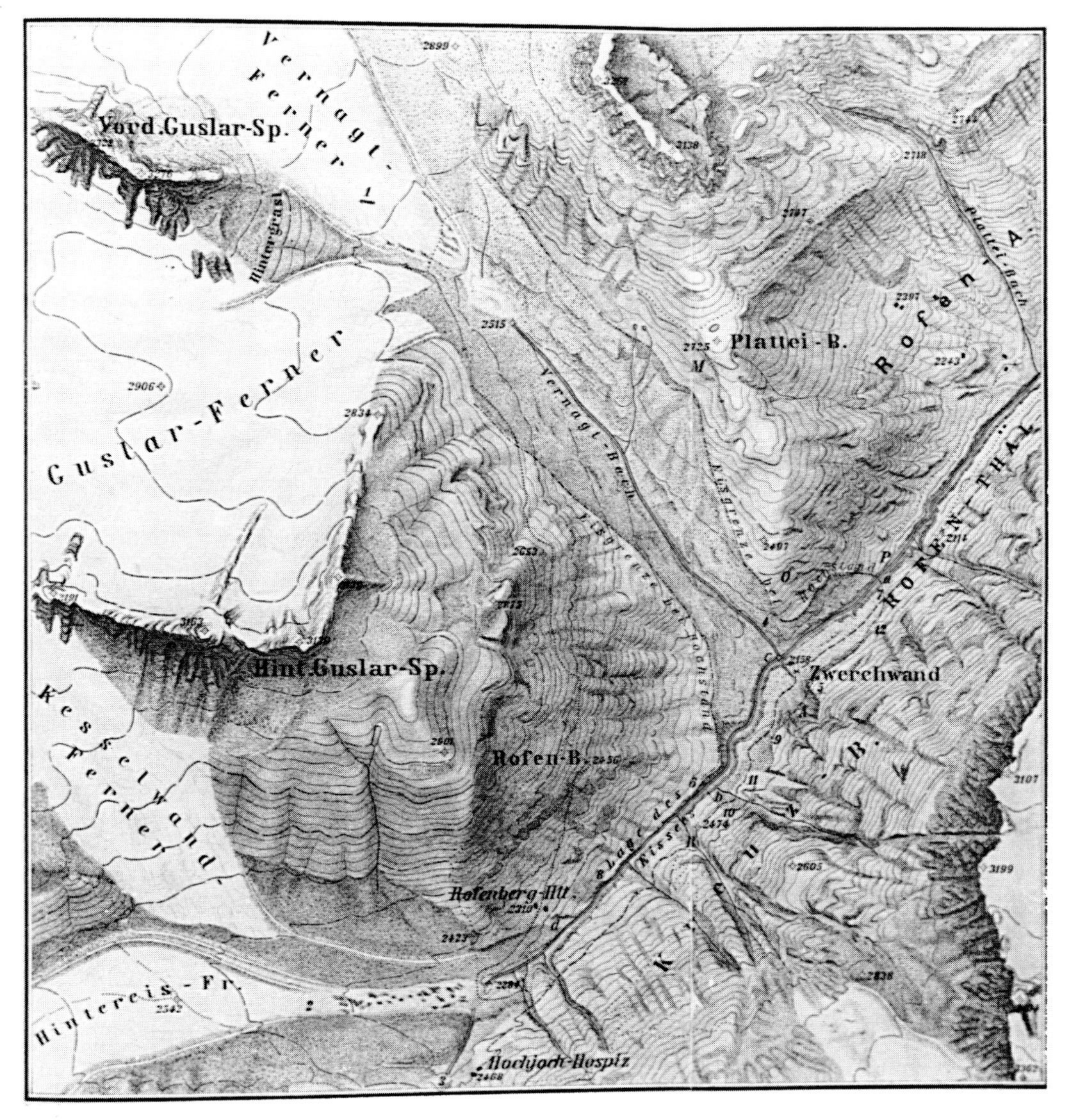

Plate 5.8 The Vernagtferner and Guslarferner in 1888. (From Richter 1892)

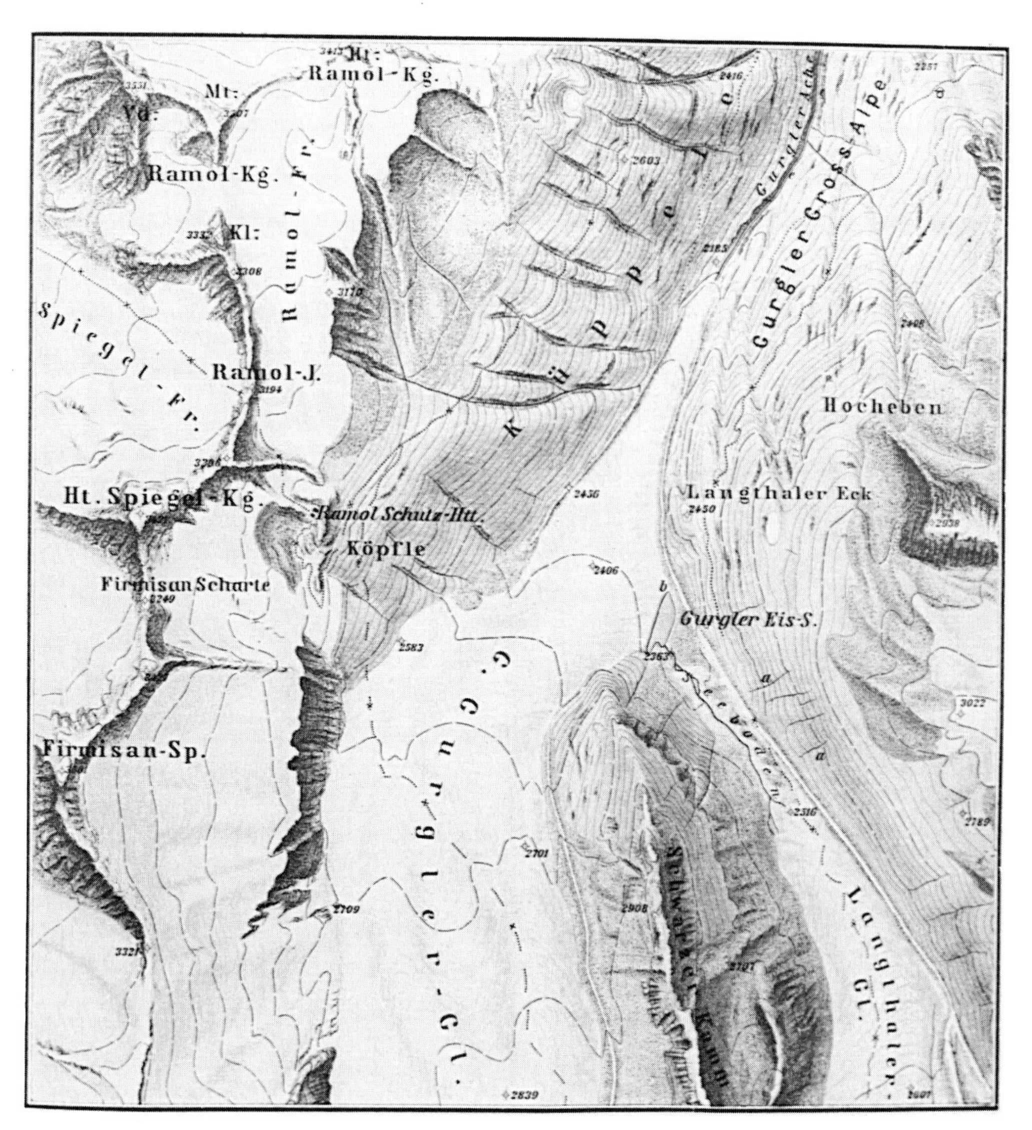

Plate 5.9 The Gurglerferner in 1888. (From Richter 1892)

Observations on the Niederjoch tongue started in 1891 (Rohrhofer 1953–4), and the Breslau section of the Alpenverein began measurements of the Gurglerferner (Plate 5.9) in 1895, though this particular set of observations was discontinued within a few years because of the difficulties of the site (Srbik 1942a). The number of tongues measured varied from year to year and was not to reach eighty until 1924 (Patzelt 1970).

In 1891 about 40 per cent of the glaciers being measured in the eastern Alps were advancing, and by 1900 about 50 per cent; this compares with only 30 per cent of the glaciers in the Swiss Alps at the beginning of the century (Patzelt 1970). The majority of the glaciers in the eastern Alps then receded for a decade before advancing once more; by 1919 about 75 per cent were waxing. Then came a major recession, interrupted only briefly between 1925 and 1927. In the Ortler and Stubaier Alpen accumulation might

Table 5.4 Annual changes in the surface height at various altitudes of the Hochjoch, Guslar and Vernagt glaciers between *c.*1893 and 1940

	Changes in height (m)					
	Hochjochferner		*Guslarferner*		*Vernagtferner*	
Altitudinal zone (m)	*1893–1907*	*1907–1940*	*1889–1912*	*1912–1940*	*1889–1912*	*1912–1940*
2500–600	− 2.32	− 1.50				
2600–700	− 1.62	− 1.06	− 1.55		+ 0.05	
2700–800	− 1.28	− 1.04	− 0.51	− 0.57	− 0.13	− 1.07
2800–900	− 0.84	− 0.49	− 0.42	− 0.59	− 0.53	− 0.51
2900–3000	− 0.92	− 0.40	− 0.20	− 0.39	− 0.26	− 0.07
3000–100	− 0.75	− 0.10	− 0.17	− 0.29	− 0.57	− 0.03
3100–200	− 0.47	+ 0.05	− 0.24	− 0.35	− 0.41	− 0.07
3200–300	− 0.69	+ 0.14	− 0.28	− 0.35	− 0.20	− 0.10
3300–400	+ 0.23	+ 0.09	− 0.36	− 0.36	− 1.12	− 0.33
3400–500					+ 0.04	− 0.40

Source: R. Finsterwalder (1953)

have been enough to cause enlargement, had it not been for the dry, hot summers of 1928, 1929 and 1930, with high rates of ablation (Steinhauser 1957).

From a map study of the changes in the area and height of the surface of the ice in eight representative glaciers, R. Finsterwalder (1953) was able to show that the ice had continued to diminish in volume as well as in extent since the mid-nineteenth century. An analysis of the results from glaciers, which included the Gepatschferner and the Hintereisferner, indicated that rates of lowering were similar for the periods 1850–90 and 1920–50, about 61 cm per year; between 1890 and 1920, wasting was only half as fast, about 30 cm per year. The variability of rates of lowering in time and space is indicated by Table 5.4.

The retreat of the glaciers at the head of the Spiegler Ache (Figure 5.2) in Ötztal has been traced by Srbik (1935, 1936, 1937, 1941, 1942a and b) and by Rohrhofer (1953–4). The withdrawal of the Niederjoch, Marzell, Mutmal and Schalfferner between 1850 and 1950 is portrayed in Figure 5.5. The waning of the glaciers of the eastern Alps and of the

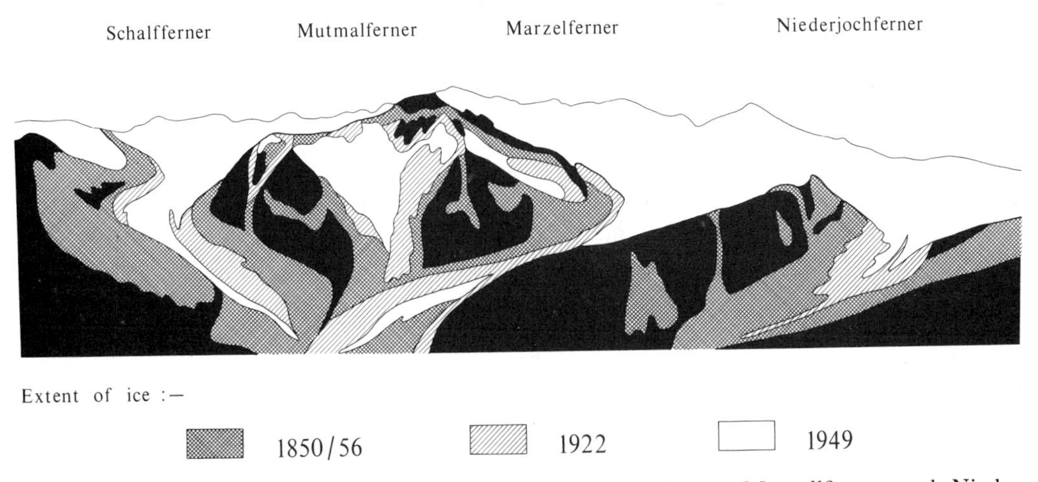

Figure 5.5 The diminishing extent of Schalfferner, Mutmalferner, Marzellferner and Niederjochferner between the mid-nineteenth and mid-twentieth centuries. (After Rohrhofer 1953–4)

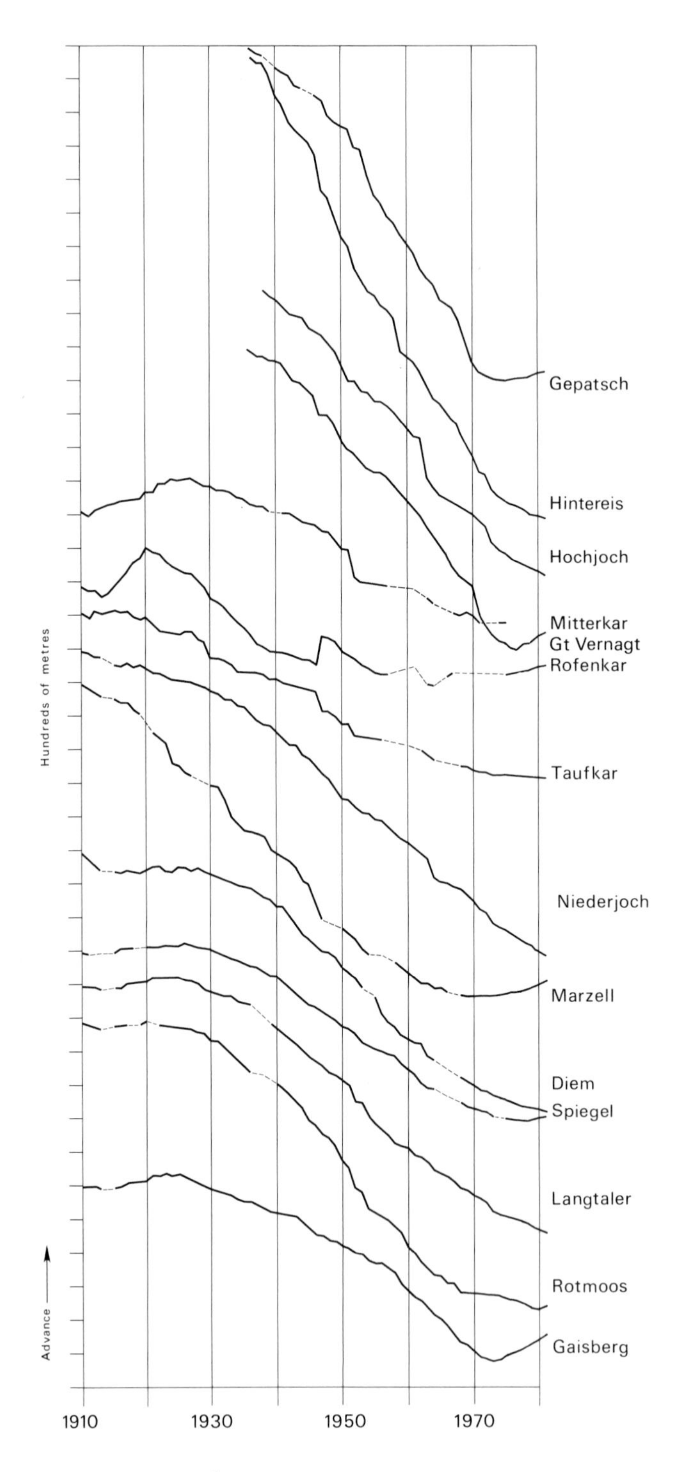

Figure 5.6 The fluctuations of the Ötztal glaciers since regular measurements began until 1982. (From figures published in *Zeitschrift für Gletscherkunde und Glazialgeologie* by Patzelt and others)

Ötztal in particular was fastest between 1938 and 1951 (Figure 5.6). It continued into the 1960s, with the wastage in 1963 and 1964 being particularly severe. Then heavy snowfall in the spring of 1965 was followed by a cool, damp summer. That same year, for the first time since 1921, less than 50 per cent of the glaciers being measured in the eastern Alps were receding (see Chapter 6 and Figure 5.6). In 1966, 25 per cent were recorded as advancing. The Rofentalferner and the adjacent Kesselwandferner in the Ötztal, which had advanced in the early 1890s and again about 1926, again responded quickly (Patzelt 1970).

5.6 Mass balance studies in the Ötztal

The cartographic studies made of the Ötztal glaciers in the closing years of the nineteenth century and the early decades of this century by Finsterwalder (1888, 1889, 1897a and b and 1928) and Hess (1918, 1922 and 1930) provided a useful basis for studies of mass balance that have been made since the Second World War.

Mass balance studies of a systematic kind began in Rofental in 1952, when Schimpp (1958, 1960) inserted a network of stakes in the Hintereisferner. This was extended by Rudolph in 1954 and kept under continuing observation, in a judiciously reduced form, from 1954 onwards (Hoinkes and Rudolph 1962a and b, Hoinkes 1970). Mass balance observations were extended to the Langtalerferner in 1962 and to the Kesselwandferner and Vernagtferner in 1965. The earlier maps were used in conjunction with new surveys by the Bavarian Academy of Sciences to produce a series of five maps of the Vernagtferner, at a scale of 1:10,000, for the years 1889, 1912, 1938 and 1969. These accompany *Fluctuations of Glaciers 1965–70* and *Fluctuations of Glaciers 1970–1975* in the series being published by the International Commission on Snow and Ice by IAHS and Unesco. They allow the variations in the positions of the ice fronts and the surface contours to be compared in detail in the successive periods between the maps.

Of the first twelve years of observations of the Hintereisferner, all except two revealed negative budgets, the exceptions being 1954/5 and 1959/60. The mean specific balance for the twelve-year period was -48 g/cm^2, that is an overall lowering of about half a metre annually. During this period of predominant loss between 1952 and 1964, more than twice as much was removed by ablation as was added by accumulation, and the average area of net nourishment exceeded the average area of net wastage by 47 per cent. The average loss for the period was very greatly influenced by two years of high budget deficits, 1957/8 and 1958/9. Temperature and radiation values were high in the summer of 1957, removing not only the snow of the previous winter but also exposing to ablation the firn layers of at least five years. There were few falls of snow or cloudy days to afford protection. The summer of 1959 had much lower radiation and lower average temperature, but mass loss was again high as there had not been much snow the previous winter, and a large area of low albedo (low reflectivity) was consequently revealed. Over the two years 1957/8 and 1958/9 net ablation removed seven times as much mass as was gained by accumulation. The area of net accumulation was only a half that of the area of net ablation.

A sequence of years with a predominantly positive mass balance came after 1964/5, when winter snowfall was above average and the following summer was cool and cloudy with not more than five days in a row having sunny weather. At the end of the ablation

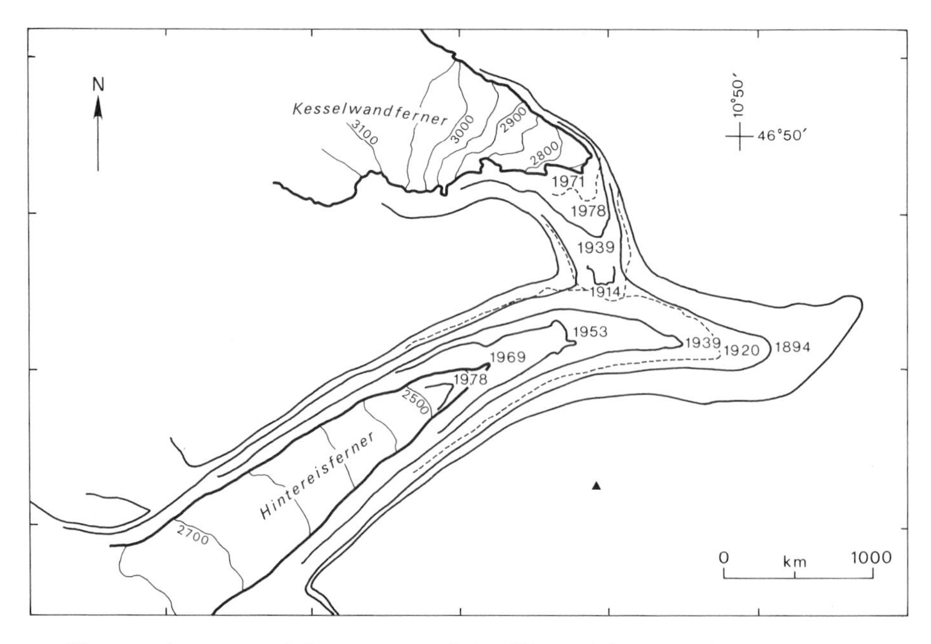

Figure 5.7 Changes in extent of the tongues of the Hintereisferner and Kesselwandferner since 1847. (From Kuhn *et al.* 1985)

season in 1965 the old snowline was much lower than usual and the area of winter snow surviving was three times the area of the glacier surface from which it had been removed. Positive mass balances occurred on the Hintereisferner in each of the four years until 1967/68 and then came two years with a negative balance. The six years with a positive balance, 1954/5, 1959/60 and 1964/68, more than counterbalanced the effects of twice as many years with a negative balance, for when Finsterwalder and Rentsch (1976) compared photogrammetric surveys of eight glaciers in the eastern Alps for 1950 and 1969, they found a mean rise in surface level of 0.1 m per year over the whole period. This compares with a mean annual lowering for the period 1920 to 1950 of 0.6 m. Between 1969 and 1979 (Finsterwalder and Rentsch 1981) there was again a mean rise in the

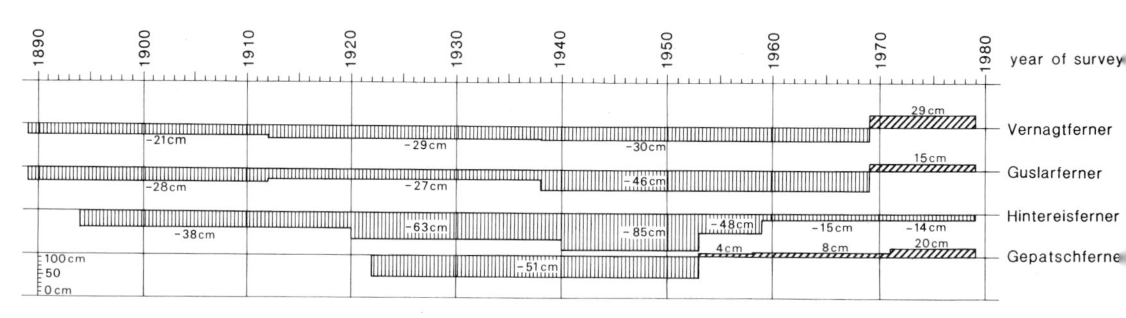

Figure 5.8 Annual changes in elevation of four glaciers in the Ötztaler Alpen, 1889–1979. (From Haeberli 1985)

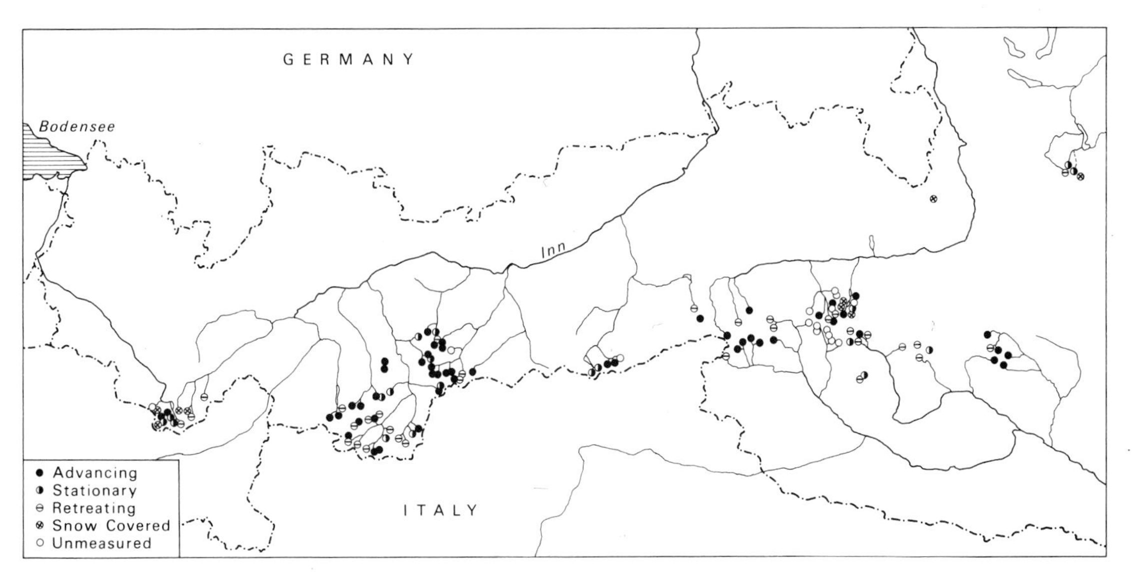

Figure 5.9 Distribution of advancing, stationary and retreating glaciers in Austria amongst those observed for the year 1983/4. (From Patzelt 1984)

surface level of all the glaciers in the sample, except for the Hintereisferner, but the rise was less than in the previous decade. The higher snowfields did not show much change, however, suggesting that the period of expansion was tailing off. But the surface of the Gepatschferner rose considerably between 1971 and 1973, although the tongue was still receding in 1977 (Finsterwalder, R. 1978).

The first three years of mass balance studies of the Vernagt were in the period of budget surplus of 1965–8. In those three years 12.6 million cubic metres of water equivalent were added to the glacier, thereby increasing its bulk by 2.5 per cent. In the succeeding six years the general tendency for ablation to exceed accumulation was resumed, but during the 1970s positive mass balances became the rule once more and the upper part of the Vernagtferner increased in volume.

5.7 The Little Ice Age in the eastern and western Alps compared

A comparison of events in the Ötztal and the Mont Blanc massif shows that the various phases of ice advance and retreat corresponded very closely in the two regions, if the evidence from the Vernagtferner is accepted as indicative of climatic conditions. The first damming of the Rofental at the end of the sixteenth century coincided with the extension of the Mer de Glace and its neighbours into the Arve valley. The formation of the second Rofensee in the late 1670s and early 1680s coincided with the flooding of the Lac du Combal, an extended Miage glacier and enlarged ice masses on both sides of the Mont Blanc massif.

The advance of the Gurgler in the early eighteenth century to a forward position in 1712 is not exactly matched in the western Alps but there were important advances elsewhere in Europe in the early 1700s, notably, as we shall see in Chapter 6, of the

Grindelwald glacier. The Gurgler remained enlarged until 1740 and during this first part of the eighteenth century the Mont Blanc glaciers were all quite extensive. The simultaneous advances of the Gurgler and Vernagtferner and many of the other Ötztal glaciers around 1770 coincided exactly with the vigorous advances of glaciers in the Arve valley to positions comparable with those they reached in the early seventeenth century.

Advances were general in both regions in the first half of the nineteenth century around 1820 and again about 1850. The advances of the Ötztal glaciers about 1850 brought tongues to within a few metres of their moraines of 1600 and 1770. They remained at or near these maxima until 1855, by which time many of the French glaciers were already waning.

By the end of the nineteenth century the broad parallelism of the behaviour of the glaciers of the eastern and western Alps was well recognized and minor discrepancies were also beginning to be appreciated. It was noticed that the advance of the Ötztal glaciers in the 1890s continued until about 1900, but by this time the Mont Blanc glaciers and their neighbours in Switzerland had been in retreat for some years. From the measurement of Klebelsberg and from diagrams showing the proportions of retreating and advancing glaciers produced by Kasser (1967, 1973) for Switzerland, Patzelt (1970) for Austria and Vivian (1975) for the French Alps, it is possible to date the glacial retreats and resurgences from one end of the Alps to the other (Figure 6.7). It is clear from the comparison that glaciers in the western Alps advance a few years before those in the east. Whereas the front of the Glacier des Bossons began to move forward in 1952, the Brenva and the Allée Blanche in 1955, and the Tour in 1960, the most sensitive glaciers in the Ötztal, such as the Rofenkar, did not begin to advance until 1965 (see Figure 5.6).

5.8 The surging of the Vernagtferner

All the evidence available indicates that the Gurglerferner responds to fluctuations in climate in phase with the other glaciers in the Ötztal and it appears that the Ötztal glaciers have in general advanced and retreated at times comparable with those of Mont Blanc. However, the only glacier in the Ötztal for which we have any detailed information before the eighteenth century is remarkable for its unconventional behaviour. Richter (1892, p. 354) noted that there was nothing very unusual about the timing of the Vernagt's advances, but he was well aware of its abnormality and indeed it was the singularity of its behaviour which decided him to choose to work on it. He knew that the Vernagt had been out of phase with the Gurgler and other glaciers in not advancing in the years about 1712, whereas both had advanced in the 1770s and about 1820. Then there had been its remarkably rapid advances in 1599, 1678, 1770 and 1845, followed by long periods of retreat and slow surface speeds. Why should its fluctuations be on so much larger a scale than those of its neighbours?

Hoinkes (1969) argued that the Vernagt was a surging glacier. Such glaciers, sometimes known as galloping glaciers, are typified by catastrophic advances occurring with some degree of regularity, during which speeds of flow are abnormally high. These advances are not always related to variations in climate. Ice accumulates for some years and then for some reason spills out of the reservoir area and pushes its way forward, shearing at the margins and crevassing chaotically. Judging by the intervals between its catastrophic advances, the Vernagt might have been expected to have pushed forward again round about 1928. This did not occur, though, as Hess (1930) pointed out, the surface velocity did increase slightly between 1924 and 1929; but surges do not necessarily occur with great regularity and it would seem likely that both topographic and climatic elements contribute to the behaviour of surging glaciers. Hoinkes (1969) emphasized the importance of the part played by topography.

Seismic investigations of the rock floor underlying the Vernagt- and Gurglerferner shows that a sill exists beneath both of them at an elevation of between 2900 and 3000 m, with basins behind about 50 m deep in the case of the Gurglerferner and about 20 m deep in the case of the Vernagtferner (Miller 1972). The main mass of both glaciers lies above the level of this sill, which is close to the height of the névéline in the quiet periods between surges. Hoinkes suggested that the glacier ice might become frozen to the bedrock at the sill at times when the ice at that altitude was thin, allowing the winter cold wave to penetrate to the bed. If the following summer were to be cool, no great quantity of meltwater would be released and the ice would remain frozen to the sill. Ice might accumulate behind the rock/ice barrier thus formed until pressure melting point was reached at the bed, when large masses of ice would suddenly be released and move rapidly down the slope towards the Zwerchwand of Rofental.

Clearly surges resulting from any such mechanism would have to depend in large part on mass balance. This point was developed much further by Kruss and Smith (1982). They pointed out that the glacier has passed through two radically different phases during the historic period, characterized respectively by cyclic surging and by shrinkage back into higher regions favourable for accumulation. Between the surge of 1845 and 1966 the Vernagtferner contracted in area from 13.8 km^2 to 9.6 km^2 and shrank to about half its former volume (Finsterwalder, R. 1972). Kruss and Smith calculate that the surging mode can only operate as long as the glacier remains large enough to create substantial basal melting, through a combination of high basal stresses and rapid flow-rates. They consider that the great retreat since 1848 resulted from a climatically led negative mass balance over the entire surface, of the order of 0.2 m per year. The glacier was particularly vulnerable to ablation because of its very great extension into lower altitudes. Should positive mass balances reoccur for a sufficiently long period the combined Vernagt–Gurslarferner will surge again.

Chapter 6

Swiss glacier fluctuations and Little Ice Age weather and climate

Snow-capped Alpine ranges appear on the horizon almost everywhere in Switzerland and routes over the major Alpine passes skirt the glaciers. Many towns and villages occupy sites that are threatened by avalanches and flooding even today, so it is no wonder that documents relating to glacier advances are abundant. In addition weather diaries and phenological records date back several centuries. The Swiss were the first to make systematic annual measurements of the changing positions of glacier fronts; these continue and are published annually by the Swiss Glacier Commission. The climatic implications of glacier oscillations, recognized in the early nineteenth century, have attracted renewed scientific attention in recent years (Pfister 1979a, 1980b).

6.1 Evidence of glacial expansion in medieval times

The strongest evidence of glacial advances prior to the sixteenth century comes from the eastern part of Valais and from the Bernese Oberland (Walliser Alpen and Berner Alpen, Figure 6.1). The Allalingletscher, on the eastern flank of the Mischabel group, is known to have advanced across Saastal, one of the Rhône's left-bank tributary valleys, on several occasions between the sixteenth and nineteenth centuries, each time holding back a lake that eventually spilled more or less violently down the lower valley. The Allalin also seems to have blocked Saastal in the thirteenth century. A document dated 13 April 1300 (Grémaud 1878) concerns the tenancy of pastures at Distelalp, near the head of the valley and upstream of the site of the Mattmarksee, the name given to the lake that is held back by the Allalin when it advances across Saastal. Two of the local farmers requested the Mayor of Visp 'to grant us tenure of the pasture in question from the glacier upwards, according to the custom of the Saaser Visp valley, so that the said men do not prevent us grazing as far down as the glacier'. This text is interpreted by Lütschg (1926), Ladurie

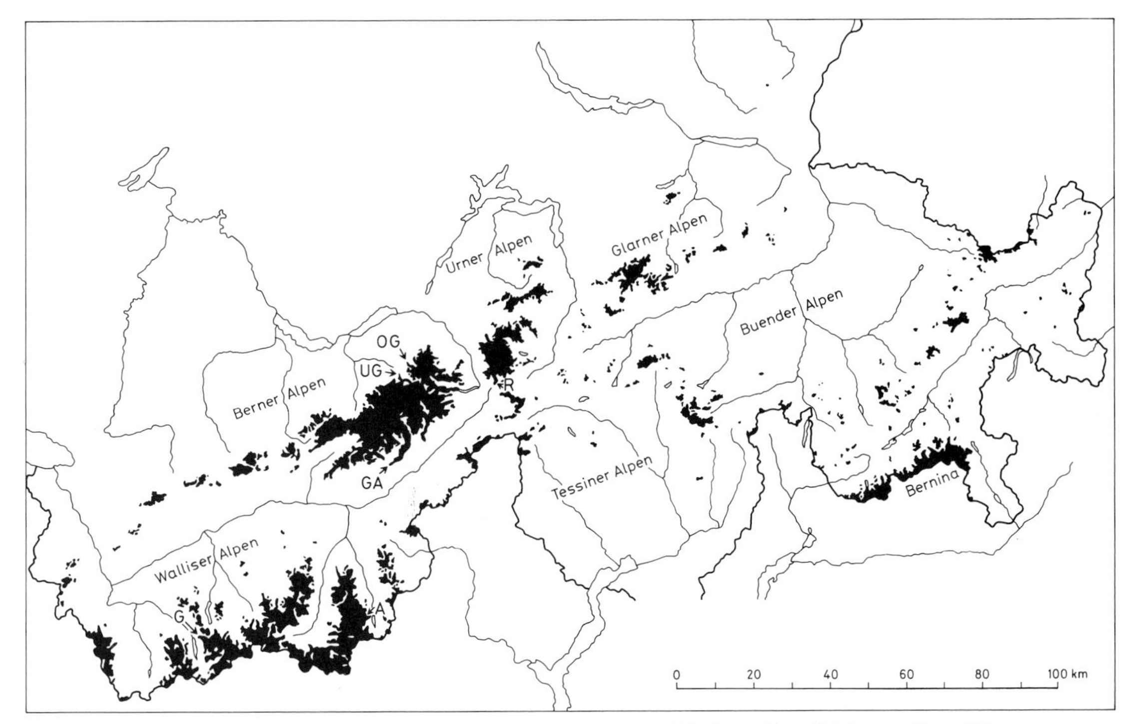

Figure 6.1 Distribution of glaciers in Switzerland. A = Allalin, R = Rhône, G = Gíetroz, GA = Grosser Aletsch, OG = Oberer Grindelwald, UG = Unterer Grindelwald

(1971) and Delibrias *et al.* (1975) as showing that the Allalin was sufficiently enlarged in 1300 to block Saastal and separate the upper mountain pastures from the lower valley of the Visp as it did in the nineteenth century. Nowadays the glacier overlooks the valley from the east, not far downstream of the dam that holds back an artificially enlarged version of the Mattmarksee that submerges the old chalets of Distelalp. If the Allalingletscher did in fact block the valley in 1300, it was much larger than it is today but probably smaller than in 1822, when the Schwarzberg, a much smaller glacier than the Allalin, extended so far down into Saastal as to bisect the Mattmarksee (Plate 6.1); in the 1300 document there is no mention of two ice dams.

If in fact the Allalin was much enlarged in 1300, the same must have been true of other Swiss glaciers, though it must be admitted that there is a dearth of supporting documentary evidence. However, it happened that the retreat of the Grosser Aletschgletscher in the twentieth century uncovered the remains of a number of larch trees still in the place where they had grown. Tree-ring counts showed that the trees were about 150 years old when they were overwhelmed by the advancing ice. Radiocarbon dating has given them ages of 800 ± 100 (B-71) and 720 ± 100 years (B-32) (Oeschger and Röthlisberger 1961, Röthlisberger *et al.* 1980). Kinzl (1932) had earlier provided evidence that the Aletsch had begun to advance well before 1385. The Valais is the driest part of Switzerland and both pasture and cropland are watered by networks of irrigation channels (*bisses*) bringing water long distances down the mountain sides from glaciers and snowfields. Many of them are several centuries old. The Oberriederin bisse is recorded as having been abandoned in 1385; its remains were exposed by the retreating ice of the Aletschgletscher as late as 1961. Lamb (1965) argues that the bisse had been constructed in the early medieval warm period and had to be abandoned when ice advanced over it in the fourteenth century.

The Unterer Grindelwaldgletscher has long been known to have advanced into woodland. Gruner (1760) wrote that 'in the central part of the glacier, all the way up on the flanks of the Vischhorn and Eiger, a good number of larch trees (*Pinus larix*) protrude from the ice. Since the wood of this species is known to harden in wet conditions, those trees may have been living there for centuries.' Gruner seems to have been correct in his assumptions, for samples of larch and other trees collected from moraines in front of the glacier have given ages of 1280 ± 150 (BM-95), 860 ± 70 (Gif-2975), 1240 ± 100 (Gif-2976), 640 ± 80 (Gif-2977) and 1060 ± 70 (Gif-2980) (Delibrias *et al.* 1975). It is not clear that the dated material came from trees found in the positions where they grew (Messerli *et al.* 1975) but they still provide strong support for an advance of the Unterer Grindelwaldgletscher well before the sixteenth century.

An enlargement of the Allalin, Aletsch and Grindelwald glaciers in the thirteenth and fourteenth centuries would no doubt have been accompanied by the enlargement of other

Plate 6.1 The dam formed by the Allalingletscher in the background, with the extended Schwarzberggletscher entering the Mattmarksee in the middle distance. This etching by Thalés Fielding, from an aquarelle by Maximilian de Meuron, dated 1822, was published in Lütschg (1926)

Alpine glaciers. Recent research indicates that this was in fact the case. The chronology of a number of moraine sequences, over twenty of them in the Valais, has been investigated by research students from the Geographical Institute of the University of Zürich.[1] The ^{14}C dates are listed by Röthlisberger *et al.* (1980) with the ages of soils from *beneath* moraines being as follows: Glacier de Fenêtre 900 ± 95 (Hv-6805), Schwarzberg 950 ± 115 (UZ-165), Glacier du Mont Miné 900 ± 75 (Hv-6800), Findelen 845 ± 225 (Hv-6791), Glacier de Corbassière 815 ± 225 (Hv-7226), Glacier du Mont Durand 790 ± 65 (Hv-6811), Glacier de Moiry 700 ± 85 (UZ-153), and Glacier du Brenay 650 ± 85 (Hv-6801). In addition a larch, 168 years old, found in place before the left lateral moraine of the Glacier de Zinal, gave a date of 920 ± 50 (B-3200). In the vicinity of the Glacier de Ferpècle, soil from beneath a left lateral moraine gave a date of 1070 ± 60 BP (Hv-7223), a juniper branch from within the moraine a date of 930 ± 220 (Hv-7222), and soil from beneath a medial moraine a date of 1045 ± 55 (Hv-6822). These results point to glaciers advancing well before the sixteenth century and probably not long after AD 1100. The positions of the moraines associated with the twelfth- and thirteenth-century advances were not very different from those reached by glaciers in the middle of the nineteenth century. Deliberate search of medieval manuscripts may well provide confirmatory evidence of this medieval precursor of the early modern Little Ice Age. At the same time it should be noticed that in 1330 the Theodulepass was not recorded as having been ice-covered (Wills 1856) whereas in 1528 the historian Aegidius Tschudi di Glarus (1538) stated that 'on its crest extends for the space of 4 Italian miles a great field of ice that never melts or disappears'.

6.2 Glacier oscillations in the Little Ice Age of early modern times

The Rhonegletscher, which is in full view of travellers over the Grimsel and Furka passes, must be one of the most closely observed and well documented glaciers in the world. In 1546, when he was riding towards the Furka, Sebastian Munster relates that he came to 'an immense mass of ice . . . about 2 or 3 pikes' length thick and as broad as the range of a strongbow'.[2] Evidently the ice reached down to the valley floor and ended in a steep front 10–15 m high and some 200 m broad. Three centuries later, in 1836, when it was sketched by a Norwegian artist, Thomas Fearnley, it still reached down to the valley floor (Plate 6.2b). On 28 July 1868 Gerard Manley Hopkins wrote in his journal (1953) that he had seen the Rhonegletscher, ending in

> a broad limb opening out and reaching the plain, shaped like the fanfin of a dolphin or a great bivalve shell turned on its face, the flutings in either case being suggested by the crevasses and the ribs by the risings between them, these being swerved and inscaped strictly to the motion of the mass. . . . We went into the grotto and also the vault from which the Rhône flows. It looked like a blue tent and as you went further in it changed to lilac.

[1] Research in Austria, particularly by Patzelt and Bortenschlager, helped to stimulate important programmes of work in Switzerland concerned with glacial and climatic history in the Late Glacial, Holocene and Little Ice Age. Research students including F. Röthlisberger, Schneebeli and Zumbühl, working under Professor Furrer at the University of Zürich, have produced a series of diploma projects and theses (listed in Röthlisberger *et al.* 1980) which, together with those from other Swiss universities, especially Bern, constitute the most detailed history of the Holocene for any part of the world.

[2] The sixteenth-century Swiss pike measured between 4.6 and 5 m.

Grønne Berg
Mose og Klipper
Mose

Plate 6.2a (on facing page—above) The Rhône glacier in 1777 by Alexandre-Charles Besson, a colleague of H. B. de Saussure (1725–1809). (From a private collection, Bern) (Photo: H. J. Zumbühl)

Plate 6.2b (on facing page—below) The Rhône in 1835 sketched by Thomas Fearnley from a site above the glacier. (Nasjonalgalleriet, Oslo, Inv. no. A.3412.) (Photo: Jacques Lathion)

Plate 6.2c (below) The Rhône in 1848 in an aquarelle by Hogard showing the series of moraines left during retreat. The outermost set has been dated to 1602, the second set to 1818, and the third to 1826. The picture indicates clearly that a fourth set was forming in 1848. (Zürich Graphic Collection, E.T.H.)

Plate 6.2d (at foot of page) The Rhône was still across the valley floor in 1862 when painted by Eugen Adam. (Zürich Graphic Collection, E.T.H. Case 648) (Photo: H. Zumbühl)

Plate 6.2e By 1912 the Rhonegletscher tongue was hanging and the valley bottom clear of ice. (Photo from Mercanton 1916)

Next year, 1869, the Swiss Alpine Club and the Swiss Academy of Sciences jointly founded the 'Gletscherkollegium', later to be called the Glacier Commission: the Rhonegletscher was chosen to be the special subject for study. Before long, measurements were being made of the rate of ablation, precipitation and firnline elevation; data were collected annually on the volume as well as the area of the glacier and the position of the tongue was mapped every month, winter and summer, from 1887 to 1910, on a scale of 1:5000. This research programme, so thorough for its time, culminated in the publication of 'Mensurations au Glacier du Rhône 1874–1915' (Mercanton 1916). One of its diagrams (Figure 6.2), showing the changes in the position of the terminus from 1602 until 1914, based on a collection of fourteen descriptive and pictorial records, is still being used in modern syntheses such as that of Furrer *et al.* (1980) and Holzhauser (1982).

The oscillations of the Allalingletscher over the last four centuries have attracted particular attention because of the repeated flooding caused by the spilling of the Mattmarksee, the lake held back by rocks and ice when the glacier advances across

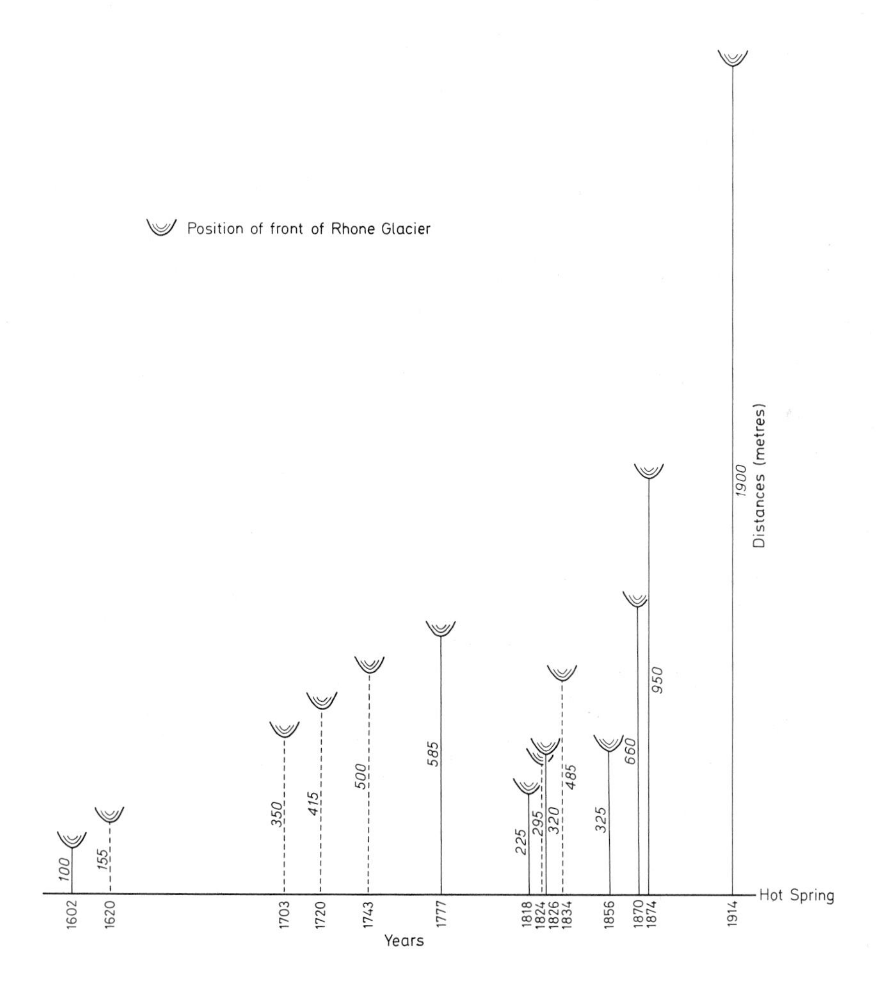

Figure 6.2 Variation in position of the front of the Rhonegletscher between 1602 and 1914. (From Mercanton 1916)

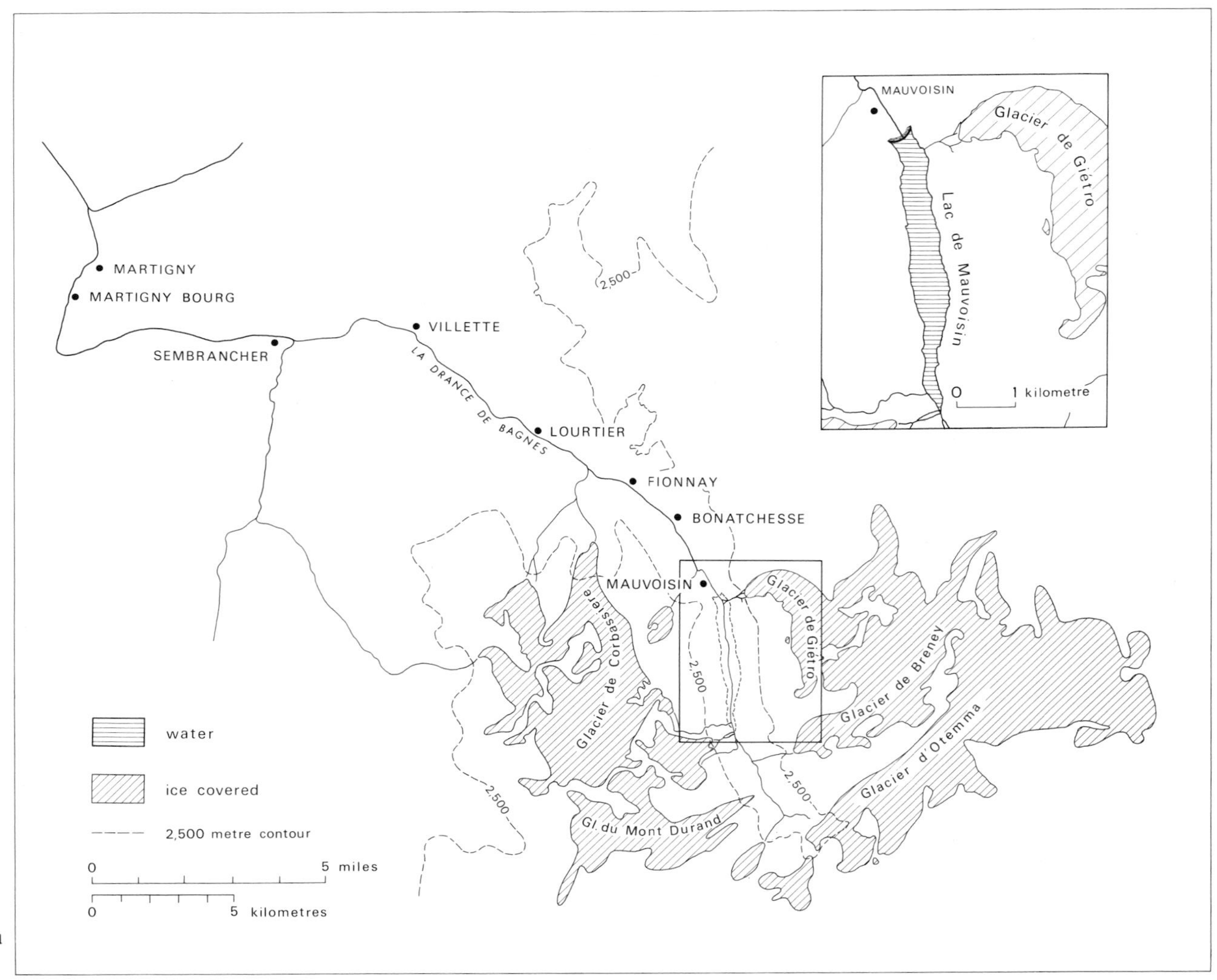

Figure 6.3 The Glacier du Giétroz and its setting

Saastal (Mariétan 1965). In early modern times, according to Lütschg (1926), who collected together many of the relevant documents,[3] the valley was first blocked in 1589. In 1633 the lake broke through the ice dam and in Saas 'half the fields were buried in debris and half the inhabitants were forced to emigrate and find their miserable bread in some other place' (Ruppen *et al.* 1979).

In the Val de Bagnes, another left-bank tributary valley of the Rhône, similar catastrophes were associated with the Glacier du Giétroz (Figure 6.3). A flood in 1595 penetrated the galleries of newly opened silver mines, overwhelmed the thermal baths at Bagne, destroyed over 500 buildings and drowned 140 people. A similar outburst occurred in 1640. Particularly well documented is the flood of 1818; the two preceding summers had been very cool and the Giétroz, like most of the other glaciers in the Alps at the time, began to advance. Blocks of ice from the tongue fell over a rockstep (which the glacier overlooks at the present day from above the Mauvoisin hydro-electric dam) and amalgamated to form an ice cone stretching across the valley floor (Bridel 1818a, b, c, Sion archives Fasc. 1. R.21.28). When it was reported to Sion in April 1818 that a lake had formed behind the ice, the authorities sent the canton engineer, Venetz,[4] to deal with the situation and artists to record the scene (Plates 6.3 and 6.4). By the middle of May the ice barrier was a kilometre long and 130 m high, holding back a lake 2 km long and 55 m deep. Starting on 11 May, Venetz, with local volunteers, worked day and night to dig a trench 200 m long across the ice. Eventually, in spite of late snowfalls and further great falls of ice, water began to escape along the trench on 13 June and about 6 million cubic metres, about a third of the lake's volume, had drained away before the ice dam collapsed at 4.30 in the afternoon of 16 June. Then the rest of the lake rushed down the valley as a wall of water 30 m high, reaching up as much as 100 m in narrow gorges, tearing away the soil here, depositing vast spreads of gravel there. It almost caught an inquisitive English visitor (his mule was washed away). It carried away forty-two chalets at Bonatchesse, thirty at Brecholey and fifty-seven at Fionnay. At Lourtier, the highest permanent village in the valley of the Drance, sixty buildings were destroyed. At Chamsec, where fifty-eight buildings were ruined, a man aged 92 saved himself by climbing an oak tree growing on a debris mound that had been left by the flood of 1595. Sawmills, fulling mills, flour mills and an iron works were engulfed as the water sped down to Martigny. At 6 p.m. it reached Martigny Bourg, flooding the streets to a depth of over 3 m, and went on to inundate the floor of the Rhône valley, spreading a layer of silt over the marshes of Guersay (Guercet). Before midnight, debris was being washed into Lake Geneva. Had Venetz and his men not released much of the water before the ice dam broke the damage would no doubt have been even greater, possibly exceeding that of May 1595 (Tufnell 1984).

There were ice falls in the Val d'Hérens in that same year, 1818, which took six years to clear; ice from the Weisshorn destroyed the village of Randa in December 1819 (Sion archives T.P. 135/1). In the Val de Bagnes, the Glacier de Corbassière continued to

[3] Lütschg was unaware of some of the safeguards of more modern methodology. He included a number of extracts from Ruppen's 'Die Chronik des Thales Saas' but very great caution must be adopted in using chronicles of this kind (Bell and Ogilvie 1978).

[4] Ignace Venetz (1788–1859) of Visperterminen came from an old-established Valaisanne family. He turned early to the study of natural science and mathematics. As an engineer, first working for his own canton of Valais and later for Vaud, much of his professional life as well as his private interest was in glaciers and glacier-fed rivers. He played a leading role in the foundation of glaciology and glacial geology, finding time for original research despite the demands of his professional career and the need to support a large family (Mariétan 1959).

Plate 6.3a The Glacier de Giétroz (2) spilling over (3) into the upper Val de Bagnes. The dam is still incomplete; the Vallon de Torembec (5) is not yet flooded. (From Bridel 1818b)

Plate 6.3b The Giétroz dam in May 1818. Venetz and his men are at work on the trench, while members of the public observe from a safe distance. (From Bridel 1818a)

Plate 6.4 Ignace Venetz about 1815, painted in oil by Laurent Ritz. (Musée de la Marjorie, Sion)

advance down its valley until 1822 and the threat of renewed blockage of the Mauvoisin valley persisted until 1824. But for the rest of the century there was no serious threat until the years 1894 to 1899. Then, the retreat of the ice at the head of the Val de Bagnes allowed a lake to accumulate between the Otemma and Crête Sèche glaciers which violently spilled over each summer between 1894 and 1899, causing floods that destroyed the bridges downstream. Eventually engineers excavated a trench in the ice and rock

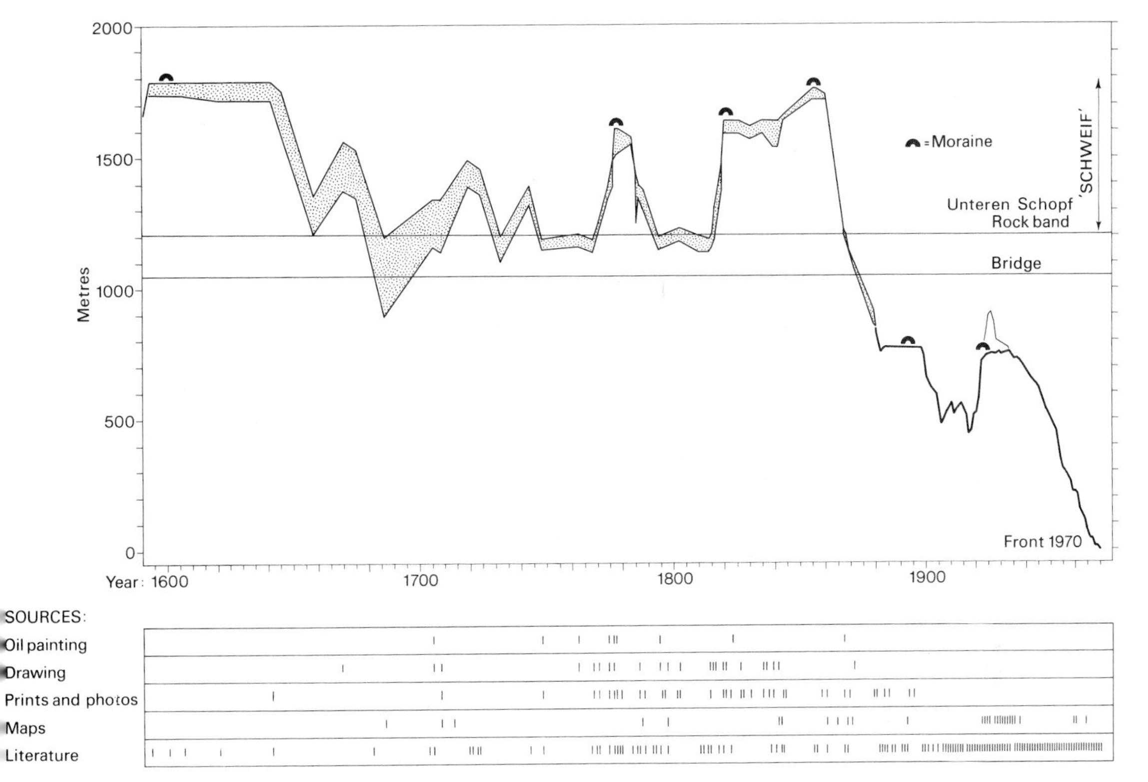

Figure 6.4 The frontal positions of the Unterer Grindelwaldgletscher, 1590–1970. (After Messerli *et al.* 1978)

barrier to allow the water to drain away harmlessly.[5]

The sequence of the expansions and contractions of many Swiss glaciers has been studied in recent years by Aubert (1980) and Holzhauser (1978, 1980). The most significant study in relation to Little Ice Age climate is that of the Grindelwaldgletscher (Messerli *et al.* 1975, 1978). In tracing the fluctuations of the Oberer and Unterer Grindelwald over the period 1590 to 1970, Zumbühl (1980) used as evidence not only written documents and maps but also relief drawings, sketches and paintings collected from libraries in half the countries of Europe. In all, he examined 300 views of the Unterer Grindelwald and 186 of the Oberer, dating from 1640 to 1900. He had to date the illustrations and establish the artists' viewpoints precisely. Some masterpieces turned out to be quite unsuitable for the purposes of the glacier historian. Particularly accurate representations were the oil paintings of Caspar Wolf (1735–83) and a watercolour of 1820 by Samuel Birmann (1793–1847).

Both the Oberer and Unterer Grindelwaldgletscher were more extensive between 1600 and 1870 than they have been since (with the possible exception of a few years around 1685). Although the upper glacier was found to respond up to a decade earlier than the lower one, the two generally behaved in a similar sense and the fluctuations of the Unterer

[5] The practical advantages of obtaining detailed histories of sensitive and dangerous glaciers such as the Allalin and Giétroz, and of keeping a strict watch on them, were unhappily underlined by the disastrous collapse of ice from the Allalin onto contractors' buildings erected below it while the hydro-electric dam was being built in 1965. There was much loss of life (Vivian 1966). The tongue of the Giétroz now hangs far above the waters held back by another dam in the Val de Bagnes, the Mauvoisin, and is the subject of especially strict surveillance by the Swiss Glacier Commission.

Grindelwald (Figure 6.4), which can be traced more accurately, hold for the Oberer (though during the period 1960–80, when the Oberer Grindelwald was advancing, the Unterer was retreating, a difference in the response time of the two glaciers greater than had previously been recorded).

From about 1590 to 1640, the Unterer Grindelwald reached its maximum extent, some 600 m forward of the rock band called the Unterer Schopf. This natural marker remained under the ice until 1870, though it was very nearly exposed by a retreat towards the end of the seventeenth century. In 1720 and again in 1743 the ice advanced to reach 100–300 m below the Unterer Schopf, retreating between and after these dates almost as far as the marker. Another advance between 1770 and 1780 brought it forward about 400 m; there it formed a moraine, since removed by erosion, which appears in several oil paintings of the time. By 1794 the ice had melted back to the Unterer Schopf. Twenty years later it was expanding and it continued to do so until 1820–2 (Plate 6.6c). Another advance brought it further forward in 1855–6 than it had been since the seventeenth century. Then came rapid thinning and retreat which has continued to 1980, except for a halt from 1882 to 1898, a minor readvance between 1915 and 1925, and a pause from 1925 to 1935. The total withdrawal from 1856 to 1970 amounted to 1800 m.

Systematic measurement of the frontal positions of the Grindelwaldgletscher began in 1880 as part of the Swiss Glacier Commission's national programme.[6] Figure 6.5 shows the variations in position of the ice fronts of the Grindelwald and other glaciers in various parts of Switzerland which have continuous records since 1890. Differences in detail, such as the rapid retreat of the Grosser Aletsch in the 1960s and 1970s when the Oberer Grindelwald and the Allalin were advancing, are probably to be attributed to differences in altitude, aspect and size, and to differences in amount of moraine cover and topographic setting of the snout. Figure 6.6 shows the fluctuations of the Unterer Grindelwald since 1600 in comparison with those of the Rhône and Fiescher and of twenty-three other Swiss glaciers. In spite of differences in cumulative mass balance values, the behaviour of the termini has been rather consistent over the long term (Reynaud 1980, 1983).

The Swiss example of regular recording of glacier fronts was followed in Austria, France and Italy, so that a comparison can now be made of glacier-front behaviour over most of the Alps for almost a century. The phases of advance and retreat correspond very closely throughout the region, as Richter (1891) recognized long ago and Ladurie (1971) and Bray (1982) have confirmed (Figure 6.7).

The record of glacier fluctuations extending back over the last four centuries and more would seem to afford opportunities for extending the climatic record back beyond the beginning of the instrumental period. The position of a glacier terminus fluctuates according to the cumulative mass balance and the dynamics of the glacier, which in turn depends on its shape and the form of its bed. The mass balance depends primarily on the

[6] A general account of the work of the Swiss Glacier Commission is given in Portmann (1975, 1976, 1978, 1980, 1981). The first fifteen annual reports were produced by F. A. Forel (1841–1912), who took a leading part in the organization of the Commission and the direction of its researches. The programme was ambitious in its aims from the first, as well as outstanding in the degree of its continuity, and has latterly become much more sophisticated. The first two reports were published in *Echo des Alpes* 17 and 18 (1881, 1882), reports 3 to 44 in *Annuaires du Club Alpine Suisse* 18 to 58 (Bern 1883–1924), and report 44 onwards in *Les Alpes*, starting with no. 1 in 1925. Since 1954 the collection of data and preparation of annual reports has been undertaken by staff of the Laboratory of Hydrology and Glaciology of the Federal Institute of Technology of Zürich.

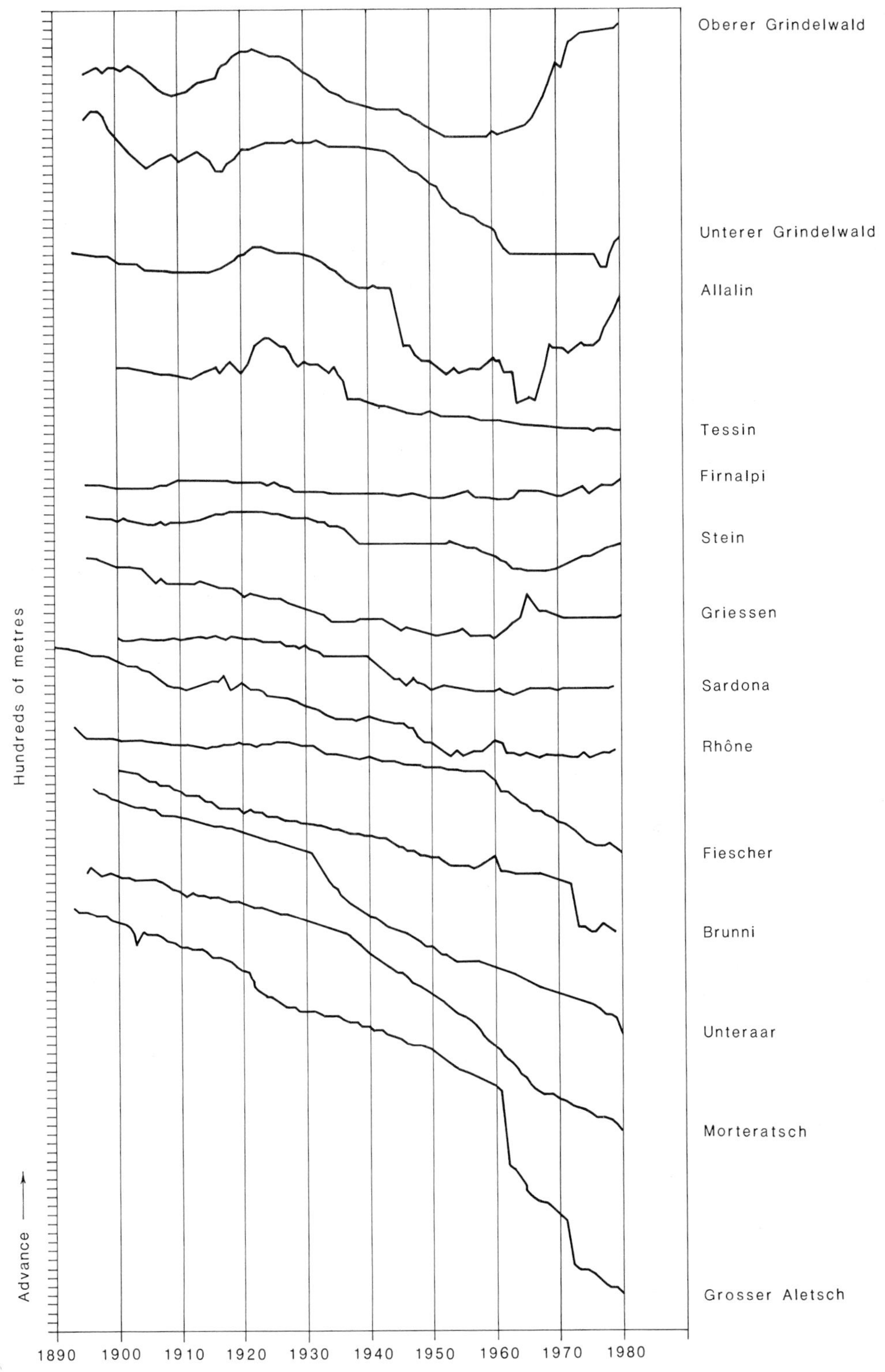

Figure 6.5 Advances and retreats of Swiss glaciers in various parts of the country since *c.*1895, based on figures published in the Annual Reports of the Swiss Glacier Commission.

Plate 6.5a The Unterer Grindelwald painted by Joseph Plepp (1595–1642). The exact date is unknown. The descending glacier (A) is pushing everything before it with a great noise. The river (B) streams out from under the ice. The houses (C) have had to be abandoned because of the glacier. (Photographed and published by Zumbühl (1980) as K1.11, p. 198)

Plate 6.5b The front of the Lower Grindelwald in 1762, painted by Johann Ludwig Aberli (1723–86). (Photographed and published by Zumbühl (1980) as K13.1, p. 205)

Plate 6.5c Samuel Birmann (1793–1847) visited the Unterer Grindelwald in 1826 and portrayed the glacier with well-developed siracs. (Original in the Kunstmuseum, Basel. Photo: H. Zumbühl. Published Zumbühl (1980) as K60, p. 193)

Plate 6.5d Grindelwald and the Unterer Grindelwaldgletscher photographed in 1885 by Jules Beck (1825–1904). (Published by Zumbühl (1980) as K139, p. 247)

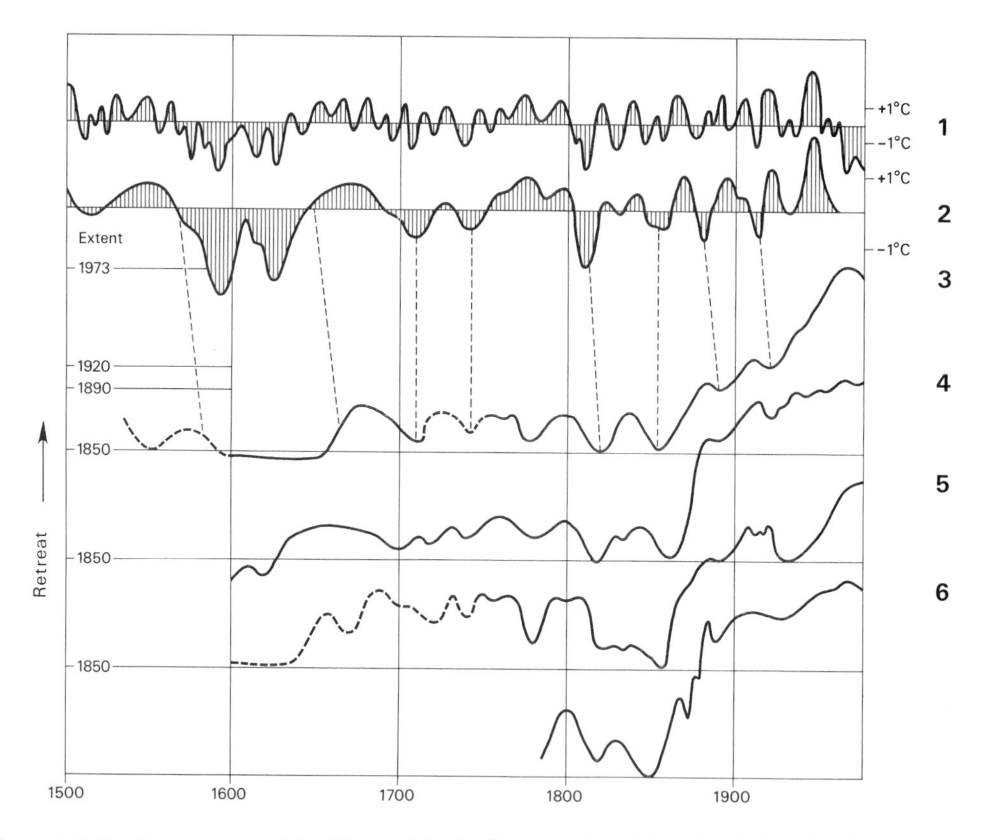

Figure 6.6 The fluctuations of the Rhône (4), the Unterer Grindelwald (5), the Fiescher (6) and the average of twenty-three Swiss glaciers (3) (Furrer *et al.* 1980), compared with tree-ring density curves (1 and 2) (Picea at Launens, Berner Alpen). It can be seen that if the density curves are shifted a few years forward (to the right), there is a clear correspondence between high tree-ring density (low summer temperature) and ice advance; the time lag depends on the size and conformation of the glacier. (From Röthlisberger *et al.* 1980)

accumulation of snow and the losses by melting. In the case of a single glacier, then, the climatic controls are the amount of snow falling on the glacier in the winter season in relation to the ablation losses caused by insolation and warm, moist winds in summer.

A comparison of cumulative mass balance curves from all the observations that have been made in the Alps shows that trends of increase and decrease have been in concert (Figure 6.8). Studies of mass balance on the Rhône and other glaciers have shown that a very large part of the variance can be explained in terms of a few climatic variables (Reynaud 1980, 1983). Most important is the mean daily temperature for July and August; this accounts for 58 per cent of the variance. June precipitation accounts for 16 per cent, total precipitation between October and May for 5 per cent. Thus the cumulative mass balances of Alpine glaciers depend primarily on summer temperature and secondly on the amount of precipitation in June and over the winter and spring.

Hoinkes (1968), who had demonstrated a good correlation between the fluctuations of Swiss glaciers and the five-year running means of deviations of summer temperatures and precipitation from 1851–1950 means, went on to investigate the relationship between

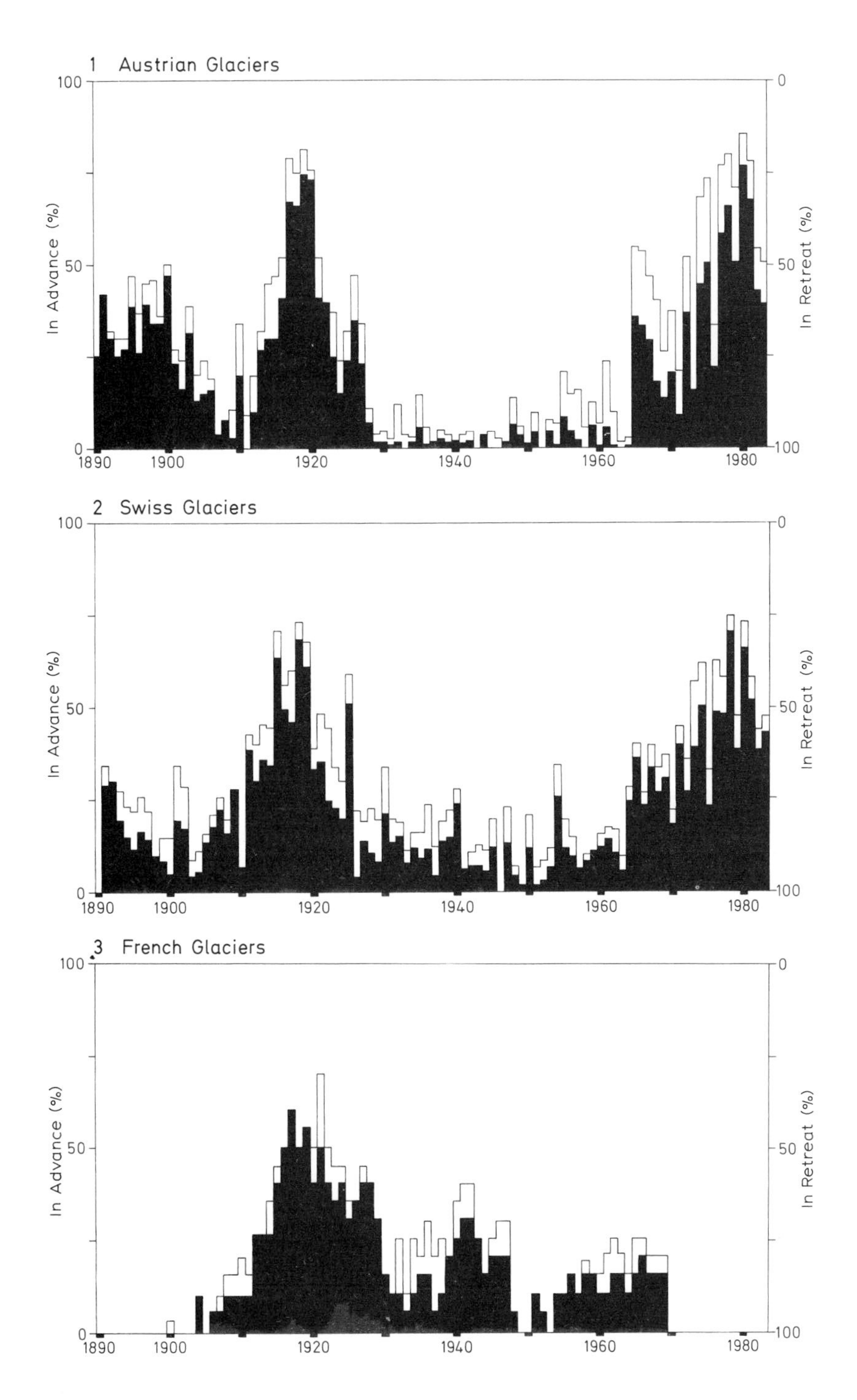

Figure 6.7 A comparison of the behaviour of glaciers in Austria, Switzerland and France since 1890, in terms of the percentage of those observed each year which were found to be advancing, retreating and stationary

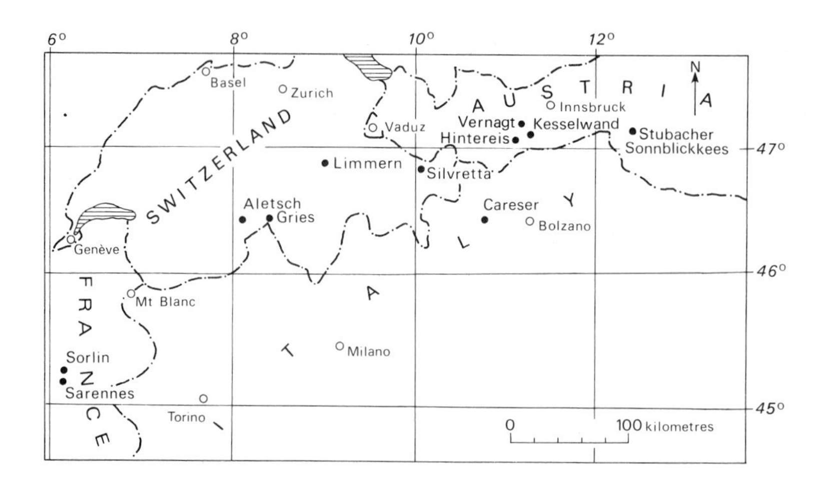

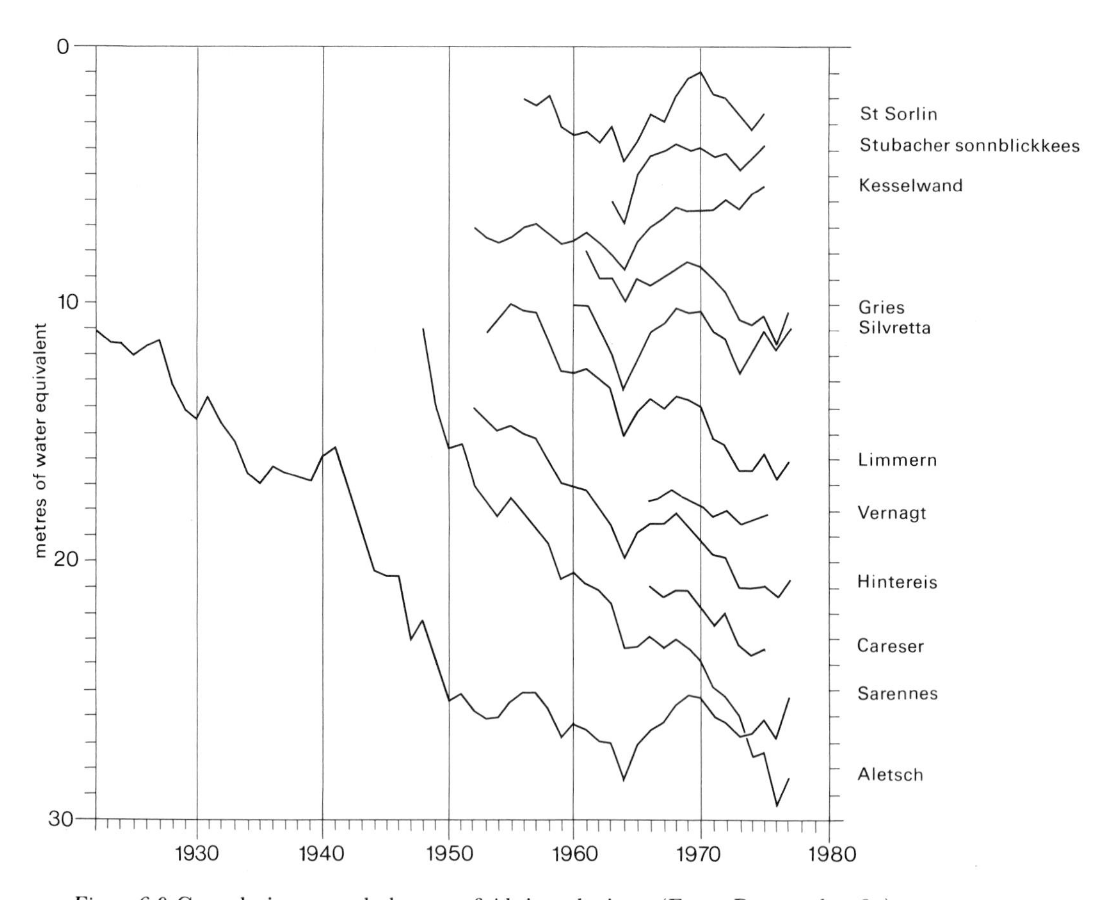

Figure 6.8 Cumulative mass balances of Alpine glaciers. (From Reynaud 1980)

glacier fluctuations and meso-scale weather situations. He showed that for the period 1953–65 there was a negative correlation between the mean specific mass budgets of the Grosser Aletsch and the Hintereisferner on the one hand and the deviations from the 1951–60 average height of the 500 mb surface over Munich and Payerne from May to September on the other (the latter being a useful indicator of the prevalence of cyclonic or anticyclonic conditions over the Alpine region). Hoinkes considered that well-defined deviation patterns of the 500 mb surface would be related to *Grosswetterlagen*, by which is meant 'fairly persistent synoptic situations which determine the form and sequence of weather events for a period of several days or even weeks',[7] and identified five types of *Grosswetterlagen* likely to have a strong influence on the nourishment or shrinkage of glaciers (Table 6.1).

Table 6.1 Grosswetterlagen with a strong influence on glacier nourishment

Favourable influence	
W1	Oceanic type in winter – winter precipitation
F4	Cold spells in spring – solid precipitation and/or late starting ablation system
S6	Monsoon-type in summer – reduced radiation, increased albedo
Unfavourable influence	
S7	Fine weather in September – long ablation season
S5	Anticyclonic weather in summer – high radiation and temperature, low albedo

Source: Hoinkes (1968)

Hoinkes made an initial assumption that a single day with favourable *Wetterlagen* would compensate for one day with unfavourable *Wetterlagen*. The algebraic sum of daily frequencies (W1 + F4 + S6) − (S5 + S7) over a year then gives an annual index and the indices can be summed over periods of years and compared with observed glacier fluctuations. The period of stationary or advancing glacier fronts in the 1890s corresponds to an average annual index of + 100 for the period 1890–4, whereas for 1947, when all the Swiss glaciers were receding, the value of the index was −27. Hoinkes concluded that for a general advance of the Swiss glaciers the index would have to exceed 90 for several years.

6.3 Swiss climate during the Little Ice Age

Ignace Venetz, the engineer from Visperterminen who had experience of coping with the hazards presented by the Giétroz and Allalin glaciers, addressed the Société Helvétique des Sciences in 1821 (Venetz 1833). Probably for the first time, he used glacier enlargement and diminution as evidence to show that temperature 'rises and falls periodically but in an irregular manner'. That the climate of Switzerland was becoming more rigorous had been conjectured at a meeting of the society in 1817, a year after its members had turned their attention to the study of glaciers (Mariétan 1959). They already had a number of weather records at their disposal. The Economic Society of Bern

[7] Hess and Brezowsky (1952) published a daily catalogue of *Grosswetterlagen* for mid-Europe from 1881 to 1951. Since then, details have been published monthly by Deutscher Wetterdienst, Zentralamt Ossenbach.

had set up a network of weather stations with uniform instrumentation in 1759, and regular observations of temperature and pressure had begun at Basel in 1755 and were started at Geneva in 1768. The climatic record can be extended back beyond the middle of the eighteenth century by utilizing a whole range of other kinds of evidence (Pfister 1975, 1977, 1978a, b, c, 1979a, b, 1980a, b, 1981). As a result of a systematic search of Swiss libraries and archives, Pfister collected together 27,000 documentary records relating to past climate, including 70,000 daily weather observations. The frequency of such records increases towards the present; between 1525 and 1550 there were about 30 records per year; from 1550 to 1657 information is available for nearly every month, and after 1659 for every month. After the middle of the eighteenth century the quantity of data becomes so large that Pfister limited his attention to the more readily quantifiable material, not only instrumental measurements but also the monthly number of rainy and snowy days noted in weather diaries, phenological records,[8] and comments on snow cover, snowfall on Alpine pastures, floods and dates of lakes freezing. These data have been brought together to provide a coherent story of changing weather patterns in Switzerland over the last four centuries.

Diaries of the weather have been especially valuable sources. An instance is the daily record kept by Wolfgang Haller; his daily entries allow estimates to be made of the temperature and of the frequency of sunny and rainy days near Zürich in 1545–6 and 1550–76. In the same area, regular observations were made by a parson, Heinrich Fries, from July 1683 to October 1718. From 1721 to 1738 a Winterthur baker, Rudolph Reiter, noted the weather hour by hour not only through the day but also through much of the night.

The first systematic observations of precipitation in Switzerland were made by Johan Jacob Scheuchzer; his records, starting in 1708 and continuing to 1731, are the earliest in central Europe. Hans Jacob Gessner, who like Scheuchzer was a Zürich clergyman, observed the rainfall from 1740 to 1746 and also from 1750 to 1754 (Pfister 1978b). In Basel a professor of law, Johann Jacob d'Annone, kept a diary of instrumental readings from 1755 to 1804.

Pfister (1980a) has made particularly effective use of phenological data in reconstructing past climate and weather. The growth patterns of plants in Switzerland, as in other

[8] Phenological calendars were used thousands of years ago in both China and the Roman Empire (Hopp 1974). The term 'phenology' was introduced by a Belgian botanist, Charles Morren, in 1853, but it was the Swedish botanist Carolus Linnæus (Carl von Linné) who initiated modern phenology and phenological networks. In his *Philosophia Botanica* he outlined methods for compiling annual plant calendars of leaf opening, flowering, fruiting and leaf fall, together with climatic observations 'so as to show how areas differ'. Phenological networks involve the co-operation of regular observers working in selected site areas and recording the dates of occurrence of specific growth stages or phenophases of chosen indicator plants. These networks can provide important supplements to climatological networks.

The time of occurrence of phenophases of many plants are to a large extent controlled by local climate. Phenological observations of selected indicator plants are recognized as being of assistance in forecasting or estimating the occurrence of succeeding phenological events in the same or other species. Maps of long-term averages of occurrence of phenophases can be used to identify zones of similar climate in the same way as meteorological maps. However, individual phenological observations must be interpreted with caution, because plants respond to the full range of environmental factors and are affected by such matters as soil type, exposure and topography in addition to micro-meteorology. Observations should therefore be considered in relation to data from neighbouring sites.

If maps of long-term phenological averages can be related to meteorological conditions, it follows that if meteorological conditions change this will be reflected in the timing of phenophases. If phenological series of adequate length are available for a period covered by good meteorological observations there is a possibility that recorded phenophase data from the past can be decoded in terms of their dominant meteorological control.

temperate latitudes, respond mainly to temperature. Most of the records relate to fruit trees, grapes and field crops. The dates of the vine harvest are especially well documented and significant; in Switzerland they are to be found in local police records.[9] Grape yields depend mainly on summer weather, being high when July and August have been hot and dry, low after cool and rainy summers (Primault 1969). For the derivation of factors for converting phenological information into climatic values, Pfister considers that corresponding data over a period of 15 to 20 years is needed, covering a wide range of weather conditions.[10] The statistical treatment has to take into account the influence of altitude and exposure; an increase in altitude of 100 m corresponds to a delay of 3.6 days in the opening of vine flowers (Becker 1969). Departures from mean values are the significant features of phenological records and it is important to know whether the mean for a particular species has varied through time for other than climatic reasons, such as the introduction of new varieties.

Studies of this kind require not only a knowledge of plant ecology but also involve the application of modern methods of historical analysis to a wide range of data (Bell and Ogilvie 1978). Some well-known European weather compilations[11] have been shown not to be verifiable (Ingram *et al.* 1981) and partly as a consequence of this there has been a tendency to discount prematurely the possibility of climatic fluctuations having been influential in human affairs (Parry 1978 and see 12.3.5). Data verification involves not only the selection of reliable evidence and the rejection of dubious material but also the identification of the time and place of an observation and, when descriptive terms are used, an appreciation of the linguistic conventions current at the time of writing (Ingram *et al.* 1981). The characteristics and shortcomings of instruments and the peculiarities of the places where they were housed have to be taken into account. Pfister (1981) had

[9] Possibilities offered by vine harvest dates were recognized by a Swiss scientist, Dufour, in 1870 and have been put to good use since especially by Ladurie (1971) and Ladurie and Baulant (1980).

[10] Most of the data used by Pfister resulted from vineyard share-cropping agreements which were held by public institutions and authorities. The division between landlord and tenant being known, the landlord's portions are listed and so provide useful series. Complications result because the acreage is not always known. Pfister compiled four regional series from different parts of the Swiss Plateau, each compiled from a variety of local series of unknown acreage. These four show correlations between 0.56 and 0.76. The main series, computed from the regional series, is highly correlated ($r = 0.87$) with a fifth series compiled from the several local series for which the acreage is known. It is recognized that yields can be affected by changes in techniques of cultivation, manuring, pruning and so on, or by a change in the variety grown. Climatic factors will be important in explaining short-term fluctuations.

[11] The first of these sources was a compilation by Hennig (1904) published by the Royal Prussian Meteorological Institute. It exhibited many of the common weaknesses of historical works produced by people with little understanding of the need for identification of the sources of all individual items of information and of the value of source analysis. Bell and Ogilvie (1978), in their discussion of the nature and use of weather compilations, noted that while Hennig distinguished between information he had gathered from newspaper articles, town guides and travelogues and references taken from historical sources, he was wholly uncritical and drew his information almost entirely from secondary sources and unreliable editions. Illustrative of the results of his approach is his statement that there were many hot dry summers between AD 988 and AD 1000. He cited eighteen references in support of the conclusion, but the two earliest and most nearly contemporary of these record only one single hot summer. All the other sources quoted are much later, many of them seventeenth- and eighteenth-century compilations. The idea of a succession of hot summers arose from Hennig's own summary of unreliable information.

Some more recent workers, such as Britton (1937), claimed to distinguish authentic from dubious material but still used a high proportion of unreliable sources, including that part of the 'Historia Croglandensis' which has been shown to be a forgery made in the interests of claiming rights to disputed land. Britton did exhibit a far more critical attitude to some other matters, demonstrating for instance the way in which events recorded by contemporary medieval writers may be magnified in the course of time. He traced the way in which the record of a severe storm flood which inundated some villages in 1099 was transmuted over time into an extraordinary storm in which the Goodwin Sands were formed.

special dating problems in Switzerland because, although the Gregorian calendar was adopted by the Catholic cantons in 1583, the majority of the Protestant cantons did not follow suit until 1701, and some continued to use the Julian calendar for varying times into the eighteenth century.

In order to handle his data effectively Pfister coded it for computer processing, distinguishing five main categories: daily non-instrumental weather observations, miscellaneous observations from chronicles and annuals, precipitation values, monthly frequency counts of various kinds, and temperature measurements. He distinguished between precise reports and interpreted data. A ten-day interval was taken as the smallest reporting unit, except in the case of phenological material, when allowance was made for individual days. Some records could not be coded and were stored verbatim, but for the most part the documentary information was quantified in such a way that temperature and rainfall could be compared with 1901–60 mean values to produce thermal and wetness indices.

Computer listing allows data from various sources referring to the same time period to be rapidly assembled and compared. For the winter of 1572–3, for instance, which was the most severe winter of the last 500 years, Wolfgang Heller's record of twelve days of snow in December is supplemented by Savion's report of Lake Geneva having frozen. There follow observations from independent sources of milder weather about the middle of January in Bern and Zürich and break-up of the ice on the lake about the same time (Pfister 1981).

The phenophases, the various stages in the development of plants, have been related to monthly temperatures using stepwise multiple regression. It has been found, for example, that if the sweet cherry flowers more than two weeks earlier than the mean date of flowering, either February and March have been warmer than average or else January was extraordinarily warm and February and March somewhat warmer than average. If flowering is three weeks late, April has been about 5 °C cooler than the mean for 1901–60). The date of the appearance of the first vine flowers usually proves to depend on May's temperature but a very warm April can also give early flowering. Distinguishing conclusively the timing of a weather abnormality is not easy but additional information, even if it is only a remark in a diary, say about unusual warmth at a particular time, can provide a vital clue and allow the timing to be determined more closely.

Vine harvest dates have generally been taken to reflect mean summer temperatures (Ladurie 1971, Ladurie and Baulant 1980) but Pfister's (1980a) analysis shows that temperatures in spring and early summer are much more important than those in August and September: the correlation coefficients he gives are May − 0.54, June − 0.59, both with 0.001 significance, and July − 0.34 with 0.03 significance. He found no significant relationship between harvest date and temperature in April, August and September. These findings are supported by the discovery that vine harvest dates are more strongly correlated ($r = +0.74$) with indicators of weather conditions in the early summer, notably the dates of tithe auctions[12] which are dependent on the time when the grain

[12] Until the nineteenth century the right to collect tithes in certain parts of Switzerland was sold by auction to tithe farmers. They undertook to collect the tithes, being allowed to retain a small proportion of the proceeds. The dates of the auctions and the inspections which preceded them depended upon the ripening of the grain. Mean dates were shown to be a function of altitude, a difference of 100 m causing 4.6 days' delay. Pfister found forty-two series of auction dates covering the period 1611–1825 which he aggregated into a unified series. The advance or delay of the auction by seven days was found to correspond to a deviation of 1°C for the 1901–60 average temperature.

Table 6.2 Indicators used for the construction of thermal indices for individual months

Month	*Cold*	*Warm*
December–February	Uninterrupted snowcover Freezing of lakes	Scarcity of snowcover Signs of vegetation
March	Long snow cover High snow frequency	Sweet cherry first flower
April	Snow cover Frequent snow	Beech tree leaf emergence Tithe auction dates
May	Sweet cherry flower	Tithe auction dates (± 0.6 °C) Vine 1st flower (± 1.2 °C)
June		Tithe auction date (± 0.6 °C) Vine full flower Vine last flower
July		Vine yields (± 0.6 °C) Coloration/maturity of first grapes
April–July		Vine harvest dates (± 0.6 °C)
August		Vine yields (± 0.6 °C) Tree-ring density (± 0.8 °C)
September		Vine quality Tree-ring density (± 0.8 °C)
October	Snow cover Snow frequency	Reappearance of spring vegetation
November	Long snow cover Snow frequency Freezing of lakes	No snowfall Cattle in pastures

Source: Pfister (1981a)

ripens, than with tree-ring densities, which reflect temperatures in late summer.

Table 6.2 gives an impression of the range of the different kinds of indicators that can be used in estimating thermal indices over the last 450 years and the standard errors of temperatures derived from the phenological data.

While the dates of individual phenopauses can give an indication of conditions in an individual month, the pattern of phenopauses over a longer time period can give additional information. Thus, a lengthening of the interval between phenopauses indicates below-average temperatures in the course of the year, as compared with the 1901–60 means. Delay of phenopauses indicates that at least four months between March and July had temperatures below the 1901–60 averages. Advance of all phenopauses indicates that two or more of the spring months as well as June temperatures were above the modern average and that July temperatures were average or higher.

Pfister's archival studies (1980a and b) have enabled him to derive precipitation as well as thermal decadal indices for Switzerland over the period 1525 to 1820 (Figure 6.9). Each month has been given negative values if it was cold and dry, positive values if warm and wet, and zero values if it approximated to the 1901–60 means. Monthly values are weighted by factors of 1 to 3 according to the magnitude of the departures from the mean, a factor of 3 indicating extreme conditions by twentieth-century standards. Then both

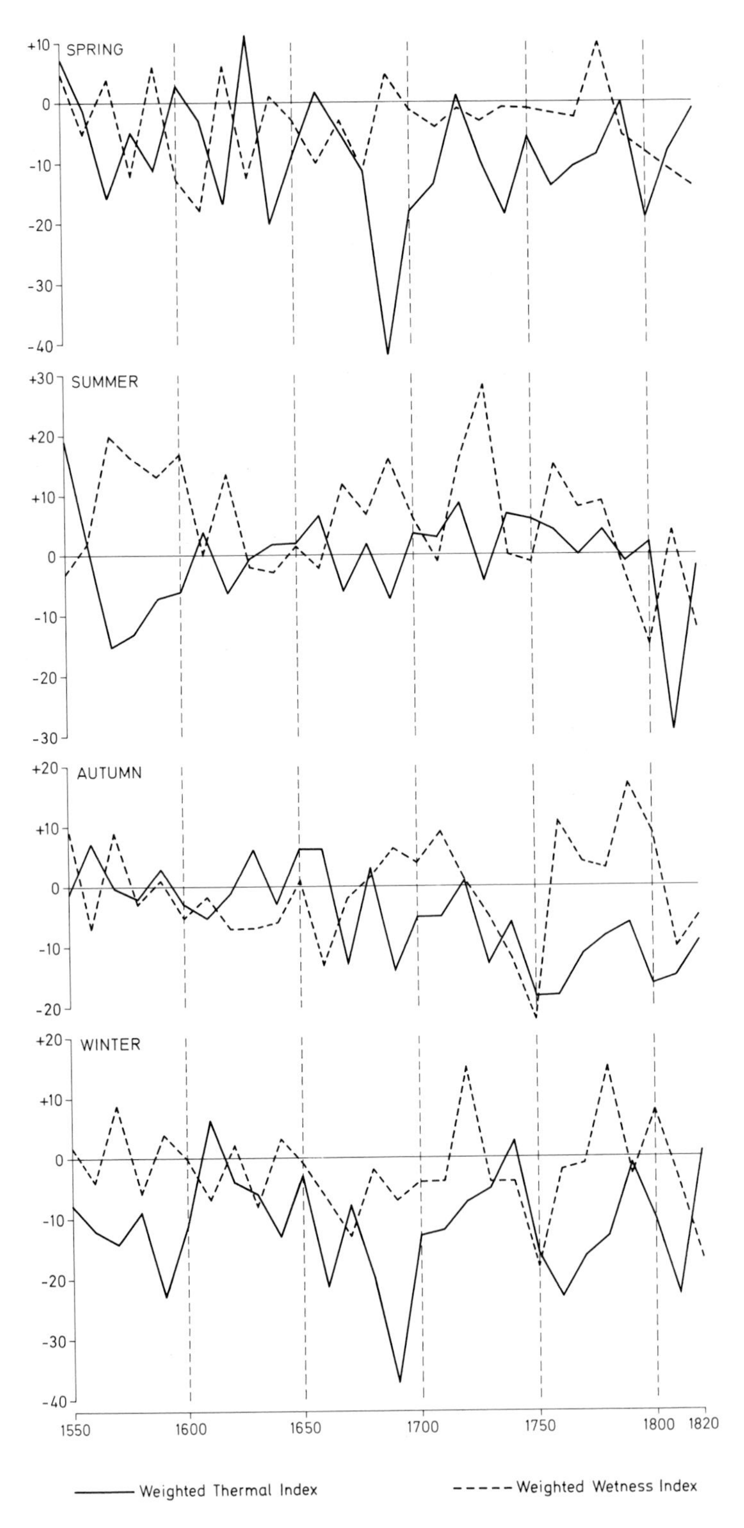

Figure 6.9 Seasonal weighted thermal and wetness decadal indices for Switzerland, 1550–1820, constructed from data in Pfister (1981, pp. 238–45)

precipitation and thermal indices for each season are obtained by summing three monthly indices to give values that range from −9 to +9.

The rings of trees near the upper limit of forests provide another source of information about past temperatures (Schweingruber *et al.* 1978a, b, 1979). The density of the summer wood, as measured by X-ray methods, varies with July to September temperatures. In the early stages of the Little Ice Age summer temperatures were about 2 °C lower than those of the late nineteenth and early twentieth centuries. In the intervening period conditions varied in much the same way that Pfister's results present in more detail.

The period between 1570 and the end of the sixteenth century was on average cooler and wetter in all seasons of the year than the preceding half century (Table 6.3). The percentage of cold winters doubled and of wet winters trebled. Between 1525 and 1564 there was no winter in which snow persisted in the Swiss lowlands for more than eighty days, yet there were four such winters in the decade 1564–74. In Zürich, the proportion of total precipitation in the form of snow rose from 44 per cent in 1550–63 to 63 per cent between 1564 and 1576. Autumn and springtime were cooler as well, but the chilling was less marked than in winter and summer. Between 1525 and 1569 there were twice as many warm summer months as cold, but between 1570 and 1600 half the summer months are rated as having been cold. The summer of 1573 was a particularly bad one and, in general, extreme conditions were more pronounced than they are today. For instance, the summers of 1588 and 1596 were extremely wet, with rain falling on 77 out of the 92 days of June, July and August in the Lucerne area.

Pfister's studies allow the changing character of individual months to be traced from

Table 6.3 Warm, cold, wet and dry months, 1525–69 and 1570–1600

	Percentage of months which were undoubtedly:			
	Warm	*Cold*	*Dry*	*Wet*
Summer				
1570–1600	28	47	12	38
1525–64	36	18	18	24
Difference	− 8	+ 29	− 6	+ 14
Autumn				
1570–1600	24	19	13	18
1525–69	24	7	20	11
Difference	0	+ 12	− 7	+ 7
Winter				
1570–1600	12	44	16	18
1525–69	15	24	9	6
Difference	− 3	+ 20	+ 7	+ 12
Spring				
1570–1600	18	33	16	17
1525–69	20	24	16	16
Difference	− 2	+ 9	0	+ 1
Year				
1570–1600	20	36	14	23
1525–69	24	18	16	15
Difference	− 4	+ 18	− 2	+ 8

Source: Pfister (1981)

1550 to 1820. As Figure 6.9 shows, the spring months were colder and drier throughout most of the period than in the first half of this century, the coldest spring decades being the 1640s, 1690s and 1740s; only the 1550s and 1730s were warmer than the 1901–60 mean. Whereas May's temperature and precipitation were near the early twentieth-century mean and April, though drier, was no cooler except slightly between 1525 and 1720, March was persistently colder throughout the whole period, except for about twenty years at the beginning of the seventeenth century. In the 1640s, 1690s and 1760s, March temperatures were particularly low, 2 °C below the 1901–60 mean; together with persistent dryness these low temperatures are believed to indicate the prevalence in early spring of northerly winds and blocking anticyclones over north-central Europe.

The high frequency of cool, wet summers in the last three decades of the sixteenth century as compared with 1525 to 1569 mainly involved June and July; June remained generally cool and wet right through the seventeenth century in spite of the fact that summers in the middle decades were not much different from those of the early twentieth century. July was frequently wet but not markedly cool in the first half of the eighteenth century and then, after 1760, it was cool as well as wet, especially between 1810 and 1820, when July temperatures, 1.2 °C below the 1901–60 mean, were largely responsible for the overall coolness of the summers.

In the three decades before and after 1600 the cool dry springs and wet summers were followed by autumns with mean temperatures similar to those of the first sixty years of this century. Then autumn seems to have become cooler, especially in the 1750s and 1760s and the early years of the nineteenth century. The autumns of the 1730s, 1740s and 1750s were wet as were those of the early nineteenth century.

While cool, wet summers were features of certain decades in the Little Ice Age, winters throughout the period were on the whole cold and dry. The decade with the coldest winters in Switzerland, as in much of the rest of Europe including Iceland and Scandinavia, was from 1690 to 1700. In London more days with snowfall were recorded in that decade than in any other before or since (Manley 1969). The coldest winter of the decade in Europe generally was that of 1694/5. In Switzerland the Bodensee (Lake Constance) froze over completely with a layer of ice thick enough to bear fully laden wagons (Lindgren and Neumann 1981). For an ice cover over the entire lake surface a temperature anomaly of at least −4°C is needed (von Rudloff 1967). Never since that year has Lac de Neuchâtel frozen over so abruptly. The winter of 1784/5 was also extraordinarily long and severe, with snow lying in Bern for 154 days, 60 of them between the beginning of March and end of May. The corresponding figures for the hard winter of 1962/3, it might be noted, are 62 and 12 (Pfister 1978a), though the city was warming itself to a much greater degree by this time.

In years of heavy snowfall, when the snow cover persisted well into the late spring on upper Alpine pastures, winter crops were heavily damaged by a snow mould (*Fusarium nivale*). This was notably the case in 1785 and also in 1757, 1770 and 1789 (Pfister 1978a). No such outbreaks have occurred in the last 120 years; *Fusarium nivale* was, it seems, a feature of the Little Ice Age in Switzerland.

The end of the Little Ice Age in Switzerland was marked by a run of warm, dry summers in the 1860s and 1870s. This was the golden age of Alpine mountaineering when so many Swiss peaks were climbed for the first time. Over the century before 1860 the summer months had 2.5 more rainy days than in the century following, a statistically significant difference. Temperatures in summer in the century after 1860 rose by only

Table 6.4 Differences in mean seasonal temperature between 1755–1859 and 1860–1965

	(1755–1859) – (1860–1965)
Spring	+ 0.3
Summer	+ 0.1
Autumn	+ 0.4
Winter	+ 0.6

Source: Pfister (1978c)

0.1 °C but the warming in the other seasons was much greater, as Table 6.4 shows, with a rise in mean winter temperature of 0.6 °C.

The Little Ice Age in Switzerland can be seen in perspective as having been the outcome of a climatic fluctuation involving cold winters extending into March, with wet summers and short growing seasons, as compared with the period before 1560 and since 1860. When it is examined in much more detail it is seen to have been made up of a number of minor fluctuations each lasting some decades. There was, for example, the period between 1560 and 1630, with cool, wet summers and cold winters and springs; this came to an end when both summers and winters became warmer and drier. Even the minor fluctuations were not usually homogeneous. The 1690s as a decade stand out as having been cold at all seasons with heavy precipitation in spring, summer and autumn. The 1720s included hot summers in 1723, 1727 and 1729, when there were booms in wine production in spite of the frequent thunderstorms. Following warm summers in the years 1759 to 1763 there came a sequence of fourteen cool summers and then seven warm ones. The last of these was followed by the long, hard winter of 1784/5. Switches in the characteristics of individual months could be abrupt; June was rather cool and wet from 1760 to 1800 and again in the decade 1810 to 1819, but in 1820 June was remarkably warm, 4 °C above the 1901–60 mean. In fact, it is not easy to make general statements about the climate of the Little Ice Age; the weather was variable as it is today, but there were more frequent cold winters and wet summers than there have been since 1860 and these were often bunched together.

What can we say now about the climatic conditions associated with the advances and retreats of the Swiss glaciers as they are exemplified and indeed represented by the Unterer Grindelwaldgletscher? The Unterer Grindelwald reached its furthest forward position right at the end of the sixteenth century, following three decades when temperatures in all the seasons were low and precipitation in the form of snow was more abundant than usual. Within this period, clusters of cold wet summers occurred in 1560–4, 1569–79 and 1585–97 (Figure 6.10).

Glacier retreat in the 1640s followed the arrival of warmer summers after 1630. A readvance starting about 1686 preceded the frigid decade of the 1690s, though this may well have provided the momentum that carried the glaciers forward until 1718. A retreat back to the Unterer Schopf coincided with the hot summers of the 1720s (which were experienced in England as well as Switzerland). A slight readvance of the Grindelwald about 1740 seems to have been associated with increased summer wetness, and then from 1756 to 1768 the glacier front was back near the Unterer Schopf. Quite suddenly it then advanced rapidly for ten years, stimulated by abundant snowfall and low spring and summer temperatures in 1767–71. The retreat that followed, from 1784 to 1794,

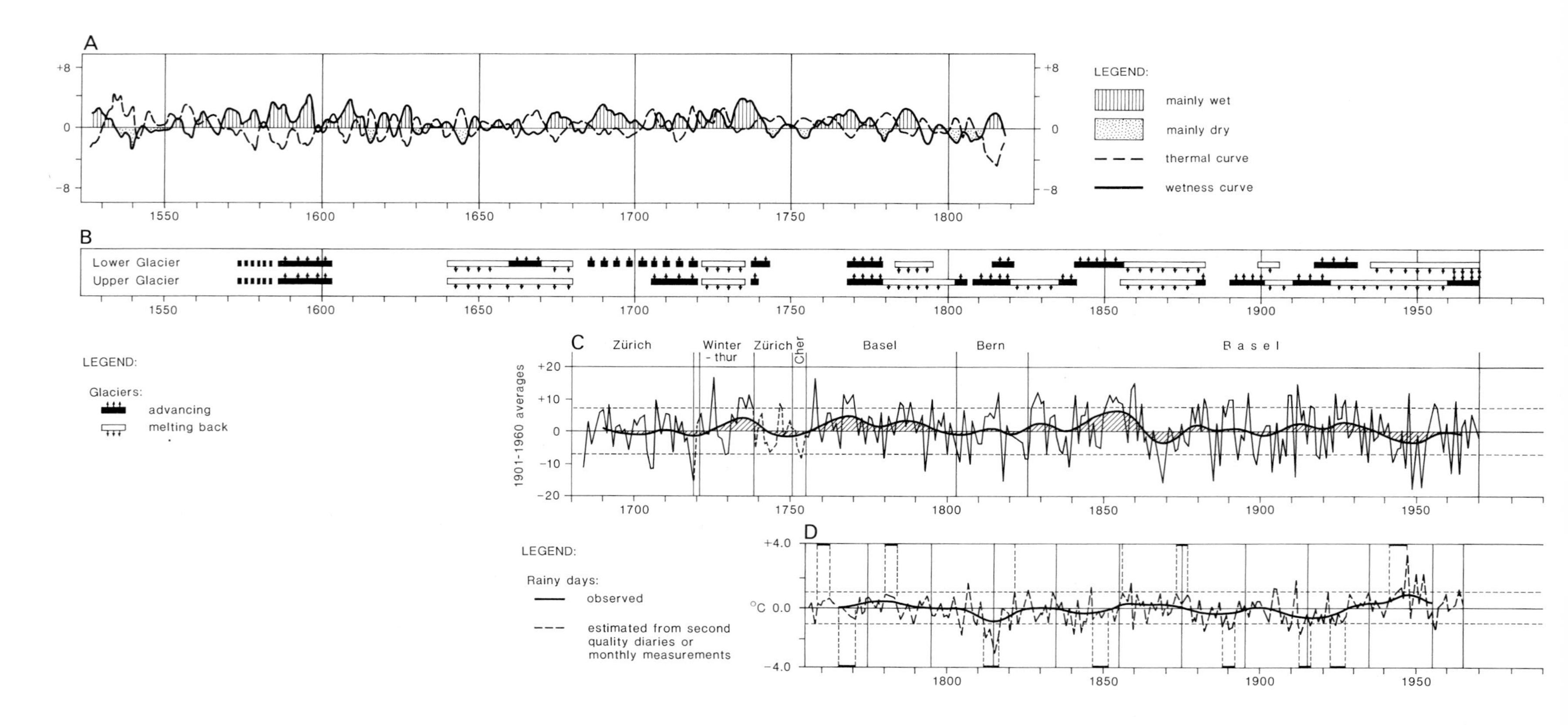

Figure 6.10 Movements of the terminal snouts of the Grindelwald glacier (b), in comparison with five-year moving averages of weighted thermal and wetness indices for the summer months (June, July and August) (a), number of rainy days in the summer months as departures from 1901–60 averages (c), and summer temperatures in Basel as departures from 1901–60 averages (d). (From Pfister 1980a)

immediately followed a sequence of four warm summers; then for twenty years the glacial terminus was back again at the Unterer Schopf.

An advance starting in 1814 carried the Unterer Grindelwald further forward in 1820 than it had been for 170 years, while the Oberer Grindelwald almost reached the 1600 moraine. This expansion was generated by a long sequence of cold, wet summers and cold autumns, starting with that of 1812 and continuing to 1817. The expansion came to a sharp halt with the hot summer of 1820 but the ice remained in an advanced position and its tongue crept forward another 100 m between 1840 and 1855. This was associated with low temperatures in all seasons between 1847 and 1851, together with high precipitation as indicated by numbers of rain-days at Basel (Figure 6.10). In 1891 the Grindelwaldgletscher began to retreat; within six years the Unterer Schopf was visible and by 1880 the tongue was a kilometre behind the 1856 moraine. Temperatures were above average in all seasons for the period from about 1860 to 1875 but not by a remarkable amount. However, according to Pfister (1980b), the deficit in precipitation was greater than had been known since 1680 and probably since 1560.

The glacier maintained its position for the last twenty years of the nineteenth century, with the aid of a cooling that was particularly marked in all seasons from 1887 to 1890 but which was not, it seems, accompanied by much increase in precipitation. The retreat was resumed for a few years until the cold summers preceding and during the First World

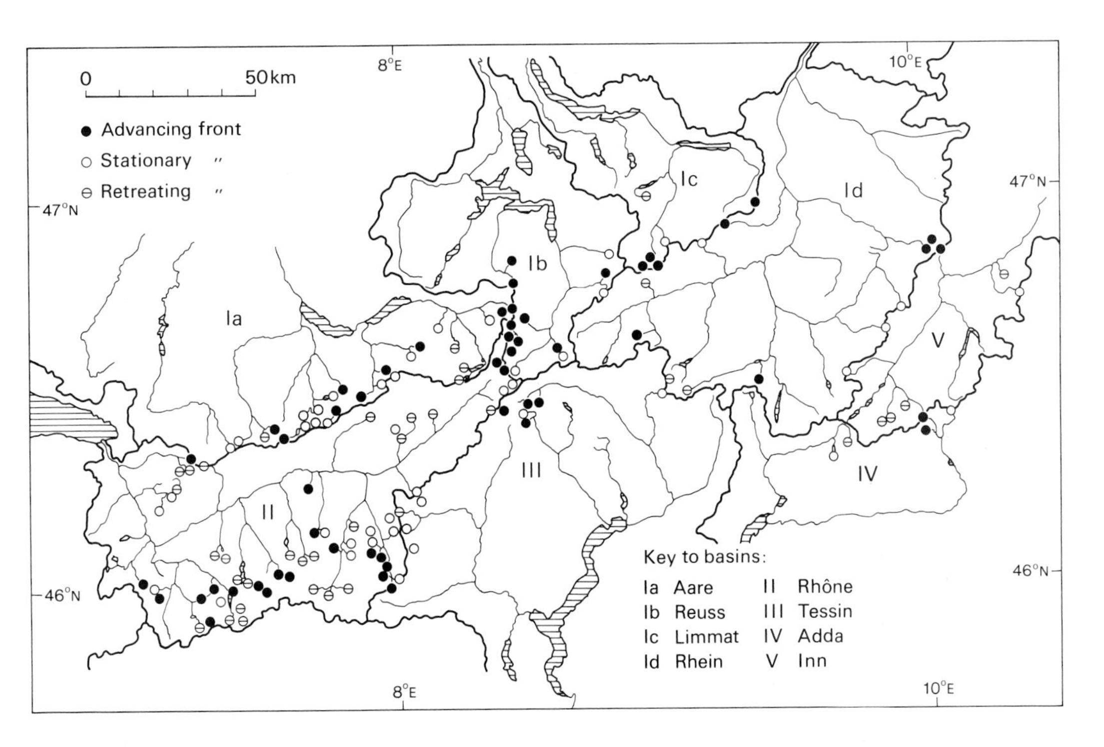

Figure 6.11 Distribution of advancing, retreating and stationary Swiss glaciers amongst those with fronts observed for the year 1974/5. (From the 96th Annual Report of the Swiss Glacier Commission)

War took effect and the ice pushed forward almost to the 1890s moraine. There the Unterer Grindelwald remained more or less stationary for the rest of the inter-war period, rather anomalously it would seem, for the Oberer Grindelwald retreated over 200 m and the Stein and other glaciers much more rapidly in the same period, presumably in response to above-average temperatures in all seasons from 1927 to 1935. Eventually, in 1940, the Unterer Grindelwald followed the example of the majority of the other Swiss glaciers and Alpine glaciers generally by retreating. It shrank steadily by an average of about 40 m annually until 1964 in response to a succession of warm dry summers culminating in the decade 1943 to 1952.

While the 1960s and 1970s saw the Oberer Grindelwald advancing in the company of the Aletsch and about 40 per cent of the other Swiss glaciers (Figure 6.11), the Unterer Grindelwald lagged behind again until it also began to advance in the early 1980s. In general, glacier tongues were less decisive in their movements than they had been between 1930 and 1960, in the sense that some were advancing while others were retreating, as had been the case in the years around 1920 (Figure 6.7). However, the glacier tongues in the 1980s are all far higher up their valleys than they were sixty years earlier. The Little Ice Age seems to have receded into the historical past; the climbing huts of the Victorians look down on glaciers from which a little of the glory has departed.

Chapter 7

The Little Ice Age in Asia

7.1 The Little Ice Age – a global event?

The last hundred years have seen European glaciers retreat from the advanced positions they had occupied in the middle or towards the end of the nineteenth century. The question arises as to whether the Little Ice Age, on which we now look back across the more genial twentieth century, was a global phenomenon. It is always tempting to seek for universals, and they are easier to promote when the data are imprecise than when they are firmly established. However, the fact that glaciers in Europe reached their most advanced positions at times which vary between the late sixteenth century and mid-nineteenth century in the Alps and the mid-eighteenth century in Scandinavia provides good reason to pause and take heed of the possibility that Little Ice Age synchroneity is unlikely to exist globally. On the other hand, the almost universal retreat of montane glaciers in the course of this century is sufficiently well established and dramatic to encourage an attempt to look into the whole question. No critical review of the assumptions and evidence has so far been attempted elsewhere.

In many parts of the world deliberate measurements have only recently started and historical records are lacking, so the data available are very uneven in amount and quality. Of the first three IAHS/Unesco volumes published on the fluctuations of glaciers (Kasser 1967 and 1973, Müller 1977), the first contained only European material and even the third had data for only one Himalayan glacier, yet the Himalaya contain by far the most substantial area of ice outside the arctic and antarctic regions, and variations in the volume of glacier runoff there are directly important to the economy of densely settled areas adjacent to the mountain chain. Regional coverage was further improved in the fourth volume (Haeberli 1985) but it still contained data for only 17 glaciers in the whole of China, compared with 120 in Austria. European historical records are more plentiful and more informative than those of other continents; elsewhere, greater reliance inevitably has to be placed on moraine dating.

A number of sample areas are examined in the following review, which is not exhaustive but is believed to represent the state of knowledge as it existed in the early 1980s. In certain areas where studies so far made have been in the nature of reconnaissances, new findings will no doubt modify the picture we have at present.

7.2 The USSR

The principal glacierized regions of the USSR almost encircle the country, as can be seen on Figure 7.1. Of the total ice-covered area, 78,000 km², three-quarters are on the arctic islands and Novaya Zemlya, for which there is little information before the twentieth century. The Siberian glaciers also became known mainly in this century and those on the

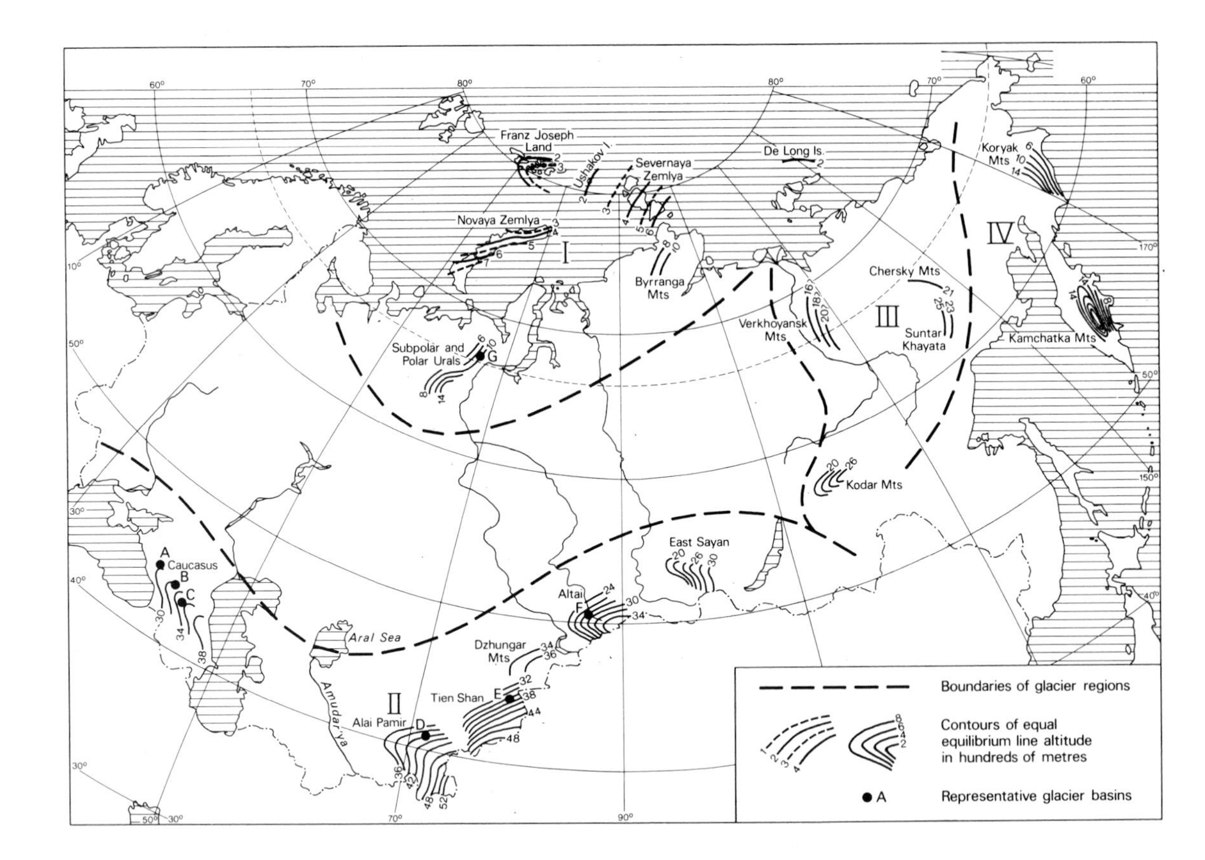

Figure 7.1 Glacierized areas and snowline elevations in the USSR. Glaciers mentioned: E = Tsentral'nyy Tuyuksu, F = Aktru, G = Igan. (After Grosswal'd and Kotlyakov 1969)

Pacific coast within the last forty years. Attention is therefore concentrated here on the glaciers of the Urals, the Caucasus and central Asia.

7.2.1 THE URALS

The glaciers scattered along the 2100 km length of the Urals are all quite small and together their total area is only 28 km^2. They lie on the eastern, leeward side of the crestline and many are in cirques 700 m to as much as 1000 m below the regional snowline (Grosswal'd and Kotlyakov 1969). They depend for their nourishment on westerly winds driving the snow and concentrating it in leeside hollows. They were close to equilibrium, their tongues being almost stationary in the second half of the nineteenth century, but negative budgets and retreats have predominated since then. Changes in the net balance of the glacier called Lednik Instituta Geografii exemplify a general trend. The values given in Table 7.1 are estimates for which the basis is not indicated by Grosswal'd and Kotlyakov (1969). From Table 7.1 it seems that wasting increased steadily from 1900 onwards, though glacier termini were stationary in 1905–10, 1921–30 and 1946–50. The Igan and Obruchev glaciers in the Polar Urals were also wasting until the mid-1960s, but there had been increased precipitation in the early 1950s and mass balances became positive in the late 1960s and early 1970s (Figure 7.2).

Table 7.1 Estimated annual net balance, Lednik Instituta Geografii, 1850–1963

	Acc./g/cm	*Abl./g/cm*	*Net budget*
1850–75	178	174	+ 4
1876–1900	168	166	+ 2
1901–25	170	192	− 15
1926–50	170	203	− 33
1951–63	220	288	− 68

Source: Grosswal'd and Kotlyakov (1969)

7.2.2 THE CAUCASUS

The Caucasus are 20° south of the Polar Urals and are very much higher. They lie between mid-latitude steppe and the sub-tropics. The glaciers are concentrated in the central part of the chain. According to Kotlyakov and Krenke (1979), there are 2047 glaciers; Kotlyakov (1980b) gives the ice-covered area as 723 km^2. In medieval times according to Kotlyakov and Krenke (1979), the snowline was higher and the glaciers smaller than they are now. A number of mountain passes then in use are now covered with ice but, as in the Valaisian Alps (Röthlisberger 1974), tracks leading to them can still be distinguished. Glacier advances between the thirteenth and fifteenth centuries are said to be indicated by radiocarbon dates and according to Kotlyakov (1981) lichenometric studies have permitted reconstruction of glacier history over the last 800 years. Though no mention is made of the methods used or the extent of the data, it seems that glaciers in the central Caucasus expanded between 1640 and 1680 and again between 1780 and 1830. The seventeenth, eighteenth and nineteenth centuries were cold, with heavy precipitation, and tree-ring studies point to large-scale avalanching in this period

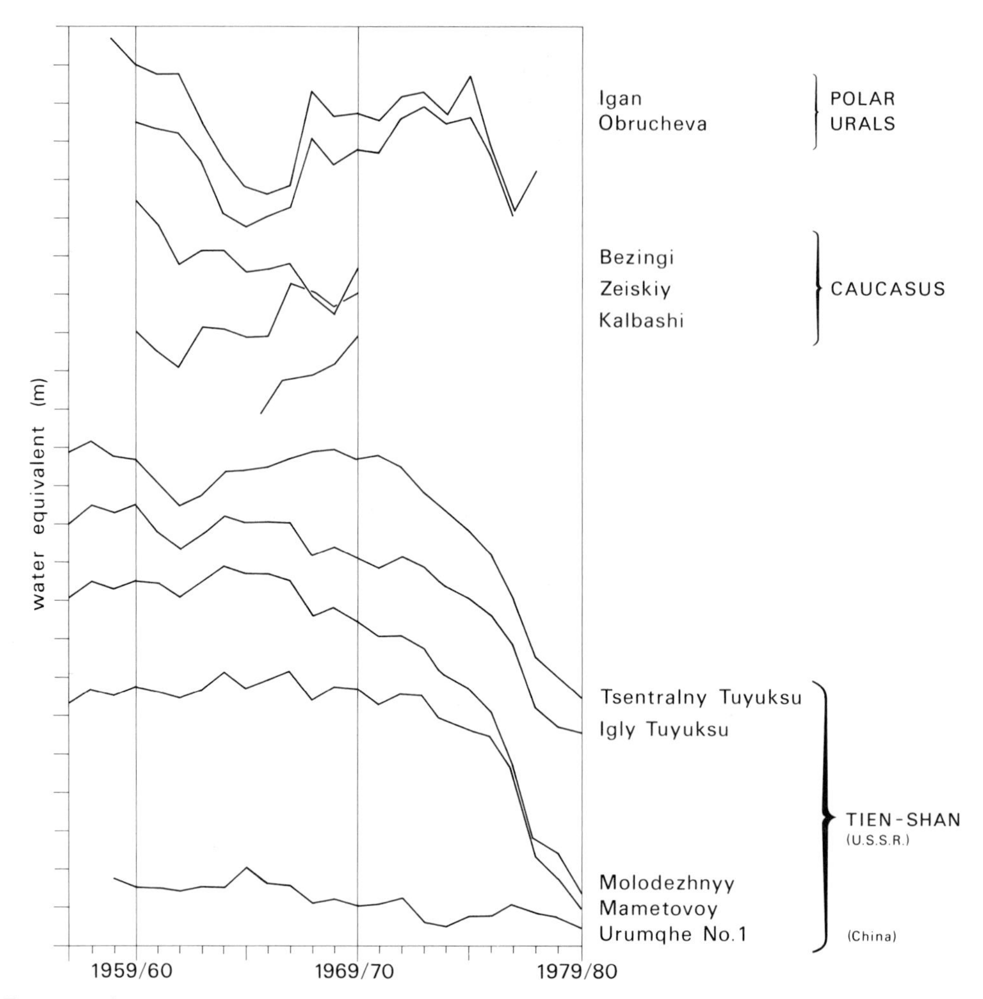

Figure 7.2 Changes in mass balance of representative glaciers in the Polar Urals, Caucasus and western Tien Shan from 1958/9 to 1979/80. (From figures published in Kasser 1973, Müller 1977 and Haeberli 1985)

(Kotlyakov and Krenke 1979). The firnline was approximately 100 m lower in the first half of the nineteenth century than in the late twentieth century and glacial advances were frequently reported in contemporary Russian literature.

Douglas Freshfield was one of the foreign travellers who ventured into the Caucasus in the nineteenth century at a time when they were scarcely known outside Russia; as Freshfield (1869) wrote in the account of his visit in 1868, they were less well known than the Andes and Himalaya. Merzbacher (1901, cited by Horvath 1975) found evidence of glacier advances in recent centuries in the central Caucasus valley where ruins lie close to the glacier tongues and local legends and songs tell of a glacier near Ushkul, probably Lednik Khalde, then six miles away from the village, having advanced and destroyed all of it but for the church. The people still held an annual festival in thanksgiving!

According to Horvath's (1975) survey, the Caucasian glaciers continued to advance until about 1849 and retreat began in 1860. Freshfield and Sella (1896, p. 53) commented that 'the movements of the Caucasian and Alpine glaciers have of late years shown a general correspondence. In 1863 the Caucasian ice was in retreat. About 1875 the tide seemed to turn, and in 1887–9 many glaciers were slightly advancing.' Since then retreat has been predominant though there have been stationary periods or slight advances in 1910–14 and 1927–33, when summer temperatures were lower and there was some increase in winter snowfall. Glaciers heavily covered in debris such as Lednik Shkheldy did not start to retreat for some 50–70 years after the rest. The ice-covered area has diminished by about a quarter and glacier volume by a similar amount (Kotlyakov and Krenke 1979). The shrinkage of the ice is said to have been more marked on the northern than on the southern slopes.

Towards the middle of the twentieth century retreat of the glaciers came to a halt. Between 1959 and 1963 firnlines were lowered by 200 to 300 m as the result of a 30 per cent increase in winter precipitation and a lowering of summer temperature by 0.3 to 0.6 °C as compared with the previous decade. In the 1960s there was a further increase in precipitation of 20–30 per cent; accumulation zones thickened by 15 m to 18 m and many glaciers advanced (Kotlyakov and Krenke 1979). Lednik Alibek, which had receded at a rate of 2–7 m per year between 1904 and 1959, advanced on average 2.2 m per year between 1960 and 1968. Irregular increases in mass balance continued until 1970 (Figure 7.2) and large-scale photogrammetric surveys of the El'brus glaciers have shown that expansion continued into the 1970s with the Bol'shoy Azau advancing 27 m and growing thicker and wider than it had been in the 1960s. However, in 1979, of twenty-six glacier tongues measured in the central Caucasus only six were still advancing; the rest were retreating (Kotlyakov 1981). Palaeoclimatic studies in the Caucasus continue to be intensified with Khar'kov University investigating past climatic conditions by using X-rays to measure the varying density of tree rings.

7.2.3 THE PAMIR

The glaciers of the Pamir, 'the roof of the world', were surveyed by Horvath (1975), using a translation of a comprehensive account by Zabirov (1955) and more recent sources. The snowline is above 4800 m over 60 per cent of the region. The ice-covered area occupies only 11 per cent of the mountains but this amounts to 8041 km^2 and the meltwater, most of which goes into the Amudar'ya, is vital for irrigation and for maintaining the level of the Sea of Aral. Because of their usefulness the glaciers of the Pamir have been receiving much attention in recent years, with programmes including mass balance studies, flow measurements and monitoring of frontal positions.

The northwestern is the most heavily glaciated part of the Pamir; southeastward the snowline rises to reach a maximum of 5200–40 m on the peaks of the Shakhdarinskiy Khrebet, including Pik Lenina, 7134 m, Pik Karla Marksa, 8726 m, and Pik Engel'sa, 6510 m, and then falling again. The best-known glacier, the Lednik Fedchenko, 71 km long, is one of the largest valley glaciers outside the sub-polar regions and the largest in the USSR. A major source of nourishment of this complex system is the névéfield of the Khrebet Akademii Nauk, which intercepts moisture-laden westerly airstreams.

Regular observations of the Fedchenko, the best-known system, began in 1959 and are being extended to monitor the differing behaviour of its various components. These include surging tributaries such as Lednik Medvezhiy and glaciers fed by large-scale avalanches possibly associated with the earthquakes to which these ranges are subject. Recession has predominated in the twentieth century, though there were minor advances of some glaciers about 1914, between 1927 and 1935, and again between 1946 and 1958. The budgets of a number of other Pamir glaciers were positive between 1959 and 1964 when many of them advanced (Kotlyakov 1968), though some were retreating again by the late 1960s. In spite of the prevalence of earthquakes in the Pamir and the great altitude of the glaciers, their general behaviour, involving retreat through much of this century, though with short-lived advances particularly around 1930 and early 1960s, is comparable with that in Alpine Europe.

7.2.4 THE TIEN SHAN

The best-known glaciers of the Tien Shan are those in the Zailiyskiy Alatau, the most northerly of its ranges, which stretch along the 43rd parallel near Alma Ata and reach up to 5000 m. As precipitation decreases the snowline rises from 3200 m in the west to 4300 m in the east. The glaciers, covering 500 km^2, are concentrated in the upper basins of the Bol'shaya and Malaya Almatinka and other large rivers that flow into Lake Balkhash.

Glacier observations, beginning in 1902, have been concentrated on the Tsentral'nyy Tuyuksu in the Almatinka basin. This glacier has retreated in the course of the century, except between 1902 and 1923 when it was stationary or advanced slightly. Makarevich (1962) suggested that this advance was attributable both to earthquakes and to a period of high precipitation in the late nineteenth century which also caused a rise in the level of Lake Balkhash. The Zhangyryk and possibly the Bogatyr' glacier on the northern slopes of the neighbouring Kungey Alatau also advanced about this time. During the next thirty-four years the Tsentral'nyy Tuyuksu lost 3 per cent of its total area and along a profile near the snout the ice was lowered 53 m between 1905 and 1958 (Table 7.2). The terminus advanced again about 1939–40 and temporarily reoccupied an area of 18000 m^2.

Table 7.2 Retreat of the Tsentral'nyy Tuyuksu, 1923–61

	Total retreat (m)	*Mean annual retreat (m/y)*
1923–37	113	7.8
1937–47	170	16–17
1947–53	30	5
1953–56	18	6
1956–59	25	8
1959–60	11–13	11–13
1960–61	16–17	16–17

Source: Makarovich (1962)

Mass balance observations began on four glaciers in the late 1950s and were extended to five more in 1965. The results show short periods of positive balance both in the late 1950s and late 1960s and Lednik Partizan gained over 5 m of water equivalent between 1964 and 1976. But the general trend of the majority of the nine glaciers was strongly downwards between 1970 and 1980, with some having water equivalent losses of as much as 5 or 6 m (Reynaud *et al.* 1984 and Figure 7.2).

Twentieth-century recession affected the Bol'shealmatinskiy glaciers in the Bol'shaya Almatinka basin, some 10 km from the Zailiyskiy Alatau, and also the glaciers of the Talgar basin. Short periods of stabilization or slight advance interrupted retreat in the mid-1930s, late 1940s and early 1950s (Horvath 1975). A significant increase in precipitation and lowering of summer temperatures began in the 1950s (Kotlyakov 1968), leading to periods of positive mass balance in the following decade. Comparison of air photographs taken in 1977 with those taken in the International Geophysical Year of 1957 not surprisingly reveal that frontal changes have not all been in the same direction (Kotlyakov 1981)

7.2.5 ALTAY AND SAYAN

The Altay and Sayan Mountains prolong the great ranges of the Pamir and Tien Shan (Figure 7.1). The snowline rises from 2300 m in the north and west to 3400 m in the southwest and 3500 m in the east. The range of conditions is wide because of the location of the mountains at the junction of three climatic regions, Mongolian, central Asian and west Siberian.

Glaciers in the Altay have recent moraines within a few kilometres of their snouts which have been attributed to the Little Ice Age by Revyakin and Revyakina (1976). The glaciers are thought to have been in advanced positions in the early nineteenth century. In 1842 Chikhachev observed that there were extensive snowfields in the eastern Altay. Whether glacial retreat, which has predominated since then, had begun or not is unknown. Intermittent observations of the termini of the Katunskiy, Akkemskiy and Sapozhnikov, which began in the 1880s, show that wasting was particularly swift in the mid-twentieth century. Some small glaciers have disappeared and the majority have lost 30–60 per cent of their areas. Valley glaciers, however, survived better, diminishing by only 5–13 per cent in area.

Investigations made by the Academy of Sciences show that retreat in this century has persisted into the 1970s (Figure 7.3), with recession most apparent since the middle of the century, especially amongst the glaciers on the southern slopes of the Altay. It is possible that retreat will not continue unbroken, for latterly positive mass balances have been measured on the five Aktru glaciers (Kotlyakov 1981).

7.3 The Himalaya

The Himalaya extend for 2000 km from the Hindu Kush and Karakoram and the sources of the Indus in the west to the headwaters of the Dikang-Brahmaputra in the east (Figure 7.4). The main east–west arc is composed of several parallel ranges, the Lesser Himalaya rising to 4500 m and the Greater Himalaya further north to over 5500 m. Ten of the thirteen mountains on the globe rising to over 8000 m, including Everest, are amongst the giant peaks. To the north lies the high plateau of Tibet and central Asia. The ice-covered area has been estimated by Wissman (1959) to be 33 times that of Europe, and though Vohra (1980) believes this may be an overestimate it remains the largest such area outside the polar regions.

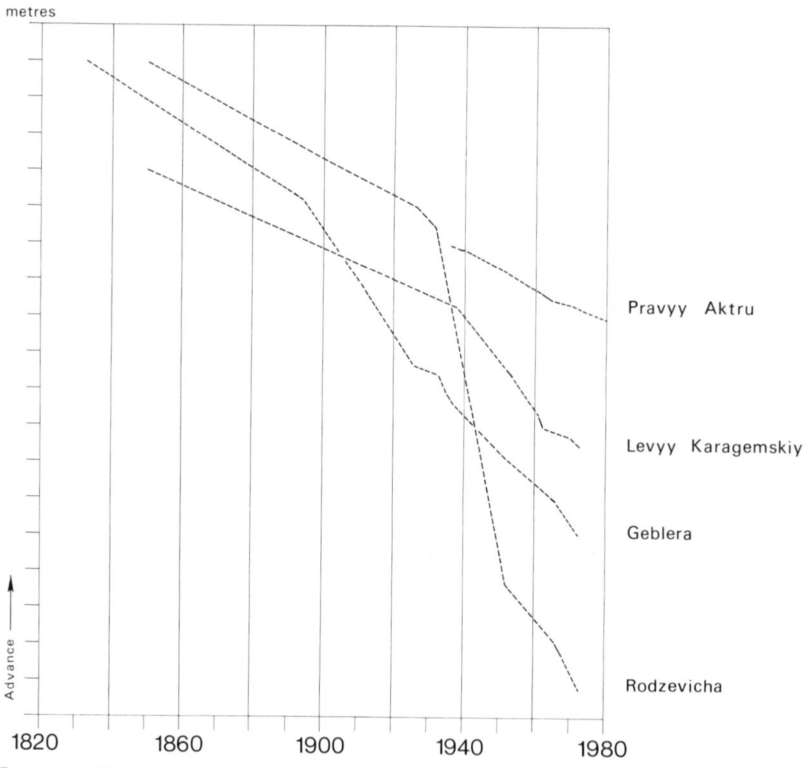

Figure 7.3 Retreat of Altay glaciers in the nineteenth and twentieth centuries. (From figures published in Revyakin and Revyakina 1976 and Haeberli 1985)

The Himalaya divide the monsoon lands to the south from the deserts and steppes of the Asian interior. Most of the precipitation is brought by the southwest monsoon. South of the main watershed glaciers are much larger than to the north, on account of the greater snowfall (Plate 7.1) and terminate at much lower levels reaching down in some cases into the forest zone. In Sikkim, in the east, the firnline is at a height of 5500 m and the treeline reaches an elevation of 3900 m; on the Nanga Parbat massif in the northwest the firnline comes down to 4700 m and the treeline is at 2000 m. Meltwater from the glaciers is used locally for watering the fields of mountain villages and it also contributes to the flow of the great rivers used for power production and irrigation in India and Pakistan.

Practically no documentary records are known relating to the extent of the glaciers before the nineteenth century. However, a Hindu scribe is quoted by Smythe (1932) to the effect that

> He who thinks of Himachal [the Himalayan snows] though he should not behold him, is greater than he who performs all worship in Kashi [Benares]. And he who thinks on Himachal shall have pardon for all sins, and all things that die on Himachal, and all things that on dying think on his snows, are freed from sin. In hundred ages of the gods, I could not tell thee of all the glories of Himachal, where Siva lived and where the Ganga falls from the foot of Vishnu like the slender thread of a lotus flower.

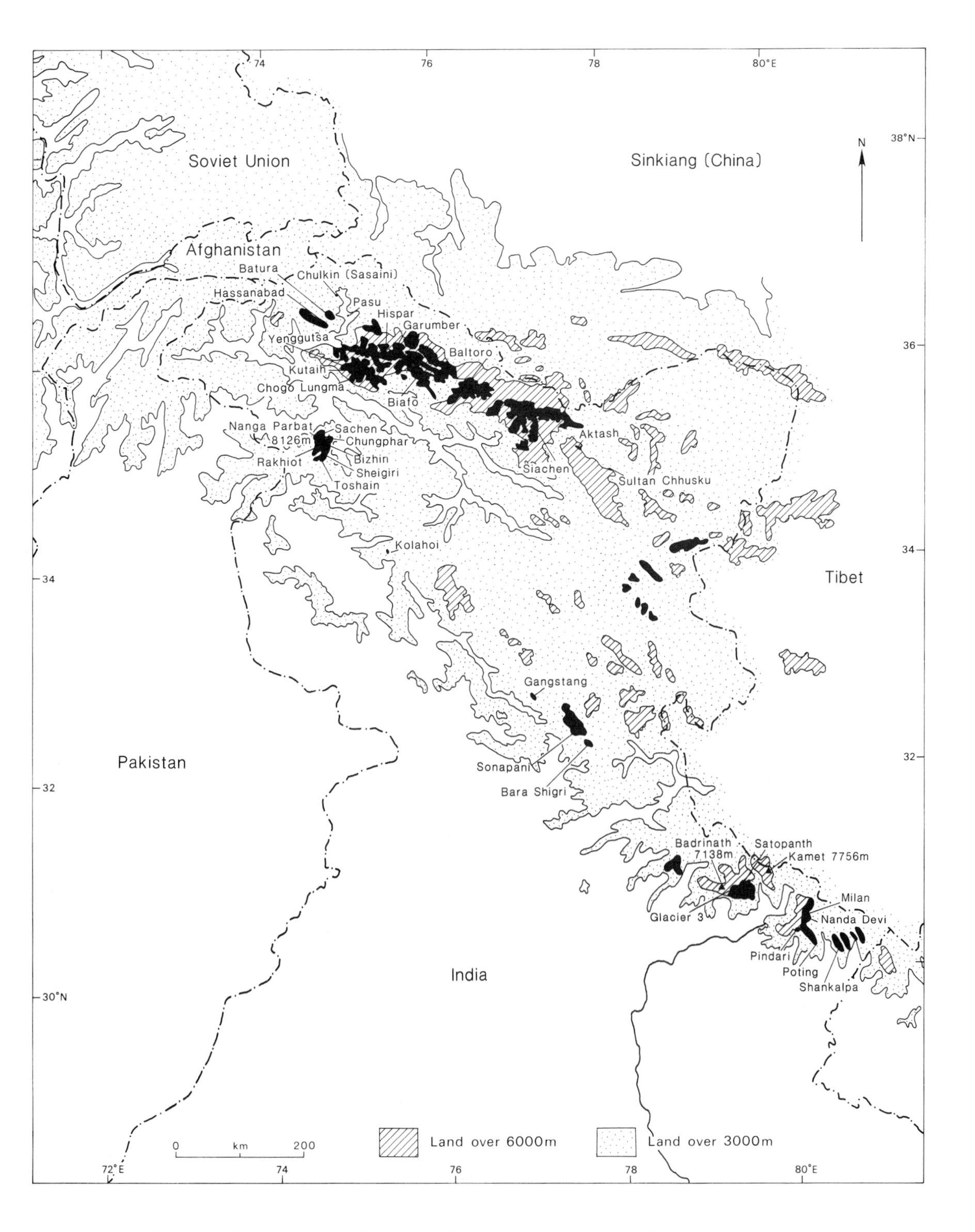

Figure 7.4 The western Himalaya with Nanga Parbat and Karakoram

Plate 7.1 The Himalaya between Mt Everest and Sikkim (Paiku Hu, 29.0°N, 86.5°E), 6 October 1984. Kathmandu is in the bottom left and La-Sa in the top right of this hand-held photograph taken from the Shuttle, which covers an area of about 75,000 km^2. Note the contrast between the well-watered zone south of the range and the aridity of the Tibetan Plateau. (NASA Astronaut Linhof)

This quotation is not merely a consolation to glaciologists; the glacier source of the Ganges evidently plays an important part in Hindu mysticism. Thousands of pilgrims visit the temples near the sources of the Ganges every year, although the temple of Gangotri is nowadays 30 km from the true source of the Ganges at Gomukh. One may wonder if the glacier source of the Ganges was much nearer to Gangotri when the temple was built and whether some documents or traditions may yet throw light on the subject. Certainly many glaciers have tracks or trade routes running along their lateral moraines and even crossing the ice, but no written records refer to them. Access by long approach roads is time-consuming and commonly dangerous on account of landslips and avalanches; political boundaries present problems; where maps exist, and they are few, they are often inaccurate. Monitoring of glaciers is consequently expensive and has to be confined to a short snow-free season between June and September.

Vivid descriptions of some of the Himalayan glaciers are to be found in the published and unpublished papers of explorers and climbers who gradually unravelled the topographical relationships of the ranges. Much of this work was accomplished by officers of the Survey of India such as Godwin-Austin, but the remarkable contributions of travellers such as Sir Martin Conway and the redoubtable wife-and-husband team of Fanny Bullock Workman and William Hunter Workman with their many expeditions around the turn of the century deserve special mention. The single-handed achievements of Aurel Stein and Kingdon Ward were on an heroic scale. A strong thread of Alpine experience ran through the early approaches to the glaciers. The Workmans were accompanied on their 1908 Hispar expedition by Calciati and Koneza, surveyors recruited through the good offices of Charles Rabot (Chapter 4). The three Schlagintweit brothers of Munich, Rudolf, Herman and Adolf, despite their youth already well known for their glacial work in the Ötztal (Chapter 5), were engaged by the East India Company in the 1850s, on the recommendation of Alexander von Humboldt, to carry out a geomagnetic survey of India. Sadly, Adolf, the one most concerned with glacial matters, was murdered at Kashgar in August 1857. His brothers produced four volumes recording the results of the expedition's work out of the twenty that had been projected. They recovered Adolf's journals and 'a considerable number of his drawings and collections' (Schlagintweit and Schlagintweit 1860–6, vol. 4, p. 565) but failed to publish them. Eventually they found their way into the Bavarian State Library where they rested in oblivion until R. Finsterwalder (1937) recognized their potential value, especially that of the paintings with the detail of the glaciers recorded by the hand of an experienced glaciologist. Kick (1960, 1967, 1969) has since made use of the written material but it seems likely that the journals have not been fully exploited.

At the beginning of the twentieth century the British Alpine Club, recognizing their importance as indicators of climatic variations, initiated the collection of records relating to glacial oscillations in all parts of the world (Freshfield 1902). After the founding of the Commission International des Glaciers, Freshfield, as the British member of the Commission, drew the attention of the Trigonometrical Survey of India to the importance of collecting data on the secular movements of the principal Himalayan glaciers (Holland 1907). There were difficulties arising from the distance of the glaciers from permanent stations as well as from 'the extreme complexity of the departmental system in Calcutta' (Freshfield 1902), and eventually the Geological Survey of India undertook responsibility for the work (Holland 1907). The first twelve glaciers to be selected were visited in 1906; sketchmaps were made, observation stations set up and photographs taken (Hayden

1907, Walker and Pascoe 1907, Cotter and Coggin-Brown 1907). Detailed accounts of glaciers and groups of glaciers appeared from time to time from 1907 onwards in the Records of the Geological Survey, some officers spending their short leaves extending the observational network (e.g. Auden 1935) and well-known travellers such as Smythe (1932), Shipton (1935) and Ward (1934) brought back information about glaciers not previously described. After a period of reduced activity at the time of partition, a new stimulus was given by the International Geophysical Year of 1957 (Tewari 1971). Lynam (1960) planned his 1958 expedition to the Kulu-Lahul-Spiti watershed partly because the Survey of India mapsheet 524, depending heavily on nineteenth-century surveys, was known to be accurate as far as the main valleys were concerned but to have side valleys like that of the Bara Shigri sketched in with 'more imagination than accuracy'.

7.3.1 KENCHENJUNGA-EVEREST

When he visited the Kenchenjunga area in 1899, Freshfield disparagingly referred to the glaciers as they were delineated on the official maps as no more than 'a few worms crawling about the heads of valleys'. Today the maps of the Mount Everest area are markedly better than those elsewhere and Royal Geographical Society maps of 1961 and 1975 on a scale of 100,000 were available to Müller (1970 and 1980), together with a large number of good-quality photographs taken in connection with climbing expeditions, when he compiled an inventory of 1936 glaciers in the region. The largest glaciers are on the southern side of the main mountain ranges where precipitation is higher than to the north, three-quarters of it being derived from the southwest monsoon which blows from June to mid-October. Evidence is accumulating that on the peaks and ridges of the Nepalese, eastern Himalaya, precipitation reaches a maximum at altitudes above 5000 to 7000 m, especially on south-facing slopes, with the floors of the main valleys becoming increasingly arid towards the north.

The proportion of the glacier surfaces within the accumulation zone (the accumulation area ratio or AAR) is on average about 0.41, a remarkably low value considering that for equilibrium a value of between 0.60 and 0.75 is usually required. The situation may find an explanation in the extraordinarily high relief which promotes very large-scale avalanching, adding to accumulation at the higher levels, and also the production of large masses of rock debris that mantle many glacier tongues and protect them from ablation.

The Mount Everest region is one of the few areas in the Himalaya where moraine sequences have been examined in sufficient detail to give any useful information about glacier front fluctuations in recent centuries (Figure 7.5). In the Imja Khola basin of the upper Khumbu, Müller (1980) distinguished a sequence of fresh-looking moraines which he thought had been left quite recently, lying within older and much larger moraines, lichen-covered and partly vegetated, which form ramparts bulging out of side valleys to partition the main valleys. These older features, called Dughla by Müller and Thukhla by Fushimi (1978), were taken to belong to the Little Ice Age on a basis of lichen size and a radiocarbon date of 480 ± 80 (B-174). Several kilometres downvalley there are still larger and older moraines (see Chapter 10). The topographical relationship of the Dughla to these Pheride moraines is comparable with that of Little Ice Age moraines to those of the Egeson and Daun stages of the Late Glacial in the Alps.

Mayewski and Jeschke (1979) found specific data on the terminal positions of only five glaciers amongst the many hundreds in the region. Of these only the Zemu and

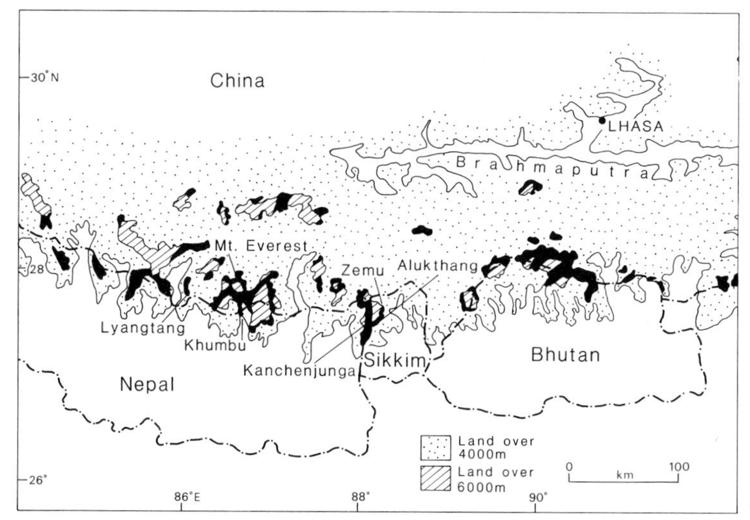

Figure 7.5 The Everest area

Alakthang had as many as three measured points between 1860 and 1968. The Zemu, the largest glacier in the eastern Himalaya, originates on the precipitous eastern slopes of Kenchenjunga (8536 m) and flows north and then east. Moraines on the north side of the tongue, extending down the valley for at least 600 m below the terminus, are covered with trees which provide opportunities, as yet unexploited, to obtain a minimum age for the moraines.

The Zemu was first visited in 1891 by the political officer in Sikkim, Claude White, and a photographer from Calcutta. They found the snout of the glacier, ending in a cliff about 150 m high, was at a height of 4200 m. Eight years later, Freshfield (1903, p. 205) spent two days making a plane-table map (which does not seem to have survived) of the lower part of the glacier, 'a huge grey, billowy stream' flowing towards him, the moraines not arranged in lines but all over the place, 'giving some excuse to the official cartographer who refused to recognise the glacier below'.

> A little plain hard by was covered with the traces of the recent passage of yaks. We had touched the 'high level' route connecting Sikkim through the Talung valley with Tibet. . . . This particular route serves chiefly for the transport of salt and timber; salt out of Tibet, timber into it. It crosses the Zemu glacier above its snout, much as the route of the Gries Pass in the Alps crosses the glacier of the same name. Ascending steeply we followed the faint tracks, soon lost, which led into the glen or hollow on the south side of the Zemu glacier and then, bearing to our right, climbed several grass-grown moraines until we found ourselves close to the edge of the retreating ice.

The Geological survey of India, recognizing the need to record the changing positions of the glacier termini and urged by Freshfield, sent La Touche (1910) to map the position of the Zemu snout. He did so in 1909, finding little evidence of change over the preceding decade. The Zemu was scarcely visited again until 1965 (Raina *et al.* 1973) when it was reported to reach down to 4260 m, having retreated 440 m in the course of the century. It ended in an ice cliff reduced to 65 m in height, with moraine covering most of the tongue.

The Alekthang glacier in the same region, according to a Major Sherwill, had by 1909 retreated half a mile since 1861. Mayewski and Jeschke (1979) present a diagram showing the glaciers of the Kenchenjunga–Everest area retreating in the latter half of the nineteenth century and then retreating more slowly or remaining stationary in the twentieth century, but the observational base is not very secure. Both the Zemu (Bose *et al.* 1971) and the Jungpu (Meyewski and Jeschke 1979) have thinned recently, as had the Khumbu (Miller 1970). Higuchi *et al.* (1980) reported that during 1960–75, 90 per cent of the glaciers free of debris had retreated, but it should be noted that this estimate is based on the change in altitude of termini since Müller's (1970) report. The Zemu has been mentioned as one of the glaciers, and there are many others, where ablation is limited by debris covering the tongue.

For the region to the east of Sikkim we have very little information, though Kingdon Ward (1934) makes some mention of the Ka-Gur-Pa glacier on the Mekong-Salween divide.

> Stretching down the glacier valley for half a mile beyond the snout is a lateral moraine, its summit and far side covered with trees, while the flank facing the glacier is almost bare below or clothed with plants trying to establish themselves. The summit of the moraine is about 350 feet [100 m] above the glacier and shows a sort of step structure as though there had been periodic fluctuations in the retreat of the ice.

He concluded that the retreat had been due to a change in precipitation.

7.3.2 GARWHAL OR KUMAON

The Garwhal area, some 700 km west of Everest–Kenchenjunga, contains some of the largest glaciers in the Himalaya, descending in some cases to as low as 3600 m. The Ganges rises in the complex massif topped by Kedarnath (6940 m), Badrinath (7138 m) and Kamet (7756 m) which is ringed by the great Hindu shrines of Badrinath, Kedarnath and Gangotri associated with the main sources of the river. To the south lies the great glacier-girt group of Trisuli (7120 m), Nanda Devi (7816 m) and Nanda Kot (6861 m). Though the main peaks were surveyed in the 1870s, the surveyors were ordered not to waste time on the uninhabited areas. Exploration of the Badrinath range between Alaknanda and Gangotri remained to be one of the objects of Smythe's 1932 expedition and Shipton (1935) entered the sanctuary of Nanda Devi in 1934.

Thousands of pilgrims visit the shrines each year and some see the ice from which the Ganges issues, but so far as is known they have left no records. Shipton (1935) recorded a tradition that 'many hundreds of years ago' when there was no high priest at Kedarnath temple, the high priest of Badrinath used to hold services at both temples in the same day, a story reminiscent of the Alpine tradition of the priests who crossed Mont Blanc from Val Veni in six hours to say mass in Chamonix. Local people believed the direct route from Badrinath to Kedarnath to be 4 km; Shipton took two days to cover the distance, which he estimated to be 40 km.

Meyewski and Jeschke (1979) plotted the frontal fluctuations of seven glaciers in Garwhal and found that over the preceding century all except No. 3 in the Arwa valley had retreated (Figure 7.6). They were able to establish that six more glaciers had retreated between 1910 and 1920 and that two others had done so in the mid-1960s. Their assertion that all of the glaciers had been in retreat since 1850 remains unsubstantiated.

The Milam glacier, flowing 16 km southeast towards Milam village and nourished in large part by tributaries from the eastern slopes of the Nanda Devi–Trisuli ridge, is second in scale only to the Gangotri amongst the Kumeon glaciers. According to local tradition recorded by Cotter and Coggin-Brown (1907), the ice was in a forward position and reached the site of the village 'a thousand years ago'. In 1906 the terminus was about a mile from the village but 'according to Rai Kishen Singh, Behader of Milam, known to science as "AK"[1] the explorer of Tibet, the ice cave 52 years ago was about 800 yards in advance of its present position'.

Sir Richard Strachey spent some time in the Milam valley in 1848 and measured the speed of flow of the ice at several points before crossing the glacier into western Tibet. He followed the well-established trade route used for the exchange of salt and borax of Tibet for the grain of India. Strachey (1900) left an excellent account of his observations, finding the people of Milam intelligent and 'decently educated', even having some knowledge of Hindu literature, and was therefore disposed to accept their statement that the ice formerly extended to a *rask* or fortified wall 'which is now several hundred yards below the terminus'.

The Milam, together with the Skunkalpa and the Poting, were amongst the twelve glaciers originally selected for monitoring by the Geological Survey of India. Accordingly Cotter and Coggin-Brown made plane-table sketches of the positions of their tongues and the points from which they had taken photographs. They noted that the existence of a whole complex of lateral moraines indicated that retreat could not have been continuous.

The nineteenth-century retreat of the Milam is recorded in more detail than that of other glaciers in Kumaon, though there is also some information to be gleaned from the retreat of the Pindari. This is a 5 km-long glacier, fed principally from the slopes of the Nandakhat–Nandakot ridge. Strachey visited it in 1847 and found 'the distance between the ice caves and the median moraine is about 2 miles'. A Colonel J. W. A. Mitchell wrote in 'a book kept in the Phurkia dâk bungalow for the purpose of recording observations of the movement of the glacier' that it 'appears to have retreated about 100 yards since I visited it in 1884'. It is not known whether this book still exists. Other material of this sort may well survive in private papers. Cotter and Coggin-Brown (1907) noted that 'the moraine terminates about a mile from the ice cave'. So the retreat here was about a mile (1.6 km) from 1847 to 1906.

During the twentieth century, observations in Garwhal have been made more frequently. Between 1906 and 1957 the Milam receded 617 m (Jangpani and Vohra 1962) and a further 23 m between 1963 and 1964 (Kumar *et al.* 1975). Observations of the Skunkalpa give no indication of advance (Cotter and Coggin-Brown 1907, Jangpani and Vohra 1962, Kumar *et al.* 1975). Auden (1935), spending his leave from the Geological Survey of India making the first map of the tongue of the Gangotri glacier in 1935, was impressed by evidence of its wasting and retreat over the preceding decades, but when Ross saw it the following year he could find no sign of any change in its position. A retreat of 600 m took place between 1936 and 1967 by which time some of the tributaries had

[1] 'AK' or Ria Krishen Singh Rawat was the greatest of the remarkable Indian explorers known as the Pundits, who worked for and were trained by the Survey of India. In the mid-nineteenth century, Europeans and other foreigners were not allowed to enter Tibet, but exceptions were made for certain groups of semi-Tibetans, amongst them the Bhotias living in the Johar valley at the head of which Milam lies. Accordingly, certain members of the Rowat clan were selected for intensive training at the Survey headquarters at Dehra Dun. They were not only instructed how to find directions by the compass, latitude by the sextant, and height by taking the temperature of boiling water, but also how to measure distances by counting paces – 2000 paces to the mile, with 31.5 inches to the pace. AK was thus a man who might be expected to give a reliable estimate of distance, though it has to be admitted that he must have been relying on memory in this particular case.

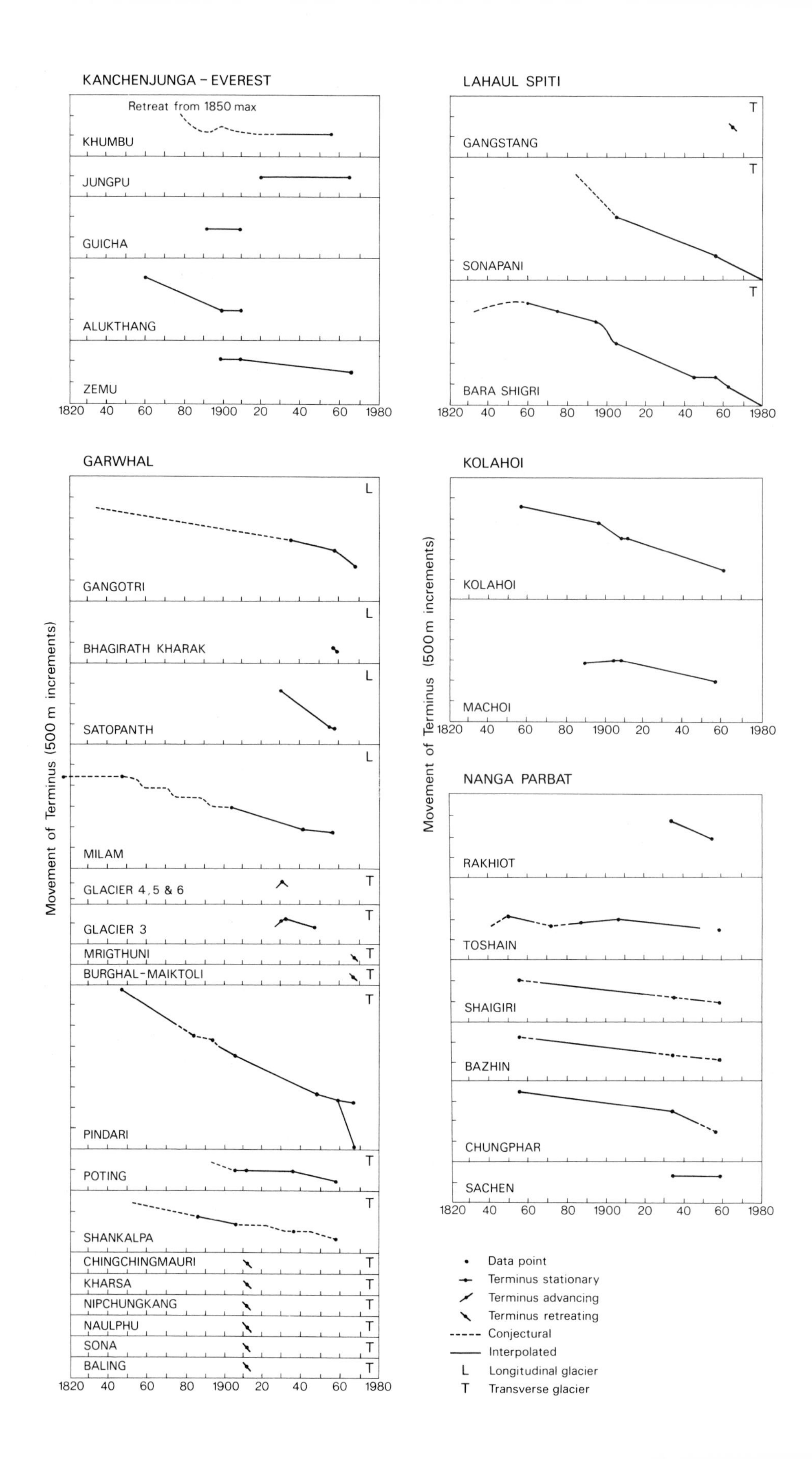

KANCHENJUNGA - EVEREST
Retreat from 1850 max
KHUMBU
JUNGPU
GUICHA
ALUKTHANG
ZEMU
1820 40 60 80 1900 20 40 60 1980
LAHAUL SPITI
GANGSTANG
SONAPANI
BARA SHIGRI
GARWHAL
GANGOTRI
BHAGIRATH KHARAK
SATOPANTH
MILAM
GLACIER 4, 5 & 6
GLACIER 3
MRIGTHUNI
BURGHAL-MAIKTOLI
PINDARI
POTING
SHANKALPA
CHINGCHINGMAURI
KHARSA
NIPCHUNGKANG
NAULPHU
SONA
BALING
Movement of Terminus (500 m increments)
KOLAHOI
MACHOI
NANGA PARBAT
RAKHIOT
TOSHAIN
SHAIGIRI
BAZHIN
CHUNGPHAR
SACHEN
L
T
Data point
Terminus stationary
Terminus advancing
Terminus retreating
Conjectural
Interpolated
L Longitudinal glacier
T Transverse glacier

retreated more than the main glacier and had separated from it (Tewari 1971). No records of even slight advances in the nineteenth and twentieth centuries have been forthcoming from any of the Garwhal glaciers that have been measured, though there are recessional moraines, so far undated, indicating that there must have been at least short halts during this period. However, Gilbert and Auden (932–3) recorded that

> Himalayan glaciers are for the most part retreating, but some are advancing in northern Garwhal. One glacier debouching into the Arwa valley has advanced so far that the valley is in danger of being choked by it. Were it to advance two hundred or three hundred yards further the valley stream will be dammed and as the valley is flat for several miles, a large lake will be formed. The bursting of the dam will be disastrous for Badrinath and other villages in the Alaknandra valley and might even result in serious floods far down in the plains.

This advance attracted the attention of the Public Works Department because of the possible danger to Badrinath and to bridges and buildings on the heavily used pilgrim route (Gunn 1930). In the event, the glacier advanced no further and a retreat of 199 m took place between 1932 and 1956 (Tewari 1971). This is the only one of the small glaciers in the Arwa valley known to have advanced in the 1930s, though some thought that the Satopanth was also advancing.

7.3.3 LAHAUL-SPITI

The transition zone between monsoon and arid climates crosses the Lahaul–Spiti region (Lynam 1960). Egerton (1864) found 'a good deal of rain and heavy clouds' and the air was damp and heavy. In Spiti the weather continued cloudy and drizzly, but the clouds were less constant and higher, while the air was delightfully pure and comparatively dry. Accordingly ice is much more extensive in the southern part of the Lahaul–Spiti region than in the northern part of the Sutlej basin. However, we have information about the fluctuations of only three glaciers, the Sonapani (Kurien and Munshi 1962), the Gangstang and the Bara Shigri, literally 'the debris-covered glacier'.

The records of the Bara Shigri are remarkably long and detailed. Egerton (1864), crossing the glacier in July, found the ice extending far across the valley of the Chandra.

> The path lay across the Chota Shigri, a vast moraine left by some former glacier which, from some cause or another (perhaps the increase in temperature asserted by some Himalayan travellers) has melted away. . . . Beyond the Chota Shigri is a comparatively open piece of ground from which we ascend slightly onto what appears another moraine, like the Chota Shigri, but of much greater extent, being from its source to the Chundra river about four miles long, and in breadth some two miles.
>
> This is, in fact, the Shigri or Great Glacier, as you soon find out from walking on it. . . . Little streamlets are everywhere trickling from the surface, exciting your wonder how there should be water on top of this mass of apparently porous rubbish, till suddenly opens before you a rent in the mass, the walls of which are clear green ice,

Figure 7.6 Advances and retreats of Himalayan glaciers since 1820. (After Mayewski and Jeschke 1979, omitting glaciers known to surge, or for which information is minimal, and with the curve for the Bara Shigri altered)

> and you see that you are travelling over an enormous glacier, coated with dirt and gravel, and sprinkled with huge rocks and boulders. . . . If you follow the glacier down to the river (at no slight risk of breaking your neck) you see a large torrent issuing from beneath the glacier itself to join the Chundra river and eating away the glacier till it forms a promontory jutting out between the torrent and the river, with the perpendicular wall some 150 feet high. . . . It seems impossible to do much to improve the road over this treacherous element. You cannot turn its flank, for below is the river and above is a chaos of crevasses.

According to Mayewski and Jeschke (1979) the Bara Shigri advanced across the Chandra valley about 1860 and dammed back a lake. This lake was mentioned by Egerton as being 'a mile or two northwards from a bend in the Chandra river'. Calvert (1873, cited by Hughes 1982) had to cross the Bara Shigri as did Colonel Tyacke (1893, cited by Hughes 1982). From Puti Runi to the next camping ground, Korcha, is nearly 10 miles. During this march the Shigri glacier is crossed. Calvert made no mention of a lake, though he went into some detail about crossing the glacier on his way to Spiti. Tyacke (1893) was more specific, writing that 'The Shigri is remarkable for the fact that some five and forty years since, it burst from the mountains above, and bringing down millions of tons of debris with it, completely dammed up the Chandra River, which remained so dammed for some months before it burst through the barrier'. The effects were quite clear when Egerton walked through in 1863

> From the Shigri, for four miles up the river, you see unmistakable signs of the river having been dammed up by the glacier. Throughout this distance the bed widens out to some 1500 to 2500 feet. The bottom is quite flat and is covered with a deep deposit of sand and gravel; and, strange to say, for these few miles there is no rock or boulder to be seen, though below the glacier the river is full of them. The natives say that the river was pent up for eleven months.

(These features were mapped by Pascoe 1926.) It seems possible that the tongue advanced to dam the river more than once. According to Egerton (1864),

> It is known that twenty-seven years ago this glacier first burst out of the mountains above and . . . formed a huge dam, extending right across the river Chandra for many months. . . . I believe no large river in the region has been seriously invaded by a glacier in the last twenty-five years, and according to the Schlagintweit brothers, snow is decreasing and the occurrence of such an event becomes yearly less probable.

There seems then to have been a glacial maximum around 1848 and possibly also in the 1830s, but by 1864 the glacier had stated to recede. Egerton's forecast of its future behaviour was correct. The position of the Bara Shigri has been recorded on several occasions; in 1906 by Walker and Pascoe (1907), in 1945 by Krenck and Bhawan (1945), in 1956 by Dutt (1961) and in 1963 by Skrikantia and Padhu (1963). Mean annual retreat rates have been estimated to have been 60 m from 1890 to 1906, 20 m from 1906 to 1945, up to 28 m from 1955 to 1963, and 6 m from 1963 to 1980, giving a total retreat of 2.6 km over a period of ninety years (Mackley and McIntyre 1980). These values cannot be regarded as precise because the ice margin is very difficult to identify at all accurately on account of the huge amounts of debris spreading from the glacier surface to moraines bordering it.

Adolf Schlagintweit sketched a glacier on the left side of the Chandra valley which he called the 'Bhoru Nag' in June 1856, and in the same month a glacier north of Shinko La (5097 m) in the Num Kum group. (He generally adapted place names used by the local people.) The sketches have never been compared with present-day photographs (private communication, W. Kick, 19 August 1984).

7.3.4 KOLOHOI

Mount Kolohoi (5024 m), on the northeast side of the Vale of Kaşhmir, was climbed and mapped by Neve and Mason in 1912. Neve (1907) had first visited the area in 1887 and in 1912 he opined that the glacier tongue had retreated about a quarter of a mile over the preceding twenty-five years and more than a mile since the Topographical Survey first mapped the area in 1857. Odell (1963) reported that when he visited Kolohoi in 1961 the ice had receded another half mile. It was on this basis that Mayewski and Jeschke (1979) indicate a recession of 800 m from 1857 to 1912 and 800 m more from 1912 to 1961. The Machoi glacier, also on Koloahoi, after advancing a little around 1900 (Odell 1963), retreated 457 m between 1906 and 1957.

7.3.5 NANGA PARBAT

The Nanga Parbat massif (Figure 7.4) carries sixty-nine glaciers with a total area of 302 km^2 (Kick 1980). The northern side is the most heavily glacierized, the reasons being the greater extent of the accumulation areas overlooking the Indus headwaters and exposure to winds bringing precipitation from the northwest in winter. Above 4500 m, annual snowfall is estimated to exceed 8000 mm as compared with less than 120 mm below 2500 m. Because of the extreme steepness of the southern slopes the tongues of some of the fifteen glaciers, all heavily encumbered with morainic debris, reach down to 2800 m in grazing areas close to villages such as Tashing and Rupal. Adolf Schlagintweit visited the area in September 1856 and Finsterwalder (1937), comparing one of his paintings with a map made in 1934, showed that the ice cover had diminished over the interval. Schlagintweit's notes and sketches show that the Tashing or Chungpar glacier advanced to reach a maximum in 1850, damming back a lake, the water of which had escaped before 1872 when Drew (1875) reported that the ice, though thinner, still blocked the valley. The Chungpar retreated by about half a kilometre between 1856 and the mid-1930s.

The Sachen glacier was painted by Adolf Schlagintweit in 1856 (see Plate 7.2). He described the lateral moraine as standing 18–28 m high above the adjacent lake. In 1958, Kick found the same moraine towering 180 m above the lake and judged that the lateral moraine must have been built up greatly after 1856, probably at the time that the glacier reached its maximum extension which was, according to the Survey of India triangulation records, around 1900. Since then the tongue has scarcely changed its position though its surface had dropped about 10 m by 1958 (Plate 7.3).

Knowledge of more recent conditions on Nanga Parbat is based mainly on German photogrammetric studies using ground photographs. Rakhiot, in the north, surveyed for the first time by Finsterwalder in 1934, was resurveyed by Pillewizer in 1954, while six other glaciers in the east and south, also originally surveyed by Finsterwalder in 1934, were resurveyed in 1958 by Loewe (1961) and Kick. The Sachen glacier was almost

Plate 7.2 The Sachen glacier, Nanga Parbat with Sango Sarr Lake, painted by Adolf Schlagintweit in 1856. (Photo: Wilhelm Kick)

stationary between 1934 and 1958, but most of the other tongues receded. The Rakhiot retreated 450 m but the surface was lowered along only 10 per cent of its length. The retreat of the Chungpar, Bizhim, Sheigiri, Toshain and three unnamed tongues was less than 20 m per year. Three other unnamed glaciers were stationary. No glaciers were found to have advanced between 1934 and 1958. Pillewizer (1958) deduced that increased supplies of névé from the north side of Nanga Parbat had already replaced the post-1934 loss in the upper and middle parts of the glaciers, indicating that the tongue was set to advance. A survey made in 1985 revealed that the Rakhiot had advanced about 200 m since 1954 (Gardner 1986).

7.3.6 KARAKORAM

Separated from the rest of the Himalaya by the Indus valley, the Karakoram, including the massifs of Rakaposhi, Haramoshi and Batura Mustagh, give rise to several of the largest glaciers outside the polar regions, some over 50 km long, including the Hispar, Biafo, Baltoro and, largest of all, the Siachen. They are fed by a multitude of transverse glaciers flowing in directions at right angles to the grain of the country down to the main longitudinal valleys. Estimates of the ice cover range from 28 to 37 per cent (Goudie *et al.* 1984) and of area from 13,500 to 15,000 km^2. The snowline is at about 5100 m in the south, 5600 m in the main Karakoram ranges, and 4700 to 5300 m in the north (Shi Ya-Feng *et al.* 1980).

Plate 7.3 The Sachen glacier and its moraines with Sango Sarr in 1958. Compare Plate 7.2. (Photo: Wilhelm Kick)

The diurnal, seasonal and altitudinal contrasts in climate are even more marked here than elsewhere in the Himalayan region. At the higher elevations precipitation may exceed 1000 mm while below 4000 m the glaciers extend down onto semi-arid valley floors with mean annual precipitation at Skardu (2288 m) only 160 mm and at Leh (3514 m), 83 mm.

Mayewski and Jeschke (1979) were able to find records of the fluctuations of seventy-four glaciers in the Karakoram, though for nineteen of them there was only a record of the position of the terminus at a single time and an indication as to whether it was advancing or retreating. Further information was added in the course of the Royal Geographical Society International Karakoram Project, 1980, when the snout position of seven glaciers in the Hunza valley were surveyed and their positions related to those found previously (Goudie *et al.* 1984).

The glaciers of the Karakoram are unusually susceptible to surging (Figure 7.7). A good deal of attention was attracted to this characteristic by the flood hazard created by the Chong Kumdum glacier damming the Shyok valley in 1928 (Ludlow 1929–30). It was known that great floods had been caused by the collapse of ice dams in the past and Mason (1930) brought together details of Indus floods since 1780 and also took up the question of which glaciers were prone to repeated, rapid advance. He was inclined to assume that all the ice-dammed lakes were the outcome of such advances. Hewitt (1969) questioned whether this was always the case. Some of the glaciers mentioned by Mason, such as the Aktash which advanced 2.5 km in five weeks in 1935–6 and the Hassanabad

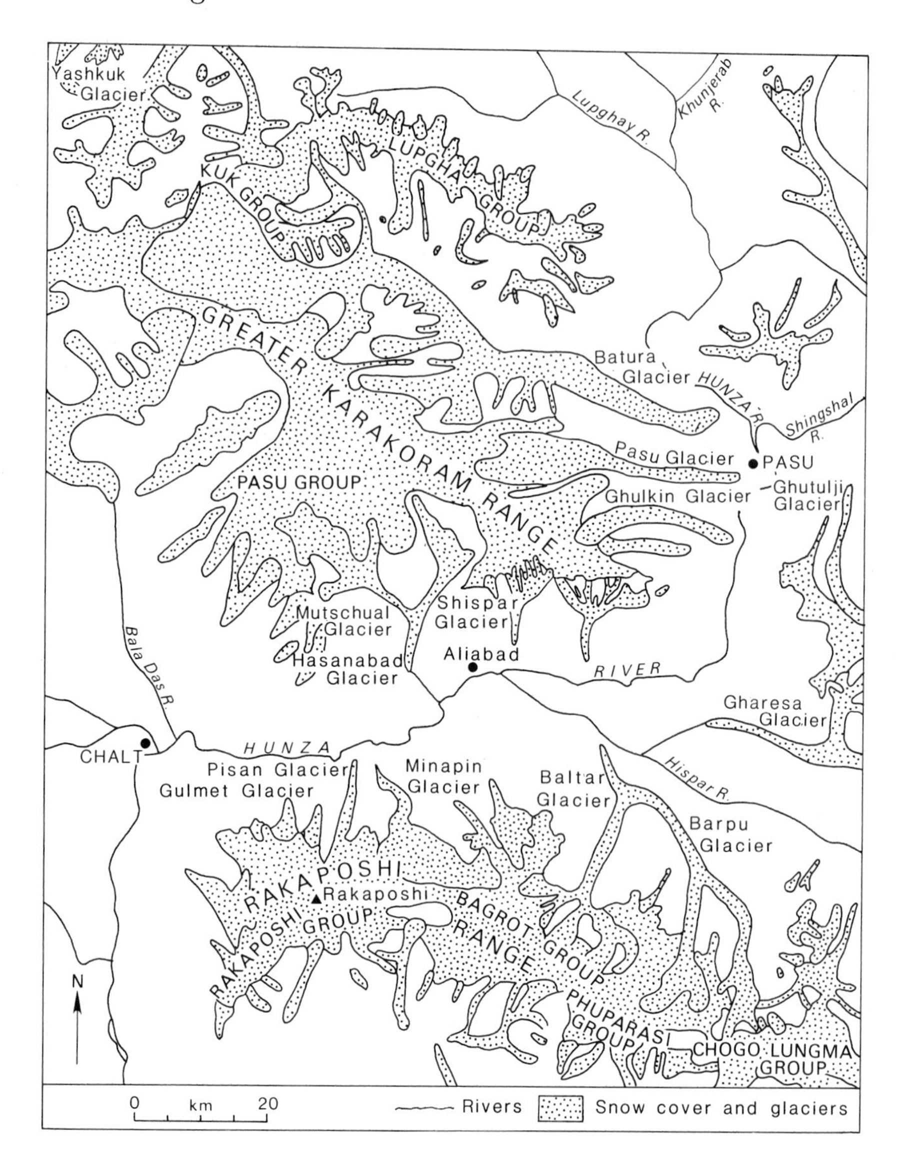

Figure 7.7 The Karakoram glaciers. (After Goudie *et al.* 1984)

which advanced 9.7 km in ten weeks in 1890 (according to Hayden 1907), certainly do surge. So do the Garumbar, Yengutsa, Kutiah and Sultan Chhusku which are amongst those listed by Mayewski and Jeschke. None of these is of much value for building up a picture of climatically controlled glacier fluctuations. On the other hand, some of the flooding could have been caused by glaciers advancing in response to climatic fluctuations. Furthermore, it is significant that over a period of over four decades in the middle of this century no major glacier dams are known to have formed (Hewitt 1982). Then, in 1978, a lake with an area of 6 km was held back by ice in the upper Yarkand.

The fluctuations of the Chogo Lungma, ranking amongst the largest icestreams in central Asia, are known in more detail than is usually the case because of its nearness to the village of Arandu. Citing the Shogar chronicle, which, as he mentioned, has not yet received the critical attention of historians, Kick (1962) argued that the village had been continuously occupied since the twelfth or thirteenth century. A watchtower that survived into the nineteenth century is believed to have dated back to the sixteenth century. According to local inhabitants the glacier was 2 km from the village at the beginning of the nineteenth century. Vigne wrote (1842) that it was advancing in 1835 and in 1861 Godwin-Austen (1864) found the terminus only 400 m from Arendu with the tongue abutting 'against the mountainside the whole way down'. Over the next two years the ice retreated, but when the Workmans (1905) and Oestreich went up the valley in 1902 the terminus was only 315 m from the village (Figure 7.6). However, the ice was thinning and the glacier had melted back leaving a natural routeway for 30 km between the valley side and a massive lateral moraine. It should be noted that the diagram given by Mayewski and Jeschke (1979) does not accord with Kick's account. It seems likely that the glacier had attained its greatest volume between 1865 and 1900 but that the lower part of the tongue was protected by debris so that the retreat of the terminus was much delayed. Between 1900 and 1913 the front advanced 200 m to bring the ice only 110 m from the village; the local people were alarmed, but it advanced no further. By 1954 the terminus had retreated about 100 m, the ice was 30–60 m below the crest of the lateral moraine, and its velocity was about half what it had been in 1903. The front retreated 200 m between 1954 and 1970 and a further 70 m between 1970 and 1979 (Best *et al.* 1981).

A similar record of nineteenth-century advance comes from the 58-km long Biafo glacier of the central Karakoram. It enters the Braldu valley at right angles and terminates opposite a prominent rock buttress. Godwin-Austen found the ice wedged against the rock buttress in 1861, completely blocking the Braldu valley. In early 1892 it was a quarter of a mile back and Conway (1894) reported that it retreated a further quarter of a mile in August of that year. As the frontal moraine it left behind was vegetated, wasting must have started several years earlier. Seven years later the Workmans (1908) found the tongue hardly reaching into the Braldu valley. It advanced again by 1902 according to Pfannl, pushing the Braldu river across into a narrow bed, but by 1905 the snout was low and fissured and retreat had recommenced (Mason 1930). It may be worth mentioning that Neve (1907), Longstaff (1910) and also Conway and Godwin-Austen reported that glaciers in Kashmir were advancing in the first few years of this century. However, when the Workmans returned to the Biafo in 1908 they found it in much the same relatively shrunken state as that in which they had seen it in 1899. On the other hand, de Filippi (1912, 1932), on the Duke of Abruzzi's expedition of 1909, wrote of

> the marvellous spectacle of the Biafo icestream, 300 feet high. . . . In its invasion of the Braldoh valley it has pushed the river up against the left wall of the valley . . . . It was only on our return journey when we ascended the left side of the Braldoh valley on the Skoro La road when we clearly saw the river flowing under the open sky through a narrow gap between the valley wall and the steep front of the glacier. The latter showed no trace of frontal moraine.

In 1922, Featherstone (1926) found the glacier still right up against the river, forcing it into the valley side and causing great landslides. Desio (1930) reported the front was

180 m from the Braldu in May 1929. Auden (1935) supervised the making of a map of the Biafo snout on 1 and 2 June 1933, which shows the snout double-nosed and 134 m from the buttress at the closest. He was much impressed by the marked variations in the position of the ice front from one season to another and expressed some uncertainty as to whether there had been much change from 1861 until the time of his own visit. At the same time he pointed to recent thinning, shown by the exposure of a band of bare unconsolidated debris 30 m wide and diminishing in width upglacier, that had also affected small hanging glaciers on Skoro La (5070 m) which were 150 m higher in July 1933 than they had been in July 1909. He noted that although the Biafo was no shorter than in 1861 it was narrower and thinner than it had been in 1909 and 'many of the lateral and hanging glaciers in Baltistan appear to be in retreat'. Hewitt (1967) observed the Biafo front continuously from September 1961 to May 1962, when it stood in the middle of the Braldu valley, and recorded fluctuations during that winter of up to 20 m, scarcely enough to invalidate observations of the kind mentioned above. There seems to be enough evidence to indicate that the Biafo and its neighbours were in advanced positions in the middle decades of the nineteenth century and then retreated. A sharp advance in the first few years of this century has been followed by retreat, but over the last 120 years the changes have been quite modest.

The Batura, on the northernside of the Karakoram, flows NNW to SSE for 59 km towards the Hunza river. To the south lies the great wall of mountains topped by Batura Mustagh (7795 m). The ridge to the north has peaks ranging from 4000 to 5000 m and none over 6000 m. The total area of the Batura was 285 km^2 in 1975 and the accumulation area ratio 0.5, much of the ice and snow being supplied by avalanches. 'Nowhere else have I heard such uninterrupted avalanche thunder as in this part of the Karakoram. The incessant row provides the solution of the problem of how this long valley with its small névé fields gets its fodder', wrote Visser (1934). Furthermore, ablation is checked by moraine covering much of the tongue. The terminus was in full view of the ancient Silk Road, just as it is now from the Karakoram Highway linking Pakistan and China. Like the Biafo, its advances are constrained by the opposing valley wall. According to the Batura Glacier Investigation Group led by Shi Ya-Feng (1979), which worked there in 1974–5, about 200 years ago the Batura advanced into the Hunza valley, penetrating 2 km below the present terminus and overrunning five canals built across Holocene moraines by the people of Pasu village, so that it seems that the Batura reached a position comparable with an earlier Holocene advance and then retreated.

The Batura still reached down onto the floor of the Hunza valley from 1880 to 1930 (Mason 1930). On a map he made in 1885, Woodthorpe noted that 'the glacier reached the Hunza river which passes round the snout'. The ice had withdrawn somewhat by 1909 when Egerton saw it but reported that it still threatened to block the valley. It was described by the Vissers in 1925 as pushing its way across the river, 'blocking the whole width of the valley'. In the 1970s the inhabitants of Passu and Khailur told Chinese glaciologists (Zhang Xiangsong and Shi Ya-Feng 1980) that in the autumn of a year between 1910 and 1930 the glacier advanced to block the river, though the water could still pass under the ice, and in the spring of the following year it withdrew. Between 1944 and the visit of a German–Austrian expedition in 1954 the terminus retreated 300 m and by the time it was mapped in 1966 on a scale of 1:10,000 in connection with the construction of the Karakoram Highway it had gone back another 527 m. In 1966 the ice front began to move forward again and by 1974, when the Batura Glacier Investigation

Group (1979) made a photogrammetric survey of the terminus, the central ice cliff had advanced 90 m. It came forward another 10 m the following year and the Investigation Group forecast that the glacier would continue to advance into the 1990s.

The recent histories of the Pasu and Gulkin glaciers, flowing parallel and to the south of the Batura, were investigated by the International Karakoram Project (Goudie *et al.* 1984). Both were a few hundred metres further forward in 1913, when Mason saw them, than when they were visited by Woodthorpe in 1885. Price Wood found the Pasu had retreated 800 m between 1885 and 1907, so a marked advance must have taken place between 1907 and 1913. Both glaciers have retreated in the course of the twentieth century; the Pasu has been more or less stationary in the 1970s whereas the Gulkin is known to have advanced at least 200 m between 1974 and 1980.

The fluctuations of the positions of the tongues of the longitudinal glaciers of the Karakoram, apart from those that have surged, have been very modest in relation to their great length. The fragmentary nature of the historical record and the different years and seasons when observations have been made complicates the identification of regional trends and the detection of anomalous behaviour. It is unfortunate that it is the longitudinal glaciers and some of those which surge that have attracted the most attention, rather than the transverse glaciers that are generally more sensitive to climatic fluctuations. Despite these difficulties it is possible to discern important differences between the Himalayan glaciers and those of the Karakoram.

7.3.7 HIMALAYAN SUMMARY

Throughout the Himalaya, glacial retreat has predominated since about 1880 though, as moraine patterns indicate, this retreat has not been continuous (Figure 7.8). Current small fluctuations can be identified by careful monitoring of sensitive glaciers such as that carried out on fourteen small glaciers in eastern Nepal between 1970 and 1976 (Ikegami and Inoue 1978). Six of them were found to be retreating, three advancing and one fluctuated irregularly (Fushimi and Ohata 1980, Fushimi *et al.* 1981). The Gyajo glacier advanced in 1970, retreated, and then in 1976 advanced again, forming a push moraine that partly obliterated the 1970 moraine. Ageta *et al.* (1980) made it clear that temperature plays the decisive role in the mass balance of the small, debris-free glaciers of Nepal, for it is temperature that determines whether the monsoon precipitation falls as rain or as snow.

Ono (1984, 1985) has shown how a chronology of advances and retreats can be derived from a study of the detailed morphology and stratigraphy of glacier forefields under Himalayan conditions. In front of the Yala glacier in the Langtang Himal of Nepal he distinguished six till sheets, differentiated by their surface fluting and texture. Each till sheet is bounded by a moraine similar in form but higher than other morainic ridges running across the till sheets which, it is argued, are annual push ridges formed by advances at the beginning of the post-monsoon seasons and left behind as the ice melts back again the following year. By analogy with push moraines formed in 1976 by the Gyajo glacier, Khumbu Himal (Fushimi and Ohata 1980) and assuming that the time span required for readvance is proportional to the height of the moraines bounding the till sheets, it is estimated that the last important advanced position of the Yala was attained in 1815 and that since then minor readvances have taken place in 1843, 1867, 1887, 1903,

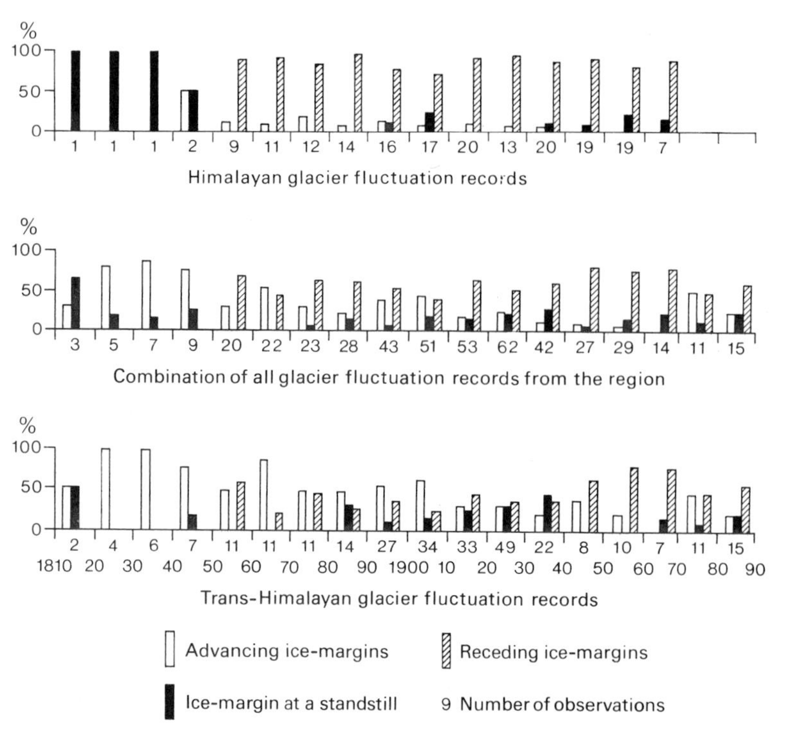

Figure 7.8 Comparison of fluctuations of Himalayan glacier termini with those of the Trans-Himalay since the beginning of the nineteenth century. Note that the majority of the observations were made between 1850 and 1960. (After Mayewski *et al.* 1980)

1921, 1953 and 1976. An earlier undated Little Ice Age extension of the Yala occupied a position similar to that reached by the glacier in 1815.

The termini of Karakoram glaciers are known to have retreated from advanced positions reached in the mid- or late nineteenth century and then to have swelled and advanced again in the 1890s and in the first decade of the twentieth century. Between 1920 and 1940, when the density of the record was greater than before or since, more glaciers were known to be stationary or advancing than were known to be retreating; retreat in the subsequent period was halted or reversed in the 1970s.

The advances of the Karakoram glaciers between 1890 and 1910 have been associated by Mayewski *et al.* (1980) with strengthened monsoonal airflow. Precipitation over lowland India was lower rather than higher than average in this period. In the Karakoram there were very few precipitation gauges but it is known that lake levels in nearby Tibet were high about the turn of the century and snowfall may have been heavy (de Terra and Hutchinson 1934).

7.4 China

The major glacierized areas of China are in the northwest of the country, with the largest areas of ice being in the Tien Shan, Kunlun Shan (Plate 7.4) and Himalaya (Figure 7.9).

The mean annual precipitation on the mountain ranges is 300–1000 mm which is in strong contrast to the intervening basins where it is no more than a centimetre or two. The snowline increases in elevation towards the south, from 2900–3400 m in the Altai, 3600–4400 m in the Tien Shan, 4400–5100 m in the Qilian Shan, 4600–6200 m in the Kunlun Shan, and 4600–6200 m in the northern Himalaya. In the southeast, the glaciers of the eastern Hengduan Shan and the eastern Qilian Shan, fed by the southeasterly Pacific monsoon, are of a maritime temperate type; those in the south, the Himalaya,

Plate 7.4 Northwestern Tibet and the Kun Lun from the Shuttle in October 1984. In this photograph of the Chopanglik region, Hotien and Min-Feng at the southern edge of the Tarim basin are in the top left-hand corner of the pleateau of Tibet with lakes Aksu Chin in the bottom left, Ya-hsi-chr-Hu at top right and Tsu-chia-Brh-Hu bottom right. (NASA Astronaut Hasselblad)

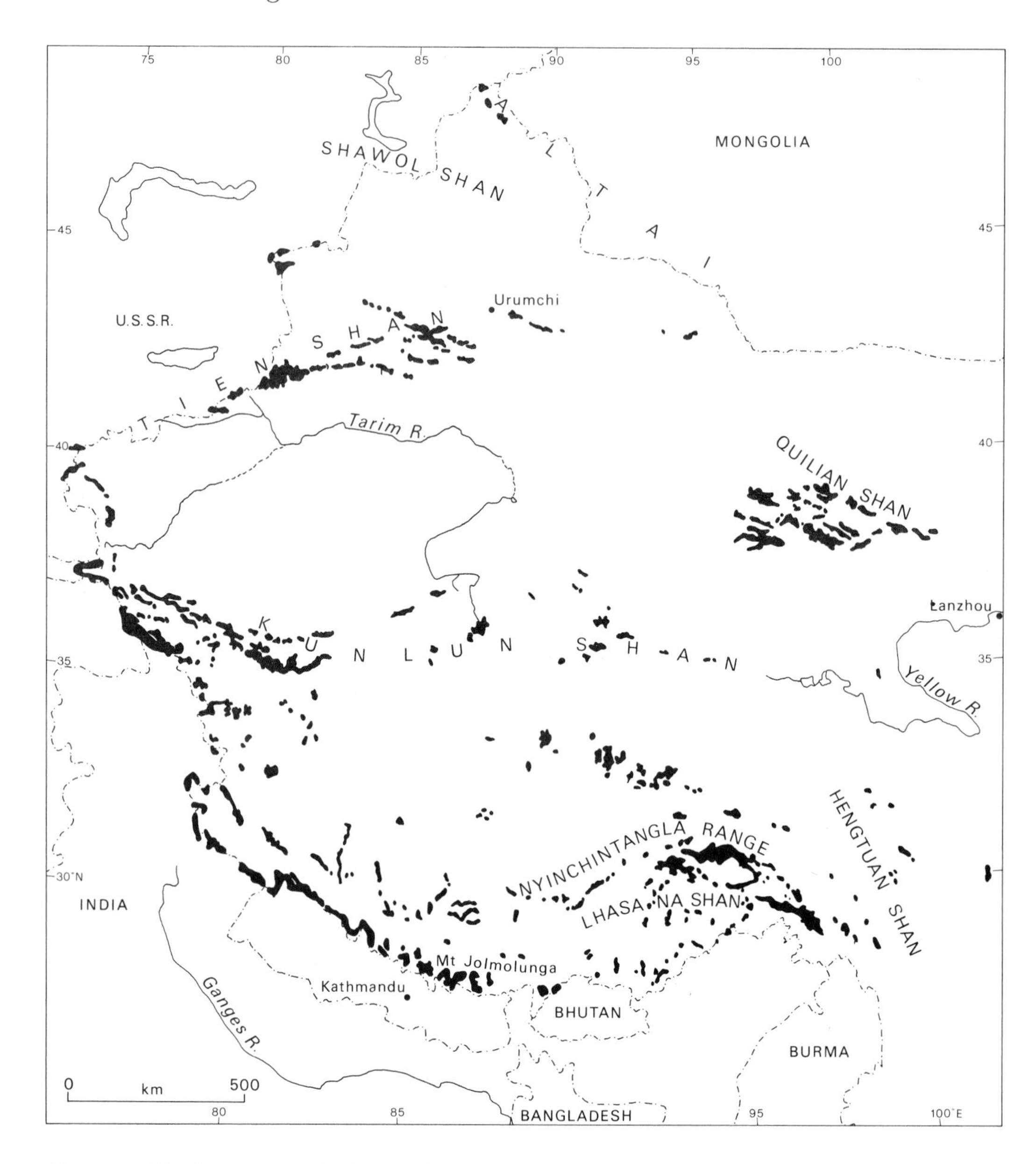

Figure 7.9 Glacierized areas in China. (After Shi Ya-Feng and Wang Jintai 1979)

Nyenchintangla and western Hengduan Shan, fed by the southwest monsoon, are cold continental as are those of the Karakoram, Kunlun Shan, Tien Shan, Altai and western Qilian Shan which are fed by snow from westerly airstreams.

Research on glaciers and climatic history in China has received increased attention since 1958 (Shi Ya-Feng *et al.* 1980, Zhang Xiangsong *et al.* 1980, 1981). A glaciological research station established in 1959 at the head of the Urumqi valley in the Tien Shan has begun to produce results in recent years (Shi Ya-Feng and Zhang Xiangsong 1984). A glacial inventory is being prepared and climatic data are being assembled (Hsieh Tze-chu and Fei Ching-shen 1980, Shi Ya-Feng *et al.* 1981).

Irrigation in the oases at the foot of the Tien Shan, Kunlun Shan and Qilian Shan is dependent on water from snowfields and glaciers. Meltwater from the Qilian Shan flows to the Kansu Corridor and is used for irrigation as well as supporting fisheries. Monitoring of glacier behaviour in these regions is considered all the more important because of falling groundwater levels in the Urumqi region and the degradation of the environment in Kansu which has led to reduction in other surface and groundwater supplies (Wang Wenying 1983).

Plate 7.5 The growth of the lichen on these rocks is being measured in order to calibrate a lichen growth curve for the Urumqi region

Temperatures in China are believed to have been lower from the fourteenth until the mid-nineteenth century than they have been in this century, with the coldest period being around AD 1700 (Shi Ya-Feng and Wang Jingtai 1979). On the Tibetan Plateau tree-ring studies indicate that cooler conditions beginning in the second half of the sixteenth century still persist (Li Chi-chun and Cheng Pen-Hsing 1980). In south Xizang (Tibet), the moraines of the Kanbukoa glacier near Yadong, as dated by lichenometry, mark advanced glacial positions about 1818, 1871 and 1885; calibration of the growth curve depends on the growth of *Rhizocarpon geographicum* over a period of less than forty years since boulders free of lichen were exposed in a mudflow in 1940 (Plate 7.5). Another set of Little Ice Age moraines in front of Glacier No. 1 at the head of the Urumqi valley and close to the glaciological research station have yet to be closely dated. The principal moraines were 600, 500 and 300 m from the tongue in 1984. At the Little Ice Age maximum the area of this glacier was 2.5 km^2, more than 35 per cent greater than its present extent, and the snowline was some 80 m lower (Shi Ya-Feng and Zhang Xiangsong 1984 and Plate 7.6).

General glacier retreat between the mid-nineteenth and mid-twentieth centuries is believed to have been caused by a rise in temperature (Shi Ya-Feng and Wang Jingtai 1979). The warmest five years of the 1940s had a mean temperature 0.5 to 1 °C higher

Plate 7.6 Glacier No. 1, the most extensively studied glacier in China, August 1985. (Photo: J. Stevenson)

Plate 7.7 Holocene and Little Ice Age moraines in front of an unnamed glacier, near Glacier No. 1, August 1985

than the mean value of the last hundred years. The tongues of the longest glaciers retreated several hundreds of metres and some several kilometres. The retreats in the interior of Tibet were less than elsewhere. The Musart in the Tien Shan retreated 750 m between 1909 and 1959, the Azda in southeast Xizang 700 m between 1933 and 1973, and the Skyany, northeast of Qogir Feng (better known as K2, 8611 m) retreated 5.25 km between 1937 and 1968. The glaciers in western Tibet, western Qilian Shan and eastern Tien Shan generally withdrew less than 10 m per year. Some debris-covered tongues were stable. The Rongbuk on the north side of Everest, surveyed in 1921 and again in 1959 and 1966, remained the same length but the ice thinned.

Glacier shrinkage was general in the 1950s and 1960s. From 1956 to 1976 all twenty-two glaciers observed in the Qilian Shan were retreating; eight termini in the eastern sector withdrew an average of 12.5 to 22.5 m a year, and fourteen in the western section at only 1.1 to 2.7 m a year. Shrinkage slowed towards the end of the period (Shi Ya-Feng *et al.* 1980). The fragmentary data from the Tien Shan suggest a similar pattern there. The Musart glacier near the old Silk Road to India withdrew at a rate of 15 m a year from 1909 and 1959. The glaciers on Mount Mystagh and Mount Kungur to the east of the Pamirs retreated at a rate of 1.7 to 3.7 m a year between 1907 and 1960. The Azar in the southeast retreated rapidly in the 1950s. In fact, wherever information is available, retreat seems to have been dominant from the 1930s until 1960.

Since 1960 the situation has changed somewhat. In the Qilian Shan average temperatures were 0.8 to 1.3 °C lower in the decade 1967 to 1976 than in the previous decade and precipitation at three stations above 3000 m increased by 3 per cent over the same period, from 345 to 355 mm. According to Shi Ya-Feng and Wang Jingtai (1979) this fall in temperature and rise in precipitation extended over western China. On the Tibetan Plateau, temperatures in the 1960s were everywhere 0.7 °C lower than in the 1950s and precipitation increased by about 5 per cent. The retreat of the Musart glacier slowed to 2 m a year between 1962 and 1978. Glaciers in the Nyenchintangla Range are still retreating but this is not expected to continue; the Aza glacier has a kinematic wave passing down its tongue. Ten glaciers in the western part, east of the Yarkand river, were advancing, the Chuanschuikon at an average rate of 15.5 m a year from 1968 to 1976. During the 1960s montane lakes such as the Yangchayung rose and flooded roads built in the previous decade. Many of the glaciers of the Amne Machin Mountains, a 200-km long extension of the Kunlun Shan, were advancing between 1966 and 1981. In 1981, of the forty glaciers in this range fifteen were known to be advancing and twenty-three were stationary or had variations within the limits of the photogrammetric techniques used for the glacier inventory (Wang Wenying 1983). In the Qilian Shan, mass balances on four glaciers were positive from 1974–7 and though many tongues were retreating, rates of recession were reduced. Thus the Shui Guan Ho No. 4 glacier in the eastern Qilian Shan receded by 16 m a year between 1956 and 1975 but only 6 m a year between 1976 and 1978. Two small glaciers here were already advancing by 1981. Comparison of air photographs of the 1950s and 1960s with recent Landsat imagery reveals that in the 1970s glaciers were advancing not only in the Qilian Shan and other eastern areas but also in the Pamir, Karakoram and Himalaya. In the Tien Shan over the period 1962–78, of thirty-eight glaciers observed, 13 per cent were advancing, 26 per cent were stationary and 61 per cent were retreating. Glacier No. 1 showed negative net balances predominating in the late 1960s and early 1970s with positive balances more recently. The Keqkar glacier advanced in 1980 onto grass-covered moraine (Ersi 1985). The greatest advances measured by Chinese glaciologists in recent years have been in the Karakoram, on the frontier with Pakistan.

Chapter 8

The Little Ice Age in North America and Greenland

Glaciers are scattered from 38° 15′N in southern California to the arctic islands of northern Canada, from the maritime coastal mountains of the northwestern United States to the Colorado Front Range and the mountains of Wyoming and from Alaska and the Yukon to the eastern Canadian Arctic between 60° and 90°W (Figure 8.1). The 490 tiny glaciers of the Sierra Nevada, none more than 2 km^2 in area, nestle in deep cirques or beneath north-facing cliffs, well below the orographic snowline (Raub *et al.* 1980). In the Pacific Mountain systems of Alaska and the Yukon, ice covers about 102,340 km^2, some 18 per cent of the glacierized area of the northern hemisphere apart from Greenland (Field 1975). The many thousands of glaciers in North America include representatives of nearly all the topographical and physical types.

The earliest recorded observations of North American glaciers were made along the southern part of the Alaskan coast where we have La Pérouse's account of Lituya Bay in 1787 and Captain George Vancouver's of Glacier Bay in 1794. In North America, as in the Himalaya, exploration and glacier observation went hand in hand. The longest documented records of terminal position begin in the 1880s when glaciological studies were initiated by such outstanding scientists as Reid (e.g. 1897, 1915), Russell (1897) and Tarr and Martin (1914). Their work was eased by the arrival of railways in the west and valuable records were left by government survey teams and boundary commissions as well as by individual observers. In 1907 the Canadian Alpine Club was founded by A. O. Wheeler with glacier observations as one of its objects (Wheeler 1931). His photographs of glaciers in Banff National Park and elsewhere provide important links in tracing the fluctuations of the Drummond, Peto and other glaciers over the last century.

American glacial studies tended to languish in the early twentieth century, apart from Cooper's in the Glacier Bay area, until in 1926 W. O. Field decided to take up Reid's investigations of the 1890s. In 1940 he joined the staff of the American Geographical Society, his office becoming an informal headquarters for glaciologists for the next three decades, and a more formal one when one of the World Data Centres for glaciology was established there before moving to the US Geological Survey at Tacoma. Many key glaciological investigations were initiated by the Geographical Society or the Geological Survey. Since the 1940s, research has proliferated and gained in sophistication with much of the impetus coming from universities and centres attached to them, notably the Quarternary Research Centre at Seattle, the Institute of Arctic and Alpine Research at Boulder and the Institute of Polar Studies at Columbus, Ohio.

Wheeler's series of observations of the Yoho glacier and others between 1901 and 1931 were exceptional. For the most part, regular observations did not begin until two or three decades later as will be apparent from a perusal of the Unesco volumes on glacier fluctuations where the bulk of the American entries appear for the first time in the 1970s. Many of the glaciers appearing in these entries were selected for their accessibility and are therefore not necessarily representative. Only a small proportion of the North American glaciers have been named, let alone having their characteristics recorded scientifically.

In Canada, the compilation of an inventory was under way by 1968. Topographical maps of the whole country are available on a scale of 250,000 but only a selection of the 1:50,000 sheets have been completed and the quality of the coverage varies considerably. By 1980, 33,000 glaciers had been identified out of an estimated total of about 100,000 and complete data as requested by Unesco was available for only 1500 (Ommanney 1980). The published inventory data for the USA in 1980 covered only the Sierra Nevada (Raub *et al.* 1980).

The Canadian Water Resources board had adopted methods of assessing volumetric changes of glaciers using ground-based photogrammetry as well as aerial survey. Photographic collections, more especially that assembled by Austin Post at Tacoma, provide invaluable resources for future glacier studies. Detailed mass balance studies have been initiated in both the USA and Canada in recent years (e.g. Müller 1962, Meier and Tangborn 1965, Østrem 1966).

Recent moraine sequences, dated by radiocarbon, tree-ring and lichenometric methods, abound in North America (e.g. Bray and Struik 1963, Osborn and Taylor 1975). Methodologies were at first imprecise; Cooper (1937) counted the growth rings of two or three trees in the Glacier Bay area and on this basis made an estimate of the time of recession there. Lawrence's (1950) studies in southeast Alaska were influential in extending the use of dendrochronology to date moraines, so that by 1967 Viereck was able to assemble data on fifty-one glacier forefields. Heusser (1957) had already come to the conclusion, on a basis of the North American data then available, that conditions had favoured glacial expansion in the mid-sixteenth, early and mid-seventeenth, early, mid- and late eighteenth centuries, and in some regions in the early twentieth century. In each century advances were greater in some areas than others and not all areas were affected at any one time. It is now quite clear that glacial expansion took place within the last few centuries in both maritime and continental environments. An attempt is made here to present some of the data about the regional pattern of fluctuations over North America in recent centuries, concentrating on those areas for which most information is available (Figure 8.1).

8.1 The Cascades and Olympics

These mountains, close to the northwest seaboard of the USA and receiving a very high

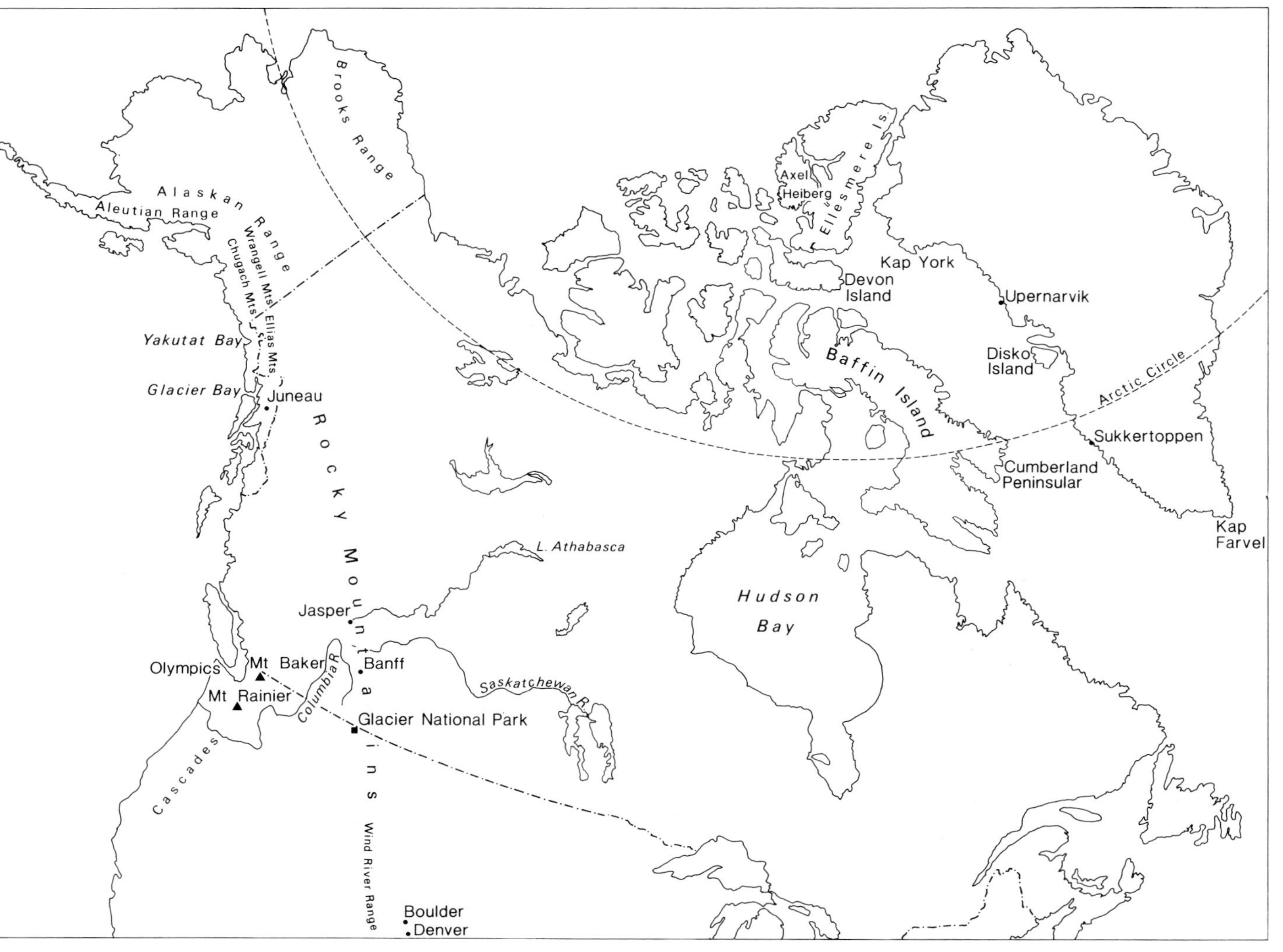

Figure 8.1 Key to glacierized regions of North America mentioned in the text

annual snowfall, support well over a thousand glaciers, most of them quite small, but together constituting a very high proportion of the total ice in the conterminous USA. The glaciers of Mt Rainier, 4392 m, are the largest. This ancient volcano is the focus of one of the oldest National Parks in the country and has been visited by naturalists and scientists since the mid-nineteenth century. Records of the position of the terminus of the Nisqually (Figure 8.2), the largest of Rainier's glaciers, go back to 1857 and are the longest in the USA. This glacier and its moraines have attracted attention for several decades (e.g. Harrison 1954, 1956a, Johnson 1960, Meier 1963, Veatch 1969, Sigafoos and Hendricks 1961, 1972).

Porter (1981) established lichen growth curves on andesite and granodiorite surfaces of

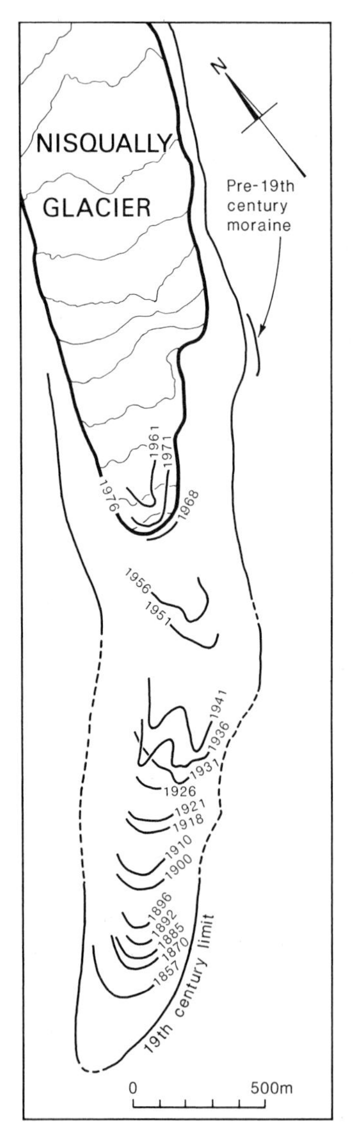

Figure 8.2 Nisqually glacier frontal positions. (From Porter 1981)

known age, taking care to avoid unusual local conditions near streams or waterfalls or where snow lies for long periods. Measurements of the diameters of thalli were made on fully exposed surfaces of known age, the largest circular or near-circular thalli being measured to the nearest millimetre. Twin curves for the two rock types were obtained, spanning the last 120 years. After an initial period of very rapid increase in size the growth rate of *Rhizocarpon geographicum* thalli was found to diminish progressively. The growth rates were faster on the andesite than on the granodiorite. The ages of lichen-bearing surfaces older than 1857 had to be estimated by extrapolation of the curves on the assumption that the decreasing rates inferred from the data persisted into the past. Although this assumption could be invalid, Porter argued that data from the Italian Alps, where the growth rate of *Rhizocarpon geographicum* is similar, indicates that differences in age extrapolated from growth rates plotted as both linear and exponential functions are not significant and that for the first four or five hundred years probably do not exceed the error resulting from field measurement. If such measurements can be taken to the nearest millimetre, the accuracy of values obtained can be taken to range from ± 3 to ± 5 years for surfaces 120–200 years old, and ± 10 to ± 20 for surfaces 200–400 years old. This careful study, with its attention to the degree of uncertainty of the results, appears to be much more accurate than tree-ring dating of moraines could be or most other lichenometric work so far published (although there is always some uncertainty as to the exact length of time required for lichen growth to be initiated after the retreat of the ice and the stabilization of the moraine it has abandoned).

Other glaciers on Mt Rainier also have well-marked moraines (Plate 8.1), a number of which have been dated using Porter's (1981) lichenometric curve supplemented by tree-ring dating for the older moraines (Burbank 1981). Periods of moraine stabilization, when ice fronts receded after readvances, were dated to 1519–28, 1552–76, 1613–24, 1640–66, 1691–5, 1720 and 1750. The sixteenth-century dates were obtained from tree-ring counts on one set of moraines only, whereas those from the seventeenth century were identified on two sets of moraines by lichenometry as well as tree rings. Burbank stressed that the interval between moraine stabilization and the establishment of trees may have been underestimated and the ages of the moraines may consequently be greater than indicated by these dates; certainly the ecesis periods, known to have been as great as 100 years, varied greatly from place to place so that these results are less precise than those based on lichen growth. More reliably dated, more recent moraines gave ages of 1768–77, 1823–30, 1857–63, 1880–5, 1902–3, 1912–15 and 1924. The synchroneity of the fluctuations of the glaciers on Mt Rainier as they retreated from their late eighteenth- and early nineteenth-century maxima was taken to suggest a common response to summer ablation conditions. After 1915 the rate of recession more than doubled, the extent of withdrawal varying according to local conditions.

Calculated mass balance variations based on monthly temperature and precipitation records back to 1909 for Longmire on the southwest flank of Mt Rainier, and extended further back by correlation with other Washington stations with longer records, are in agreement with observed glacier behaviour since 1850. Comparison of three-year running means of net balance at Mt Rainier with dates of moraine stabilization reveals a clear correlation with decreasing net balance. Confirmation of these results is provided by observations of standstills of the Nisqually glacier around 1875 and 1900 and a thickening of the ice for several years after 1932 and again after 1945 (e.g. Harrison 1956a, Veatch 1969).

Plate 8.1 Moraines left by the retreat of the Emmons glacier on Mt Rainier, July 1970

Temperature data from Longmire, like that from California, Oregon and Washington, indicate no long-term rise in mean annual temperature since about 1850. The lichenometric data presented by Porter (1981) and Burbank (1981) suggest that a rise in mean annual temperature of the order of 1 °C took place earlier, between 1770 and 1850, and that this initiated and sustained the major recession of the last 150 years. The dominantly negative mass balances calculated for the period 1850 to 1890 were associated with thinning of the glaciers; retreat of the termini was not great at that time, but the tongues were made susceptible to rapid retreat from 1910 to 1940, when ablation season temperatures rose and precipitation diminished. The extreme sensitivity of the Olympic glaciers to small climatic changes was demonstrated by Tangborn (1980) who reconstructed mass balance values for the South Cascade glacier and the neighbouring Thunder Creek glacier from the late nineteenth century onwards and calculated that mean summer temperature would only need to have been 0.5 °C lower or winter accumulation 10 per cent greater for the mass balances of the two glaciers to have remained continuously positive during the twentieth century (Figure 8.3).

Four moraine sequences in the Dome Peak area (48°20′N, 1°21′W) were investigated by Miller (1969) using tree rings and ash layers for dating (Figure 8.4). He recognized that the true dates of moraine formation were likely to be several decades earlier than tree-ring counts. Glaciers were found to have come further forward in the last few centuries than at any time since the Fraser Glaciation of 10,000 years ago. The outermost moraine of the Chickamin glacier he dated to the thirteenth century on a basis of trees

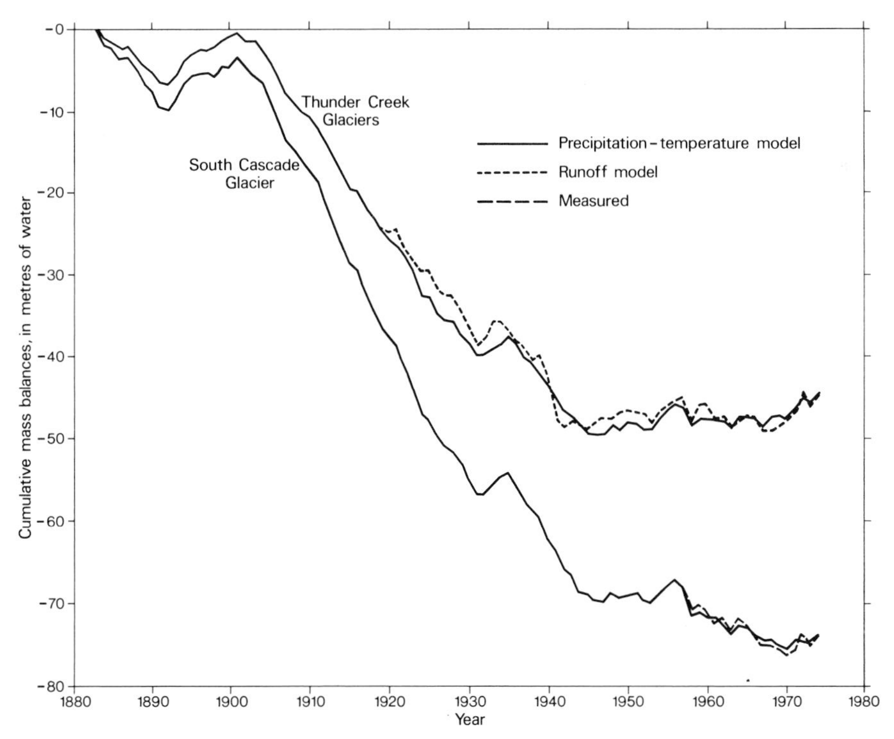

Figure 8.3 Mass balance values calculated for South Cascade and Thunder Creek glaciers, 1885–1975. (From Tangborn 1980)

which 'began growing more than 680 years ago', but the moraine could very well be much older than this (Figure 8.4). The outermost moraines of the South Cascade glacier were dated to the sixteenth or seventeenth century and those of the Le Conte and probably the Dana to the sixteenth century. The ratios of the areas of these little glaciers in the 1960s to their areas at their Little Ice Age maximum extensions are similar (Table 8.1).

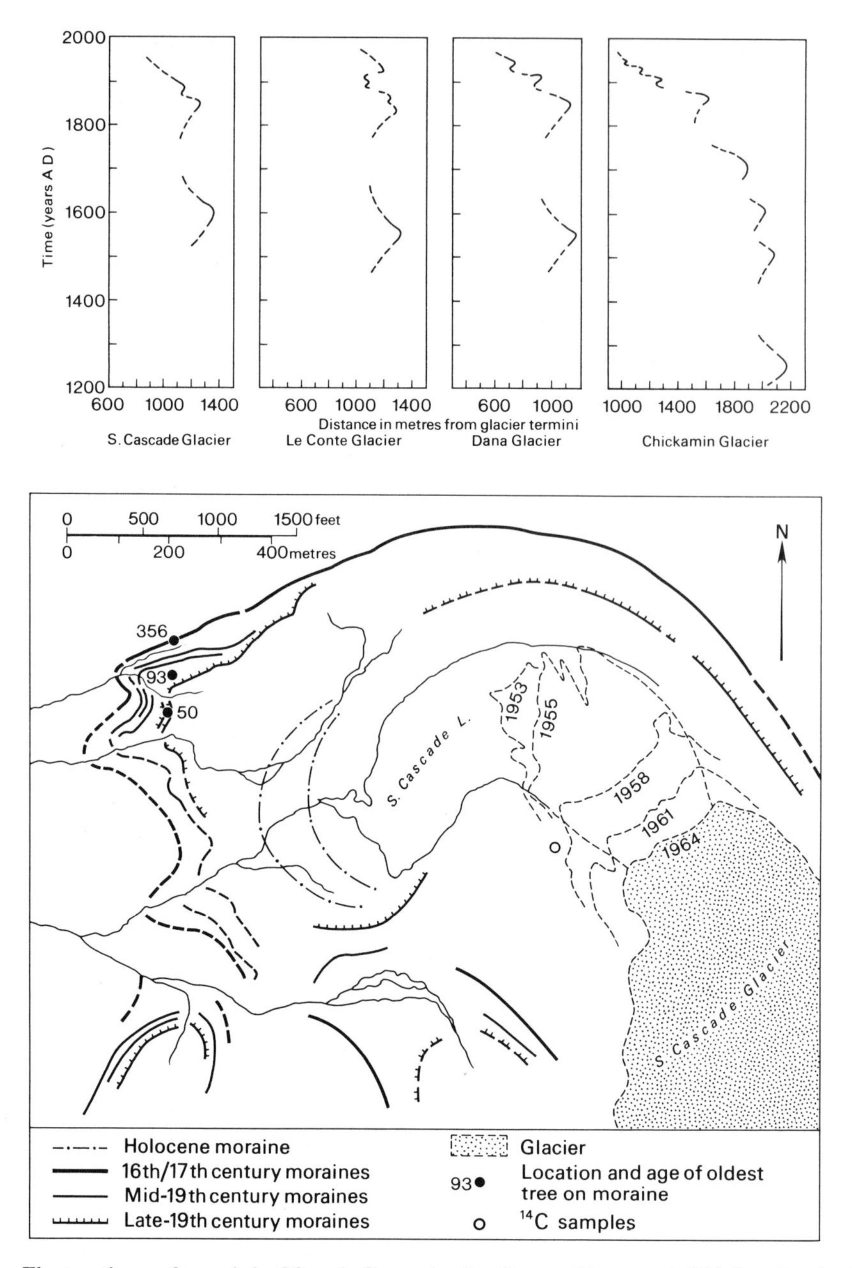

Figure 8.4 Fluctuations of termini of South Cascade, Le Conte, Dana and Chickamin glaciers and moraines and retreat stages of the South Cascade glacier. (From Miller 1969)

Table 8.1 Comparison of glacier area and terminal altitude of Dome Peak glaciers during the Little Ice Age maximum and in 1963/4

	South Cascade	*Le Conte*	*Dana*	*Chickamin*
Area (km) Little Ice Age Max.	4.15	2.98	3.99	7.80
Area (km) 1963	2.72	1.58	2.46	4.87
Ratio 1963/Max.	1/1.5	1/1.9	1/1.6	1/1.6
Altitude of terminus (m) at Max.	1490	1340	1270	1100
Altitude of terminus (m) in 1964	1615	1829	1768	1525

Source: Miller (1969)

Sixteenth- and seventeenth-century advances were followed by retreat and then by renewed but smaller advances in the nineteenth century, some of which were multiple. Volcanic ash from Mt Mazama helps to confirm the dating. For instance, the 'O' layer, dated by radiocarbon to 400 BP (Wilcox 1965), is present within the South Cascade moraine dated to the sixteenth century but is absent from within the moraines attributed to the nineteenth century. A seventeenth-century advance was also identified on Mt Olympus when Heusser (1957) made a reconnaissance of the Blue and Hoh glaciers and suggested that the earliest datable Little Ice Age advance of the Blue glacier occurred about 1650; this estimate depended on the ages of only two trees, an Alpine fir (*Abies lasiocarpa*) and a hemlock (*Tsuga mertensiana*) growing on a moraine remnant. An advance about 1850 tilted trees on Blue glacier moraines which, except in one place, obscure moraines dating to the seventeenth century. For the Hoh, the most extensive recent expansion also seems to have been that of 1850.

8.2 Montana and Colorado

In the relatively dry front ranges of the Rockies, small cirque glaciers, now fed almost entirely by wind-drifted snow (Alford 1974), are bordered by moraine fields. Benedict (1965, 1973) made careful and detailed lichenometric studies in the Indian Peaks region, northwest of Denver, but did not produce different growth curves according to the rock sub-strates nor did he indicate the degree of uncertainty of his lichen ages. However he did suggest the likelihood that an interval of fifty years had elapsed before lichens became established on moraines in the Front Range. He concluded that *Rhizocarpon geographicum* colonized the oldest of the moraines examined about AD 1750 and that they marked an ice advance about 1700 greater than any other in the Holocene history of the region (see Chapter 10). The Arapaho cirque, one of the best known, has an outer moraine attributed to the seventeenth century and two inner moraines, probably deposited in quick succession, perhaps dating to about 1820 and 1850.

The Agassiz and Jackson glaciers in Glacier National Park, northwest Montana, 80 km apart on either side of the Continental Divide, appear to have reached their most forward positions in the Little Ice Age about 1860 (Carrara and McGimsey 1981). No prominent moraines mark the limit of that advance, possibly because it was short-lived. Within the limit, trees were less than a century old and diminished in age towards the glacier; outside the limit they were older. It was concluded that the equilibrium line, as calculated from estimated accumulation area ratios at the nineteenth-century maximum, had been depressed by 180 m in the case of Jackson and 300 m in the case of the Agassiz glacier. The Agassiz, in 1979, was merely a patch of stagnant ice covering 0.75 km^2 as compared with an area of 3.38 km^2 in the mid-nineteenth century. Retreat, judging from tree-ring counts, was slow until 1910; on a US Geological Survey map of 1906 the Agassiz glacier was still 2.8 km long and projected outside its cirque into the forest. Between 1917 and 1926 it shrank back at a rate of more than 40 m a year. Over the next sixteen years, in a period of increased summer temperatures and decreased precipitation, the rate of retreat more than doubled and glacier shrinkage in these continental areas was much more drastic than in the maritime west (Johnson 1980). Since then retreat has slowed down but has not ceased.

8.3 The Canadian Rockies

The largest glaciers in the Rockies are in the front ranges near the boundary between Alberta and British Columbia where peaks rise to over 3000 m and there are several extensive icefields (Figure 8.5 and Plate 8.2). Well-developed, fresh moraines, many of them unvegetated, occur within one or two kilometres of present glacier tongues. Larger glaciers generally show nested sequences of moraines, but many of the steeper glaciers have only one complex moraine formed by repeated readvances (Kearney and Luckman 1981).

Moraines in the forefields of twelve glaciers in Banff, Jasper and Yoho National Parks and the Robson Provincial Park were dated by Heusser (1956) using tree rings. He obtained the fullest record for the Robson glacier, identifying readvance or recessional moraines of which the oldest was dated to 1782 and others to 1801, 1864, 1891 and 1907. Luckman (1977) used written records, photographs and tree rings to prepare a growth curve for lichens on the moraines of the Mt Edith Cavell and Penstock Creek glaciers in Jasper National Park. On the Angel glacier moraines, where Heusser (1956) had cored thirty-three trees, Luckman cored an additional seventy-five trees, many of them older. In a third of the trees his cores failed to reach the middle and these were not used for dating purposes. The dates he obtained were earlier than Heusser's, partly because he assumed that the interval between stabilization of the moraines and colonization by trees was 15–30 rather than 10 years, but he stressed that the ages he obtained were minima values:

Angel glacier moraines

Heusser (1956)	1723	1783	1871	1901
Luckman (1977)	1700–10	1715–25	1851–6	1881–9

He used his lichens to date the moraines of six other glaciers and emphasized that, because of the short calibration period and the shape of his growth curve, the ages he obtained by this method were also minima.

The available data from the Middle Canadian Rockies (Table 8.2) have been assembled by Luckman and Osborn (1979). Major advances seem to have occurred in the late seventeenth to early eighteenth, early to mid-nineteenth, and late nineteenth to early twentieth, centuries. The dates for the moraines dated by lichenometry are dependent on extrapolation of a growth curve that is not regarded as definitive. Apart from these, the

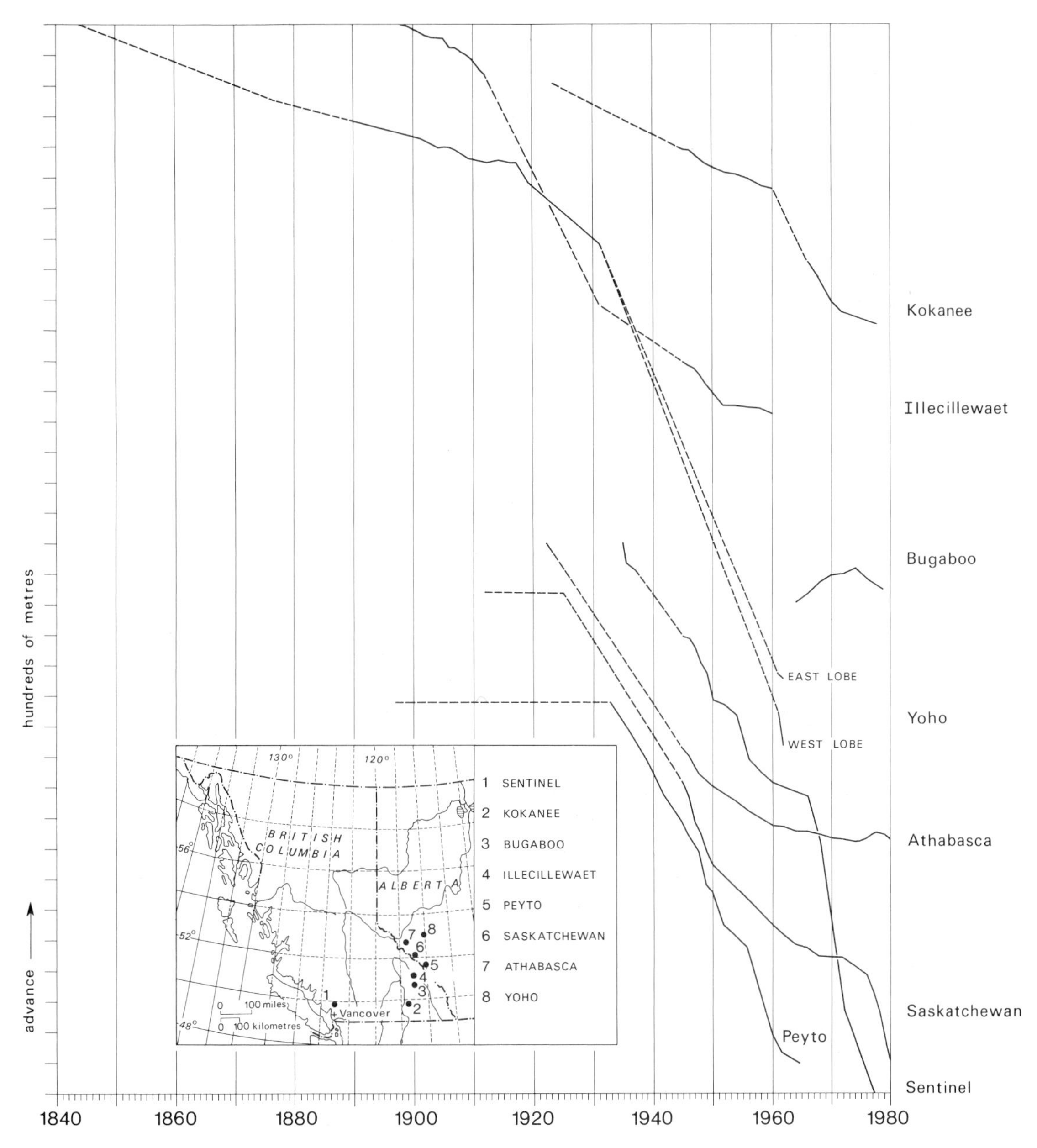

Figure 8.5 Glacier retreat in the Canadian Rockies. (Derived from figures published in Kasser 1973, Müller 1977 and Haeberli 1985)

earliest moraine date is for the outermost moraine of the Bow glacier, dated by the rings of an Engelman spruce (*Picea engelmannii*) to 1669 or earlier. The closely spaced multiple moraines of the Columbia, Athabasca, Peto and Bow glaciers mark nineteenth-century advances to positions comparable with those reached in preceding centuries, differences between the patterns of Little Ice Age moraines in front of the various glaciers probably depending on geomorphic setting. Since the late nineteenth century, recession has been the rule (Brunger *et al.* 1967).

In 1858, Dr Hector of the Palliser expedition discovered the Lyell glacier and described how 'after crossing shingle flats for about a mile, we reached a high moraine of perfectly loose and unconsolidated materials, which completely occupies the breadth of the valley, about 100 yards in advance of the glacier' (Thorington 1927). This, one of the earliest descriptions of a glacier in the Canadian Rockies, shows that modern recession from the mid-nineteenth century maximum had already begun.

Measurement of glacier fronts started soon after the opening of the Canadian Pacific Railway to passenger traffic in 1886. The following year, George Vaux, a leading Quaker businessman from Philadelphia who was also a dedicated mineralogist and supporter of the Philadelphia Academy of Sciences, visited the Rockies with his three children, William, George and Mary. They stayed at the newly opened Glacier House Hotel and went to see the Great Glacier of the Illecillewaet. Later, William Vaux Jr was to comment in his monograph *Modern Glaciers* (1907), 'during railway construction the glacier was doubtless often visited by those stationed on the work but no records were made until July 17th 1887 when our party passing through, roughly mapped the tongue and made a photographic record of the conditions as they existed'. When the Vaux returned in 1894 and noticed that the glacier front had retreated they realized the value of their 1887 photographs and became so interested in the study of glaciers that they intitiated the first measurement of glacier front positions in Canada (Cavell 1983). The family was in a position to establish a routine which involved Mary and her father spending most of the

Plate 8.2 The Columbia icefield with its Holocene and Little Ice Age moraines, July 1970

Table 8.2 Dated moraines of the Little Ice Age (Cavell) advance in North America

Site	*Outermost moraine*	*Inner moraines*	*Source*	*Technique*
Mount Robson	1782	1801, 1864, 1891, 1907	Heusser 1956	D
Columbia glacier	1724	1842, 1871, 1907	Heusser 1956	D
Dome glacier	1870	1900, 1919	Heusser 1956	D
Athabasca glacier	1714	1841, 1900	Heusser 1956	D
Saskatchewan	1807	1854, 1893	Heusser 1956	D
Southeast Lyell	1840	1855, 1885, 1894	Heusser 1956	D
Freshfield	1853	1881, 1905	Heusser 1956	D
Peto glacier	1711	1861, 1880, 1888	Heusser 1956	D
Bow glacier	1669	1847, 1894	Heusser 1956	D
Yoho glacier	1844	1877, 1890s	Bray and Struik 1963	D
Wenkchemna	1906–25		Gardner 1978	P & DO
President's Peak	1714	1832	Bray 1964	D
Scott glacier	1780		Shafer 1954	D
Hector glacier	*c.* 1660		Brunger *et al.* 1967	D
Drummond glacier	1570–1660		Nelson *et al.* 1966	D
Cavell glacier	1705	1720, 1858, 1888	Luckman 1977	D
Penstock Creek	1765	1810, 1876, 1907	Luckman 1977	D
Fraser glacier	1595	1725, 1880, 1915	Luckman and Osborn 1979	L
Dungeon Peak	1620	1780, 1880, 1905	Luckman and Osborn 1979	L
Redoubt South	1705	1890, 1905	Luckman and Osborn 1979	L
Redoubt North	1630	1860, 1890, 1920	Luckman and Osborn 1979	L
Bastion Peak	1705	1885, 1915	Luckman and Osborn 1979	L
Turret glacier	1530	1602, 1735, 1895	Luckman and Osborn 1979	L

Source: Luckman and Osborn (1979)
Notes: D = denrochronology;
P = old photographs;
DO = documentary sources;
L = lichenometry

summer in the Rockies with George and William visiting as often as business would allow (Plates 8.3, 8.4 and 8.5). Inevitably they came into close contact with Wheeler, Dominion Surveyor working in the Selkirks. Already in 1899 he noted with approval that the Vaux surveys of the Illecillewaet and Asulkan had become 'properly systematised'. Their most detailed studies were of these two glaciers but they also photographed and observed the Yoho, Victoria and Wenkchemna glaciers, establishing standard photographic points and measurement procedures. Their observation points were later adopted by Sherzer (e.g. 1905) and Wheeler (e.g. 1910 and 1933). These simple reconnaissance studies are the only ones from Canada dating back to the time when the glaciers had yet to retreat far from their most forward positions in the nineteenth century. All the available evidence goes to show that recession has been predominant ever since. The withdrawal of the Athabaska and Columbia glaciers was accelerated, when terminal lakes formed in front of them, the Athabaska losing about two-thirds of its 1870 volume in the subsequent century (Kite and Reid 1977). The frontal recessions of a selection of glaciers with the longest records in British Columbia and Alberta are shown in Figure 8.5.

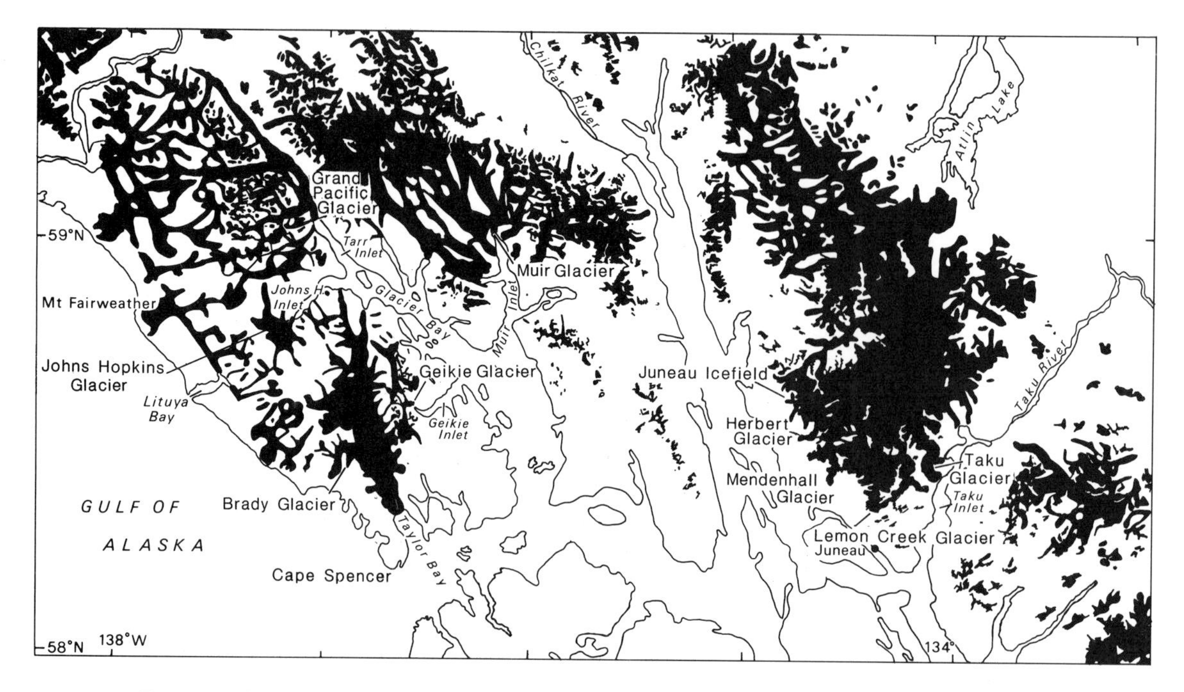

Figure 8.6 Glacier Bay in Alaska

8.4 Alaska and the Yukon

Many of the earliest observations of glaciers in North America relate to the coastal glaciers of Alaska which gained the attention of explorers and pioneer scientists. Nevertheless, pre-twentieth-century data are sparse in comparison with the size of the country and the range of environmental conditions. Much of what does exist was collected together by Field (1975) without being fully analysed.

The region is subject to earthquakes. Rapid glacier advances at the beginning of the century after a series of severe shocks centred near Yakutat Bay led to the notion that so much snow and ice had been shed onto the glaciers that anomalous advances had been caused (Tarr and Martin 1914). This hypothesis was seriously questioned by Miller (1958) who pointed to the evidence of exceptionally heavy precipitation in the 1870s and 1880s and argued that diastrophism was a relatively minor factor. The ocurrence of a major earthquake (8.4–8.6 on the Richter scale) with an epicentre immediately south of the heavily glaciated Chugach mountains in 1964 provided an excellent opportunity to test the earthquake advance theory (Post 1965). Extensive air surveys of the glacier in Alaska and Canada were available for comparison with air photographs taken after the earthquake and little significant avalanching was found to have been caused. It can be safely concluded that earthquakes have not been responsible for the anomalous behaviour that characterizes some of the best-known Alaskan glaciers. Surging is a much more important factor, especially in the St Elias, Wrangell, Chugach, Alaska and Aleutian ranges (Post 1969).

However, surging is not the whole explanation for the behaviour of the glaciers in, for example, the Taku and Glacier Bay areas (Figure 8.6). The Juneau Icefield, covering about 1820 km^2 of the most heavily glaciated sector of the Coast Range, is drained by a dozen or so valley glaciers. It has been studied intensively and a chronology of glacier advance and retreat worked out in some detail (e.g. Miller 1964, 1965, 1970, Field 1975). Ten of the outlet glaciers receded from 1870 maxima. Lawrence (1950, 1951, 1958) and others counted the rings of trees growing on the larger moraines and showed that there had been an important advance of the glaciers at about the time when they were first seen. A detailed study of the forefield of the Lemon Creek glacier was made as part of the Juneau Icefield Project (Heuser and Marcus 1964, and Figure 8.7, Table 8.3).

Cores were taken from Sitka spruce (*Picea sitchensia*) growing on the remnants of trimlines and moraines and it was assumed from the ages of the oldest spruce growing on a moraine (dated from air photographs) that the time taken for their establishment was ten years. The pattern of halt and readvances in the long-term recession corresponded to that of the large Juneau outlet glaciers except the Taku and the Hole-in-the-Wall. The Herbert, for instance, retreated only very slowly between the mid-eighteenth and the late nineteenth centuries and then withdrew rapidly before slowing down again in the early twentieth century. The most rapid retreat was in the middle of the twentieth century. All

Plate 8.3 William Vaux photographing Illecillewaet glacier, which was already in retreat, in August 1898 from 'observation rock'. (From the Vaux family collection of the Whyte Museum of the Canadian Rockies)

Table 8.3 Changes in the area and terminal position of Lemon Creek glacier, Alaska, *c.*1750 to 1958

Dated terminus	*Cumulative area loss (%)*	*Recession rate (m/y)*
Max. *c.*1750		
1759	0.8	
1759–69	1.5	12.5
1769–1819	2.4	3.5
1819–91	3.3	3.8
1891–1902	8.0	61.4
1902–19	10.2	4.4
1919–29	11.7	7.5
1929–48	21.4	32.9
1948–58	25.1	37.5

Source: Heusser and Marcus (1964)

the Juneau outlets without exception were expanded in the mid- to late eighteenth century (Table 8.4).

The Taku, the major trunk glacier draining the southeast of the Juneau icefield, was in an enlarged state, blocking the headward end of Taku inlet when Vancouver first saw it in 1794:

> A compact body of ice extended some distance nearly all round . . . from the rugged gullies in the side were projected immense bodies of ice that reached perpendicularly to the surface of the ice in the basin, which admitted of no landing place for boats but exhibited as dreary and inhospitable an aspect as the imagination can possibly suggest. (Vancouver 1798)

He mentioned that 'the basin' was about 13 miles from the mouth of the inlet, indicating

Table 8.4 Dates of the outermost moraines of the Juneau icefield outlet glaciers

Glacier	*Recession date*	*Source*
Norris	1750s	Lawrence 1950
Taku	1750s	Lawrence 1950
Twin Glaciers	1775–7	Cooper 1942
Twin Glaciers	1777	Field 1932
Lemon Creek	1750	Heusser and Marcus 1964
Gilkey	1783	Heusser and Marcus 1964
Mendenhall (middle)	1765–9	Lawrence 1950
Mendenhall (east)	1786–8	Lawrence 1950
Herbert	1765	Lawrence 1950
Eagle	1785	Lawrence 1950

Source: Viereck (1967)

Plate 8.4 (on facing page—above) Mary Vaux at the foot of the Illecillewaet glacier, 17 August 1899. (From the Vaux family collection of the Whyte Museum of the Canadian Rockies)
Plate 8.5 (on facing page—below) The smooth tongue of the retreating Illecillewaet glacier in the summer of 1902. (From the Vaux family collection of the Whyte Museum of the Canadian Rockies)

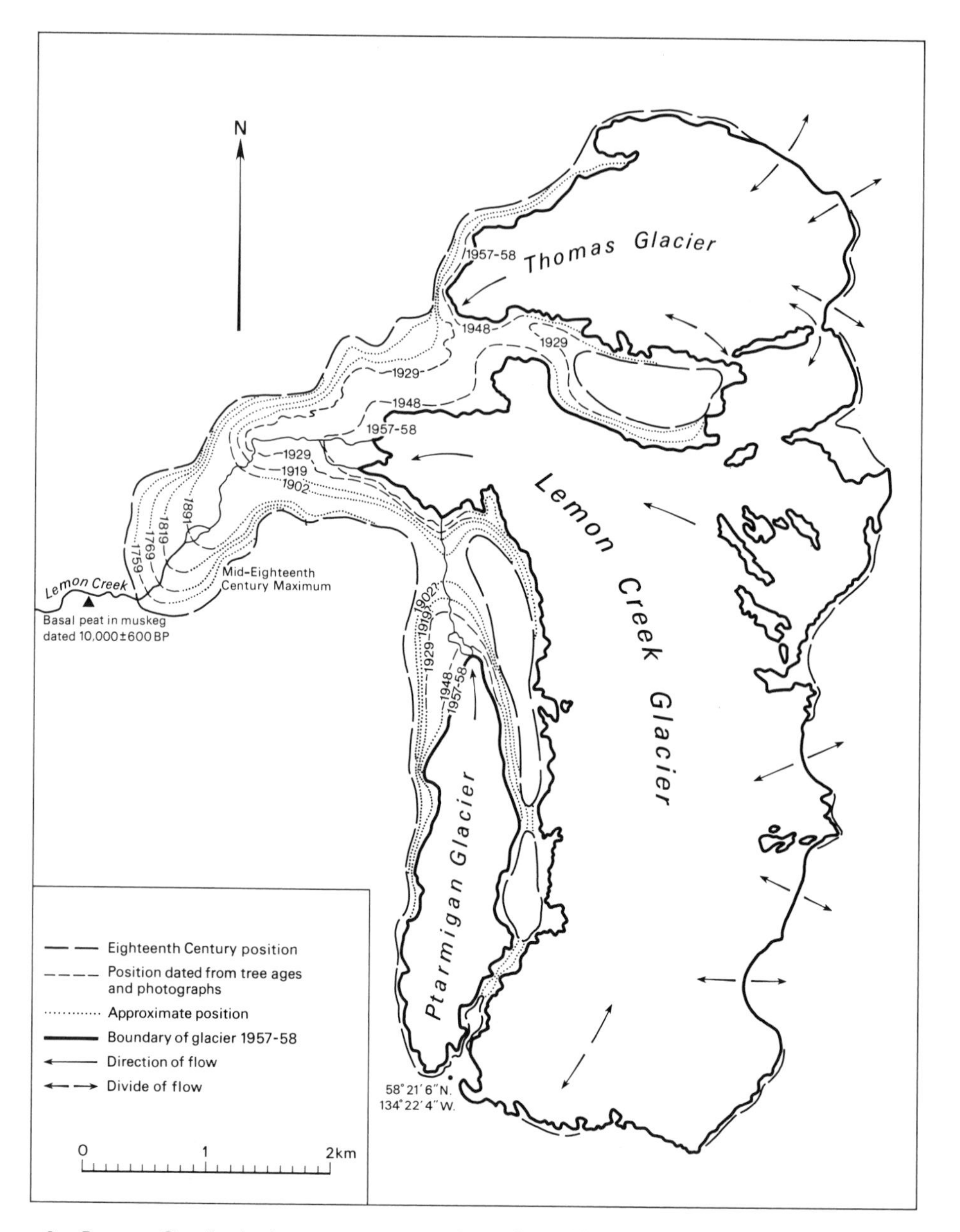

Figure 8.7 Lemon Creek glacier retreat stages since the mid-eighteenth-century maximum. (From Heusser and Marcus 1964)

that the ice-front was not far from Taku Point. This agrees with the local Thlingit accounts of an ice barrier which prevented access to the interior valley of the Taku river 'before the White Men came'. In the nineteenth century, however, the ice had withdrawn and Taku inlet and the valley at its head were being used by the Indians to cross the Coast Range to the Atlin area. In the 1870s to 1890s the route was also used by prospectors (Miller 1964). By the end of the century, Klotz (1899) found only the Foster glacier discharging directly into the sea and noted that 'the gullies are not now so generally filled with ice'. But already around 1890 the Taku had begun to advance again. Between 1890 and 1952 the front had advanced 5550 m and by 1961 a further 600 m. In the early 1960s, 80 per cent of the surface of the Taku was above the firnline compared with 60 to 70 per cent of the surfaces of the Norris and Mendenhall glaciers which were retreating. This circumstance may help to account for the strongly positive regime and the anomalous behaviour of the Hole-in-the-Wall, but it is probably much more to the point that they are tidewater glaciers. The Taku front was in water 100 m deep in 1890 and was calving actively until 1937. Deposition of sediment from the Taku and Norris glaciers and by the Taku river caused so much shoaling that by 1941 exposed sediments fronted most of the Taku terminus. The average advance rate of over 100 m a year of the 1890–1952 period dropped, the front advancing only about 75 m between 1964 and 1968.

Ice in Glacier Bay, a broad inlet about 160 km west of Juneau, has had an even more violently anomalous history (Figure 8.6). In 1794, the Bay was 'terminated by a solid, compact mountain of ice, rising perpendicularly from the water's edge and bounded to the north by a continuation of the united, lofty, frozen mountains that extended eastwards from Mount Fairweather' (Vancouver 1798). From tree-ring evidence, Cooper (1937) argued that by the time Vancouver saw it the glacier had already retreated 10 km from a position near the mouth of Glacier Bay, and placed the beginning of the retreat at between 1735 and 1785. Since Vancouver saw it the glacier has almost disappeared and Glacier Bay has been enlarged from less than 10 km in length to over 100 km at the head of Tarr and Johns Hopkins Inlet and 86 km at the head of Muir Inlet. The water area of the bay has been enlarged by about 130 km^2, i.e. by an amount comparable with the combined area of all the glaciers in the Swiss Alps. In Muir Inlet the ice was receding at about 400 m a year in the late 1970s, the terminus moving through a cross-section that had contained ice 1100 m thick in the 1890s. Recession was equally fast in the northwest arm of Glacier Bay, which split in the 1880s into the Grand Pacific glacier in Tarr Inlet and the Johns Hopkins glacier in the Johns Hopkins Inlet. The Grand Pacific had waned so that it was near the head of its fiord between 1925 and 1935, but since then has readvanced over 2 km. Meanwhile the Johns Hopkins reached the head of its fiord about 1930 and has since readvanced 1.8 km. The Hugh Miller and Geikie glaciers in the southwest part of Glacier Bay receded to land in the first half of the twentieth century and are now about 2 km and 3.5 km away from tidewater. Both were receding at about 60 m a year in the late 1970s (Field 1979).

The Brady glacier, 520 km to the east of the Fairweather Range, discharges both north into Reid Inlet at the landward end of the Glacier Bay and southward into Taylor Bay about 30 km west of the entrance to Glacier Bay. When Vancouver's party visited Taylor Bay, Lieutenant Whidbey was ordered to survey it on 10 July 1794. Setting out from Cape Spencer, despite being much inconvenienced by 'immense numbers of pieces of floating ice', he reached the side of Taylor Bay which he found took

> nearly a north direction for about three leagues to a low pebbly point; from which, five miles further, a small brook flowed into the sound and on its northside stood the ruins of a deserted Indian village. To reach this station the party had advanced up an arm about six miles wide at its entrance but had decreased to about half that width and there further progress was now stopped by an immense body of compact perpendicular ice, extending from shore to shore, and connected with a range of lofty

> mountains that formed the head of the arm, and, as it were, gave support to this body of ice.

Klotz (1899) compared Vancouver's chart with that of the 1894 Boundary Commission and concluded that the front of the Brady had advanced 5 miles (8 km) in the intervening period when the glaciers at Taku inlet and Glacier Bay were retreating. Derksen (1976) does not agree that the Brady glacier advanced at all at this time, arguing that Vancouver's chart is distorted in detail though it reproduces the regional coastline roughly. He identified the locations mentioned in the report in such a way as to suggest that in 1794 the glacier front was quite close to its 1970 position. This interpretation rests essentially on his identification of the 'low pebbly point', 5 miles SSE of the terminal position in 1794, with an undated and probably Holocene moraine 5 miles SSE of the present terminus. Derksen surveyed the area in detail, dating the climax of ice advance to 1876–88 on tree-ring evidence at trimlines, and accepted Bengtson's (1962) conclusion that a forest incorporated in till 24 km north of the present terminus was destroyed by advancing ice about 685 ± 40 BP (UW-14). He accordingly visualized the most recent major advance of the Brady glacier as extending over the period 700 BP to the late nineteenth century.

Lituya Bay, to the west of the Fairweather Range, was first visited and surveyed by La Pérouse (1787) in the course of his scientific expedition of 1786. He found it

> disturbed by the fall of enormous masses of ice, which frequently separate from five different glaciers. . . . It was at the head of this Bay that we hoped to find channels by which we might penetrate to the interior of America. . . . At length, having rowed a league and a half-mile we found the channel terminated by two vast glaciers. . . . Messrs Lelangle de Monte and Daglet, with several other officers attempted to ascend the glacier. With unspeakable fatigue they advanced two leagues, being obliged at the extreme risk of life to leap over clefts of great depth, but they could only perceive one continued mass of ice and snow, of which the summit of Mt Fairweather must have been the continuation. . . . I sent M. Momeron and M. Bernizet to explore the eastern channel which terminated like this at two glaciers. Both these channels were surveyed and laid down in the mouth of the Bay.

Klotz (1899) compared La Pérouse's chart with the map of the area made by the Canadian–Alaskan Boundary Commission of 1894 (Figure 8.8) and demonstrated that a substantial advance of the Lituya glacier had taken place in the course of the nineteenth century.

Violent retreats and advances have occurred not only in the Glacier Bay region but also elsewhere in Alaska, notably in the Prince William Sound area, and so dates of glacial maxima may appear in any century, eighteenth, nineteenth or twentieth (see Tables 8.4 and 8.5).

Apart from surging glaciers, it is tidewater glaciers that have produced the major anomalies (Field 1979). Glaciers reaching down to sea-level are generally nourished by firnfields situated well above the average elevation; shifts in the positions of their snowlines can trigger rapid movements of the fronts (Mercer 1961). Wasting, affected by calving at the terminus as well as sea water ablation, is greatest when the calving cliff is longest and ends in deep water. The consequent instability allows many large icebergs to break away from the tongue, and sediments deposited from the ice are spread thinly in

Table 8.5 Dates of maximum advances, Prince William Sound area

Glacier	*Date of maximum*	*Source*
Columbia	1914–22	Cooper 1942
Meares	1957 +	Viereck 1967
Harvard	1957	Viereck 1967
Bury	1898	Tarr and Martin 1914
Harrison	1957	Viereck 1967
Tebenkoff	1875–85	Viereck 1967
Ultramarine	1880–90	Viereck 1967
Nellie Juan	1860–85	Viereck 1967
Taylor	1865–75	Viereck 1967
Falling	1875–85	Viereck 1967

Source: Viereck (1967)

deep waters as swift recession proceeds. The rate of ice flow is maintained right to the terminus and may even increase towards it as there is little frontal resistance. If the ice discharge is greater than can be supplied from the accumulation area, the glacier surface will be rapidly lowered, the snowline will migrate upwards and the accumulation area will be reduced, thereby diminishing the ice cover inland. Reduction of volume resulting from the feedback effect is largely independent of climate. On the other hand, if the terminus reaches shallow water, deposition promotes further shallowing and material accumulates at the base of the ice cliff, forming a barrier protecting the terminus from tidewater. As deposition continues, reduction in the rate of terminal ablation may initiate an advance. As this proceeds, a moraine-outwash barrier is pushed ahead until equilibrium is established between the flow of ice and the melting of the terminus. Conversely, very rapid recession occurs when the terminus recedes off a bar into deep water, as is currently happening at the Columbia ice front (Meier *et al.* 1979, Weller 1980). The behaviour of the ice in the Glacier Bay and Taku regions can be explained adequately by the operation of these controls. It is evident that when one is attempting to discern the climatic sequence for the coastal and near-coastal ranges of northwestern America, the records of advance and retreat of the surging and tidewater glaciers have to be ignored.

In the western Chugach mountains, the head of the Gulf of Alaska, the firn limit along the coast is below 1000 m, rising inland to 1500 m in interior valleys. The ice cover, with an area of 10,400 km², is second only to that of the St Elias mountains in Alaska. The whole crest of the range is covered with interconnected névéfields of which the largest is the Bagley Icefield. The first appraisal of the voluminous and scattered evidence (Field 1975) indicates that with few exceptions non-tidal and non-moraine covered valley glaciers have receded 1–2 km from positions occupied within the last two and a half centuries.

There is evidence that many of them have been receding from nineteenth-century maxima, but this recession has not been unbroken. The glaciers of the lower Copper river and adjacent basins, for example, are all retreating from moraines formed when they were up to 5 km longer. Some of these moraines have been dated. The outer moraine of the Martin River glacier seems to have been formed by 1650 (Field 1975) and the Miles was in an advanced position in 1880 (Tarr and Martin 1914). The Grinell advanced in 1904 and many of the glaciers in the area, including the Childs, Allen, Henry and Grinell, advanced about 1910. The early twentieth-century advances were of crucial interest to the operators of the Copper River and Northwestern Railway which was put through the valley between 1906 and 1910. It passed between the Miles and Childs glaciers. During construction both advanced, decreasing the distance between them by a third until the Childs was less than 1500 m from the bridge carrying the railway over the river. It was feared that advance of the Grinell might continue long enough to break up the stagnant ice in front of its active tongue. The railway track crossed 400 km of this stagnant ice which was thinly covered with ablation moraine and with alder thickets and cottonwood trees growing on it. Worse still, the Grinell might push forward over the 500 m separating it from the river. The railway passed over the Allen glacier for 8 km; fortunately the advance of the Allen in 1912 'was short-lived and did not communicate appreciable motion to the stagnant ice on which the track was laid' (Field 1932). The Grinell retreated between 1911 and 1931 and further retreat of several hundred metres had taken place by the time it was seen by a party from the American Geographical Society in 1966. There do not seem to have been any observations of the Grinell since that date, but there is known to have been a minor advance of the Allen in 1966 and of the Childs in 1968 (Field 1975).

The same story of recession during the last century comes from the Kenai mountains which form the backbone of the Kenai peninsula where glaciers of a wide range of morphological types occupy a total area of 200 km^2. Most of the valley glaciers are accessible and much of the area has been mapped on a scale of 1:63,360. All the glaciers have receded from nineteenth-century positions to extents varying from about 100 m to 2.3 km. Recent moraines abound. 'Pre-1750' moraines occur within a kilometre of the Sargent Icefield snout. Inside there are moraines dated to 1830 and smaller ones dated to 1890 and 1950. The Tebenkof glacier, the largest in the Blackstone area, has an outer moraine dated 1875–85 by Viereck (1967) from which it had receded 250–350 m, according to Tarr and Martin (1914), and a further 300 m by 1935 (Field 1975).

The glaciers of the coastal ranges of Alaska were in advanced positions in the eighteenth and nineteenth centuries and have receded since then. This recession has been interrupted by minor advances between 1909 and 1914 and around 1935. There were less widespread advances in the late 1940s and again in recent decades. Glaciers with higher névéfields have receded least, it seems, and have been most apt to show small advances, while those with sources close to the firnline have receded most. A composite of mean annual temperature pieced together from available weather bureau records, from 1895 onwards, shows a net rise of 1 to 1.5 °C from the late 1900s to 1960. But this was complicated by a series of fluctuations with troughs in 1895, 1920, and 1935 of which the first was much the most marked. The rise in temperature in the early twentieth century was not continued in the 1940s and 1950s (Hamilton 1965). It is not within the compass of the present work to attempt to pick out regional variations in this general scheme. Indeed it is doubtful whether there is yet enough accurately dated evidence to do so satisfactorily. It is, however, worth commenting on the general lack of firm evidence of moraine formation in the sixteenth and seventeenth centuries in the coastal belt. Traces of the earlier part of the Little Ice Age are probably better preserved in the drier interior ranges.

Moraine sets in the White River valley on the northern flanks of the St Elias and Wrangell mountains were dated by Denton and Karlén (1977), using lichenometry, to between 1500 and the early twentieth century. The work was based on measurements of the largest thalli of *Rhizocarpon geographicum* (possibly with some *Rhizocarpon superficiale* included as they are difficult to differentiate) and the growth curve was based on tree-ring dating of surfaces together with some radiocarbon dating. Full details were given of the control surfaces. The growth curve was rated as only 'reasonably accurate' but 'affording a valuable reconnaissance tool'. Omitting data from glaciers known to surge, four sets of moraines were identified with lichens having maximum diameters of 37–45 mm, 27–29 mm, 15–16 mm and 6–10 mm. The geomorphological descriptions seem to make the attribution of these moraines to the last four centuries acceptable. Denton and Karlén

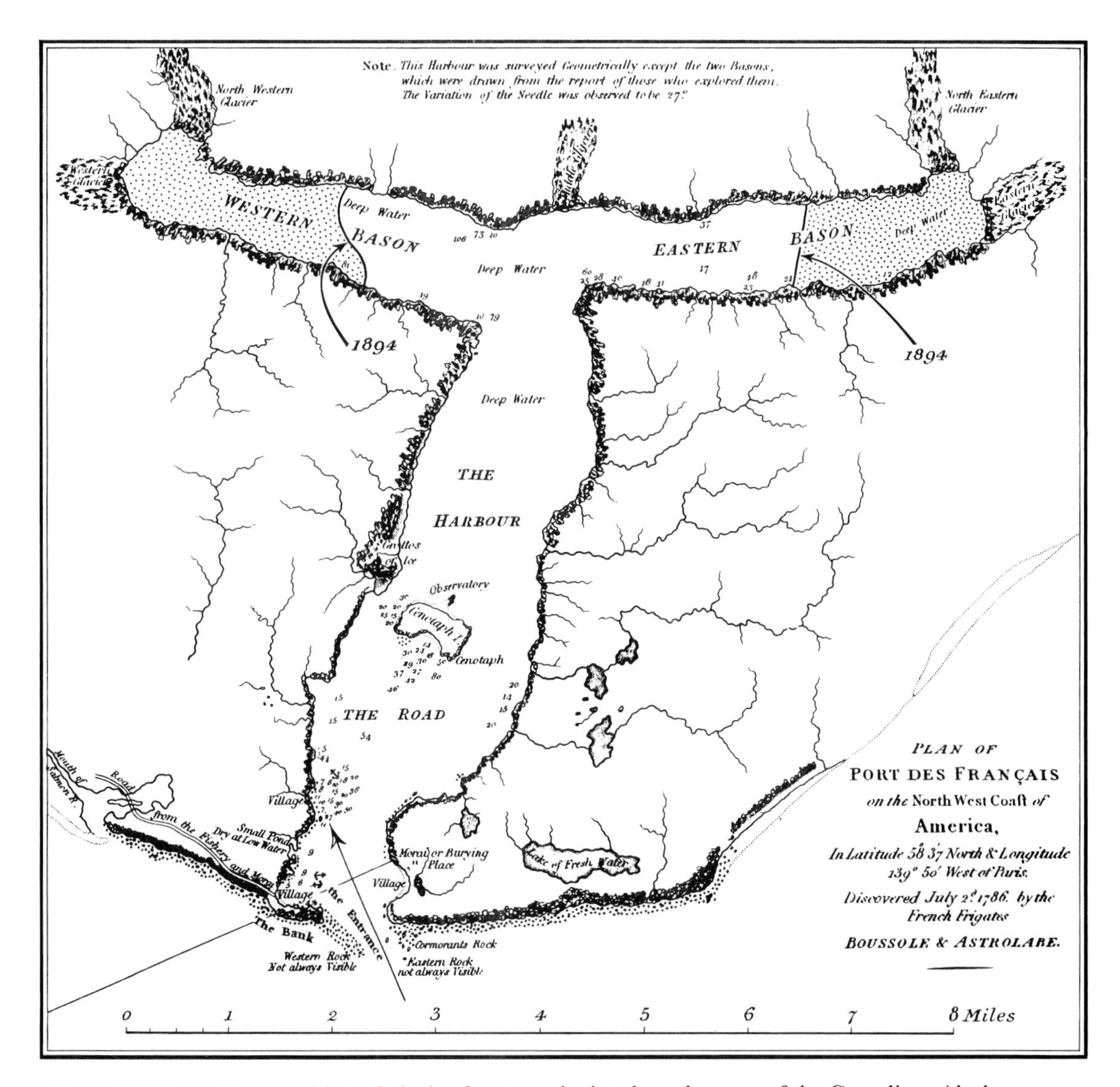

Figure 8.8 Lituya Bay: position of glacier fronts as depicted on the map of the Canadian–Alaskan Boundary Commission of 1894 superimposed on La Pérouse's map of 1799

suggested that there had been fluctuating glacial expansion between 1500 and the twentieth century but they did not attempt to itemize these expansions chronologically, and it would not seem that their growth curve was sufficiently accurate either to do so safely or to confirm that the outer moraines were caused by glacier expansion as early as 1500.

The outer moraine of the Gulkana glacier in the Alaska Range has been tentatively dated to 1580 and those of three others to 1650 (Reger and Péwé 1969) but these results depend upon comparison of the largest lichen thalli measured at a number of stations with the ages of the oldest trees near by. The resulting growth curve is strikingly different from that obtained by Denton and Karlén (1977) from the White River valley which also lies in the interior of the southern mountain belt and has a similar climate.

The Brooks Range forms a barrier roughly 1000 km long and 200 km wide which arcs east to west across northern Alaska between latitudes 67° and 70°N. In the centre of this Arctic range, mountain peaks reach 1200 to 1800 m in the south and 2100 to 2400 m in the north. In the Franklin and Romansof mountains, with a combined ice cover of 260 km^2, most of the glaciers lie above 1500 m and include small icecaps, cirque glaciers, and valley glaciers up to 8 km long. At Anaktuvuk Pass in the north-centre of the range the mean annual temperature is about − 10.1 °C and mean annual precipitation about 280 mm, of which 75 per cent falls as snow. A lichenometric curve for the central Brooks Range was set up using measurements of the largest diameter of *Rhizocarpon geographicum* sp. and the faster-growing *Alectoria miniscula/pubescens* (Calkin and Ellis 1980).

Major periods of moraine stabilization for fifteen cirque glaciers in the Kilich river area were dated at 750, 320 and 70 BP. This work was later extended to other areas in the Brooks Range (Haworth *et al.* 1983) and similar results obtained. More than half of the glaciers investigated in the west and central parts of the range had especially well-defined moraines dated to 475–325 BP. These dates cannot be taken as very accurate for the growth curve was calibrated by radiocarbon dates of which two were less than 400 BP, as well as dendrochronologically dated control points and dated surfaces from an Eskimo village abandoned between 1873 and the 1880s and gold placer mines believed to have been abandoned soon after 1901. Calkin and Ellis estimate their curve to be accurate to within 20 per cent. It may be speculated that some of the moraines in the interior ranges include members deposited in the sixteenth or seventeenth centuries, while in the coastal areas with heavier precipitation such moraines could have been overwhelmed by later advances. There is no doubt that the little glaciers in the Brooks Range and other interior mountains have receded in the present century. Comparison of present conditions with those shown in photographs from 1901 and 1911 show general recession and thinning throughout the area.

Records from a thermal borehole in the Ogoturuk Creek area in northwest Alaska indicate that mean ground surface temperature has risen 2 to 2.5 °C during the last 100 years, while similar records from the Barrow area indicate a temperature rise of 4 °C since 1850, with about half the increase occurring after 1830. There are indications from both sites that recent cooling has penetrated the upper parts of the profiles (Péwé 1975).

8.5 Baffin Island

The present-day glaciation limit for the area of Baffin Island around the northern part of the Barnes icecap is between 0 and 300 m above the surface of the plateau on which the icecap lies. Bradley (1972), using an observed mean lapse rate of 0.5 °C per 100 m, argued that a mean summer cooling of 1.5 °C would lower the limit below a substantial part of the land surface. Ives (1962) had suggested that extensive lichen-free areas represented the ice cover of the Little Ice Age and that while only 2 per cent of the region is now glacierized, it has recently been as much as 70 per cent ice-covered. The reliability of lichen-kill zones as indicators of Little Ice Age snow cover has been questioned (Koerner 1980) but there is no doubt that the topography and geographical position of Baffin make it an area peculiarly sensitive to changing climate (Williams *et al.* 1979) and it may be accepted that glacial extension here at high levels in recent centuries was on a much larger scale than in other parts of North America so far discussed. It seems probable that the area mapped by Andrews *et al.* (1970) may be taken as giving a reasonable indication of the extent of snow cover in recent centuries, whether the zone represents that in which lichen was killed by glacial extension or that in which lichen sparsity is due to high incidence of extensive seasonal snow cover.

In the Cumberland Peninsula, the easternmost part of Baffin Island, surmounted by the Penny Icecap, recent moraines have been dated by lichenometry (Miller 1973). The forefields of forty-six glaciers were examined and the lichens on 100 moraine crests. Advances were found to have culminated around 750, 380 and 65 years BP, but the lichen growth curve, dependent on radiocarbon data, was regarded as a first approximation and its reliability was rated at about 20 per cent.

In the Watts Bay area of southwest Baffin Island, relative age units have been defined using both lichenometry and weathering-rind data (Dowdeswell 1984) and the classification of the data was confirmed by discriminant analysis (Dowdeswell and Morris 1983). Moraines of three age groups were identified at the margins of the Grinell icecap and a nearby cirque glacier. The two youngest sets of recessional moraines, formed during temporary slowing of more general retreat, were judged to have been formed not more than 130 years ago and the older ridges some sixty years earlier. Moraines about 100 years old were also identified near the Terra Nivea icecap, about 40 miles east of Watts Bay (Müller 1980), in northern Baffin near the Barnes icecap (Andrews and Barnett 1979) and also in the Pangnirtung area (Davies 1980). The recession of the last century which has been so marked in the northwest USA, western Canada and Alaska, has evidently also taken its toll of the glaciers in the eastern Canadian Arctic.

8.6 Mid- to late twentieth-century glacier fluctuations in North America

In North America there was a change from marked glacial recession towards stability or advance about 1945. The Palisade glacier in the Sierra Nevada is known to have been increasing in thickness in 1947. At Tatoosh Island, near the Straits of Juan de Fuca, a trend towards decreasing temperature and increasing precipitation was under way by 1943 and glacier growth in the Cascades and Olympics followed (Hubley 1956). By 1953–5, of seventy-seven glaciers investigated in this region, fifty were advancing at rates of 3 m to 100 m a year. Of the remainder, all but one was showing clear signs of increased thickness. The first glacier known to advance was the Coleman on Mt Baker (Harrison 1956b). The tongue of the Nisqually advanced 100 m between 1961 and 1971. Inland, the glaciers in the Wind River Range were swelling at high levels in the early 1950s, although the fronts themselves were still retreating (Harrison 1956a). Further north in Banff National Park, the Athabasca and Saskatchewan slowed down their precipitate retreat after 1950 as did the smaller Drummond glacier (Nelson *et al.* 1966). The expansion of the 1950s observed in the Rockies, Cascade and Olympic mountains did not show up in the Juneau area, although there was a decrease of the rate of retreat of the Lemon Creek glacier beween 1948 and 1958.

In 1961 the state of 475 glaciers in western North America was surveyed from the air (Meier and Post 1962, and Figure 8.9). Detailed studies have shown that the accumulation area ratio of glaciers, that is the ratio of the area above the equilibrium line to the total area, is closely related to specific net budget or mass balance. For a glacier with a linear increase in net budget with altitude and a symmetrical distribution of area about median altitude, an accumulation area ratio (AAR) of 0.5 would indicate that the glacier was in equilibrium and neither advancing nor retreating. AARs can be readily calculated from air photographs taken towards the end of the summer season. It is also often possible to judge from air photographs whether glaciers are advancing, retreating or stationary (La Chapelle 1962). Advancing glaciers are characterized by convex tongues and many crevasses; retreating glaciers commonly terminate in a fine feather edge and have fewer crevasses, while stagnant tongues often have concave surfaces and are frequently more or less covered with ablation moraine.

Meier and Post examined 960 glacier tongues appearing on oblique air photographs taken in 1961. A consistent pattern was revealed (Figure 8.9). The area along the Pacific coast from the Kenai Peninsula to latitude 53° north in British Columbia had AARs of 0.6 and the glaciers appeared healthy. Over a large area of eastern Alaska, the southern half of the Coast Range of British Columbia and the Monashee mountains, the AARs ranged from 0.5 to 0.6. It was believed that values in the eastern part of the Alaska Range and the Wrangell Mts were probably similar but little data were available because of weather conditions. The glaciers of the Canadian Rockies, the Cascade Range and Montana had AARs of 0.25 to 0.5. In the western part of the Alaska Range, the Rocky Mountains in Wyoming, and the remnant glaciers of Idaho, AARs were less than 0.2.

Where AARs were more than 0.5, more than half the glaciers appeared to be active or advancing; where they were less than 0.2, all were retreating or stagnant. At the time of the 1961 survey, glaciers in the Cascades such as the South Cascade had negative mass balances once more, but further small increases of mass balance were to take place in the 1970s. Small glaciers like the avalanche-fed Vesper in the northern Cascades are much more sensitive to climate fluctuations than the more massive glaciers in the Rockies (Dethier and Frederick 1981).

The lowering of temperature which affected western North America was also influential in the eastern Canadian Arctic (Bradley and Miller 1972, Bradley 1973). Seasonal means of temperatures on Baffin Island showed a decrease of as much as 2.1 °C for the June–August ablation season for the period 1960–9 and an increase of as much as 2 °C for the accumulation season September–May. Winter precipitation increased markedly everywhere with surprising consistency considering that local topographic features cause total precipitation amounts to vary widely from place to place. The lower summer temperatures were apparently caused by increasing advection of cool air from the north and east and a concurrent decrease in warmer air from the west and southwest.

A single decade was sufficient for the snow cover to respond to the new conditions. Comparison of air photographs showed that while snowbanks had been diminishing

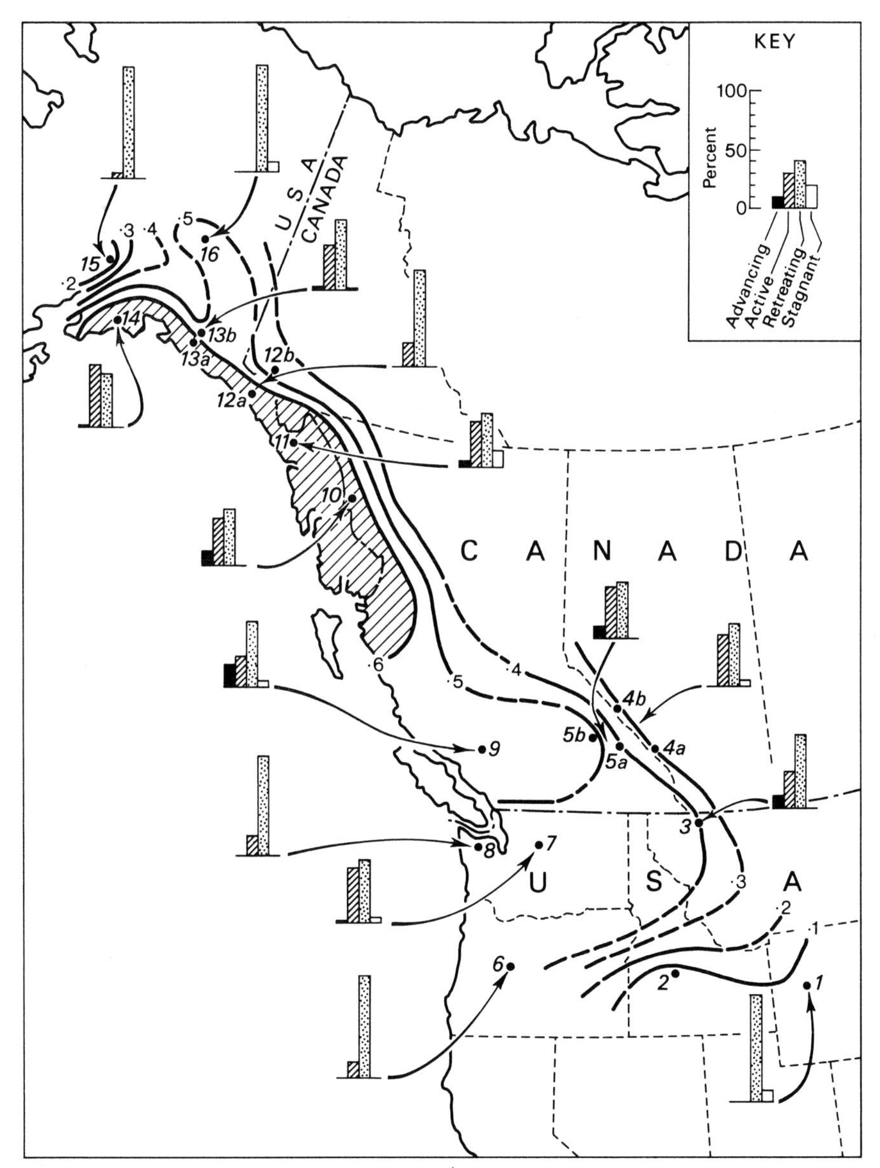

Figure 8.9 Distribution of values of accumulation area ratio (AAR) over western North America in 1962 in relation to glacier advance and retreat. (From Meier and Post 1962)

between 1949 and 1960, in the next ten years permanent snow cover increased markedly. Snow patches less than 10 m in diameter doubled in size; larger snow patches, more than 50 m in diameter, expanded less than the smaller ones. In the early 1970s, permanent snow was lying in many areas that had been snowfree at the same seasons in the 1960s. Extension of snow over lichen-covered surfaces suggested that a decade of climatic deterioration had been sufficient for the snow cover to become as extensive as it had been forty years earlier. Glaciers, because of their longer response times, continued to retreat between 1960 and 1971.

On Devon Island an abrupt transition to cooler summers took place from 1962–4. Annual degree-days totals here are highly correlated with mass balance values. It has therefore been possible to reconstruct the mass balance for the northwest of the icecap back to 1947–8 (Bradley and England 1978). It emerges that the mass of ice lost was six or seven times greater in the period 1947–8 to 1962–3 than it was in the succeeding decade. Icecap growth here is, however, limited by low values for precipitation. An occasional warm summer can wipe out the cumulative gain of a succession of years with positive mass balances. A fall in temperature in this high Arctic environment has therefore a much less noticeable effect on glacier expansion than it has in regions with higher precipitation (Figure 8.10).

The White glacier, 80°N on Axel Heiberg Island, switched from a strongly negative balance in 1960 to a slightly positive balance in 1961 (Müller 1962). Snow beds and small glaciers were reported to be expanding on northern Ellesmere Island by Hattersley-Smith and Serson (1973) but Bradley (personal communication, November 1984) reported that the northern Ellesmere Island icecap shrank by 11–14 per cent between 1972 and 1983 and mass balance measurements showed a net loss of about 0.4 m in thickness. Differences in nourishment have caused noticeable variations in response in nearby

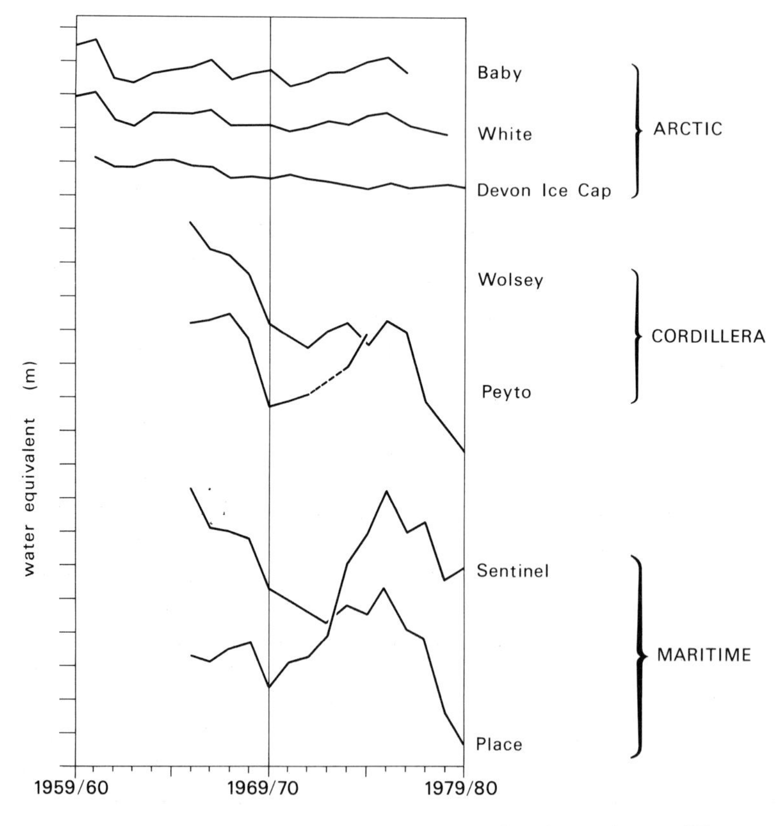

Figure 8.10 Cumulative mass balance values of selected glaciers in arctic, cordilleran and maritime North America. (Derived from figures published in Kasser 1967, 1973, Müller 1977 and Haeberli 1985)

glaciers. In the Borup Fjord area of northwest Ellesmere, the Carl Troll glacier advanced 20 m between 1950 and 1959 and retreated 50–70 m by 1978 when it was showing signs of expanding again. The Webber glacier advanced 60–150 m during the same period, 1959–78. Its accumulation comes from a large icefield nourished by superimposed ice and drifting snow, while the Carl Troll depends mainly on accumulation in small cirques (King 1983).

Recent seasonal and interannual variations of global snow and ice cover have been observed since 1967 by means of satellite sensing (Kukla and Gavin 1980). The mean extent of snow and ice in the northern hemisphere peaked in 1973, and snow was again very extensive in 1976. Global cooling at surface levels amounted to 0.5 °C between 1958 and 1970; subsequent warming gave 1980 and 1981 values approximately 0.1 °C higher than those of 1958 and 1959. There was again appreciable cooling between the northern springs of 1981 and 1982 (Angell and Korshover 1983). At present it is not known whether the cooling trend since mid-century has come to an end or not. On the Grizzly glacier, a sub-polar cirque glacier in the Brooks Range, positive mass balances since 1977 switched to negative in 1981–2 and the July temperature rose from 1.6 °C in 1981 to 4.3 °C in 1982 (Calkin *et al.* 1983). It remains to be seen whether this is a significant indicator or whether it was just a minor oscillation.

8.7 Greenland

When early settlers, including Erik the Red, succeeded in establishing permanent settlement in Greenland towards the end of the tenth century AD the climate was relatively benign. Their settlements persisted for several hundreds of years and, indeed, survived long after contact with Scandinavia had broken. From archaeological evidence it is known that Herjolfsnes remained inhabited until after 1480, though it is clear that the climate had deteriorated by 1350. This is shown by the fact that costumes from burials were penetrated by tree roots until those of the 1350s and subsequent decades, which remained in a remarkable state of preservation, indicating that not long after burials of this time the ground a metre or two deep had become frozen throughout the year (Norlund 1924, Hovgaard 1925, Gad 1970).

The isolation of the Greenland settlements and their eventual demise were the outcome of both climatic deterioration and economic and political factors. Drift ice discouraged shipping and in addition the attention of the authorities in Scandinavia was increasingly diverted to continental affairs. The Norse farmsteads dwindled and eventually became deserted in mysterious circumstances; Greenland was left to the Inuit.

Until the eighteenth century the positions of the ice edge and the extremities of the glaciers remained unrecorded. Then, in 1729, Governor Paars wrote to the king of Denmark describing a visit he had made to the edge of the Inland Ice in the Godthåb district, mentioning that the ice edge was near to a waterfall. The only possible location for the waterfall is in Austmannadalen, at a position well forward of the present ice edge, where a trimline marks the extreme limit of ice advance in recent centuries (Bobe 1936). Weidick's (1959, 1963) careful studies of all the historical evidence relating the extension of the ice in east Greenland show that the eighteenth-century expansion was general. A trader called Peder Olsen Walloe made a reconnaissance of southernmost west

Greenland, setting off in the summer of 1751 from Frederikshåb in an umiak (a large Inuit canoe) towards Julianehåb. 'The Greenlanders say', he wrote in his diary,

> that the ice increases every year, which is mostly recognisable from the fact that tracks where the Greenlanders used to go hunting are now quite overridden and covered by ice, and, as far as may be concluded from their simple chronometry, the change that has taken place in a score of years is very considerable. (Weidick 1959)

Detailed confirmation comes from Otho Fabricius, the incumbent missionary at Frederikshåb from 1768 to 1773:

> the ice spreads out more and more every year . . . the experiment has been tried of erecting a post on the bare ground a good distance from the ice, and the next year it was found to be overtaken by it. So swift is this growth that present day Greenlanders speak of places where their parents hunted reindeer among naked hills which are now all ice. I myself have seen paths running up towards the interior of the country and worn in bygone days but now broken off at the ice, which confirms the Greenlanders' statement. The ice advances especially in the valleys and where these reach the sea and the heads of fjords (I mean the inner ends of the fjords) it becomes so dominating as to have great floes hanging over the water. (Weidick 1959)

It was not only the Greenlanders' reindeer hunting paths that were disappearing under the ice. In 1765, an Icelandic pastor, Egill Thorhallesen, was sent to Godthåb and made summer expeditions to both north and south for two years before joining the Godthåb mission. He wrote a brilliant description of the region (Gad 1970), in which he mentioned that the Greenlanders 'also speak of other ruins, some of which are still to be seen and some are already in under the ice which has laid itself over the entire hinterland, indeed over the highest mountains and filled the valleys between them'. Thorhallesen was not too concerned with the extent of the ice advance which had overwhelmed some of the old Norse ruins. He reported that he had no doubt that Icelandic farmers could still make a livelihood for themselves if they were settled on the fjords of southwest Greenland.

Glacier advances about 1600 and 1750 were inferred by Beschel (1961) from his studies of the moraines of the two Tasiussaq glaciers in the Sukkertoppen area. His pioneering study was influential in initiating the use of lichenometric dating but it was not really sufficiently well based to establish that there was an advance about 1600. The historical evidence demonstrates clearly that the glaciers of West Greenland advanced in the mid-eighteenth century, as did those in Norway. It is not known when these advances began nor yet whether they were succeeded by a general retreat. There could have been a stand-still until the beginning of the nineteenth century, though this is perhaps unlikely in view of the contemporary recession in Norway. For the period between the mid-eighteenth and mid-nineteenth centuries little information about fluctuations in the position of the ice is available except for one glacier, the Sermitsiaq in Tasermiut, which was reported to be advancing in 1833 (Weidick 1959).

From the mid-nineteenth century onwards much more information is available, especially after 1877, when a commission was appointed for geographical and geological investigations in Greenland (Kommissionen for Ledelsen af de geografiske og geologiske Undersøgelser i Grønland) and the appearance of the journal *Meddelelser om Grønland.* Pictorial records also survive. The best of the early charts and maps for glaciological purposes are those drawn by reindeer hunters and compiled by Rink in the mid-nineteenth century. Though often distorted in outline, they reveal good knowledge of detail. Like Jan Møller's pictures, they are safely housed in the Royal Library collection in Copenhagen (Weidick 1959)

Weidick (1968) reviewed the evidence for the fluctuations of both the Inland Ice and the local glaciers in western Greenland. Compiling data on some 500 ice lobes, he plotted the fluctuations of 135 lobes between Kap Farvel in the south and Upernavik in the north for which information is available for more than three decades, providing detailed lists of all his sources. The results show very marked consistency of behaviour although a few anomalies are revealed, especially amongst calving glaciers. The oscillations of 94 per cent of the Inland Ice lobes were in phase despite their wide geographical distribution and differences in sub-glacial topography, size and dynamics. Predominant retreat since the mid-nineteenth century was interrupted by halts or readvances around 1880 and 1920. In some places in the south the readvance of 1880 covered nearly all the trim zone; further north, in Disko Bay, it brought glaciers generally to positions near their historic maxima. The lobe Pakitsup ilordlia in Qingua kujatdleg was exceptional in advancing beyond the earlier maximum over a vegetated area. The 1920s readvance was on a smaller scale but tended to be more marked in the north than the south (Davies and Krinsley 1962). Afterwards a general retreat set in that was fastest in the decades before 1940. Since the historic maximum, total retreat of the majority of the lobes of the Inland Ice has been of the order of a kilometre. Most of those with larger fluctuations, and all of those that have oscillated through more than 5 km, are calving lobes.

The most numerous records of the behaviour of local glaciers come from the Sukkertoppen area and Disko Island. Like the lobes of the Inland Ice, most of these local glaciers reached their maximum recent extents before the nineteenth century, perhaps as early as 1750. The majority were close to their maximum positions until the mid-nineteenth century, although a few receded earlier leaving large areas of dead ice. The general reactivation of the fronts between 1880 and 1890 involved advances of lesser magnitude than the earlier ones. The subsequent retreat was interrupted by minor advances between 1915 and the mid-1920s, which in many cases sufficed only to stabilize moraines. The general retreat of the mid-twentieth century was fastest between the 1920s and 1940s.

Seven glacier forefields in the Ikamiut kangerdluarssuat mountains, north of Sukkertoppen (65°24′N, 52°52′W), were mapped and moraine ridges dated by lichenometry in 1978 (Gordon 1980, 1981). In addition, investigations of the environment of two outlet glaciers of a small icecap were made. Air photographs on a scale of 1:10,000, taken by the Danish Geodetic Institute in 1968–9 provided a base on which ice fronts were mapped in 1978. As the linear growth rates of the lichens had to be estimated from the maximum size of thalli inside the ice margin as it appeared on 1942 air photographs, the results must be regarded as tentative. Photographs taken of lichen thalli in 1978 against a scale rule should allow better calibration in the future. Estimates made of ice retreat from moraines about 1745, 1850, 1885, 1930 and 1944 are in agreement with documentary and other data tabulated by Weidick (1962). Evidence of a more extensive glaciation before 1745 was found on two of the forelands. The most prominent end moraines and trimlines were formed about 1850. Between 1945 and 1968/9 the nine glaciers receded for distances between 180 and 570 m. Recession of three of them continued until 1978, no lichens or plants growing near their fronts. Six glaciers advanced between 20 and 158 m between 1968 and 1969; patches of moss were growing right up to the ice margins.

Elsewhere in Greenland, some glaciers have become stationary and some have advanced in the last few years. Weidick's (1968) graphs show a generally reduced rate of retreat or stability after 1940 and there were signs of readvance in Disko Bay by 1960. The valley glacier Sermikavsak, on Upernavik Island, was advancing in the 1970s (Gribbon 1979).

The most striking feature of the Little Ice Age record from Greenland is the synchroneity of the oscillations of the Inland Ice and the local glaciers. The major trends distinguished by Weidick appear also to have affected North and East Greenland. The historical data for northern Greenland between Kap York and Nioghalvfjerdsfjorden was examined by Davies and Krinsley (1962), using for comparison accounts by Chamberlain and Salisbury written after the Peary auxiliary expeditions of 1894 and 1895. All the glaciers had retreated markedly though by 1960, out of the 203 glaciers considered, 160 were stationary. Much less analysis has been attempted of the scattered data available for the east coast.

The uniformity of trend displayed in Weidick's (1968) graphs strongly suggests the dominant control of temperature. Minor readvances of the Sukkertoppen glaciers are explicable in terms of periods of low summer and autumn temperatures in the 1880s and low summer temperatures in the 1890s, and low temperatures at all seasons in the 1920s, recorded at the meteorological stations at Godthåb (64°10′N and 51°33′W) and at Sukkertoppen. Between 1875 and 1920, periods of climatic deterioration were short and so the general recession was broken by only short periods of enlargement. The rapid recession between 1920 and 1942 was associated with long ablation seasons associated with higher spring, summer and autumn temperatures. A slight decrease of mean summer temperature between 1942 and 1960, shortening the ablation season, was followed by climatic deterioration in the Arctic during the 1960s and 1970s. In West Greenland, summer temperatures decreased and there were colder and drier autumns. For 1968–73 the mean temperature of 3.4 °C at Godthåb was the lowest 5-year mean since records began in 1875. Both winter and summer precipitation increased sharply in the 1950s and 1960s so that the effects of lower temperatures were reinforced by higher precipitation.

Glaciological and hydrological investigations have been stimulated in recent years by the opportunities for developing hydro-electric power resources. By 1981, mass balance measurements were being made on eight glaciers between Søndre Strømfjord and Kap Farvel and the Geological Survey of Greenland was mapping and making annual measurements of the glacier margins. Work had also started on compiling information with a view to producing an atlas showing all West Greenland glaciers on a scale of 1:250,000.

Chapter 9

Glaciers in low latitudes and the southern hemisphere

These are amongst the least known of the world's glaciers. All – or nearly all – of them have retreated in the course of this century but information about the timing of their fluctuations in earlier centuries is sparse. There are a few historical records concerning the glaciers of Latin America and it is conceivable that others will be forthcoming from Iberian, Vatican or other archives. For Africa, New Guinea, New Zealand and the sub-Antarctic Islands no such hopes can be entertained (Figure 9.1).

The regimes of equatorial and tropical glaciers differ in many respects from those of middle and high latitudes. Precipitation near the equator usually occurs in two seasons of the year and in some cases is large enough for glaciers to extend down into the forest zone. In the vicinity of the tropics precipitation occurs during a single season and is usually limited in amount. Because of the aridity, ice equilibrium lines and snowlines rise to very high levels, especially on the lee side of mountain ranges. Ablation in such cases mainly involves sublimation, the direct transition from ice to water vapour, as a result of strong solar radiation inputs throughout the year. In more humid areas, the build-up of clouds in the afternoon results in glaciers more commonly being sited on western than east-facing slopes.

9.1 East Africa

Of the three glacierized mountains in East Africa, Kilimanjaro (5895 m) and Mt Kenya (5199 m) are both volcanoes. Ruwenzori, on the border between Uganda and Zaire, is a horst, an uplifted block of crystalline rock overlooking the western arm of the Great Rift Valley. Kilimanjaro rises from rather arid plains and has a higher snowline than either of the other two. On both Kenya and Ruwenzori, accumulation dominates from March–May and September–November, ablation in the intervening months. The four

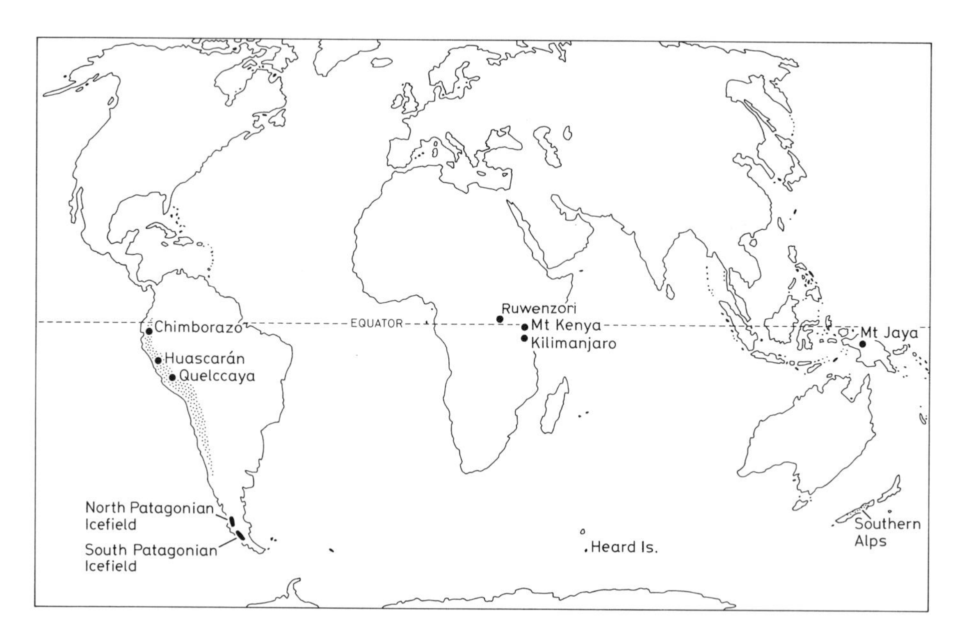

Figure 9.1 Distribution of glaciers in low latitudes and the southern hemisphere

seasons are reflected in the firn profiles on Ruwenzori, though their impress is limited by their short duration and the lack of strong seasonal contrasts (Temple 1968).

The East African glaciers, especially those on Mt Kenya, have been the subject of a book by Hastenrath (1984). He provides a remarkably detailed presentation of the evidence of the changing state of all the glaciers since the nineteenth century, with full photographic and cartographic as well as documentary inventories. Most remarkable are his photographs of the glaciers which he took from a hot air balloon in 1974.

Sets of moraines, only very sparsely vegetated, in front of all the glaciers mark more advanced positions of the ice in recent centuries. The moraines in front of the Lewis glacier on Mt Kenya, which is the most accessible and closely studied of all the East African glaciers, are multiple, with two or more crests up to about 50 m apart (Charnley 1959). They lie at a height of between 4270 and 4420 m (Figure 9.2). Between 1899 and 1974 the area of the glacier diminished from 0.63 km^2 to 0.31 km^2 and over the same period the elevation of the snout rose by 130 m (Hastenrath 1975). Shrinkage was at a maximum in the early part of the century, slowed down in 1934, and almost ceased in the 1960s. After 1963, though the tongue continued to retreat, there were signs of thickening of the upper glacier. This came to an end in 1974 and over the next four years the volume of the Lewis glacier is estimated to have diminished by 1.2 million cubic metres, i.e. by one-sixth, and a continuation of current conditions seems likely to result in a further drastic reduction in size (Bhatt *et al.* 1982).

The story of the Ruwenzori glaciers is very similar in site of the fact that their alimentation seems to depend more on moisture derived from the Atlantic than is the case on Mt Kenya. In the late nineteenth century it is clear that the ice cover was substantially greater than now; since then it has continuously diminished with the exception of the years 1959–62, when the glaciers advanced slightly or receded only slowly (Figure 9.3, Whittow *et al.* 1963, Temple 1968). The larger ice masses have become fragmented, the smaller ones have disappeared and little ice will be left on Ruwenzori by the end of the century if trends of recent decades persist.

The glaciers of Kilimanjaro, all of which are on the western Kibo cone and within the crater, have shrunk continuously since the end of the last century, those on the eastern crater rim disappearing entirely.

The shrinkage of the East African glaciers over the last century seems to reflect important changes in the climate of the region. It is known that Lake Victoria and other lakes in East Africa were much enlarged in the 1870s as a result of higher rainfall than the present. It is also known that in late 1961, about the time when the glaciers began to advance or retreated more slowly for a few years, the lakes rose sharply and then remained at high levels for some years.

Between 1920 and 1949, according to Sanson (1952), rainfall decreased by about 150 mm on Mt Kenya, though in the neighbourhood of Kilimanjaro and Ruwenzori it increased by 75 mm. Mean annual temperatures exhibited oscillations between 1915 and 1963 with interannual amplitudes of about 2 °C and with perhaps a slight cooling

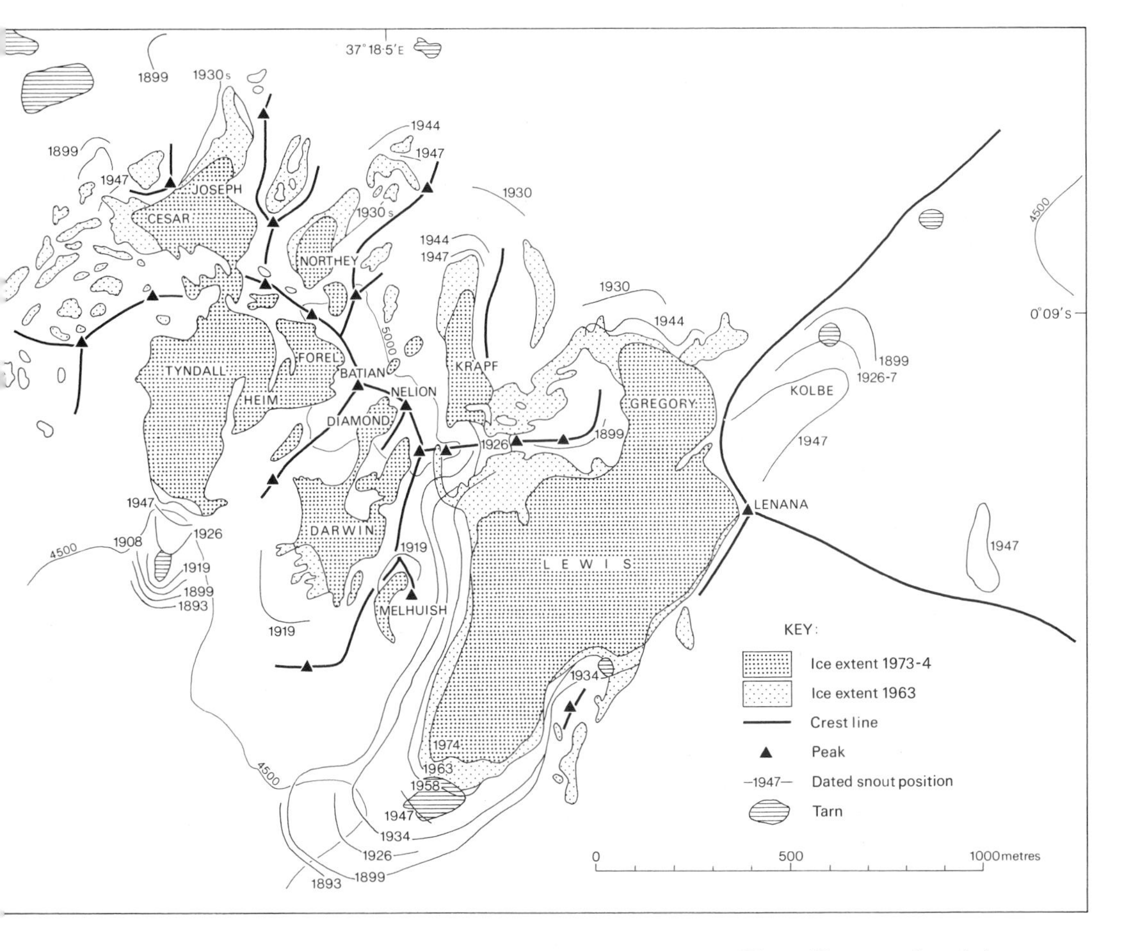

Figure 9.2 Changing extent of glaciers on Mt Kenya, 1893–1974. (From Hastenrath 1984)

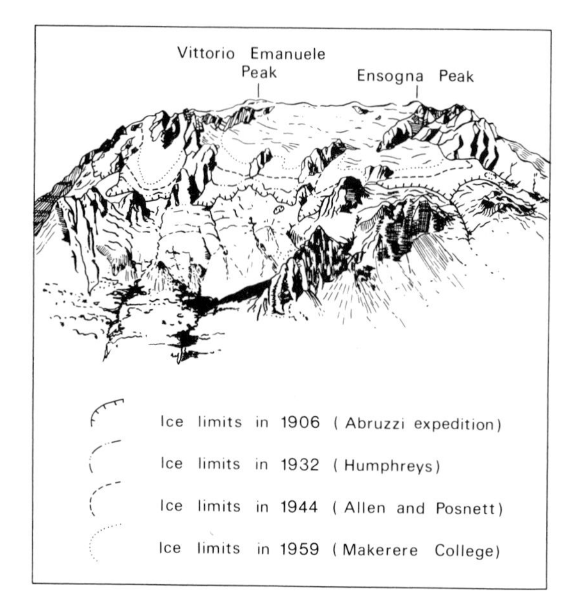

Figure 9.3a The retreat of the ice on the eastern side of Mt Speke, Ruwenzori, Uganda, 1906–59. (From Whittow *et al.* 1963)

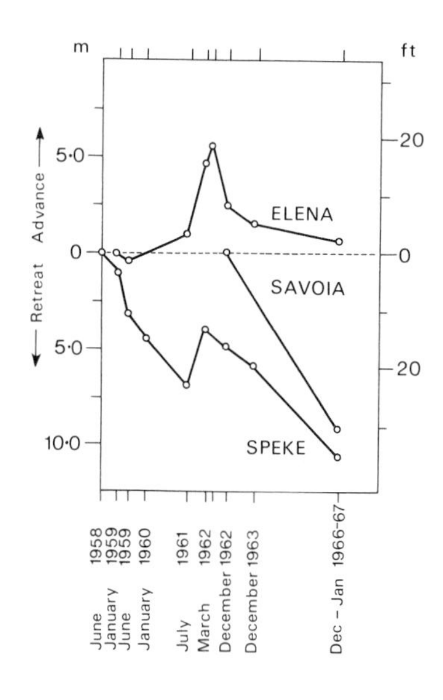

Figure 9.3b Fluctuations of the termini of the Speke, Savoia and Elena glaciers between June 1958 and January 1967. (From Temple 1968)

between 1933 and 1963 (Whittow *et al.* 1963). Taking a longer view, Hastenrath (1985) interprets the numerical modelling by Kruss (1984) of the Lewis glacier as indicating that the retreat since 1890 has been caused by annual precipitation diminishing in the two decades after 1883 by something of the order of 150 mm, with an accompanying decrease in cloudiness; he also finds evidence that temperatures in East Africa have been a fraction of a degree higher in the twentieth than in the latter part of the nineteenth century and that there was a particularly marked rise of temperature in the 1920s.

The moraines bordering the East African glaciers still offer opportunities for learning more about the fluctuations of the ice in recent and earlier centuries. Attention has of late been directed to the record provided by the ice itself. A core 11 m long, penetrating to bedrock, was extracted from the Lewis glacier in 1977 (Davies *et al.* 1979). It lacked annual banding but was characterized by peaks in the concentration of particulate material. These are believed to represent periods of ablation, each lasting some years or even decades. A layer of ice and firn above the highest dirt band probably represented the last period of net accumulation, starting in the early 1960s; the dirt band beneath, it was thought, might represent ablation and the concentration of particulate material over the period since about 1865. Other cores of a similar length have since been obtained (Hastenrath and Patnaik 1980, Thompson 1981, Thompson and Hastenrath 1981) but the information provided in the form of ice and firn horizons, oxygen isotope profiles and micro-particle content is difficult to interpret.

9.2 New Guinea

Several peaks of the Merauke range, the western half of the great Cordillera which runs 2000 km from east to west through the centre of New Guinea, rise above 4500 m. Three in Irian Jaya, Mt Mandala (Juliana 4640 m), Mt Idenburg 4717 m and Mt Jaya (Carstensz 4883 m), carry glaciers. The small icecap on Mt Trikara (Wilhelmina 4730 m), further east in Papua New Guinea, disappeared entirely at some stage between 1943 and 1972 and all the little glaciers scattered along the central cordillera have greatly diminished in the course of this century (Peterson *et al.* 1973). Mt Jaya (4°5'S, 137°10'E), part of a huge block of Miocene limestone, carries the largest area of ice; five separate glaciers on its horseshoe-shaped margins had a total area of 6.9 km^2 in 1972. Amongst these glaciers, the Meren and the Carstensz (discovered as late as 1913) were the object of concentrated attention by Australian expeditions in the early 1970s.

The glacier snouts reach down into valleys that appear to have been abandoned by ice in quite recent times, leaving behind a series of moraines. On the youngest ones with sharp crests and unstable slopes, plant colonization is at a very early stage; as no organic matter has accumulated between valley walls and lateral moraines, these can be presumed to have formed within the last century or two. The retreat on Mt Jaya over the last century or so has been traced from air photographs, dated cairns and expedition records (Figure 9.4 and Table 9.1). Extrapolation of retreat rates from the period 1936 to 1974 suggests that recession from the most recent moraines may have begun 120 to 150 years ago. This conclusion is supported by numerical modelling of the retreating glaciers (Allison and Peterson 1976). Such observations as are available from elsewhere are sufficient to confirm that the very marked recession of the Carstensz glacier is typical of all the glaciers in New Guinea.

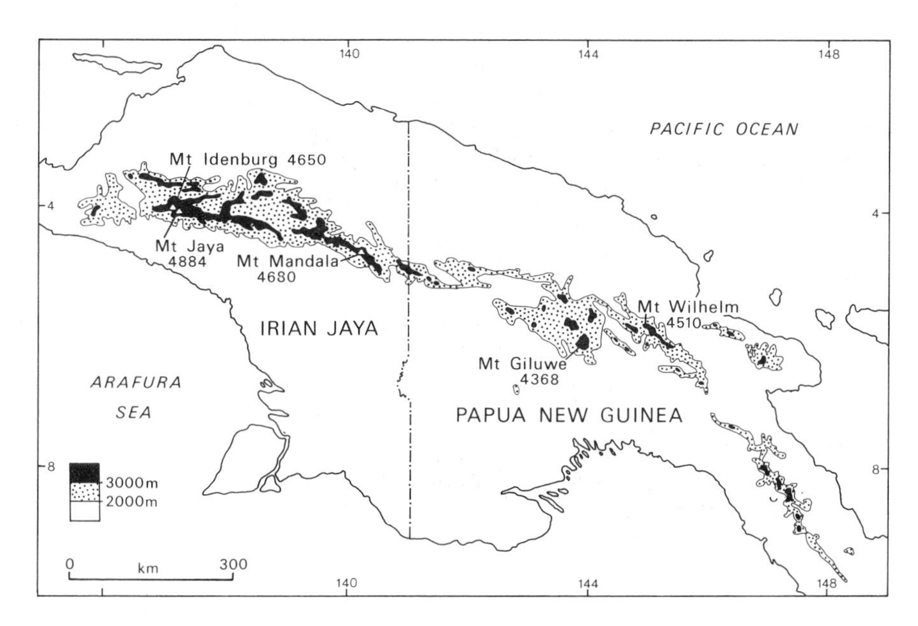

Figure 9.4 The mountains of New Guinea

Table 9.1 Rate of retreat of the Carstensz glacier, New Guinea, 1850–1974 (metres per year)

1850–1936	24
1936–42	54
1942–62	24
1962–72	37
1972–73	15
1973–74	100

Source: Peterson *et al.* (1973)

The outstanding feature of Mt Jaya's climate is its constancy through the year and the high humidity; it is only 80 km from the Arafura Sea. The mean diurnal temperature range for much of 1972 was only 2.7 °C at 4250 m and 3.4 °C at 3600 m, while the range of temperature for 10-day means at 3600 m was less than 1.5 °C. On most days cloud builds up in the middle of the day and then, in the afternoon and evening, rain falls at low altitudes and snow on the peaks (Allison and Bennett 1976). Total precipitation measured in the Meren area in 1972, an unusually dry year, was about 2440 mm, and it is thought that values between 2800 and 3300 mm are probably typical of the area as a whole, with the highest monthly totals between December and March. Hardly surprisingly, no significant stratigraphy was noted in 10 m cores taken from the firnfields of the two glaciers.

The observed retreat of the Carstensz glacier could be accounted for by a warming at a rate of 0.6 °C per century starting in 1830–50, with a thirty-year lag before significant retreat began (Allison and Peterson 1976). This corresponds to the warming of the tropics calculated by Mitchell (1961). Retreat might also be associated with increased cloud cover and reduced radiation inputs. A change in precipitation alone is unlikely to have been responsible for the recession; it has been calculated that a decrease in mean annual precipitation of about 1 m would have been required to account for it (Allison and Kruss 1977). A minor but significant and unusual role may have been played by decreased albedo of the ice caused by increased colonization by 'cryovegetation', mainly algae, some black, some highly coloured, which is unusually prominent on these glaciers (Kol and Peterson 1976).

9.3 South America

Glaciers are distributed along the eastern and western ranges of the Andes in Venezuela, Colombia, Ecuador, Peru, Bolivia and northern Chile. Maps and photographs show that the glaciers on the Pico Bolivar of Venezuela have retreated dramatically in the twentieth century, some thinning by as much as 100 to 150 m and the ice-covered area diminishing by as much as 80 per cent (Schubert 1972). A reconnaissance expedition to the Sierra Nevada de Santa Marta of northern Colombia in 1939 (Cabot 1939) found a striking sparsity of glaciers and a profusion of glacial deposits, indicating rapid recent wasting which continued until 1969, by which time approximately a third of the ice present in 1939 had gone (Wood 1970).

In Ecuador, some of the glaciers are situated on great volcanoes aligned along fault zones on the western flank of the Cordillera Oriental and the eastern flank of the

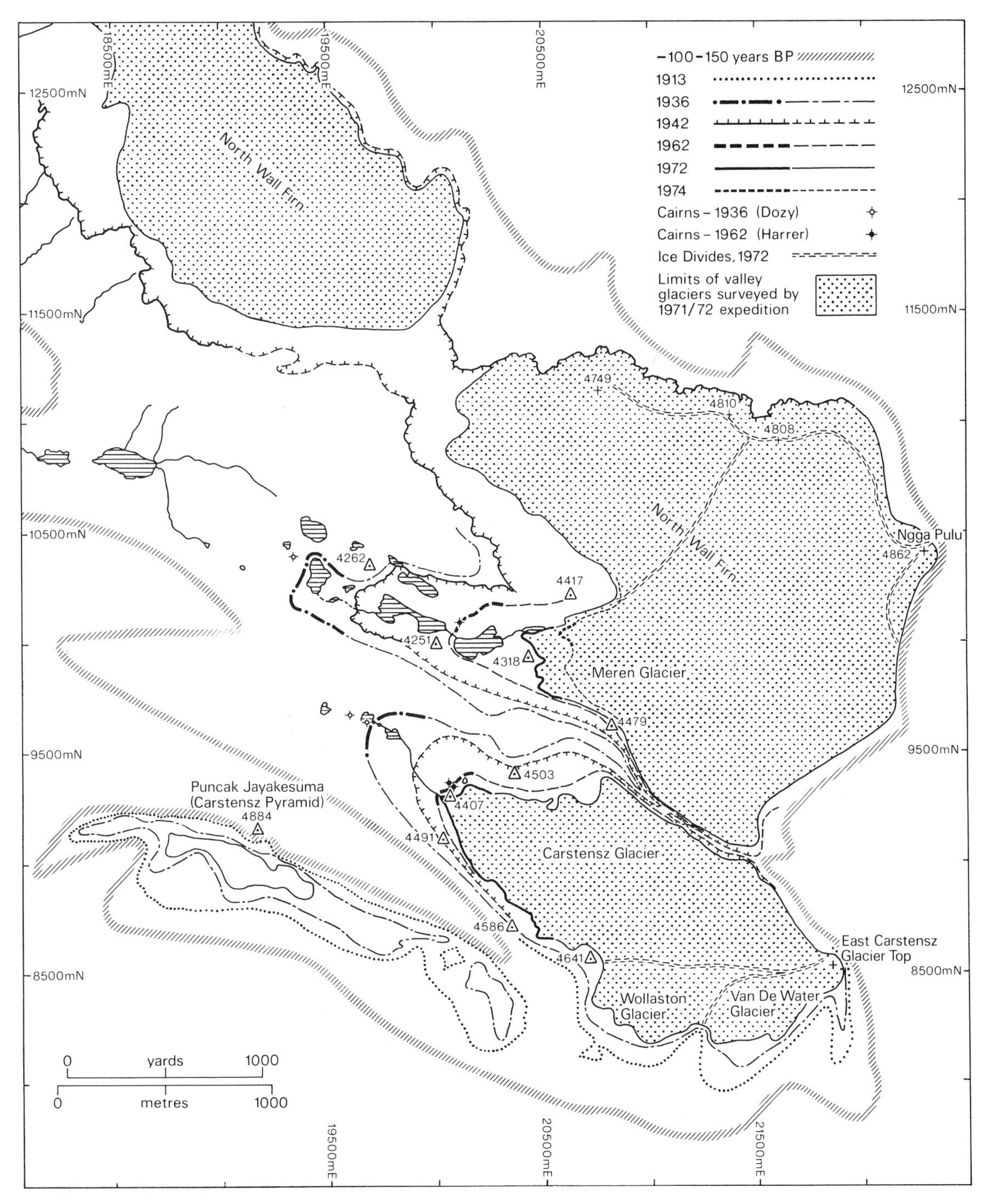

Figure 9.5 Changing extent of the ice on Mt Jaya (Carstensz), New Guinea. (From Allison and Peterson 1976)

Cordillera Occidental. As the Atlantic is the main source of precipitation, snowlines are lower on the east than the west. Mercer (1967) and Hastenrath (1981) have gathered together the scattered information available. It includes a few documentary descriptions in the Quito archives, dating from the early years of Spanish rule, which suggest that in the sixteenth century conditions were probably somewhat more severe than at present, with permanent snow on the crests of Pichincha, Corazon and other mountains now free of perennial snow (Figure 12.1). No information has been found for the period between 1580 and 1730. Later in the eighteenth century, records of the French Academy's geodetic mission of 1735–43 provide reliable data. Velasco, who lived in Ecuador from his birth, about 1727, until 1767, and von Humboldt, who travelled through Ecuador in 1802, left some accounts. The German geologists Reiss and Stubel in their studies (1886) of volcanoes between 1870 and 1874 made careful observations of contemporary glaciers. They were accompanied by the Ecuadorian artist Troya, some of whose pictures were used to illustrate the geological publications that eventually emerged; the originals in Leipzig do not seem to have survived. Whymper (1892), climbing in Ecuador in 1880, published several sketches of glaciers. Meyer's (1907) interest was centred on the glaciers and his writings are the most important early source of information. His photographs, forty-three of them, were published with an explanatory text in English (1908). They show that the largest glaciers on the northeast side of Chimborazo, 6272 m, and about 1°25′S, had all receded by the beginning of the twentieth century from fresh terminal moraines girdling the mountains at heights between 4000 and 5200 m. Air photographs taken in 1960 and 1977 of the Reschreiter glacier on Chimborazo show the ice bulging above the crests of its lateral moraines but by 1983 the surface of the ice was lower (Clapperton 1983).

Topographic maps based on air photographs have been produced in recent years by the Instituto Geografico Militar. The maps and photographs were used by Hastenrath (1981) in his 1974, 1975 and 1978 field studies and for assembling glacier records for individual mountains. He indicates that the ice-covered area is an order of magnitude smaller now than it was at the Little Ice Age maximum and that the regional snowline has risen intermittently over the last 250 years by about 300 m. He estimates that the volume of the Caldera glacier on El Altar, 5319 m, diminished by 50 million cubic metres between 1870 and 1900, and by a similar amount since then. He argues that such a shrinkage would have required a diminution of mean annual precipitation by about 500 mm, a change of albedo of about 15 per cent, or a reduction of cloudiness by one- or two-tenths. As he found in East Africa, the glaciers seem to be less sensitive to changes in precipitation than to changes in other climatic parameters.

The Peruvian glaciers are the largest between the tropics but many of them are in remote, unmapped places. An initial inventory and mapping of those in the Cordillera Blanca by the three Österreichischer Alpenverein expeditions between 1936 and 1939 found the glaciers strikingly different from those in the Alps, being characterized by gigantic, paper-thin cornices and abundant icicles. They were in advanced positions in the 1870s (Spann 1948); subsequent retreat was especially rapid after 1890 until 1910 when a readvance took place. The ice limit on Huascuran rose 430 m in altitude between 1886 and 1909 and then descended 150 m between 1909 and 1932, and rose again by 500 m over the next ten years (Broggi 1943). The glaciers near the head of the Laguna Perron in the Huandoy group shrank greatly between 1932 and 1947. The Atlante glacier in the northeast of the group is near to a mine-working and more is known about its variations than any other glacier in the range. Sievers (1914) had been told in 1909 that the firnline in the Cordillera Blanca had risen about 50 m since 1895. In 1932, Kinzl (1949) visited the glaciers and heard that a small advance had taken place about 1920, closing the entrance to one of the mines. The recession that followed averaged about 10 m a year between 1939 and 1948 but slowed to 2 m a year between 1948 and 1957 (Smith 1957). Systematic monitoring of glaciers in the Cordillera Blanca was prompted by disasters with serious loss of life resulting from avalanching and the overflowing of moraine-dammed lakes (Kinzl 1970, Lliboutry *et al.* 1977).

The glaciers in the Cordillera Blanca are typically fronted by well-marked systems of Little Ice Age moraines which are set apart from older moraines in many cases. Their number and size vary from one glacier to another, some forming long crescentic loops 50 m high (Clapperton 1972) and most of these hold back proglacial lakes. Lliboutry *et al.* 1977 noted signs that many glaciers had advanced towards these moraines again in the early twentieth century. Recession after 1932 involved the appearance of pools of water which united and enlarged to form supraglacial lakes, which themselves expanded until they reached frontal moraines. Kinzl's maps and photographs of 1948–50 disclose that ten big lakes had appeared between Quebrada Ulta, above Carhuas, and Quebrada Shallap, above Huaraz. Some of these lakes enlarged swiftly. The Laguna de Safuna Alta did not exist in 1932, was only a pool in 1950, but by 1969 a body of water 4.85 million cubic metres in volume had collected and was threatening a small town 40 km downstream. Glacier recession and the consequent formation of proglacial lakes behind morainic dams caused a great increase in large and sudden floods of liquid mud (*aluviones*), not all resulting from the bursting of moraine-dammed lakes, but the largest from huge rock and ice avalanches, notably the 1962 aluvion which overwhelmed Ranrahivca drowning 4000 people and that of 1970 which again devastated Ranrahivca and also the neighbouring town of Yungaya, causing 15,000 deaths. A similar disaster had occurred in 1725 but no other had been reported until 1938, when the mid-twentieth century retreat was in full swing. A commission intended to control the lakes of the Cordillera Blanca, set up in 1941, succeeded in gradually lowering the levels of thirty-five of the most dangerous lakes by means of pipes and concrete-lined trenches, so effectively that an earthquake in 1971 failed to cause any outbursts from the lakes which had been treated.

The glaciers in the Cordillera Huayhuash, a spectacular range 30 km long, east of the Cordillera Blanca, are steeper, thinner and more crevassed than those in the Alps. The largest are about 4 km long. In 1909 Sievers (1914) found that many had evidently retreated about 150 m from massive, fresh moraines. They were still in much the same positions in 1939 and Kinzl (1955) considered that, like others in the Cordillera Blanca, they had probably advanced in the 1920s. Between 1936 and 1954, although the larger glaciers showed little change, the smaller ones noticeably shrank (Mercer 1967).

The glaciers in the Andes Centrales to the north and south of the trans-Andean railway have receded since the middle of the nineteenth century in a spectacular fashion. When Hauthol (1911) visited the area in 1908 he found the glaciers receding but inferred they had advanced briefly but sharply in 1886–7. Some peaks such as Paragate had small glaciers in 1917, but these had shrunk by 1923 and disappeared completely by 1942. The snowline is estimated to have been at about 4600 m in 1862, 4900 m in 1923, and 5100 m in 1940 (Mercer 1967).

The Quelccaya icecap offers a peculiarly good opportunity to obtain palaeoclimatic information from ice cores. With an area of 55 km^2 and a central summit at 5650 m rising 400 m above the snowline, temperatures are so low, less than −3 °C, that percolation is inhibited, and the relief is low enough to allow the ice to remain little distorted. Net radiation is 'for all practical purposes zero' and so no energy is available for evaporation or melting. Under present conditions the mean annual net balance would seem to be about the same as mean annual accumulation (Thompson 1980). Though the annual temperature range is small, precipitation is seasonal with the Atlantic, 1800 km away, being the main original source of moisture, and thunderstorms recycling the water *en route*. The ice is well stratified on account of the seasonal snow accumulation.

Preliminary cores 15 m long revealed a periodic variation of micro-particle content, oxygen 18/16, and radioactivity, believed to be annual (Thompson *et al.* 1979, Thompson 1980). Two cores 155 m and 163 m long penetrating to bedrock were therefore extracted from the Quelccaya summit area in 1983 (Hastenrath 1985), in the hope that they would provide a key to regional atmospheric conditions over a long time period and possibly register the occurrence of El Niño conditions in the eastern Pacific, 400 km away. They have proved to be extremely informative, covering a period of over a thousand years. Dating dependent upon a combination of annual dust layers, micro-particulate concentrations, conductivity values and identification of ash attributable to the eruption of Huaynaputina (16°35′S, 70°52′W) in the spring of 1600, is rated as accurate to +2 years back to AD 1500. Decadal temperatures, inferred from 18O records, were generally low between 1530 and 1900. The details of temperature fluctuations at Quelccaya, plotted as departures from the 1881–1975 mean compare remarkably closely with those of the northern hemisphere as compiled by Groveman and Landsberg (1979). Values of conductivity and micro-particle concentration were higher than those obtaining during the fourteenth, fifteenth and twentieth centuries throughout most of the 1530–1900 period. This increase in dust deposition was probably due to increased wind velocities across the high altiplano. The wettest conditions of the last thousand years occurred between 1500 and 1720; in contrast, 1720 to 1860 was very dry. The Little Ice Age thus stands out in the Quelccaya record as an important climatic event which evidently affected this part of the southern hemisphere as clearly as the northern hemisphere (Thompson *et al.* 1986).

The glaciers in the widest part of the Andes in Bolivia and northern Chile have received relatively little attention until recently except for mapping carried out by the Österreichischer Alpenverein in the late 1920s. The Illampu–Anconhuma–Casiri area of the Cordillera Real, northwest and southeast of the La Paz railway, between 15°50′ and 16°60′S, was surveyed in 1928 and the altitudes of most terminal moraines were determined. The glaciers were quite close to moraines but there had been some shrinkage. Parts of the Bolivian section were little known until the 1950s when a British expedition mapped the snouts of fifty-one glaciers. Both retreating and advancing tongues were found and 'the fronts of many glaciers were reaching past old terminal moraines and covering plants that were at least ten years old' (Melbourne 1966). Jordan (1985) mapped 1775 glaciers in sixteen mountain ranges of the Eastern Cordillera of Bolivia, using 1:70,000 air photographs taken in July/August 1975, thereby providing an accurate base from which future change can be measured.

9.4 Extra-tropical southern hemisphere

The glaciers of the southern hemisphere outside the tropics, apart from those of Antarctica, are situated in the Andes, New Zealand and the sub-Antarctic islands. Their regimes are very much associated with the southern ocean and the westerlies.

9.4.1 SOUTHERN ANDES

Some of the Andean peaks in northern Chile are high enough to carry ice despite the aridity in the vicinity of Capricorn, but tiny glaciers like the one on Llullaillaco (24°43′S, 6723 m) between 6000 and 6500 m, are little more than sheets of ice with no crevasses or signs of movement. Like the Quelccaya icecap, they either accumulate or ablate over the whole surface area. It is reported that they showed no signs of recession in the 1940s and 1950s (Lliboutry *et al.* 1958) but as they have no moraines, less recent changes would be difficult to detect.

The arid zone reaches as far as 30°S, to the northern limit of winter cyclonic depressions. Glaciers up to 12 km long occur around 32°S (Corte and Espizua 1981), for instance on the slopes of Aconcagua (7040 m). Many glaciers in the Mendoza–Santiago region were retreating rapidly in 1945 (Mercer 1962, Corte and Espizua 1981); little seems to be known about their previous fluctuations. The best known, the Nevado Universidad and the Juncal Sur, are surging glaciers. Diminished flow in the San Juan Mendoza and other Argentinian rivers fed by the snow and ice of the high Cordillera led to publication of requests to alpinists to donate photographs taken before 1962, the year in which the area was covered by air photography, to facilitate the study of volumetric changes of ice on the Argentinian side of the Andes (Bader 1973). An inventory of the Argentinian glaciers from 28° to 35°S is in progress at the Institute of Snow and Ice at Mendoza (Corte and Espizua 1981, Cobos and Boninsegna 1984, Aguado 1984) and south of 35°S by Rabassa (Rabassa *et al.* 1978).

An analysis of the fluctuations of glaciers in the upper Atuel river basin in Argentina between 34°20′S and 35°20′S was made by Cobos and Boninsegna (1983). The Cordon Limite between Argentina and Chile, rising above 4000 m, supports glaciers on both sides. Striking changes over the last century or more have been reasonably well documented by travellers in the region. A base map published in 1947 was updated in 1961 and air photographs were flown in 1948, 1955, 1963 and 1970. On the western, Chilean slopes, the Ada glacier was observed in 1858 to reach down to 1797 m; by 1882 the front had risen to 1929 m and by 1980 to 2500 m above sea-level. The glaciers on the Argentinian side depend on redistribution of snow by southwesterly winds. In the past, ice from several tributary valleys reached down into the valley of the Atuel. Now ice is confined to cliff glaciers in sheltered ravines and only the Corto and Humo glaciers enter the Atuel valley. The situation in 1895 was observed by Hauthol (1895) who reported that ‘the present glaciers clearly indicate not only that until a short time ago they were of much greater extent but also that they are now losing volume and receding at a great rate’. At that time the lowermost kilometre of the Humo tongue consisted of dead ice covered in moraine. It was observed by Groeber (1947, 1954) in 1914 and again in 1937. Between his visits the glacier had retreated markedly; and he later estimated that between 1914 and 1947 the front receded 3200 m, involving the loss of 0.5 km^3 of ice. The retreat of

Table 9.2 Recession of the Humo glacier, Argentina, 1914–82 (metres)

1914–47	– 3200
1948–55	– 300
1955–63	– 160
1963–70	– 300
1970–82	– 150

Source: Cobos and Boninsegna (1983)

the Humo and its neighbours continued at least until 1982 (Table 9.2). All the other glaciers surveyed by Cobos and Boninsegna had also dwindled rapidly during the twentieth century, with the single exception of a glacier in the Laguna valley which advanced 1400 m between 1970 and 1982.

Air photographs flown in 1945 revealed much more ice in Argentina south of 40°S than had been indicated on earlier maps (Mercer 1967). High avalanche cones and reconstituted glaciers were found to be common between 43° and 45°20′S, signs of recent recession being widespread and comparable in extent with those in the Rockies. Most information is available for the glaciers of Cerro Tronador (41°10′S, 71°53′W) on the border between Chile and Argentina. The gently sloping accumulation area surrounding the steep summit of Tronador ends in cliffs over which the ice avalanches, giving its name 'Thunderer' to the mountain. Some of the reconstituted glaciers below are easily accessible. Little Ice Age advances brought the tongue of Rio Manso on the southeast side to the Holocene moraine (see Chapter 10.6). Lawrence and Lawrence (1959) found evidence from tree rings suggesting that the latest Rio Manso moraines were formed when the glacier was in an advanced state from the early eighteenth century to 1795, in 1809–21, 1832–4 and about 1847. Tables presented by Rabassa *et al.* (1978b) show that over the periods 1942–53 and 1953–70, three of the main Tronador glaciers were retreating at mean rates of something like 10 m annually. However, an advance, of the Rio Manso glacier, which the Lawrences thought was probably still continuing, had brought the front in the late 1950s into forests dating from the 1920s, and figures given by Rabassa *et al.* indicate readvances of the Tronador glaciers in the 1970s (see Table 9.3).

A great many independent glaciers and two icefields lie polewards of 46°S. The northern Patagonian icefield, the Hielo Patagonico Norte, extending from 46° to almost 48°S, with a length of 130 km and a width of 75 km, has an area, according to Landsat imagery, exceeding 4400 km^2 Its outlet glaciers on the eastern side have all shrunk during the twentieth century; those originating on or near Cerro San Valentin (4058 m),

Table 9.3 Fluctuations of Mt Tronador glaciers, Argentina, 1942–77 (metres per year)

Castano Overo		*Alerce*		*Frias*	
1942–53	0	1942–53	– 11.1	1942–53	– 16.9
1953–70	– 1.8	1953–69	– 9.9	1953–70	– 9.8
1970–4	+ 8.2	1969–70	– 16		
1974–5	– 15	1970–5	– 6.8	1970–6	+ 47.4
1975–6	– 12	1975–6	+ 18		
1976–7	+ 9	1976–7	+ 6	1976–7	+ 32

Source: From figures in Rabassa *et al.* (1978)

the highest peak in the southern Andes, have heavy debris covers and horizontal rather than vertical shrinkage has predominated. The whole area is dominated by cyclonic depressions moving in from the west. The outlets on the western, more maritime sides of the icefield are almost free of surface debris and the 1945 air photographs showed them to be much healthier than those on the east. Many of these tongues end in tidewater and access from the land is difficult. The climate is ideal, not only for glacier nourishment but also for the growth of rain forest so dense as to be almost impenetrable without the aid of machetes (Heusser 1960). Only the moraines of the San Rafael glacier, which currently calves into the 140 m-deep Laguna San Rafael (Plate 9.1), have so far been investigated. The distribution of lateral moraines indicates that the front has been much wider and more extensive in the past. The outermost moraines of recent centuries were dated by the Lawrences (1959) to 1882 on tree-ring evidence and the ages of trees outside them were taken to show that the San Rafael had not been more extensive than in 1882 for more than 500 years. The tongue extended 8 km across the Laguna when the first bathymetric survey was made in 1875. It was touching the edge of the Laguna in 1776 when Father José García Alsue observed large icebergs on the water, but John Byron made no mention of seeing icebergs during his voyage of 1742, and neither did Antonio de Vea in 1675 when he wrote of the glacier 'which extends from the beach to the land inshore' (Müller 1959). The Lawrences inferred from the scatter of information that a large advance had taken place between 1742 and 1776; such an advance would have obliterated evidence of any less extensive Little Ice Age event. This observation is not otherwise substantiated. Retreat has not been continuous since 1882. An inner moraine was stabilized by 1910, according to the tree rings; retreat was rapid between 1910 and 1935 and the 6-km front

Plate 9.1 The calving front of the San Rafael glacier in 1986. (Photo: Charles Harpum)

was then more or less stationary until 1959, when a sharp readvance was observed to be taking place and trees established in the 1950s were overthrown (Müller 1959). Recent retreat has also affected all the independent glaciers between the northern and southern icefields as well as many of those in the area to the east of the icecap between Lago Buenos Aires and Lago San Martin. A glacier flowing east towards Rio Meyer had receded only a short distance from its end-moraines when it was photographed in 1897, although it had obviously thinned considerably (Hatcher 1903). By 1945, the snout was 2 or 3 km from the end-moraine (Mercer 1967). A small glacier on Cerro Hermoso, showing signs of recent retreat on the 1945 air photographs, was reported by Magnani (1961) to be up against its terminal moraine. The scatter of data available shows that the recent behaviour of the San Rafael glacier has been in line with that of small independent glaciers in the region.

The southern Patagonian icefield, the Hielo Patagonico Sur, extends 360 km from 48°15′S to 51°20′S with an average width of 40 km. Major outlet glaciers run down to the sea in the west and many of the magnificent fjords of southern Chile are often choked with floating ice. The climate is extremely maritime; at Evangelistas, south of the icefield, mean monthly temperatures range between 4.4 and 8.7 °C; mean annual precipitation is 2620 mm. Mean annual precipitation on the icefield has been estimated to be about 7000 mm and drifting in blizzards with winds up to 200 kph occurs especially in summer (Shipton 1962). The topography is large-scale, the region remote and, below the level of the ice, very densely forested. Many of the western glaciers were scarcely known at all before the 1949 air photographs were taken. Calving glaciers like the Bruggen, ending in Fiordo Eyre, subject to rapid and anomalous advances, have attracted most attention. The Bruggen, which had a river flowing from it across a lowland when it was seen from the *Beagle* in 1830, had advanced to the opposite side of the fiord by 1926 (Agostini 1941), retreated 3 km by 1945, and readvanced 5 km by 1962 (Mercer 1967).

In the north, five western tongues, Ofhido Norte, Ofhido Sur, Bernado, Tempano and Hammick were investigated in 1967–8 (Mercer 1970). An excesis period of seventy years was suggested by the discrepancy between the age of a tilted but living cyprus (*Pilgerodendron uvifera*) and the ages of trees on the outermost recent moraine of the Tempano. All the tilted *Nothofagus* trees found proved to have rotten centres and so were useless for dating purposes. Radiocarbon dating of Holocene moraines shows that the two Ofhido tongues have not entered the Fiordo Ofhido for several thousand years. *In situ* tree trunks, rooted in peat and buried in outwash, occur outside the recent moraine sequence. The centre part of a 100-year-old tree has been dated to 800 ± 95 years BP (I–3827), the eleventh to fourteenth century AD according to the kind of corrections proposed by Klein *et al.* (1983), when the trees were overwhelmed by a glacial advance. When visited by Mercer, the Ofhido Norte ended in a proglacial lake bounded by a massive, multiple end-moraine. Outside this, but within 200 m of it, there are three older moraine ridges. Mercer examined sections of the trees on each moraine and, assuming the growth rings to be annual, found the ages to be 45, 50 and 105 years. He concludes that, allowing for 70 years for the establishment of trees, the moraines were formed between 1790 and 1850. Air photographs show that the ice front reached the base of the multiple, undated end-moraine in 1945. Mid-twentieth-century retreat was swift, for the terminus had retreated as much as a kilometre by 1958. The main tongue of the Hammick ends 12 km from the Fiordo Tempano, less than 50 m above sea-level. Moraines of this glacier were dated in the same way, suggesting that the glacier reached its maximum recent extent about 1750 and readvanced about 1840. The ice margin in 1968 was 50 to 200 m from an afforested end-moraine, with trees 150 years old on the outer face and 60 years old on a smaller moraine superimposed on the inner face. The terminus had retreated 50 to 60 m between 1945 and 1968. The calving fronts of the Tempano and Bernado also seem to have reached their greatest extents of recent centuries in the eighteenth century, with moraines formed between 1750 and 1800. In 1968, all the glaciers on the western side of the southern icefield were receding but were much closer to their recent maxima than those on the eastern side (Mercer 1970).

In general, glaciers on the eastern side of the southern icefield appear to be less well nourished than those to the west and to have taken part in the general shrinkage of the twentieth century which is well established for Patagonia (Bertone 1960, Magnani 1961). Many of the large outlets calve into the great piedmont lakes, San Martin, Viedma and Argentino (Plates 9.2 and 9.3). There are more travellers' reports on the east side of the icefield than the west. The best-known glacier, the Moreno, calving into Argentino, which has repeatedly attracted the attention of glaciologists, is notorious for its anomalous behaviour (Nichols and Miller 1952, Lliboutry 1953, Liss 1970). Since 1899 it has repeatedly dammed a southern arm of Lake Roca, flooding pastureland and forest. In 1981 it was advancing into forest on the north side, though the southern arm of the lake was not dammed.

Plate 9.2 The South Patagonian icefield from the west, with Lago O'Higgins, Lago Viedma and Lago Argentino seen from left to right on the far side. (G. M. Grechko and Y. V. Romanenko, from Salyut-6, 10 March 1973)

Plate 9.3 The southern Patagonian icefield seen from the east with Lago Argentino. The Moreno glacier was almost closing the southern arm of Lago Argentino when this handheld photograph was taken from the Shuttle on 31 October 1985

The only major investigations of the eastern glaciers have been those initiated by the American Geographical Society together with the Museo Argentino de Ciencias Naturales 'Bernadino Rivadavia' in 1948. A series of expeditions starting in 1949 (Nichols and Miller 1951) and ending in the late 1960s (Mercer 1965, 1968, 1970) visited major outlet glaciers and studied a substantial selection of moraines. The data collected provide a valuable insight into the fluctuation history of the Patagonian glaciers, but it should be recognized that the work was essentially exploratory. Few of the 350 outlets of the eastern margin have been carefully examined (Bertone 1960). The Little Ice Age expansion may have begun by 1600 or earlier; a ^{14}C date of 390 ± 85 (I-989) from the Cerro Norte suggests an advance at some time between the fourteenth and seventeenth centuries. Mercer (1965) identified a possible maximum on the west side of the Upsala, the largest glacier in South America, calving into the Bahia Upsala, an arm of Lago Argentino, in 1600. Tree-ring dating is hampered by a lack of exact knowledge of excesis times which Mercer considered to have great local variation. If he was correct in using an interval of fifty years on the west side of the Upsala tongue, then the ice retained the position it had reached by 1600 until 1760. The recession which followed was broken by an early nineteenth-century readvance succeeded by further substantial shrinkage. The outer recent moraine of the Adela glacier in the valley of the Rio Fitz Roy appears to have stabilized by 1690. Photographs of the glaciers taken in 1931 show the ice only slightly further forward than in 1959 and suggest that the kilometre of recession from the nearest moraine may have begun in the nineteenth century. Adopting an excesis time of about a hundred years, Mercer (1965) reckoned that the moraines of the Dos Lagos glacier in the Canon Cerro Note were ice-free by 1760, 1780, 1820 and the late nineteenth century. Nichols and Miller (1951) tentatively suggested a brief expansion between 1870 and 1880 on the south side of the Ameghino which was the greatest for several centuries. Since then the glacier has shrunk 100 to 150 m vertically. Apart from the Moreno, all the eastern glaciers have shrunk or receded more than those on the west of the icefield. Some of the small glaciers, for example those between Cerro Mellizo Sur and Brazo Norte Occidente of the Lago San Martin, had lost a third to a half of their areas by the 1960s (Mercer 1965).

In a review of the glacial history of southernmost South America, Mercer (1976) summarized Little Ice Age history, finding glacial advances culminating in the seventeenth, eighteenth and nineteenth centuries and that the great majority of glaciers taking part in these advances had since receded. Details of dating must be considered tentative in view of the reliance on one method of dating, the small numbers of trees sampled and uncertainty about the excesis period. Perusal of the latest account of the southern icefield (Martinic 1982) indicates that no further investigations of note had been made. Heusser (1961) is, it seems, probably correct in finding a general correspondence between the patterns of fluctuations in Patagonia and in Alaska.

9.4.2 NEW ZEALAND

The most substantial data on southern-hemisphere glacier fluctuations come from New Zealand. The two islands are situated 33° to 47°S and 167° to 178°30′E, between the high pressure cells of the sub-tropics and the low pressure zones of the Southern Ocean. Weather and climate are governed by a procession of anticyclones separated by troughs of low pressure and depressions, nearly always moving from west to east (Trenberth 1976). The Southern Alps of South Island, running southwest to northeast, present a barrier to the westerly winds and, except in the far south of the island, precipitation declines from west to east.

Though North Island has many peaks rising above 1200 m, only Ruapehu, a massive volcano near the centre of the island rising to almost 2800 m, is high enough to support glaciers. On South Island, glaciers are scattered along some 400 km of the Southern Alps between latitudes 40°54′S and 44°53′S (Figure 9.6). Between the tiny outlying glaciers on Mt Rolleston in Arthur's Pass in the northeast and those on Mt Elliot and Mt Pyramid in the southwest, there are over 250 glaciers, the largest of which are in the Mt Cook region, around 43°35′S. Of the ten main trunk glaciers, the Spencer, Burton, Franz Josef and Fox flow to the west of the main divide, and the Tasman, Hooker and Mueller, the Godley, Classen and Murchison descend to the east.

No systematic inventory of the glaciers of New Zealand has yet been published, though a preliminary study of snowline variation in the Southern Alps has appeared (Chinn and Whitehouse 1980). Exploration of the larger glaciers began in 1862 with the remarkably detailed investigations of Von Haast which were supported by planimetric diagrams of the fronts of the Godley, Classen and Tasman glaciers (Figure 9.7) and many field sketches, some of which were later interpreted in water colour by John Gully (Paul 1974, Gellatly 1985a) (Plates 9.4 and 9.5). The first photographs were taken in 1867 when 'Mr

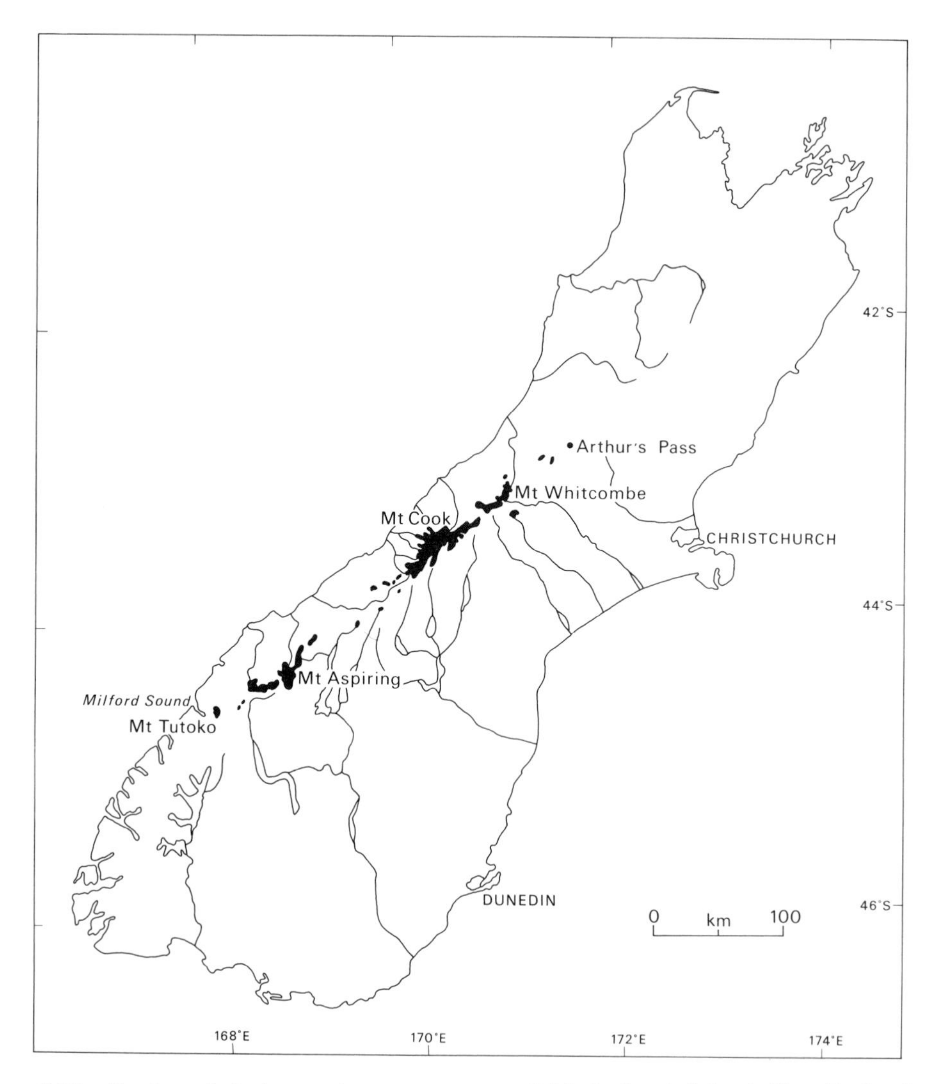

Figure 9.6 Distribution of glaciers and permanent snowfields in South Island, New Zealand

A. P. Sealey made some long expeditions with his camera' (Harper 1896). Serious surveying and mapping made good progress in the 1890s with the explorations of C. E. Douglas and his assistant A. P. Harper (1896) on the eastern side of the Southern Alps, and of T. N. Broderick on both sides (Figure 9.8). Douglas and Harper's (1895) fieldbooks form part of an usually informative repository of written records and illustrations which allow the fluctuations of the New Zealand glaciers to be traced through the late nineteenth and into the early twentieth century (Gellatly 1985a). Identification of phases of glacier expansion earlier than the second half of the nineteenth century has to be based on moraine dating.

As early as 1893 Harper had recognized important differences between the glaciers flowing to the east and to the west:

Plate 9.4 View from Mt Cook range to the Tasman and Murchison glaciers. A watercolour by Julian Von Haast (1822–87), made on 12 April 1861. This workmanlike sketch by the Provincial Geologist was used as a basis for the painting in Plate 9.5. (From Alexander Turnbull Library Collection, Wellington, New Zealand. Ref. no. 52096½)

Plate 9.5 Watercolour of the Tasman and Murchison glaciers painted by John Gully on the basis of Julian Von Haast's sketch. (From Alexander Turnbull Library Collection. Ref. no. 51360½)

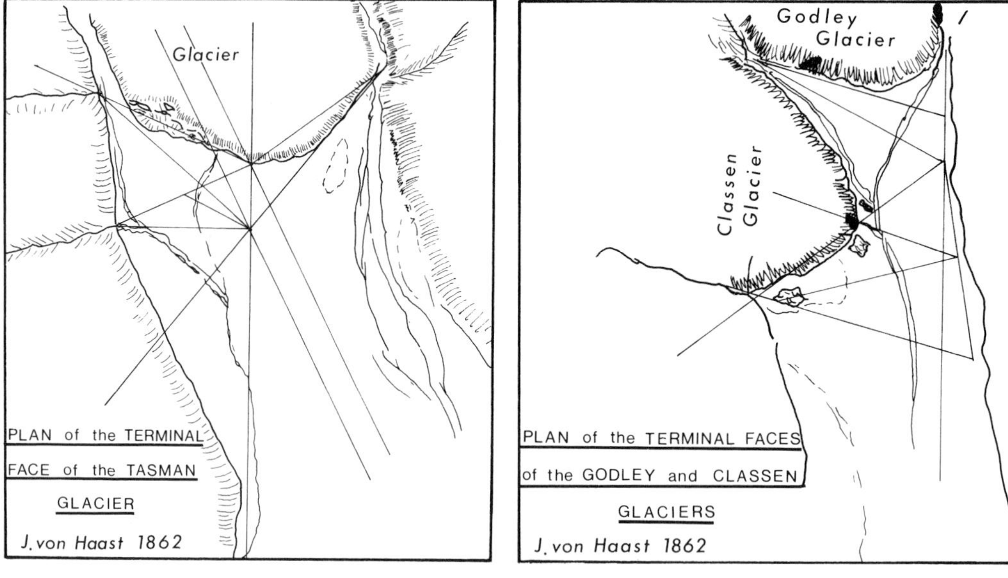

Figure 9.7 Sketchplans of the tongues of the Tasman, Godley and Classen glaciers made by J. Von Haast in 1862. (Copy by Dr A. Gellatly)

> The great glaciers on the east side of the watershed are chiefly flat, hummocky ice . . . covered roughly speaking for a quarter of their length with a considerable quantity of surface moraine, formed of detached masses of rock and debris of all kinds, with which the Swiss moraines cannot be compared for roughness and extent: it is lifted in heaps or hillocks of 50 feet or more above the general level of the glaciers [Plate 9.6]. The old lateral moraines too are most marked, especially on the Hooker [Figures 9.8 and 9.9]. The western glaciers on the other hand, are very different in character, being practically icefalls to within a short distance of their lower ends [Plate 9.7]. They are also almost entirely free of surface moraine, a fact easily accounted for, since the strata in the rocks dips steeply towards the coast, thus presenting a smooth surface on the west, and on the east a broken face, which under the action of weather and climate sends down vast masses of debris onto the glaciers. (Harper 1893, p. 37)

Harper noticed that the western tongues reached lower altitudes than the eastern ones, the Fox 'being the lowest of all the glaciers' and 'like the others on this side having tree-firns and bushes growing almost on the moraine, and in some cases overhanging the ice'.

Plentiful moraines and moraine remnants east of the continental divide provide evidence of repeated fluctuations but its interpretation has not been straightforward. An initial linear lichen growth curve (Burrows and Lucus 1967, Burrows and Orwin 1971) was employed to date several sequences (Burrows 1973, 1975, 1977) and it was concluded that there had been regular, well-marked periods of glacier activity occurring at least once every hundred years since the twelfth century (Burrows and Greenland 1979). Differences in soil development on moraines dated only forty years apart according to this curve, together with the unusual steepness of the curve itself, led to suspicions of its inaccuracy (Burrows 1980, Birkeland 1981, Burrows and Gellatly 1982).

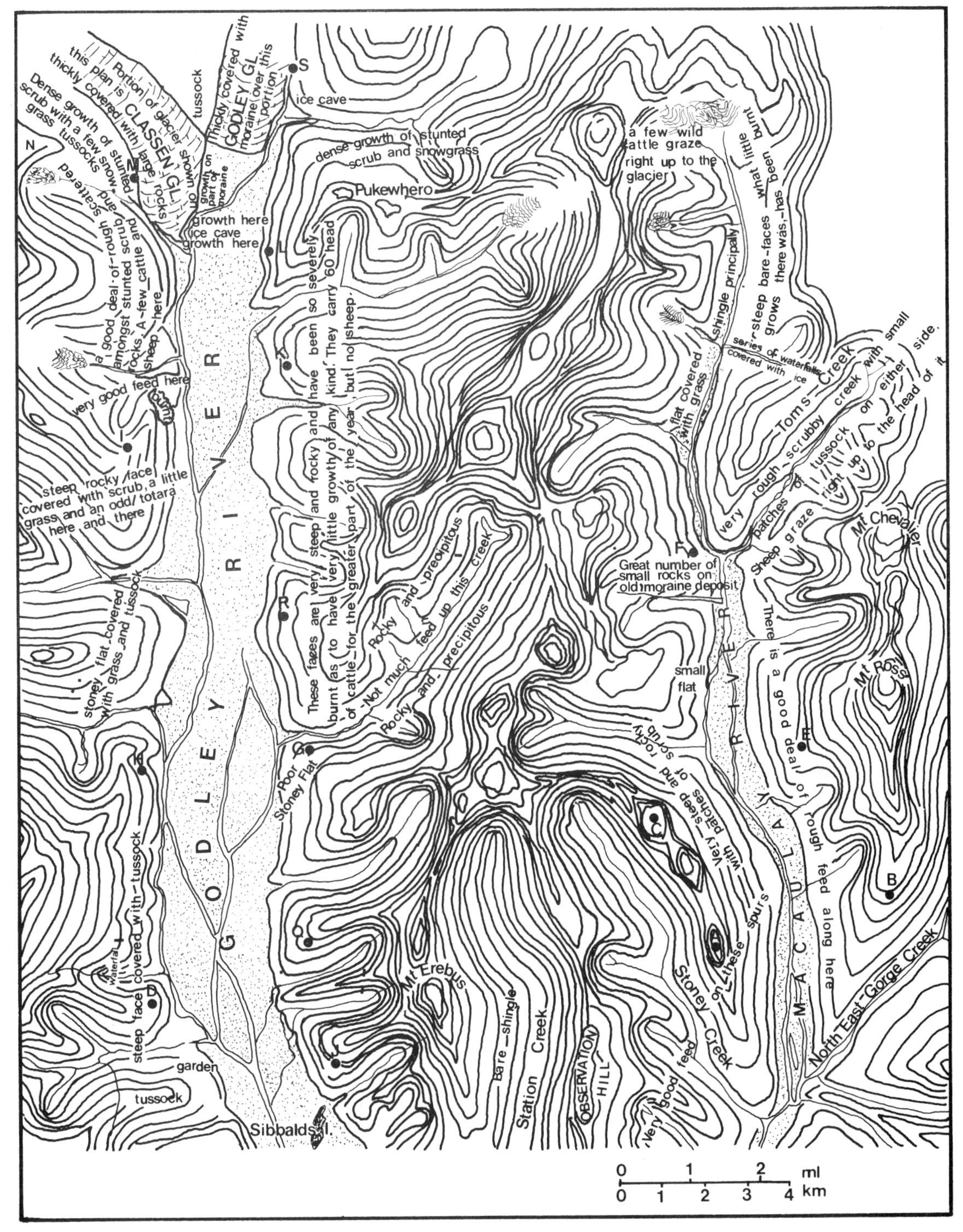

Figure 9.8 Broderick's map of the tongues of the Godley and Classen glaciers and their environs made in 1888. (Copy by Dr A. Gellatly)

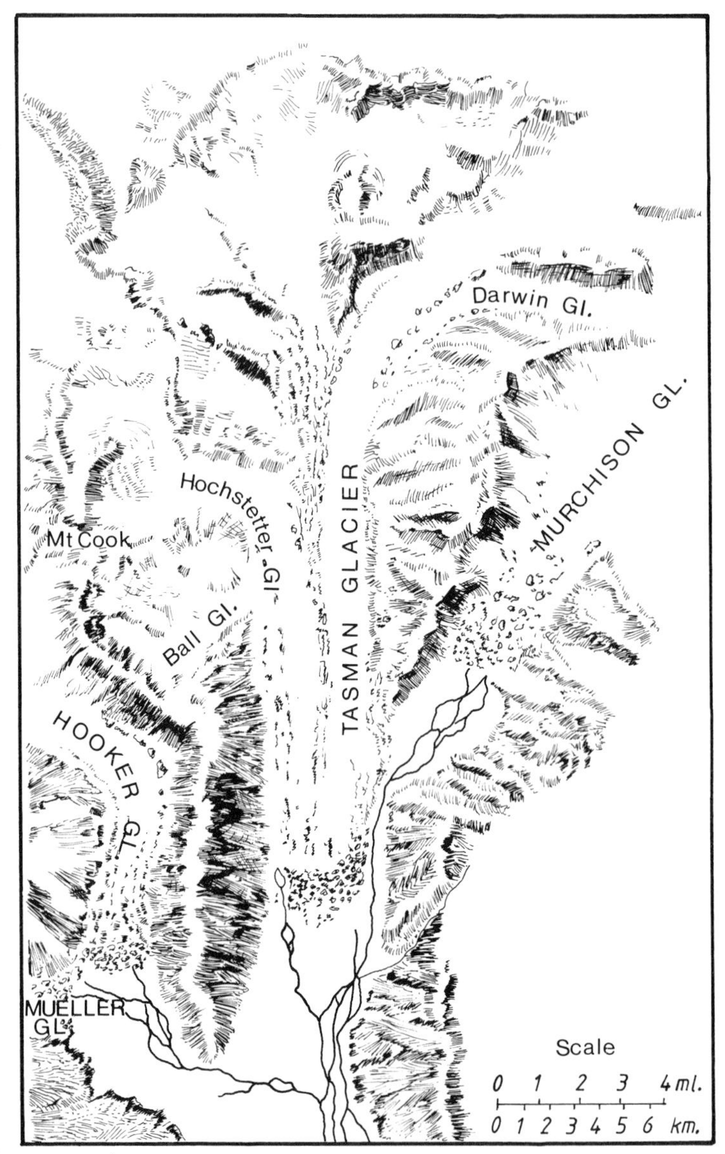

Figure 9.9 Green's map of the Great Tasman glacier together with its tributaries and part of the Hooker glacier made in 1884, with additions from Dr von Lendenfeld's map. It was first published in the *Proceedings of the Royal Geographical Society*, 1884. (Copy by Dr A. Gellatly)

Three important control points on the Mueller forefield have been redated, further control points have been identified, and a new curve has been set up for the Mt Cook area (Gellatly 1982, 1983, and Figure 9.11). The new, non-linear curve is based on a combination of five historically dated surfaces all post-1860 and nine surfaces approximately dated by weathering-rind thickness. Rates of rind growth on fine-grained sandstone were calibrated by Chinn (1981) using surface rocks on radiocarbon-dated landslides. He considered his dates to be within 20 per cent of actual values, though it may be noted that three of the ten radiocarbon dates on which he based his curve were less than 400 BP, within the time-range when such values are not unambiguous (Klein *et al.* 1982).

The forefields of the Mueller, Hooker, Tasman, Murchison, Godley and Classen glaciers have been remapped and the moraine dating substantially revised. Separate peaks of glacial expansion occurring about AD 1140, 1350, 1640 and 1845 have been recognized as a result. In view of the dependence of the lichen growth curve on weathering-rind thickness, these dates must be regarded as approximate. However, the convergence of results from different lines of relative dating is striking, with lichen size, weathering-rind thickness and degree of soil development varying accordingly. The earlier part of the chronology is substantiated independently by radiocarbon dating (Table 9.4). All of these dates refer to organic material providing a minimum age for an earlier advance and a maximum age for a later one. Size frequency histograms of lichen populations growing on the various dated surfaces and difference in modal rind thicknesses suggest that the recognition of four main phases of advance is justified. Discriminant analysis of the data might well be used to test its statistical validity (Dowdeswell and Morris 1983). Meanwhile it seems reasonable to accept the chronology proposed by Gellatly (1982, 1983, 1984, 1985a, b) pointing to Little Ice Age advances, including a well-marked precursor about the fourteenth century, as being the most accurate available for glacier fluctuations east of the continental divide. A large number

Plate 9.6 The moraine-covered surface of the Mueller glacier, typical of the glaciers east of the watershed. The Stocking glacier can be seen hanging on the mountainside on the far side of the Mueller in the centre of the photograph. Photographed in 1928. (Alexander Turnbull Library, Radcliffe Collection. Ref. no. 7550½)

Table 9.4 Radiocarbon support for lichen chronology

Radiocarbon date yr/BP	*Laboratory no.*	*Moraine locality*	*Description*	*Calendar age*
634 ± 48	NZ-711	Tasman glacier	Wood between till layers	AD 1240–1385
664 ± 57	NZ-4774	McCoy glacier	Plant roots and leaves between till layers	AD 1250–1395
650 ± 60	NZ-4015	Collin Cambell glacier	Wood between till layers	AD 1255–1400
520 ± 60	NZ-4016	Collin Cambell glacier	Wood between till layers	AD 1330–1430
537 ± 42	NZ-1413	Cameron glacier	Wood from soil on outwash buried by till or outwash	AD 1325–1425

Source: From data in Burrows and Gellatly (1982) and Klein *et al.* (1982)

of the moraines remain to be dated and it may yet become possible to obtain a more detailed picture of the several advanced phases which have occurred within the last millennium.

The fluctuations of eighteen glaciers on the densely vegetated and steep western versant of the Southern Alps were investigated by Wardle (1973) using both tree rings and lichens to supplement early historical sources. Difficulties arise with both these techniques. *Rhizocarpon geographicum* occurs only locally. It is excluded from moraines associated with the Franz Josef and Fox by more rapidly growing lichens and by mosses and has not been found on surfaces known to have been exposed for less than thirty-five years. Moreover it 'includes in addition to the usual yellow form, a dull greenish-yellow more diffuse form which seems to be faster growing'. Wardle (1973) used maximum lichen diameters to estimate age in the absence of other evidence. He plotted the diameters of the largest thalli of *Rhizocarpon geographicum* found against the approximate ages of moraines derived from other evidence, but did not have a properly constructed growth curve at his disposal and so could not use lichen size as an independent dating tool. The number of woody plants with well-defined annual rings growing on the Westland moraines is limited. *Dracophyllum traversii*, with the clearest rings, is uncommon on moraines and does not appear in the early stages of plant succession. Some species of *Olearia* have clear rings as young shrubs but not as older trees. The main trees at low altitudes are *Weinmannia racemosa*, in which the central rings are invariably decayed in trees over 300 years old, and *Metrosideros umbellata* which has indistinct rings and wood too hard to penetrate with an increment corer. Two Podocarps, *Dacrydium cupressium* and *Podocarpus ferrugineus*, grow on the moraines of the Fox and Franz Josef but the evidence that growth rings are annual is not conclusive. Fortunately *Nothofagus menziesii*, which colonizes the moraines of the Hooker Range, has a life-span of 600 years, tolerably distinct rings and is easily cored. The growth rings of sub-alpine trees and plants can reasonably be regarded as annual (Wardle 1973) but the seeding period is extremely uneven. Very coarse moraine can remain bare for a century while seedlings can be found on finer moraine still underlain by ice. Wardle concluded that seedlings of *Olearia, Dracophyllum, Nothofagus, Weinmannia* and *Metrosideros* are likely to appear in favourable sites within five to ten years but *Podocarpus ferrugineas* and *Dacrydium cupressinum* do not enter the succession on wetland moraines during the first century of exposure.

Despite all the difficulties Wardle was able to obtain thirty incremental cores providing minimum dates for moraines and produced a chronology for the Westland glaciers. He

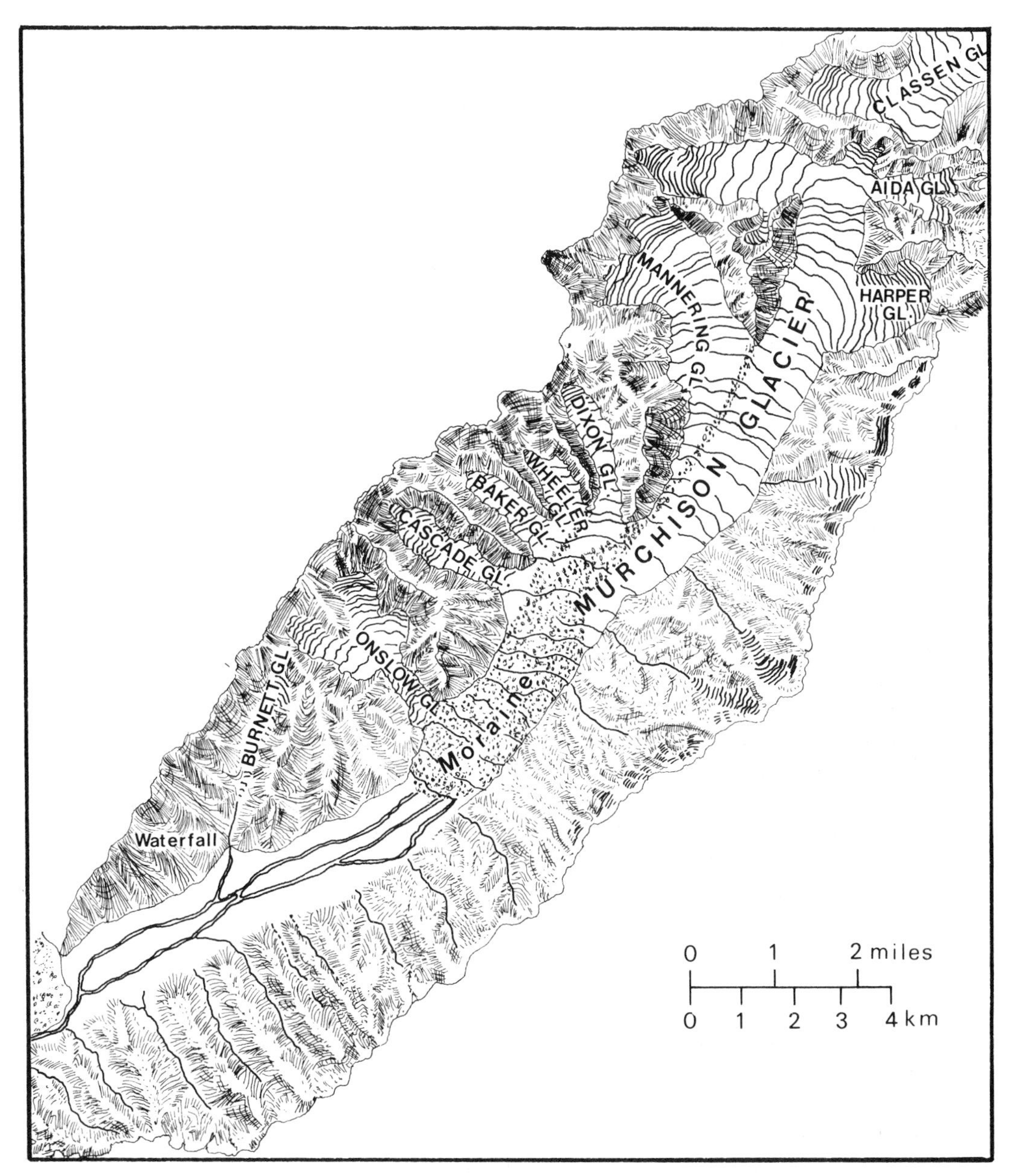

Figure 9.10 The Murchison glacier in 1892 from the Government Survey of the central portion of the Southern Alps of New Zealand, with additions by A. P. Harper *et al.*, first published by the Royal Geographical Society in 1893. (Copy by Dr A. Gellatly)

attributed the multiple moraine sequences which he found to a succession of glacial advances occurring during the last four centuries, of which the largest were those in the seventeenth and eighteenth centuries. The number of moraines formed in recent centuries differs from one forefield to another and compound moraines with crests of two or more ages are not uncommon. Wardle found few moraines which he could not confidently date as either younger or much older than 350 years. He recognized advanced phases of the Westland glaciers in the mid- to late eighteenth century, early to mid-nineteenth century

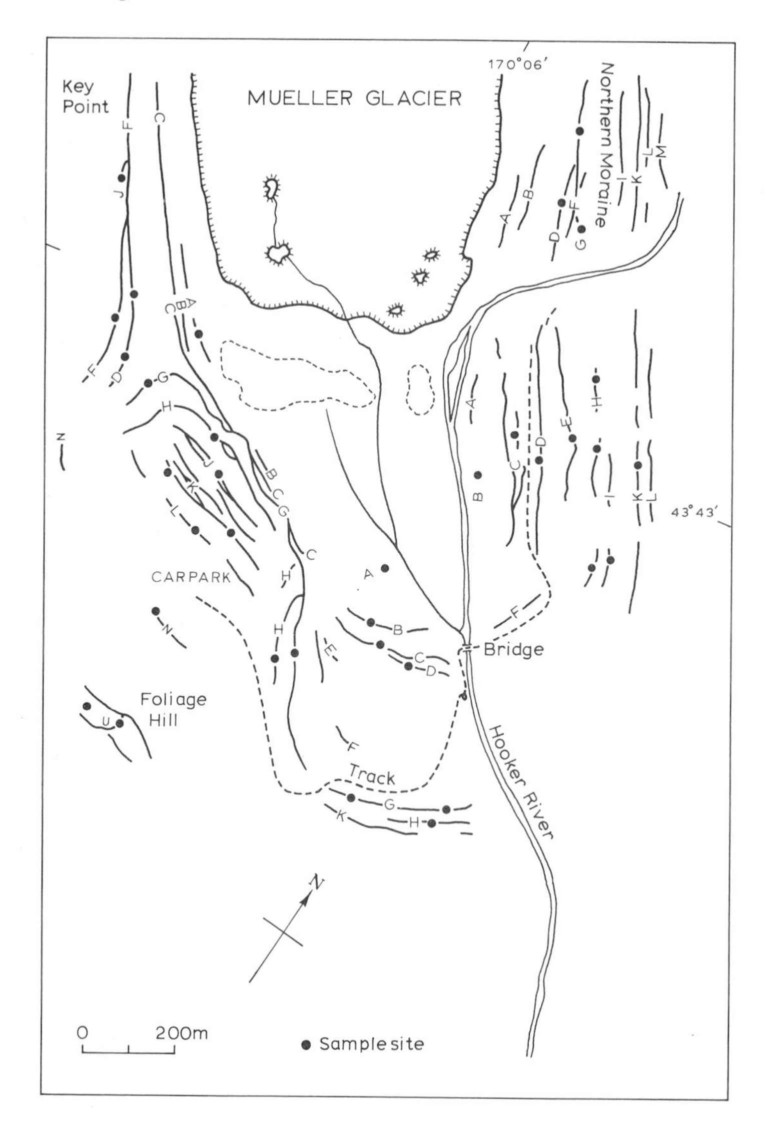

Figure 9.11 Forefield of the Mueller glacier, South Island, New Zealand. Age of moraine rock surfaces from lichens and weathering rinds; A 100, B 135 ± 35, C 340 ± 88, D 580 ± 150, E 840 ± 280, F 1150 ± 300, G 1490 ± 385, H 1830 ± 476, I 2160 ± 562, J 2540 ± 660, K 2940 ± 765, L 3350 ± 870, M 3790 ± 960, N 4200 ± 1090, U 7200 ± 1870. (From Gellatly 1984)

and late nineteenth century. His results from the Westland glaciers conform reasonably well with those of Gellatly from the east side of the divide.

The recorded history of glacier fluctuations in Westland National Park has been summarized by Wardle (1973) and of the Mt Cook National Park by Gellatly (1985a). The present whereabouts of manuscripts, early maps and photographs are given and some account is provided of the observations made by early scientists and travellers. The first observers found the glaciers in advanced positions (Figure 9.7 and Plate 9.7). Von

Haast, the Canterbury Provincial Geologist who explored the Mt Cook area in 1861 and 1862, wrote of the Godley glacier: 'from the fact that several older moraines densely clothed in sub-alpine vegetation were already half buried in the present terminal moraine of the glacier, it was clear to me that the glacier, after a period of retreat, was now advancing' (Von Haast 1879). Between Von Haast's discovery of the Tasman in 1862 and his second visit in 1869 the snout advanced half a mile. The Westland glaciers were also advancing; early photographs show the Franz Josef further forward in 1885 than in 1875. Evidently small advances were widespread in the final decades of the nineteenth century. Since then retreat has predominated, though this has been broken by a number of small readvances and the six main trunk glaciers on the east side of the divide were all within one or two kilometres of their late nineteenth-century moraines in 1981 except for the Godley which had withdrawn about four kilometres (Gellatly 1985a).

Rapid twentieth-century wastage, swiftest since 1930, has caused not only the retreat of the fronts of trunk glaciers but also the complete disappearance of some small ice bodies. The fluctuations of the Franz Josef are the best known for the historic period (Plates 9.7, 9.8, 9.9 and 9.10, Figure 9.12). Small advances, each insufficient to redress previous retreat, took place in 1900–9, 1922–34, 1947–50 and 1965–8 (Ṡara 1970). Between early 1984 and mid-1985 'its terminus . . . advanced more than 200 m' (*New Zealand Herald*, 5 July 1985); this latest advance, it was suggested, may be attributable to an increase in precipitation associated with more frequent cool southwesterlies in the early 1980s (Plate 9.10).

In the absence of high altitude meteorological data and only limited mass balance studies (e.g. Thompson and Kells 1973), climatic interpretation of the ice fluctuations remains controversial. It has been suggested that precipitation plays the dominant role in controlling the fluctuations of the Franz Josef (Suggate 1950, Hessell 1983). Salinger and Gunn (1975) pointed out that mean temperatures in New Zealand were low between 1900 and 1935 and that warming was experienced from 1935 to 1970. Detailed evidence has since been presented that annual temperature has risen over 1 °C between the early 1860s and the present, amelioration in New Zealand continuing into the second half of the twentieth century at a time when temperatures were falling in the northern hemisphere (Salinger 1979, 1982). Although this analysis has been disputed by Hessell (1983), the evidence of warming is convincing and is supported by independent data including the diminishing incidence of years with exceptionally heavy snowfall and also dendroclimatological data pointing to increasing warmth. It seems that twentieth-century retreat of the glaciers in New Zealand has been mainly caused by rising temperature and longer ablation seasons (Gellatly and Norton 1984).

Full explanation of glacier oscillations must, of course, take into account precipitation as well as temperature. An explanation for the fluctuations of the Stocking glacier on the east side of the divide, presented by Salinger *et al.* (1983), took both factors into account (Plate 9.6). Changes in the extent of the glacier were related to climatic variables represented by smoothed monthly temperature values from Hokitika and precipitation data from Otira. Principal component analysis showed variance of these values could account for 83 per cent of the changes in glacier front position; the relationship with temperature was found to be closer than with precipitation.

Plate 9.7 The Franz Josef glacier as portrayed by Sir William Fox in 1872. The terminus was then about 1000 metres wide and the ice was probably advancing. The rock on the right of the picture is the 'Sentinel Rock', later used as a convenient observation point. (Alexander Turnbull Library Collection. Ref no. 69165½)

Plate 9.8 The Franz Josef glacier in about 1905. The front had withdrawn considerably since 1872, though the ice was still not very far from Sentinel Rock. (Photo: A. C. Graham, Westland National Park Collection)

Plate 9.9 By the end of the 1930s the tongue of the Franz Josef glacier had thinned and a lake lay on low ground which was ice-covered in 1903. Note the rapid growth of vegetation on moraines exposed during the twentieth century. (Photo: M. C. Lyson, *c.*1940, Westland National Park Collection)

Plate 9.10 The Franz Josef glacier during its advance in 1985. (Photo: David Norton)

9.4.3 THE SUB-ANTARCTIC ISLANDS

It is not proposed to discuss the marginal fluctuations of the Antarctic ice-sheet. The east and west sheets together cover about seven times the area of the Greenland icecap and the lag between climatic change and ice marginal shifts is probably lengthy, longer than the duration of the change itself. However, it may be noted that an ice core 473 m long, extracted from the summit of Low Dome, has provided a well-dated isotopic (O18/O16) temperature record showing a warm period from AD 300 to 1000 followed by well-defined, quite rapid cooling which, after a partial recovery between AD 1400 and 1600, reached a minimum between 1790 and 1850. The temperature then increased to about the same level as before AD 1000 (Morgan 1985). Most of the continent is so cold, even in summer, that it is doubtful whether slight changes in temperature would affect the economy of the ice sheet unless it were to be accompanied by changes in precipitation or rates of wastage. No records of any significant changes in the extent of the ice sheet itself are available for the last few centuries. Alpine glaciers such as those in the Dry Valleys of the Ross Sea area are believed to be useful as climatic indicators only when used to integrate climatic change over periods of some thousands of years (Chinn 1981).

The most detailed and lengthy record from the sub-Atlantic was produced when volcanic eruptions in 1969 and 1970 ruptured the ice cover on Deception Island, exposing annual ice layers formed over the period since 1680 (Orheim 1972, 1977). The layers were clearly delineated by summer dust surfaces and the stratigraphic interpretation was made relatively easy by the great width of sections, one 50 m and the other 100 m high. As the sections were in the accumulation area, there is no reason to suppose that the mass balance record they provided has been affected by earlier volcanic events. The main effect of volcanism here is the deposition of pyroclastic material which is buried by snow the following winter and then ceases to affect surface albedo. This record provides a valuable indicator of regional climatic trends, since mean summer temperatures for 1944–67 for Deception Island show a strong positive correlation with temperatures at Orcadas in the South Orkneys, Argentine Island, Port Stanley, Punta Arenas and Grytviken.

Orheim compared the Deception Island series with mass balance series from Scandinavia (Liestøl 1967), the Alps, the Polar Urals and northwest USA (Kasser 1973)

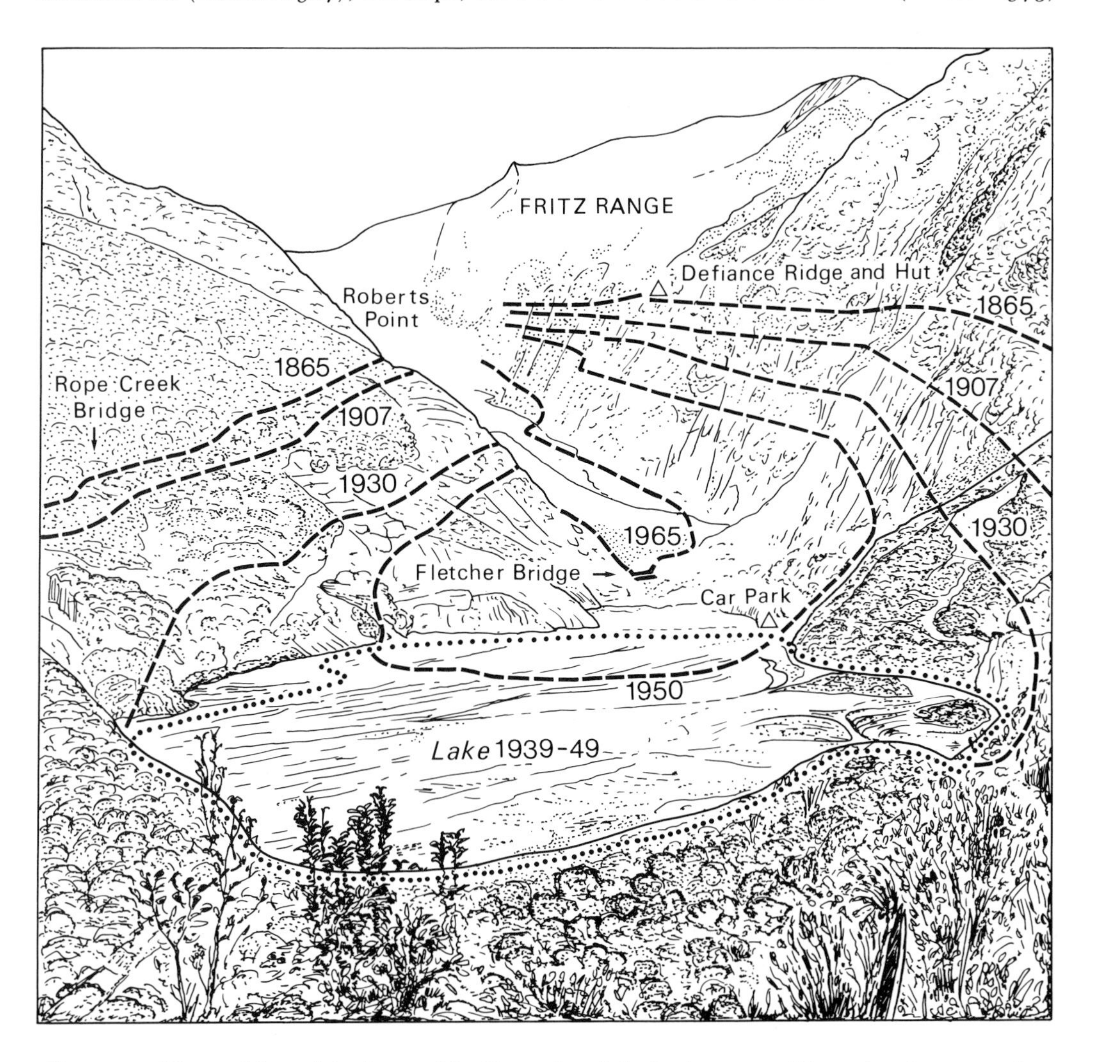

Figure 9.12 The position of the front of the Franz Josef glacier between 1865 and 1965 as seen from Sentinel Rock. In 1865 the whole of the foreground would have been covered by ice. (Drawn from annotated photograph in Westland National Park Collection)

and also with measurements from cores taken from Antarctic sites including the South Pole and Byrd Station. No relationship of any kind was found with the Antarctic series, probably because the mass balance values for Antarctica are almost entirely controlled by variations in accumulation while the Deception Island values are determined largely by summer warmth. Comparison of the Deception Island series with series from the northern hemisphere revealed that short-term variations in the mass balance were inversely correlated during the last few decades and, Orheim considered, probably over the last 150 years. Spectral analysis suggested that this relationship was caused by an anti-phase cycle with a period of about ten years. The cycles also occurred in the oldest part of the Deception Island series but were then apparently in phase with those of the northern hemisphere. This effect could be due to a gradual error in dating caused by missing years in the Deception record.

Long-term variations, so far as they can be safely identified from the data, appear to be in phase. There seems to have been a sharp decrease in mass balance on Deception Island between 1870 and 1880. Although this was both preceded and followed by long periods of more positive mass balance, the mass balances of the century after 1870–80 were significantly smaller than for the preceding hundred years. This pattern parallels the

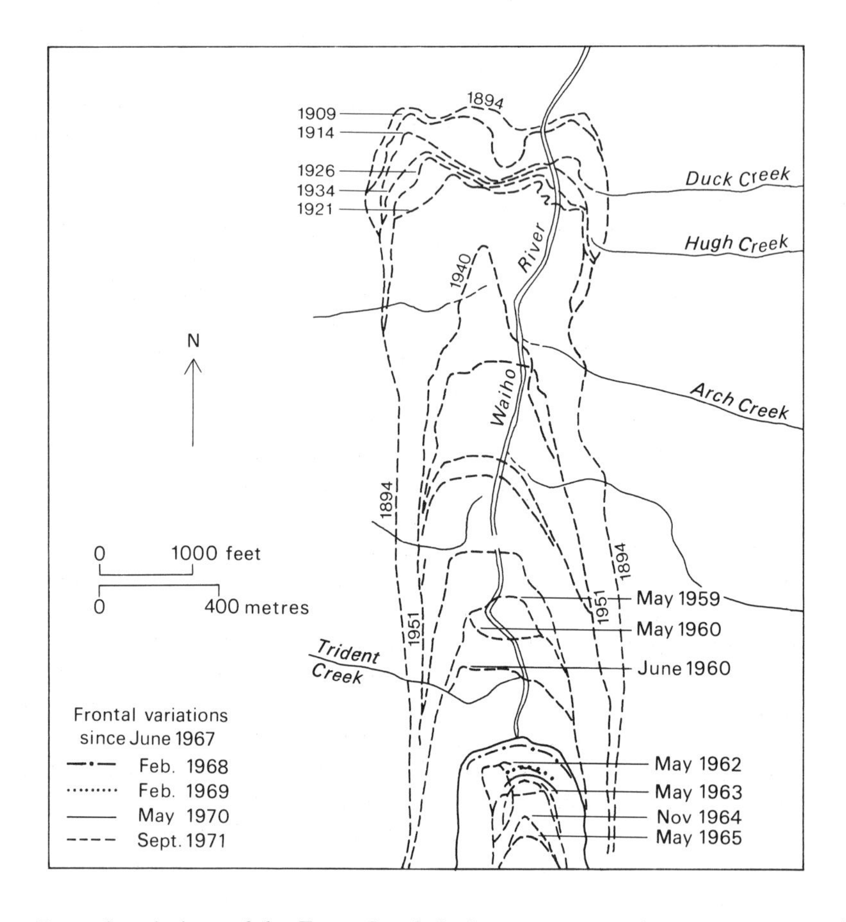

Figure 9.13 Frontal variations of the Franz Josef glacier, 1894–1971. (From Kasser 1973)

overall trend in the northern hemisphere and in low latitudes and also in New Zealand and the southern Andes, with glacier shrinkage predominating after 1850.

In South Georgia in 1958, according to Mercer (1962), every glacier 'had a series of moraine ridges close to its snout, where the hummocky relief and scarcity of vegetation contrast with the older ground beyond'. Many glaciers had evidently reached more advanced positions in the last few centuries and had left behind multiple lateral and terminal moraines more than 300 m outside the margins of the land-based glaciers. The outermost of these moraines are more weathered and more densely covered with vegetation than those nearer the ice. Behind the terminal moraine of the Heaney glacier in St Andrews Bay, the top layer of peat, buried by till, gave a ^{14}C date of 155 ± 45 (SRR 738) (Clapperton *et al.* 1978). This provides some support for a Little Ice Age origin for the till, although it cannot be taken as providing a date more precise than that (Klein *et al.* 1982). The only glacier not calving into the sea for which details were available to Mercer (1962) had thinned by 60 m and retreated by 300 m between 1912 and 1958. However, many of the glaciers on South Georgia were thought by Clapperton *et al.* (1978) to have had positive mass balances in the early twentieth century; instrumental records point to a climatic deterioration which led to glacial advances between 1924 and 1936.

The scattered information available from other sub-Antarctic islands was surveyed by Mercer (1962). The glaciers on Kerguelen were reported to be retreating when the gunship *Gazelle* called there on a voyage of 1874–6. Many glacier fronts were mapped during the voyage of the *Curieuse*, 1912–14, but this work is not known to have been repeated. In 1931, the outlets of the main icefields were in marked and apparently recent retreat. At Port au Français, 49°12′S, 70°12′E, temperatures have increased steadily by 1.5 °C between the mid-1960s and mid-1980s (Allison and Keage 1986).

More detailed information is forthcoming from Heard Island in 53°05′S, 73°30′E. The climate is cool and wet throughout the year and 80 per cent of the island is covered with ice. Heard Island is made up of two volcanic cones. The larger one, Big Ben, which is still active, rises to 2750 m and supports fifteen or more rapidly moving glaciers. Typically, they widen and steepen as they approach the sea and terminate in ice cliffs. A few piedmont glaciers on the eastern and southern sides, notably Brown and Stephenson glaciers, are now land-based with lobate tongues.

The smaller cone, on the northwest side of the main mass of the island, forms the Laurens Peninsula which rises to 706 m. The ice here is thinner and does not reach the coast but terminates at about 200 to 350 m above sea-level.

Little or no change in the general appearance of the Heard Island glaciers seems to have been reported between the visits of the Challenger expedition in 1874 and the Gauss expedition of 1902, nor again between 1902 and the Banzare expedition of 1929. This is probably not just the result of the inspections having been cursory but of the circumstance that many of the glaciers are calving into the sea. It was evident from trimlines that glaciers had been 30 to 90 m thicker not many decades before (Allison and Keage 1986).

Rather detailed knowledge of recent fluctuations is to be attributed to the enthusiasm of Dr Budd, who was appointed medical officer to the island in 1954. In spite of the incidence of snow or rain on 200 days in the year he took every opportunity to return to the island and make rapid surveys of the glaciers and wildlife (Budd 1970, Budd and Stephenson 1970, Radok and Watts 1975). It seems that a slight recession of the ice which occurred between the setting up of a station on Heard Island by the Australian Antarctic Research Expedition in 1947 and its abandonment in 1955, was accentuated between 1955 and 1963. It was at this time that the Brown and Stephenson glaciers lost their coastal cliffs and became land-based. The Winston glacier, a little to the south, retreated a mile up its lagoon between 1947 and 1963 when Budd surveyed the glaciers encircling Big Ben. In 1965 Budd returned again and found some indications that the glaciers were enlarging. Two years later on his next visit he found that the Winston and Vahsel in particular had expanded well outside the area they had occupied in 1954. The Little Challenger glacier had spread laterally and had developed sea cliffs 20 m high.

This readvance of the 1960s, which was marked (though it did not affect all the glaciers on the island), was still continuing in 1971, according to Radok and Watts (1975). They suggested that glacier fluctuations on the island were controlled by temperatures varying in accordance with the latitude of depression tracks affecting the region (Parkinson and Cavalieri 1982). Comparison of air photographs taken in 1980 with those of 1947 shows that recession, which had been general over the period as a whole, had been most marked since about 1970. Tidewater glaciers have thinned; the steep glaciers on the north side of Big Ben have not retreated noticeably but those on the east have receded several hundreds of metres. Small icecaps and glaciers on the Laurens Peninsula had shrunk so much by 1980 that it was evident that they would disappear entirely if the trend were to continue for long. The shrinkage seems to be in response to warming since the mid-1960s, which has been recorded on Kerguelen and which appears to have been associated with a northerly shift of depression tracks (Allison and Keage 1986).

Chapter 10

The glacial history of the Holocene

10.1 Development of a Holocene glacial chronology

The conventional view of the Pleistocene, until a few years ago, was that there had been four great glacial periods separated by warmer interglacials and broken by interstadials, the whole lasting for the million years of the Quaternary era. The last major glaciation, ending with the withdrawal of the Wurm–Weichsel–Wisconsin ice sheets, was succeeded by a period of gradually rising temperatures during the Post-Glacial, to reach an 'optimum' followed by a 'deterioration'. The Little Ice Age of the last few centuries, which it seemed could be regarded as the culmination of this deterioration, was succeeded by a warming in the early decades of this century. The Boreal, Atlantic, sub-Boreal and sub-Atlantic sequence was illuminated and refined by palaeobotanists using pollen analysis, first in Europe and then elsewhere, allowing a more detailed succession to be elaborated (e.g. Godwin 1956, Terasmae 1967).

The rise in temperature that caused the melting of the Late Pleistocene ice sheet allowed a poleward migration of vegetation belts. Regions of Europe that had been tundra were covered with closed forests and in central North America the limit of closed forest continued to shift north across Keewatin until 5500 BP (Figure 10.1), and then retreated (Nichols 1974). In Siberia too the coniferous taiga expanded well to the north of its present limits (Kind 1967). This Hypsithermal, Altithermal or Climatic Optimum has been viewed either as occupying a warm interval from the Boreal to the onset of the sub-Boreal, as those who introduced the terms seem to have intended, or as having started with the withdrawal of the major ice sheets (Mercer 1972).

Latterly the simple model of fourfold glaciation during the Pleistocene has been undermined and the onset of glacial conditions has been shown to have occurred well over a million years ago (Einarsson *et al.* 1967, Shackleton and Opdyke 1973). Not only has the simple picture of the Pleistocene been discarded in favour of a much more complicated

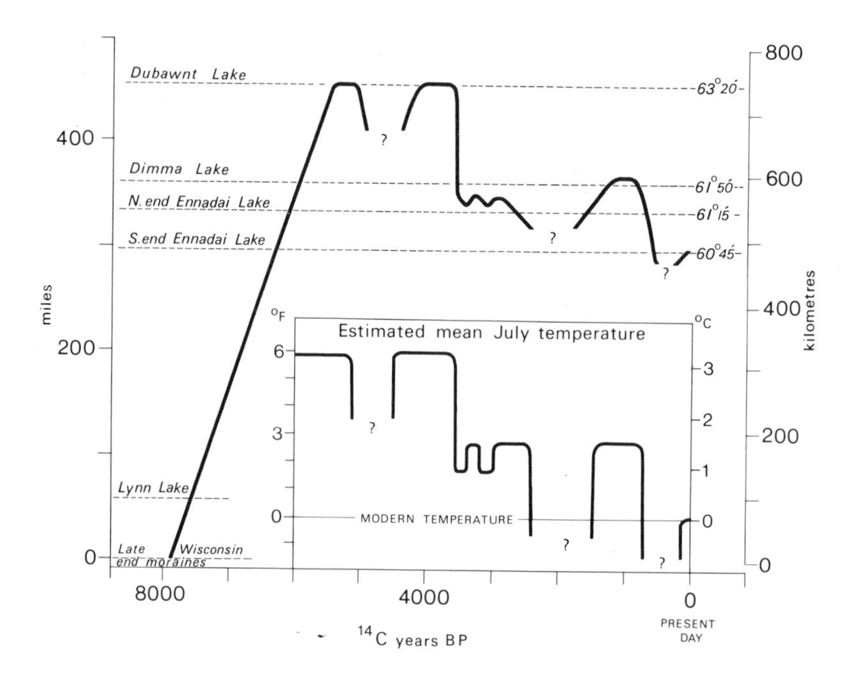

Figure 10.1 A tentative reconstruction of the position of the northern limit of continuous forest along the 100 W meridian in central Canada during the Holocene. It is mainly based on radiocarbon-dated pollen diagrams from the Ennadai and Lynn Lake areas. Inset: the departures from modern mean July temperatures have been calculated from the varying distance from Ennadai of the forest limit. (After Nichols 1974)

model (Figure 10.2), but the course of the Holocene has also come to be recognized as more eventful than was once thought. Holocene moraine systems have been shown to indicate that the glacial advances of the last few centuries are by no means the only ones to have occurred since the withdrawal of the Pleistocene ice sheets.

A chronology of the Holocene was proposed by Porter and Denton (1967) which summarized the glacial fluctuations then known for the North American cordillera over the last 4000 years and also outlined the global sequence. Their chronological framework consisted of about fifty radiocarbon dates. They found that glacial advances were widespread about 4600 BP and that a major glacial expansion had culminated between 2800 and 2600 BP, as well as there having been a cool period in recent centuries. There was evidence in Washington State and British Columbia of an initial period of glacier growth during what was regarded as the Hypsithermal, but they suspected this apparent anomaly might be accounted for by glacial surges.

A few years later, Denton and Karlén (1973a, b) put forward a generalized model of glacial fluctuations over the last 6000 years (Figure 10.3a). It was based on a field programme which had involved them in mapping and dating moraine systems fronting fifty-one glaciers, forty of them in Swedish Lappland and eleven in southern Alaska and Yukon Territory. They concluded that the Holocene was punctuated by at least three important phases of alpine glacier expansion of which the two most recent, the Little Ice Age of the seventeenth and eighteenth centuries AD and that of 3300 to 2400 calendar years BP, were approximately equal in magnitude. Between these two they recognized a lesser phase of glacier expansion in western North America between 1250 and 1050 calendar years BP (AD 700–900). Another glacial expansion of moderate extent and less widely documented was found to have occurred between 5800 and 4900 calendar years BP, and in Sweden an expansion around 8000 BP. Each of these phases of glacial advance involved a number of minor oscillations of ice fronts giving several moraines. In the intervals between glacial expansion, treelines were elevated.

Viewed as a whole, the climatic events of the Holocene were seen by Denton and Karlén to describe a broad pattern

> with periods of glacier expansion about 600 to 900 years in duration, separated by intervals of contraction up to 1750 years in duration. The intervals of glacial expansion may occur about each 2500 years with the most intense Holocene phases peaking about 250 to 350, 2800 and 5300 years BP. This pattern can be extended into the past, for the expected timing of Late Wisconsin events, based on projection of the Holocene climatic scheme, corresponds closely with the Cary, Cary/Port Huron, Port Huron, Allerod, Younger Dryas, pre-Boreal and Cochrane-Cockburn stadials and inter-stadials. Moreover, the Roman Empire–Middle Ages and Little Ice Age sequence shows marked similarities with the Allerod-Younger Dryas, thus strengthening the argument that they are analogous. The only major difference is that during the Late Wisconsin these climatic fluctuations were superimposed on an earth undergoing an Ice Age, whereas during much of the Holocene they affected an earth that had recovered from an Ice Age. (1973b, p. 202)

A very different view of the incidence of glacial advances put forward by Heuberger (1974) attracted less attention. It was based on very detailed and long-continued investigations of moraine systems in the Alps (e.g. Kinzl 1932, Aario 1945, Heuberger and Beschel 1958, Mayr 1964). The model portrayed in Figure 10.3b involves several periods in the Post-Glacial when glaciers are known to have reached or passed their Little Ice Age positions, seven occurring within the last 9000 years and six within the last 7000 years. It was recognized that the individual advances were not exactly contemporaneous,

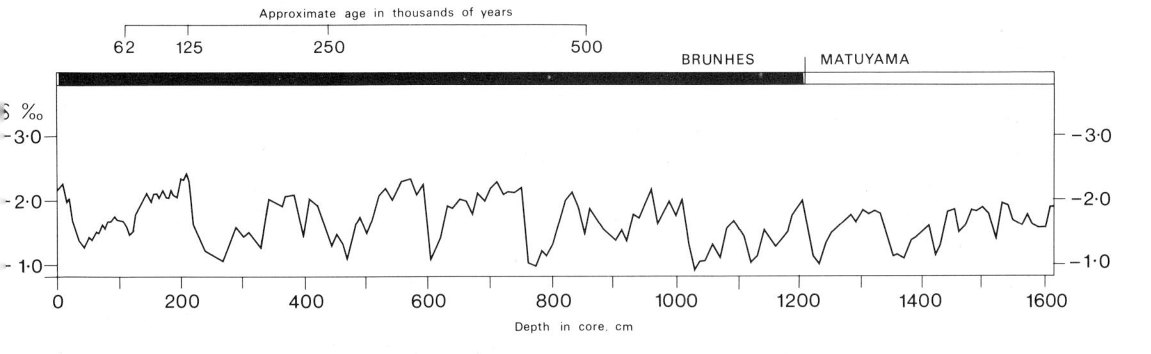

Figure 10.2 The fluctuations in isotopic composition through depth, and therefore through time, exhibited in the oxygen isotope record from core V28–238. They are attributable to changes in ocean isotope composition caused by variations in the mass of the northern hemisphere ice sheets and thus record the history of glaciation in the Pleistocene. (After Shackleton and Opdyke 1973)

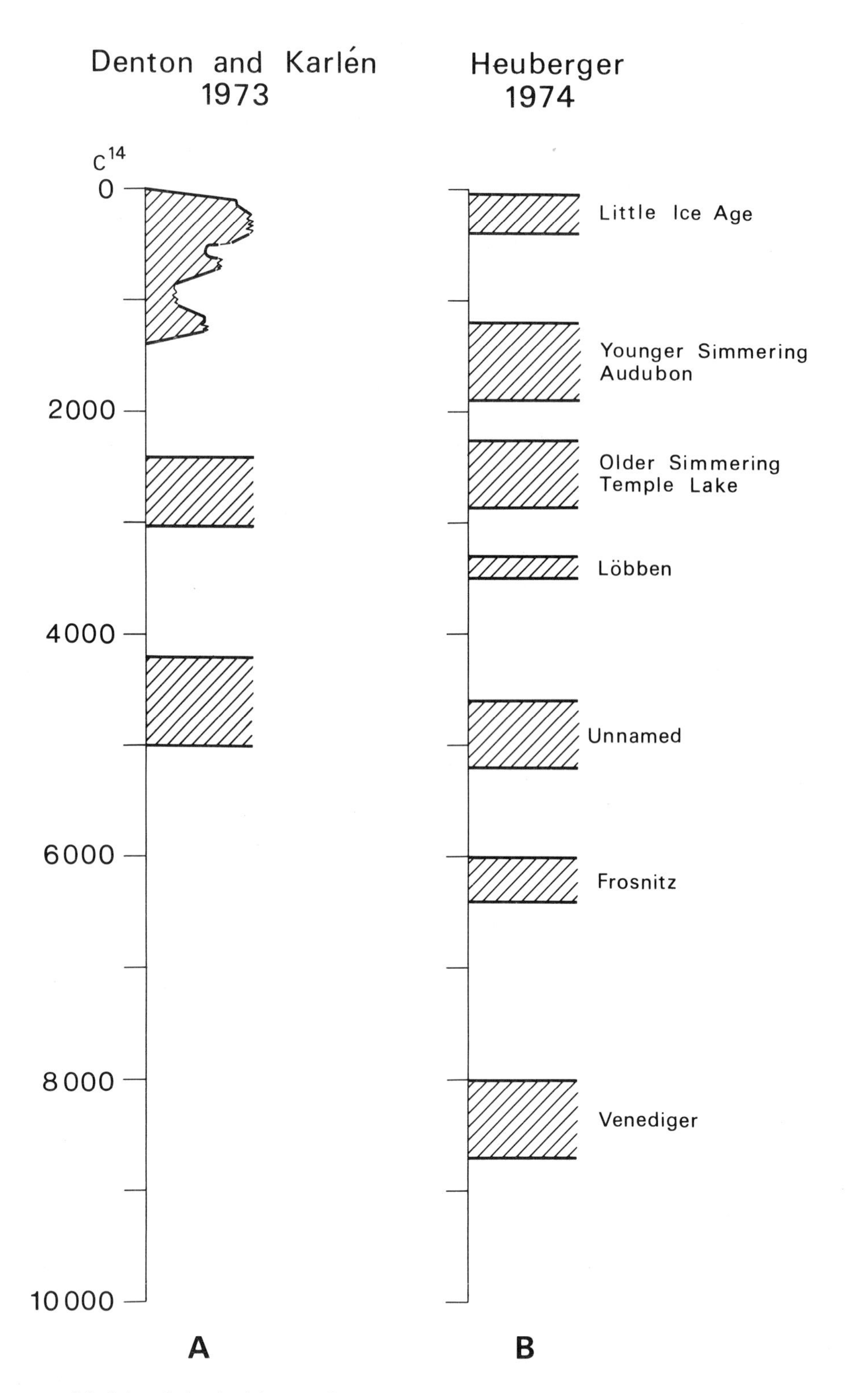

Figure 10.3 Models of the incidence of glacial episodes during the Holocene: (a) according to Denton and Karlén (1973b), (b) according to Heuberger (1974)

but Heuberger claimed a general correspondence of the main phases of advance and retreat, not merely in the Alps but in mountain areas worldwide. Furthermore, he concluded that the Alpine glaciers had never entirely melted away after the last major glacial period and that the Hypsithermal was too short to have left a decisive mark on the history of Alpine glaciation. Patzelt (1974), from studies of soils, moraines, pollen and ^{14}C dating, distinguished eight phases of Holocene glacial advance in the eastern Alps. Of these, the Venediger and Schlaten were similar in magnitude to the Little Ice Age, and Löbben and Frosnitz were probably slightly greater. Some of these Alpine phases correspond in time to advances in North America, New Zealand and elsewhere; others do not, as far as one can tell, though they could turn out to do so (Figure 10.4b).

The fluctuations of the Alpine glaciers in the Holocene appear to have occurred in response to quite small fluctuations in temperature. The snowline and treeline variations in height were equivalent to a range of mean summer temperatures of 1.3 to 1.6 °C, judging from the rise in the snowline and the accompanying rise of summer temperatures at Alpine observatories between the 1840s and 1940s.

The question which concerns us here is whether or not the Holocene glacial advances and retreats were contemporaneous in different parts of the world. The behaviour of glaciers within the last several centuries was broadly synchronous. Was this synchroneity, as Benedict (1973, p. 597) has inquired, an anomaly, or is the apparent lack of synchroneity at earlier times a reflection of errors in dating and interpretation of Holocene deposits? The question was again raised by Andrews *et al.* (1975) in the course of a discussion arising from the lack of correlation, so far as could be seen, between the sedimentary units associated with cirques in the San Juan Mountains and other Holocene chronologies, notably those of Denton and Karlén (1973b) and of Benedict (1973) for the Colorado Front Range. The analysis of worldwide radiocarbon dates by Bryson *et al.* (1970) had indicated globally synchronous climatic shifts during the last 10,000 years occurring with considerable rapidity, and Easterbrook (1974) had also concluded that in spite of differing response rates the main glacial phases after 11,000 BP were synchronous. The proliferation of studies using a variety of techniques differing in precision has increased the detail of climatic sequences for individual regions, but it has complicated rather than eased the problem of determining whether or not the glacier variations have in fact been globally synchronous.

10.2 Europe

10.2.1 THE EASTERN ALPS

Many detailed studies have been made in the eastern, central and western Alps, using pollen analysis and dating moraines from organic matter beneath, within or overlying them. Patzelt (1974, 1977), a notable contributor to our understanding, worked in the Venediger and was able to refer to work by Mayr (1964, 1968) in the Stubaier Alpen, Bortenschlager (Bortenschlager and Patzelt 1960) and Heuberger (1966) in the Ötztal, and Zoller (Zoller 1960, 1966, Zoller and Kleiber 1971, Zoller 1972) in eastern Switzerland.

Patzelt demonstrated that phases of glacial advance show up distinctly in certain pollen profiles, the proportion of arboreal pollen decreasing and non-arboreal pollen increasing, but that this is only the case in profiles from bogs near the ice or at about the elevation of

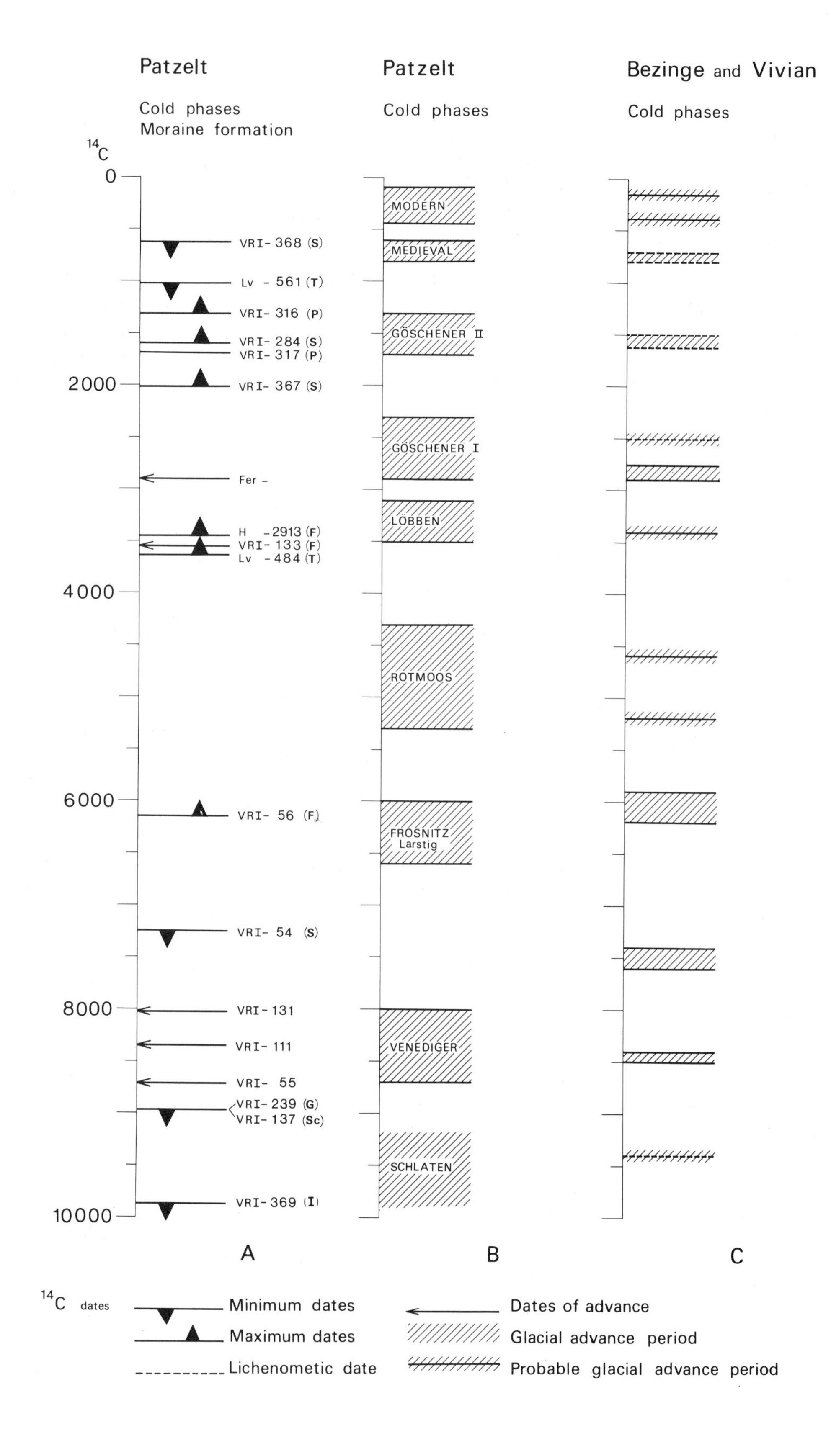

Patzelt
Cold phases
Moraine formation
Patzelt
Cold phases
Bezinge and Vivian
Cold phases
14C
0
2000
4000
6000
8000
10000
VRI-368 (S)
Lv - 561 (T)
VRI-316 (P)
VRI-284 (S)
VRI-317 (P)
VRI-367 (S)
Fer -
H -2913 (F)
VRI-133 (F)
Lv -484 (T)
VRI-56 (F)
VRI-54 (S)
VRI-131
VRI-111
VRI-55
VRI-239 (G)
VRI-137 (Sc)
VRI-369 (I)
MODERN
MEDIEVAL
GÖSCHENER II
GÖSCHENER I
LÖBBEN
ROTMOOS
FROSNITZ
Larstig
VENEDIGER
SCHLATEN
A
B
C
14C dates
Minimum dates
Maximum dates
Lichenometic date
Dates of advance
Glacial advance period
Probable glacial advance period

the timberline, between 2100 and 2300 m. Dating of moraines varies in precision. For some moraines only a maximum or minimum age can be given; others are quite closely bracketed. Moraine assemblages differ from one glacier to another and a large number have to be examined to obtain a complete chronological sequence.

The phases of glacial advance identified by Patzelt in the eastern Alps were as follows:

(a) The Schlaten, comparable in scale with the Little Ice Age of recent centuries, occurred probably about 9400 BP. (As there is no consensus as to which correction curve should be adopted, ^{14}C dates are given here and not sidereal, calendar dates.) A sample of *Pinus cembra* (stonepine) from immediately above a lateral moraine of the Schlatenkees in the Venedigergruppe gives a minimum age of 8970 ± 130 BP (VRI–137, see Figure 10.4) but neither beginning nor end of the phase has been closely dated.

(b) The Venediger, occurring between about 8700 and 8000 BP, saw the deposition of lateral and terminal moraines in the Venedigergruppe which are still well preserved and lie about 300–400 m outside the Little Ice Age moraines. It was characterized by two main advances (Patzelt 1974) and may have been even more complex, for pollen profiles suggest repeated strong depressions of the treeline about this time. The dating is controlled by three radiocarbon dates of 8720 ± 150 (VRI-55), 8340 ± 130 (VRI-111) and 8040 ± 120 (VRI-131) from samples representing peaks in the non-arboreal pollen profile.

(c) The Frosnitz or Larstig advances, between 6600 and 6000 BP, are marked by non-arboreal pollen increasing by up to 50 per cent. A lateral moraine of the Frosnitz glacier in the Venedigergruppe buried *Pinus cembra* with a maximum date for its growth of 6130 ± 130 (VRI-56). The local treeline was depressed about 200 m. The moraines are outside those of the Little Ice Age but are not large, suggesting that the advance was not long sustained. Mayr (1969) dated an advance of the Glacier du Tour in the Mont Blanc massif to about 6400 ± 100 BP, and an analysis of wood in till-like material in front of the Fernauferner gave dates of 5590 ± 80 and 5590 ± 130 BP (Mayr 1968).

(d) The Rotmoos in the Ötztal, dated by Patzelt (1974) as between 5300 and 4300 BP, is a climatic deterioration indicated by a sharp non-arboreal pollen maximum but without support from geomorphological evidence in the Austrian Alps.

(e) The Löbben, occurring between about 3500 and 3100 BP, saw glaciers in the Vendigergruppe and Stubaier Alpen advance 100–150 m outside the Little Ice Age moraines, with outwash gravels burying peat dated to 3440 ± 60 BP (H-2913). Zoller's date of 3620 ± 85 (LV-484) for an advance of the Tiefengletscher (Col de Furka, Switzerland) would roughly correspond to this.

Figure 10.4 Holocene phases of glacial advance in the eastern Alps with a chronology from the western Alps for comparison: (a) key ^{14}C dates for moraines from the eastern Alps pointing to the cold phases shown in (b) and described by Patzelt (1974). Cold phases shown in (c) were identified by Bezinge and Vivian (1976a) by ^{14}C dating of fossil wood near glaciers in Valais. In some cases two of them correspond to a single cold phase identified by Patzelt. (F) = Frostnitzkees, (Fer) = Fernau, (G) = Gschnitz moraine, (P) = Pasterze, (S) = Simonykees, (Sc) = Schlatenkees, (T) = Tiefen

(f) Göschener I, comparable in extent with glacial advances of recent centuries, covers the period 2900 to 2300 BP, when the treeline was lower and non-arboreal pollen prevalent. In the Stubaier, the Fernau glacier drove its snout into a bog, disrupting the stratigraphy and shoving pieces of organic material into a pile, called a 'moorstauch moraine', dated to after 2900 ± 60 by analysis of wood beneath. Environmental conditions show no marked change for three centuries after 2900, according to an intensively studied profile in front of the Fernau glacier, and then there was heavy avalanching, a sharp decrease in *Picea excelsa* and a rise in *Alnus viridis* pollen. Glacial silt deposits indicate that the glacier remained in an advanced position until some time before 2050 ± 80 BP.

(g) Göschener II was a period of expansion from roughly 1700 to 1300 BP, that is from the second to the sixth centuries AD, with silts from an advance of the Fernau glacier dated to 1660 ± 90 BP (Mayr 1968). The treeline was depressed and at Simonykees a moraine with a maximum date of 1600 BP (VRI-284) has been recognized, banked against another one dated to over 7220 ± 140 BP (VRI-54).

Patzelt (1974) emphasized that this sequence and the time limits of individual cold phases were likely to be subject to modification as a result of further work. Some events, notably the Schlaten, Rotmoos and Lobben, were not then very clearly defined. But the evidence, obtained from seventy pits excavated at various sites in the eastern Alps and eight radiocarbon-dated profiles, continues to carry considerable weight.

10.2.2 THE CENTRAL AND WESTERN ALPS

In the central and western Alps a great deal of the evidence for Holocene glacial fluctuations is in the form of moraines dated from the wood of trees killed by advancing ice. Such wood has been found in considerable quantity within moraines above the present treeline. Some samples have tree rings which indicate increasingly severe conditions during the final years of tree growth (Holzhauser 1984, Furrer and Holzhauser 1984). Measurement of the varying density of the wood in long sequences of tree rings, using X-ray methods, provides independent indices of summer temperature. The positions of the tongues of large glaciers such as the Aletsch and Gorner are believed to have responded to lower temperatures within 20 or 30 years, the smaller glaciers within 4 or 5 years.

Bezinge and Vivian (1976a), as Figure 10.4c indicates, distinguished nine phases of glacial advance in the western Alps and four phases of probable advance. Of the total number, twelve correspond closely to those distinguished by Patzelt. Zoller (1977) also found a close correspondence between the Holocene cold phases in the central Alps and those in the east, though he shows discrepancies in the timing of cold phases in the millennium 3500 to 2500 BP.

Probably the most thorough investigation into Holocene glacial and climatic relationships in Switzerland has been carried out by a large group of researchers at the Geographical Institute of Zürich under the direction of Professor Furrer. (The results appear in various issues of *Physische Geographie*, available from the University of Zürich.) Amongst its first fruits were studies by Schneebeli (1976) and Röthlisberger (1976) of the Val de Bagnes, Zermatt, Ferpècle and Arolla areas in Valais. A strong basis for this work was again provided by wood found in quantity in moraines. Between 1970 and 1975, thirty pieces were discovered within 300 m of the Zmuttgletscher alone. The numbers of tree rings in trunks overridden by ice advances give minimum values for the length of preceding periods between the ice advances. Pits 4 m deep were dug in the sediments and soil layers in sites between moraine ridges, and sections exposed in moraines were examined, yielding much detailed information about the sequence of events. In front of the Findelengletscher, for instance, eight soil layers lying one above the other and separated by morainic material were found to have been revealed by erosion of a lateral moraine complex. The soils provide a chronology of glacial oscillations over the last 4500 years, and the lateral moraines a much fuller sequence than has yet been yielded by the terminal moraines (Figure 10.5). The fossil wood found in moraines has allowed not only ^{14}C dating of glacial advances in the course of the Holocene, notably those of the Aletsch (Holzhauser 1984), but also the production of a dendroclimatological curve for the last 9000 years (Röthlisberger *et al.* 1980).

The curve shown in Figure 10.6 summarizes the fluctuations over the last 8500 years in Valais, the known glacier front positions for 1850, 1890, 1920 and 1976 being used to provide a scale against which the earlier oscillations can be compared. Alongside are plotted ^{14}C dates obtained by Röthlisberger and other workers with an indication of whether they point to maximum or minimum ages for the moraines. This provides an impression of the wealth or poverty of the data on which the chronology is based. The data for the last 3000 years are so abundant that the dates for this period are plotted on a larger scale (Figure 10.7).

The frequency of fluctuations more than 3000 years ago is now known to be greater than is shown on Figure 10.6. A survey of the results of Swiss investigators up to 1982 (Gamper and Suter 1982, and Figure 10.8) identifies in particular a number of short glacier oscillations, indications of which were lacking in Figure 10.6, about a forward position corresponding to the Lobben advance period in the eastern Alps. The increasingly lengthy record showing rapid switches between cold and warm conditions as indicated by moraine dating, is supported by evidence of alternating phases of solifluction and soil formation, and especially clearly by the dendroclimatic data assembled by Renner from fossil wood in the St Gothard area and by Bircher in Saastal. The longest warm phase in the Holocene identified in Switzerland by Gamper and Suter (1982) lasted only from 4300 to 3600 BP; another between 6300 and 5700 BP is rather less well established. The pollen record shows broad fluctuations in climate corresponding to those distinguished by Patzelt in the eastern Alps; further detailed examination of fossil wood, especially by X-ray densitometry, may well provide an even more detailed picture of early Holocene climatic history in Switzerland.

The need for further detailed investigations has been underlined by the results of studies in the upper Val d'Aosta (Porter and Orombelli 1985). Here retreat of the Rutor glacier in recent decades has revealed a peat bog sandwiched between tills. The top of the bog has been dated to 6275 ± 75 and 6270 ± 85 BP (UW-587 and UW-467) and the base to 8395 ± 125 BP (UW-468) indicating that the glacier lay upvalley from the peat bog and was less extensive than in the mid-1980s during a warm phase lasting over 2000 years. This unusually precise evidence of the extent of withdrawal during the earlier Holocene conflicts with the chronology presented in Figure 10.8 which shows an advance phase comparable with that of 1850 between 7000 and 8000 BP. Several Valaisian glaciers advanced sufficiently during this period to overwhelm trees growing behind the present glacier limits (Röthlisberger *et al.* 1980). Further study of fossil wood specimens using

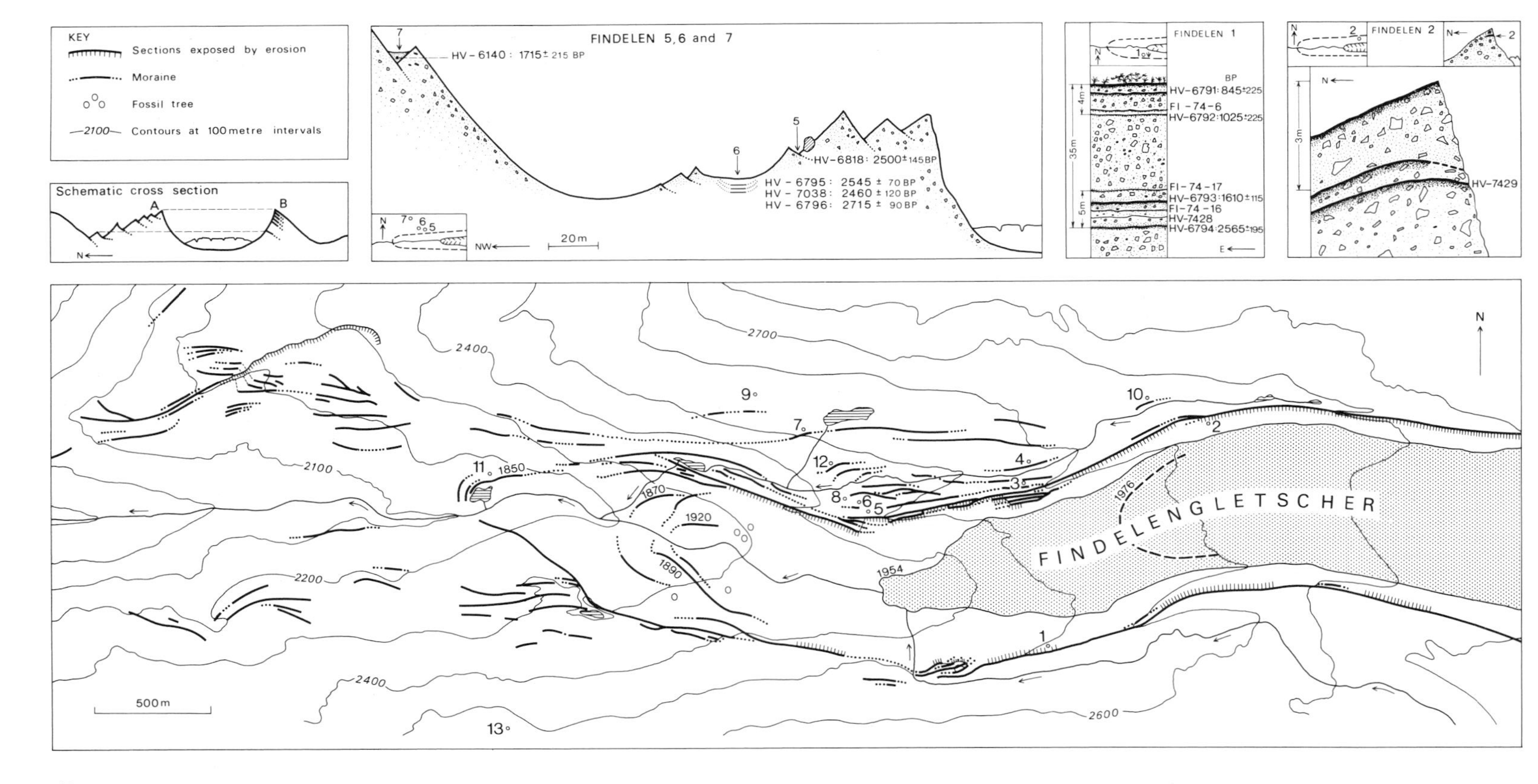

Figure 10.5 Findelengletscher: lateral and terminal moraine fragments. Numbers refer to sections examined in detail, three of which are shown above the map, together with a schematic cross-section. At Findelen 1, on the south of the glacier, alternating tills and fossil soils indicate a sequence of warm and cold phases. (After Röthlisberger 1976)

X-ray densitometry together with a continued quest for field evidence of the extent of other glaciers in the Alps in the earlier Holocene is clearly necessary.

10.2.3 THE ALPS AS A WHOLE

Throughout the Alps the Holocene was characterized by repeated glacial advances at intervals of 1000–1500 years. In some, possibly all cases, they were multiple, lasting several centuries, and they were comparable in extent to the Little Ice Age. The advances were quite sudden and rapid with each main advance involving a number of minor fluctuations. The similarity between the scale of the various Holocene advances resulted in the later ones, notably those of the Little Ice Age, obscuring the earlier events. It is only the intensity of recent investigations and the happy circumstance that fossil wood has been found in greater abundance than elsewhere that has allowed the fuller story to be revealed.

10.2.4 SCANDINAVIA

In Scandinavia, there does not seem to have been as clear a break as there was in the Alps between Late Pleistocene glaciation and Holocene advances (Figure 10.9). Ice marginal deposits belonging to the Younger Dryas, and dated to between 11,000 and 10,000 ^{14}C years BP, have been traced around the coast of Norway, everywhere within about 50 km of the sea (see, for example, Sollid and Sørbel 1975), and then across southern Sweden and Finland. The greatest thickness of the ice at that time was in the northern part of the Gulf of Bothnia.

In Gotland, Mörner's (1980) studies of carbonate deposition in lakes point to extraordinarily rapid rises in temperature about the time of the Younger Dryas/Boreal transition, through as much as 8.4 °C in the course of a century around about 10,000 BP, followed by a slight cooling and then renewed warming at 9650–9250 BP when the Gulf of Bothnia became clear of ice. Within little more than 1000 years, a large section of central Lappland had been abandoned by the ice, judging from Karlén's (1979) accounts. They are based on radiocarbon dates of 9000 to 8500 BP from samples of peat and wood fragments in the area and a date of 9640 ± 130 BP (LU-176) for basal organic sediment from Lake Vuolep Allakasjaure near Abisko. However, two marked readvance phases, the Ornes, 9900–9800 ± 50 BP, and Skibotn, 9600–9500 ± 150 BP (and possibly a third about 9400 ± 250 BP), were identified by Corner (1980) in the Lyngen–Storfjord area of northern Norway.

Holocene moraines in northern Scandinavia are above the limit even of birch woodland and so it has not been possible to use wood fragments for dating them. Glacier advances have been identified from moraines dated by lichenometry (Karlén 1973) and radiocarbon analysis of associated soils.

In the Kebnekaise area of northern Sweden glaciers overlook complex frontal moraine systems as much as 1.5 km wide. Karlén (1973) mapped fifty-three of these systems in detail and found that they consist of steplike series of upturned drift bodies banked against each other in imbricate fashion, each moraine belt marking an individual glacial advance (see Figure 10.10). Adopting Benedict's lichenometric techniques to find the relative ages of the moraines, he measured the diameters of *Rhizocarpon geographicum* and *Rhizocarpon alpicola* thalli in sample areas of 500 m^2. On moraines known to date from a

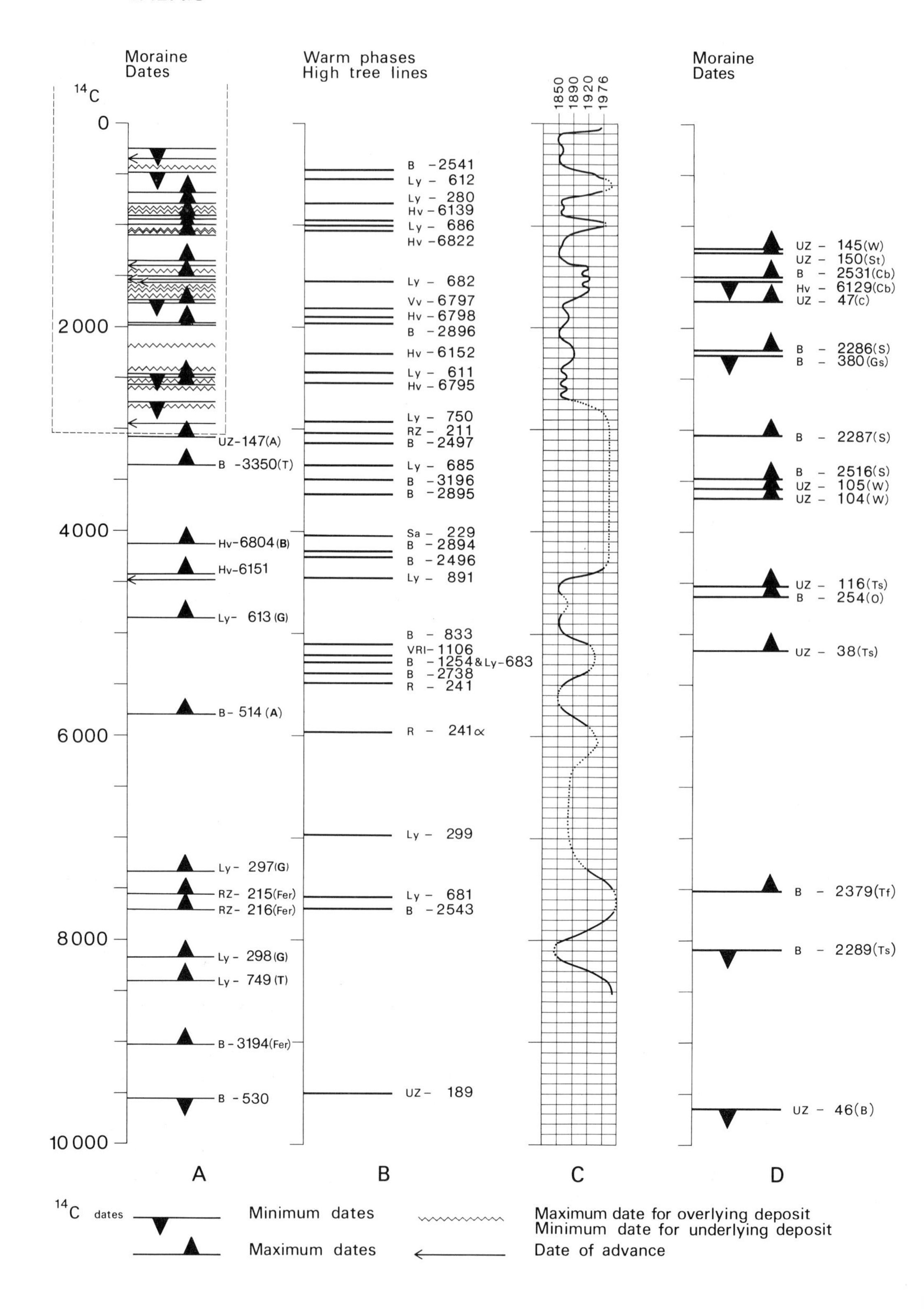
WESTERN ALPS
VALAIS
Moraine Dates
Warm phases High tree lines
Moraine Dates
14C
0
2000
4000
6000
8000
10000
1850
1890
1920
1976
UZ-147(A)
B -3350(T)
Hv-6804(B)
Hv-6151
Ly- 613(G)
B- 514(A)
Ly- 297(G)
RZ- 215(Fer)
RZ- 216(Fer)
Ly - 298(G)
Ly - 749(T)
B - 3194(Fer)
B - 530
B -2541
Ly - 612
Ly - 280
Hv - 6139
Ly - 686
Hv - 6822
Ly - 682
Vv - 6797
Hv - 6798
B - 2896
Hv - 6152
Ly - 611
Hv - 6795
Ly - 750
RZ - 211
B - 2497
Ly - 685
B - 3196
B - 2895
Sa - 229
B - 2894
B - 2496
Ly - 891
B - 833
VRI - 1106
B - 1254 & Ly - 683
B - 2738
R - 241
R - 241α
Ly - 299
Ly - 681
B - 2543
UZ - 189
UZ - 145(W)
UZ - 150(St)
B - 2531(Cb)
Hv - 6129(Cb)
UZ - 47(C)
B - 2286(S)
B - 380(Gs)
B - 2287(S)
B - 2516(S)
UZ - 105(W)
UZ - 104(W)
UZ - 116(Ts)
B - 254(O)
UZ - 38(Ts)
B - 2379(Tf)
B - 2289(Ts)
UZ - 46(B)
A
B
C
D
14C dates
Minimum dates
Maximum dates
Maximum date for overlying deposit
Minimum date for underlying deposit
Date of advance

glacial advance that culminated in 1916 the largest diameters were 21 mm throughout, in spite of variations in altitude and rock type from one moraine to another of this age. Lichens were also measured on many mining tips of known age from the seventeenth century and on a moraine below Nipalsglaciären which had overrun a buried soil, giving radiocarbon dates of 2320 ± 160 BP (ST-3811) and 2460 ± 90 BP (I-6854). These dated surfaces were used to construct a lichen growth curve. Lichenometric dates relate to the times when moraine surfaces were uncovered and stabilized and therefore record times when glacier retreat was beginning.

Karlén found that the Holocene moraines fall into four groups:

1 The oldest moraines were in front of only three glaciers. The maximum thalli diameters of up to 410 mm occurred on moraines in front of Björlings glacier where, it seemed, there were at least two and possibly as many as five advances around 7000 BP (Figure 10.10).
2 In front of four glaciers were moraines with maximum thalli diameters of 256 to 285 mm indicating glacial advances in two episodes between 4800 and 3900 BP.
3 In front of thirteen glaciers maximum lichen diameters of 140 to 210 mm indicated moraines dating to advances between about 2800 and 2200 BP.
4 The youngest moraines, with maximum lichen diameters of 21 to 100 mm, were the outcome of Little Ice Age advances (see Chapter 3).

The diagram summarizing Karlén's results does not seem to provide altogether conclusive proof of four glacial advance phases; in a 1976 résumé of his lichenometric work he mentions intermediate advances around 3300, 2000 to 1800, 1600, 1300 to 1000, and 700 to 400 BP (Figure 10.11a). Clearly a great deal depends on the reliability of the lichen growth curve and the assumption that lichens grow steadily even if the climate itself is fluctuating.

Karlén (1979) went on to examine the evidence for glacier variations in the Svartisen, Okstindan and Saltfjell areas of northern Norway. He obtained samples for ^{14}C dating from seven moraines and lichenometric data from 125 moraines. He constructed a lichen growth curve on a basis of maximum thalli diameters on mining tips and radiocarbon dates of soils and buried peat. He concluded that glacial expansions took place about 2800, 1900, 1500 to 1300, 1100 and since 600 ^{14}C years BP. He considered that dates from peat or soil from Okstindan could indicate glacial expansion at some time around 6000 and 4500 BP.

Innes (1984) is critical of some of the assumptions underlying Karlén's lichen growth

Figure 10.6 Holocene glacier fluctuations in the Valais, western Alps. Key ^{14}C dates in (a) point to moraine ages. Dates for fossil wood in (b) indicate warm phases when treelines were higher than at present. Holocene glacier fluctuations in (c) are based on these data sources and are related to known extensions of glaciers between 1850 and 1976 (Röthlisberger 1976, Schneebeli 1976). (d) gives additional dates on fossil wood and high treelines in the western Alps. Vivian and Peretti collected most of the wood and published the dates in *Radiocarbon* 1973, 1974 and 1975. Column (a): (A) = Allalin, (B) = Brenay, (C) = Corbassière, (Fen) = Fenêtre, (Fer) = Ferpècle, (F) = Findelen, (G) = Gorner, (MD) = Mont Durand, (mm) = Mont Minet, (T) = Tsijiore Nouve, (Tr) = Trient, (Z) = Zmutt. Column (d): (B) = Alpe Palü Bundnerland, (Tf) = Trift, (O) = Oberaar, (W) = Witenwasserengletscher, (S) = Stein, (Cb) = Cambrena, (St) = Stelliboden, (Gs) = Göscheneralp and (Ts) = Chiera

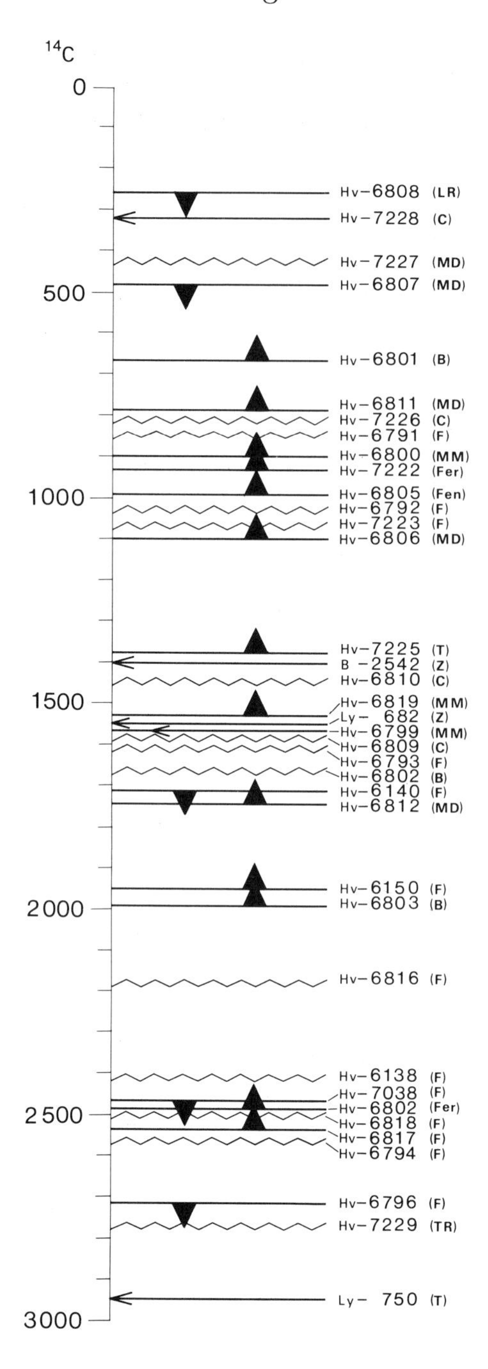

Figure 10.7 Enlargement of inset on Figure 10.6 showing ^{14}C dates in Valais from 3000 BP to the present. The symbols and abbreviations are as for Figure 10.6.

curves, particularly at Blockfjellbreen in Svartisen, and is consequently unconvinced by the conclusions he has reached elsewhere in northern Scandinavia (see Chapter 3). This is important because his Scandinavian work is an important part of Karlén's contribution to the paper he published with Denton (Denton and Karlén 1973b) on Holocene climatic fluctuations worldwide.

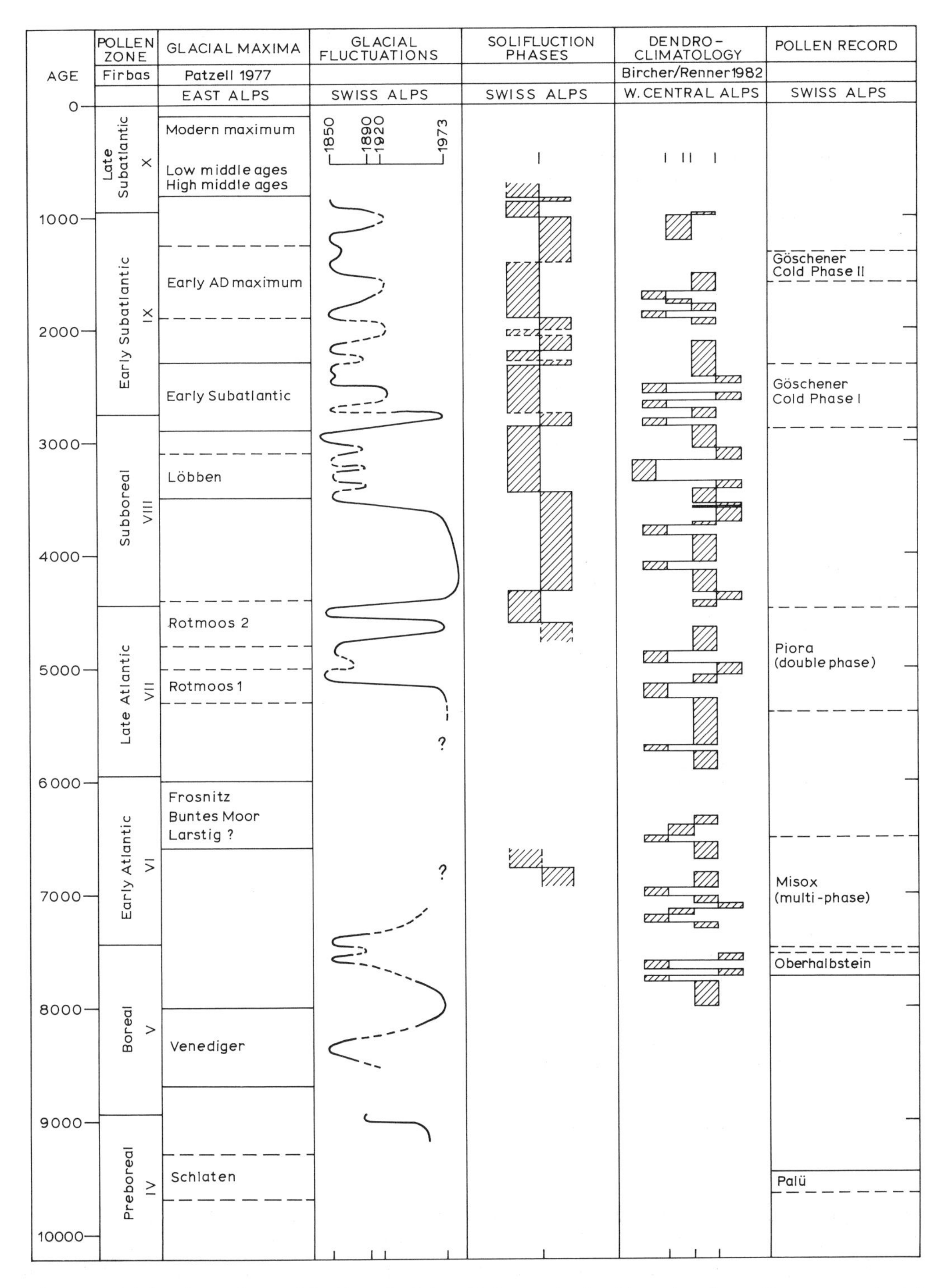

Figure 10.8 The chronology of the Postglacial or Holocene in the Alps, showing pollen zones, glacial fluctuations, solifluction alternating with soil formation, tree-ring and pollen indicators of temperature. (From Gamper and Suter 1982)

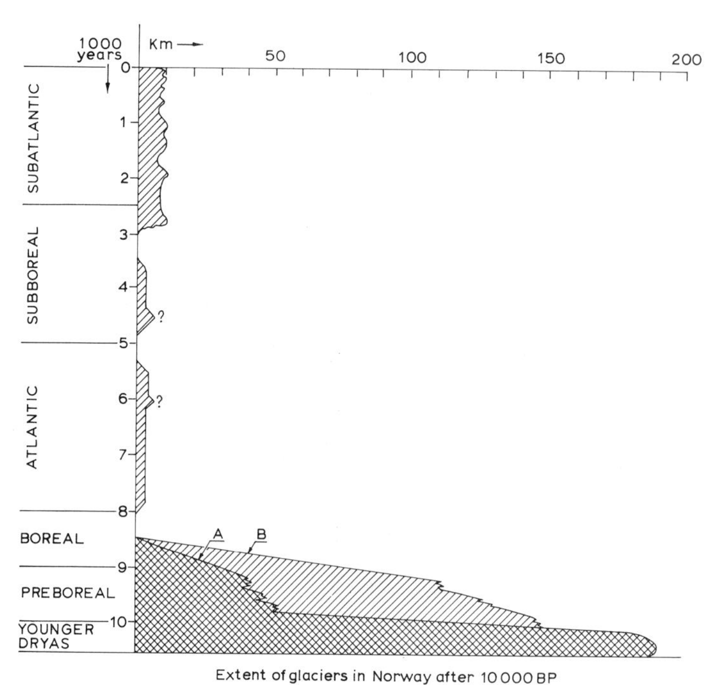

Figure 10.9 Retreat of the Norwegian ice sheet and subsequent Holocene glacier fluctuations.
(a) glaciers in the fjord districts of western Norway;
(b) glaciers in other parts of the country. (From Andersen 1980)

To supplement the lichenometric chronology of the moraines, Karlén attempted to use the sediments deposited in a small glacial lake as a record of the fluctuations of a glacier feeding it (Figure 10.11c). He argued that the silt supply would vary with glacier size. The post-glacial sediments in the lake basin he chose, the Vuolep Allakasjaure, are one or two metres thick. Cores from them contain organic matter which shows up on X-ray photographs and can be measured by combustion. It also allows ^{14}C dating. The results of the study suggested the sequence of events in Table 10.1.

Table 10.1 Dates of glacier events derived from lacustrine sediments deposited in Vuolep Allakasjaure

9345 BP	deglaciation
9200–7500 BP	warming conditions
7500–7300 BP	sudden increase in glacial activity
7300–7100 BP	warming
7100–7000 BP	short recurrence of glacial activity
7000–6000 BP	warm with lowest glacial activity of the Holocene and possibly complete melting of glacier
5900–4000 BP	glacial activity increases in a fluctuating manner (Figure 10.11c)

Source: Karlén (1976)

These results are no more than tentative and clearly require substantiation by studies of other cores before reliance can be placed on them. The dating is uncertain because the organic matter content of the cores was so low that samples believed to be of the same

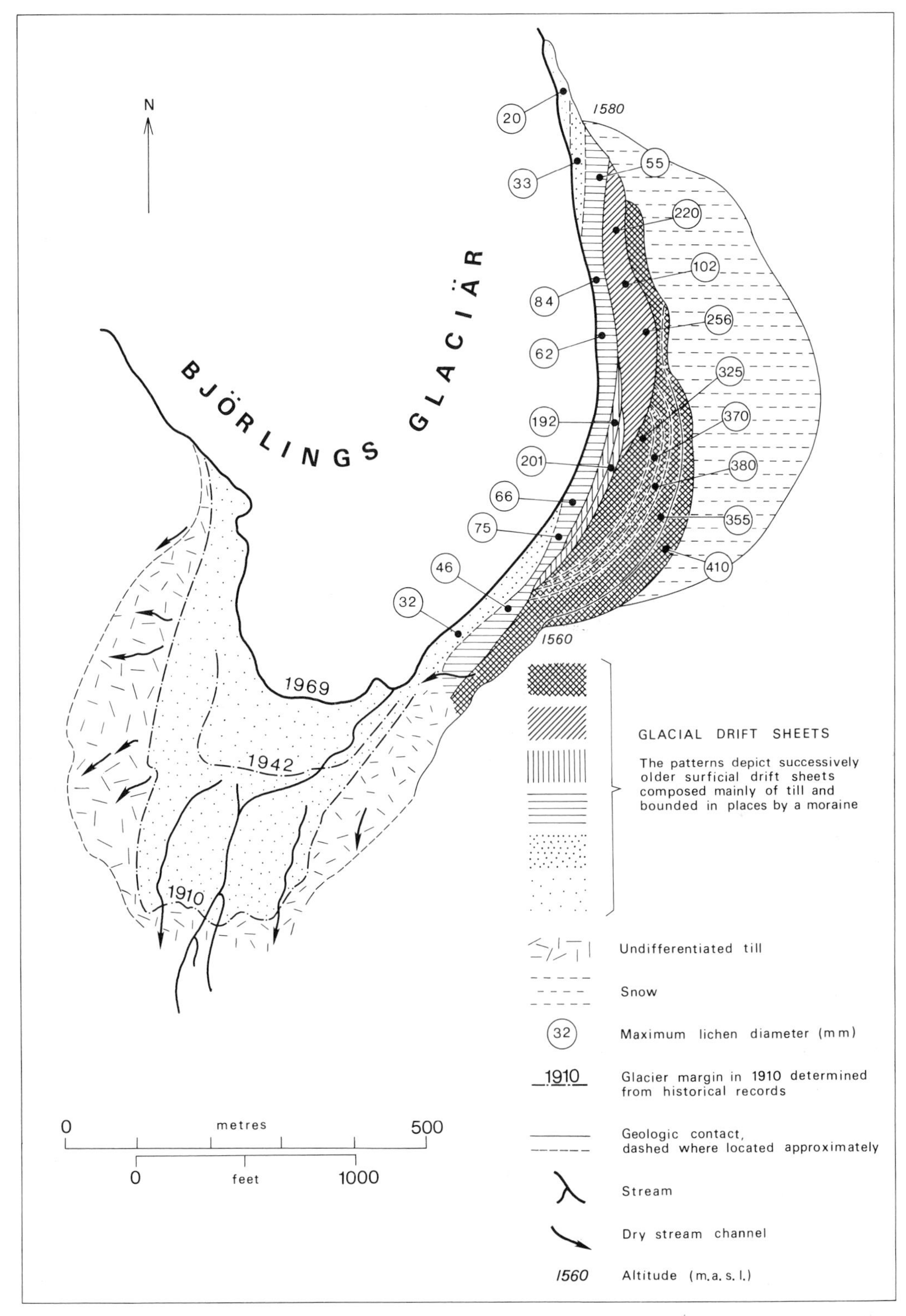

Figure 10.10 The moraines of the Björlings glacier. They include some of the oldest set dated by lichenometry in Scandinavia. (From Karlén 1973)

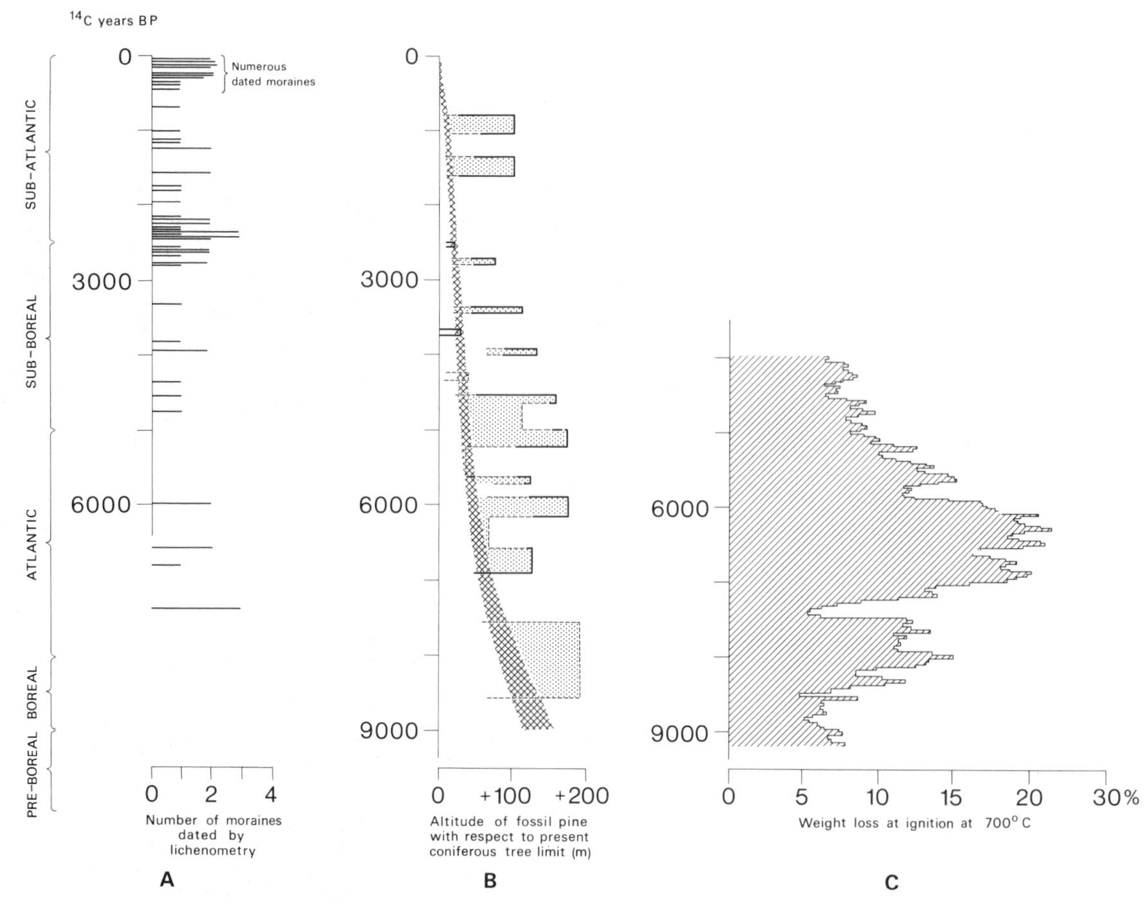

Figure 10.11 Holocene glacier history in Sweden according to Karlén (1976). (a) summarizes findings from lichenometry; (b) shows the treeline evidence superimposed on land uplift curve (from Donner 1969); (c) gives organic content of lake sediment from Vuolep Allakasjaure

horizons but from different cores were amalgamated to provide enough material for dating purposes.

Karlén also brought together ^{14}C dates for pinewood collected above the present treeline in Lappland, Jämtland, Härjedalen and Dalarna and, taking into account the number of tree rings in the complete trunks, put forward a general scheme for the incidence of warm phases in the Holocene (Figure 10.11b). *Pinus silvestris* colonized Scandinavia as the Pleistocene ice sheet melted away, reaching the Torneträsk area about 9000 BP. The height of past treelines since then has been affected not only by climate but also by post-glacial rebound, amounting to 260 m since 9000 BP near the Gulf of Bothnia, and to about 80 m since 7500 BP in the Swedish mountains (Donner 1969). The main reason for the height of the treeline appearing to decrease before 3000 BP, according to Lundqvist (1969), was this uplift of the land, whereas after 3000 BP climate change must have been responsible. Karlén therefore took the effect of rebound into account, plotting the heights of dated wood samples above the present treeline and comparing these heights with the curve of glacial rebound. He was thereby able to distinguish about a dozen periods with the last 9000 years when the treeline was high, with pines growing well above the altitude of the present treeline. This meant that summers were warmer than now as early as 8000 BP as well as around 6000 and 5000 BP. Again there is evidence of frequent and rapid changes of climate represented by shifts in the treeline through 100 to 175 m.

Problems of interpretation arise, for it is not always easy to determine the height of the present treeline and Karlén was obliged to use the general coniferous treeline as a datum in a study of specifically fossil pine. The ^{14}C dates are obtained from the wood of trees that grew from seed germinating many years, even a century or two, earlier. It is the climatic conditions at the time of germination which are probably critical; once established, trees may survive long after temperatures have fallen again. Karlén himself points out that although single trees are known to have germinated above the general treeline during the 1920s and 1930s, the altitudinal limits of pines do not seem to have increased markedly in Scandinavia as a result of twentieth-century warming. Nevertheless, he sees the fluctuations of the treeline in the Holocene as having been through 2000 m, implying variations of summer temperatures of at least 1.5 °C, a figure similar to that deduced by Patzelt for the Austrian Alps.

Østrem's (1964) studies of the age of organic matter contained in the ice cores of certain moraines still provide much of the information about the middle Holocene in Scandinavia. The technique he uses is fraught with difficulties and uncertainties. Much of the fine debris in the ice consists of mineral material; samples taken from Kebnekaise contained so much old carbon in the form of graphite that they were useless for dating purposes. Some of the organic matter also seems to be older than the ice containing it; Østrem points out that humic material from a recent snowbank gave an apparent age of 1000 years, and organic matter from the surface of Gråsubreen in the Jotunheim ranged in age from a few hundred to a thousand years (Figure 10.12). Thus although the ice cores in the end-moraines of Gråsubreen gave ^{14}C dates of 6770 ± 270, 4190 ± 80, 4060 ± 170, 3780 ± 150, 2600 ± 100 and 1300 ± 100 BP, a correction of a thousand years has to be subtracted. When this is done the corrected values do not correspond very well with the Swedish ice advances as derived from Karlén's lichenometry. Barsch (1971) was not convinced by Østrem's explanation of his ice-cored moraines as till-covered snowbanks; the decrease in age of the moraine ridges as distance from the glacier increases, though explicable in terms of overriding known to have occurred elsewhere in Scandinavia and Switzerland, still presents problems.

Griffey (1975, 1976) obtained a date of 4790 ± 120 (Birm–493) for an organic horizon in the inner end-moraine of Austre-Okstindbreen but rejected the possibility that it might be older than a better vegetated, more rounded moraine downvalley which had been dated to 2000 ± 110 (Birm–492). Alexander and Worsley (1973) had shown that old peat could become incorporated in much younger moraines and Griffey explains his date of 4790 BP in terms of this kind of anomaly. However, Page (1968) had identified an advance in Glomfjord, north of Svartisen, as having culminated about 4550 ± 170 (Gak–1445) and was inclined to attribute the outer Glommen moraines to a moist climatic phase at the Atlantic/sub-Boreal transition, involving a fall in the firnline of about 200 m and either summer temperatures 1.2 °C lower than now or winter precipitation 600 mm greater. He cited a body of literature dating back to the beginning of the century, claiming to recognize Atlantic glacial advances in Nordland, Dovre and Hardangervidda. Page's work has not gained general acceptance but in view of the more recent work in the Alps, showing advances there about 5000 BP, and results obtained by various other people in Scandinavia, the Glommen moraines deserve renewed attention.

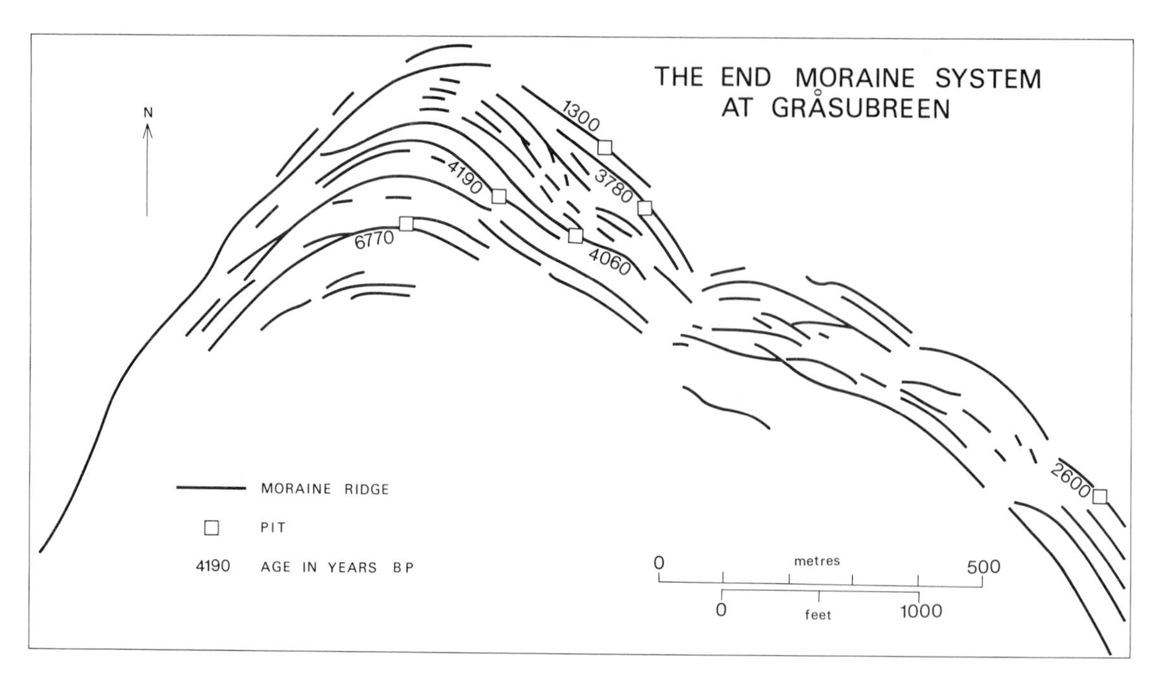

Figure 10.12 The moraines of Gråsubreen. (From Østrem 1964)

Matthews and Dresser (1983) have obtained eighteen radiocarbon dates from a single horizon of a humo-ferric podzol buried beneath the outermost Neoglacial end-moraine of Haugabreen in southern Norway. The dates range from 485 ± 60 near the surface to 4020 ± 70 BP near the base of the horizon, indicating that its development probably began about 5000 to 4500 BP (when botanical evidence shows that climate in Scandinavia was deteriorating) and that organic accumulation continued until the soil was buried beneath the end-moraine of advancing ice approximately 230 calendar years BP. In the interval between 4500 BP and the present, it is concluded no other advance of Haugabreen reached as far downvalley as in the Little Ice Age.

Mörner's (1980) study of lake carbonates in Gotland indicated a fall of temperature of 3.7 °C at the sub-Boreal/sub-Atlantic transition, about 2500 BP. Glacier advances may well have occurred about this time. Ackert (1984) mapped a series of ice-cored lateral moraines in the Tarfala valley of Swedish Lappland and found them to be composed of material from deflected talus tongues. They were evidently formed by a glacier advance greater than the last one in 1910. From a comparison of the Tarfala moraine patterns with those mapped by Karlén (1973), Ackert concludes that the Tarfala lateral moraines and the talus tongues from which they originated were formed between 2700 and 2400 BP and that they point to a lowering of temperature of 0.9 °C. The outermost moraines were continuous with the frontal moraines of Storglaciären dated by Karlén (1973) whose chronology was accepted by Ackert. The cooling of 0.9 °C was considered by Ackert to be a minimum value (personal communication, 5 March 1985).

The Okstindan Research Group of Reading University found evidence of advances of the Okstindan glacier, at 66°N near the Swedish border, between 3000 and 2500 BP and again between 2000 and 1600 BP (Griffey 1975, 1976). The dates were obtained from organic lenses of podsol profiles beneath till and basal peats revealed by digging pits in moraines which lie outside those formed in the Little Ice Age.

Worsley (1974) had shown that Engabreen advanced about 1600 BP but suspected that this was a surge; it seems more likely that it was responding to a climatic fluctuation that caused other glaciers in northern Scandinavia to advance about the same time.

Although a substantial number of Holocene moraines in northern Norway and Sweden have been shown to have been formed earlier than the Little Ice Age, few have been discovered in southern Norway. In Jotunheimen, Matthews and Shakesby (1984) obtained values for the relative age of moraines using lichenometry and a Schmidt hammer to assess the degree of weathering. They estimated that over 90 per cent of the moraines examined were formed in the Little Ice Age as against less than 40 per cent from a similar sample size in northern Norway and Sweden. It seems quite possible that the status of the Little Ice Age in relation to its Holocene predecessors differed in northern and southern Scandinavia.

10.2.5 ICELAND

Even the large icecaps in Iceland are commonly believed to have disappeared completely between 8000 and 2500 BP (Thórarinsson 1958, Björnsson 1979), but very little direct evidence of the extent of Icelandic glaciers during the Holocene has so far been found. Tephra from the 1362 eruption of Öræfi (Thórarinsson 1956, 1964) on the moraines of Svínafellsjökull, Skaftafellsjökull, Kvíárjökull (see Figure 2.15) indicates that they were formed before the Little Ice Age; how much earlier is not clear.

Excavations in the forefield of Skalafellsjökull, a non-surging outlet of Vatnajökull, show that organic sediments were accumulating in the zone between the present front and the outermost Little Ice Age moraine of 1887 throughout the period 5710 ± 90 BP to 1370 ± 90 BP, and hence that the glacier must have been smaller before Öræfi's eruption than it has been since. The earlier date does not provide a close approximation to the beginning of this prolonged period of ice withdrawal. The tephra above the organic sediments probably dates from 1362, strongly suggesting that the Little Ice Age advances were the most extensive during the Holocene. Similar evidence that Eyjabakkajökull was smaller than during the Little Ice Age between 6940 ± 50 BP and AD 1362 is less significant since this is a surging glacier. No evidence was found in either case of Holocene advances before the Little Ice Age but the onset of solifluction at Skalafellsjökull after 2820 ± 40 BP gives an indication of climatic deterioration (Sharp and Dugmore 1985).

Direct evidence of pre-Little Ice Age glacial activity comes from an examination of the relationship between dated tephra layers and the lateral moraines of Sólheimajökull. Eystrheidi, the outermost moraine, formed of till deposited between 1200 and 1400 years ago, overlies loess which was deposited above an older Holocene till more than 3000 years old (Maizels and Dugmore 1985). This till, dated to before 3100 BP and probably after 3500 BP, was laid down at a time when the glacier was less extensive than at its Holocene maximum, which occurred some time after 7000 BP and before 4000 BP. However, the results of studies of neighbouring glaciers such as Gígjökull, Steinholtsjökull, Seljavallajökull and Lambafellsjökull suggest that the behaviour of Sólheimajökull may have been anomalous; the neighbouring glaciers all reached their greatest extents with the last few centuries when, for some reason, Sólheimajökull failed to cover as large an area as it had earlier in the Holocene. (Private communication from Andrew Dugmore, May 1986.)

However this may be, the results of the investigation of the Sólheimajökull moraine sequence have clearly refuted the possibility of the Icelandic icecaps having been absent during much of the Holocene.

10.2.6 SPITSBERGEN

Holocene advances in Spitsbergen were most extensive in the Little Ice Age when ice fronts overrode earlier Holocene moraines. Many of the frontal moraines lie directly on early Holocene marine terraces indicating their asssociation with the most extensive ice cover in at least the last 8000 years. However, certain smaller cirque glaciers were unable to breach the early Neoglacial moraine during subsequent Little Ice Age advances. At one such glacier on western Prins Karls Forland, more than ten ice-cored moraine crests are preserved, of which only the innermost few are lichen-free and hence believed to have been formed in the Little Ice Age. The outer crests contain a diverse range of maximum *Rhizocarpon geographicum* thalli 35–60 mm in diameter and show incipient soil development. In some areas where the thalli diameters are smaller, about 25–27 mm, lichens have been killed presumably by later continuous snow cover and moraines show signs of reactivation (personal communication from G. H. Miller, 10 September 1979). Miller tentatively suggests 'an early Neoglacial advance at least 60 mm *R. geog.* years ago, a mid-Neoglacial advance about 25 mm *R. geog.* years ago and a complex LIA advance terminating in the early 1900s'.

10.3 ASIA

10.3.1 HIMALAYA

Efforts to unravel the Holocene history of the Himalayan glaciers have been delayed by difficulties of access, political restrictions, lack of adequate maps and air photographs and the sheer size of the region (Figure 7.4). Investigations have recently been aided by space imagery and new series of topographical maps of the Nepalese Himalaya, and extended by Chinese glaciologists working on the Tibetan side of the ranges.

Work has been concentrated in the Mt Everest region and especially on the moraine of the Khumbu glacier on the southwest side of the mountain in the Sapte Kose drainage basin. Valleys in this system, which extends across the Himalaya, typically contain two moraine complexes, of which the older, though as yet undated, is generally taken to relate to the last glacial maximum. These inner and outer moraine sets are more easily distinguished on images of the Tibetan valleys than those in Nepal. Two advances causing moraine deposition on the Tibetan Plateau have been dated at 2980 ± 150 and 2920 ± 100 (Zhen and Li 1982). Some of the Nepalese valleys such as the Dudh Kunda contain evidence of at least four advanced stages during the Holocene (Williams 1983).

The present equilibrium line altitude rises northward at 71 m per km across the Himalayan range. It is estimated that in the Late Pleistocene the line was depressed by about 850 m below the present level in the Nepalese valleys but by only about 400 m on the Tibetan Plateau. Depression during the Holocene was also apparently less in Tibet than in Nepal, probably as a result of decrease in precipitation in cooler periods.

The subdivision of the inner moraine complex of the Khumbu glacier has been the subject of considerable disagreement between different investigators, each of whom has

Table 10.2 Tentative correlation of Holocene glacial deposits in Khumbu region, Mt Everest

Khumbu Müller (1958)	*Khumbu Iwata (1976)*	*Khumbu Benedict (cited Williams 1983)*	*Khumbu Fushimi (1977)*	*Dudh Khunda Williams (1983)*
	Lobuche I	Pumore	Thuckla 6	Yuligolcha I
Recent	Lobuche II	Tsola II	Thuckla 5	Yuligolcha II
	Lobuche III	Tsola I	Thuckla 4	Yuligolcha III
Dhugla			Thuckla 3	
	Thukla	Dhugla II	Thuckla 2	Tamba
		Dhugla I	Thuckla I	

Source: Williams (1983)

developed his own nomenclature (Müller 1958, Iwata 1976, Benedict 1976, Fushimi 1977, Williams 1983; see Table 10.2). The stratigraphic relationships of some of the radiocarbon samples which have been collected are somewhat uncertain, according to Williams, but the dates from lateral moraines obtained by Müller, and from outwash terraces probably connected with different parts of the inner moraine complex by Benedict, fall into two groups indicating glacial expansion around 1200 BP (1150 ± 80, B–173, Müller 1958, and 1155 ± 160, I–6728, Benedict 1976) and in the Little Ice Age (550 ± 85, I–6642, 530 ± 165, I–672 and 480 ± 80, B–174). In front of the nearby Kyuwo glacier, Fushimi (1978) collected a sample of wood from the oldest moraine of the inner set which gave a date of 410 ± 100 (Gak–6807), and of charcoal from the youngest moraine of the outer set which gave a date of 1200 ± 100 (Gak–6808). He therefore attributed the inner moraine of the older Khumbu complex to the Holocene expansion around 1200 BP, thereby implying that all the moraines of the inner complex relate to several phases of the Little Ice Age. The large number and complexity of moraines in, for example, the Dudh Kunda valley and its tributary the Yuligocha (Williams 1983) suggest that there were several more Holocene phases.

Röthlisberger has examined the moraines of sixteen glaciers along the length of the Karakoram and Himalaya, comparable in size with glaciers in the European Alps. Avoiding glaciers in the arid zone and those known to be subject to surging or affected by earthquakes, he collected sixty-eight samples of palaeosoils and *in situ* tree trunks overriden by ice and covered with till (Röthlisberger and Geyh 1985a). The soil material gave two sets of radiocarbon dates; the older set was given by material consisting mainly of lichen fragments, the younger set by humic acids, roots and *in situ* wood. It was argued that the lichen dates indicated the time when soil began to form after the ice had retreated; the other dates referred to the time when soil formation ceased as readvance of the ice began (Geyh *et al.* 1985).

An account of early Holocene conditions in the Num Kum area of Kashmir was given in a preliminary survey by Röthlisberger and Geyh (1985a), but the full results of Röthlisberger's studies in the Himalayas, which he made with a view to comparing the timing of glacier advances in the northern and southern hemispheres, appear in his book (1986) *1000 Jahre Gletschergeschichte der Erde* (see Preface). Large lateral and terminal moraine lobes of the Rantac and Tarangoz glaciers in the Suru valley (34°N, 76°E) mark their Late Pleistocene positions, the onset of which Röthlisberger dates to 19,490 ± 1630 BP. Present-day glacier termini lie 2–4 km further upvalley. Three soils in the right lateral

NORWAY

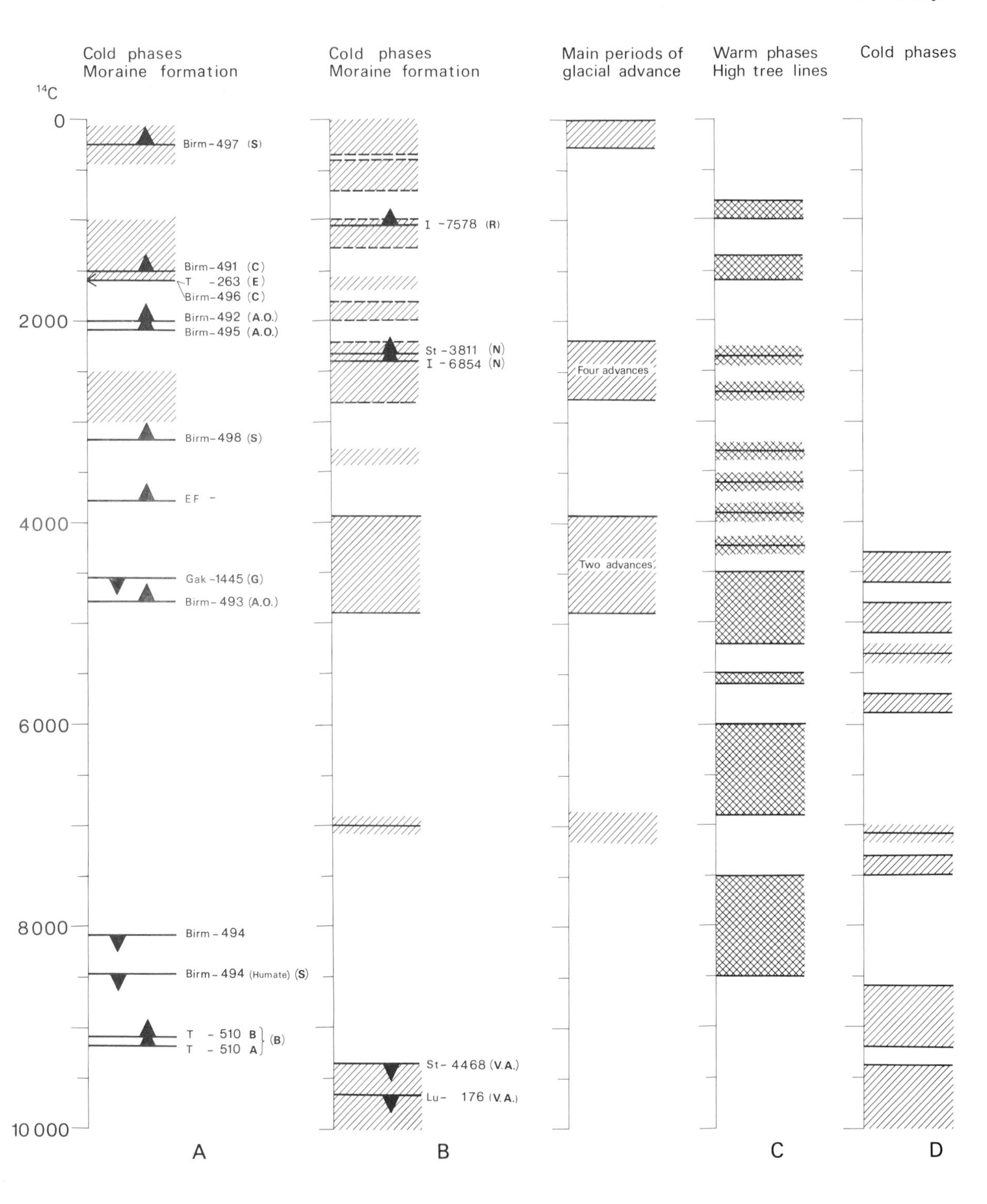
SWEDEN
Vuolep Allakasjaure
Cold phases Moraine formation
Cold phases Moraine formation
Main periods of glacial advance
Warm phases High tree lines
Cold phases
14C
0
2000
4000
6000
8000
10 000
Birm-497 (S)
Birm-491 (C)
T -263 (E)
Birm-496 (C)
Birm-492 (A.O.)
Birm-495 (A.O.)
Birm-498 (S)
EF -
Gak-1445 (G)
Birm-493 (A.O.)
Birm-494
Birm-494 (Humate) (S)
T -510 B
T -510 A
(B)
I -7578 (R)
St-3811 (N)
I -6854 (N)
St-4468 (V.A.)
Lu- 176 (V.A.)
Four advances
Two advances
A
B
C
D
14C dates
Minimum dates
Maximum dates
Lichenometric date

LAPPLAND

Warm phases
High tree lines

SWEDEN

Jämtland,
Härjedalen and
Dalarna

Warm phases
High tree lines

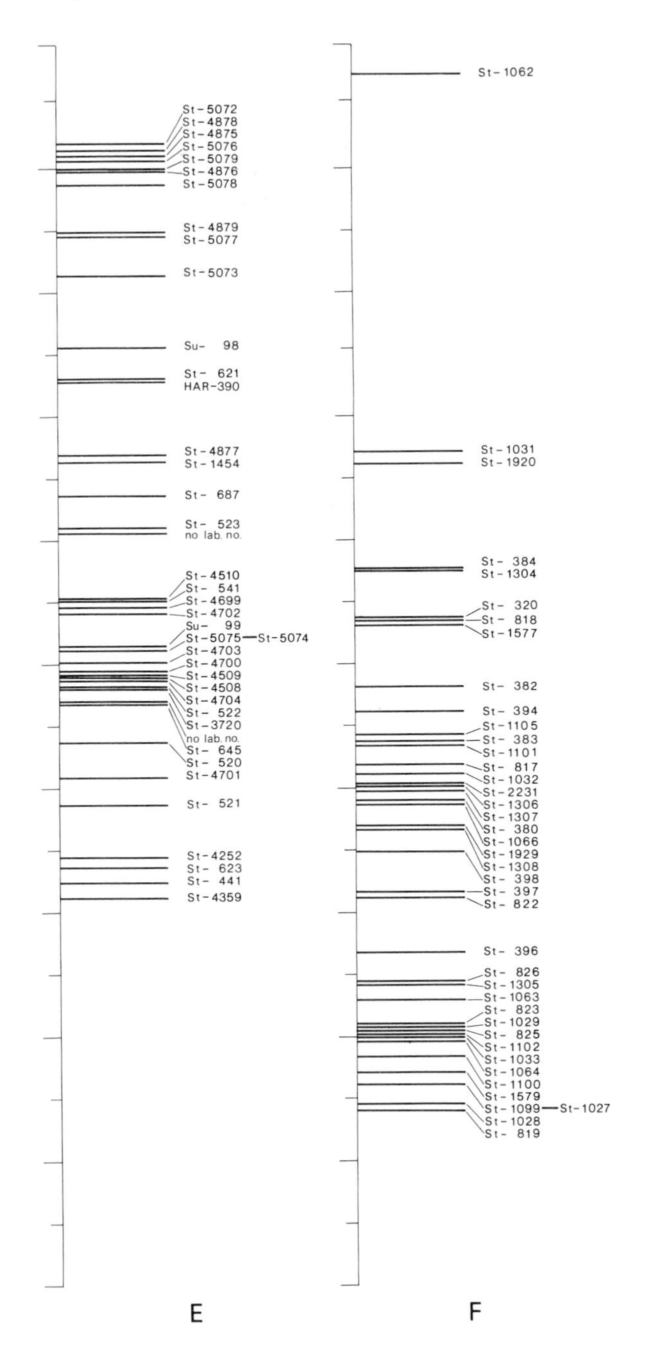

Glacial advance period

Warm phases

Figure 10.13 Holocene glacier and treeline fluctuations in Scandinavia. (a) key ^{14}C dates pointing to glacier advances in Norway plus duration of advances since 3000 BP according to Griffey (1975, 1976). The advance phases of the early Holocene in Norway are not shown. (b) ^{14}C dates and details of glacial advance phases in Sweden (Karlén 1976) alongside high treelines indicating warm phases and based on data displayed in (e) and (f) (Karlén 1973). (d) cold phases indicated by character of lacustrine sediments. (e) ^{14}C dates on fossil wood found above the present treeline in Lappland. (f) ^{14}C dates on fossil wood from above the present treeline in northern Sweden (d, e, f: Karlén 1976). (AO) = Austre Okstindbreen, (C) = Corneliussenbreen, (B) = Balsfjord, (E) = Engabreen, (G) = Glomfjord, (S) = Steikvassbreen, (N) = Nipalsglaciären, (VA) = Vuolep Allakasjaure, (R) = Årjep Ruotesjekna

moraines of the Tarangoz give dates of 13,900 to 12,750 BP, indicating retreat. Then the glacier readvanced, coming forward so far in the early Holocene, about 7400 BP, as to override Late Pleistocene moraine. There were other readvances, one before 3280 and another about 2230 BP, but they were on a smaller scale.

Röthlisberger and Geyh (1985a) identified the main Holocene phases of glacial expansion in the Himalaya as having been around 7400 (2 dates), 4900–4600 (5 dates), 3700–3100 (12 dates), 2700–2100 (14 dates), 1700–1500 (10 dates), 1200–950 (2 dates), 800 (2 dates) and Little Ice Age, 550–100 BP (9 dates). He did not consider he had identified all the glacier advances in the Holocene. He recommended that all dates earlier than 7000 should be considered tentative. The advances of 3700 to 3100 BP and 2700 to 2100 BP were, he thought, probably multiple. The sequences for the Alps and the Himalaya correspond so closely in number and spacing that, although there are uncertainties in the absolute chronology derived from the radiocarbon analyses, they can reasonably be regarded as synchronous (Figure 10.14).

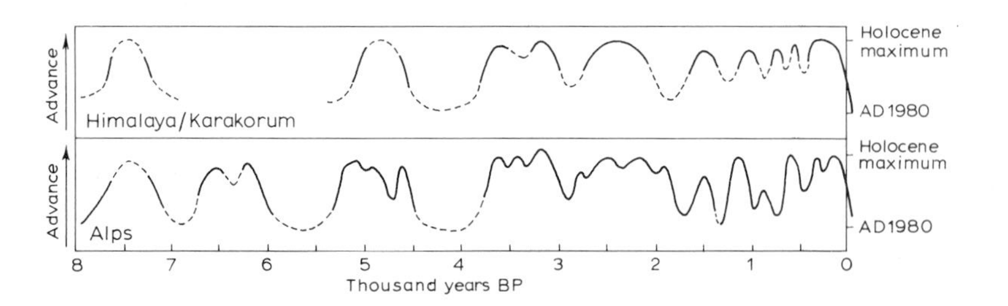

Figure 10.14 A comparison of Holocene glacier fluctuations in the Himalaya/Karakoram with those in the Alps. It is likely that not all the Himalayan fluctuations have been recognized and dated. (From Röthlisberger and Geyh 1985)

10.3.2 CHINA

During the Late Pleistocene maximum, valley glaciers in China were up to six times longer than today; the climate was drier but temperatures were 6 to 8 °C lower. By the middle of the Holocene, the Tien Shan watershed seems to have been free of glaciers (Shi Yafeng and Wang Jingtai 1979). A decrease in rainfall and temperature in the Urumqi valley is indicated by a section of lake deposits which accumulated between 7320 ± 200 and 3950 ± 150 BP, showing *Picea schrenkiano* pollen being followed by *Artimesia* and *Ephedra* (Shi Yafeng and Zhang Xiangsong 1984).

Over the last 3000 years there have been at least two fluctuations. A terminal moraine lies 2 km down the valley from the terminus of the Atza glacier in southeast Xizang (Tibet); wood from the associated lateral moraine, which rises 100 m above the glacier, has given a date of 2980 ± 150 BP (Li Jijun 1978, unpublished and cited by Shi Yafeng and Wang Jingtai 1979). The Rongbude moraine, 2.2 km north of the Rongbuk glacier of Everest, and the Turgbilichi moraine in the Muzart valley of the Tien Shan, 12 km downvalley from the present glacier, are thought to be of similar ages. These advances were considerably greater than those of the Little Ice Age.

A later cold period about 2000 years ago, marked by an advance of the Rougon glacier near Yigun, southeast Xizang, has been dated 1920 ± 40 and 1500 ± 85 BP. There followed a warmer interval which was accompanied by large-scale recession of glaciers in western China and temperatures a few degrees higher than in the twentieth century.

10.4 North America

Substantial progress has been made in unravelling the course of the 10,000 years of Holocene glacial history in the Rocky Mountains of Colorado, Wyoming, Montana, southern Alberta and British Columbia, the Cascades of Oregon and Washington, the mountains of Alaska and Yukon and Baffin Island. As more detailed studies have been made in these widely separated areas, the number of Holocene advances recognized has tended to increase, but there is no general agreement at the present time on terminology and questions of synchroneity.

10.4.1 THE ROCKY MOUNTAINS OF COLORADO AND WYOMING

Shortly after Matthes (1941) had first identified Holocene moraines in the Sierra Nevada, Hack (1943) drew attention to the moraines around Temple and Miller Lakes in the southern Wind River Mountains of Wyoming. These moraines were regarded as having been formed in the last few thousands of years, since the end of a mid-Holocene Altithermal, and for many years little more attention was paid to them. Then, in the early 1960s, Richmond (1960a, b, 1962, 1965) pointed out that the moraines represented glacial advances not only in the last few centuries (the Gannett Peak deposits) but also at two other Holocene stades which were called the Temple Lake 'A' and 'B' events, said to have occurred from 3100 to 2800 BP and 1800 to 1000 BP respectively. He proceeded to apply the Wind River Range terminology to Holocene deposits in the Rocky Mountain National Park in Colorado.

A glacial advance which Benedict (1968) dated by lichenometry to 1900 to 1000 BP he called the 'Arikaree'. As this term had been pre-empted in the literature, the alternative name 'Audubon' was suggested by Mahaney (1972); this name has stuck. In view of the similar ages of the Audubon and Temple Lake B advances, Birkeland *et al.* (1971) put them in the same class. However, Benedict (1973) pointed out that the Audubon and Temple Lake moraines are geomorphologically distinct. The Audubon moraines, widely recognized for instance in Wyoming by Birkeland and Miller (1973), are often overlain and partly obscured by Gannett Peak deposits.

The type Temple Lake moraine in Wyoming was described by Moss (1951) as post-Pinedale and pre-Altithermal (Figure 10.15) and he was of the opinion that it had been emplaced by a Late Pleistocene pulsation. In contrast to this view we find Richmond (1960a, b, 1964, 1965) describing the Temple Lake moraines as early Neoglacial. He equated them with moraines in the La Sal mountains of Utah which were regarded as 2800 ± 200 years old on the basis of a radiocarbon date obtained from hearthsites downstream.

Currey (1974) re-examined the Holocene glacial sequence in Temple Lake valley (Figure 10.16) and concluded that 'the type Temple Lake moraine may represent two episodes of substadial rank. Highly significant from the point of view of Neoglacial chronology is the probability that the unit mapped originally as "Cirque (Neoglaciation)" moraine almost certainly represents at least two episodes of stadial rank' (Currey 1974,

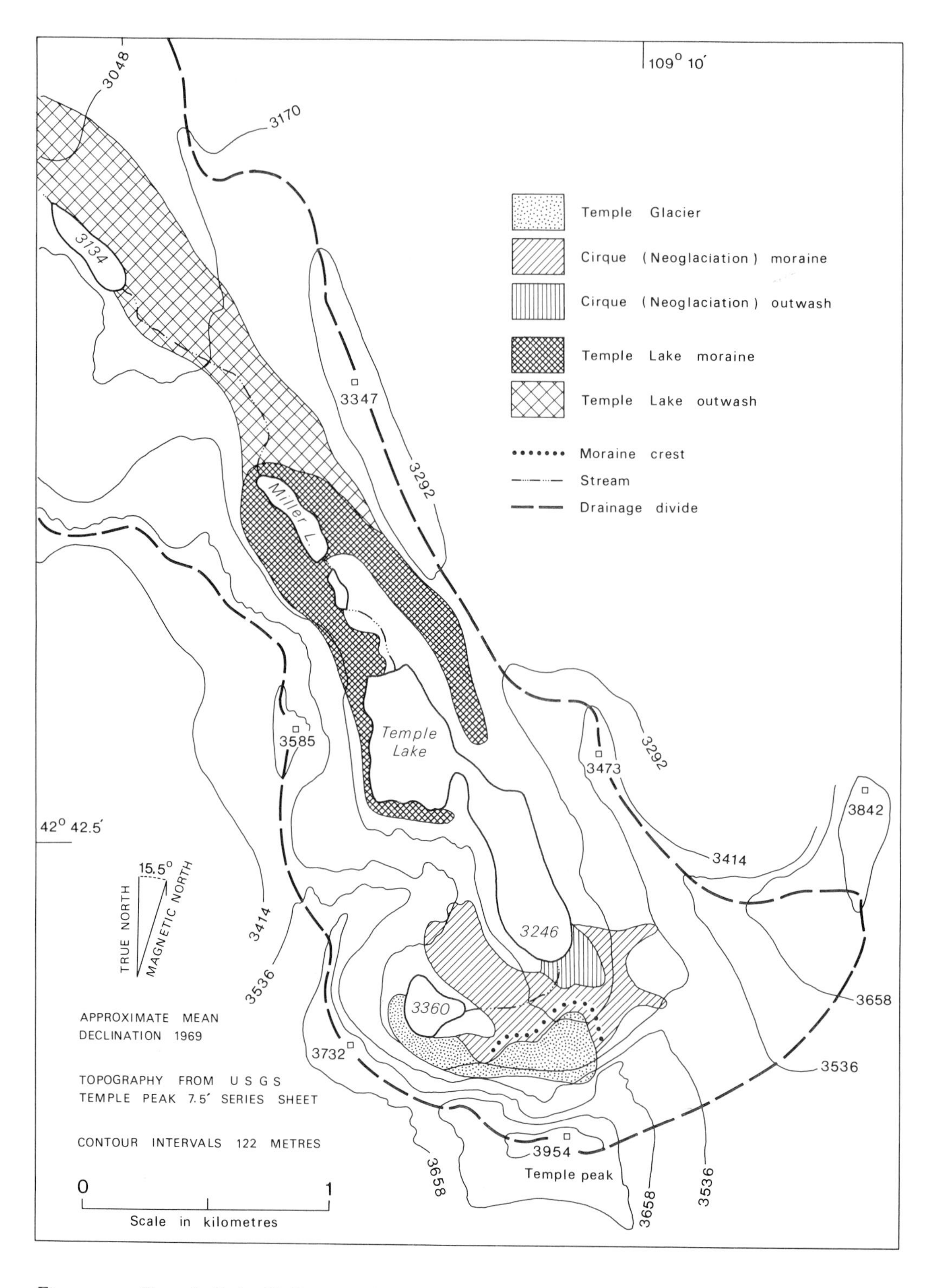

Figure 10.15 Temple Lake Valley moraines, Wyoming, including the type Temple Lake moraines, as first mapped by Moss (1951). (From Currey 1974)

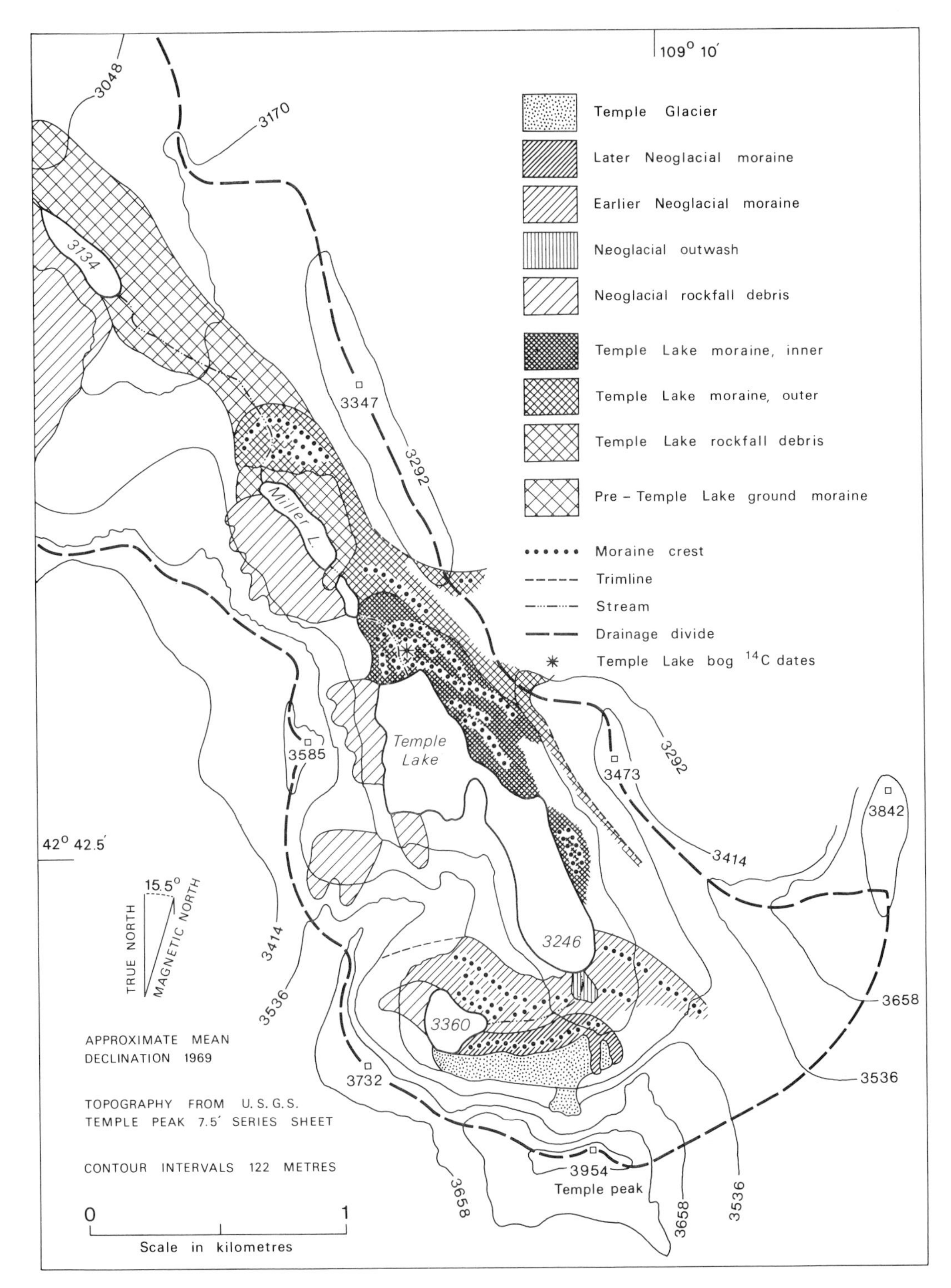

Figure 10.16 Temple Lake Valley Holocene moraine sequence and rockfall deposits as mapped by Currey (1974). This illustrates the need for careful geomorphological analysis as well as accurate dating if a complete sequence of Holocene glacial deposits is to be identified. (From Currey 1974)

p. 297), thus agreeing so far with Richmond's and Birkeland's interpretation. However, radiocarbon dates were obtained from a bog between two ridges of the inner moraine set of the type Temple Lake moraine; one from a depth of 189 to 198 cm above glacial outwash gave a date of 6500 ± 230 BP (GX–3166D); the other from a depth of 138 to 168 cm, which included wood fragments and conifer needles, gave a date of 3520 ± 155 BP (GX–2748) and seems to represent a time when the site was warmer and drier than at present. The original type Temple Lake moraine, Currey therefore concluded, was much older than Richmond had supposed and must be attributed to a time preceding the Altithermal. In fact, as Currey (1974, p. 298) pointed out, the type Temple Lake moraine is older than the latest limiting date, 6170 ± 240 BP, for the last stage of the Pinedale glaciation in Colorado (Richmond 1960) and for the Rockies generally (Richmond 1965):

> The type Temple Lake would appear to be at least in part correlative with the Pinedale IV moraines reported by Graf (1971) for southern Montana and the Pinedale IV moraines reported by Kiver (1972) for northern Colorado; it may be correlative with the period of multiple glacial advance which is reported by Currey (1969) to have occurred between 6000 and 7000 years ago in the Sierra Nevada.

There seems to be no doubt that we have evidence here of at least one pre-Altithermal Holocene glacial advance and at least two post-Altithermal advances of which the last was that of Gannett Peak. A number of studies have pointed to there having been more numerous fluctuations. For instance C. D. Miller (1973), using lichenometry to establish a chronology, found evidence in the northern Sawatch Range of Colorado for three main Neoglacial phases of rock glacier development, though no evidence for the Gannett Peak.

Conflicting conclusions have been reached from studies in the Fourth of July Cirque of the central Colorado Front Range. Mahaney (1973), making use of Benedict's (1967) growth rate curve for *Rhizocarpon geographicum* to assign ages to glacial and periglacial deposits there, pointed to three Neoglacial advances amongst which he included Temple Lake as well as Audubon and Gannett Peak. However, Williams (1973) strongly disputed Mahaney's findings; she distinguished four stratigraphic units in the Fourth of July Cirque and considered that three of them preceded the Altithermal, allocating two to the Pinedale and one to the Bull Lake (i.e. Early Wisconsin glaciation). Mahaney (1973) rejected Williams's chronology, emphasizing that his own conclusions are based on a combination of criteria including degree of weathering, morphology and soil characteristics as well as lichenometry. But he concedes that the Altithermal on the Colorado Front Range may be more complicated than was previously supposed.

Benedict (1973), reassessing the local chronology of cirque glaciation in the Colorado Front Range, took the view that 'despite an unusually complete and well-preserved sequence of glacial deposits and more than half a century of study the history of Holocene glaciation in the southern and central Rocky Mountains is still poorly understood'. He rejected seven of the ten radiocarbon dates listed by Birkeland *et al.* (1971, chart B) and reinterpreted two of them which were his own. He argued that interpretation of Holocene glacial sequences had been hampered by the assumption that glacial advances in high-altitude cirques can be correlated directly with alluvial deposition in far-distant lowland, and he also criticized reliance on certain relative dating techniques. His own work had relied heavily on lichenometry but he had utilized in addition a wide range of morphological and other data, and in presenting a preliminary radiocarbon controlled sequence for the Colorado Front Range in the Holocene (see Figure 10.18b) he indicated

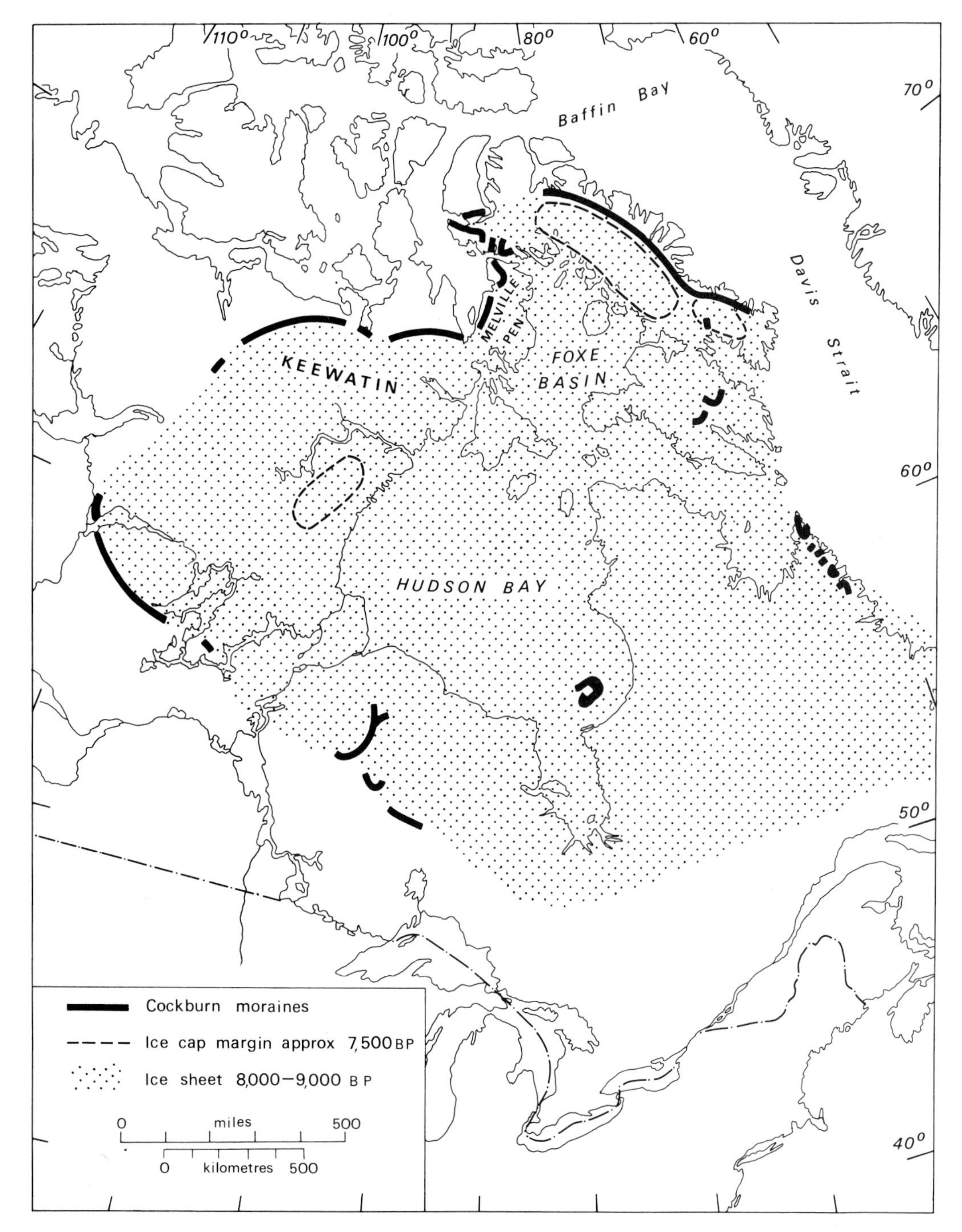

Figure 10.17 Relict ice in the Baffin area about 7500 BP compared with ice sheet at Cockburn moraine stage only a few centuries earlier. (After Falconer *et al.* 1965)

Figure 10.18 Holocene glacier fluctuations in North America. (a) White River and Skolai Pass, Alaska and Yukon, key ^{14}C dates and ages of glacial advance. Note the clear agreement with intervening warm phases indicated by high treelines during the last 4000 years (Denton and Karlén 1977). (b) Colorado Front Range Holocene phases of glacial advance, dated by lichenometry with some ^{14}C support (Benedict 1973, Currey 1974). (c) Cumberland Peninsula, Baffin Island. Moraines formed in the last 4000 years are dated by lichenometry (considered to be ± 20 per cent accurate). ^{14}C dates are from marine shells alive when the sea was warmer than at present (Miller 1973).
(G) = Guerin,
(Gf) = Giffin,
(L) = Lemon Glacier,
(LL) = Llewellyn Glacier,
(M) = Mendenhall Glacier,
(N) = Natazhat,
(R) = Russell,
(RH) = Rusty and Hazard,
(SS) = Seven Sisters,
(D) = Temple Lake

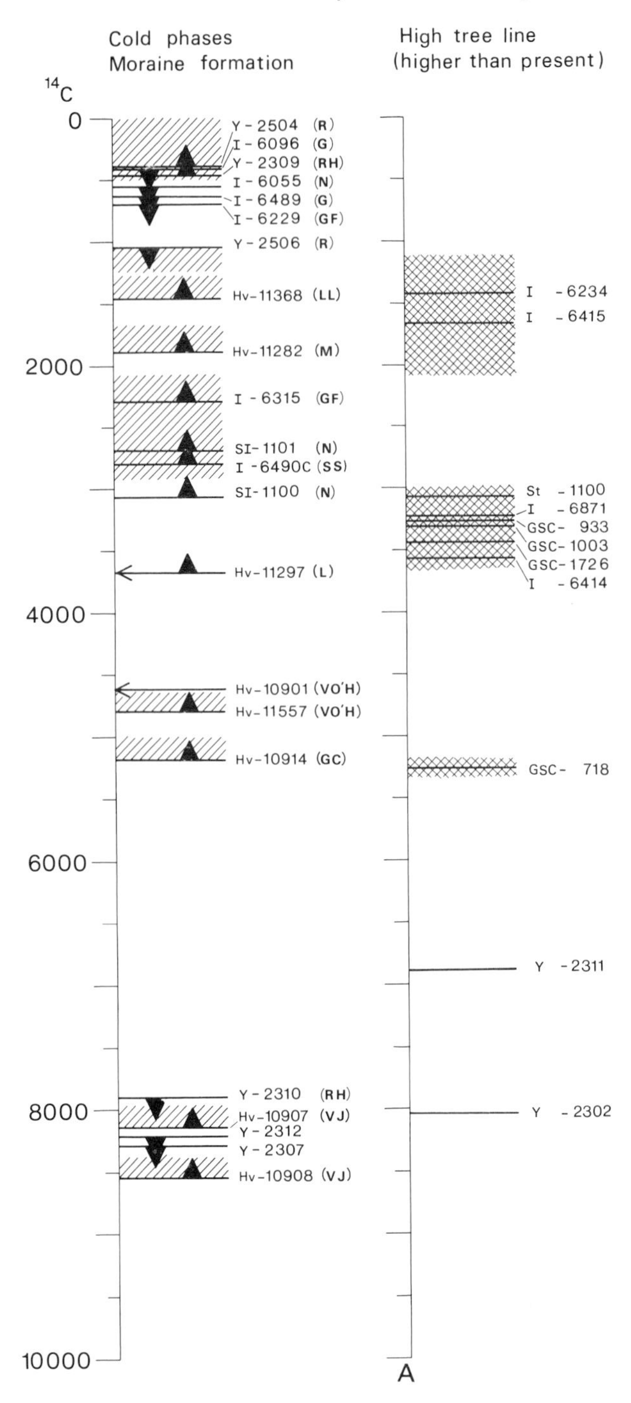

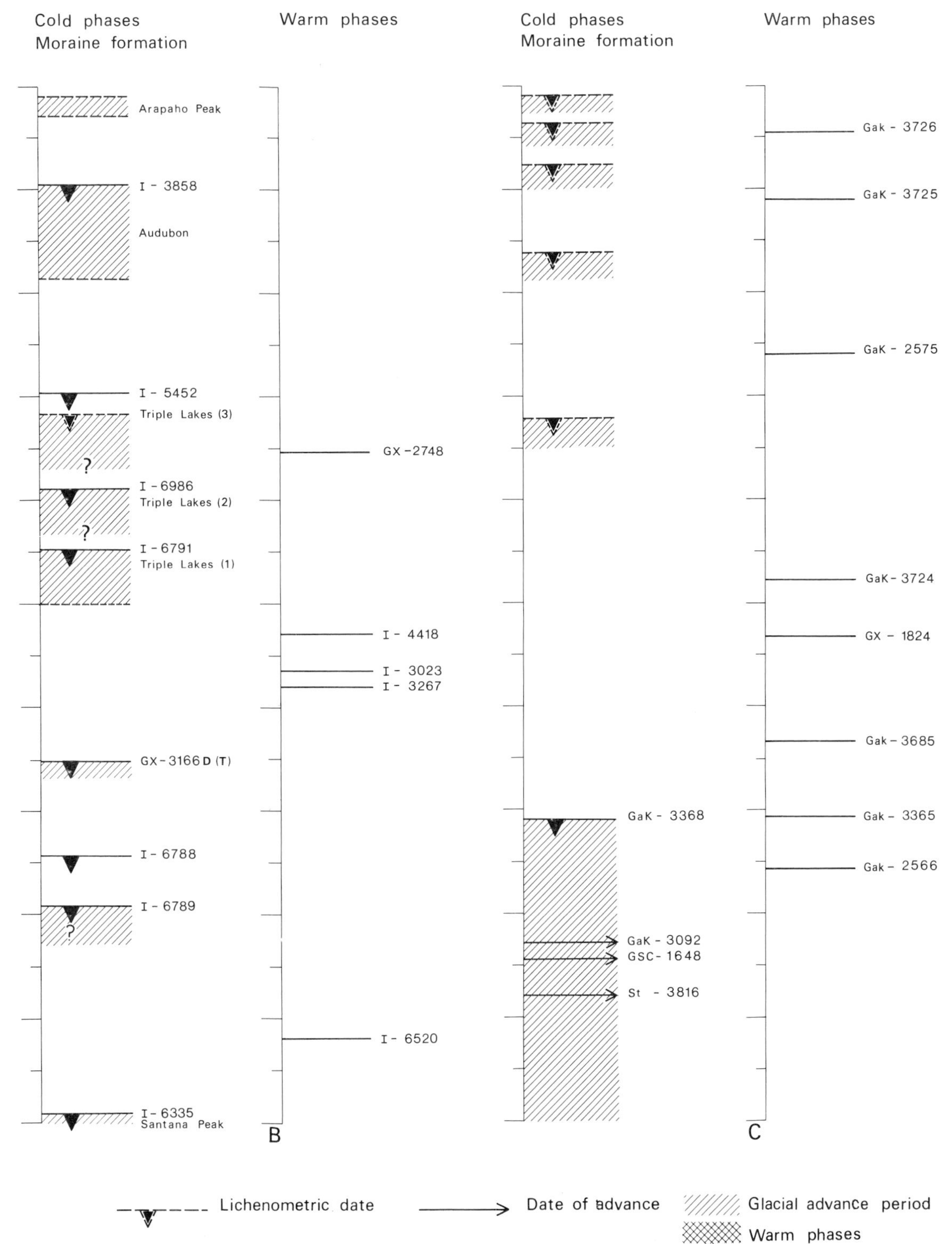

COLORADO
Front range
Cold phases
Moraine formation
Warm phases
Arapaho Peak
I - 3858
Audubon
I - 5452
Triple Lakes (3)
I - 6986
Triple Lakes (2)
I - 6791
Triple Lakes (1)
GX - 3166 D (T)
I - 6788
I - 6789
I - 6335
Santana Peak
GX - 2748
I - 4418
I - 3023
I - 3267
I - 6520
B
CUMBERLAND PENINSULA
Baffin Island
Cold phases
Moraine formation
Warm phases
GaK - 3368
GaK - 3092
GSC - 1648
St - 3816
Gak - 3726
GaK - 3725
GaK - 2575
GaK - 3724
GX - 1824
Gak - 3685
Gak - 3365
Gak - 2566
C
Lichenometric date
Date of advance
Glacial advance period
Warm phases

clearly and meticulously the points where his limiting radiocarbon dates were obtained. The phases of glacial advances he tentatively identified are as follows:

(a) Sanatana Peak moraines, commonly paired terminal moraines lying in many Front Range valleys at about 3340 to 3410 m, are at a higher level than those of the 'late stage' of the Pinedale. At Caribou Lake, organic matter above outwash gives the Sanatana Peak advance a minimum age of 9915 ± 165 BP (I–6335), though it could have begun much earlier. A sample of wood collected 20 cm above the organic matter suggests that by 9200 ± 135 BP the treeline had risen to at least its present elevation.

In the past the Sanatana Peak moraines have been equated with the early Temple Lake moraines and Benedict considers this correlation 'not unreasonable'. But Currey's (1974) more recent limiting date from the Temple Lake moraines of 6500 ± 230 appears to be as closely related as Benedict's to the period of moraine formation and therefore the Temple Lake–Sanatana Peak correlation, resting mainly on the conclusion that they are both pre-Altithermal, is not very convincing. Close spacing in time of the Sanatana Peak advances has yet to be proved.

(b) An advance around 8000 BP (Figure 10.17b) is uncertain. There is a 'possibility that snowbanks may have expanded at this time and had begun to diminish in size by 7900 ± 130 BP (I–6789)'. The snowbank expansion is based on sedimentary evidence from a bog at the foot of the Sanatana Peak deposit; no moraines of this age have been dated.

From 7500 to 6000 BP, during the Altithermal maximum and for the next thousand years as well, Benedict finds no evidence of perennial snowbanks surviving in the Front Range. In that 2500 years, Bezinge and Vivian (1976a, b) identified three phases of glacier advance in the Alps and Röthlisberger and Schneebeli (1976) at least two. Currey's (1974) date of 6500 BP suggests a cold phase in Wyoming also.

(c) The Triple Lakes moraines are not readily distinguished from the Sanatana Peak moraines and have been confused with them on occasions in the past. They are situated only a little higher, at elevations between about 3415 and 3475 m in the Arapaho cirque. They are stable, being without ice cores and covered with herbs, shrubs, krummholz spruce and fir trees. Though they were correlated with the Temple Lake moraines by Richmond (1960a, 1965) and by Benedict himself (1968, pp. 84–5), in view of later work Benedict considers this correlation erroneous and believes them to be the outcome of three or four advances between 5000 and 3000 years ago. The onset of cold conditions followed the formation of the A1 horizons of soils at archaeological sites above the present timberline, dated to 5800 ± 125 (I–3267), 5650 ± 145 (I–3023), and 5300 ± 130 (I–4418). They were affected by frost action shortly before 5000 BP; the ice advanced and basal peat overlying a lateral moraine in the Arapaho cirque gave a date of 4485 ± 100 BP (I–6791).

Another advance about 4000 BP is marked by a terminal moraine overlapped by pro-glacial lake sediments dated 3805 ± 100 BP (I–6986). Then, by 3000 BP, a non-glacial interval had begun 'perhaps on a par with the Altithermal' (p. 592). Currey's dating of a warm phase in the Temple Lake area at 3530 BP supports this. The youngest Triple Lakes moraine in the Indian Peaks area was colonized by *Rhizocarpon geographicum* no later than 2850 BP (Benedict 1968, p. 85), and there is a date of 2945 ± 95 BP (I–5452) for the end of the Triple Lakes outwash deposition in the Fourth of July bog. Following these Triple Lakes advances there was an interval when the ice retreated, soils formed and cavernous weathering took place.

(d) The Audubon moraines and rock glaciers are mostly ice-cored, bouldery and unstable, with a sparse cover of vegetation. Soils are weakly developed and the Audubon moraines differ from the Triple Lakes moraines in lacking a loess mantle and having much less lichen, a 10 to 40 per cent cover as compared with 80 per cent. The size of the lichen patches, with *Rhizocarpon geographicum* thalli diameters of 42 to 71 mm, suggests that the Audubon advances occurred between 1850 and 950 BP (Benedict 1968, p. 82). There were two or possibly three advances in this time. Glacial ice remained in the Arapaho Cirque until at least 955 ± 95 BP (I–3858) and may have persisted until the Little Ice Age.

(e) The Arapaho Peak moraines would seem to be equivalent to the Gannett Peak stage of the Wind River mountains and thus belong to the Little Ice Age. Lichen covering of boulders is sparse, less than 5 per cent, with maximum thalli diameters of 20 mm. Although there is no radiocarbon control, the moraines are likely to be between 250 and 100 years old.

Benedict's (1973) assessment of Holocene chronology in the Colorado Front Range seems to be the most complete and convincing presently available, though it will no doubt be amplified and corrected. He argued that this record represents changes in regional climate, this being demonstrated by its correspondence with well-dated pollen records. The remnant of the disintegrating ice sheet lay too far to the north to have had an important effect on temperature trends in the southern Rocky Mountains. Because of their small size, the many Front Range glaciers would have responded rapidly and sensitively to rather minor climatic fluctuations. Ablation season temperatures have probably always been of great importance in controlling glacier nourishment here, as they are known to have been since the nineteenth century (Waldrop 1964), but because of a four- to eightfold concentration of snow in the cirques (Outcalt 1964) relatively small variations in snowfall could have a great effect and changes in winter wind direction and velocity play a determining role in mass balance history. This suggests that the course of Holocene glacial events in the southern Rocky Mountains might not have been identical with those further north.

The distribution of Late Holocene lithofacies in the Indian Peaks region of the Front Range supports the notion that while similar increases in cloudiness and lowerings of temperature in summer were common to the Triple Lakes, Audubon and Arapaho Peak phases, it was differences in prevailing winter climate, including snowfall and wind drifting, which were responsible for the differing character and amplitude of these three phases (Olyphant 1985).

10.4.2 CASCADES

Several regionally distributed tephra layers bracket Holocene glacial deposits in the Glacier Peak district of the northern Cascades and adjacent regions to the east and northeast. Glacier Peak erupted repeatedly in the Late Pleistocene. Two thick pumice beds (1–1.5 m thick) with seven thin layers between them, have been identified as far as a thousand kilometres downwind (Porter 1978). The youngest of these layers, B, was deposited about 11,250 BP; the oldest, G, less closely dated, has been placed at around 12,500 BP but may well have been younger (Beget 1984). The Glacier Peak area subsequently received ash from Mt Mazama between 7100 and 6700 BP, which left a tephra layer, O, 2–10 cm thick (Mack *et al.* 1979, Beget 1984). Tephra layer Yn, erupted from Mt St Helens about 3400 BP, was less widely distributed but the fallout extended northwards across the Glacier Peak area into Canada. Layer Wn, also from Mt St Helens, has an approximate age of 450 BP dependent on dendrochronology. These various tephras are usually readily recognizable in the field on a basis of their sediment characteristics such as grain size, phenocryst content and colour (Table 10.3). On well-drained moraine crests the Mazama ash generally oxidizes to strong brown and the Mt St Helens Yn to pale yellow, though both are pale or light-coloured in bogs and Mt St Helens Wn is white in most localities according to Waitt *et al.* (1982). Confirmation of identification by petrographic characteristics (Smith and Leeman 1982) depends on collection of samples for laboratory analysis, which has unfortunately not been customary among all those using tephrochronology.

Table 10.3 Characteristics and age of tephra from Glacier Peak, Mazama and Mt St Helens

Tephra	*Colour and grain size*	*Source*	*Age (yrs BP)*	*Reference*
Wn	white sand	St Helens	450	Mullineaux
1974				
Yn	yellow to white sand	St Helens	3400	Mullineaux
O	strong brown	Mazama	6900	Mullineaux
B	lapilli pumice	Glacier Peak	11250	Porter
1978				
G	lapilli pumice	Glacier Peak	12000	Porter

Source: Mullineaux (1974) and Porter (1978)

Since deglaciation in the Glacier Peak area took place before the Late Pleistocene eruptions, the occurrence of Glacier Peak tephras within cirques downwind from the volcano provides a convenient basis for differentiating Late Pleistocene and Holocene glacial deposits. At the end of the Pleistocene, glaciers retreated 20 to 40 km upvalley from their end moraines, but subsequently readvanced to deposit moraines just below or within a few kilometres of cirques situated near the crests of the Cascades. In the Glacier Peak area, deposits of the Carne Mountain Advance were buried by Glacier Peak tephra and therefore antedate 11,250 BP (Beget 1984). Further south, near Snoqualimie Pass, the Pleistocene moraine sequence in the upper Yakima valley ends with the Hyak moraines which have a minimum age of 11,050 BP from wood incorporated in stratified drift behind the type Hyak moraine (Porter 1976). However, while Porter (1978) inferred an age of between 12,000 and 11,000 BP for the Hyak Advances and also for the Rat Creek Advances which left type moraines at Stevens Pass, due south of Glacier Peak, Waitt *et al.* (1972) suggest that the Hyak and Rat Creek Advances may have been as early as 13,000 BP. They estimate that the Rat Creek age snowline was at about 1800 ± 50 m, that is, roughly 650 ± 50 m below the present snowline. This degree of snowline depression contrasts strongly with their estimates of less than 100 m depression associated with Holocene glacial advances.

An Early Holocene advance, the White Chuck Advance, left moraines typically overlain by Mazama ash and themselves overlying or lying just upvalley from Glacier Peak tephra deposits. These moraines contain comminuted pumice lappili, probably because the Holocene glaciers ploughed over and incorporated parts of Glacier Peak tephra layers. The type White Chuck moraine on White Chuck Cinder Cone, to the east of Portal Peak, must have been emplaced after 11,250 BP and before 6700 BP. Woody charcoal with growth rings visible, found about a metre below the surface of the moraine, has been dated to 8350 ± 50 BP (USGS-1070) and 8380 ± 90 BP (W-4277). This is interpreted as approximately dating the glacial advance (Beget 1984). Other cirque moraines bracketed between Mazama and Glacier Peak ash are found in Harmony Peak and Pumice Peak cirques, also in the Glacier Peak district of the north Cascades, and in each case they lie several hundred metres downvalley from later Holocene moraines (Beget 1984). The Brisingamen moraines in the Enchantment Lakes basin are also covered with Mazama ash which is separated from the moraine surface by grus and organic debris which has not been dated. These moraines are much less weathered than the Rat Creek moraines and it has been estimated that they were deposited in the course of a glacial expansion between 10,000 and 7500 BP when the snowline was at about 2410 m, about 55 m below the present snowline (Waitt *et al.* 1982). Waite also reports the existence of similar moraines overlain by Mazama and underlain by Glacier Peak tephra in many cirques east and southeast of Glacier Peak.

Glacier Peak erupted in the Mid-Holocene between 5500 and 5100 BP and coignimbrite ash-falls more than a metre thick covered ridges near the volcano, while contemporary mudflows and pyroclasts filled upper Dusty Creek to a thickness of 300 m (Beget 1984). Subsequently, the Dusty Glacier advanced over this fill and deposited moraines high on the flank of Gemma Ridge. This 'Gemma Peak Advance' was smaller than the Little Ice Age (Streamline Ridge) advances which followed later, but evidence of its occurrence was preserved because subsequent erosion cut a 300 m-deep gorge through the Dusty Creek assemblage, leaving the Gemma Peak moraines perched high on the valley sides, safe from erosion by the later glacial advances, which never filled up the canyon although they were more extensive. This advance is tentatively correlated with a Mid-Holocene advance of the South Cascade glacier, near Dome Peak, 25 km north of Glacier Peak. This was dated by the root of a tree which was apparently overridden and killed about 4900–4600 BP (Miller 1969). No Mid-Holocene advances have been dated in the Enchantment Lakes basin but there are signs that the Little Ice Age (Brynhild) glacier here overrode a less extensive moraine set with a different orientation (Waitt *et al.* 1982).

The majority of the cirques in the north Cascades seem to display a compound Little Ice Age moraine set and an outer, generally smaller, moraine dating from the Early Holocene; the evidence of intermediate advances lies buried beneath the more extensive advance and retreat of recent centuries.

10.4.3 CANADIAN CORDILLERA

The Late Wisconsin valley glaciers in the southeastern Canadian Cordillera had retreated close to or within present glacier limits by about 10,000 BP or earlier (Luckman and Osborn 1979). Dates of 9660 ± 280 (BGS–465) and 9600 ± 300 BP (BGS–490) were taken from basal peat between ice stagnation features on Tonquin Pass and beneath tephra on Castleguard Meadow on the southern margin of Columbia Icefield. Both overlie non-glacial deposits, and were taken within 1000 m of contemporary glaciers at similar elevations, suggesting that post-Wisconsin valley glaciers had receded close to present glacier limits.

Air photographs of Banff, Jasper and Yoho National Parks (Luckman and Osborn 1979), Upper Elk Valley (Ferguson 1978) and Waterton Lakes National Park (Osborn 1982) show that although much the most common deposits are the youthful, undissected moraines of the last few centuries, at some sites there are also older moraines with rounded crests, some of which have been heavily eroded. Thirty-five such moraines have been identified in Banff, Jasper and Yoho National Parks and seventy-five in Glacier National Park, Montana (Osborn 1982). The early Holocene expansion responsible for these features, termed the Crowfoot Advance by Luckman and Osborn (1979), cannot yet be dated closely. Ash from Mt Mazama, now Crater Lake, Oregon, dated to 6600 BP by Westgate (1975) and to nearer 6900 BP by Mack *et al.* (1979), has been found on eleven out of the twelve moraines inspected in the field in Banff, Jasper and Yoho National Parks and on thirteen out of seventeen moraines examined in Waterton Lakes Park and Glacier National Park, Montana (Osborn 1982). No underlying datable material has been found.

Treeline studies by Kearney (1981) indicate that relatively cool episodes with lowered treelines occurred before 8500 BP and again between 8000 and 7500 BP, although, like other workers using proxy climatic indicators, he found that the Early Holocene as a whole was warmer and drier than the present time. He placed the transition to cooler, wetter conditions between 6000 and 5700 BP. Kearney considered the warmest and driest part of the Holocene to have been the period 7500 to 5900 BP and similarly Ferguson (1978) places it in the interval 8300 to 5700 BP. Osborn (1982) takes the view that the Crowfoot Advance was no later than 8500 BP when, according to Kearney, the Hypsithermal began. Kearney and Luckman (1981) suggest that the Crowfoot Advance probably occurred either before 9700 or between 9200 and 8500 BP. It will be recalled that the White Chuck Advance in the Cascades would seem to have occurred at about this last date.

Crowfoot moraines occur more frequently in the southernmost part of the Canadian Rockies than further north, where a higher proportion were wiped out by Little Ice Age advances. A few cirques in Waterton Lakes Park, southernmost Alberta, and Glacier National Park, Montana, have only Crowfoot moraines and none from the Little Ice Age. Holocene glacial fluctuations in the Rockies of southern Canada and Montana have been on a modest scale, all the moraines lying within one or two kilometres of the present glacier fronts.

The only indication so far found of glacier expansion in the middle of the Holocene, between the Crowfoot and Little Ice Age advances, has been a rock glacier in the headwaters of Lost Horse Creek in Banff National Park, which cannot, unlike its neighbours, be assigned to either the Crowfoot or recent advances (Luckman and Osborn 1979). Even as lately as 1982, knowledge of Holocene glacial and climatic history in the southern Canadian Rockies has been described as 'quite deficient' (Osborn 1982). Osborn (1984) has noted that the Bugaboo glacier in the Purcell mountains had begun to advance by about 2400–2200 BP according to radiocarbon dates of wood near the base of moraine, with possible readvances culminating about 1100 BP and towards the end of the last century. It is possible that more detailed field examination, especially of exposures in lateral moraines, may produce yet more satisfactory evidence of the course of glacial history in the Mid-Holocene.

Moraines in cirques in the Shuswap Highlands in the North Thompson river basin, north of Kamloops, which were deglaciated by about 11,000 BP (Fulton 1975) have also been found to bear Mazama ash and therefore to be older than 6600 BP (Duford and Osborn 1978). These Dunn Peak moraines are known to be younger than Late Wisconsin. According to Duford and Osborn, one of them has a minimum age of 7390 ± 250 BP (GX–4039) on charcoal found in the moraine and they could thus correspond with the White Chuck Advance. Their interpretation, it might be noted, is disputed by Alley (1980) who argues that the moraines were formed after 7985 ± 125 (I–9162) and before 4320 ± 95 (I–9161). The later date, from charcoal in a regosol behind the moraine, is not accepted by Duford and Osborn (1980) as relevant. In one of the cirques, a moraine downstream of the type Dunn Peak moraine, also bracketed between the dates 11,000 and 6600 BP by Duford and Osborn (1978), may represent either an older Holocene event or a standstill during the Late Wisconsin retreat.

In the Coast Mountains of British Columbia, in the Garibaldi National Park, glacier expansion between 6000 and 5000 BP is indicated by tree stumps in place, overridden by the ice and dated 5260 ± 200 (Y–140 bis), 5300 ± 70 (GSC–2027) and 5590 ± 140 BP (GSC–760). No sign of this episode was found by Ryder and Thomson (1986) during their survey of the Waddington Range to the northwest of Garibaldi between 50° and 51°N. Here, an advance of the Tiedmann glacier in the centre of the Waddington Range between 2940 and 2250 BP had already been recognized by Fulton (1971). Following systematic mapping from air photographs, the most promising sites, including the Tiedmann forefield, were visited by helicopter. Fragments of older laterals lying outside the Little Ice Age moraines were identified in several sites and a general phase of glacier expansion between 3300 and 1900 BP is suggested by Ryder and Thomson although only the Tiedmann is believed to have reached the Holocene maximum at this time. The advance of the Tiedmann began before 3345 ± 115 (S–1470) and culminated around 2355 ± 60 (S–1471). The Gilbert glacier started to advance before 2220 ± 70 (S–1459), probably reaching a maximum after 2040 ± 40 (S–1572), although the evidence here is less clear-cut. The timing fits in with the expansion between 3200 and 2300 BP recognized by Alley (1976a and b) at Battle Mountain. Most of the Waddington glaciers seem to have reached their Holocene maxima during the Little Ice Age.

10.4.4 THE ALTITHERMAL

As compared with ice sheets in Europe, the Laurentide ice sheet was very persistent, continuing to occupy the Hudson Bay catchment until about 7500 BP (Figure 10.17). The succession of events identified in northwest Europe by Blytt and Sernander is not necessarily directly applicable to North America. After all, the succession is based on the macro-fossil remains of arboreal plants that immigrated to Scandinavia over various routes and arrived at various times depending on the refuge areas from which they had come and the distances they had travelled. Futhermore, northwest Europe lies to the leeward of an ocean that was receiving abundant meltwater until the eighth millennium BP, whereas much of North America, especially the west, received its weather from the Pacific Ocean which was not chilled to the same degree by Arctic meltwater.

In North America the most striking feature of the Holocene climatic succession has generally been taken to be a period lasting a few millennia when temperatures were somewhat higher than they are now. The Altithermal or Hypsithermal is often taken to have occurred around 6000 years ago. However, there is no consensus as to the timing of this Altithermal and indeed no certainty that there ever was such a single, clearly identifiable warm period in the Holocene.

In the far north of Alaska, Calkin and Ellis (1982) found pollen changes, poplar and spruce tree expansion, peat formation, thaw lake, pingo and permafrost histories as well as archaeological evidence in the Brooks Range all pointing to a period of thermal maximum starting as early as 10,000 BP and lasting until 4000 BP. However, Mid-Holocene warming, which saw the advance of alder north of the Brooks Range (Tedrow and Walton 1964), had been rated as 'a relatively minor event' compared with early Holocene warming between 10,000 and 8300 BP (Hopkins 1967). In northwest Canada, the pollen record from the Tuktoyaktuk Peninsula shows it was occupied by spruce from 10,000 to 6000 BP; by 4000 BP the landscape had reverted to tundra which persists into the present. In the central and eastern Canadian Arctic, in contrast, where the ice sheet persisted for so long, Andrews *et al.* (1981), from the application of transfer functions to sequences of peat and lake sediments deposited after the disappearance of the Laurentide ice sheet, estimated that July temperatures reached a maximum, 2 to 3 °C warmer than at present, between 4000 and 3500 years ago.

Nichols (1975) deliberately selected peat profiles from four sites along the Canadian boreal forest–tundra ecotone (transition) in order to detect climatic changes causing ecotonal displacements, adopting Bryson's hypothesis that the mean summer position of the Arctic Front controls the northern limit of forest. With rates of peat accumulation of as much as 1 cm per three years, sampling was possible at intervals of less than a decade in some profiles, allowing considerable precision in the detection of the changes. The oldest sediments, dating back to 6200 BP, showed a spruce forest in existence, representing a climate with summers substantially warmer than those of today. These altithermal conditions persisted until 4800 BP with a possible cool interval between 5600 and 5500 BP and a maximum of summer warmth between 5300–5200 and 4800 BP. The cooling brought tundra as far south as it is now for 300 years. There was a recovery and then a brief cooling about 4200 BP followed by a peak of summer warmth about 4000 to 3900 BP. Fires which were frequent after 4000 BP became widespread and synchronous over a period of a century or two around 3500 BP, possibly marking the summer expansion southwards of cold dry arctic air masses over the northernmost forests which were changed to tundra until about 3300 BP. Then warmer summers are indicated by the regeneration of woodland, but by 3000 BP a prolonged cooler period was causing renewed southward expansion of the tundra. This was accentuated about 2500–2100 BP when sand incorporated in the peat is believed to have been blown in by the wind as the tundra advanced still further south.

In Ontario, Warner *et al.* (1984) working on Manatoulin Island found the Altithermal to be 'a bipartite period from about 10,500 to approximately 8000 BP when climate was probably drier and warmer than at present, followed by a period between 8000 and about 5000 BP when the climate was humid and warmer than at present'. Osborn (1982), from a variety of evidence, deduced a more complicated sequence of events, concluding that in the southern Rocky Mountains of Canada 'there is some consensus that prior to 7500 BP one or more periods of relatively cool/wet climates occurred and that a shift towards cool/wet climates at about 6000 BP ended the warmest, driest part of the Holocene, with further cooling at about 3000 BP'.

Pollen and macro-fossil data from the Pacific Northwest of the United States suggest that the climate was warming by 12,500 BP (Barnosky 1984). Lake levels were dropping; by 10,500 BP ice had gone from the lower and middle altitudes, and the period from 10,000 to 7000 BP was the driest of Holocene times. Modern temperate steppe reached its greatest extent before 8500 BP and the most xerophytic part of the Holocene, with July temperatures higher and precipitation lower than now, was before 6000 BP. To explain the late arrival of closed arboreal communities in the western Olympic Peninsula of Washington State, Heusser (1978) was inclined to invoke greater summer warmth and dryness before 3000 BP.

We thus have indications from North America of generally warmer conditions than those of the present from as early as 10,000 BP until as late as 3000 BP. As more detailed studies are made in sensitive areas, cool interruptions of the warm conditions are revealed and there are signs that at times the climate was drier and at other times wetter. The problem is to distinguish the pattern in time and space of these variations and to attempt to understand their causes and effects and the implications for the future. The glacial record can help to provide a solution because of the short response times of small glaciers and the opportunities afforded to make use of dendrochronological and tephrochronological methods in conjunction with ^{14}C dating and moraine stratigraphy in the same areas. One can perhaps refer to a Neoglacial period following an Altithermal, so long as one recognizes that they are broad categories; many glaciers did not melt away completely in the Altithermal and the beginning of the Neoglacial is imprecise in time and may vary from one region to another.

10.4.5 THE ST ELIAS MOUNTAINS OF ALASKA AND YUKON TERRITORY

The conclusions reached by Denton and Karlén (1973b) on Holocene climatic variations were based in part on fieldwork carried out between 1962 and 1970 on the northeast flank of the Wrangel and St Elias mountains in Alaska and the Yukon. The area provides evidence in the form of moraine systems, and two volcanic ash deposits covering much of the area assist in the correlation of Holocene events. Changes in altitude of the spruce treeline provide additional information.

During the last Wisconsin glacial maximum the Macauley–Kluane glaciers flowing north and northwest from the St Elias mountains coalesced to form a piedmont glacier. This began to disintegrate about 13,660 BP and the glaciers had already retreated to their present positions by 12,500 to 11,00 BP (Denton 1970, Denton and Stuiver 1966). However, Holocene moraine systems fringe all the present glaciers and extend as much as ten kilometres in front of their tongues. Some of the more recent advances overran earlier ones; piecing together a complete sequence involved examining several moraine suites in order to discover the old moraines that happened to have been preserved. On the eastern side of the mountains there were only Little Ice Age moraines whereas on the western side Little Ice Age moraines nested behind older Holocene deposits. The difference between the two sides seems to be the result of the western glaciers having been thickly covered with debris, so that massive Holocene ice-cored moraines were formed which the Little Ice Age advances failed to overrun, whereas the eastern glaciers were cleaner and probably deposited smaller Early Holocene moraines which were subsequently destroyed by meltwater or obscured by Little Ice Age advances. Indeed, Denton (1974) noted that no evidence for events comparable with the Younger Dryas advances of the Scandinavian ice sheet or the Cochrane–Cockburn advances of the Laurentide had been found in the St Elias mountains and mentions the possibility that moraines of these ages could have been overrun by Little Ice Age advances and destroyed or concealed.

The presence of spruce wood dated 8020 ± 120 BP (Y–2302) in muskeg alluvium and drift indicates that the treeline was by this time no more than 183 m below the present treeline of 1250 m. By 5250 ± 130 (GSC–718) the spruce treeline was at 35 m above the present level. A spruce log near Giffin glacier, in alluvium 76 m above the present treeline, gave a date of 3580 ± 95 (I–6414), and together with similar data (Figure 10.18a) indicates a period considerably warmer than the present. Another date of 3050 ± 55 (SI–1100) was given for wood found 10 to 15 m above the spruce line of the present day.

Cooling seems to have set in by 3300 BP, for between this date and 2400 BP glaciers advanced to form massive moraines fringing the Seven Sisters glacier, the northern tongue of the Russell glacier and the Guerin glacier, drift fronting the Natazhat glacier and loess near the Kaskawash glacier. All of these glaciers expanded more at this time than they are known to have done at any other time in the Holocene. The moraine of the Seven Sisters glacier is well dated by silt from an organic silt layer immediately below the outermost moraine, giving an age of 2780 ± 90 BP (I–6490C), and a maximum age for the moraine of the Natazhat is given by organic sediments beneath it with a date of 2675 ± 85 BP (SI–1101). Loess deposition attributed to the advance of the Kaskawash buried grass which has been dated to 2640 ± 80 (Y–1435). The treeline again rose between 2100 and 1230 BP when a spruce 1775 ± 90 (I–6091) was killed by White River ash dated to 1230 BP, and then came another glacial advance from about AD 700 to 900. Only the Giffin glacier reached its maximum Holocene extent then. Warmer conditions are indicated by a higher treeline for nearly 500 years, and then moraines dated by organic debris mark the Little Ice Age advances that followed.

The best-dated studies of Holocene glacial variations in the Alaska–Yukon area are those in the White River valley and Skolai Pass area (Denton and Karlén 1977) (Figure 10.18a) and in the northeastern St Elias mountains where the same authors undertook earlier fieldwork (1973b) and also summarized (1973b, 1977) the work of others such as Sharp (1951), Borns and Goldthwaite (1966), Denton and Stuiver (1966) and Rampton (1970). The accordance of views of workers in this region is striking and it will be of interest to see whether the results of studies made by Röthlisberger in the region concur (Röthlisberger and Geyh in press).

10.4.6 BROOKS RANGE

The glaciers in the Brooks Range (see Chapter 8), although they are cold-based, sub-polar and slow-flowing, have left a complex morainic record (Ellis and Calkin 1979, 1984, Calkin and Ellis 1980, 1982, Calkin *et al.* 1985), marking their sensitivity to changes in the balance between snow accumulation and ablation. Solar radiation accounts for some 60 per cent of current ablation.

Deposits fronting more than ninety glaciers have been examined and mapped by Calkin, Ellis and their associates. They conclude that at least seven periods of expansion have occurred in the last 5000 years. The chronology of advance and retreat is hence much more complicated than Porter and Denton (1967) indicated. The moraines are 1000 to 2000 m above the uppermost level of spruce forest and well to the north of its poleward limit in zones of tundra and continuous permafrost. Dating is therefore heavily dependent on lichenometry. Calkin and Ellis (1980) presented a lichen growth curve which is based on *Rhizocarpon geographicum*, s.l. and to a lesser extent upon *Rhizocarpon eupetraeoides/inarense* and *Alectoria minuscula/pubescens*. The *Rhizocarpon geographicum* curve is controlled by historical and dendrochronological data, and also by reference to direct measurements of Alectoria thalli which grow seven times faster. In all, thalli dimensions of six different species were employed, their differing growth rates being taken into account (Ellis and Calkin 1984). The linear phase of *Rhizocarpon geographicum* is calibrated to 1300 BP by radiocarbon dates from Atigun Pass but its further extension is dependent upon measurements of *Rhizocarpon* obtained in other areas with similar climates, controlled by radiocarbon dates between 1500 and 9000 BP (e.g. Andrews and Barnett 1979). In these circumstances, Calkin and Ellis assign a subjective ± 20 per cent age reliability to their growth curve.

The central Brooks Range was more or less clear of ice by about 11,800 to 10,500 BP, the final Late Pleistocene maxima having culminated just beyond the mountain front about 13,000 to 12,500 BP (Hamilton and Porter 1975). Indications of early Holocene advances are sparse. Numerous moraines dated lichenometrically to between 5000 and 300 BP typically cluster in narrow belts less than 50 m wide encircling the deposits left by recent frontal retreats. Evidently there have been several Holocene advances, all to a similar distance. Calkin and Ellis (1982) found lichenometric evidence of moraine stabilization at about 4400 ± 900, 3500 ± 700, 2900 ± 600, 1800 ± 400 (1150 ± 230, 800 ± 160, added by Calkin *et al.* 1985b), 390 ± 90 BP. Evidence of recent oscillations is more voluminous but earlier advances are represented by only a few moraine fragments (Calkin and Haworth in press). Only four instances have been found of Holocene moraines approximately assigned to the period before 6000 BP and it is conceded that 'we may be a thousand years off on these old 7000 to 8000 lichen ages'. But Calkin is convinced that the moraines concerned, situated well up in their cirques, are 'older than the ones with 4500 year lichen but clearly not as old as Hamilton's youngest Pleistocene' (private communication, 1986). He takes the view that extension of investigations would be likely to reveal further evidence of early Holocene fluctuations, although deposits from the earlier millennia are now buried beneath Neoglacial or Little Ice Age moraines.

Calkin and Ellis have demonstrated convincingly that glaciers in the Brooks Range expanded by about 50–60 per cent of their present sizes on several occasions but, it has to be noted, that the various clusters attributed to the last 3000 years are not clearly separated and that error bars overlap. The dating of moraine stabilization at about 1150 BP is in conformity with evidence from the Golden Eagle glacier near Atigun Pass. Here, glacial retreat has revealed dead moss with a ^{14}C age indicating that it was killed by advancing ice about 1120 ± 180 BP. However, the maximum diameter of lichens preserved under the ice by the same advance suggests a preceding period of lichen growth, during which the glacier was less extensive than at present, which lasted for some 1500 years. The crudity of the dating has to be borne in mind when the chronology of Holocene advances in the Brooks Range is being compared with that in other regions. An advance postulated about 1120 ± 180 BP (BGS–614, Calkin and Ellis 1981, Ellis and Calkin 1984), though based on only one ^{14}C date, is supported by lichenometric evidence from fifteen glaciers in the central Brooks Range and four further east. Nevertheless, additional dating techniques rigorously applied, might well modify and enhance the reliability of the chronology that has been proposed.

10.4.7 BAFFIN ISLAND

While small cirques and valley glaciers on Baffin provide the most information relevant to changes over the last four thousand years, data on more remote events comes from the margins of ice sheets. The most detailed records come from the Cumberland Peninsula and the Pangnirtung area.

The Cumberland Peninsula, the easternmost part of Baffin, is surmounted by the Penny icecap, the glaciers of which reach down to the sea in several fjords. Smaller icecaps descend to about 200 m asl and cirque glaciers are restricted to elevations above 600 m on the east coast and above 1200 m nearer the Penny icecap.

The maximum of the Last Glaciation occurred much later than on the Canadian mainland, where the last major advances of the Wisconsin took place between 28,000 and 14,000 BP. On the Cumberland Peninsula the Laurentide ice and local glaciers reached their maxima about 8000 BP (G. H. Miller 1973). The best evidence comes from Okoa Bay where shells on a raised glacio-marine delta give ages of 8290 ± 170 BP (Gak–3092) and 8769 ± 35 (St–3816) (Figure 10.18c). The shells were all of types living in cold conditions. Andrews and Miller (1972) estimate that the Holocene equilibrium line altitude was 350 m below present on the coast of the Cumberland Peninsula and 450 m lower near the Penny icecap, 80 km inland. The explanation for the anomalous behaviour of the ice on Baffin would seem to be that the Laurentide ice blocked storm tracks until the Early Holocene, when final disintegration allowed cyclonic depressions to bring increased amounts of snow to the island (Mercer 1956, Andrews and Ives 1972).

The extension of the ice on the Cumberland Peninsula to its furthest limits about 8000 years ago corresponds in time to the Cockburn advance (Falconer *et al.* 1965) of an ice sheet centred on Foxe Basin and Hudson Bay and occupying a very large part of northeast Canada. The moraine systems of the advance can be traced along the northeast coastal zone of Baffin Island, round to the west of Melville Island, and across northern Keewatin for a distance of over 3000 km. This ice sheet expanded between 9000 and 8000 BP and then shrank again, the ice wasting away extraordinarily quickly in the southwest to disappear from Keewatin by 7000 BP (Craig and Fyles 1960; see Figure 10.17). It was suggested by Andrews *et al.* (1974) that east Greenland glacial history has followed a similar course to that of Baffin with the key factor in both areas being fluctuations in the amount of snow.

The delayed wasting of the Laurentide ice sheet on Baffin Island has resulted in the distinction between moraines of Late Wisconsin and Holocene age being even less apparent than in western Norway. After about 8000 BP the local ice bodies also melted away. By 7100 BP they were no larger than at present and by 5500 BP there was probably less ice on Baffin than there is now. Species of molluscs, *Chlamys islandicus* and *Mytilus edulis*, now found only in the south of the region, reached their most northerly positions before 7590 ± 170 (Gak–2566) and remained there until at least 4810 ± 110 (Gak–3724). Maximum temperatures according to Carrara and Andrews (1972) and G. H. Miller (1973) were probably reached between 7000 and 5000 BP. No signs of any major readvance or standstill of the retreating ice have been found although the recession of the

Greenland ice sheet in the Søndre Strømfjord area, on the other side of Davis Strait, was interrupted by seven halts or slight readvances between 10,000 and 6500 BP (Ten Brink and Weidick 1974)

About 5000 BP on Cumberland Peninsula 'the Hypsithermal amelioration reversed and a series of climatic oscillations between relative glacial and non-glacial modes began. This period of new glacial growth and climatic oscillation is recorded in suites of Neoglacial moraines' (G. H. Miller 1973, p. 569) except in the Merchants Bay area where only one set of moraines, dating from the last few hundred years, is to be seen (Locke 1980, Andrews 1982a, b). The dating of late Holocene events presents substantial difficulties and much reliance has to be placed on lichenometry. A linear growth curve first established by Miller and Andrews (1972) was modified slightly to fit a new radiocarbon calibration in 1973. It must still be regarded as a first approximation; Miller regards its reliability as ± 20 per cent, a point of considerable importance when one is attempting to make comparisons with events elsewhere. Although Miller considers the useful time range of *Rhizocarpon geographicum* to include the whole of the Neoglacial, the effective limit might be nearer 3000 years. Furthermore, any ice-cored moraine is unstable and its age, as indicated by the size of the largest lichen thalli, may in fact refer to a time long after the moraine's original formation, when it was stabilized, possibly at the onset of a warm period.

Miller's studies in the Okoa Bay area of Cumberland Peninsula were sufficiently numerous to reduce some of the possible inaccuracies. They involved the examination of valleys below forty-six glaciers and measurement of lichen on over 100 moraine crests. Five periods of glacial advance were recognized (Figure 10.19). Only three glaciers had outer moraine crests with thalli diameters taken to represent calendar ages of 3200 ± 600 BP. An advance around 1600 BP was marked by moraines around six glaciers. Other advances culminated 750 years ago and during the Little Ice Age (see Chapter 8.5). Evidence from radiocarbon-dated organic material gives some support to the lichenometric findings (Figure 10.19). A basal peat section indicating the onset of warmer conditions provided a date of 1670 ± 90 BP (Gak–2575) which fits in well with lichens starting to grow on the stabilized moraines about 1600 BP. Relatively warm summers are indicated by the formation of peat for which six dates have been obtained between 1670 and 1000 BP. Then a date of 850 ± 110 BP from the top of a peat layer buried by sand indicates the onset of cooler conditions about the time of the medieval advances in the Alps.

Investigations by Davis (1985) in the Pangnirtung area of the Cumberland Peninsula revealed a very similar chronology with three ice advances between about 3000 and 2000 BP and another three within the last 1200 years (Table 10.4).

Andrews and Barnett (1979) examined the moraines of the western and southern margins of the Barnes icecap in the northern half of Baffin Island and found a striking similarity in the size distribution of thalli with those on moraines in the Cumberland Peninsula. They suggested that moraines were formed along the western margin of the Barnes icecap between 3100 and 2800, around 2100, 1500, 1000 and 750 BP. Andrews and Barnett concluded that their data indicated that, within the resolution of the dating methods used, the expansion and contraction of the little independent glaciers on the Cumberland Peninsula were synchronous with events along the 150 km length of the western margin of the Barnes icecap where several of the glacial lobes surged at about the same times as independent glaciers in the region expanded, all of them presumably

Table 10.4 Holocene glacier advances in Cumberland Peninsula, Baffin Island

Maximum R. Geographicum diameter (mm)	*Name of advance*	*Age in ^{14}C years BP*
< 12	Cumberland 1	< 100
< 17–26	Cumberland 2	200–400
28–32	Cumberland 3	500–650
39–46	Pangnirtung	900–1150
68–69	Kingnait 1	1900–2000
75–80	Kingnait 2	2200–2400
95–100	Snow Creek	2900–3100
105		3250

Source: Davis (1985)

being affected by the same climatic fluctuations. Heusser (1983), from his studies in the Prince William Sound, also concluded that surging had accompanied more widespread expansion of glaciers in Alaska (see Chapter 5.8).

Complex moraine assemblages elsewhere in the eastern Canadian Arctic appear to be the outcome of Holocene oscillations. A preliminary survey suggests that a very similar situation to that in the Cumberland Peninsula exists in Labrador (McCoy 1983). Several moraine stages have been recognized in the Borup Fjord area of northern Ellesmere Island and King (1983) suggested on the basis of a number of radiocarbon dates that these may have occurred about the same time as the Little Ice Age, Göschener I, Frosnitz and Venediger advances in the eastern Alps.

10.5 Equatorial glaciers

The most detailed Holocene record of an equatorial glacier is for the Carstenz glacier on Mt Jaya in Irian Jaya, Indonesia (Hope and Peterson 1975). Evidence of multiple glaciation is emerging here with advances dated to before 2900 BP, about 2500 BP and two or three later expansion phases including the Little Ice Age. No evidence has been found of the presence of ice on any of the Papua New Guinea mountains between about 7000 and 5000 BP; the treeline on Mt Wilhelm seems to have been as much as 200 m above the present level from 8300 to 5000 BP, suggesting that temperatures were then 1 to 2 °C above the present level. Following the last ice advances of the Late Pleistocene, deglaciation was probably under way by 15,000 to 14,000 BP in New Guinea as it was also on Ruwenzori in central Africa (Livingstone 1967). In New Guinea, deglaciation was interrupted at various times and the treeline did not rise until after 10,000 BP, suggesting that the climate was colder (or drier) than at present during most of the retreat period. The summit of Mt Wilhelm (4510 m, 5°47′S, 145°10′E) was not clear of ice until about 9000 BP (Peterson *et al.* 1973).

On Mt Kenya, the ice seems to have retreated from Late Pleistocene moraines by 12,500 BP. Later moraines indicate a complicated subsequent history of advance and retreat (Baker 1967). Mahaney (1985), using relative dating methods, found no evidence of early Holocene advances. However, Johansson and Holmgren (1985) obtained bracketing dates for a non-organic sediment taken to be of glacial origin, in Naru Moru

Tarn at the head of the Teleki Valley, which gave ages of 6070 ± 225 BP and 4135 ± 70 BP; these dates provide a minimum age for the moraine damming the tarn. A cooler interval on Mt Kenya had earlier been discerned by Coetzee (1967) at 6000 to 5000 BP, on a basis of pollen evidence, and a minimum age of 6277 to 5425 BP was obtained for a moraine at an altitude of between 4100 and 4200 m in the Hobley valley by Perrot (1982). Karlén (1985), using methods similar to those he had employed in Sweden (see above, pp. 312–14), investigated the proglacial lacustrine sediments of Oblong Tarn below the Cesar glacier. Arguing that non-organic silt layers represent periods of enhanced glacial activity, he concluded from the dates of sediment accumulation, controlled by four radiocarbon dates, that such activity was most intense on Mt Kenya from 3200 to 2900, 2350 to 1700 and 600 to 50 BP.

In Mexico, Heine considers that glaciers were very limited in extent during the Late Pleistocene continental glaciation between 30,000 and 13,000 BP, because of the prevailing aridity. His studies (1975, 1976, 1983), involving tephrochronological methods, indicate multiple glaciation in Mexico at the end of the Pleistocene and in the Early Holocene. Glaciers on Popocatepetl, Pico de Orizaba and Malinche left behind moraines about 12,000 BP, one on Malinche being dated to 12,060 ± 165 BP (Hv-4244). The same three volcanoes, together with Nevado de Toluca, also display threefold moraines dated to between 10,000 and 9000 BP. Higher moraines dated to between 3000 and 2000 BP have been identified on Pico de Orizaba and Popocatepetl and in the first half of this millennium rock glaciers formed on Nevado de Toluca and Malinche.

In Peru, the Quelccaya icecap readvanced about 4 km between 11,460 and 11,00 BP (Mercer and Palacios 1977). This compares with an advance after 11,400 BP in New Guinea (Hope *et al.* 1976) but precedes the European Younger Dryas by about 500 years. By 10,000 BP, a single date suggests, it had retreated again and between 2700 and 1600 BP was smaller than it is today. The most recent expansion of the icecap took place between 600 and 300 years ago.

10.6 South America

The most heavily glacierized region in the southern hemisphere is in Patagonia, an area larger than Scandinavia and hardly known to the outside world until the twentieth century (see 9.4.1). In this remote area and difficult terrain between 39° and 53°S, Mercer has dated a series of ice advances since the Last Glaciation. Many of his key sites are near the margin of the Hielo Patagonico Sur, which is the more southerly of the two Patagonian icefields. Other sites are distributed over the area between the Chilean Lakes in the north and Tierra de Fuego in the south (Mercer 1965, 1968, 1970, 1976).

Mercer suggested that the last major glacial episode of the Pleistocene, the Llanquihue, was over before 12,000 BP. Radiocarbon dates for the main deglaciation of Seno Otway, near the Straits of Magellan, 12,460 ± 190 BP (I–3512), and for the disappearance of ice from Lago Ranco in the Chilean Lakes, 12,200 ± 400 (GX–2935), are similar although these areas are 1400 km apart. Dates of 13,200 ± 320 (GX–4169) and 13760 ± 295 (GX–4170) for a readvance of ice on the western shore of Lago Llanquihue which culminated round about 13,000 BP, together with a date of 11,245 ± 245 for the deglaciation of Rio Baker, show that the transition from full glacial to full interglacial conditions in the region occurred within 2000 years, between 13,000 and 11,000 BP

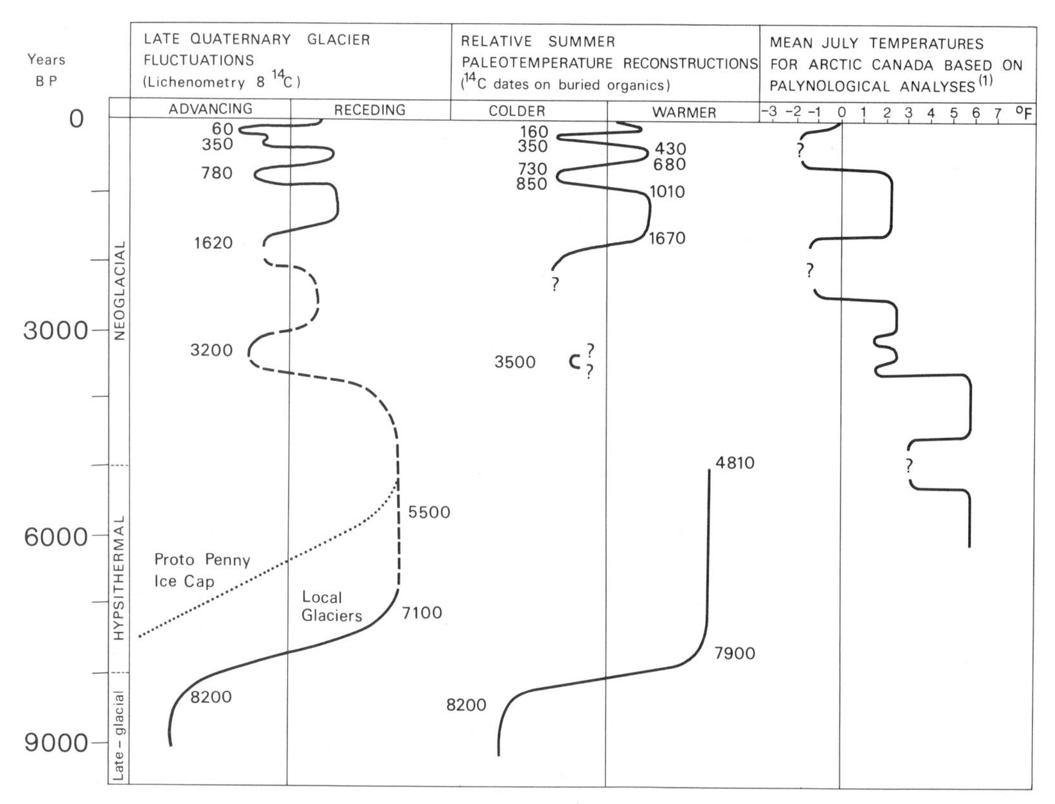

Figure 10.19 Miller's (1973) model of glacier fluctuations on Baffin Island; palaeotemperature indications derive from buried organics; Nichols's (1967) mean July temperature curve based on palynology. (From Miller 1973)

(Mercer and Ager 1983). By 11,000 BP Glaciar Tempano, a western outlet of Hielo Patagonico Sur, had already receded within its present limits. The compressed peat resting on till, revealed by the retreat of the terminus between 1945 and 1968, gave a date of 11,070 ± 760 (I–3825, Mercer 1976). In view of this evidence, Mercer is not inclined to accept a minimum date of 10,000 ± 140 (I–2209) for the shrinkage of Glaciar Moreno on the east side of Patagonico Sur as contradicting the early onset of Holocene conditions in Patagonia. He concluded that 'after 13,000 BP, glaciers in Patagonia shrank rapidly and without interruption until about 11,000 BP, at a time when ice still covered most of Canada and Scandinavia, they were within their present borders' (Mercer 1976, p. 160).

Mercer's conclusion conflicts with that of Heusser (1974), based on pollen analysis, that temperatures in Chile were still 4 °C below those of the present day between 11,000 and 10,000 years ago. The discrepancy is not to be explained in terms of a lag between deglaciation and biotic migration, for the botanical evidence indicates rising temperatures after 12,000 BP followed by a reversal to more severe conditions from 11,000 to 10,000 BP, concurrent with the Younger Dryas stage in Europe (Mangerud 1970). Heusser reckoned that the Holocene began in Patagonia about 10,000 BP. His earlier work was substantiated by a temperature and precipitation record for Alerce in southern Chile for the past 16,000 years, obtained using regression equations to relate pollen taxa from surface samples to temperature and precipitation and applying the results to a radiocarbon-dated pollen sequence in a lake core (Heusser and Streeter 1980). According to this, mean January (summer) temperature was within 2 °C of the present value of 15.2 °C for much of the past 16,000 years, but at the beginning of the period it was definitely lower. This early cool interval was followed by a warming which reached a peak about 11,300 BP. This was followed by another cool period which was becoming pronounced about 10,000 BP and followed by a warm episode peaking between 9410 and 8600 BP. After the second warm peak, cooling set in and successive minima were reached between 4950 and 3160 BP, between 3160 and 800 BP and during recent centuries.

Mercer defined the Altithermal as 'the local post-glacial interval that began when the climate first became warmer than it is today and ended when the climate became colder'. Assuming that Patagonian glacial fluctuations mainly reflect changes in temperature, he concluded that the Altithermal, or Hypsithermal as he called it, began shortly before 11,000 BP and ended at the start of the first Neoglacial advance, perhaps about 5500 BP. Again, his view is in conflict with Heusser's pollen data.

Radiocarbon dates upon which Mercer placed his Patagonian Holocene chronology are plotted in Figure 10.20 (Mercer 1965, 1968, 1970). Many moraines remain to be dated and when more is known about the ages of the Punta Bandara moraines, near the shores of Lago Argentino, east of Glaciar Moreno, it should become clearer whether or not there was a Patagonian equivalent of the Younger Dryas. Certain moraines, which Mercer reckons at present to be 'Late Glacial' have a minimum age of 6740 ± 130 (I–2200) and could turn out to be Holocene. Multiple moraines such as the Pearson 1 Group, near the Upsala glacier, could conceivably have been formed by more than one ice advance. The area involved is enormous, the research done so far has been remarkably revealing, but a good deal remains to be done before we can be sure that all the main Holocene glacier fluctuations have been identified.

As it stands, Mercer's model for Patagonia has a precocious and lengthy Altithermal followed by three Late Holocene advances. These culminated between 4600 and 4200, between 2700 and 2000 BP, and during the last three centuries, the first being the greatest and the last the smallest. If fluctuations on a smaller scale than those of 4600 to 4200 occurred in the middle Holocene, evidence for them has yet to be found; studies of lateral moraine sequences may shed more light on the subject.

To the north of Patagonia in the sub-tropical Andes of Argentina, the glacial chronology of the Holocene is very different (Stingl and Garleff 1985). There are no signs of important Late Glacial and Early Holocene advances but large advances between 6000 and 4500 BP were followed by rapid shrinkage until about 3000 BP. In this semi-arid region of the Rio Atuel, Stingl and Garleff suspect that the Mid-Holocene advances were in response to increased precipitation rather than cooling. It must also be borne in mind that the southern Andes are tectonically active and that some of the glaciers are liable to surge.

10.7 New Zealand

The glaciers which filled valleys in the Southern Alps in the final Otira phase of the Pleistocene reached down to below present sea level in Westland and to the mountain front overlooking the Canterbury Plains on the east of the Main Divide (Burrows and

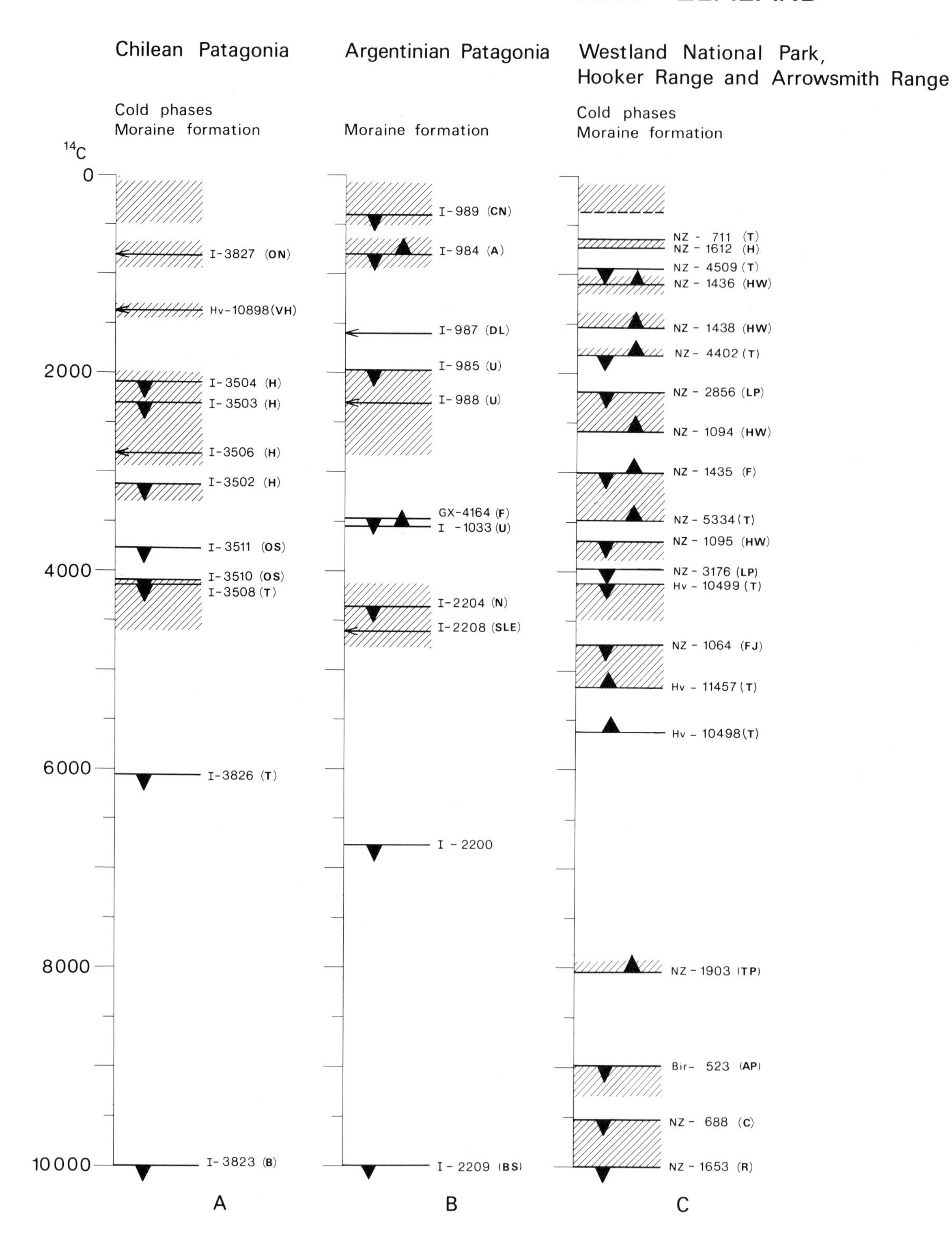

SOUTH AMERICA
NEW ZEALAND
Chilean Patagonia
Argentinian Patagonia
Westland National Park, Hooker Range and Arrowsmith Range
Cold phases
Moraine formation
Moraine formation
Cold phases
Moraine formation
14C
0
2000
4000
6000
8000
10000
I-3827 (ON)
Hv-10898 (VH)
I-3504 (H)
I-3503 (H)
I-3506 (H)
I-3502 (H)
I-3511 (OS)
I-3510 (OS)
I-3508 (T)
I-3826 (T)
I-3823 (B)
I-989 (CN)
I-984 (A)
I-987 (DL)
I-985 (U)
I-988 (U)
GX-4164 (F)
I -1033 (U)
I-2204 (N)
I-2208 (SLE)
I - 2200
I - 2209 (BS)
NZ - 711 (T)
NZ - 1612 (H)
NZ - 4509 (T)
NZ - 1436 (HW)
NZ - 1438 (HW)
NZ - 4402 (T)
NZ - 2856 (LP)
NZ - 1094 (HW)
NZ - 1435 (F)
NZ - 5334 (T)
NZ - 1095 (HW)
NZ - 3176 (LP)
Hv - 10499 (T)
NZ - 1064 (FJ)
Hv - 11457 (T)
Hv - 10498 (T)
NZ - 1903 (TP)
Bir- 523 (AP)
NZ - 688 (C)
NZ - 1653 (R)
A
B
C
14C dates
Minimum dates
Maximum dates
Lichenometric date
Date of advance
Glacial advance periods

Gellatly 1982). Subsequent retreat was broken by at least two readvances (Suggate and Moar 1970, Burrows 1974). According to Burrows (1979), the last stadial of the Otira glaciation began soon after 14,800 ± 230 BP and culminated before 13,400 ± 150 BP. The Rakaia glacier had lost 40 km, or 60 per cent of its length, before 11,900 ± 200 BP (Burrows 1977). In the period 12,000 to 9000 BP, glaciers readvanced, some of them extending a third as far down the valleys as they had reached at the Otiran maximum (Porter 1975, Burrows 1977).

The Waiho Loop, 9 km downvalley from the present front of the Franz Josef glacier and 14 km upvalley from the last Otiran moraine, was probably formed by a readvance about 11,450 ± 200 BP (NZ–4234) (Wardle 1978), although it could be somewhat older than this. East of the Divide, the Birch Hill moraines have been correlated with the Waiho Loop, but only on grounds of their setting being similar. The existence of well-marked moraine systems interrupting the valleys east of the Divide, such as the Rakaia, clearly demonstrate the complexity of Holocene history but the number of moraine sets differs from one valley to another.

Local name systems were established in a number of valleys where field investigations were concentrated by Burrows, his co-workers and research students. Attempts were made to compare the sequences and correlate them (Burrows and Russell 1975, Burrows 1975, 1977, 1980). However, correlation was handicapped by the sparsity of radiocarbon dates and also because some of the available dates could not be unambiguously related to specific moraine surfaces, and on reflection Burrows judged it would have been better to have waited until more information had come to hand (Burrows and Gellatly 1982).

Subsequently, interpretation of the moraine records was facilitated by additional radiocarbon dates becoming available (Wardle 1978, Burrows and Gellatly 1982) and by the employment of rock weathering-rind thicknesses measured on the sandstones east of the Divide (Chinn 1981, Birkeland 1981, Gellatly 1984). By 1982 Burrows and Gellatly were able to list thirty-two radiocarbon dates closely associated with glacial deposits for the period between 12,000 and 500 BP and to compare Burrows's results on the east side of the Divide with Wardle's on the west. Gellatly (1982) presented a radiocarbon-controlled chronology for the Mt Cook region which included her own results together with those of Burrows (1980, 1983), Burrows and Greenland (1979) and Burrows and Gellatly (1982).

Gellatly, Röthlisberger and Geyh (1985) extended the detailed examination of the forefield of glaciers in the Mt Cook region, paying particular attention to superposed and accreted till units in lateral moraines according to the methods employed by Röthlisberger and his colleagues in the Swiss Alps (10.2.2). Deep gullies created by

Figure 10.20 Holocene glacier fluctuations in the southern hemisphere: (a) key ^{14}C dates from moraines in Chilean Patagonia and advance phases identified by Mercer (1968) and Röthlisberger (1986). (b) key ^{14}C dates for moraines in Argentinian Patagonia and glacial advance phases recognized by Mercer (1976). (c) key ^{14}C dates for moraine formation in New Zealand and glacial advance phases recognized by Wardle (1973), Burrows and Gellatly (1982) and Gellatly *et al.* (1985). Column (a): (B) = Bernado, (H) = Hammich, (ON) = Ofhido Norte, (OS) = Ofhido Sur, (T) = Tempano, (VH) = Ventisquero Huemal. Column (b): (CN) = Cerro Norte, (A) = Adela, (DL) = Dos Logos, (U) = Upsala, (F) = Frias, (N) = Narraez, (SLE) = San Lorenzo Este, (BS) = Punta Bandara. Column (c): (T) = Tasman, (HW) = Horace Walker, (FJ) = Franz Josef, (LP) = La Pérouse, (AP) = Arthur's Pass, (R) = Rakaia, (C) = Cameron Valley, (TP) = Taruahuna Pass, (F) = Fettes.

erosion and slumping have cut back into lateral moraines, permitting access for field study and sampling to determine their genesis and construction (Figure 10.21). Twenty-four additional radiocarbon dates from the Tasman and Horace Walker sequences have enabled a chronology to be established as complete as those that have been obtained for the European Alps. Of particular importance has been the recognition and exploitation of the fact that micro-residuals consisting of lichen material in buried soils can be used to provide radiocarbon dates which are believed to correspond closely to the time at which the underlying surface first came into existence (Geyh *et al.* 1985), and which are consequently much more precise chronological indicators of the time of moraine stabilization after ice retreat than are radiocarbon dates obtained from the rest of the organic material in buried soil horizons in these environments. In the Mt Cook region, as a result, it has been possible to distinguish significant time differences in rates of soil formation and plant colonization depending on the nature, especially the coarseness, of the soil's parent material. Gellatly (1985) has found that on particularly bouldery substrates the establishment of a plant cover can be delayed for as much as five centuries.

Gellatly *et al.* (1985) present detailed chronologies from the lateral moraines in the Tasman valley to the east of the Main Divide and from the Horace Walker moraines to the west, and show that the periodicity of glacial response has been similar on both sides. Moraines of the Tasman glacier were constructed about 5000, 4500–4200, 3700, 3500–3000, 2700–2200, 1800–1700, 1500, 1100, 900, 700–600 BP and during the Little Ice Age of the last few centuries. The Horace Walker chronology is less complete. Moraine stabilization occurred after about 4600, 3700 and 2800 BP and glacial advances are dated about 3000, 2500–2000, 1800 and 1400 years BP.

Terminal moraines within Mount Cook National Park, east of the Divide, had previously been dated, using weathering rinds, to about 7000, 4200, 3800, 3400–3000, 2600–2100, 1800, 1500, 1150, 840 BP and recent centuries. These dates and the spacing between them correspond surprisingly well to the radiocarbon dates derived from the macro-organic constituents of soils in the near-vertical moraine sections exposed further

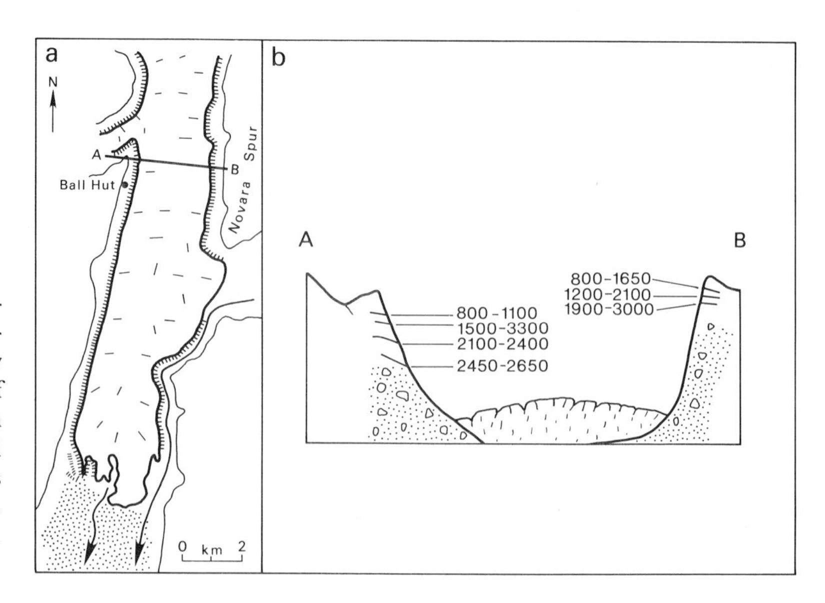

Figure 10.21 Radiocarbon-dated lateral moraine section, Tasman glacier, New Zealand: (a) position of sample sites in the Tasman valley; (b) superimposed lateral moraine sequences with approximate dates BP. (Based on Gellatly *et al.* 1985)

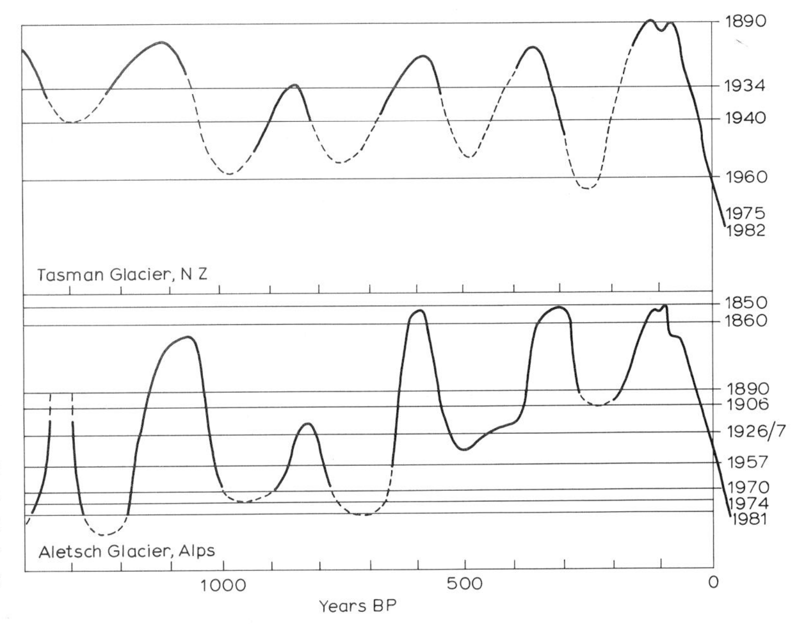

Figure 10.22 The fluctuations of the Tasman glacier, New Zealand, over the last 1500 years compared with those of the Grosser Aletsch dated by Holzhauser (1982). (From Gellatly *et al.* 1985)

upvalley (Gellatly 1984), and the whole sequence is very similar to that obtained from Westland National Park. Gellatly *et al.* (1985) concluded that Holocene glacial advances in New Zealand occurred between 11,500 and 10,500, around 8000, 5000, 4500–4200, 3700, 3500 to 3000, 2700 to 2200, 1800–1700, 1500, 1100 and three times in the present millennium.

Whereas in the European Alps, Little Ice Age advances were comparable in scale to ice advances earlier in the Holocene, in New Zealand glacial advances of the last few centuries have been smaller than their Holocene predecessors. In New Zealand the intervals between advances have diminished in duration towards the present. No equivalent to the Schlaten (about 9500 BP) has yet been closely dated in New Zealand, though there were advances between 14,000 and 9000 BP one or more of which might turn out to be comparable, nor is there an equivalent of the Frosnitz. The New Zealand glacial advances in the latter half of the Holocene might be seen as coinciding with those of the European Alps; on the other hand, the Alpine glacial advances in the Holocene as a whole could just as well be seen as corresponding to the intervals between the advances of the New Zealand glaciers. In spite of the new abundance of radiocarbon dates from both areas there is still this degree of uncertainty which may or may not disappear when more dates on wood from the Westland moraines are forthcoming.

10.8 Summary

The conclusions that emerge from the regional studies at the stage they had reached in 1985 are assembled in Figure 10.23. The methods used to date the events presented there vary in precision according to the techniques used, the soundness of their calibration and the closeness of bracketing values. Some phases of advance shown as having lasted several

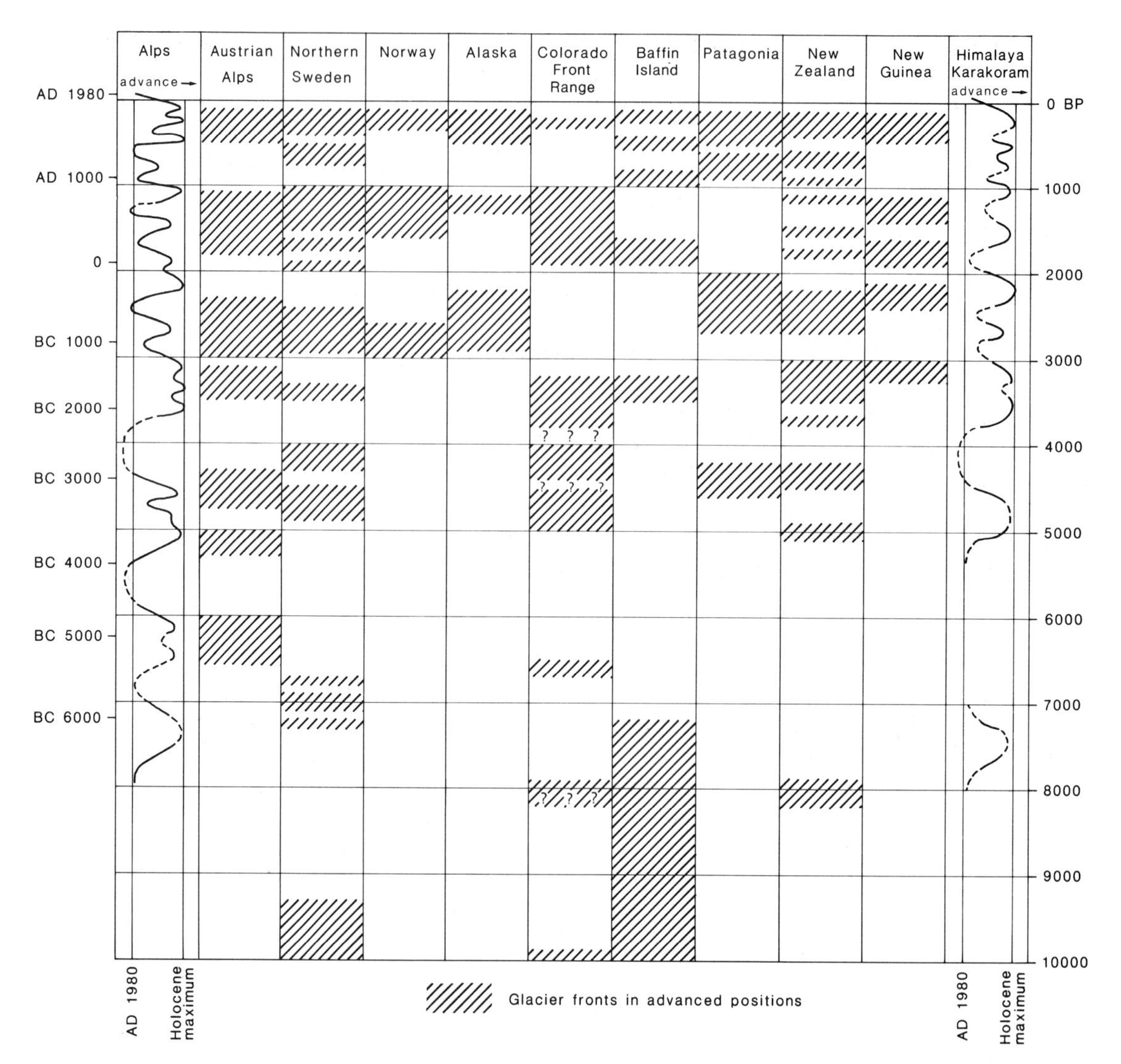

Figure 10.23 Regional summaries of glacier expansion phases. Some advance phases shown as having been lengthy may in fact have been imprecisely dated and consist in fact of one or more shorter phases. The diagram suggests a lack of synchroneity, except in the Little Ice Age and possibly around 2500 BP and also 3200 BP, but this impression cannot be taken as conclusive because of the varying quality and quantity of the data. (Based on: Alps – Röthlisberger and Geyh 1985; Austrian Alps – Patzelt 1977; northern Sweden – Karlén 1976; Norway – Griffey 1976; Alaska – Denton and Karlén 1977; Colorado Front Range – Benedict 1973 and Miller 1973; Baffin Island – Miller 1973; Patagonia – Mercer 1976; New Zealand – Burrows and Gellatly 1982 and Gellatly *et al.* 1985; New Guinea – Hope *et al.* 1976; Himalaya/Karakoram – Röthlisberger and Geyh 1985)

centuries may be made up of several shorter advances or have been imprecisely dated. Advances in some regions may not yet have been recognized, especially where subsequent advances have been of similar or greater extent and lateral sections of moraines have not been examined.

The regions where the chronology seems to be most reliable exhibit some features in common. Phases of glacier expansion in the Holocene are shown as having lasted about as long as the intervals between them. The alternations are more closely spaced in the latter Holocene than in the Early Holocene, though this may be a function of the greater sparsity of data with increasing time. The Little Ice Age and the subsequent retreat of the glaciers is common to all the regions; in those best studied, advances appear to have taken place at intervals of roughly a millennium or less, with near synchronous advances about 2500, 3200 and 4300 BP.

Further light was thrown on the nature of Holocene glacial and climatic fluctuations with the publication of the full results of Friedrich Röthlisberger's examination of moraine sequences in both the northern and southern hemispheres (Röthlisberger 1986). In this extended study only glaciers of similar size were selected for investigation. Comparability of results was further enhanced because Röthlisberger himself collected all the samples using the same geomorphological criteria, and all dating was carried out in the same laboratory. Some of Röthlisberger's results from the Alps, Himalaya and New Zealand have already been mentioned in this chapter but the complete work also embodies data from North and South America.

Röthlisberger concluded that phases of extension and shrinkage have occurred synchronously in both hemispheres (see Table 10.5) and presented a diagrammatic summary of his results together with a curve for Scandinavia drawn up by Karlén and based in part on his unpublished data (Figure 10.24). Röthlisberger pointed to the generally good agreement beween the regional curves for the period between 11,000 and 3000 years BP and proposed that the lags in timing and even contradictions for the period

Table 10.5 Phases of glacial extensions during the Holocene (years BP)

					South America		
	Alaska/Yukon	*Alps*	*Scandinavia*	*Himalayas*	*Tropics*	*Southern*	*New Zealand*
350–100	√	√	√	√	√	√	√
900–500	√	√					
(2 phases)							
700–1050							
(2 or 3 phases)							
1700–1500	√	√	√	√	√	√	√
c. 2200	√	√	√	√		√	√
c. 2700		√		√		√	√
c. 3200	√	√	√	√	√	√	√
c. 3700	√	√		√			√
c. 4600	√	√		√	√	√	√
c. 5200	√	√	√			√	√
c. 6300	√	√	√		√		
c. 7500	√	√	√	√	√		√
c. 8400		√				√	√

Source: Röthlisberger (1986)

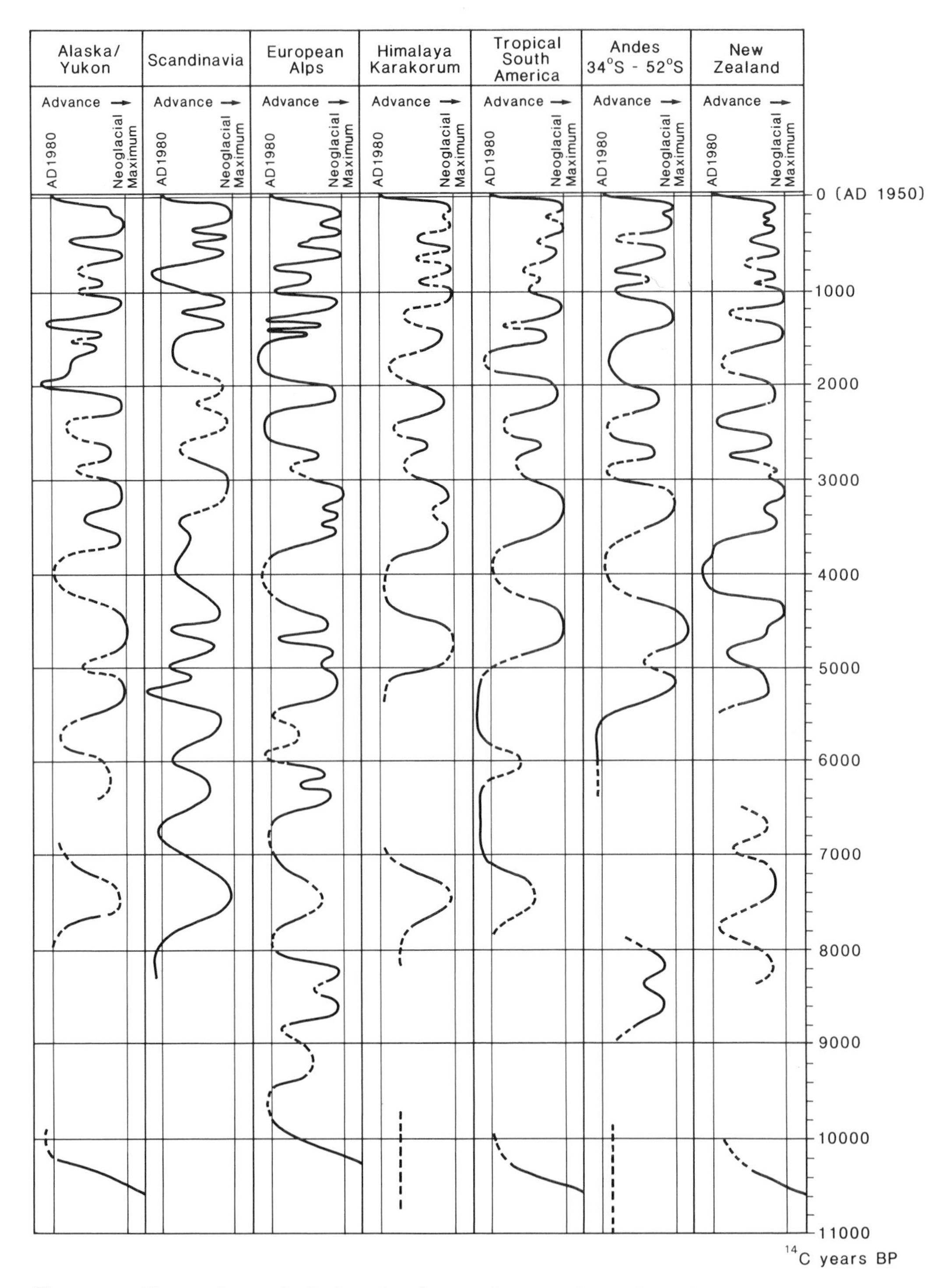

Figure 10.24 Fluctuations of glaciers in the northern and southern hemispheres during the Holocene. (From Röthlisberger 1986)

after 3000 BP 'are due more to variations in sample material than to climatic conditions and may be explained by the large number of data, the concomitant broad deviation and variation in the ^{14}C content of the samples'. He suggested that the accuracy levels of ^{14}C had been reached and that better resolution could only be obtained in future by stratigraphical analysis of soil horizons in lateral moraines or by dendrochronological analysis. He found that advances between 3000 and 1600 years BP had been generally smaller than those between 1600 and 100 BP. In the northern hemisphere advances during the last 600 years generally reached maximum Neoglacial positions but in the southern Andes the phase culminating around 4600 BP was markedly more extensive, a finding in agreement with that of Mercer (1970, 1976). In New Zealand, advances around 4400, 3100, 1500 and 1100 BP were greater than those of more recent centuries. The increased knowledge of glacial behaviour in the southern hemisphere tropics provided by Röthlisberger's dating of fluctuations in the Cordillera Blanca is especially noteworthy in relation to the demonstration by Thompson *et al.* (1986) that the Little Ice Age affected the Quelccaya icecap as clearly as glaciers in the northern hemisphere.

Some of Röthlisberger's time series are, he concedes, better based than others. The best substantiated, those for the European Alps and New Zealand, are in good accordance. Data from the period before 5000 BP are much sparser than for more recent millennia. All in all, Röthlisberger's work, though it cannot be taken as conclusive, provides substantial support for the view that periods of glacial enlargement have occurred at intervals throughout the Holocene, and that, like the Little Ice Age, the earlier phases were globally synchronous, affecting both hemispheres. The implication is that the dominant controls have been changes in global temperature, regional variations in precipitation causing some of the departures from the general pattern.

Chapter 11

Little Ice Age and other Holocene phases of glacier advance: a consideration of their possible causes

11.1 Overview of Little Ice Age

Glaciers on every continent have expanded in the last few centuries; the Little Ice Age was a global phenomenon. The multiple moraines that were left behind indicate that a number of individual pulses were involved.

In the European Alps the main advance was preceded by another of comparable magnitude in the decades around AD 1300. The subsequent retreat probably did not reduce ice extent to that of the medieval warm period and this early advance phase may therefore be regarded either as a forerunner or as a part of the Little Ice Age proper. Evidence of thirteenth/fourteenth-century advances in Scandinavia and North America is not well substantiated; for the Caucasus it is not explicit; but for the Himalaya and New Zealand it seems to be better founded.

The beginning of the main advances which brought ice fronts near to their Little Ice Age maxima is best dated in the Alps, where glaciers are known from historical records to have reached advanced positions by 1600 and where X-ray densitometry of tree rings shows a sharp decline in summer temperatures about 1570 (Holzhauser 1982). Lichenometric studies, which cannot be regarded as conclusive, give some support to the view that Little Ice Age glacier advances in northern Scandinavia were either contemporaneous with those in the Alps or took place even earlier. In southern Scandinavia, on the other hand, the historical evidence indicates that glaciers failed to expand before the latter part of the seventeenth century. Sea ice is known to have become more extensive around Iceland about the middle of the century but, as in southern Scandinvia, evidence of glaciers advancing is lacking until the last decade of the seventeenth century. Manley has suggested (in a personal communication, 1978) that chilling of the Norwegian Sea may have steered depressions southwards, inhibiting precipitation and delaying ice advance. At the same time it should be pointed out that the glaciers in southern Scandinavia and Iceland to which historical records refer are outlet glaciers of large ice masses; it is therefore conceivable that icecaps were building up in the course of the seventeenth century and spilled down the marginal valleys only towards the end of the century.

Elsewhere in the world, historical records relating to the beginning of the Little Ice Age are lacking and reliance has to be placed on determining the ages of moraines. Measurement of maximum lichen diameters and the ages of trees growing on the moraines has given results which accord well with those from the Alps, in North America, on Mt Rainier in Washington State, in the Canadian Rockies and in Alaska; they are at least credible if not undoubtedly reliable. The time of onset of the Little Ice Age in South America is more uncertain, except for the evidence from Quelccaya.

In any single region, the number of moraines deposited and surviving varies from one glacier to another according to local topography and conditions such as the amount of debris released into glacier tongues. None the less there is a striking consistency in the timing of the main advances. In Europe they have been dated to around 1600 to 1610, 1690 to 1700, in the 1770s, around 1820 and 1850, in the 1880s, 1920s and 1960s (Figure 11.1b). Where information is available it appears that this timetable was also followed in many parts of the world outside Europe, with familiar dates appearing in the moraine records from Canada and Alaska over the last four centuries, and from the Caucasus, Himalaya and China from the mid-nineteenth century onwards.

Some exceptions to these generalizations include the absence from the Norwegian record of evidence of advances in the 1850s and, on a smaller scale, the fact that there was no separate forward pulse in the 1820s in the Venediger, Austria, where advance was uninterrupted between 1810 and 1850, although a separate pulse was recorded in the Ötztal at that time. The absence of indications of minor advances in the later nineteenth century and twentieth century in the Himalayan region east of the Karakoram is almost certainly due to the focusing of attention on the less sensitive tongues of the major longitudinal glaciers; the transverse glaciers and the small glaciers in the heads of tributary valleys to either side are likely to respond more sensitively, as has been shown in the 1970s and 1980s by Japanese studies in Nepal and also by Chinese work in the northern Himalaya and elsewhere.

The repeated advances of ice in the Alps between 1600 and 1850 brought glacier tongues forward to very similar terminal positions time after time (Figure 6.4). Their maximum extent was not always reached at the same time even in a single locality; in the case of the Mont Blanc massif, for instance, the Mer de Glace was largest in 1644 (Figure 4.3), the Trient in 1780 (Figure 4.4). In Norway the situation is different. Moraines formed there since the mid-eighteenth century are well separated one from another, with successively younger moraines towards the heads of the valleys (Figure 3.8) reflecting the time of greatest extension, about 1750, and later advances each smaller than its predecessor. The Lemon Creek glacier, reaching down from the Juneau Icefield in Alaska (Figure 8.7), has retreated more than 2 km since its mid-eighteenth-century maximum; its recession is comparable in distance and, if the dating of the moraines is anything like correct, in time with several of its neighbours (Table 8.4) and, it happens, with Nigardsbreen, an outlet glacier of Norway's Jostedalsbreen. Far down the Pacific coast at 48°20′N, in the Dome Peak area of the Olympics, terminal moraines of the South Cascade and nearby glaciers dated to the sixteenth/seventeenth centuries, mid-nineteenth and late nineteenth centuries are all clustered near together, with the younger members of the

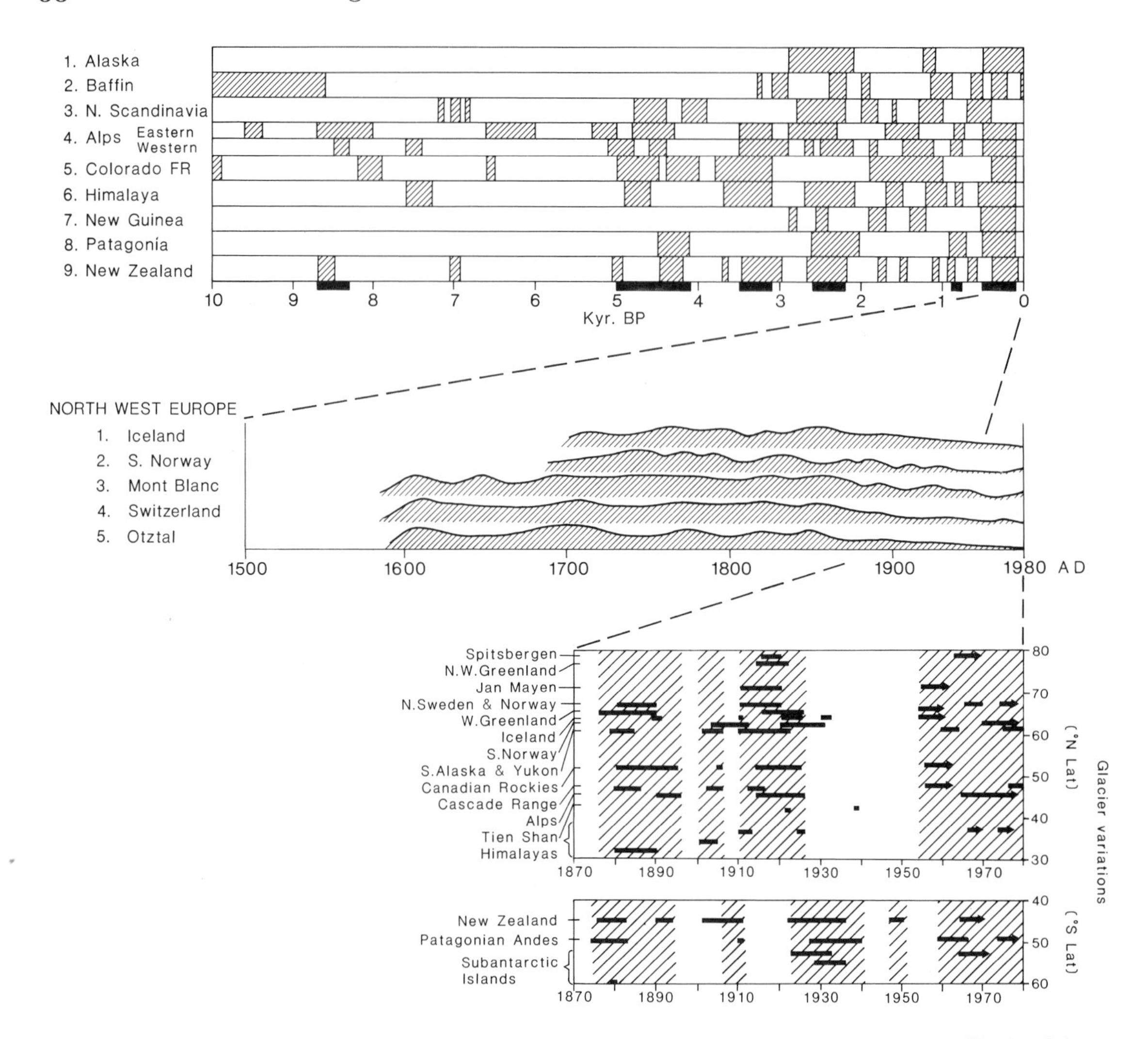

Figure 11.1 Summary diagram of the main phases of ice advance in the Holocene. (Derived from data in previous chapters and after Porter 1981, with indications when advances appear to have been widely synchronous)

clusters successively closer to the ice (Figure 8.4). This pattern is entirely reminiscent of the forefield of the Lower Grindelwald in Switzerland; the difference is that the dating of the Grindelwald moraines can be confidently accepted, whereas strong reservations must be felt as to the reliability of the dating of the Dome Peak moraines, based as it is on dendrochronology (see Chapter 8.1).

There is a hint here that the more northerly glaciers in both Europe and North America advanced late and retreated at quite an early stage in the Little Ice Age, whereas the more southerly ones advanced early and subsequently advanced repeatedly to similar positions, eventually waning decisively in the latter half of the nineteenth century and the first half of the twentieth century. However, Iceland does not conform with this general picture, a number of the Icelandic glaciers, possibly a majority, reaching their greatest

Little Ice Age extent (and, as we have seen, their greatest extent in the Holocene) towards the end of the nineteenth century.

In the southern hemisphere, the advances of the New Zealand glaciers east of the Alps left terminal moraines dating to 1650 and 1885, spaced less than 50 metres apart in the cases of the Mueller (Figure 9.7), Classen and Tasman glaciers (Gellatly 1982). The frontal moraines of the western glaciers, such as the Fox and La Perousse, are somewhat more widely spaced probably on account of their rapid turnover and greater sensitivity.

The common behavioural feature of nearly all the world's glaciers is their recession over the last hundred years or more. The exceptions are relatively few and can usually be explained in terms of special circumstances of some kind, such as the occurrence of unusually heavy debris cover on ice tongues or exceptional accumulation conditions, as on Folgefonni in southwest Norway. The scale and ubiquity of the recession is comparable with that of the advance that seems to have taken place in so many parts of the world around 1600. These events can be seen, at this distance in time, as bracketing the Little Ice Age.

11.1.1 HALTS IN THE RECESSION FROM LITTLE ICE AGE MAXIMA

The recession of the late nineteenth and twentieth centuries, which in the early 1980s was still in progress, was interrupted by minor small readvances, halts or decelerations of the retreats of ice fronts at intervals of a few decades (Figure 11.1c). These interruptions were remarkably synchronous worldwide in the 1880s, 1920s and 1960s.

The minor advances of glaciers which took place in the 1960s and in many places persisted into the 1970s (Table 11.1) are better recorded than those of earlier times and

Table 11.1 Localities where glaciers have advanced (a) and/or mass balances have been positive (b) in the 1960s and/or 1970s and/or 1980s

		Chapter and section
Iceland	a	2.3.3, 2.4.3
Norway	ab	3.1.3, 3.2.3, 3.3
Mont Blanc	a	4.5
Ötztal	ab	5.5, 5.6
Switzerland	ab	6.3
Urals	b	7.2.1
Caucasus	a	7.2.2
Pamir	ab	7.2.3
Tien Shan	b	7.2.4
Altay and Sayan	b	7.2.5
Karakoram	a	7.3.6
Nepal	ab	7.3.7
China	ab	7.4
North America	ab	8.6
Greenland	a	8.7
East Africa	a	9.1
Ecuador	a	9.3
Argentina	a	9.4.1
New Zealand	a	9.4.2
Heard Island	a	9.4.3

mass balance studies provide an insight into the accumulation/ablation and climatic conditions responsible for them (Reynaud *et al.* 1984).

Patzelt (1985) brought together data pointing to the high proportion of glaciers in the Alps observed to be advancing in the 1960s and 1970s, showing the association with increased precipitation as well as lower annual temperatures. The rate of shrinkage of glaciers diminished and some increased in volume, but the proportion of monitored Alpine glaciers with advancing termini exceeded 50 per cent for only a few years (Figure 6.7) In the early 1980s retreat was dominant and it seems that the small fluctuation of the 1960s and 1970s was superimposed on continuing retreat.

In Scandinavia, as in the Alps, glacier fronts retreated more slowly than they had for several decades, or even advanced in the 1960s and 1970s, especially in the more maritime areas; again mass balance studies indicate that an increase in precipitation was involved. In central Asia too, the years around 1970 saw glaciers slowing their retreat or advancing, but here a fall in temperature was a more prominent feature than any increase in precipitation. Fragmentary information from Ruwenzori in East Africa and about individual glaciers in extra-tropical South America, New Zealand and the sub-Antarctic islands points to short-lived advances during the period between about 1960 and the mid-1980s.

The behaviour of North American glaciers generally conformed with that of glaciers elsewhere, with an important exception. The slowing down of the retreat or the advance of glaciers in the Sierra Nevada, Cascades and Olympics began a decade or more earlier, in the years around 1950. Another anomalous area is New Zealand, where interruptions of the glacier recession that has persisted since the early part of the century have not generally been very apparent and the cooling recorded in the northern hemisphere for two or three decades after 1945 did not take place.

The fluctuation that interrupted glacier retreat in most parts of the world in the 1960s and 1970s seems to have been comparable in scale with those of the 1880s and 1920s. However, glaciers the world over by 1960 were smaller than they had been in 1920, and in 1920 they were smaller than they had been in 1880.

11.2 Overview of Holocene glaciations

Periods of glacier expansion, each lasting a matter of centuries, have occurred from time to time in the course of the Holocene in all parts of the world. Many if not all of these events were multiple, indicating the existence of smaller-scale glacial and presumably climatic fluctuations within the larger ones and thus their similarity to the Little Ice Age.

The sequence of Holocene glaciations is best known from the European Alps, where opportunities for dating organic material, especially tree fragments in moraines, are good and where there has been an exceptional concentration of investigations. Conclusive proof of the occurrence of a number of individual phases of glacier expansion is also available from Scandinavia, the Caucasus, the Himalaya and many regions in North America as well as South America, New Zealand and the sub-Antarctic islands, but the dating in all these regions is either sparser or less reliable than that in the Alps.

The rapidly increasing volume of information about the times at which glaciers expanded during the Early, Late and even Mid-Holocene is as yet insufficiently precise to affirm conclusively that glaciers of similar size expanded synchronously or, indeed, that they did not (Figure 11.1a). Bracketing ^{14}C dates are frequently lacking or too wide apart, while lichenometry, especially when carried out in a number of different ways, some of them unstated, and weathering-rind studies, are not precise enough to demonstrate interregional relationships conclusively. Detailed comparisons have been made of the timing of glacial episodes in the Karakarom/Himalaya and in the European Alps (Figure 10.14); a more restricted comparison has also been made of the fluctuations of the Tasman and Aletsch glaciers (Figure 10.22). In both cases the similarity in the recurrence intervals of glacier responses is as strong an indicator of their synchroneity as their dates.

In the Alps, the positions reached by glacier termini at different times in the Holocene were remarkably close to each other, as they were in northern Scandinavia and the Himalaya. The same appears to be the case in Baffin Island and the Brooks Range of Alaska.

Holocene glacial episodes were not everywhere of comparable extent. In Greenland, the simultaneous Little Ice Age advances of the margin of the inland ice and of local glaciers seem to have been greater than any forerunners in the Mid- or Late Holocene. In Iceland too, the Little Ice Age expansion was the only Holocene advance that had been recognized, until 1985 when evidence was found at Sólheimajökull of moraine formation between 1400 and 1200 BP and also earlier in the Holocene. Elsewhere in Iceland and also in Spitsbergen, further evidence of Holocene glacial history may yet be discovered in lateral moraine sections.

In the southern hemisphere, glaciers advanced further in the middle Holocene than at later stages. In mid-latitude South America, moraines dated to between 4600 and 4200 BP are furthest downvalley; those formed 2700 to 2000 BP are in intermediate positions and the Little Ice Age moraines are closest to the present ice fronts. Nearer the tropics a much greater expansion than any since, and probably attributable to greater precipitation than at present, seems to have taken place at some time between 6000 and 4500 BP. In New Zealand too the earlier Holocene advances on the eastern side of the southern Alps were on a somewhat larger scale than those of the Little Ice Age.

There seems to be a contrast between the Holocene behaviour of mid- to high-latitude glaciers in the northern and southern hemispheres. In both Patagonia and New Zealand, Little Ice Age advances were on a smaller scale than those of the middle Holocene, whereas in Iceland, Greenland and Spitsbergen, data at present available point to Little Ice Age advances, and notably those of the nineteenth century, as having been the greatest to have taken place since the end of the last glacial period.

11.3 Possible causes of Holocene glacial phases

There is a substantial body of literature concerned with the causes of climatic variation on a scale of a few hundred years, that is on the scale of the Little Ice Age and its forerunners. Many workers have concentrated their attention on one possible cause, more or less ignoring the rest, whereas it is very likely that several factors are involved. Explanations advanced fall into two main classes, those which rely on internal adjustments within the atmosphere–ocean system and those invoking external factors to account for changes in the mean temperature of the globe.

Before considering the strength of any of the current hypotheses it must be borne in mind that changes in the general circulation of the atmosphere are likely to be complex. A change in global mean temperature, far from being uniform, may be amplified, diminished or even reversed in particular areas. Thus, while the twentieth-century warming trend that was so well marked up to the 1940s was followed by significant cooling in the northern hemisphere, proxy data confirm the view of Salinger (1976) that, in New Zealand, warming continued into the 1970s (Gellatly and Norton 1984).

Despite the complexities, there is now substantial evidence that mean global temperature has fallen to the extent of 1 or 2 °C for periods of several hundred years over large portions of the globe, not only in the later part of the Holocene but on several occasions over the last 10,000 years. It would be unsatisfactory to devote a whole volume to the Little Ice Age and its earlier equivalents without mention of the causation of such climatic oscillations. It is not attempted here to provide a full explanation but only to sketch the main outlines of the factors that appear to be of major importance.

11.3.1 PROCESSES INTERNAL TO THE CLIMATIC SYSTEM

A number of processes involving feedbacks from one part of the climatic system to another are related to the occurrence of cooler periods. The most important of these are concerned with the interaction of the atmosphere and ocean.

Bjerknes (1968) believed that the increase in ice cover that occurred during the Little Ice Age was insufficient to have changed planetary albedo and that there was not, therefore, any radical change in the earth's radiation balance involved. He fastened his attention first on the advance of arctic sea ice in the Icelandic sector (see Chapter 2.1), noting that the Gulf Stream and North Atlantic Drift kept the Norwegian coast ice-free during the period of glacial enlargement. Very reasonably he believed that the glacial advances of the last few centuries and the extension of sea ice in the northeast Atlantic must have stemmed from the same cause.

A consideration of the basic conditions of wind-driven ocean currents shows that water transport is divergent when the wind-field is cyclonic and convergent when the wind-field is anticyclonic. It follows that under steady-state conditions meridional geostrophic water movement is poleward beneath cyclonic and equatorward under anticyclonic wind systems. The relative warmth of Icelandic waters during the first half of the twentieth century was dependent on the northward movement of geostrophic water transport in the open ocean south of Iceland. The north-flowing branches of the Gulf Stream were meteorologically supported during the first part of the twentieth century. During this period annual average pressure maps show cyclonic wind stress over this part of the Atlantic, occasional anticylonic wind conditions being insufficient to reverse the flow. The situation in previous centuries was very different. Lamb and Johnson (1959) constructed annual pressure maps for January for the period 1790 to 1829 which reveal that there was less cyclonic wind stress south of Iceland than in the twentieth century and consequently it may be presumed less northward water transport and therefore colder water in the northeast Atlantic, North Sea and Norwegian Sea. Lamb (1979) has since shown, using both early instrumental records and proxy data, that the ocean surface around the Faroe Islands and between the Faroes and Iceland was up to 5 °C colder on average between 1675 and 1704 than in the first half of the twentieth century. Bjerknes rated this temperature anomaly and the associated lowering of the freezing level in the precipitation-producing maritime air masses as 'probably the most important factor in maintaining the European Little Ice Age'.

In the northwest Atlantic, Lamb and Johnson's maps indicate more cyclonic wind stress and thus more northerly transport of warm water in the period 1790–1829 than in the period 1900 to 1939. Although none of this water reached any coast, Bjerknes judged that it would have increased the moisture carried by occasional southeast winds over Labrador and southeast Greenland and so increased snowfall and promoted glacier growth there.

Bjerknes envisaged that the centre of the oceanic anticyclonic gyre and therefore the area of maximum transport of the Gulf Stream, now situated near Bermuda, would have been further down the Gulf Stream in the Little Ice Age (i.e. further northeast). This idea is again supported by the Lamb and Johnson pressure map which shows the west Atlantic trough in January extending further south than at present, exposing the western Sargasso Sea to cyclonic instead of anticyclonic wind stress. They used sea temperature data collected by British naval vessels to sketch the January distribution of positive and negative temperature anomalies for the period 1790–1829. A positive anomaly appears in the Sargasso Sea, bordering abruptly the 'slope water' between the Gulf Stream jet and the American coast. This tends to confirm Bjerknes's interpretation of ocean–atmosphere relationships in the Atlantic sector.

Bjerknes traced these situations back to secular persistence of low-index atmospheric circulation which favoured anomalies in ocean temperature, causing not only spread of sea ice in the northern Atlantic but also favouring glacier growth on the land areas around the North Atlantic. He surmised that glacier growth in western North America could be attributed to low-index split of the Aleutian Low, causing increased precipitation on the mountains from Alaska to the northwestern United States. He also attributed glacial expansion in the Pyrenees, Alps and Atlas mountains to increased winter precipitation associated with intensification and westward recession of the 500 mb trough over Europe.

Bjerknes did not offer any further explanation for the persistence of mid-latitude low-index circulation patterns in the northern hemisphere. He based his chronology mainly on Scandinavian data from Ahlmann (1953) and was unaware of the more widespread incidence of Little Ice Age conditions outside Europe and North America. He did not envisage the occurrence of earlier Holocene events of similar duration and magnitude, although he mentioned 'a moderate readvance' in the pre-Roman era, nor did he take into account the possible relevance of extended snow as apart from ice cover and was hardly in a position to do so at the time when he was writing.

Increased precipitation has no doubt had a part to play in these glacier advances but it is known that temperatures were lower over periods of decades during the Little Ice Age in North America (see Chapters 8.1, 8.4, 8.5 and Figure 11.2) and in the Alps (6.3, 11.2) and it is argued here that the immediate causes of Little Ice Age advances were reduced ablation and increased solid precipitation.

Before the coming of satellite imagery it was impracticable to measure continental snow cover, but such mapping has been underaken by the National Oceanic and Atmospheric Administration on an annual basis since 1966 (Matson and Wiesnet 1981). During the period of record, Eurasia experienced two periods of large increase in snow cover, in 1971–3 and 1976–7. The running mean for Eurasia between 1966 and 1980 showed an upward trend. In the winters of 1972, 1973, 1977 and 1978 over 50 per cent of the Eurasian landmass was snow-covered. The record from Eurasia includes a particularly sharp increase in autumn snow cover in 1976, when October snow cover was

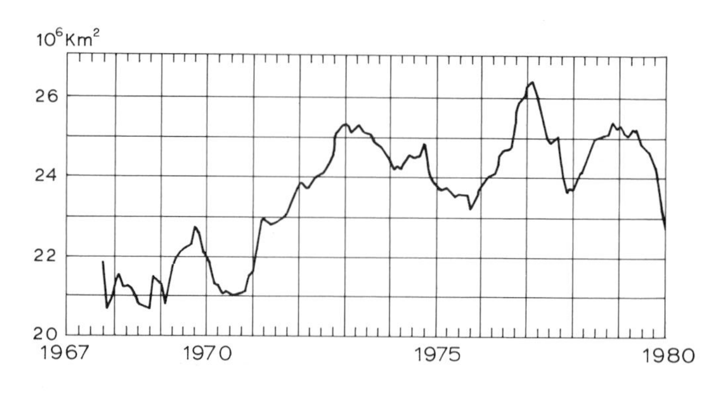

Figure 11.2 Twelve-month running means of northern hemisphere monthly mean snow cover. Data are plotted on the last month of the twelve-month period. (From Matson and Wiesnet 1981)

19.7 × 10^6 km² compared with a ten-year average of 9.6 × 10^6 km². The snowline advanced from 60° to 50°N in an area bordered by longitudes 30°E and 80°E in one week. This gave a high (45 per cent) zonal albedo value. The snow cover in Eurasia dominates the running mean for the northern hemisphere shown in Figure 11.2, though it may be noted that the average winter snow cover in North America rose steadily from 1973 to 1979, and that during the severe North American winter of 1978 nearly 70 per cent of the continent was snow-covered. It is significant that hemisphere snow cover and therefore albedo value rose during the cold period of the 1970s. In the northern winter of 1971–2 there had already been a rapid increase of snow and ice cover, estimated by Kukla and Kukla (1974) as about one-sixth the difference between present-day conditions and those at the end of the last glacial maximum. Such variations in snow cover in the 1970s and the variations in global albedo associated with them indicate the likelihood of even greater increases in snow-cover and global albedo under Little Ice Age conditions, as the evidence presented on Switzerland (6) and Baffin Island (8.5 and 8.6) suggests, and consequent modification of the radiation balance on a scale that Bjerknes did not consider.

Weyl (1968), who was also concerned with the interaction between ocean and atmosphere, suggested that changes in salinity distribution in the oceans controlled both Little Ice Age and Pleistocene sea ice distribution, which in turn affected atmospheric conditions and hence the extent of glaciation. He emphasized the importance of the marked difference in salinity between the North Atlantic with 37.4/K and the North Pacific with 34.5/K and suggested that the high salinity of the North Atlantic depends on the state of water vapour flux (westwards) across the isthmus of Panama. He believed that complete blocking of this flux could be effected by changes in pressure distribution and that it would only take 600 years of such blocking to remove the salinity contrast between the two oceans. Reduction of water vapour flux across Panama, leading to reduced salinity in the North Atlantic, would cause sea ice to extend there and so lead in turn to atmospheric cooling. Lower surface salinity in the north would cause a reduction in the formation of bottom water and so a reduction in the thermohaline part of the northern advection of both heat and salt. This would allow the sea ice to spread further south and enhance the cooling. Eventually this positive feedback would influence the whole deep-water circulation, ultimately leading to increased salinity deficit near the Antarctic and so to the spread and thickening of sea ice there. Wiley considered that in the case of the Little Ice Age the whole process had been halted by an increase in water vapour flux across Panama, which prevented the southern hemisphere from being affected and so prevented the development of full glacial conditions.

A similar mechanism is discussed by Broecker *et al.* (1985) in an attempt to explain variations in carbon dioxide (CO_2) content of the atmosphere in Late Quaternary times. They postulate changes in the rate of exchange between polar surface waters and the rest of the ocean in terms of changes in the production of deep water. During peak glacial times, faunal, chemical and isotope studies suggest that the production of deep water in the North Atlantic was greatly reduced. Is it possible, they inquire, that evidence of brief warm periods represents periods during which the glacially weakened northern Atlantic deep-water sources was rejuvenated? Equally, one may consider the possibility that, during the Holocene, deep-water production in the North Atlantic may have slackened from time to time, with accompanying reductions in the release of heat to the atmosphere over the northern Atlantic. But why should such oscillations occur at intervals of a few thousand years? There is evidence that small changes in CO_2 due to changes in circulation in the top kilometre of the ocean, particularly in high latitudes, may be greatly amplified, to lead within a few decades to large decreases or increases in atmospheric CO_2 and temperature (Lal and Revelle 1984). Evidently it will be necessary to understand the climate of the deep oceans before a full understanding of changes in the atmosphere can be achieved. If rather modest periodic variations in the rate of deep-water formation, with periods of 50 to 200 years, are capable of causing variations in surface temperature of 0.1 to 0.5 °C, then this internal variation could have played a part in causing the measured climatic fluctuations which occurred during the past century. Calculations made by Watts (1985) imply that variations in the rate of bottom-water formation could have been responsible for global temperature variations of the magnitude of those which have actually occurred on time scales ranging from a few years to several centuries.

Weyl postulated that slight warming in the sub-tropics caused by variations in earth–sun geometry might have led to wind shifts in low latitudes and so accounted for the key changes in water vapour flux across Panama. In recent years there has been renewed interest in the climatic effects of the variations in the earth's orbit, first discussed by Milankovitch, and the major climatic changes in the Quaternary are now generally attributed to the variations in the earth's orbital parameters (see Weertman 1976, Hays *et al.* 1976, Imbrie and Imbrie 1980, Ruddiman and McIntyre 1981). Kutzbach (1983) has argued that increased incident solar radiation in June, July and August around 10,000 BP would have increased the strength of the summer monsoon, bringing rain to lands in the northern tropics and thereby explaining Early Holocene high lake levels there. However, such perturbations are too widely spaced in time to provide an explanation for the Little Ice Age and similar Holocene glacial advances which, as we have seen, occurred at intervals of the order of about two thousand years.

Botanical evidence from North America indicates relatively warm dry conditions in the Early Holocene, probably associated with increased dominance by the sub-tropical Pacific anticyclone, and cooler wetter conditions with increased storminess after 5000 BP under the influence of polar Pacific cyclones (Heusser *et al.* 1985). It is argued that these circulation patterns are attributable to the incidence in the northern hemisphere of high solar radiation at the time of the summer solstice in early millennia and low summer radiation thereafter; the broad characteristics of Holocene climate are seen as having been controlled by changes in seasonal solar radiation intensity caused by variations in the earth's orbital parameters, notably precession. However, it is further argued that, during the Early Holocene, glaciers were inactive apart from 'isolated instances of minor glacial advance'. But the White Chuck moraines (Beget 1981, 1983, and 10.4.2) were outside those formed later in the Holocene and, although there are as yet rather incomplete data about glacial expansion in North America around 8000 BP, there are increasing indications that here, as in the Alps, there were other advances in the Early Holocene comparable with those in the Little Ice Age. For instance relative age dating of a small moraine in Idaho is supported by ^{14}C evidence of periglacial activity between 10,000 and 7500 BP (Butler 1983, 1984). Nor, it seems, is Early Holocene climatic cooling confined to the northern hemisphere; it is also recognized in New Zealand (Gellatly *et al.* 1985 and see 10.7).

Heusser *et al.* (1985) themselves take the view that assessment of the patterns of temperature and precipitation for short time periods of the order of centuries on the basis of pollen analysis is not justified. It has also been suggested recently that the response of forest growth to cooling of the order of 2 °C is likely to be delayed by up to two hundred years, producing gradual forest changes in response to abrupt temperature changes and reducing the amplitude of response to brief climatic events. Delayed responses would imply that fossil pollen deposits may not be able to resolve climatic changes of the order of one hundred to two hundred years in length or to record brief climatic events (Davis and Botkin 1985). Orbital control seems likely to have been responsible for the broad sweep of environmental change through the Holocene, but does not account for the increasingly complex picture of climatic fluctuations which is emerging with more detailed field investigations in many regions.

It is possible that the current circulation pattern of the atmosphere and oceans as we know it is not the only one compatible with existing boundary conditions. Lorenz (1968, 1976) introduced the idea of intransitivity, according to which there can be alternative climatic states and the circulation pattern may flip from one to the other without external forcing. Broecker *et al.* (1985) see such oscillations as having occurred.

Such alternations, possibly involving deep-water production in the northern Atlantic being turned on and off, are possible, according to Broecker *et al.* (1985), but if they are to be substantiated more evidence is required than is available at present. In the meantime we must continue to seek for a more deterministic explanation for the Little Ice Age and its predecessors, while accepting that Holocene climatic change was characterized by atmosphere–ocean interactions, leading to or led by snow and ice albedo feedback.

11.3.2 PROCESSES EXTERNAL TO THE CLIMATIC SYSTEM

11.3.2.1 Geomagnetism

Geomagnetism and climatic change have been linked together by a number of investigators. King (1974) pointed to a remarkable similarity between the pattern of the earth's magnetic field over the northern hemisphere, shown by isopleths of magnetic intensity, and mean height contours of surfaces of constant atmospheric pressure which define the shape of the circumpolar vortex of the upper westerlies. Both patterns have a bipolar form, with one end in the Canadian Archipelago and the other in northeast Siberia (Figure 11.3). King suggested that the earth's magnetic field positively influences the average tropospheric pressure system at high latitudes and also that the average pressure system moves west as the magnetic field rotates. The magnetic declination at London and Paris was zero in 1600 and the different magnetic field configuration may

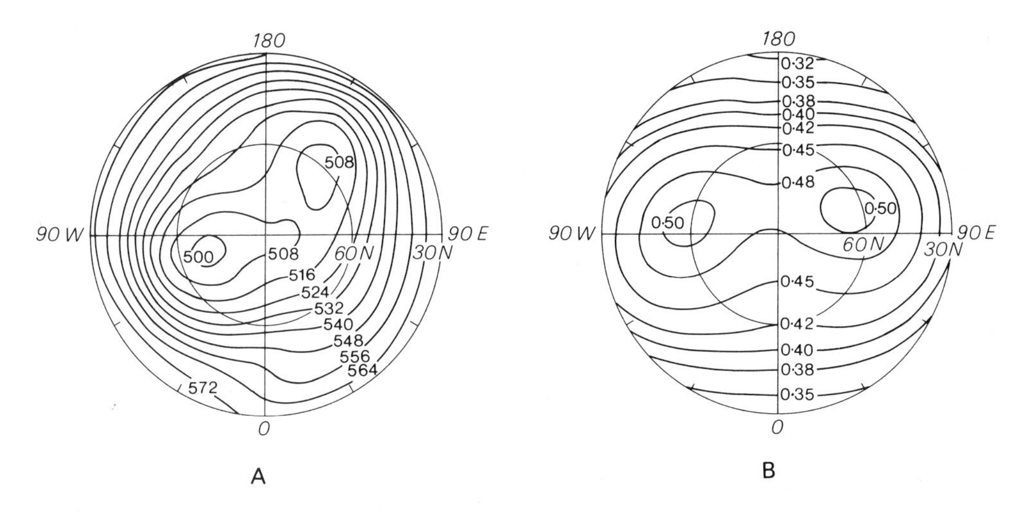

Figure 11.3a Northern hemisphere polar plot showing the average height in decimetres of the 500 mb level for January

Figure 11.3b Northern hemisphere polar plot showing contours of constant magnetic field strength at 400 km, 1965. (From King 1974)

perhaps have controlled the climatic pattern of the Little Ice Age. He did not demonstrate the existence of a causal relationship but pointed to the need for further research in this area.

Wallin *et al.* (1971), following a study of remnant magnetism in deep-sea cores over the last 2.2 My, identified a correlation between variations in magnetic intensity and climatic change, with high magnetic intensity indicating cold climates, and concluded that 'magnetism may modulate climate to some degree by the ability of the earth's magnetic field somehow to provide a shield against corpuscular radiation'. Other authors have concurred, but there are many indications that if it is an important factor it is operating amongst several others, and probably on a longer timescale than the Little Ice Age.

11.3.2.2 Variations in solar radiation

It is evident that changes in climate over time are likely to be related to variations in ground-level solar intensity, which could be due either to changes in solar output or to changes in the transparency of the atmosphere. Calculations suggest that a decrease of 1 per cent in the radiation output of the sun would be enough to lower global average temperature by 1 or 2 °C (Eddy 1980). Sunspots appear to be useful indicators of other forms of solar activity such as active regions, flares, intense magnetic fields, prominences, coronal form and solar wind (Harvey 1980). It was for long thought that the sun is a constant star with regular repeating behaviour, these characteristics being demonstrated firstly by the record of regular sunspot periodicities, that is by the eleven-year sunspot cycle, and secondly by measurements of radioactive output or solar constant. However, both these propositions appear to have been falsified.

The possibility that there were very few sunspots for a prolonged period before 1716 was raised as early as 1887 by a German astronomer, Gustav Sporer, though an English astronomer, William Derham, had written in 1711 that

> there are doubtless great intervals sometimes when the sun is free, as between the years 1660 and 1671, 1676 and 1684, in which time Spots could hardly escape the sight of so many Observers of the Sun, as were then perpetually peeping upon them with their Telescopes in England, France, Germany and Italy and all the world over (quoted Eddy 1976).

Maunder, superintendent of the solar department of Greenwich Observatory, summarized the evidence for such gaps in 1890 and 1894, pointing out that the occurrence of a sunspot minimum which was prolonged would have important implications for solar–terrestrial relations, but his work was ignored. Eddy (1976) analysed the historical data critically and came to the conclusion that the prolonged absence of sunspots between 1645 and 1715, known as the Maunder minimum, was to be taken seriously (Figure 11.4). It has been established that at the present time the rotational speed of the sun is inversely correlated with sunspot numbers (Sakurai 1977, Herr 1978). Drawings by early astronomers suggest that solar rotation was anomalously slow before the Maunder minimum and anomalously fast during it (Eddy *et al.* 1976). Nor was the Maunder minimum the only one. It seems that there was another prolonged minimum in the seventh and early eighth centuries, and maxima in the twelfth and early thirteenth centuries and in the late fifteenth and early sixteenth centuries.

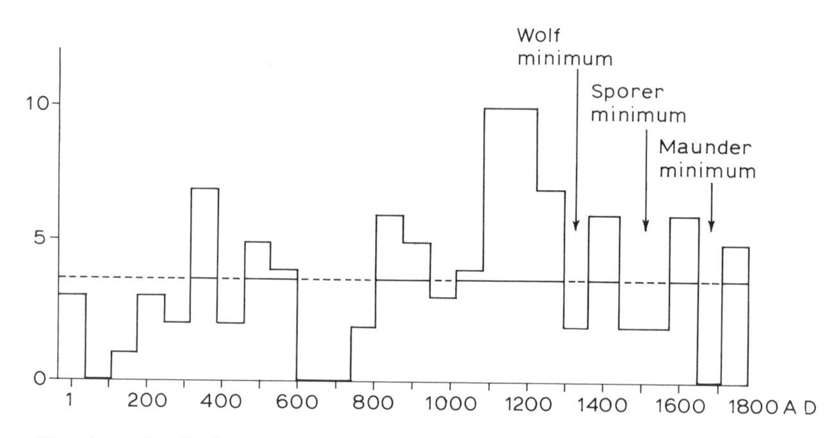

Figure 11.4 Number of naked-eye oriental reports of sunspots and aurora. (From Eddy 1977)

Changes in atmospheric ^{14}C can also be attributed to a variable sun. It can be shown that periods of low solar activity promote high ^{14}C activity in the upper atmosphere. The connection with solar variation comes about because ^{14}C is produced by the interaction of neutrons with atmospheric nitrogen and is therefore dependent on cosmic ray intensity. Interplanetary cosmic ray flux is believed to be constant on geological timescales (Oeschger *et al.* 1970) but changes in cosmic ray intensity are caused by changes in the magnetic field strength of the solar wind. These cause changes in the deflection of galactic cosmic rays and so in cosmic ray flux in the upper atmosphere. The situation is complicated further because variations in ^{14}C values are also caused by climatically triggered change in the ^{14}C mixing rate between the atmosphere and the oceans, as well as by changes in geomagnetic field intensity which modulate incoming cosmic ray flux. Lal and Venkatavaradan (1970) and Suess (1970) concluded that ^{14}C activity variations on a scale of about ten years could be explained by variations in climate, while the trend caused by changing geomagnetic dipole strength, which can be deduced from thermoluminescent magnetism of archaeological ceramics, is important on a scale of 100,000 years (Bucha 1970).

The ^{14}C value of tree rings of known age was measured by Stuiver and Quay (1980). The residual ^{14}C variations, after the removal of long-term trend caused by changes in magnetic field intensity, were assumed to have been caused by ^{14}C production rate changes. Using a carbon reservoir model, Stuiver and Quay calculated the production rate changes and compared them with sunspots and auroral records and found strikingly good agreement. They found major increases in ^{14}C production rate during the 1282–1342 Wolf minimum, the 1450–1534 Sporer minimum, and the 1645–1715 Maunder minimum. They suggested that increased rotation rate of the sun, associated with a change in turbulence in the convective layer, inhibits transport of the magnetic field from the deep interior to the surface and could cause a reduction of solar wind which ultimately results in high ^{14}C production. The residual ^{14}C variations shown in Figure 11.5 are, it will be noted, on the timescale of 100 to 10 years. The coincidence of low solar activity and high ^{14}C production with periods of low global temperature suggests the possibility that variations of solar output were responsible both for changes in atmospheric ^{14}C values and for global temperature changes, all the more because a medieval maximum of solar activity coincides with the medieval warm period. Nevertheless it must be acknowledged that while there are striking coincidences in timing, they are not perfect; the Maunder minimum did not cover the whole of the Little Ice Age, for example. Other difficulties still remain and it would seem that satisfactory interpretation of the ^{14}C record demands more precise understanding of the carbon cycle of the oceans than we have at present (Lal and Revell 1984).

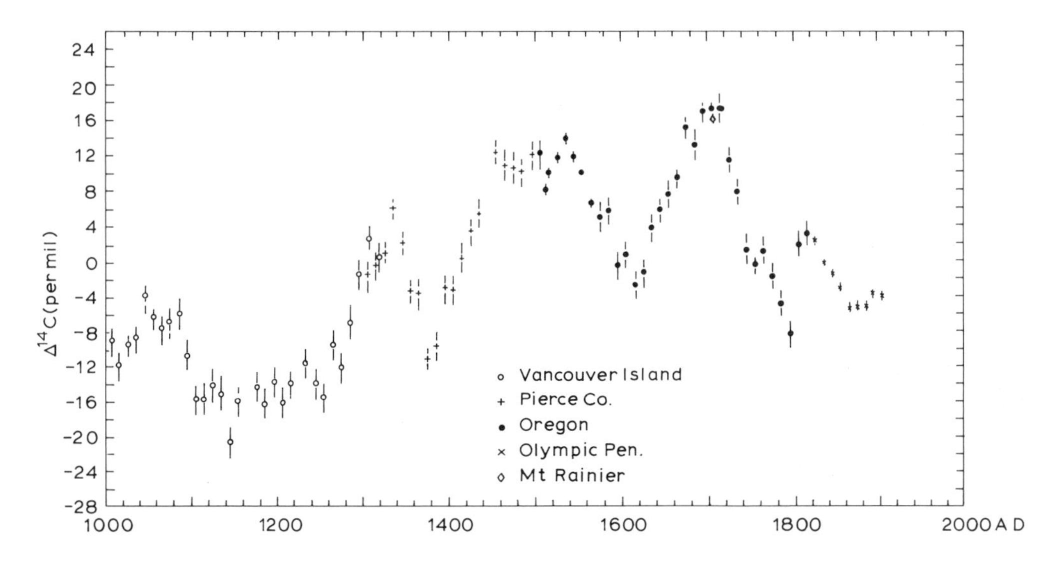

Figure 11.5 Residual ^{14}C variations after removal of the long-term trend attributed to change in geomagnetic dipole strength, based on data from tree rings from Oregon, Washington and British Columbia. (From Stuiver and Quay 1980)

The precise ways in which solar variability might affect climate are uncertain. In a survey of solar variability as a contributory factor to Holocene climatic change, Harvey (1980) summarized the voluminous literature concerned with the possible influence on planetary albedo and climate of solar corpuscular radiation or solar wind, variations in stratospheric total ozone and variations in the solar optical and infrared output. Ozone in particular is created photochemically by solar ultraviolet radiation and could influence climate by modulating the amount of solar radiation entering the troposphere. Harvey concluded that it might be possible to explain global temperature changes of the order of 1–3 °C in terms of solar constant variations of 1 per cent or less, but that while neither solar constant nor ozone changes may be capable alone of explaining climatic change, the total effect of the two together could be adequate to explain the known fluctuations of climate during the Holocene.

11.3.2.3 Volcanism

Major volcanic eruptions inject large quantities of silica micro-particles and gases into the stratosphere. The consequent effects on atmospheric transparency and incident radiation have been warranted to be sufficient to explain climatic changes on a variety of long and short timescales (Bray 1974a, Lamb 1970, Bryson and Goodman 1980) and specifically invoked as determining factors in the patterns of recent glacial advance (Bray 1974b, Porter 1981). A great deal of historical documentation relates to the sort of catastrophic eruption which creates worldwide or hemisphere-wide dust veils in the high atmosphere, which dim the sun and the moon, brighten the background sky and create superb sunsets. Lamb (1970), in a major pioneering work, built a dust veil index which has frequently been used for comparison with climatic and glacial records.

There seems little doubt that volcanic activity influences climate but the extent of this influence is controversial. Indeed Rampino *et al.* (1979) even argue that, on the longer timescale, global cooling may cause volcanism. On shorter timescales, difficulties are encountered, such as the absence of large-scale glacial advances after some large eruptions, including Krakatoa, in the late nineteenth century. They concluded that volcanically induced cooling has a timescale of only one to three years and is only on the same sort of scale as background variations in temperature, that is of the order of less than 1 °C.

Many instances of small but significant falls in surface temperature have been discerned following large volcanic injections into the upper atmosphere (Lamb 1970, Mitchell 1971, Mass and Schneider 1977) and their occurrence has been explained by theoretical studies (Pollack *et al.* 1976). Bradley and England (1978) presented evidence that the eruption of Mt Agung (18°S, 115°E) in 1963 was responsible for a marked change in the climate of the American high Arctic and that this change had a significant impact on glacier mass balance in the region. Newell (1981) estimated that in the tropics tropospheric temperature decreased up to 1 °C following the Agung eruption, with most cooling between August 1964 and August 1965. He suggested that this cooling could have been due to absorption either of near infrared or visible radiation by stratospheric aerosols. In the first case, cooling would have been direct, as there would be less energy available in the troposphere for absorption by water vapour and carbon dioxide; in the second case, cooling would have been indirect as there would be less energy available for evaporation at the sea surface and therefore less available for latent heat evaporation. On the other hand, not all recent large eruptions, for example Bezymyannyy 1956 (Self *et al.*

1981), have been followed by a discernible decrease in surface temperature, and Landsberg and Albert (1974) found no convincing evidence for cooling after several major nineteenth-century eruptions.

Some of the apparent contradictions are beginning to be resolved as a result of the very large amounts of scientific data collected during recent eruptions, especially that of Mt St Helens in 1980. More extensive and precisely based compilations of volcanic eruptions have been made (Newhall and Self 1982, Simkin *et al.* 1981), and independent data sources have been exploited in the form of acid layers in ice sheets (Hammer 1980, Hammer *et al.* 1980, 1981) and frost rings in trees (LaMarche and Hirschboeck 1984).

Lamb (1970) intended his dust veil index (DVI) to provide a measure of the impact of volcanic eruptions on the atmosphere. He sifted through documentary records of some 250 eruptions and took into account not only the qualitative assessments of observers but also, where available, information on changes in radiation or temperature, volume of volcanic ejects, or a combination of these. Adjusting his values to give a DVI of 1000 for Krakatoa in 1883, he found that total world frequency of eruptions of this magnitude or greater since AD 1500 averaged five per century, but between 1900 and 1965 there were scarcely any, giving 'the appearance of a world-wide wave of volcanic activity' about the time of the Little Ice Age (Lamb 1972, p. 423).

The occurrence of temperatures below freezing leaves a permanent anatomically distinct record in tree rings. Frost early in the growing season causes changes in the early-wood part of a tree ring; frost late in the growing season causes damage to the small thick-walled cells of the latewood. The remarkable coincidence of frost-ring dates within two years of the eruptions of Krakatoa in 1883, Mt Pelée and La Soufrière in 1902, Katmai in 1912 and Agung in 1963 suggested that the influence of atmospheric veil effects provided the setting for widespread and severe frost damage (LaMarche and Hirschboeck 1984).

Data from bristlecone pines at seven locations in the western USA make it possible to identify 'notable frost-ring events' which caused widespread latewood damage. These data were compared with Lamb's geographically weighted Dust Veil Index (Figure 11.6), according to which nineteen major volcanic events occurred in the period 1500 to 1963. Ten of these were followed by notable frost events in the same year or within one or two years afterwards. In ten other cases, however, a frost occurred without a preceding volcanic event and seven volcanic events had no associated frost event. The statistical probability of the relationship between frost events and volcanic events was tested on the assumption that the probability of such events within three-year periods is equal to the observed average frequency. The observed number of joint occurrences was found to be six times that expected by chance. LaMarche and Hirschboeck (1984) suggested that the anomalous circulation over western North America in the years following great eruptions could have involved southerly displacement of the general westerly flow and/or more frequent developments of an upper-level trough accompanied by occasional outbreaks of unseasonably cold air from higher latitudes, synoptic situations typical of wintertime occurring in late spring and early autumn.

There is not a one-to-one relationship between volcanic eruptions and frost events. Frost rings could be associated with quite other causes; there have been frost events unrelated to dust veils, as in 1941, and conversely notable eruptions with no associated frost rings in the USA. There is also the likelihood that the eruption chronology is not adequately specified or complete; Lamb warned that the reliability of the volcanic record diminished with age.

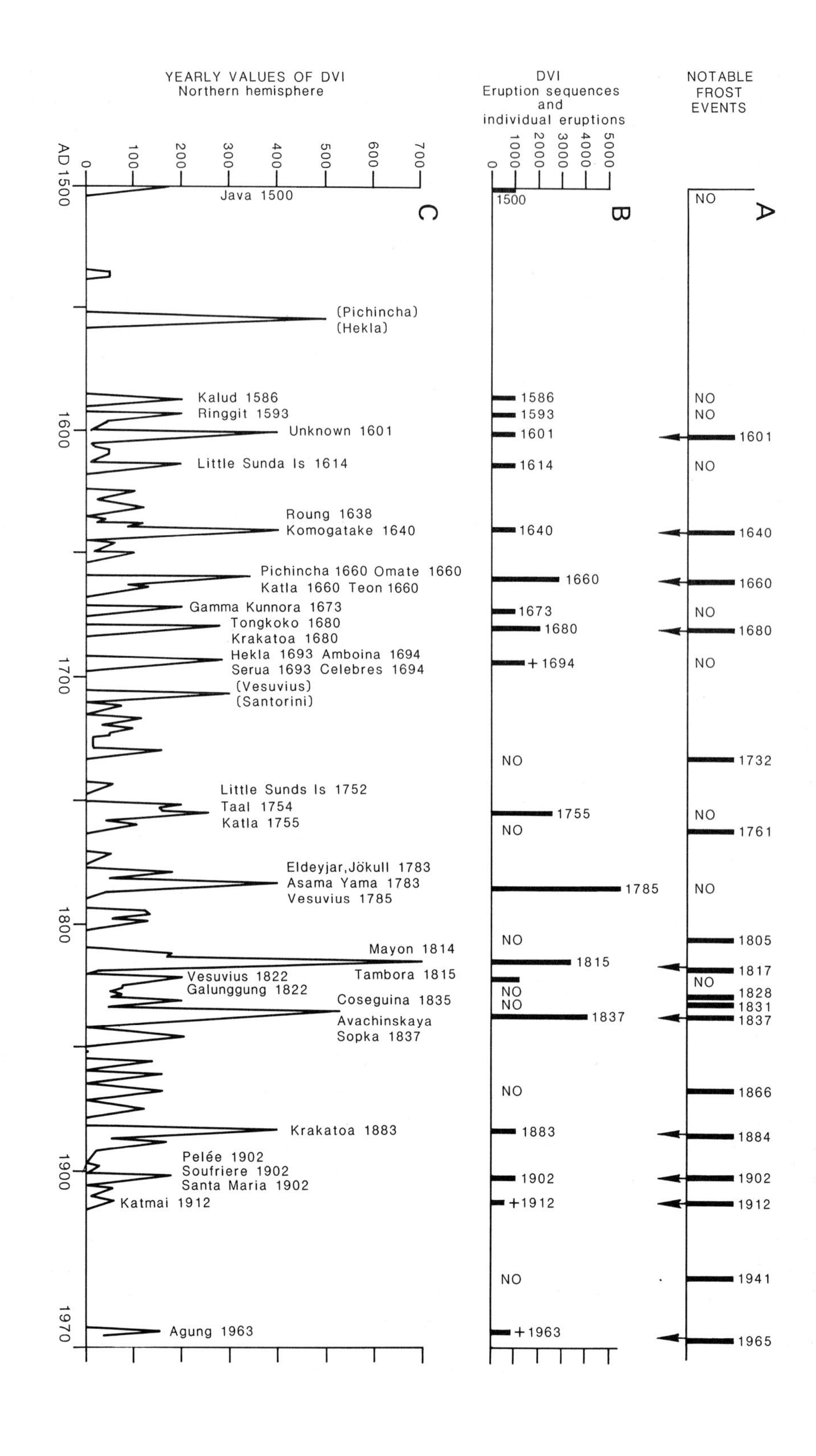

YEARLY VALUES OF DVI
Northern hemisphere
DVI
Eruption sequences and individual eruptions
NOTABLE FROST EVENTS
A
B
C
AD 1500
1600
1700
1800
1900
1970
0
100
200
300
400
500
600
700
1000
2000
3000
4000
5000
Java 1500
(Pichincha)
(Hekla)
Kalud 1586
Ringgit 1593
Unknown 1601
Little Sunda Is 1614
Roung 1638
Komogatake 1640
Pichincha 1660 Omate 1660
Katla 1660 Teon 1660
Gamma Kunnora 1673
Tongkoko 1680
Krakatoa 1680
Hekla 1693 Amboina 1694
Serua 1693 Celebres 1694
(Vesuvius)
(Santorini)
Little Sunds Is 1752
Taal 1754
Katla 1755
Eldeyjar,Jökull 1783
Asama Yama 1783
Vesuvius 1785
Mayon 1814
Tambora 1815
Vesuvius 1822
Galunggung 1822
Coseguina 1835
Avachinskaya
Sopka 1837
Krakatoa 1883
Pelée 1902
Soufriere 1902
Santa Maria 1902
Katmai 1912
Agung 1963
1500
1586
1593
1601
1614
1640
1660
1673
1680
+1694
NO
1755
NO
1785
NO
1815
NO
NO
1837
NO
1883
1902
+1912
NO
+1963
NO
NO
NO
1601
NO
1640
1660
NO
1680
NO
1732
NO
1761
NO
1805
1817
NO
1828
1831
1837
1866
1884
1902
1912
1941
1965

Bryson and Goodman (1980) examined the listing of 6000 historically recorded eruptions on file at the Centre for Climatic Research at the University of Wisconsin. They noted in particular that more eruptions of all magnitudes were recorded in the early twentieth century and during the 1970s than during the period 1925 to 1955. From 1945 to 1970, annual eruption numbers doubled and aerosol optical depth also doubled, strongly suggesting that the increase in ejecta in the stratosphere might be responsible for the lowering of temperature during these decades. As previous chapters have shown, glacial advances during the 1960s and 1970s were widespread globally and it would seem quite likely that earlier periods of advance within the Little Ice Age could be explained in the same way. Bryson and Goodman made use of a much longer data source than Lamb's, extending back 10,000 years, and they considered it represented no more than 1 per cent of the volcanic events during that time.

Newhall and Self (1982) introduced a qualitative Volcanic Explosivity Index (VEI) using volcanological data only, assigning indexes of 0 to 8 to over 8000 eruptions. This VEI is primarily a function of volume of ejecta and eruptive column height and expresses the potential to eject dust and gas into the atmosphere. However it contains no correction for latitude or for elevation of the source vent, both of which are important from the climatic point of view. While volcanic dust injected into the lower stratosphere in low latitudes is gradually spread around the whole globe, dust originating in high latitudes does not spread in significant quantities below about 30° latitude in the hemisphere of origin and tends to maintain its concentration mainly between latitudes 60° and 90° (Lamb 1972, p. 216). A temperature record from 1755 to 1977 based on data from 2384 stations was used by Self *et al.* (1981) for comparison with the best-documented eruptions from Newhall and Self (1982). They evaluated the temperature data hemispherically and for each of six 30° zones of latitude. Fluctuations in temperature which could be associated with volcanism did not stand out as being greater than the general background fluctuations when there were no eruptions recorded, and this was true of all the zones considered. Large eruptions like Tambora, Krakatoa, and Santa Maria 1902 were found to be associated with small temperature decreases of the order of 0.2 to 0.5 °C on a hemispherical scale for periods ranging from one to five years. The maximum change in temperature after some eruptions appeared to lag by up to three years, going from equatorial to polar latitudes. Decreases in temperature of up to −1.5 °C occurred within individual high-latitude zones. After the Agung eruption in 1963 there was a fall in temperature of 0.8 °C north of latitude 60°N, whereas after other eruptions no fall in temperature was apparent.

It has often been suggested that a series of major eruptions might have a cumulative effect on temperature but the closely spaced series of eruptions between 1881 and 1889 including Krakatoa, and between 1902 and 1903 including Mt Pelée, La Soufrière and Santa Maria, were found to have had no greater effect on temperature than a single large eruption. Just as surprisingly, eruptions ranging from moderate to very large seem to have had essentially similar effects on temperature. This could be a reflection of the

Figure 11.6 Frost rings in bristlecone pines in the western USA and major volcanic eruptions: (a) dates of notable frost ring events; arrows indicate associated eruptions. 'No' indicates absence of major frost events at time of major eruption; (b) dust veil index (Lamb 1970) and dates of eruption or eruption sequences of 1000 or greater. 'No' indicates apparent absence of major eruptions corresponding to frost ring events shown in (a); (c) integrated yearly dust veil index, with names of volcanoes and dates of major eruptions. (From LaMarche and Hirschboeck 1984)

composition of the material ejected by the different volcanoes. Smaller, sulphur-enriched explosions such as those of Agung in 1963 and Fuego in 1974 may have ejected material relatively rich in SO_2 and HCl, compared with others such as Krakatoa and Santa Maria which ejected larger quantities of silicic dust (Rampino and Self 1982).

Detailed studies of the Fuego eruption in 1974 and that of Mt St Helens in 1980 have led to significant changes of view about the nature of atmospheric hazards caused by volcanic explosions (Rose *et al.* 1983). It is estimated that the mass of sulphur injected into the atmosphere by Fuego was one or two orders of magnitude higher than the mass of silicates injected and that this eventually contributed to the formation of a sulphate layer in the stratosphere capable of persisting for several years (Murrow *et al.* 1980). The Mt St Helens eruption released very large quantities of highly fragmented silicic ash particles but these were removed from the atmosphere very rapidly by a process of particle aggregation. This process is not yet fully understood but probably involves moisture or electrical forces. The importance of aggregation, which is most likely to affect finer particles, was not recognized earlier partly because such aggregates are generally disrupted on impact and partly because grain size studies of tephras have mostly been concerned with larger particles. The mass of small silicates which reach and remain in the stratosphere for any length of time may be small compared with the mass of sulphur even in the case of eruption of highly explosive magma poor in sulphur, such as that of Mt St Helens.

The Fuego eruption released much more sulphur than that of Mt St Helens, although it was substantially smaller, mainly because of the high sulphur content of the Fuego basalt. The 1982 eruption of El Chicon in Mexico was similar in magma volume to that of Mt St Helens but this also released much more sulphur. Monitoring of some of these recent eruptions has revealed that many times more sulphur is released in explosive eruptions than would be indicated by the mass of exploded lava. For instance, the mass of sulphur released by Mt St Helens between 20 May and 21 December 1980 was several times that supplied by the mass of magma erupted. It appears that a magma body about three times the volume of that erupted is necessary to account for the observed release of volatiles. It is concluded that shallow magma bodies which are the source of eruptions must contribute the remainder of the sulphur. This means that estimates of volatiles released to the atmosphere based on volumes of erupted magma will be far too low.

We must now accept that the primary atmospheric effects are caused by volcanic sulphur which is converted to sulphuric acid to become the dominant aerosol. Its main effect seems to be cooling of the lower troposphere by back-scattering of solar radiation. The quantity of sulphur injected into the stratosphere is more important from the climatic point of view than the overall magnitude of eruptions. We have already noted the incomplete nature of the historic record and now it seems that in any case those large explosions which are most likely to be recorded are not necessarily the most important. This provides a possible explanation for the lack of frost rings registered against certain historic eruptions by LaMarche and Hirschboeck (1984).

An extremely valuable source of information about the amount of sulphate in the atmosphere in the past is available in the form of measurements of the acidity of annual layers in cores extracted from polar ice sheets. Volatiles entering the stratosphere are oxidized photochemically and dissolve in water droplets to form sulphuric acid which is eventually washed out in precipitation. After eruptions the acidity of snowfall therefore increases significantly. Seasonal variations in the value of 18 O provide a sequence of couplets of high and low value which mark annual increments of deposition (Bradley

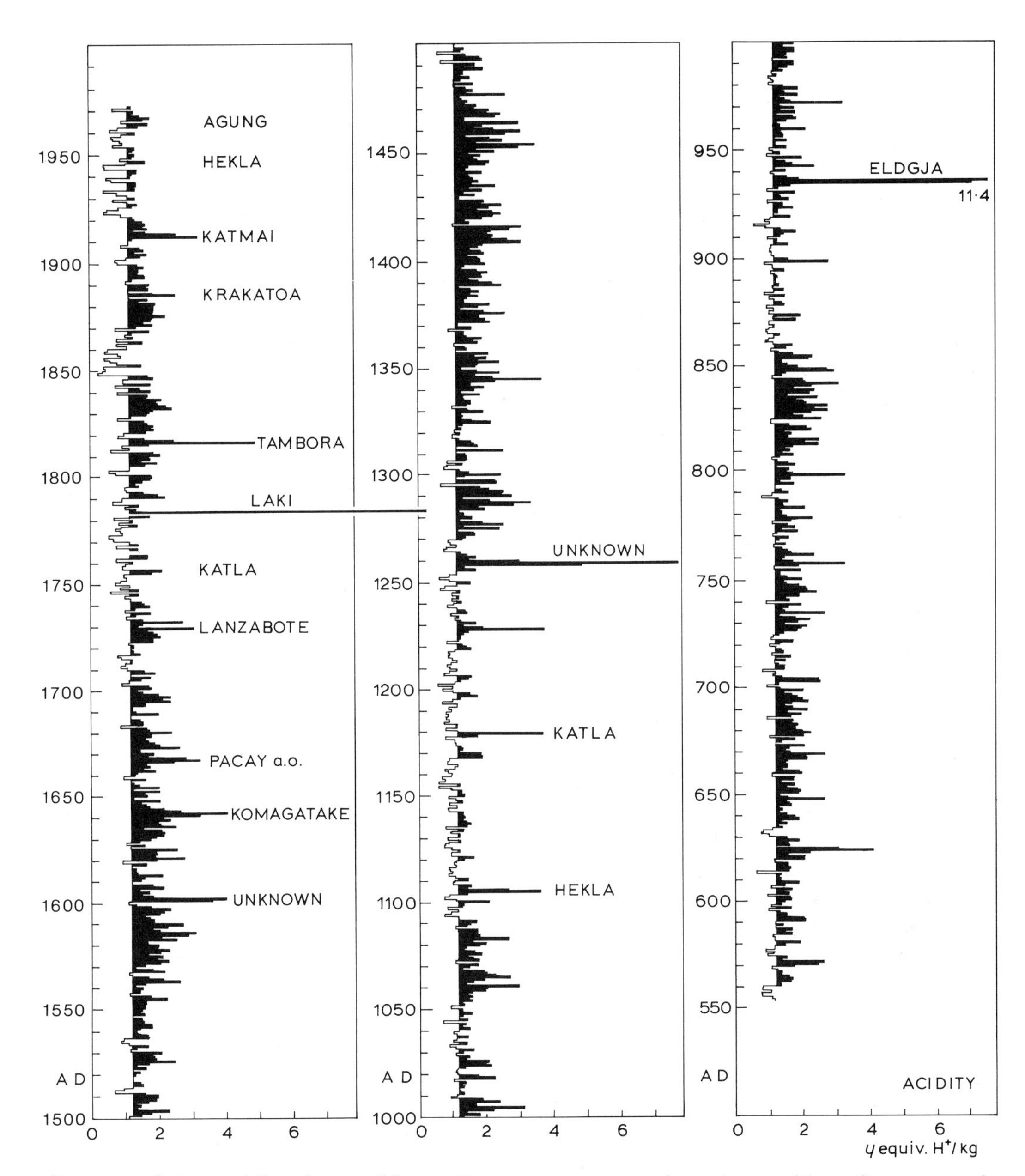

Figure 11.7 Mean acidity of annual layers from AD 1972 to 553 in an ice core from Crete, central Greenland. Acidities above the background are 1.2 ± 0.1 equivalent H per kg ice due to fallout of volcanic acids, mainly H_2SO_4 from eruptions north of 20°S. The ice core is dated with an uncertainty of ± 1 year over the past ninety years, increasing to ± 3 years at AD 553. (From Hammer *et al.* 1980)

1985, pp. 136–46). These allow accurate dating down a core, so long as it is taken in a locality where wind scour of snow is not severe and melting and refreezing does not take place. Acidity variations from layer to layer can be measured by their electrical conductivity (Hammer 1980).

A 404 m-long core from Crete, central Greenland (71°N, 37°W), has been dated to an

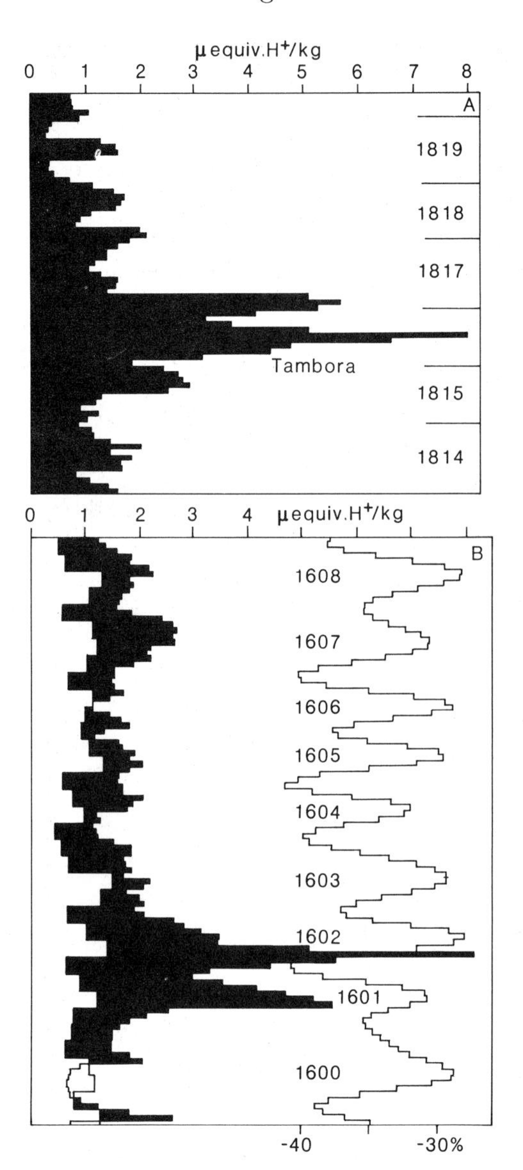

Figure 11.8 The acid fallout found in annual layers in the Crete core (black curves) from two major eruptions: (a) the Tambora eruption in Indonesia, which began in 1815, the fallout in Greenland peaking in 1816, 'the year without a summer'; (b) the fallout, which lasted until the end of 1602, from an eruption in AD 1600 or early 1601 at an unknown locality. (From Hammer *et al.* 1981)

accuracy of ± 1 year through the last 900 years and ± 3 years back to AD 553 (Hammer *et al.* 1980, 1981). Individual volcanic events show up clearly (Figure 11.7). It is possible to pick out the effects of well-known eruptions such as Agung, Katmai, and Krakatoa but it is immediately clear that a very large number of eruptions are recorded in the ice which were previously unrecorded. One of these is illustrated in Figure 11.8b. Many eruptions have no doubt taken place in isolated localities. Astronauts now report eruptions seen from spacecraft and from their counts it has been estimated that one-quarter to one-third of the eruptions that occur have not been brought to the notice of the Smithsonian Institute (private communication from Gordon Wells, NASA, 1985) (Plate 11.1).

Estimates can be made of the total sulphate production of recent eruptions of which the

Plate 11.1 The volcano Iti Boleng, on the island of Andonora, Indonesia, photographed in eruption by the Shuttle crew in the autumn of 1983. This eruption was otherwise unreported. (NASA Astronaut Hasselblad)

source is known, using the distribution patterns of fallout from bomb tests. This makes it possible to compare the relative output of ash and sulphate aerosols and reveal, for instance, that while silicate ash was produced by Tambora, Krakatoa and Agung in the ratio of 150:20:1, sulphate production was in the ratio 7.5:3.1 (Rampino and Self 1982). Calculations of the magnitude of volcanic eruptions in terms of global acid fallout are most satisfactory for those in high northern latitudes. Some estimates may be out by factors of two or three. Some acid could be neutralized before deposition or wind patterns especially favourable to deposition in Greenland could lead to over-estimation.

The Crete core shows clearly that the impact of volcanism has varied significantly over time. Periods of low volcanic activity are marked in black on Figure 11.9. Fifty-year average values of the Crete acidity profile are compared with three sets of northern hemisphere proxy climatic data and a generalized temperature curve based on the three together. There is a good general correlation between periods of below-average volcanic activity and above-average temperature. The quietest period volcanically was from AD 1100 to 1250, that is in the medieval warm period. The most active period volcanically came between AD 1250 and 1500 and between AD 1550 and 1700, suggesting that it had an important role in the causation of the Little Ice Age. This impression is considerably strengthened if the periods of glacial expansion within the last century or so are viewed against the isotopic records from ice cores (Porter 1981, 1986, and Figure 11.10).

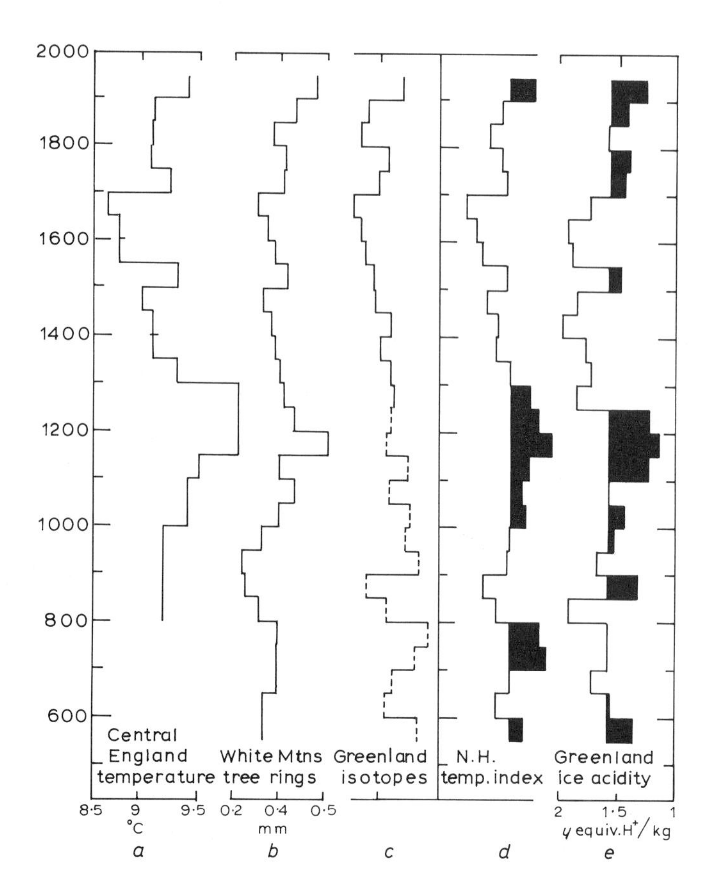

Figure 11.9 The correlation between temperature variations in the northern hemisphere and volcanic activity: (a) central England temperature estimates; (b) tree-ring width in the White Mountains of California; (c) long-term isotopic variations in Greenland; (d) a northern hemisphere temperature index (no scale), formed by combination of (a)–(c); (e) Greenland acidity from the Crete core profile, with 50-year average values. Warm periods are shown in black. (From Hammer *et al.* 1981)

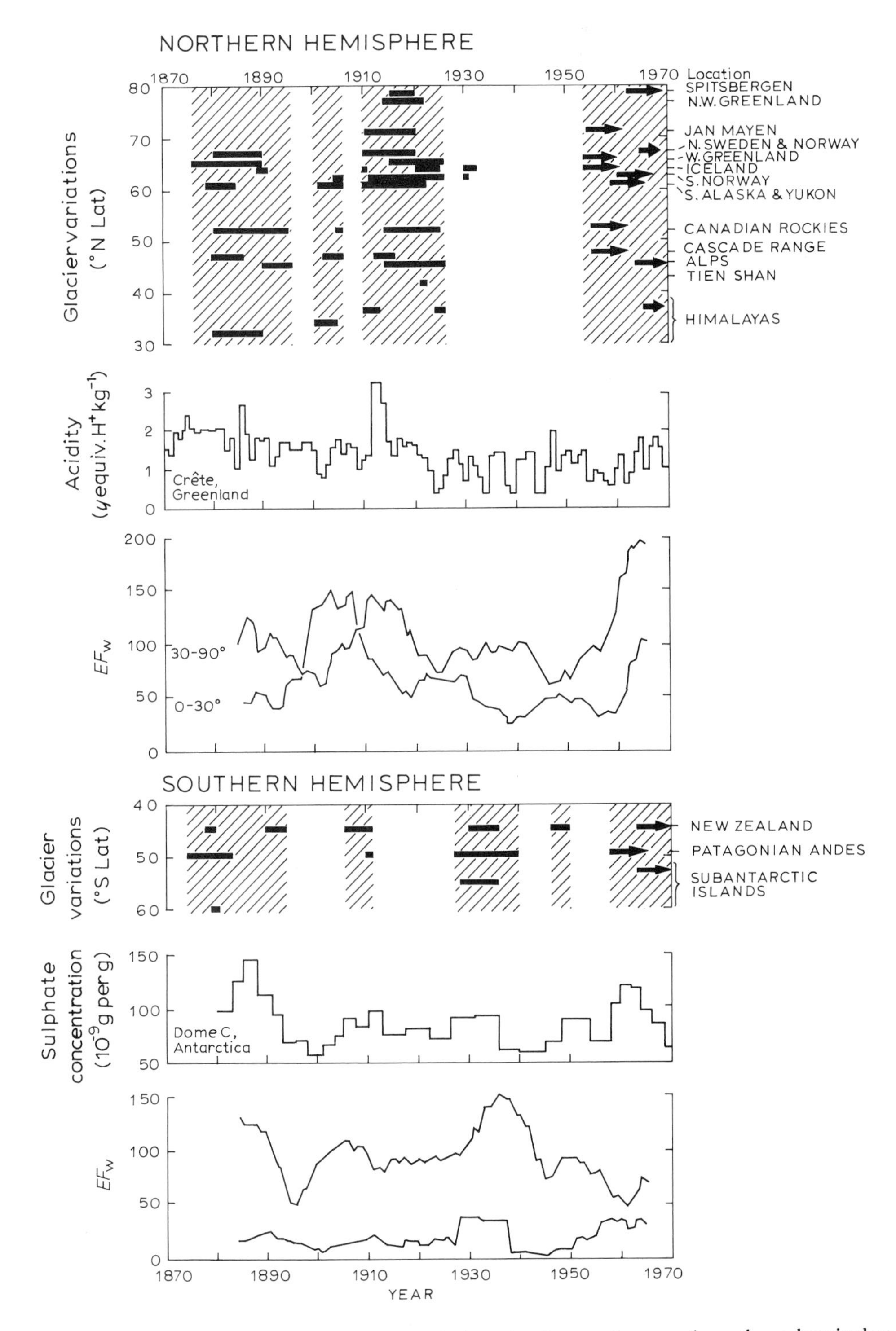

Figure 11.10 Comparison of recent glacier variations in the northern and southern hemispheres, with acidity profiles from polar ice sheets (from Hammer *et al.* 1980 and Delmas and Boutron 1980) and frequencies of recorded eruptions (from Hirschboeck 1980). Bars represent intervals within which glacial advances culminated. (After Porter 1981)

11.4 Assessment of the causes of Little Ice Age and earlier Holocene glacier advances

Global climates in the course of the Holocene have been influenced by orbital forcing. In the Early Holocene this involved greater inputs of solar radiation than at present in June, July and August, that is in the northern hemisphere summer and southern hemisphere winter. The difference in the seasonal inputs of solar radiation may help to explain the extension of vegetation zones into higher latitudes in the northern hemisphere about 10,000 years ago and, conceivably, the greater relative scale of earlier Holocene glacial advances in the southern hemisphere than in the northern hemisphere (11.2).

In the course of the Holocene, solar orbital parameters moved towards those of the present day. In any case, alternative explanations to orbital forcing have to be found for the shorter-term fluctuations, each lasting only a few centuries, that were involved in the glacier phases of the Holocene. Solar variability (11.3.2.2), which may well have been on this timescale could, it seems, have lowered temperatures globally to an appropriate extent. This in turn would have caused continental snow cover and sea ice to extend and no doubt there would have been consequent changes in atmosphere/ocean circulation patterns, atmospheric CO_2 concentrations and the distribution of global precipitation of the kind considered by Weyl, Broecker and Bjerknes (11.3.1).

Volcanism offers an attractive possible explanation for fluctuations of glaciers on a decadal scale within the cold phases, though its effectiveness has yet to be substantiated conclusively (11.3.2.3). The occurrence of longer-term variations in the incidence of volcanic eruptions, causing clustered injections of aerosols into the high atmosphere over centuries, cannot be altogether discounted.

Chapter 12

Consequences of the Little Ice Age climatic fluctuation

Some of the consequences of the impact of the Little Ice Age have already become apparent in earlier chapters but the range of effects has been much wider than has so far been indicated. Even a sketch, as this must be, of the variety of impact on the natural and human environment of series of temperature changes over 1 or 2 °C, such as that which made up the Little Ice Age, has to be extremely selective.

12.1 Physical consequences

12.1.1 SEA-LEVEL

The relationship between the build-up of great ice sheets and the drop in worldwide sea-level is well known. The possible contribution of small glaciers to oscillations of sea-level has received little attention. A rise of level of 10 to 15 cm appears to have taken place during the last 100 years (Gornitz *et al.* 1982). A part of this is attributable to thermal expansion of sea water; the rest has generally been accounted for by the melting of polar ice. But recent studies suggest that the mass balance of the Antarctic ice sheet, which represents 85 per cent of the global ice cover, is probably not diminishing and may be building up, and that the Greenland ice sheet, representing 12 per cent of the total ice, is close to balance (Meier 1984). Estimates based mainly upon long-term volume change data and the results of hydrometeorological mass balance models (such as that of Tangborn 1980, see Chapter 8) have been used to provide data concerning the contribution to the oceans of glacier ice other than that of Antarctica and Greenland (Meier 1984). The smaller glaciers seem to account for one-third to one-half of the observed rise in sea-level, though it must be noted that the error band is large. If this estimate is correct, we must assume that in the course of the Holocene, small-scale oscillations have been superimposed on the general rise known to have occurred because of the shrinkage of the large ice sheets. Some models taking this into account have been put forward, such as that of Fairbridge (1976, Figure 5).

We must note, however, that the old idea of a worldwide, uniform sea-level rise following the last glacial maximum, the eustatic rise, is now known to be untenable. The rise in sea-level due to melting has not been uniform over the globe. Ocean floors, continental shelves and land masses respond to varying extents to ice and water loads and, moreover, the shape of the ocean surface itself is neither uniform nor unchanging (Clark *et al.* 1978). In detail, sea-level curves differ according to their location and relative proximity to ice sheets (e.g. Inman 1983). Ice melt has been of predominant importance during much of the Holocene, so that the most common set of sea-level curves shows a rapid rise of about 1 m per century between 10,000 and 6000 BP, followed by a much more gradual rise up to the present. It is upon such curves that small fluctuations due to the Little Ice Age and similar-scale events must be assumed to have been imposed. It has been suggested that sea-level around parts of the North Atlantic may have been slightly raised by melting during the medieval warm period so that levels in the thirteenth century may have been about 50 cm higher than around AD 700 (Tooley 1978, Lamb 1980).

Small changes in sea-level are likely to be most influential on low coasts, such as those of Lancashire, the Wash or the Netherlands, but such coasts are also sensitive to changes in sediment supply, tide height, vegetation and storminess. It seems likely that there may have been an increase in storminess in the North Atlantic sea area soon after AD 1200, that is at the time of the appearance or reappearance of sea ice around Iceland (see 2.1). Very severe storms, causing surges, lead to abnormally high tides on low-lying coasts. Some, but by no means all such events, are recorded in historical documents because of damage caused or because of erosion or alteration by massive transport of sand (Lamb 1985). Detailed study of the stratigraphic record of storm incidence is now under way, but many difficulties of correlation, dating and interpretation remain to be solved (Tooley 1985). At present, the obviously close correlation between climate, sea-level and coastal changes can only be demonstrated in a general way.

12.1.2 SNOWLINES, SNOWPATCHES AND RATES OF EROSION

During the Little Ice Age, snowlines were lower than at present and snow cover lasted longer into the spring. We noted in Chapter 6 the historic evidence for lengthy snowlie in Switzerland and in 8.5 field evidence of extensive Little Ice Age snow cover on Baffin Island. In the Andes of Ecuador, the high mountains, such as Corazon, are known to have been perennially snowcapped in the 1500s and there is also documentary evidence for similar conditions around 1740 and into the nineteenth century. The snowline in the Western Cordillera rose almost 300 m between the mid-eighteenth century and 1975; that in the Eastern Cordillera by about 250 m (see Figure 12.1 and Hastenrath 1981). Similar figures come from many other mountain areas.

Areas without present-day snow or ice cover were also affected. While the theoretical mid-twentieth-century snowline in the Ben Nevis area of Scotland was at about 1600 m (Manley 1949), travellers in Scotland in the seventeenth and eighteenth centuries referred to permanent snow cover on the Cairngorm summits at around 1200 to 1300 m, the difference in height implying that temperatures were 2 to 2.5 °C lower than those of the mid-twentieth century (Lamb 1985). John Taylor (1618) visited Deeside in about 1610 and recorded that 'the oldest man alive never saw but snow on the tops of divers of these

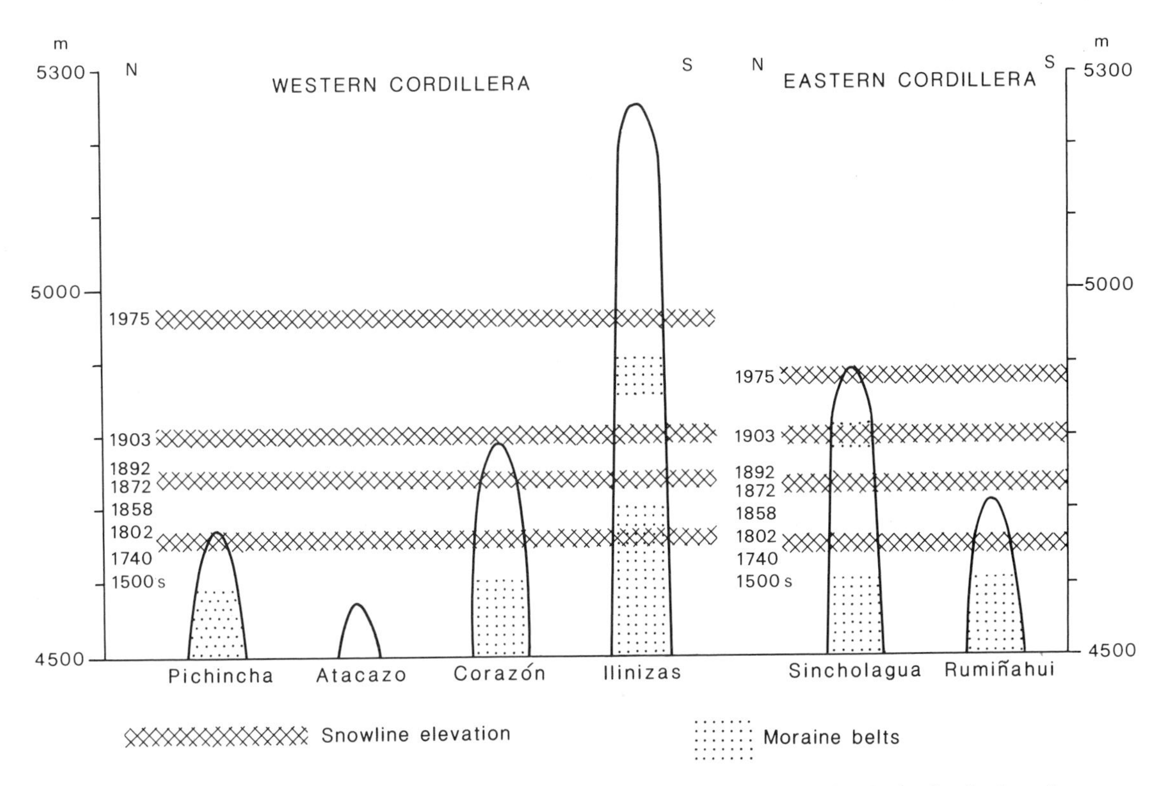

Figure 12.1 Schematic diagram to show snowline variations between 1500 and 1980 in the Andes of Quito, Ecuador. Cross-hatched bands show snowline elevation and dotting indicates moraine belts. (From Hastenrath 1981)

hills, both in summer as well as in winter'. Bishop Pococke who, it will be recalled, had visited Chamonix in 1741 with Wyndham (see 4.2), also noted lying snow when he went north of Fort William in July 1760 and wrote that 'we had the high mountain Ben Nevis to the south on which the snow lies in holes fronting the north the whole year' (quoted Manley 1971). Thomas Pennant (1771) visited not only the Cairngorms, which he described as 'naked summits of surprising height . . . many of them topped with perpetual snow' but also Ben Wyvis in Ross-shire, which reaches only 1045 m. He found that 'snow lies in the chasms of Ben Wyvis in the form of a glaciere throughout the year'. Semi-permanent snowbeds are still to be found in north- and northeast-facing gullies of Ben Nevis (1343 m), though these have not survived all the warm summers of the twentieth century (Manley 1969), but they are not found on Ben Wyvis, nor are the Cairngorm summits capped throughout the year. It is clear that snow was more widespread and longer-lasting in Scotland during the Little Ice Age and the possibilities have even been raised that glaciers may have formed in the corries of the Cairngorms (Sugden 1977). This has not been established but Lamb (1977) calculated that a fall of only 1.3 °C in mean annual temperature in the Scottish lowlands, a fall similar to that shown by the central England record during the Little Ice Age, combined with cold unstable winter weather, would have been sufficient to have brought the snowline down to 1200 m on the mountains and allowed glacier ice to form in some shaded corries (Lamb 1977, p. 526).

If snowlines were lower, snowpatches more permanent and larger and glaciers more extensive in the Little Ice Age, glaciers may have moved somewhat faster and eroded more actively. We have seen that pulses of glacier advance associated with changes in climate on scales of decades within the last few centuries are commonly represented by individual moraine units or even, in exceptional circumstances, by till sheets (Ono 1984). Such geomorphological evidence strongly suggests changes in rate of process.

The Arapaho glacier, in the Colorado Front Range, is a cirque glacier which lost over 30 per cent of its area and thinned considerably in the last century (Waldrop 1964). Lichen data suggest that there were at least three and not more than five intervals of deposition in the Little Ice Age. Debris within the cirque comes partly from rockfalls and avalanches falling on to the ice surface and partly from sub-glacial erosion. As it is possible to distinguish clasts from these sources on a basis of degree of rounding, polish and striations, it has proved possible to estimate their relative contribution to deposition (Rehais 1975). A maximum of 70 per cent of the present glacier load is derived from glacier erosion and the denudation rate in the cirque is estimated to be 95 to 165 mm per 1000 years at the present rate. During the Little Ice Age (Gannett Peak Stage) it is estimated that 88 per cent of the debris was of sub-glacial origin and the rate of erosion may have been of the order of 4920 to 8160 mm per 1000 years. If this denudation rate is corrected for difference in extent of ice cover and the possibility that part of the terminal moraine could be of Audubon age (1850 to 950 BP) the rate for Gannett Peak time still comes out at 1260 to 2040 mm per 1000 years. Though these figures are necessarily tentative estimates, they indicate a considerable increase in erosive power during the Little Ice Age.

The contribution of rockfall and avalanche debris formed a lesser proportion of the whole during the Little Ice Age, but none the less the rate is estimated to have been 290 to 485 m^3 per year during the Gannett Peak Stage as against only 30 to 50 m^3 per year during the twentieth century. The implication is clear, that glacial advance periods during the Holocene have been times of increased glacier erosion and there was simultaneously an increase in the production of debris by mass movement. This last conclusion is supported by many other studies. For example, a morphological study of the relationship between lateral moraines and talus tongues in the Tarfala valley of Swedish Lappland led to the conclusion that the glacier must have received an unusually high input of debris from the surrounding mountain slopes during periods of glacier advance (Ackert 1984).

12.1.3 MASS MOVEMENTS

A classical investigation by Rapp (1959, 1960a, b) drew attention to the importance of mass movements as principal agents of transfer of debris on steep mountain slopes, both erosional and depositional effects being produced. The frequency and magnitude of such occurrences is of obvious importance geomorphologically. There have been a number of collections of historical data, such as the chronicle of avalanches in the district of Davos (Laely 1951) and a two-volume work bringing together information about landslides and avalanches through the ages in Iceland (Jónsson 1957). Much interesting material is included, much of it obviously authentic. But Laely depended in part on local chronicles which have not been examined critically; the sources are not always clear and for the period after 1841 he provided no source references at all. Jónsson also drew his information from miscellaneous sources of all types. In both works the number of incidents increases towards the present, almost certainly because of more voluminous documentation.

In order to obtain a valid notion of the incidence of mass movements over time from documentary sources it is necessary to ensure homogeneity of the data and that the records are both contemporary and authentic. The Norwegian fiscal system had built into it a provision to respond to the incidence of major physical hazards. A petition could be made for tax and land rent relief if a farm was seriously damaged by landslides, rockfalls, avalanches, floods or ice. Crop failure was not included in the list of factors on which an appeal might be based. Only serious damage was considered to provide adequate grounds for reduction. The legal proceedings that had to be employed before relief was granted were rigorous and resulted in the accumulation of substantial official records (Grove and Battagel 1983). The incidence of physical hazards affecting farms in the valleys peripheral to the northern margins of Jostedalsbreen has been extracted from the records for Oppstryn, Nedstryn, Loen and Olden parishes. The results are plotted on Figure 12.2.

The delay between damage and appeals for relief was often brief and less than five

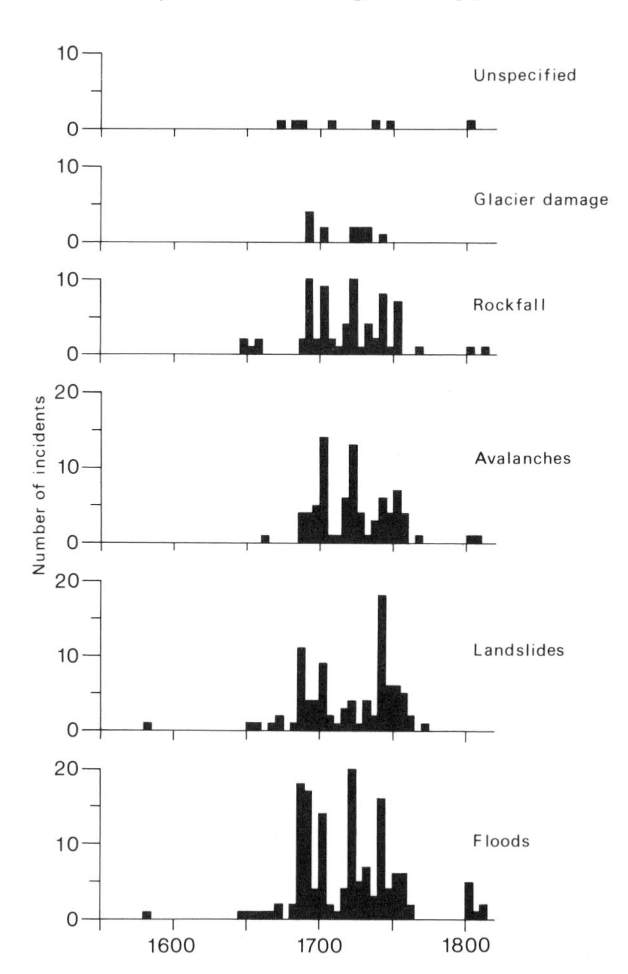

Figure 12.2 The incidence of mass movements and floods, over five-year periods, in the parishes of Oppstryn, Nedstryn, Loen and Olden revealed in *avtak* records. This diagram is not exactly the same as that published in Grove (1972), which was based on the farm history published by Aaland (1932). Reference to original manuscripts revealed that Aaland's version was not completely accurate

years in the great majority of cases. Data have therefore been plotted for five-year periods. In some instances damage was caused by a single incident, in others by a series of incidents or by several kinds of incident occurring more or less simultaneously. The incidence of damage from mass movements and floods was concentrated in the period between 1650 and 1765, during which advancing glaciers invaded the farmed land and swelled to their maximum extent (see 3.1.3 and Grove 1985). There is no lack of documents predating 1650 and no signs of diminishing prosperity (Aaland 1932). The incidence of large-scale mass movements of all sorts as well as flooding increased sharply and was concentrated in particular decades, such as the 1690s and 1740s. It is evident from the documents that the scale of damage was unprecedented. For example, Gutdal farm, which had been assessed at 4 løber 2.5 pund of butter by an official tax commission in 1667 was found by a Court of Inspection in 1693

> to have suffered damage by flooding and landslides on its home fields and utmark with more than half its pasture completely gone and with the rushing river now running through the site of their former outlying barns with the ground covered with small heaps of rock fallen from the mountains ruining a great part of the little still remaining. . . . The cattle pastures are nothing but rocks, scree and scrub swept by rockslides so that little or nothing is left of their summer pastures.

The Gutdal assessment was reduced by nearly 50 per cent to 2 løber 2.5 pund of butter. This sort of account is duplicated many times in the records, though in many cases the damage was attributed to one or two major incidents.

No petitions for relief were made in the late eighteenth century. The tax system changed after 1815, but the collection of miscellaneous incidents made by Aaland (1932), together with inquiries made in Oldendalen, makes it quite clear that the incidence of damage had diminished very markedly and that the lack of petitions was not merely a reflection of the new taxation system.

A period of frequent small-scale mass movements in southwest Norway was independently identified as occurring between 1670 and 1720 by Innes (1985), using lichenometry. Active debris flow formation he dated to the period 1790 to 1860. It is possible that some of the debris flow events in this period coincided with minor glacial advances but, as Innes points out, such flows can be triggered by intense or prolonged rain or by rapid snowmelt and so a correlation with glacial history would not necessarily be expected.

A detailed study of rockfall deposits on the Italian side of Mt Blanc has revealed a series of massive lobes terminating on the valley floors or rising against opposing valley walls and in nearly all cases lying outside the outermost Little Ice Age moraines of nearby glaciers (Porter and Orombelli 1981, and Figure 12.3). The sources of the debris are still marked by large light-coloured scars which stand out against the darker rock faces surrounding them. Lichenometric and dendrochronological dating indicate that the deposits below the Triolet glacier range in age from 150 to 600 years (4.3 and Figure 4.6). Differences in the lichen ages of adjacent boulder zones imply multiple rockfalls with similar distributions. Most of the major rockfalls for which there is evidence of age date from the Little Ice Age. The town of Courmayeur is built on the site of an enormous rockfall which seems, according to ^{14}C dates, to have been emplaced about 2500 BP. Two other major remnants of rockfall lobe are undated but assigned on general grounds to the Late Glacial between about 11,000 and 10,000 BP. Most of the dates come

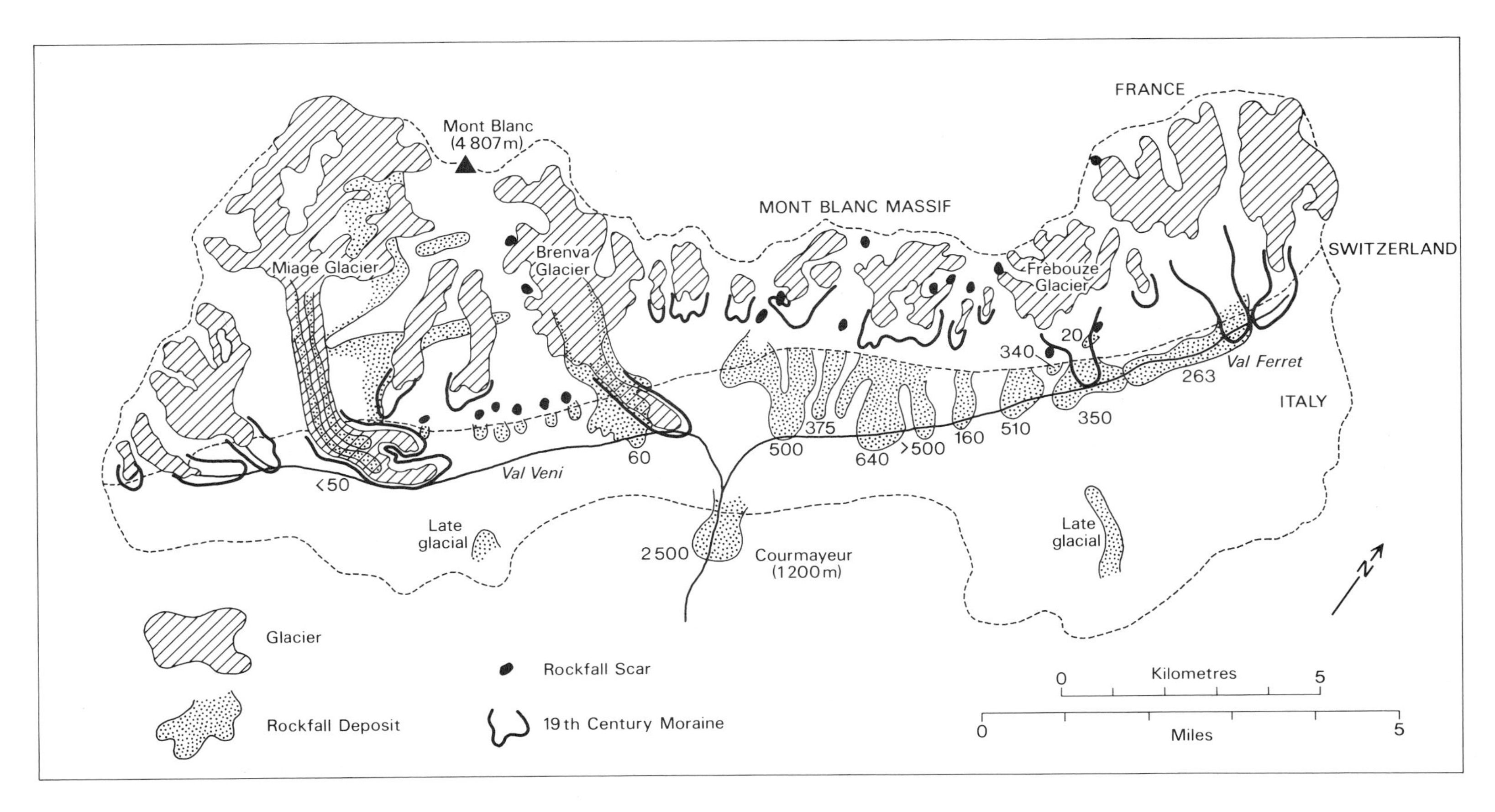

Figure 12.3 The distribution and approximate age of major rockfall deposits in Val Veni and Val Ferret, adjacent to the southern flank of Mont Blanc. (From Porter and Orombelli 1981)

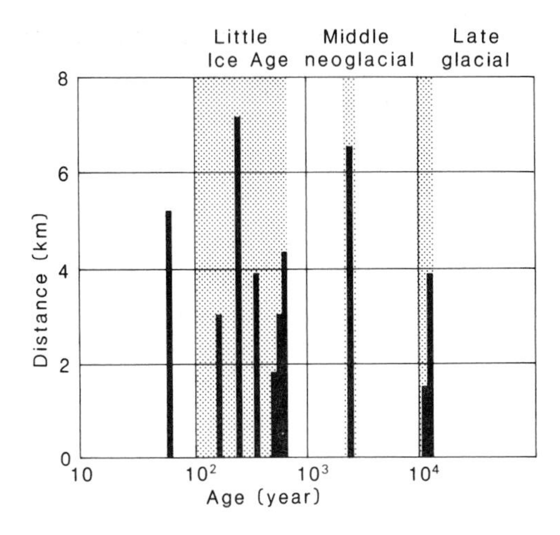

Figure 12.4 The distance travelled by giant boulders originating in rockfalls from the southern flank of Mont Blanc shown as a function of the time of deposition. (From Porter and Orombelli 1981)

from the superficial deposits in the valley and relate to the Little Ice Age (Figure 12.4). It has to be recognized that other rockfalls could have occurred between 600 and 2500 BP or between 2500 and 10,000 BP and that these could be concealed beneath the Little Ice Age deposits. Here, as in Norway, minor falls continue to take place, but the available evidence suggests that all the major rockfalls occurred during times of increased glacier activity. While Porter and Orombelli did not themselves find independent evidence of glacier expansion between 2300 and 3000 BP in the upper Val d'Aosta, two such periods of expansion were recognized in Switzerland by Gamper and Suter (1982) in their survey of Holocene glacial fluctuations in Switzerland (see Figure 10.8). During this time it seems that a well-marked period of solifluction was preceded and followed by soil formation in the western Alps.

Increased incidence of mass movements and floods was not retricted to the immediate vicinity of glaciers during the last few hundred years. It emerges from inspection of the complete set of *avtak* records from Sunnfjord Fogderi, an administrative region stretching from Jostedalsbreen to the Norwegian coast around 61°30′N, that environmental deterioration set in decisively in the last three decades of the seventeenth century. During the eighteenth century, incidents causing severe damage occurred in several groups of years; difficult conditions were widespread with increased danger from all types of physical hazard. The ten years from 1740 to 1750 brought the greatest toll. Flooding was the most prevalent form of hazard with 1743 especially disastrous, heavy rain in December triggering landslides not only in Sunnfjord but right across western Norway from Ryfylke to north of Nordfjord (Grove and Battagel 1983, and Figure 12.5).

12.1.4 FLOODS AND FLUVIAL ACTIVITY

Details included in *avtak* reports from both the Nordfjord and Sunnfjord areas show clearly that glacial streams caused both pronounced erosion in some places and increased

Figure 12.5 Farms in Norway damaged by heavy rains in December 1743 according to information assembled from *avtak* reports. The area shaded is known to have been affected by the storm; it may also have affected areas further east

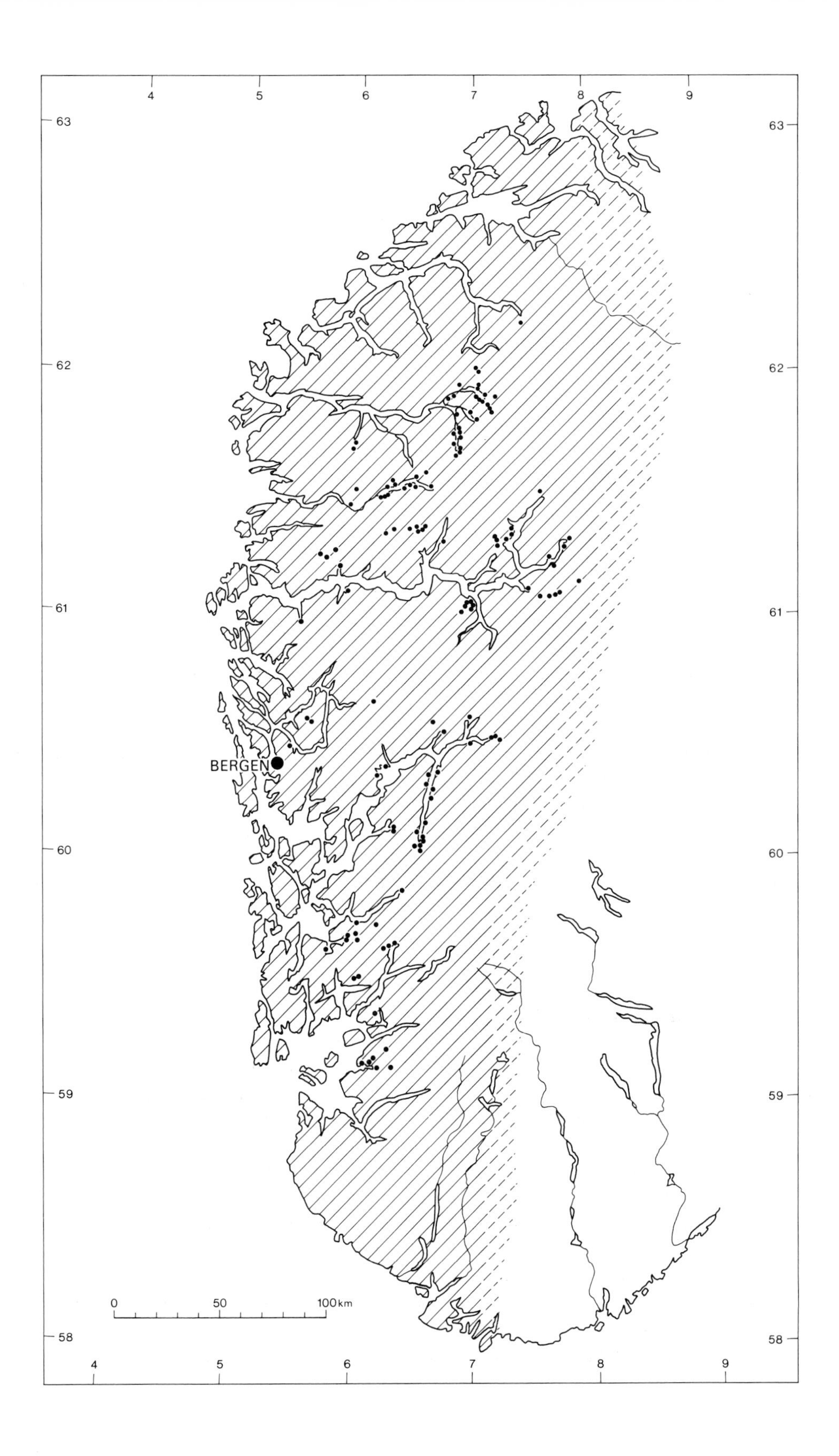

4
5
6
7
8
9
63
62
61
60
59
58
BERGEN
0
50
100 km

deposition in others. It has been claimed by some writers that aggradation of proglacial streams is especially pronounced in periods of glacial advance, while others have seen it as associated with glacial retreat. Observations made during the advance of the Glacier des Bossons between 1968 and 1975 confirm the impression given by the historical records that both initial degradation and its subsequent aggradation can occur in periods of glacial expansion (e.g. Maizels 1979). An increasingly close braided network was created as the sediment came to exceed the competence of the river to carry it. In the short term it seems evident that net aggradation or degradation may occur either during glacial advance or retreat, according to the balance between meltwater load and meltwater competence. In the longer term it might be suspected that periods of lower temperature, glacial advance, increased flooding and increased sediment transport might show up in the form of sedimentary evidence of changes in fluvial regime and sediment yield, corresponding to the Little Ice Age and previous periods of similar character during the Holocene.

Varve sequences in lakes in Banff National Park, Canada, show that sedimentation rates were generally high between the late twelfth and late thirteenth centuries and between about AD 1600 and 1950, but substantially lower in the three intervening centuries. High sedimentation rates thus correspond to periods of increased ice extent and decreased temperature, and lower sediment rates to periods of decreased ice extent and higher temperatures (Leonard 1985). Year-to-year variation in sedimentation rates seems to be controlled dominantly by summer temperature variation but, on longer timescales of decadal duration, increased sedimentation rates were found to be associated both with moraine-building periods and with those in which the ice receded very rapidly.

Studies of sequences of sediments from the seventeenth to the nineteenth century in Carparthian valleys indicate that phases of higher flood frequency lead to the formation of new valley fills. The bases of these fills are composed of gravels and sands of channel facies, with coarse pebbles in the bottom horizons. Fine overbank sediments and fills of meandering palaeochannels occur on top of these channel fills (Starkel 1983). Cyclic variations in river discharge and sediment load during the Holocene, each lasting about 2000 years, have been discerned in Europe; parallel Holocene fills have apparently been found in many European river valleys, though after about 3500 BP fill formation does not seem to have been generally synchronous, probably because of human interference and increasing input of suspended load from arable land in some areas (Starkel 1983).

In the lower Danube between Linz and Vienna, five major alluvial bodies have been recognized and dated by ^{14}C and dendrochronology on logs and *in situ* tree stumps. After 5 to 10 m of end-Pleistocene dissection, a thick fill was deposited between 9800 and 8100 BP, followed by a similar cut-and-fill cycle with alluviation between 7600 and 5100 BP Both these deposits are at similar elevations to the last Pleistocene fluvio-glacial terrace, but later fills generally form lower terraces. The subsequent depositional unit is dated about 4500 to 3000 BP and there are two other minor fills formed between about 2700 BP (750 BC) and 1100 BP (AD 850). Accumulation of the modern floodplain began shortly before AD 1400 (Butzer 1980). Generalized phases of higher flood frequency seem to have affected central Europe between 10,800 and 10,000 BP, 8500 and 7500 BP, about 6500 to 5900 BP, 5000 to 4500 BP, 2800 to 2200 BP and in the sixteenth to nineteenth centuries AD.

Some of the moister periods associated with the formation of valley fills were also periods of gullying and deposition of small alluvial fans, for example in Poland between

8400 and 7500 BP and in Scandinavia between 8400 and 7700 BP. Starkel (1983) noted also the formation of landslides in the Pennines and the Carpathians during such moist periods of the Holocene and suggested a similarity with the environment in western Norway during the Little Ice Age (Grove 1972). There appears to be some similarity between periods of fill formation in central Europe and the timing of glacial episodes during the Holocene but the match is by no means exact according to the dating which is so far available. Palaeohydrology is a new and rapidly expanding field (Gregory 1983); a clearer understanding of the relationship between formation of major valley-fill units and changes in pluvial characteristics with episodes of glacial expansion and perhaps of increased mass movement must await more detailed investigations and better dating than we have so far attained.

12.1.5 AEOLIAN ACTIVITY

Cold dry periods during the Holocene have allowed episodic dune formation in high latitudes. A chronology from northern Quebec based on 101 ^{14}C dates from dune palaeosols discloses that aeolian activity has occurred during thirteen periods since the Mid-Holocene (Filion 1984). Three major aeolian intervals are recognized at 3250–2750 BP, 1650–1050 BP, and between 750 BP and the present. Plant colonization was apparently terminated by naturally occurring fires. Many of the palaeosols contain charcoal; some of them show clear signs of having suffered frost cracking or involution before burial by wind deposits. The number of alternations between soil formation, stabilization by plant cover and active aeolian activity decreases from the forest zone towards the arctic tundra. To the north of the forest limit, dune formation evidently resulted from more intense and longer-lasting dominance of wind action. In the forest zone frequent fires caused sudden ecosystem changes in cold periods. The alternation of cool, dry phases and warmer, perhaps moister, conditions was well recorded in the forest zone. Dry, cool conditions in the tundra, especially after 3000 BP, were scarcely interrupted and so dunes remained active.

Very few dates earlier than 3300 BP have been obtained and so the attribution of aeolian processes specifically to the latter part of the Holocene must be considered tentative. The period of aeolian conditions between 3150 and 2750 BP overlaps a time of pronounced glacial expansion in the Alps; that from 1650 to 1050 BP also coincides with glacial expansion in the Alps although it is longer, and the final period corresponds to the Little Ice Age and its medieval forerunner. There seems no doubt that colder periods during the latter part of the Holocene were also periods of greater windiness and dune formation in parts of the Arctic.

12.2 Biological consequences

12.2.1 CHANGES IN PLANT DISTRIBUTION AND FOREST STRUCTURE

Pollen extracted from lake and bog sediments can provide a great deal of information about past biological assemblages. As a pollen assemblage from a particular time and place is a function of the regional flora and vegetation, which are controlled by regional climates, there has to be a relationship between palaeopollen assemblages and past climates. This relationship is complex and the business of extracting information about past climates from the pollen is not simple (Birks 1981, Chapter 9 in Bradley 1985). None the less, the method has proved to be one of the major tools by which changes in vegetation may be traced and past climatic conditions reconstructed. Here we are concerned with tracing the effects of the Little Ice Age and like periods earlier in the Holocene upon the vegetation of the time.

While in general pollen analysis has not proved capable of revealing the response of plant communities to short phases of glacial advance, climatic oscillations of this order have been shown to be clearly distinguishable in pollen profiles taken through bogs near the ice or at about the height of the timberline (e.g. Patzelt 1974, and see Chapter 8). Given certain clearly defined conditions, it has also proved possible to demonstrate that changes of the order of 1 °C in summer temperature can have well-marked consequences for the vegetation of a large region (Webb 1980). In order to reveal vegetational change within time spans of 50 to 100 years it is necessary to work in regions where plants are highly sensitive to change in climatic parameters and for pollen samples to be placed close together at intervals representing, say, 20 to 50 years and for them to be dated with more precision than is possible using radiocarbon. Investigations fulfilling these criteria are possible if there are lakes with varved sediments in areas near vegetational boundaries or where there are species-rich communities where small climatic shifts can change the competitive advantage from one species to another.

The pollen record over the last 3000 years in varves from Marion Lake, Michigan, shows an increase in the percentage of Western hemlock (*Tsuga*), pine (*Pinus*) and spruce (*Picea*) between 1430 and 1860. The increase was not enormous, but comparison with similar data from a lake 40 km to the east and with well-dated and detailed digrams from more distant sites in Minnesota and New York states showed that distinctive changes in vegetational assemblages occurred in all of them at the time of the Little Ice Age (Bernabo 1981). It has also emerged that beech trees (*Fagus*) expanded 78 km westwards in upper Michigan at the same time. It is believed that this is probably a reflection of cooler, moister conditions which came to prevail in the upper mid-west. Another study of the contents of varved sediments, from Hell's Kitchen Lake in north-central Wisconsin, also revealed an increased percentage of white pine (*Pinus strobus*) and hemlock pollen and yellow birch seed (*Betula*) between 600 and 150 BP. This has also been interpreted as an assemblage change due to moist conditions during the Little Ice Age (Swain 1978). A similar example of vegetational response is revealed by pollen studies along the prairie margin in Minnesota, which have shown the spread of maple-basswood (*Acer-Tilia*) bigwoods forest during the last 400 years (Grimm 1983).

In earlier chapters reference was made to the records of summer temperature extracted from trees in Switzerland (Chapter 6) and those of the oscillation of summer conditions to be found in the tree-ring sequences of North America (Chapter 8), as well as the oscillations of treelines in Europe and elsewhere (Chapter 10). It is thus already clear that trees provide several important sources of palaeoclimatic information, especially those growing in marginal situations, and that this is because of their sensitivity, especially in the way of regeneration, to climatic parameters. Treeline migration upward during climatic amelioration, such as that which has involved large numbers of bristlecone pine (*Pinus longavea*) establishing themselves above the former treeline in the White Mountains of California as a result of the warming since 1880 (Davis and Botkin 1985), is more rapid than the reverse migration in response to deterioration, because recession of the treeline will be delayed until trees which die are not replaced. The same applies to migration of treelines poleward; trees at the forest limit in Keewatin, for instance, are relics of a former warm period (Nichols 1976). It follows that climatic deterioration of the order of several hundred years may not appear in treeline macrofossil records. More detailed discussion of the possibilities and difficulties involved in interpreting treeline data and tree-ring sequences may be found in the survey by Bradley (1985, Chapters 8 and 9).

The ecological impact of the Little Ice Age climate on woodland near the treeline has been demonstrated by detailed studies of spruce morphology and regeneration in the Bush Lake area of northern Quebec (57°47′N, 75°45′W). Between 1389 and 1890 tree growth was generally slow, particularly narrow rings being formed in the periods 1420–35, 1480–95, 1600–10, 1690–1700, 1825–30, 1835–40 and 1855–60. The population formed a krummholz, with the individuals taking on a stunted form. In the short intervals with milder climate between 1389 and 1890 cone production was stimulated and seedlings established as, for instance, in 1640–70, 1740–50, 1800–20 and 1840–60, but the normal method of reproduction through the period of the Little Ice Age was by layering. At the end of the Little Ice Age, there was a major shift from krummholz to forest, but there was also mass mortality of old spruces with limited photosynthetic mass, which were unable to take advantage of the more favourable climate. These were replaced by seedlings which later became normal trees. It is suggested that the time lag in growth-form shifts was determined, at least in part, by the critical position of apical buds and needles relative to the height of the winter snowpack. Seedlings established during warmer periods of the last few hundred years which developed into trees probably reached the snow–air interface when winters were somewhat milder and the snow depth was greater. Winters during the Little Ice Age in this area are believed to have been windy as well as very cold, with a thin snow cover, while during the recent warming the snowpack has been on average 20 cm deeper, judging by the growth form of the trees. The shift in growth habit from krummholz to forest at the end of the Little Ice Age is apparently a large-scale phenomenon observed throughout the northern Quebec forest–tundra zone (Payette *et al.* 1985).

12.2.2 CHANGES IN FAUNAL DISTRIBUTION

Climate is one of the principal environmental controls of living organisms; temperature is the factor of greatest biological significance. The primary effects of changing climate are physiological, but indirect effects, for instance factors affecting transportation, such as ocean currents or cyclone tracks, may also be important. The relationship between animal and insect species and climate is complex and so are responses to changes in climate. In general, the most important response of a particular species is adjustment of its distributional range. As animals are more mobile than plants, their responses tend to be more rapid, especially to amelioration (Ford 1983). There is room here to give only a few illustrations of the importance of even minor climatic change to faunal distribution.

The physiology of cod (*Gadus morhua*), especially the regulation of water balance within the body, is disturbed below 2 °C. The kidneys of the cod do not function in colder water. Cod can therefore only occur briefly and in very small numbers in water below 2 °C. It is abundant in waters where the surface temperature is between 2 and 13 °C; temperatures between 4 and 7 °C are optimal for reproduction and early survival.

The changing distribution of water masses during the past few centuries in the northern seas between Scandinavia, Greenland and Scotland was discussed in Chapter 2.

It is possible to follow from fisheries' records the changes in the distribution of cod caused by displacement of water masses (Beverton and Lee 1965). The Faroese cod fisheries are known to have failed totally in 1625 and 1629 and from 1675 onwards there were no cod in the area for many years. Between 1685 and 1704 cod fishing also failed off southwest Iceland; it is to be presumed that the warm water of the Irminger current did not reach so far north as the Icelandic coast during those years (Lamb 1979). The Norwegian fisheries showed signs of decline in the seventeenth century and in 1695 fisheries failed along the whole coast of Norway south to Stavanger, except for a small cod stock cut off in the inner part of Trondheimsfjorden, which was probably living in an isolated pocket of old North Atlantic Drift water. The disappearance of cod from the whole Norwegian Sea area south past the Faroe Islands was, it seems, due to the extension of polar water below 2 °C, which was unsuitable for the survival of these fish. In 1695 cod were sparse even as far south as the Shetland Islands. Fisheries evidence, though not complete, is sufficient to indicate that polar water dominated the whole region for periods of up to 20 to 30 years during the Little Ice Age and was present for much of the time between about 1600 and 1830, and the fauna of this whole sea area was substantially affected.

More details of the susceptibility of cod distribution to changes in water characteristics are available for the period of climatic warming during the present century. During the Little Ice Age the waters around Greenland were too cold for cod except for small populations in sheltered fjords, although at the time of the Norse settlement in Greenland cod had been abundant, judging by their remains found in Viking middens. As the flow of the Irminger current extended round West Greenland after 1917, cod spread progressively northward to reach Upernavik (72°50′N) by 1933. As cod were by then spawning to the north and west of Iceland, cod eggs and larvae were carried across the Denmark Strait by the western branch of the Irminger current and then round the tip of Greenland by the West Greenland current (Figure 12.6). The temperature of the water was by now high enough to allow the fish to survive and develop. In most fish it is the egg and larval phases which are most vulnerable to adverse conditions, including temperature. There is a positive correlation between the year–class strength of Greenland cod, between 1924 and 1951, and water temperature on the spawning grounds. The build-up of population in Greenland waters was due to exceptionally good survival in a small number of exceptionally warm years. The northern limit of cod was back to 69°N at Disko Bay by 1950, with the return of cooler sea conditions. There was little recruitment to the West Greenland stocks after the cold East Greenland current became dominant during the last three decades.

Information about the distribution of animals, fish and birds is in most cases scanty, if available at all, for periods earlier than the twentieth century. Some of the complex and in many cases sensitive reactions of such organisms to minor climatic changes are discussed in Ford's (1983) study, which is concerned essentially with the twentieth-century period of climatic warming and the cooler period after about 1950. Extension of the ranges of European birds, for instance, during twentieth-century warming was well marked (Sharrock 1974). Critical factors as far as bird distribution is concerned are the severity of the winter months, depth and duration of snow cover, warmth and length of summers and availability of water. Food supply is often crucial and so the range of one species may depend upon the survival of another more sensitive species. An example is provided by the decline of puffins (*Fratercula arctica*) nesting around Britain between 1920 and 1950, during which time the fish species upon which puffins depend changed their location. As sea temperatures declined again after the 1950s, cold-water fish such as sand eels (*Ammodytes spp.*) returned and the northern colonies of the puffins were able to start increasing again.

During the warming period between 1890 and 1940, depressions were moving eastwards into higher latitudes and there were more opportunities for European migrants to be carried to Europe and Iceland in the backing winds of the northern sectors of lows. Accordingly, more European migrants were recorded during this period than between 1820 and 1890 or after 1940 (Williamson 1974). Extension of the range of European birds during the twentieth-century warming included the establishment of breeding pairs of blackheaded gull (*Larus ridibundus ridibundus*) in Iceland after 1911, swallows (*Hirundo rustica*) in the Faroe Islands and Iceland in the 1930s, starlings (*Sternus vulgaris*) in Iceland from 1941, and the fieldfare (*Turdus pilaris*) in Greenland and Jan Mayen after 1936 (Williamson 1974). Effects were by no means restricted to the higher latitudes. Despite the complication caused by human impacts on the environment it is possible to discern the spread of many species northward from southern Europe (Burton 1981). One of the most striking is the expansion of the serin (*Serin serinus serinus*), which was confined to the sunny borders of woods in the western half of the Mediterranean in the eighteenth century. By 1876 it had colonized much of central Europe and during the next fifty years spread into most of France, Germany and western Poland. It now breeds regularly as far north as northern France, the Low Countries and southern Scandinavia. Cetti's warbler (*Cettia cetti cetti*), then considered to be a non-migrating species, was confined to the

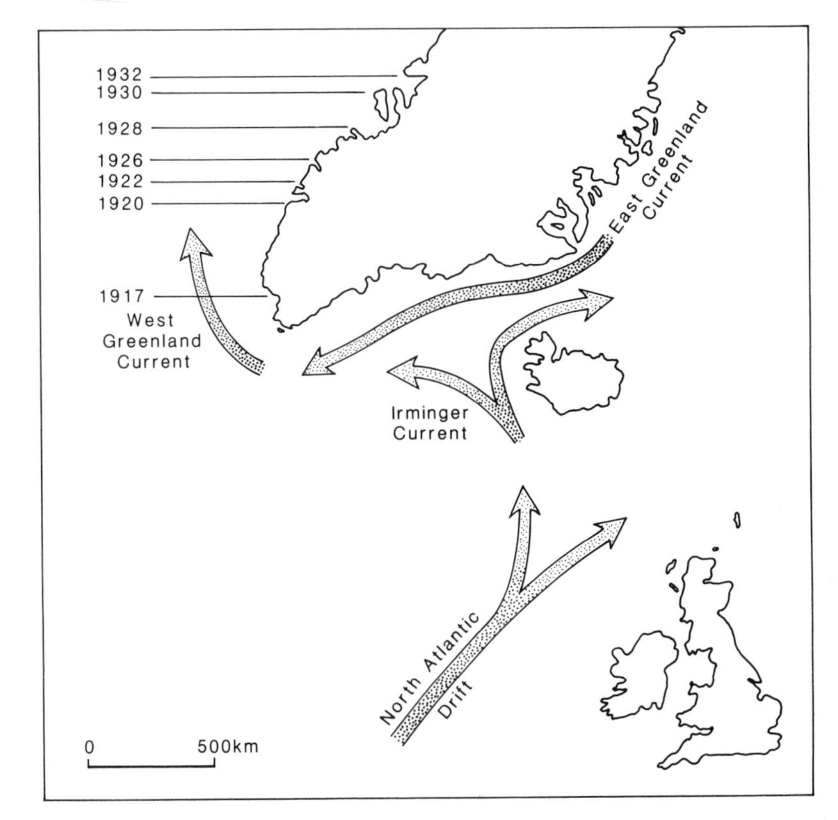

Figure 12.6 The northward spread of cod up the west coast of Greenland during the twentieth-century warming period. (From Ford 1983)

coastal parts of the Mediterranean in the early twentieth century. It was only found away from the coast in the Balkans, Italy and Iberia. Soon after 1920 it started to spread and by 1932 it had moved up the Rhône valley and in the west had reached the Seine basin. It was checked by cold winters between 1940 and 1952 and then resumed expansion into northern France, the Low Countries and northern Germany, despite temporary checks caused by winters such as that of 1962–3. By 1975, Cetti's warbler was nesting in most of the southern and eastern counties of England.

Within Britain, some of the resident fauna also made territorial gains. The great spotted and green woodpeckers (*Picus major Linnaeus* and *Picus viridis pluvius*), the nuthatch (*Sitta Europacea*), mistle thrush (*Turdus viscivorus*), the rook (*Corvus frugilegus*) and insects such as the peacock and comma butterflies (*Inachis io* and *Polygonia c-album*) have all pushed north. The blackbird (*Turdus merula*) has been especially successful both in Britain and Fenno-Scandinavia. It has colonized the Faroes and now winters in southern Iceland. Extension of the range of these and other bird species from the south has been matched by those of a considerable number of insects. Since 1850, more than fifty species of butterflies and moths from the continent have appeared in the British Isles and twenty-six of these have become permanently or temporarily established. The most striking case is that of the golden plusia moth (*Polychrysia moneta*). The first two British examples were captured in 1890; since then this moth has spread through England and Wales and into southern Scotland and eastern Ireland.

With cooling, especially of northern latitudes, there was a movement of species southwards after 1960. The snowy owl (*Nyctea scandiaca*) nested in Shetland in 1968 after a gap of sixty years. The great northern diver (*Gaveia immer*) began to nest again in Scotland after 1970. During the late 1970s and early 1980s in Great Britain more southerly based species were still moving north while, conversely, there were some northern species moving south. The simultaneous expansion of both northerly and southerly species suggests a more immediate response of fauna to deteriorating conditions than to amelioration since the end of the nineteenth century.

In view of the changes in faunal distribution known to have taken place in the last century or two, apparently as a result of minor climatic fluctuations, it is reasonable to assume that changes of distribution would have taken place on a more substantial scale in sensitive areas during the more marked phases of the Little Ice Age.

12.3 Human consequences

The impact of climatic change on society is inevitably coincident with that of many other factors, social, economic and political, which also influence the human environment. Climate is most influential in areas marginal for agriculture either latitudinally or altitudinally. It is therefore logical to consider the part played by the climatic factor in human affairs in such situations before touching on the much more controversial question of whether climatic change has had any discernible influence on more fortunately placed societies.

12.3.1 ICELAND AND GREENLAND

The case of Iceland seems to be most appropriate for initial discussion. Iceland has never gone to war with another country, civil unrest has not been of any consequence for several centuries, and it is three hundred years since the carrying of arms was abolished (Thórarinsson 1956). Despite this, life for the Icelanders was exceptionally difficult throughout the Little Ice Age period.

The evidence from these earlier centuries can be viewed more effectively in the light of the events that took place after the middle of the present century, when cooling was marked in northern latitudes. In 1965, mean annual temperature at Stykkishólmur in western Iceland fell below the 1846–1982 mean for the first time since the 1920s (Figure 12.7) and, in the same year, sea ice returned to Icelandic waters (2.1 and Figure 2.7). With modern agricultural methods, improved stock, good transport and the availability of indoor heating, Iceland is less vulnerable to environmental change than in the past, but agriculture and fishing, which are still the basis of the economy, remain highly sensitive to temperature.

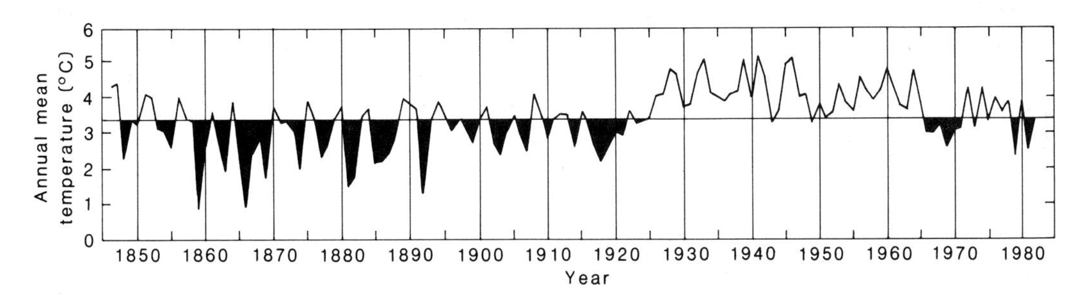

Figure 12.7 Annual mean temperatures at Stykkishólmur, 1846–1982. Values below average are shaded. (From Bergthórsson 1985)

Grass is the main crop grown. With the return of the sea ice and atmospheric conditions characteristic of the nineteenth century, there was widespread winter killing of the grass crop and its almost complete failure in some places, especially in the north (Figure 12.8 and 12.9). The yield of grass depends very much on the air temperature during both the growing season and the previous winter. Cold winters are even more effective than cool summers in restricting grass growth. Very hard frosts in late winter kill the grass; low temperatures in the growing season retard growth (Bergthórsson 1985). Prolonged snow cover, especially if there is melting and refreezing, is especially harmful. The soil may remain frozen late into the spring, thereby delaying growth; then a thaw may leave water lying on the frozen soil, killing the grass outright. In the single year, 1967, yields of hay per hectare were 870 kg lower than the average over the previous twenty-five years. Over 100,000 ha there was a decrease in production of 87,000 tonnes, worth 260 million krónur, reducing the basic productivity of Icelandic agriculture by one-fifth. This figure does not include the loss of grazing or the 'downstream' shortage of raw material for other agricultural production, which was valued that year at 2,300 million krónur (Fridriksson 1969). The year 1967 was not the only one with severe icing; 1970 and 1975 were similar in many respects.

The drop in hay yield affected livestock production, as did the productivity of the grazing land. A recent calculation of the relationship between climate, the productivity of cultivated grassland and potential livestock capacity led to the conclusion that a

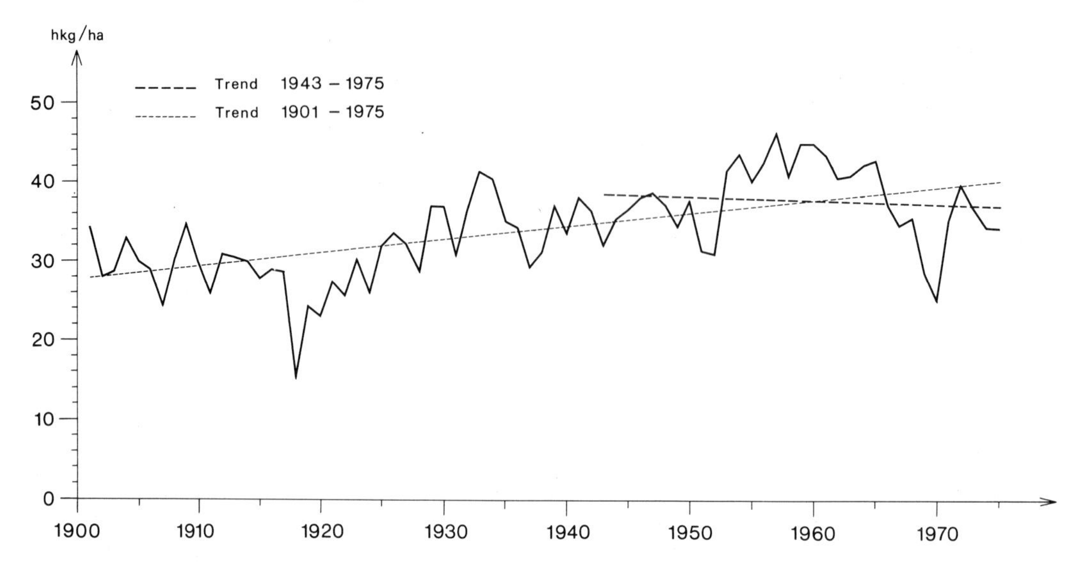

Figure 12.8 Yield of hay in 100 kg/ha for the whole of Iceland. (From Bergthórsson 1985)

temperature deviation of 1 °C from the 1901–30 normal of 3.2 °C would reduce carrying capacity by 30 per cent (Bergthórsson 1985). This estimate is in excellent agreement with the course of events between 1962 and 1982, when there was a reduction in livestock of 27 per cent per degree C below 3.2 °C. It appears that dependence on climate has only been slightly reduced despite improved farming practices.

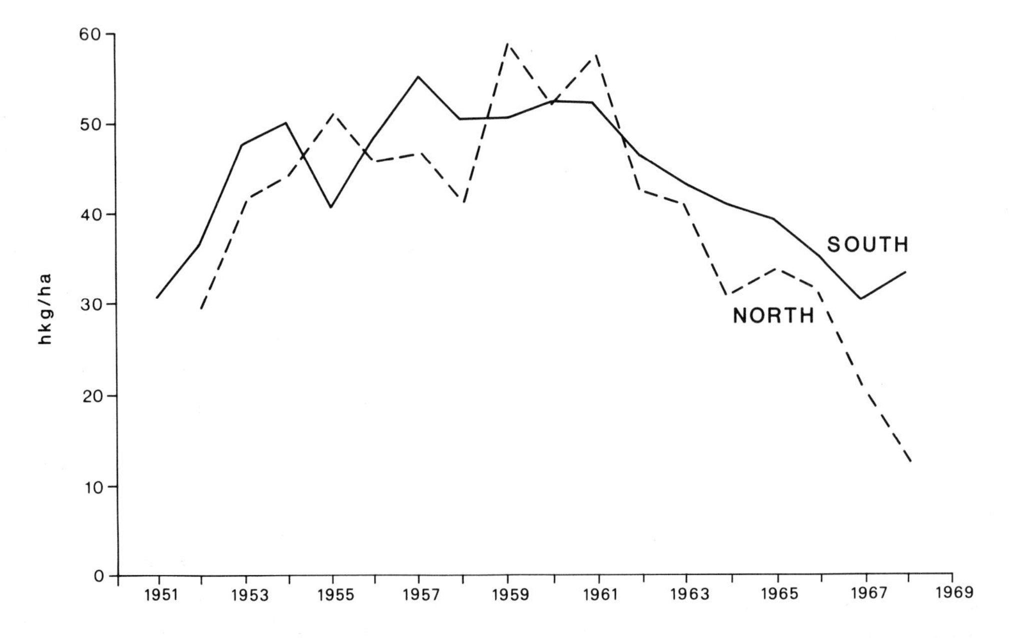

Figure 12.9 Average yield of hay during 1951–68 in districts in the north and south of Iceland. (From Fridriksson 1969)

Some protection against hard conditions is possible in a modern state, but protective measures are expensive. The cost of purchasing and then transporting hay in 1965 was 14.2 million Icelandic krónur. In 1967, government compensation loans and transport subsidies totalled 18 million krónur, and in 1968, 15.1 million krónur were paid in compensation loans and 4.7 m million krónur in hay transport subsidies. Both compensation loans and hay transport subsidies were discontinued after 1968, although adverse conditions continued. Several years in the 1970s were warmer than average, but temperatures were low again in 1978 and 1979 and in the early 1980s.

If low temperatures have an immediate effect upon grass growth, low water temperatures can equally drastically affect the distribution and migration of fish, as shown earlier in this chapter. Fish had been exported from Iceland since the fourteenth century, although the amount was limited until the introduction of decked vessels around 1890. As in the case of agriculture, the introduction of modern technology during the period of climatic amelioration since the late nineteenth century has revolutionized the industry. The Icelandic fish catch quintupled between 1920 and 1960, although variations in weather and hydrographic conditions caused violent year-to-year variations, especially in catches of herring. Larger catches increased exports and so helped to provide the funds for imports to sustain the needs of modern Iceland and its increasingly urban population.

The changes in sea temperature around Iceland after 1960 affected the fisheries fundamentally. In 1965 and 1966, the Icelandic herring catch was over 750,000 tonnes. In 1967 it was below 500,000 tonnes and by 1968 it had fallen to 150,000 tonnes (Jakobsson 1969). This was in turn largely responsible for the devaluation of the Icelandic currency (Kristjánsson 1969). Deterioration in the fishlife around Iceland as a result of falling sea temperatures was one of the major factors causing international friction with Britain over fishing rights in the Icelandic sea area. Iceland's dependence on fishing makes her peculiarly vulnerable to such changes and helps to explain and justify her firm political stance at this time.

Although grass has always been the main crop in Iceland, barley was cultivated in the centuries after the settlement. It is mentioned in the classical literature and in ecclesiastical and other documents evidence of former cultivation is found in field names, in signs of former tillage, and in finds of cereal pollen (Thórarinsson 1956). It was most extensively grown in the south and southwest and least extensively in the north, where cultivation had probably ceased before the end of the twelfth century. In the thirteenth and fourteenth centuries it diminished in the south and southwest as well, and in most places died out in the fifteenth century, although it may have been grown on the southern shore of Faxaflói into the sixteenth century. Occasional efforts to grow cereals were made in the following 300 years but regular cultivation experiments did not succeed until the twentieth century.

The obvious conclusion, that cereal cultivation dwindled and ceased for climatic reasons, has been challenged but observations made in southern Iceland show that modern fast-growing barley requires 850 degree-days from sowing to harvest (Bergthórsson 1969) and that this will be raised by 30 degree-days for every 100 mm of precipitation above 200 mm in the growing season. In 60 per cent of the years between 1873 and 1922, barley would have ripened at only one weather station, Reykjavík, while in the period 1931–60, in 60 per cent of summers it would have ripened at 22 out of 48 stations (Bergthórsson 1985). It seems that no reason other than the climatic one is required to explain the cessation of barley growing in Iceland and the decision of farmers in Vestur-Skaftafellssýsla to rely on collecting the seeds of the wild lyme grass (*Elymus arenarius*) instead (Thórarinsson 1956).

The advance of glaciers caused abandonment of farms in the eighteenth century (Chapter 2), but this was not the only reason for abandonment. Overwintering of stock depended on the collection of sufficient hay, but grass growth was affected by the many severe winters during the seventeenth, eighteenth and nineteenth centuries. Today it is reckoned that a sheep needs 3 m^3 of hay, a horse 10–15 m^3, and a cow 35–40 m^3. Eighteenth-century animals often had to do with much less and often had not sufficient to survive. Although some were slaughtered in bad years, many were kept in the hope that they would pull through the winter. The results were catastrophic (Figure 12.10). A survey of primary sources has shown that livestock died of hunger and cold in 24 out of the 36 years between 1730 and 1766 (Ogilvie 1981). In 1757 the sheriff of Dalasýsla (Figure 2.1) reported that

> conditions in this district . . . are so difficult that if I described them without furnishing clear proof they would seem incredible. . . . I should like to draw your attention . . . to the fact that just in this year 321 cows and bulls, 1292 sheep, 3209 young lambs, and 151 horses have died in this one district. Forty four people have died of hunger and wretchedness, and 15 dwellings have been deserted. With this loss of livestock and the resulting difficulties, it is a foregone conclusion that during the coming winter there will be such dearth and misery in the district that it will far exceed the dearth that has already been experienced. And it will be a pure miracle if a third of the population does not die of hunger, as the small number of livestock that the peasants have left over is so wretched that they have no milk. They do not have any fish either as the fishing off the coast the previous spring and winter was poor and they have no cattle to slaughter with which to maintain life. (Quoted in Ogilvie 1981)

As conditions worsened for farming so they did for fishing. Fish, especially cod, were the staple part of the diet and from the fourteenth century onwards had been the main export. The Icelanders only had open rowing boats and so they could not go far out to sea. Catches were never large, but those in the eighteenth century were smaller than those in the seventeenth century. The failure in the fisheries, which lasted through the seventeenth century to the eighteenth century, was most marked from around 1685 to 1704 and between 1744 and 1759. An official in Bardaströnd district in western Iceland reported in 1744 that

> in the west the fishing was poor; there were average to small catches of lumpfish but everywhere the cod catch was the worst that people could remember for many years, both during the autumn fishing last year and this spring. In most places most boats did not catch more than about 10 fish and in many places less. (Quoted in Ogilvie 1981)

The main reason for the decline of fishing was the withdrawal of the cod southward, with the arrival of colder water around Iceland, but there were also subsidiary factors. The presence of ice around the coast prevented the boats going out and stormy weather meant more boats lost at sea. In some areas the people were so dependent on fish that when the fishing failed they had to leave. Thus in Húnavatnssýsla in 1732

> instead of the usual 14 to 15 boats from Olafsvik only two could be manned this year because of the lack of labour. Many people who lived by the coast have left and people who lived further inland do not want to risk setting out to fish because of the recent failure of the fishing.

In the seventeenth century boats usually had 10 or 12 oars, but before the middle of the eighteenth century boats for six men or less came into use because of shortage of labour.

Farm desertion was widespread throughout the eighteenth century, the poorer districts in the north and east being the first to be affected. In 1735

> there was a lot of traffic of people who left their homes, especially from the north, which increased after the previous severe winter. That spring, all over the country, poor people died of hunger, in poverty and some were found like cattle out on the heath.

The movement out of the worst-affected parts of the country was too much for the more prosperous parts to cope with successfully. In 1756 the sheriff of Bardaströndarsýsla wrote that

> what is proving the ruin of our inhabitants more than anything else is the great number of people coming from the northern part of the country as they take the little we have and it already seems as though the situation will soon not be much better than in the north. (Quoted in Ogilvie 1981)

Both birth and death rates were affected in the years of scarcity, many dying of hunger and others of diseases such as dysentery and gastro-enteritis associated with malnutrition. Death rates peaked in the most severe years, nearly 17,000 dying in 1750–8. In 1757 the sheriff of Dalasýsla, describing the situation in his immediate vicinity, mentioned that because of the lack of grass and the poor harvest the previous year,

> people have become so impoverished that they would be unable to buy foodstuffs even if they were readily available. The people who have died of hunger including those who have come from other settlements number seven in all. Two farms have been deserted. (Quoted in Ogilvie 1981)

The simultaneous failure of both fishing and farming affected not only the quality of life but also the size of the Icelandic population, but demographic trends cannot be explained only in these terms. When the first census was taken in 1703, the population was 50,358. Changes in subsequent decades were characterized by slow rises to about 50,000, followed by sudden falls. The most serious of these was caused by the smallpox epidemic in 1707, which killed one-third of the population. The eruption of Laki in 1783 and its consequences reduced the population to 38,363 by 1786. There was a further drop in 1890, caused by emigration to America. This exodus was triggered by the eruption of Askja in 1875, which damaged the grass over a wide area, and by the severe ice years of 1881 and 1887, which reduced livestock so much that people were on the verge of famine (Thórarinsson 1956).

Not only did the Icelanders have to contend with the effects of volcanic eruptions and epidemics as well as climate, but also with the rigours of the Danish monopoly of trade, which lasted from 1602 to 1787. Essentials such as flour, timber and metal had to be imported and there is no doubt that the attitude of the Danish merchants added to the difficulties of the Icelanders, but it was not the primary cause of suffering.

In Húnavatnssýsla, a letter written in 1754 explained that the previous winter had

> caused great misery to the inhabitants here in the district as they have lost almost all their livestock and means of subsistence and they were not able to buy flour and foodstuffs from the merchants, let alone obtain them on credit, the monopoly's control being what it is. This year, the behaviour of the merchants has been even more absurd and contrary than in previous years; in this time of dearth they have let people travel to the trading place 2 or 3 times. Sometimes they were turned away with scornful words, on other occasions made to wait there for no reason at all as it suited the merchants, only to be finally refused unless they had goods to pay with at once. (Quoted Ogilvie 1981)

The Icelandic community survived the impact of climatic deterioration, along with the rigours of volcanic eruptions, epidemics and overseas domination, but the cost was very high. The relative importance of these influences varied over time and was also unequal from one district to another. There is no evidence that the Icelanders attempted to mitigate the effects of harsh climate by adaptation or innovation and indeed there were few opportunities to do so.

The Norse settlement in Greenland, initiated in AD 985, did not survive in spite of the fact that there were opportunities to adapt to increasingly severe conditions. By about 1350 the Western Settlement, in what is now the Godthåb District, was abandoned and by 1500 the larger Eastern Settlement, in the modern Narsaq and Julianehåb District, was also empty. Possible explanations for the evidently sudden desertion of the Western Settlement and the dying out of the Eastern Settlement have ranged from plague or the decline of trade with the outside world, through attacks by Basque pirates or by the Inuit, through to the effects of climatic change (Jansen 1972). None of these explanations is wholly convincing if taken alone.

We know from a letter written by the Pope to the Archbishop of Nidaros [Trondheim] in 1279 that Greenland was visited 'only infrequently'. The craft used had a carrying capacity of between 30 and 74 deadweight tons (Gad 1970, p. 123) and so the Greenlanders cannot have depended on imports for their basic food supply, although goods such as honey (required for the making of mead) were undoubtedly much-valued items. In any case, the decline and eventual cessation of the sealink with Scandinavia was not solely due to the increasing extent of the sea ice but had also a great deal to do with the orientation of the Danish-Norwegian state towards the Baltic. Available evidence about intentional sailings to Greenland in the late fourteenth and fifteenth centuries is extremely sparse compared with the number of documentary references to unintended landings. After surveying the documentary evidence, Gad (1970, p. 151) was left with a strong impression that it was chance that brought ships to Greenland.

The Greenlanders occupied the lowland areas around the inner shores of fjords where they could pasture their cattle, sheep and goats and perhaps attempt unsuccessfully to grow corn. All the stock had to be kept inside over the long winters, the settlers then making use of dried meat and stored dairy products. Excavations have revealed that the people, even from the farms furthest inland, were also basically dependent upon seal meat, particularly that of the harp seal (*Pagophilus groenlandicus*). Sealing during the annual migrations of the animals up the west coast in May–June, hay harvesting during the summer and caribou hunting during the autumn probably involved much of the population (McGovern 1981). In addition, walrus and polar bear were hunted in the course of annual long-distance expeditions to Nordsetur, the northern hunting ground round Disko Bay, over 800 km north of the Western Settlement. These arduous trips provided the walrus tusks and bear skins which were the basis of overseas trade. The sudden collapse and abandonment of the Western Settlement in mysterious circumstances which left a few domestic animals wandering unattended, some time between 1341 and 1364, cut off access to Nordsetur and so the source of trade goods. Norwegian visitors had little interest in purchasing any of the products of the Eastern Settlement. The lack of acceptable trading goods further diminished the likelihood of deliberate visits to Greenland.

McGovern (1981) argued that the economy always operated on a thin edge, depending on the skilful co-ordination of communal labour to make maximum use of the seasonal abundances of both terrestrial and marine resources. The necessarily tight scheduling would have been most successful in the context of high predictability of seasonal and year-to-year fluctuations in climate and hence in resources. Harp seal catches during the period of cooling which occurred between 1954 and 1974 in the Godthåb district is known to have decreased as interannual variability increased. The inner fjord station of Kapisigdlit, situated in the main north settlement area, showed the greatest increase in variability of catches between 1954–8 and 1959–74, the percentage of harp seals in the total catch falling from 30 per cent to 4 per cent and the mean annual catch also falling from around 30 per cent to just under 4 per cent. If the minor cooling of the mid- to late twentieth century had such a noticeable effect on the migration of the harp seal it can reasonably be assumed that the more prolonged and severe medieval cooling would have had even more pronounced consequences. At the same time, lengthening of the winter season and greater persistence of snow cover would have increased the byring time of the domestic animals, while the hay harvest, by analogy with experience in Iceland, would have been reduced. Moreover, deep winter snow, especially if associated with ice crusting, is known to have caused dramatic reduction and even local extinction of caribou herds in southwest Greenland (Vibe 1967, p. 163). It is quite clear that the Norse settlements in Greenland suffered severe climatic stress after the first century and a half of their existence.

With contraction of the resource base the position of the Norse farmers in Greenland became more hazardous and adequate food supplies more problematical. Malnutrition and premature deaths increased and the ability of the community to withstand hostility from outsiders was lessened. It is possible that the Western Settlement was attacked by Basque raiders, the few women and children left being taken over by the Inuit, but the discovery of a cap of a type fashionable in Burgundy after 1450 from excavations in the Eastern Settlement indicates that contacts with the Basques were not necessarily warlike. It is rather more likely that there was an attack by Inuit. According to Ivar Bardarson, who visited Iceland sometime after 1341, 'the Skrælings have the entire West Settlement; but there are horses, goats, cows and sheep, all wild. There are no people, neither Christians nor heathens' (Gad 1970, p. 141).

The increasingly harsh climatic conditions which impoverished the farming settlements allowed the successful spread of Inuit culture dependent on ice hunting. The Inuit used skins for clothing and for constructing boats that made travel along the coast relatively easy, and they also made use of a variety of equipment, including toggling harpoons which enabled them to catch the ring seal in the winter months. When sea temperatures fall, the ring seal increase in number while the harp and common seal decrease. The

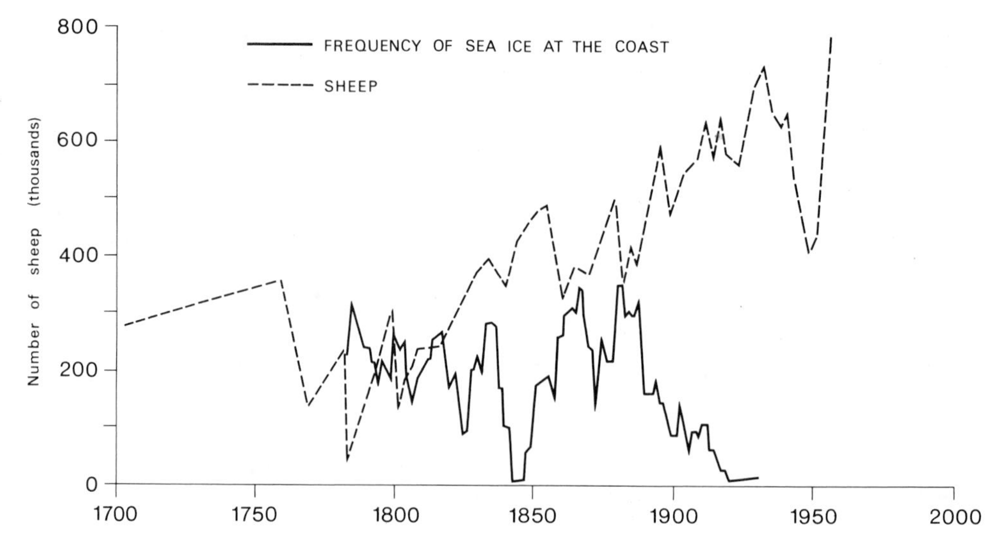

Figure 12.10 The number of sheep in Iceland and frequency of sea ice. (From Fridriksson 1969)

Norse farmers, without effective means of hunting sea mammals in winter, could not use this alternative food supply. Communal boat-drives used to force the harp seal onto the beaches or into nets were impossible when the fjords were ice-covered.

Had the Norse been able to respond to the changing conditions by shifting emphasis from cattle-keeping in the inner fjord country towards greater dependence on coastal and marine resources they might have been able to survive (McGovern 1981). While it is only possible to speculate about the exact social and cultural barriers which must have prevented such adaptations, the archaeological evidence makes it quite clear that rather than attempting any variations in their economy, the Norse persisted in their traditional arrangements and eventually succumbed to climatic stress.

12.3.2 WESTERN NORWAY

Even in Iceland and Greenland there are no grounds for viewing history from the point of view of climatic determinism, though climate has undoubtedly had a pervading influence on social life, economic conditions and political events. The impact of deteriorating climate was also felt in maritime western Norway. Although farm desertion was rare, inspection of the legal records resulting from applications for tax relief provide a picture of decreasing prosperity caused by physical damage, reduction of the area available for cereal cultivation, and poor harvests. Physical damage was especially crippling if it affected the fenced land, the 'Innmark', which was most intensively cultivated. The effects are described in a report on the state of Mere Sunde farm in Nordfjord Fogderi in 1702 which

> suffered almost indescribable damage from river bursts and landslides to both the Innmark and the Utmark for almost all of their fenced pasture, particularly round the farm buildings, has almost completely been carried away, being covered with rocks and grit, so that it cannot possibly be of any further use. . . . Further, the

occupants have had to turn about half of the above home farm, ruined by rocks and grit, into Utmark pasture and to move their fences further in. So now they have to scrape together such hay and fodder which they need for the upkeep of their wretched cattle, from up under the mountain in the scree with great toil and danger. The state of this farm was not unusual amongst those in the parishes adjoining Jostedalsbreen. (Grove 1972, Grove and Battagel 1983)

The reduction in the area of pasture lowered the number of beasts that could be kept and overwintered. The cattle were enumerated farm by farm in both 1667 and 1723. These returns cannot be regarded as precise but appear to be acceptable as estimates. Every parish in the region between Jostedalsbreen and the sea suffered a drop in the number kept between the two enumerations and in most parishes this was substantial (Figures 12.11 and 12.12). Five out of the six parishes in Nordfjord Fogderi had lost more than 20 per cent of their cattle by 1723. Of the nineteen parishes in Sunnfjord Fogderi, only six showed a drop of less than 15 per cent and eight lost more than 20 per cent of their cattle. Cattle were the mainstay of the economy and hay the crop that really mattered.

Some part of the reduction in the cattle population was certainly caused by direct damage to farmland. The incidence of physical damage was high from the 1680s to the 1750s, the decade between 1740 and 1750, when the glaciers were at their greatest extent, bringing the highest toll (Figure 12.2). The Nordfjord parishes were those worst hit but the trail of damage extended right out to the coast. The farms furthest inland from Sunnfjord were especially affected by landslides, while those further downvalley were

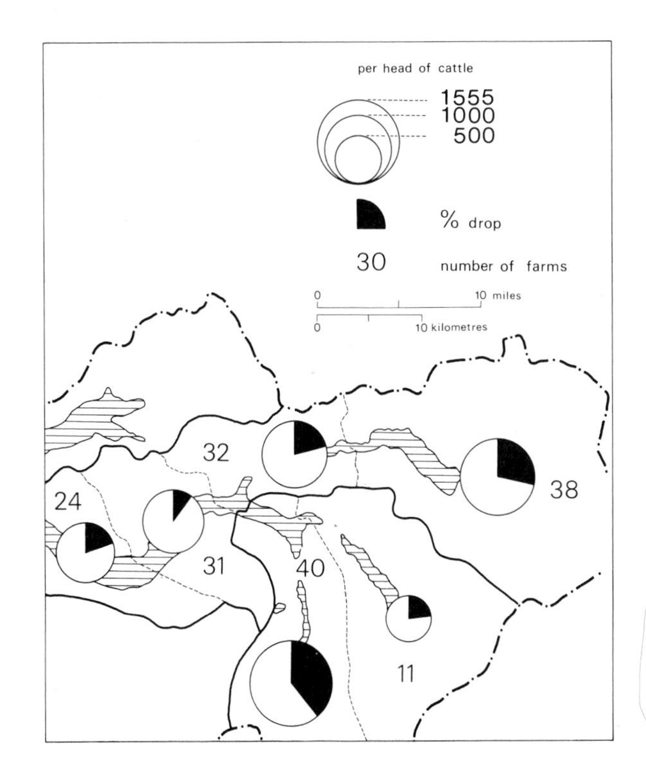

Figure 12.11 The percentage drop in the number of cattle kept between 1667 and 1723 in the parishes of Stryn and Olden Skipreider, Nordfjord Fogderi. These parishes lie north of Jostedalsbreen. (From Grove and Battagel 1983)

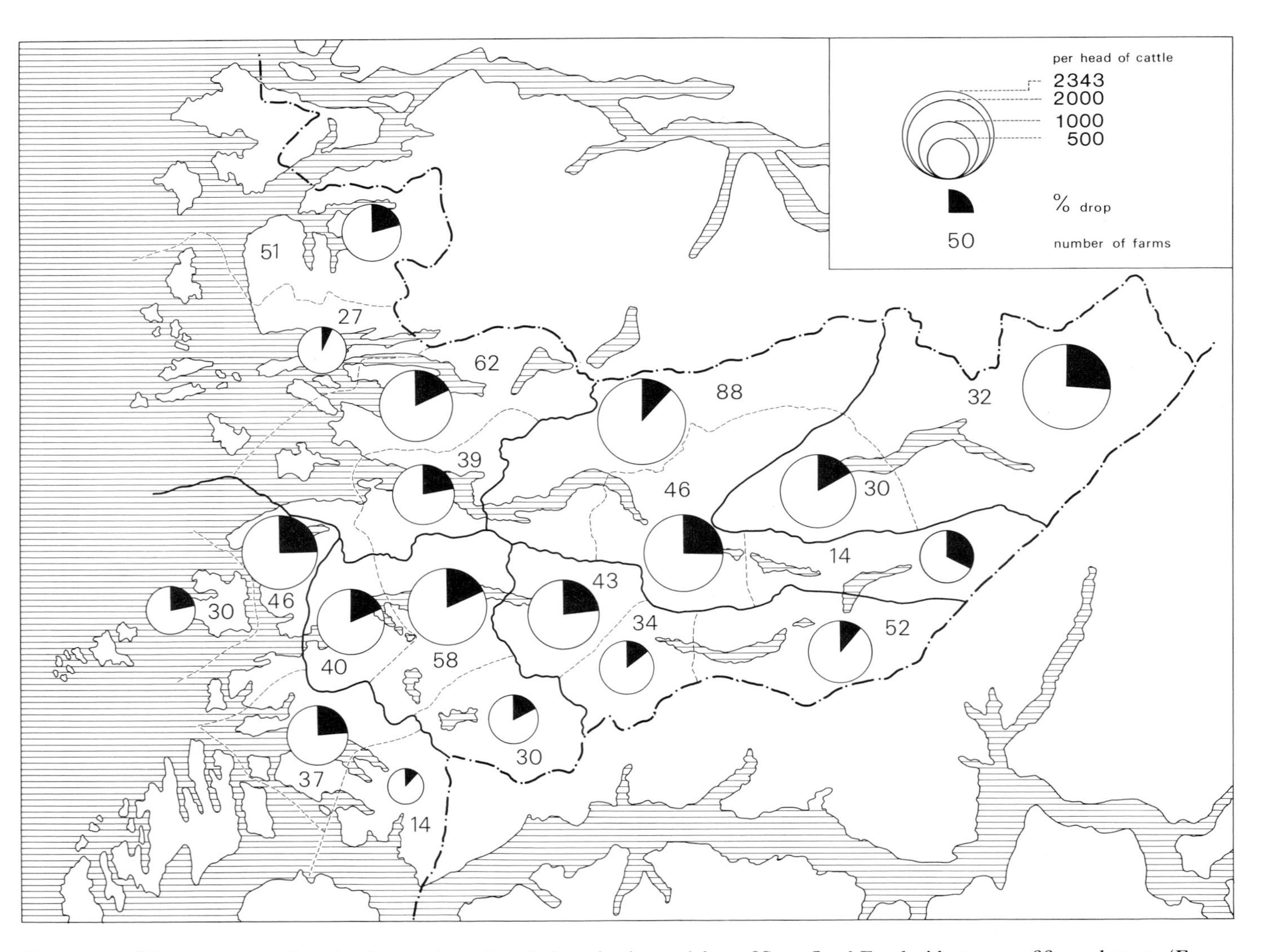

Figure 12.12 The percentage drop in the number of cattle kept in the parishes of Sunnfjord Fogderi between 1667 and 1723. (From Grove and Battagel 1983)

subjected to floods. The percentage of farms gaining tax relief as a consequence was highest in the parishes affected by landsliding. In Jølster, thirty-three farms or 38 per cent of the total were affected. In all, fifty-nine farms in Sunnfjord were sufficiently seriously damaged to be granted relief by 1763, when the last successful claim was made in the region, but there were in all 774 farms in Sunnfjord, so that although physical damage was widespread and is known to have affected the pastures it has to be concluded that much of the reduction in the number of cattle must have been due to deterioration of the grass crop. There were no obvious socio-economic reasons for the change. Agricultural techniques were still backward in the early nineteenth century (Kraft 1830), and the available evidence precludes the possibility that farmers had turned from cattle to some more profitable enterprise.

In western Norway, unlike Iceland, some cereals were being grown for home consumption in the seventeenth and eighteenth centuries, but diminution of the head of cattle involved a reduction in arable productivity. This is particularly clearly revealed in court records from the 1740s in Nordfjord. In 1744 eleven farms in Oldendalen petitioned for reduction of assessment, following damage by landslides and avalanches. Some 332 cattle had been kept on these farms in 1667 together with 22 horses. Of corn, about 75 tønner had been sown and 286 tønner had been harvested; in 1745 the farmers were found to have only 207 cattle and 18 horses plus a few sheep. They now reckoned to sow about 63 tønner of corn and harvest 238 tønner. Further reduction in arable productivity was expected. The relationship was clearly explained in the records. Thus at Muri, for instance, it was recorded that

> at the time that the matrikkel was drawn up (1667) they could keep 80 cattle and 5 horses and sow 18 tønner of corn and harvest 80 on the whole farm . . . but since that time, and particularly in 1743 there had been great and irreparable damage to arable and pasture inside the fences and also to the high pasture and hayfields outside, whereby both fodder and harvest, but particularly fodder, had been much reduced, so now only 58 cattle and 6 horses can be kept, as well as a few sheep, on the whole farm, and although sowing and harvesting of corn can be regarded as approximate to the old amount, the great damage suffered has so reduced the hay harvest and fodder that it will not be possible in future to supply all the arable with manure and part of the arable will have to be turned over to pasture.

Some of the nearby farms had already been forced to reduce their arable for lack of manure.

While inland farming communities depended essentially upon their cattle, those near the sea relied on fishing to a greater or lesser extent. In the eighteenth century, fishing also suffered (3.1.3). In 1693, an inspection was made of Mauri, on the shore of Nordfjord, and its condition was found to be wretched:

> The said farm, Mauri, has suffered irreparable damage to its Utmark and cattle pastures from a number of falls of scree . . . this farm, in times past, has time and again had increases made to its landskyld [land rent] by His Majesty's sheriffs, because it has been blessed by God with fisheries, for their land reaches them and they and many others living in the valley were able to supply themselves, but many years have now passed since the fish came into land, and they now have only the meagre growth of the soil and the produce of their cattle with which to meet the royal taxes and dues, and with which to support their wives and children.

The consequent reduction in assessment was to be further increased in 1702 when the occupants were found to be starving and

> a great deal of the arable to be sour and useless and what could be fairly good soil gives little return for it is not properly cultivated or fed with manure, because their cattle are so poor and wretched and because of the poorness of the land, the occupants are poor and wretched.

Out on the coast of Sunnfjord, two other farms were granted reductions before 1700 on account of damage to fisheries. At Vilnes in 1733, it was recorded that a reduction

> had taken place because there was an island called Lodoen, belonging to the farm which because of the good sealing there, had been formerly assessed at half a laup of butter on the whole farm but where neither now nor for a long time since, had there been any sealing.

At Stubseide, on the north shore of Stongfjorden, the farm has been highly assessed because of its fisheries but these were in decline and had not been productive for a long time.

The inhabitants of Nordfjord and Sunnfjord Fogderier suffered not only from general reduction in prosperity but also, from time to time, from actual famine conditions. Periods of shortage before the harvest were expected by sixteenth- and seventeenth-century Norwegian peasants and there were traditional expedients for coping with longer famines following really severe crop failure. The use of alternatives to the normal staples was clearly explained in Bishop Pontoppidan's (1752) treatise on Norway:

> If grain is scarce, which generally happens after a severe winter, the peasants are obliged to have recourse to an old custom as a disagreeable but sure means of preserving life. Their bread in times of scarcity, is made thus, they take the bark of the fir tree, boil it and dry it before the fire, then they grind it to a meal and mix a little oatmeal with it, of this mixture they make a kind of bread which has a bitterness and a resinous taste and does not afford that nourishment that their usual bread does. However, there are some people, that think it right not to disuse this sort of bread entirely, and even in plentiful years they sometimes eat a little of it that they might be prepared against a time of scarcity.

The inhabitants of western Norway seem to have suffered the need to resort to substitutes for the normal staples several times in the late seventeenth and eighteenth centuries, and the combination of starvation diet and disease had dire effects on their numbers. Details of population trends in Norway before 1735, the year in which annual registration of births, marriages and deaths was made obligatory for each diocese, are not available, but a preliminary examination of a sample of the early parish registers has shown that years of demographic crisis occurred between 1668 and 1677, in the 1690s, and in the period 1712 to 1719. Dyrvik *et al.* (1976) defined crisis conditions as those obtaining when the death rate reached 50 per 1000 as against a birth rate which in pre-industrial Norway fluctuated around 30 per 1000. All the coastal parishes examined suffered such conditions in both 1676 and 1695. There are indications that years of crisis tended to have a regional impact affecting some areas and not others. The sample of parishes taken by Dyrvik *et al.* was particularly deficient for the Bergen diocese, which includes the whole area from Sognfjord to Nordfjord and thus both the parishes around

Jostedalsbreen and Sognfjord Fogderi. It is likely that a more precise idea of the incidence and importance of demographic crises in maritime western Norway will emerge from further study of parish records.

A government enquiry made in 1743 and answered individually for each separate parish included a question on diseases. The priest of Innvik on Nordfjord mentioned in his answer that 'Tomantille-rod, which grows in the meadows and is serviceable for both people and animals when it is boiled and drunk in sweet fresh cow's milk, has helped many against the dysentery which has been rife this year.' Drake (1969, pp. 66–71) points to the association of the incidence of dysentery and typhus with crop failures and the contemporary eighteenth-century Norwegian opinion which attributed the prevalence of these diseases to malnutrition and the eating of infected food. This association seems to have occurred in the valleys around Jostedal in the 1740s. There were peaks in the death rate in 1741–2, 1748 and 1773, years in which the grain harvest failed over large parts of the country.

The Norwegian areas that have been considered here are not particularly high and so altitude had little to do with the incidence of crop failure in years which bear a striking resemblance to those in which damage to farmland was concentrated. But harvest failure was associated with altitude in southern Scotland.

12.3.3 SCOTLAND

In southeastern Scotland, as in much of northwestern Europe, higher-altitude arable land has commonly been used to grow oats, but there has been extensive abandonment of such land in the last millennium. Oats above 250 m in Britain are sensitive to summer warmth, exposure and summer wetness, which may be measured in terms of accumulated temperature, average windspeed and end-of-summer potential water surplus (Parry 1975). In the Lammermuir Hills, minimum levels of summer warmth for the ripening of oats have been established to be 1050 degree-days above a base of 4.4 °C, and maximum tolerable levels of the other two limiting factors 6.2 m/sec average wind speed and 60 mm potential water surplus. The zone which is climatically marginal for the harvesting of ripened oats can be identified by contouring combined isopleths of these three parameters (Figure 12.13).

Trends of accumulated temperature, the dominant limiting factor, and potential water surplus have been constructed by Parry (1975, 1976, 1978) for the period AD 1100 to 1950, using data from Lamb (1977). It is estimated that if, over the period 1250–1450, the mean temperature fell a little less than 1 °C, then summer warmth at 300 m OD in northern England would have been reduced by 15 per cent, frequency of crop failure increased sevenfold from 1 year in 20 to 1 year in 3, and the frequency of two consecutive crop failures increased 70 times. The existing chronology for this early period is likely to be improved and refined as more documentary sources are treated critically (Farmer and Wigley 1983) and more data from sources such as X-ray densitometry of tree-rings become available. The existing estimates indicate clearly that there were substantial changes in the viability of upland farming as the climatic limits of cultivation rose fluctuatingly over this early period. Figure 12.14 shows the close temporal and spatial accordance between the zone through which the theoretical limits to cultivation fell between AD 1300 and 1600 and the distribution of farmland known to have been abandoned before 1600. The same examinations carried out for the period AD 1600 to

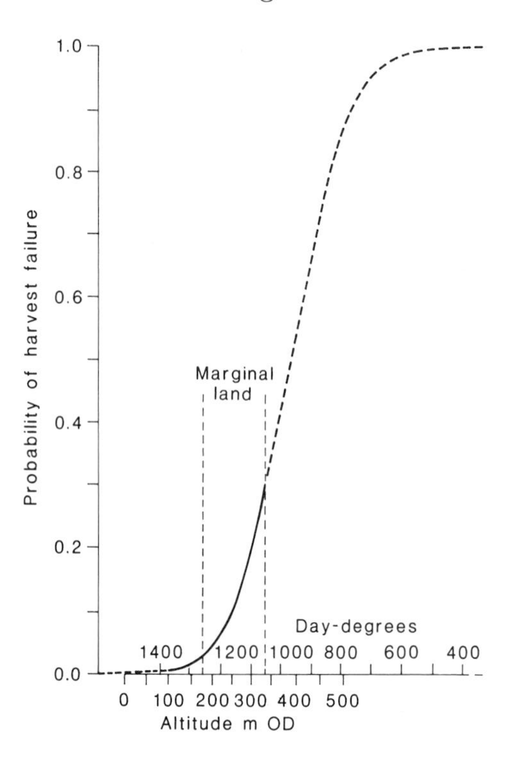

Figure 12.13 The probability of harvest failure in southeast Scotland. (From Parry 1976)

1800 provide an equally convincing result (Figure 12.15). Between 1600 and 1750, three out of an existing total of fifteen farmsteads in the Lammermuir Hills were abandoned within the zone calculated to have become sub-marginal by about 1530. A further eight were deserted in the zone which became sub-marginal between 1530 and 1600 and all but one of the remainder are in the marginal zone. Three-quarters of the 2900 ha of cultivated land which reverted to permanent moorland between 1600 and 1800 had become climatically sub-marginal before 1600. It is therefore logical to assume that there was a causal relationship here between climatic change and land abandonment.

The processes involved in land abandonment can be better understood if changes in climate are considered in terms of changes in frequency of short-term anomalous events. Such events affect the level of risk faced by the farmer. For the subsistence or semi-commercial farmer, the viability of farmland depends on its ability to sustain the household from one harvest to the next. The extent to which yield exceeds this essential minimum is less important than the likelihood of achieving the minimum in a given year, and the probability of harvest failure is more important than average yield over a number of years. Harvest failure in two successive years, leading to the consumption of the seed grain or the exhaustion of cash reserves, is likely to be disastrous.

Figure 12.14 (*on facing page – above*) Abandoned farmland and lowered climatic limits to cultivation in southeast Scotland between AD 1300 and 1600. (From Parry 1975)

Figure 12.15 (*on facing page – below*) Abandoned farmland and lowered climatic limits to cultivation in southeast Scotland between AD 1600 and 1800. (From Parry 1975)

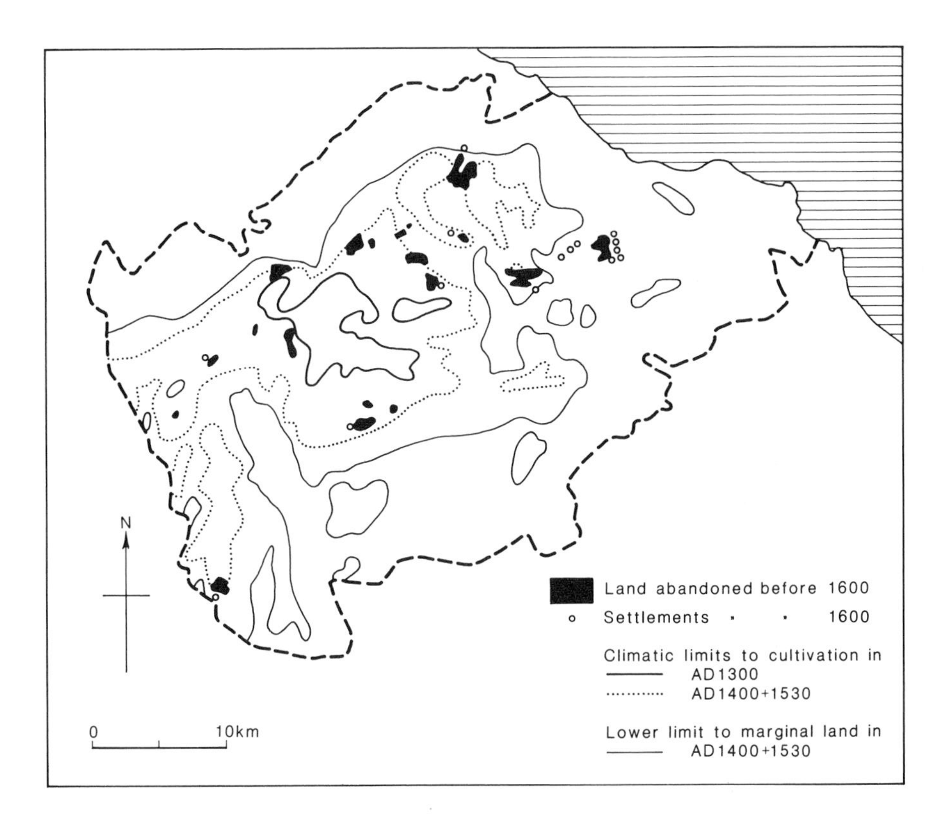
N
0
10km
Land abandoned before 1600
Settlements · · 1600
Climatic limits to cultivation in
AD1300
AD1400+1530
Lower limit to marginal land in
AD1400+1530

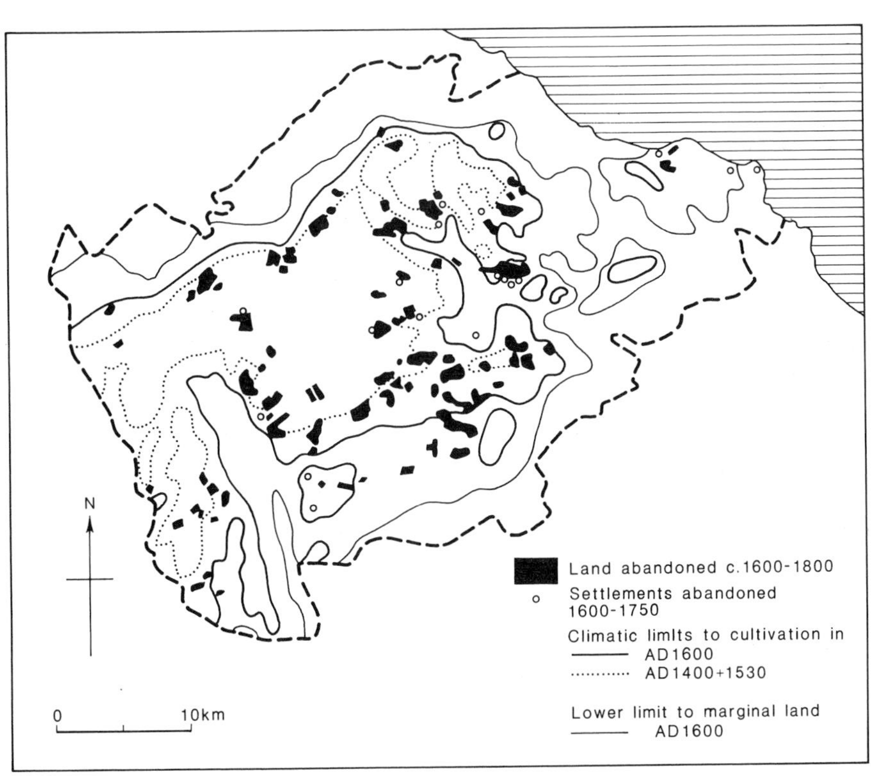
N
0
10km
Land abandoned c.1600-1800
Settlements abandoned 1600-1750
Climatic limits to cultivation in
AD1600
AD1400+1530
Lower limit to marginal land
AD1600

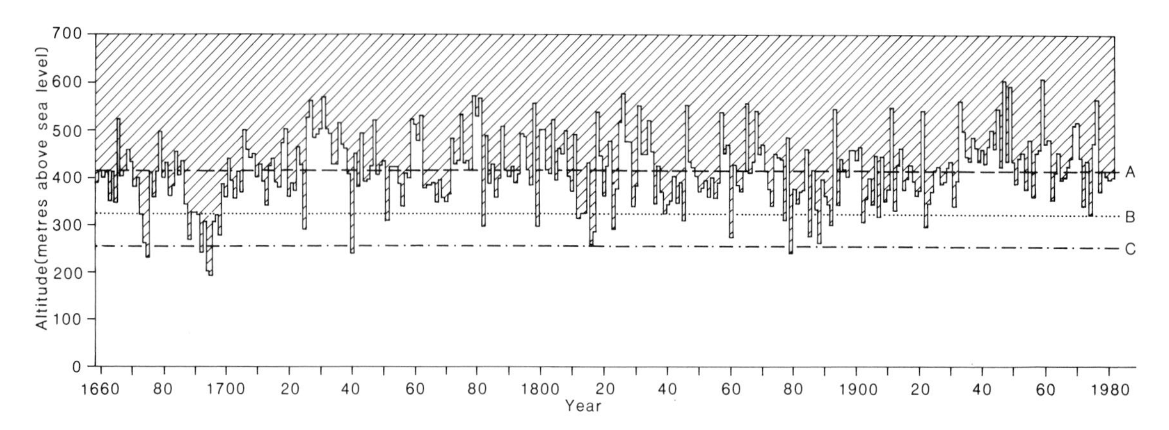

Figure 12.16 Hypothetical shift of oat crop failure with altitude in southern Scotland 1650–1981: (a) mean altitude of crop failure; (b) 1:10 failure frequency; (c) 1:50 failure frequency. (From Parry and Cater 1985)

In maritime upland areas, quite small increases in altitude commonly result in marked shortening of the growing season and a great reduction in the accumulated warmth as measured in degree-days above a certain temperature (Manley 1945). Furthermore, the variability of accumulated warmth relative to the mean increases with altitude, enhancing further the altitudinal fall in agricultural potential. Parry and Carter (1985) have evaluated the risk of crop failure resulting from low levels of accumulated temperature in the Lammermuirs, deriving a run of temperature data for the period 1659–1981 by bridging data sets from central England (Manley 1953, 1974) to Edinburgh and from Edinburgh to a network of twenty-seven stations covering the Southern Uplands. As the regionally averaged lapse rate of temperature with elevation is known to be 0.68 °C per 100 m, the height at which the minimum accumulated temperature for ripening of oats of 970 degree-days is achieved each year could be calculated. This gave the hypothetical shift of the level at which the oat harvest would have failed between 1659 and 1981 (Figure 12.16). The incidence of years with less than 970 degree-days was calculated for 10 m intervals of height and the theoretical frequency of two consecutive failures computed, assuming year-to-year accumulated temperature totals to be statistically independent. It had previously been assumed that annual totals of accumulated warmth would be normally distributed but comparison of calculated and theoretical frequencies showed that clustering of cool years results in a greater risk of consecutive failures. At the approximate level of cultivation of 340 m the observed frequency of consecutive failures is more than double the frequency that would result from random temperature variations about a mean. Frequencies of 1:10 and 1:50 were taken to delimit the upper and lower margins of the high risk zone for oats. This zone has shifted through a range of more than 85 m in altitude in the last few centuries, so that the level of 1:50 frequency for the warm period 1931–80 lies well above that of the 1:10 frequency for the cool period 1661–1710 (Figure 12.17). While in the first period the area above 280 m was sub-marginal for oats cultivation, the climatic limit of the modern period is at about 365 m. The isopleth for 1:50 frequency for 1931–80 coincides with the present limit of cultivation and the moorland edge.

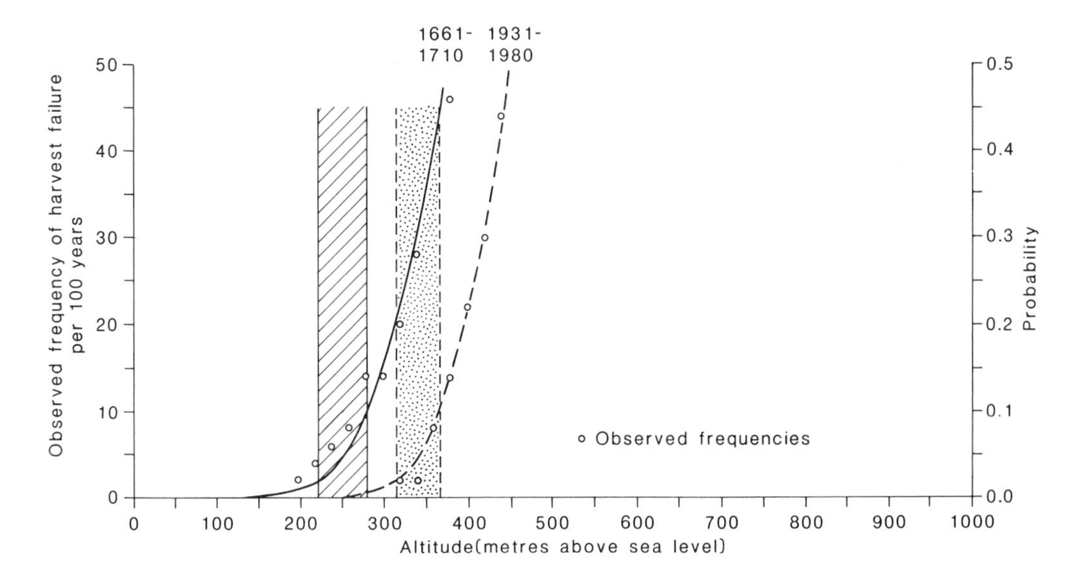

Figure 12.17 Differences in the risk surfaces of crop failure in cool (1661–1710) and warm (1931–80) periods in southern Scotland. High-risk zones are shown by shading and stippling. (From Parry and Carter 1985)

It could be expected that the effects of single isolated cool summers, such as those which occurred in 1740, 1782, 1860, 1879 and 1922, might differ from those of periods characterized by a high frequency of scattered but not successive cool summers, as between 1812 and 1817 and again between 1879 and 1892. The greatest effects might be expected to come from clusters of years with two, three or more cool summers in a row. The available historical data appear to bear this out. There was, for example, an outstandingly poor harvest in Scotland in 1740 but food shortages were made good the following year (Parry 1978). In 1782 the oats harvest in the uplands was delayed until December. In the parish of Lauder 'it was the end of December before the harvest was finished after a great part of the crop was destroyed by frost and snow. None of the farmers could pay their rents; some of them lost two hundred to five hundred pounds sterling' (*Statistical Account of Scotland 1791–99*). There were some farm bankruptcies causing amalgamation of holdings but the parish returns from the *Statistical Account* show little sign of lasting hardship in the 1790s resulting from the single poor summer in the previous decade.

Clusters of cool summers which would have caused consecutive harvest failures above 300 m occurred in 1674–5, 1694–5 and 1816–17. Above 340 m, failures would theoretically have occurred in all of eleven successive years from 1688 to 1698. A comparison of the manuscript Pont Maps dated about 1596 with those of the Military Survey of Scotland of 1747–55 shows that few of the high-level farms survived. Of thirty-two recorded about 1600, only ten survived into the next century (Parry and Carter 1985).

Though the impact of periods of frequent though scattered cold summers was no doubt less immediately marked, the effect seemed to have been as long-lasting as those of the clusters of extreme years. For example, harvest failure was calculated to have occurred in five years out of the sixteen between 1887 and 1892 at heights above 310 m, and official annual acreage returns for the region show a 5 per cent reduction in the areas under crops and grass between 1880 and 1885 and reductions of 1 per cent every five years from 1885 to 1900.

Parry's work has illuminated the relationship between the impact of brief climatic events which may be proximate causes of short-term economic events and long-term climatic change which may cause long-term changes in the resource base. The relationship beween the incidence of extremes which may be of crucial importance and changes in the mean (Wigley 1985) is central to the whole problem of the human consequences of climatic change. The correlation between historical data and events predicted on a basis of meteorological parameters and phenological requirements of specific plant types provides a logical basis for embarking upon evaluation of the importance of climatic change in relation to other factors, economic, political and social. A declining resource base would cause greater sensitivity to such other factors, a point which has not been readily acknowledged by historians.

The retreat of high-level cultivation in southern Scotland may have been triggered by political upheavals, the Black Death, the decline of the monasteries or runs of cool summers in the 1590s and 1690s, but it was the operation of these forces on a zone where marginality had been enhanced by climatic deterioration, particularly in the fourteenth and seventeenth centuries, that produced such a marked and lasting effect (Parry 1981 and Figure 12.18). The theoretical shift of the cultivation limit over the British Isles as a whole between 1300 and 1600, based on estimates of accumulated temperature, delineates an area of over 2 million hectares likely to have been abandoned for climatic reasons. About one-third of Britain's unimproved moorland was climatically viable for cereal cultivation in the early Middle Ages. The phase of widespread desertion of villages in England, 1370–1500, coincided broadly with the cooling phase associated with southward shift of depression tracks in the North Atlantic and reduced cyclonic activity. By 1500 summers may already have been 0.7 °C cooler than those of the medieval optimum.

Upland in other parts of the British Isles appears to have been abandoned at much the same time as in the Lammermuirs. Many farmsteads on Dartmoor, for instance, were left deserted in the late thirteenth and fourteenth centuries. Of the 110 medieval settlements known to have been deserted on Dartmoor, twenty-seven are at between 210 and 275 m, twenty-two between 275 and 305 m and sixty-one over 305 m (Linehan 1966). Conversion of abandoned houses for use as barns and corn driers shows that desertion was gradual but by the middle of the fourteenth century all of the settlements above 300 m had been abandoned (Beresford 1981).

12.3.4 LOWLANDS IN NORTHWEST EUROPE

Desertion of settlements in lowland England, along with the decline of agriculture, has conventionally been viewed in terms of the effects of population growth on the availability and productivity of the land, considerable emphasis being placed on soil exhaustion due to population growth outstripping the optimum determined by contemporary techniques (e.g. Postan and Hatcher 1978). Historians such as Postan (e.g. 1975) accepted the climatic chronology put forward by Brooks (1949), now superseded, according to which the early medieval period, during which population expansion took place, was cold. The relevance of climate to the desertion of villages has therefore been seen as minimal, as few such desertions took place during or soon after the supposed period of climatic deterioration. In fact, the build-up of population took place during the medieval warm period (Lamb 1965) and desertion in the following period of damper and probably cooler summers (Lamb 1977, pp. 449–61). The most frequent reasons given by contemporaries for abandonment were shrinkage of village population, shortage of seed corn, soil exhaustion and shortage of plough teams (Baker 1966). Shortage of seed and of plough teams would both have been dependent on previous harvests.

There is no question that a number of factors, social and economic, were involved in the diminishing number of villages and abandonment of arable farming, but successions of wet, cool summers and disastrous harvests, such as those of 1314–16 and 1320–1, may have had important triggering effects (Kershaw 1973). By 1500, summers were of the order of 0.7 °C cooler than the summers of the medieval optimum, and shortening of the

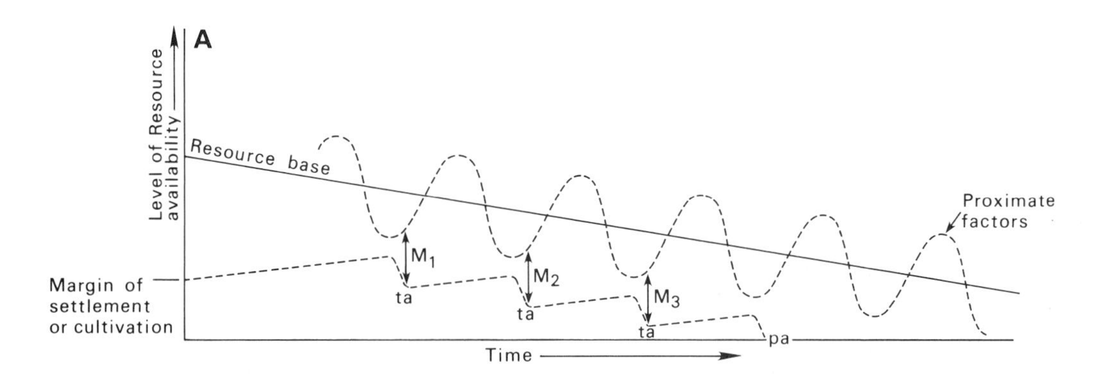

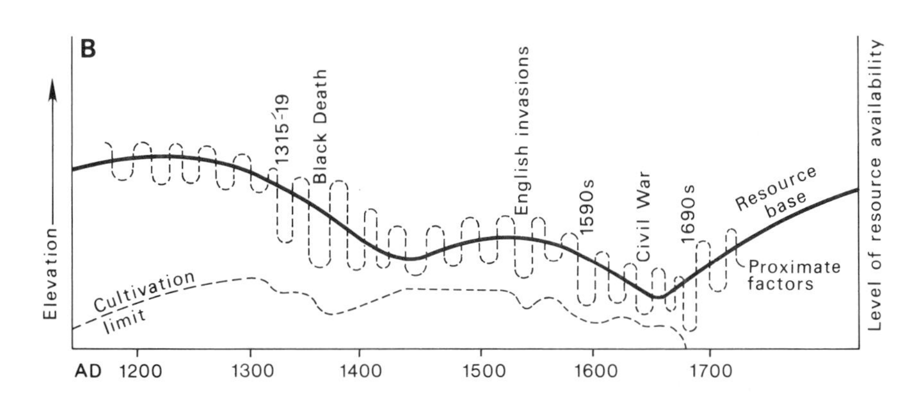

Figure 12.18 Proximate and indirect factory behind land abandonment or settlement desertion. M1-3 = marginal tolerance of resource scarcity (high frequency of crop failure, lowest average yield tolerated); ta = temporary land abandonment; pa = permanent land abandonment. (From Parry 1975)

growing season as prevailing temperature fell probably amounted to about three weeks in England by 1400 and as much as five weeks by the seventeenth century.

Increased wetness of the ground may also have had a part to play. Reduction in temperature and consequent evaporation would have caused increased wetness. Reports tell of marshes spreading and rivers increasing in volume, reaching a maximum in the fifteenth century (Lamb 1984). It has been argued that desertion of medieval villages cannot be attributed to increased wetness because deserted and non-deserted villages are to be found adjacent to one another (Beresford and Hurst 1971). But soils vary in their drainage characteristics over quite short distances, and such variations are especially important in the wetter western parts of the country. Even in eastern England, 14 km east of Lincoln, at only 21 m OD, the site of Goltho provides a case in point. The soils surrounding it are mainly on the Ragdale Series, which are especially subject to waterlogging and compaction and difficult to manage in wet years even with modern machinery. Similar conditions are found around most of the sites of the twenty-six deserted medieval settlements within a 10 km radius (Beresford 1981).

England was far from being the only place in Europe to suffer a decline in both population and agricultural production after 1300. Rural depopulation was very widespread. Utterström (1955) reinvigorated the suggestion made by Norwegian historians decades earlier that political and economic decline might have been caused by climatic deterioration. Utterström's work was severely criticized by Ladurie (1971, pp. 8–11), on the basis not only that much of his evidence was not *a priori* climatic but that many of the situations described 'can be explained equally well or better in terms of purely economic considerations' and also that 'bad years have to be regarded not as a long series but merely as episodes in short term meteorological fluctuations'. Since Ladurie wrote *Times of Feast, Times of Famine*, great advances have been made in the methodology of palaeoclimatology and our knowledge of the general course of European climate in the historic period has been placed on a sounder footing. As we have seen, the economic relevance of short-term fluctuations as well as changes in the long-term mean can be considerable.

In Denmark agricultural decline was widespread after 1340, with most of the villages in some parts of Jutland being abandoned. In northern Norway, the limits of cultivation fell at least 150 m between 1300 and 1600 and through the country as a whole about half the medieval farms were abandoned. Interdisciplinary studies made in the course of the Ødegårdsprosjekt (deserted farm project) launched in 1968 (Gissel *et al.* 1981) have revealed the great complexity of the phenomenon. In Norway, desertion was found to be greatest in high-lying, hilly, forested areas and least in low-lying agricultural districts. In most places there was a reduction in grain pollen in the late Middle Ages, but it was found open to question whether this was due to plague decreasing the population or to climatic deterioration. Ultimately, the participants came to the conclusion that their work had 'strengthened an explanation based on plague as a cause of regression' but there has yet to be a full-scale study of the location of pre-Black Death desertion in terms of degree-days and other climatic indices (Sandnes 1971, Parry 1975).

The course of events in Scandinavia in the present century points to the influence of temperature on yields there in medieval and early modern times. The rise in temperature in the twentieth century improved conditions not only for arable farming but also for horticulture and forestry. A considerable rise in wheat and rye yields has been attributed to extra summer warmth; the warm summers of the 1930s led to superb harvests. The trend was halted by the exceptionally cold summers of 1940–2 and the cooling trend of the following decade. In Lappland cultivation of wheat was abandoned in the early 1950s and of rye at the end of the decade. Arable farming was in clear decline by the 1960s, with falling yields and uncertain profits (Bourke 1984).

Many of the years of crisis mentioned earlier in this chapter were common to several of the areas discussed. In particular the 1690s, the years around 1740, and 1816–19 were all years of dearth and famine. The famine which began with the harvest of 1693 was one of the worst in western Europe since the Middle Ages, turning France into a 'great desolate hospital without provisions' and bringing 'green years' to Norway when the crops failed to ripen, and carrying off perhaps a third of the population of Finland in 1696–7 (Jutikkala 1955). In the winter of 1739–40, the major rivers of northern Europe froze over, as did the Zuider Zee and the Kattegat. The temperature in central England in January and February 1740 was more than 6 °C below the contemporary decennial average (Manley 1958). Temperatures remained well below normal in every month that year except September. The year 1740 had the lowest mean annual temperature in Manley's series for central England (Manley 1974). Most of northern and western Europe suffered severely (Bourke 1984). Potatoes were by now the staple of the Irish poor, being customarily left in the ground and lifted when they were needed during the winter. In 1739–40 the crop was frozen in the ground and destroyed and the death toll in this 'year of the slaughter' was only to be exceeded in the 1840s (Drake 1968, cited Bourke 1984).

The 'year without a summer' of 1816 and the one that followed were marked by a crisis of at least hemispherical proportions according to Post (1977), who surveyed not only the agricultural consequences but a whole range of concurrent economic and social phenomena which he saw as causally connected. He argued that it was the interaction of economic and climatic fluctuations, rather than political decisions, that led to the disruption of commodity market relationships and attendant social upheavals. Post, like Pfister (1975) who made a detailed study of the crisis from 1768–71 in the Bern region, concluded that climate is an important and neglected historic force. Many historians remain reluctant to accept this conclusion.

12.3.5 SOME PROBLEMS OF ASSESSING THE IMPACT OF THE LITTLE ICE AGE ON HUMAN AFFAIRS

Assessments of the importance of climatic change in human affairs have often been unsatisfactory, partly because climatology and history are such different disciplines and adequate communication between their practitioners has been lacking, partly because of the very complexity of the interactions of climate, biosphere and human activity. Until recently climatologists have rarely been familiar with the methods of historical research, more particularly with procedures of source validation, and have been too prone to assume that meteorological and human events occurring at the same time are necessarily causally related. It is all too easy to assume that runs of grain prices can be used as proxy indicators for variation in climatic elements, or that the climatic history of medieval or early modern Europe is expressed by the sequence of its harvests. Early work on the impact of climate on history was naïvely deterministic and not unnaturally led to a long-lasting bias against the subject amongst many of those historians who considered the matter at all. But it is beginning to be acknowledged that Huntington's (1907, 1915) thesis regarding the major role that climate plays in human affairs has never been satisfactorily refuted (Fischer 1980).

The trickle of literature on climatic change, its causes, chronology and biological consequences, has grown to a flood too great for any single person to keep up with advances in all the different fields into which it has diversified. It is hardly surprising that many historians are unfamiliar with the main themes and suspicious of methodologies so foreign to their own. They have often failed to appreciate the inexactness of early chronologies and have referred back to work long since superseded or discredited. The disinclination of historians to consider the possibility that changes in climate may play a role in historical explanation has not entirely disappeared and even some of those who are more open-minded remain sceptical. De Vries (1980), in attempting to highlight the problems of reconstructing past sequences of weather from scattered observations, refers to work by Easton (1928) despite the fact that it is known to include important errors (Bell and Ogilvie 1978). The views of both Postan (1975) and van Bath (1963) on the relative unimportance of climate as a determinant of economic recession and decline of yields were used in the same paper to counterbalance those of other historians disposed to consider climate as more influential. Yet Postan and van Bath accepted and used the outdated climatic chronologies of Britton (1937) and Brooks (1949).

One wonders how many historians appreciate the cumulative and progressive nature of scientific research! Certainly they are inclined to ignore the fact that one and the same climatic fluctuation can have more marked effects in one place than another, and even different effects in adjacent regions. A period marked by pronounced cold in northwest Europe can be accompanied by a change in rainfall amount in a part of the Mediterranean basin or eastern Europe. On the plain of the Danube, for instance, little evidence has been found of lower temperatures in the Little Ice Age; however, agriculture in the region was affected between 1550 and 1850 by conditions more humid and less continental than they were before or have been since (Czelnai 1980).

Whether a change in the length or intensity of the growing season will be of economic importance will depend partly on the geographical location and the climatic sensitivity of the area in question and partly upon the vulnerability of the socio-economic system concerned. De Vries (1980) found no significant correlation between climatic change and the level of economic activity in Holland in the seventeenth and eighteenth centuries although he discerned the influence of wet winters on farming and cold winters on canal transport. But the Netherlands are not amongst the most sensitive parts of Europe climatically, and the Dutch ecnomy in the seventeenth century was described by de Vries himself as having been 'in the unique position of being a net gainer from events which brought so much grief to so much of the rest of Europe' (Post 1980). Yet even in Holland with its flourishing grain trade, some parts of the population were affected by the incidence of extremes; for instance, the incomes of barge-owners were lowered by cold winters because they had fewer passengers and had to pay more for fodder for their horses, their largest single expense.

Increased recognition of the need for interdisciplinary understanding and communication has led not only to the appearance of volumes of review papers concerned with 'Climate and History' (Wigley *et al.* 1981, *Jounal of Interdisciplinary History*, volume 10, 1980), but also to reviews of food–climate relations on a world scale (Takahashi and Yoshino 1978, Bach *et al.* 1981). Crop–climate relationships, with which both climatologists (e.g. Lamb

1982) and historians (e.g. Post 1977, de Vries 1980) have had to involve themselves, are the specialist province of agronomists.

12.3.5.1 The complexity of weather sequences, crop responses and economic consequences
Weather affects crops directly, but the effect of a particular weather sequence may differ from one crop and one variety to another, with vulnerability to adverse conditions varying according to stage in the growth cycle. The effects of a particular season are commonly felt in a succeeding growing season, through seed quality, soil moisture and temperature at the time of sowing, and the carry-over of pests and diseases. Understanding of climatic requirements of individual crops has advanced greatly in recent decades as detailed crop–weather models have been developed and refined. These simulate response to environmental factors in terms of growth rate, development towards maturity, and harvest yield. Such models depend on the input of very detailed meteorological parameters, not merely seasonal averages. They cannot be used for analysis of pre-twentieth-century harvests because data is inadequate, but they should aid understanding and act as a disincentive to over-easy generalization about crop–climate relationships. None the less, it remains true that the primary requirements of crop plants, like their wild relations, are warmth and moisture, though not in excess, and agricultural production can be related in a general way to how these requirements are met, particularly during the growing season.

The implications of particular rainfall patterns vary from region to region, according to whether the main climatic stress on agriculture arises from deficit or excess of heat or water. However, there are other aspects of climate that cannot be overlooked, such as the incidence of sunshine necessary for ripening, of hailstorms which can cause direct damage, especially if they are associated with high wind, or high humidity which can favour fungal attacks. Snow can lead to severe lodging of cereals, though it can also play a protective role, sheltering crops from very low temperatures that would otherwise kill them outright. It is the sequence of weather conditions involving all or several of these factors in a particular year which may cause outstandingly good or bad harvest years.

12.3.6 ANOMALOUS WEATHER SEQUENCES IN PARTICULAR YEARS: 1695, 1771 AND 1816

Contemporary accounts sometimes allow the course of weather events to be traced that led to dearth in particular years. The impact upon society of adverse climatic conditions is increased if the anomalies concerned are extensive, if there are several successive poor harvests and if there is simultaneous failure of several crops.

The decade of the 1690s was one of the coldest, if not the coldest, on record in western European history. Volcanic eruptions of Hekla (Iceland) and Serua (Indonesia) in 1693, and of Aboina (Indonesia) in 1694 produced readily recognizable frost rings in trees in North America (Chapter 11 and Figure 11.5) though their traces are not prominent in the Greenland ice-core record (Figure 11.6). The weather in Europe in 1695 was so outstandingly severe that an account of it appeared in a German serial publication *Theatrum Europaeum* of 1702, which is believed to be the first detailed account of its kind (Lindgren and Neumann 1981). Finland has been estimated to have lost a third and Estonia 20 per cent of its population; references have been made here earlier to famine at this time in Norway (3.1.3 and 12.3.2) and Scotland (12.3.3). Summers had been cold and wet in the 1680s and early 1690s. The winter of 1694–5 started early and was extraordinarily hard and long in most parts of Europe, sea ice coming far south on the coast of Iceland and lakes and rivers freezing over well into the spring. Snow on the mountains persisted into the summer and then thawed in June, at the same time as rainy unsettled weather was spoiling the pastures and crops. Lowland rivers flooded heavily; crops could not be gathered in 'partly because they were retarded in growth by the wet weather and partly because they could not be dried'; salt, widely used for preserving foodstuffs, could not be made at the coastal sites where it was usually manufactured. Famine and social distress were widespread in Europe at this time, and it was in this decade that both the Icelandic and the Norwegian glaciers were advancing strongly into farmland.

In the eighteenth century, the peasants' diet in the Pre-Alps of Switzerland included bread, potatoes, milk, cheese, fruit and vegetables, which could, to a considerable extent, be used as substitutes for each other. Moreover, grain could be supplied from government stocks in times of scarcity. Despite this broad nutritional base, a major subsistence crisis brought famine in 1770–1. Every summer from 1767 to 1771 was wet. A wet autumn in 1768 was followed by a long winter in 1769–70 and an extremely snowy summer in 1771. Wet autumns affected wheat production in the lowlands, long winters damaged seeds in the uplands and depleted hay stocks. Wet summers diminished yields of hay, potatoes and grain, and snowy summers on the Alpine meadows reduced cheese production. The worst impact came in 1771, when harvests were poor for the second consecutive year. Dairy production was low because of reduced herds, shortage of pasture and poor quality of hay; seed potatoes were rare and expensive; government grainstores were empty and supplies from traditional sources such as Burgundy, Swabia and Alsace were hindered by embargoes probably resulting from a more general shortage. Grain prices escalated with harvest failures in central and western Europe in 1769 and 1771 and in northern Europe in 1771 and 1772 (Pfister 1975, 1978). This, it will be recalled, was at the beginning of a decade when the Grindelwald and other Swiss glaciers advanced rapidly.

The sequence of weather and crop reactions in New England during the summer of 1816, following the eruption of Tambora in 1815 (Figures 11.7 and 11.8), may be taken as another example of an extreme year in a period of unusually severe conditions (Post 1977, pp. 9–12). There was great cold in April, with late snows and crop growth much retarded. By mid-May it was reported that spring was 'nearly six weeks less forward than common'. Frosts in southern New England in early May were not serious because of the retardation of growth, but those in late May were a threat to growing crops. Temperature at the beginning of June was more seasonable and there was hope of a successful harvest after all, but then came a pronounced cold wave between 5 and 11 June, with piercingly cold winds from the northwest. The earth was frozen half an inch deep at Williamstown in the far south and by 11 June cucumbers and other vegetables were mostly destroyed. On 10 June came severe frost in the morning and 'ten days after the frost the trees on the sides of the hills presented for miles the appearance of having been scorched'. The effects of snow, frost and sub-freezing temperatures were harsh in the south; in northern districts they were disastrous. Most of the vegetables and Indian corn in the northeast were killed. Undaunted, the farmers replanted and after 11 June there was a month of good growing weather, but July 1816 averaged 5 °F (3 °C) below normal. This was the coldest July in American meteorological history. The hay crop suffered, reducing the number of animals that could be overwintered the following year. Wheat and rye fortunately survived better than the corn. The weather modified in late July but there was another frost on 11 August, before the harvest. It was particularly severe in northern New England and corn was again the crop mainly affected. On 28 August a further cold wave hit the whole region and a succession of frosts ended the growing season before the end of September. The corn crop had been almost totally lost, the hay crop was small, and the wheat and rye crops were fair to normal. Though the mean temperature in 1816 was 1.3 °F below the normal for the period 1816 to 1838, it was not as cold as in 1836 or 1837, the years following the eruption of Coseguina, when mean temperatures were 3 °F below the norm. But the early morning temperatures in June, July and August 1816 were the lowest recorded in the period. There is historical evidence for an advance of glaciers in the Alps, Scandinavia and Greenland, as well as moraine dates suggesting advances at this time in North America and elsewhere (11.1).

It was the low temperatures in the growing season of 1816 which made it such an outstanding year. One is led to the conclusion that simple correlations of average winter temperature, or the incidence of frost in March, against economic variables, such as grain shipments or butter prices, are unlikely to be particularly conclusive. One may also be impressed by the global hazard presented by weather conditions following unusually violent volcanic eruptions.

12.4 A future prospect

The importance of the climatic factor in history has been minimized by some twentieth-century historians and historical geographers. They have been working in the more prosperous parts of Europe and America during a period when history, like many other fields, has become more specialized and, moreover, in a period during the first half of the century when climate was somewhat more benign and dependable than it had been earlier or has been since. The broad overview has not been fashionable but such syntheses are needed. Post's (1977) study of weather patterns and the great subsistence crisis of 1816–17, including within its scope discussion of the interaction of climate and social unrest, demography, epidemic disease and trade cycles, covered both North America and northwest Europe. In view of the glaciological evidence presented in Chapters 8 and 9, it is not at all surprising to find preliminary studies suggesting that the economic and social crises affecting Europe during the Little Ice Age had parallels in Latin America (Claxton and Hecht 1978) and Mexico (Swan 1981). Pfister (1985) has shown the way forward with his meticulous and comprehensive studies of Swiss material relating to the climate of the Little Ice Age and its effects. If studies of this kind were to be made elsewhere, not only in Europe but also in China, Korea and Japan (Yamamoto 1971a, b) where comparable source material is available, and even in Latin America, the adequacy of an assessment of the part played by climate in human affairs would be much improved.

The importance of climatic trends extending over wide areas within which social, political and economic conditions differ from one part to another has been brought forcefully to our attention during the drought which has affected large parts of Africa in the 1970s and 1980s. Here it is water deficiency that has brought starvation to large numbers of people. Here, as in Europe in the subsistence crises of earlier centuries, other factors are also influential. War, maladministration, conflicts between the needs of urban and rural communities, and, not least, adverse terms of trade have had their parts to play.

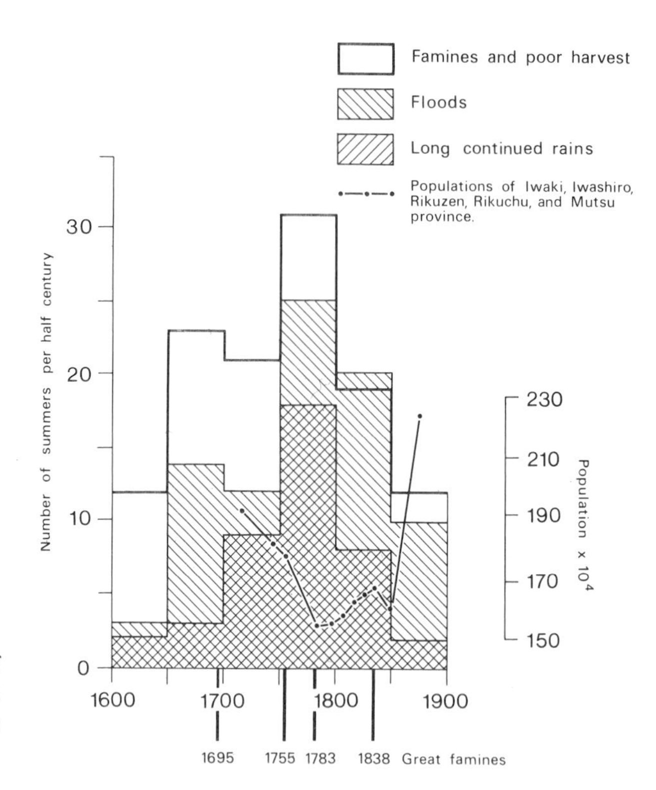

Figure 12.19 Secular changes of summer climate in Japan during the Little Ice Age. (From Yamamoto 1971a)

But their effects would not have been so severe had there not also been a downturn in rainfall. In present-day Africa, as in Little Ice Age Europe, the poorest suffer most and those least able to adapt to changing circumstances are most affected. Famine in the regions hardest hit brings increased susceptibility to disease and increased mortality. It would be impossible to argue that other economic factors have been unimportant; equally it is impossible to argue that their influence has not been much enhanced by climatic conditions. This modern analogy shows that while climatic change cannot be ignored, it never acts alone. The impact must vary with the severity of the change, the character of the environment affected and the social and economic resilience of the people involved.

Technological advance and sophistication of the economic structures of societies provide important buffers against the immediate impact of climate, but protection is not without cost. Buildings can be heated, roads and railways kept open and foods in short supply imported from elsewhere, so long as these are available. Although western Europe and North America are currently in a position of surplus food production, deficiencies exist elsewhere. The agricultural revolution since the Second World War generated by plant breeding and selection, with increased use of fertilizers and pesticides, led for a time to the assumption that the weather was no longer of importance. This complacency was shaken by the severity of the cooler years of the 1970s. It is now recognized that the more efficiently farming is organized to make optimum use of 'normal' weather conditions, the more sharply production is likely to fall if the climate deviates markedly from this norm. If crop varieties are bred to thrive in a narrow range of temperature and moisture conditions they may give poorer yields than their predecessors if the actual conditions are outside this range.

It is the climatic implications of increased carbon dioxide (CO_2) concentrations in the atmosphere that currently exercise most of the climatic community. It is possible that such man-induced changes helped to bring the Little Ice Age to an end and that the global rise in temperature since the middle of the last century has been caused by these increased concentrations. The fact remains that a small but distinct temperature fall affected large parts of the globe in the 1960s and 1970s and occurred despite increasing CO_2 concentrations. This cooling was, as we have seen, sufficient to have marked biological effects and to influence agricultural production in high latitudes and even to have some political consequences, notably in relation to Iceland. The small advances of glaciers triggered calculations of the probable consequences of continued positive mass balances of glaciers near mining sites in Canada (Fisher and Jones 1971) and the possible hazards involved for hydro-electric installations in Switzerland (Bindschadler 1980). The advance of the Findelen in the 1980s enforced the rebuilding of water offtakes for the Grande Dixence hydro-electric scheme in the Valais. A more severe episode would not only cause a number of major engineering works to be threatened by advancing glaciers but would also present a hazard to numerous recently built ski resorts. A recurrence of mass movements in the Mont Blanc massif would now affect far more people than was the case a century or two ago (Grove 1987).

A reversion to the climatic conditions of the Little Ice Age is perhaps less likely now than it would have been but for the burning of fossil fuel in enormous quantities over the last few decades. But it might be wise not to assume that a return to Little Ice Age conditions is entirely out of the question.

Bibliography

1 Introduction

Ahlmann, H. W. (1949) *Glacial Research on the North Atlantic Coast*, Royal Geographical Society Special Research Publication 1.

Bell, W. T. and Ogilvie, A. E. J. (1978) 'Weather compilations as a source of data for the reconstruction of European climate during the medieval period', *Climatic Change* 1: 331–48.

Benedict, J. B. (1968) 'Recent glacial history of an alpine area in the Colorado Front Range, U.S.A. II. Dating the glacial deposits', *Journal of Glaciology* 7: 77–87.

Bernabo, J. C. (1981) 'Quantitative estimates of temperature changes over the last 2700 years in Michigan based on pollen data', *Quaternary Research* 15: 143–59.

Beschel, R. (1961) 'Dating rock surfaces by lichen growth and its application to glaciology and physiography (lichenometry)', in Raasch, G. O. (ed.) *Geology of the Arctic*, vol. 2, University of Toronto Press, pp. 1044–62.

Birkeland, P. W. (1981) 'Soil data and the shape of the lichen growth curve for the Mt Cook area (note)', *New Zealand Journal of Geology and Geophysics* 24: 443–5.

Bradley, R. S. (1985) *Quaternary Paleoclimatology. Methods of Paleoclimatic Reconstruction*, London, Allen & Unwin.

Britton, C. E. (1937) *A Meteorological Chronology to AD 1450*, Meteorological Office Geophysical Memoir 70, London, HMSO.

Brooks, C. E. P. (1949) *Climate Through the Ages. A Study of the Climatic Factors and their Variations.* Revised edn. London, Benn.

Burbank, D. W. (1981) 'A chronology of late Holocene glacier fluctuations on Mount Rainier, Washington', *Arctic and Alpine Research* 13: 369–86.

Burke, R. M. and Birkeland, P. W. (1977) 'Re-evaluation of multiparameter relative dating techniques and their application to the glacial sequence along the eastern escarpment of the Sierra Nevada, California', *Quarternary Research* 11: 21–51.

Calkin, P. E. and Ellis, J. M. (1984) 'Development and application of a lichenometric dating curve,

Brooks Range, Alaska', in Mahaney, W. C. (ed.) *Quaternary Dating Methods*, Amsterdam, Elsevier, pp. 227–46.

Carrara, P. E. and McGimsey, R. G. (1981) 'The late-Neoglacial histories of the Agassiz and Jackson glaciers, Glacier National Park, Montana', *Arctic and Alpine Research* 13: 183–96.

Childe, V. G. (1936) *Man Makes Himself*, London, Library of Science and Culture.

Collins, D. N. (1984) 'Climatic variation and runoff from Alpine glaciers', *Zeitschrift für Gletscherkunde und Glazialgeologie* 20: 127–45.

Davis, M. B. and Botkin, D. B. (1985) 'Sensitivity of cool-temperate forests and their fossil pollen record to rapid temperature change', *Quaternary Research* 23: 327–40.

Deevey, E. S. and Flint, R. F. (1957) 'Postglacial hypsithermal interval', *Science* 125: 182–4.

Denton, G. H. and Karlén, W. (1973) 'Holocene climatic variations – their pattern and possible cause', *Quaternary Research* 3: 155–205.

Furrer, G. and Holzhauser, H. (1984) 'Gletscher- und klimageschichtliche Auswertung fossiler Hölzer', *Zeitschrift für Geomorphologie*, N.F.50: 117–36.

Gad, F. (1970) *The History of Greenland II*. Earliest Times to 1700, London, Hurst.

Gellatly, A. F. (1985a) 'Glacial fluctuations in the central southern Alps, New Zealand: documentation and implications for environmental change during the last 1000 years', *Zeitschrift für Gletscherkunde und Glazialgeologie* 21: 259–64.

Gellatly, A. F. (1985b) 'Phosphate retention; relative dating of Holocene soil development', *Catena* 12: 227–40.

Gellatly, A. F., Röthlisberger, F. and Geyh, M. A. (1985) 'Holocene glacier variations in New Zealand (South Island)', *Zeitschrift für Gletscherkunde und Glazialgeologie* 21: 265–73.

Geyh, M. A., Röthlisberger, F. and Gellatly, A. (1985) 'Reliability tests and interpretation of ^{14}C dates from palaeosols in glacier environments', *Zeitschrift für Gletscherkunde und Glazialgeologie* 21: 275–81.

Haeberli, W. (ed.) (1985) *Fluctuations of Glaciers 1975–1980*, International Commission on Snow and Ice of the International Association of Hydrological Sciences, Paris, IAHS and Unesco.

Harbor, J. M. (1985) 'Problems with the interpretation and comparison of Holocene terrestrial and lacustrine deposits: an example from the Colorado Front Range, USA', *Zeitschrift für Gletscherkunde und Glazialgeologie* 21: 17–24.

Hastenrath, S. (1984) *The Glaciers of Equatorial East Africa*, Dordrecht, Reidel.

Heine, K. (1983) 'Spät- und postglaziale Gletscherschwankungen in Mexiko: Befunde und paläoklimatische Deutung', in Schroeder-Lanz, H. (ed.) *Late and Postglacial Oscillations of Glaciers: Glacial and Periglacial Forms*, Rotterdam, Balkema, 291–304.

Hoinkes, H. (1970) 'Methoden und Möglichkeiten von Massenhaushaltsstudien auf Gletschern. Ergebnisse der Messreihe Hintereisferner (Ötztaler Alpen) 1953–68', *Zeitschrift für Gletscherkunde und Glazialgeologie* 6: 37–90.

Holzhauser, H. (1984) 'Rekonstruktion von Gletscherschwankungen mit Hilfe fossiler Hölzer', *Geographica Helvetica* 39: 3–15.

Hope, G. S., Peterson, J. A., Radok, U., and Allison, L. (eds) (1976) *The Equatorial Glaciers of New Guinea*, Rotterdam, Balkema.

Hoskins, W. G. (1968) 'Harvest fluctuations and English economic history, 1620–1759', *The Agricultural History Review* 16: 15–31.

Ingram, M. J., Underhill, D. J. and Farmer, G. (1981) 'The use of documentary sources for the study of past climates', in Wigley, T. M. L., Ingram, M. J. and Farmer, G. (eds) *Climate and History: Studies in Past Climates and Their Impact on Man*, Cambridge University Press, 180–213.

Innes, J. L. (1982) 'Lichenometric use of an aggregated *Rhizocarpon* "species"', *Boreas* 11: 53–7.

Innes, J. L. (1983a) 'Use of an aggregated *Rhizocarpon* "species" in lichenometry: an evaluation', *Boreas* 12: 183–90.

Innes, J. L. (1983b) 'Size frequency distributions as a lichenometric technique: an assessment', *Arctic and Alpine Research* 15: 285–94.

Innes, J. L. (1984) 'The optimal sample size in lichenometric studies', *Arctic and Alpine Research* 16: 233–44.
Innes, J. L. (1985) 'Lichenometry', *Progress in Physical Geography* 9: 187–254.
International Study Group (1982) 'An inter-laboratory comparison of radiocarbon measurements in tree rings', *Nature* 298: 619–23.
Jacoby, G. C. and Cook, E. R. (1983) 'Past temperature variations inferred from a 400-year tree-ring chronology from Yukon Territory, Canada', *Arctic and Alpine Research* 13: 409–18.
Jacoby, G. C., Cook, E. R. and Ulan, L. D. (1985) 'Reconstructed summer degree days in central Alaska and northwestern Canada since 1524', *Quaternary Research* 23: 18–26.
Johnson, A. (1980) *Grinnell and Sperry Glaciers, Glacier National Park, Montana – A Record of Vanishing Ice*, US Geological Survey Professional Paper 1180.
Jones, P. D., Wigley, T. M. L. and Wright, P. B. (1986) 'Global temperature variations between 1861 and 1984', *Nature* 322: 430–4.
Klein, J., Lerman, J. C., Damon, P. E. and Ralph, E. K. (1982) 'Calibration of radiocarbon dates', *Radiocarbon* 24: 103–50.
Ladurie, E. Le Roy (1971) *Times of Feast, Times of Famine: A History of Climate since the Year 1000*, New York, Doubleday.
Lamb, H. H. (1977) *Climate: Present, Past and Future*, vol. 2, London, Methuen.
Landsberg, H. E. (1985) 'Historic weather data and early meteorological observations', in Hecht, A. D. (ed.) *Palaeoclimatic Analysis and Modelling*, New York, Wiley, 27–69.
Lawrence, D. B. (1950) 'Estimating dates of recent glacier advance and recession rates by studying tree growth layers', *Transactions of the American Geophysical Union* 31: 243–8.
Manley, G. (1974) 'Central England temperatures; monthly means 1659–1973', *Quarterly Journal of the Royal Meteorological Society* 100: 389–405.
Matthes, F. E. (1939) 'Report of Committee on Glaciers, April 1939', *Transactions of the American Geophysical Union* 20: 518–23.
Matthes, F. E. (1950) 'The Little Ice Age of historic times', in Fryxel, F. (ed.) *The Incomparable Valley: A Geological Interpretation of the Yosemite*, Berkeley, University of California Press/Cambridge, Cambridge University Press, 151–60.
Matthews, J. A. (1980) 'Some problems and implications of ^{14}C dates from a podzol buried beneath an end moraine at Haugabreen, southern Norway', *Geografiska Annaler* 62A: 185–208.
Matthews, J. A. (1981) 'Natural ^{14}C age/depth gradient in a buried soil', *Naturwissenschaften* 68: 472–4.
Matthews, J. A. (1984) 'Limitation of ^{14}C dates from buried soils in reconstructing glacier variations and Holocene climate', in Mörner, N.-A. and Karlén, W. (eds) *Proceedings of the Second Nordic Symposium on Climatic Changes and Related Problems, Stockholm*, Dordrecht, Reidel, 281–90.
Matthews, J. A. (1985) 'Radiocarbon dating of surface and buried soils: principles, problems and prospects', in Richards, K. S., Arnett, R. R. and Ellis, S. (eds) *Geomorphology and Soils*, London, Allen & Unwin, 269–88.
Matthews, J. A. and Dresser, P. Q. (1983) 'Intensive ^{14}C dating of a buried paleosol horizon', *Geologiska Föreningens i Stockholm Förhandlingar* 105: 59–63.
Orombelli, G. and Porter, S. C. (1983) 'Lichen growth curves for the southern flank of the Mont Blanc massif, western Italian Alps', *Arctic and Alpine Research* 15: 193–200.
Parry, M. L. (1978) *Climatic Change, Agriculture and Settlement*, Folkestone, Dawson.
Paterson, W. S. B. (1981) *The Physics of Glaciers*, 2nd edn, Oxford, Pergamon Press.
Pearson, M. G. (1978) 'Snowstorms in Scotland – 1831 to 1861', *Weather* 33: 390–3.
Pfister, C. (1981) 'An analysis of the Little Ice Age climate in Switzerland and its consequences for agricultural production', in Wigley, T. M. L., Ingram, M. J. and Farmer, G. (eds) *Climate and History: Studies of Past Climates and Their Impact on Man*, Cambridge University Press, 214–48.
Porter, S. C. (1979) 'Glaciologic evidence of Holocene climatic change', Review papers volume, International Conference on Climate and History, Norwich, UK, 148–79.

Porter, S. C. (1981a) 'Glaciological evidence of Holocene climatic change', in Wigley, T. M. L., Ingram, M. J. and Farmer, G. (eds) *Climate and History: Studies of Past Climates and Their Impact on Man*, Cambridge University Press, 82–110.

Porter, S. C. (1981b) 'Lichenometric studies in the Cascade Range of Washington: establishment of *Rhizocarpon geographicum* growth curves at Mount Rainier', *Arctic and Alpine Research* 13: 11–23.

Porter, S. C. (1986) 'Pattern and forcing of northern hemisphere glacier variations during the last millennium', *Quaternary Research* 26, 27–48.

Porter, S. C. and Denton, G. H. (1967) 'Chronology of Neoglaciation in the North American Cordillera', *American Journal of Science* 265: 177–210.

Porter, S. C. and Orombelli, G. (1985) 'Glacier contraction during the Middle-Holocene in the western Italian Alps: evidence and implications', *Geology* 13: 296–8.

Röthlisberger, F. and Geyh, M. A. (1985) 'Glacier variations in Himalayas and Karakorum', *Zeitschrift für Gletscherkunde und Glazialgeologie* 21: 237–49.

Röthlisberger, F., Haas, P., Holzhauser, H., Keller, W., Bircher, W. and Renner, F. (1980) 'Holocene glacier fluctuations – radiocarbon dating of fossil soils (FAh) and woods from moraines and glaciers in the Alps', *Geographica Helvetica* 35, no. 5 (special issue): 21–52.

Russell, J. C. (1948) *British Medieval Population*, Albuquerque, University of New Mexico.

Ryder, J. M. and Thomson, B. (1986) 'Neoglaciation in the southern Coast Mountains of British Columbia: chronology prior to the late Neoglacial maximum', *Canadian Journal of Earth Sciences* 23: 273–87.

Sigafoos, R. S. and Hendricks, E. L. (1961) *Botanical Evidence of the Modern History of Nisqually Glacier, Washington*, US Geological Survey Professional Paper 387-A.

Sigafoos, R. S. and Hendricks, E. L. (1972) *Recent Activity of Glaciers of Mount Rainier, Washington*, US Geological Survey Professional Paper 387-B.

Stuiver, M. (1978) 'Radiocarbon timescale tested against magnetic and other dating methods', *Nature* 273: 271–4.

Van Bath, S. (1963) *The Agrarian History of Western Europe A.D. 500–1850*, London: Edward Arnold.

Weidick, A. (1967) 'Observations on some Holocene glacier fluctuations in West Greenland', *Meddelelser om Grønland 165*, no. 6.

Zumbühl, H. J. (1980) *Die Schwankungen der Grindelwaldgletscher in den historischen Bild- und Schriftquellen des 12 bis 19 Jahrhunderts*, Zürich, Denkschriften der Schweizerschen Naturforschenden Gesellschaft.

2 Icelandic glaciers and sea ice

Ahlmann, H. Wilson and Thórarinsson, S. (1937) *Vatnajökull. Scientific Results of the Swedish–Icelandic Investigations, 1936, 37, 38*, Chs 1–4 in *Geografiska Annaler* 19; Ch. 5 in *Geografiska Annaler* 20; Chs. 6–9 in *Geografiska Annaler* 21; Ch. 10 in *Geografiska Annaler* 22; Ch. 11 in *Geografiska Annaler* 25.

Bárdarson, G.G. (1934) *Islands Gletscher. Beiträge zur Kenntnis der Gletscherbewegungen und Schwankungen auf Grund alter Quellenschriften und neuester Forschung*, Reykjavík, Vísindafélag Íslendinga (Societas Scientiarum Islandica) 16.

Bell, W. T. and Ogilvie, A. E. J. (1978) 'Weather compilations as a source of data for the reconstruction of European climate during the medieval period', *Climatic Change* 1: 331–48.

Bergthórsson, P. (1962) 'Preliminary notes on past climate of Iceland', Conference on Climate of the 11th and 16th Century, 16–24 June, Aspen, Colo.

Bergthórsson, P. (1969) 'An estimate of drift ice and temperature in Iceland in 1000 years', *Jökull* 19: 94–101.

Bradley, R. S. (1985) *Quaternary Paleoclimatology. Methods of Paleoclimatic Reconstruction*, London, Allen & Unwin.

Dansgaard, W., Johnsen, S. J., Møller, J. and Langway, C. C., Jr (1969) 'One thousand centuries

of climatic record from Camp Century on the Greenland Ice Sheet', Technical Note, Hanover, New Hampshire.
Dickson, R. R., Lamb, H. H., Malmberg, S.-A. and Colebrook, J. M. (1975) 'Climatic reversal in northern North Atlantic', *Nature* 256: 479–81.
Einarsson, E. H. (1966) 'Sudurbrún Mýrdalsjökuls vid Gvendarfell. Breytingar sídustu 100 ár o. fl.', *Jökull* 16: 216–18.
Eythórsson, J. (1931) 'On the present position of the glaciers in Iceland. Some preliminary studies and investigations in the summer of 1930', Reykjavík, Vísindafélag Íslendinga (Societas Scientiarum Islandica) 10.
Eythórsson, J. (1935) 'On the variations of glaciers in Iceland. Some studies made in 1931', *Geografiska Annaler* 17: 121–37.
Eythórsson, J. (1952) 'Thættir úr sögu Breidár', *Jökull* 2: 17–20.
Eythórsson, J. (1963a) 'Variation of Iceland glaciers 1931–1960', *Jökull* 13: 31–3.
Eythórsson, J. (1963b) 'Jöklabreytingar 1961/62 og 1962/63', *Jökull* 13: 29–31.
Eythórsson, J. (1964) 'Jöklabreytingar 1962/63', *Jökull* 14: 97–9.
Eythórsson, J. (1965) 'Jöklabreytingar 1963/64', *Jökull* 15:148–50.
Eythórsson, J. (1966) 'Jöklabreytingar 1964/66', *Jökull* 16:230–1.
Eythórsson, J. and Sigtryggsson, H. (1971) 'The climate and weather of Iceland', *The Zoology of Iceland*, 1.
Fridriksson, S. (1969) 'The effects of sea ice on flora, fauna, and agriculture', *Jökull* 19: 146–57.
Gad, F. (1970) *The History of Greenland. I. Earliest Times to 1700*, London, Hurst.
Gaimard, P. (1838) *Voyage en Islande et au Groënland exécuté pendant les années 1835 et 1836 sur la corvette 'La Recherche'*, 2 vols, Paris.
Helland, A. (ed.) (1882a) 'Islændingen Sveinn Pálssons beskrivelser af islandske vulkaner og bræer', *Den Norske Turistforening, Årbog*, pp. 19–79.
Helland, A. (1882b) 'Om Islands Jökler og om Jökelelvenes Vandmængde og Slamgehalt', *Archiv for Mathematik og Naturvidenskab* 7: 200–32.
Henderson, E. (1819) *Iceland: or the Journal of a Residence in that Island, During the Years 1814 and 1815*, Edinburgh, Wayward Innes.
Ingram, M. J., Underhill, D. J. and Farmer, G. (1981) 'The use of documentary sources for the study of past climates', in Wigley, T. M. L., Ingram, M. J. and Farmer, G. (eds) *Climate and History: Studies in Past Climates and Their Impact on Man*, Cambridge University Press, 180–213.
Jóhannesson (1956) *Íslendiga saga I*, Reykjavik.
John, B. S. and Sugden, D. E. (1962) 'The morphology of Kaldalon, a recently deglaciated valley in Iceland', *Geografiska Annaler* 44: 347–65.
Jónsson, B. H. (1969) 'Sea ice in satellite pictures', *Jökull* 19: 62–8.
Kålund, K. (1872–82) *Bidrag til en historisk-topografisk Beskrivelse af Island*, 2 vols, Copenhagen, Kommissionen for det Arnamagnæanske Legat.
Keilhack, K. (1886) 'Beiträge zur Geologie der Insel Island', *Zeitschrift der Deutschen geologischen Gesellschaft* 38: 376–449.
Koch, L. (1945) 'The East Greenland ice', *Meddelelser om Grønland*, 130, no. 3.
Larson, L. M. (1917) *The King's Mirror. Speculum Regale – Konungs Skuggsjá*, Scandinavian Monographs 3, New York, American Scandinavian Foundation.
Magnússon, A. (1702–12) 'Chorographica Islandica', AM 213 8°, Universitetsbiblioteket, Copenhagen, in Lárusson, Ó. (ed.) (1955) *Safn til Sögu Íslands*, flokkur 2, vol. 1, fasc. 2, Reykjavík.
Magnússon, A. (1930) *Árni Magnússons Levned og Skrifter*, 2 vols, Copenhagen, Kommissionen for det Arnamagnæanske Legat.
Magnússon, A. and Vídalín, P. (1710) *Jardabók. Vol. VII. Ísafjardarsýsla*, Copenhagen, 1940.
Malmberg, S.-A. (1969) 'Hydrographic changes in the waters between Iceland and Jan Mayen in the last decade', *Jökull* 19: 30–43.
Nielson, N. (1937) 'A volcano under an ice-cap. Vatnajökull, Iceland, 1934–36', *Geographical*

Journal 90: 6–23.
Ogilvie, A. E. J. (1984) 'The past climate and sea-ice record from Iceland', *Climatic Change* 6: 131–52.
Ólafsson, E. and Pálsson, B. (1772) *Vice-Lavmands Eggert Olafssons og Land-Physici Bjarne Povelsens Reise Igiennem Island. Foranstaltat af Videnskabernes Sælskab i Kiøbenhavn og beskreven af forbemeldte Eggert Ólafsson*, 2 vols, Sorøe. (English edn published 1975.)
Olavius, O. (1780) *Oeconomisk Reise igniennem de nordvestlige, nordlige og nordostlige Kanter af Island*, etc., 2 vols, Copenhagen.
Olavius, O. (1964–5) *[Oeconomisk Reise.] Ferdabók: Landshagir í nordvestur-, nordur- og nordaustursýslum Íslands 1775–1777*, 2 vols, Reykjavík.
Paijkull, C. W. (1866) *En sommar på Island. Reseskildring . . .*, Stockholm (English edn published 1868).
Pálsson, S. (1945) *Physisk, geographisk og historisk Beskrivelse af de islandske Isbjerge*. Written 1794 and published by A. Helland in *Den Norske Turistforening, Årbog*, 1882–4; republished 1945 by J. Eythórsson in *Ferdabók Sveins Pálssonar*.
Price, R. J. (1982) 'Changes in the proglacial area of Breidamerkurjökull, southeastern Iceland: 1890–1980', *Jökull* 32: 29–35.
Rist, S. (1955) 'Skeidarárhlaup 1954', *Jökull* 5: 30–6.
Rist, S. (1967a) 'The thickness of the ice cover of Mýrdalsjökull, southern Iceland', *Jökull* 17: 237–42.
Rist, S. (1967b) 'Jöklabreytingar 1964/65, 1965/66 og 1966/67', *Jökull* 17: 321–5.
Rist, S. (1967c) 'Jökulhlaups from the ice cover of Mýrdalsjökull on June 25, 1955 and January 20, 1956', *Jökull* 243–8.
Rist, S. (1970a) 'Jöklabreytingar 1966/67, 1967/68 og 1968/69', *Jökull* 20: 78–82, 93.
Rist, S. (1970b) 'Jöklabreytingar 1967/68, 1968/69 og 1969/70', *Jökull* 20: 83–7.
Rist, S. (1971) 'Jöklabreytingar 1931/64, 1964/70 og 1970/71', *Jökull* 21: 73–7.
Rist, S. (1972) 'Jöklabreytingar 1931/64, 1964/71 og 1971/72', *Jökull* 22: 89–95.
Rist, S. (1973a) 'Jökulhlaupaannáll 1971, 1972 og 1973', *Jökull* 23: 55–60.
Rist, S. (1973b) 'Jöklabreytingar 1931/64, 1964/72 og 1972/73', *Jökull* 23: 61–6.
Rist, S. (1974) 'Jöklabreytingar 1931/64, 1964/73, og 1973/74', *Jökull* 24: 77–82.
Rist, S. (1975) 'Jöklabreytingar 1931/64, 1967/74, og 1974/75', *Jökull* 25: 73–9.
Rist, S. (1976) 'Jöklabreytingar 1964/65–1973/74 (10 ár), 1974/75 og 1975/76', *Jökull* 26: 69–74.
Rist, S. (1977) 'Jöklabreytingar 1964/65–73/74 (10 ár), 1974/75–1975/76 (2 ár) og 1976/77', *Jökull* 27: 88–93.
Rist, S. (1978) 'Jöklabreytingar 1964/65–73/74 (10 ár), 1974/75–76/77 (3 ár) og 1977/78', *Jökull* 28: 61–5.
Rist, S. (1981a) 'Jöklabreytingar 1964/65–1973/74 (10 ár), 1974/75–1978/79 (5 ár) og 1979/80', *Jökull* 31: 42–6.
Rist, S. (1981b) 'Jöklabreytingar 1964/65–1973/74 (10 ár), 1974/75–1977/78 (4 ár) og 1978/79', *Jökull* 31: 37–41.
Rist, S. (1982) 'Jöklabreytingar 1964/65–1973/74 (10 ár), 1974/75–1979/80 (6 ár) og 1980/81', *Jökull* 32: 121–5.
Rist, S. (1983) 'Jöklabreytingar 1964/65–1973/74 (10 ár), 1974/75–1980/81 (7 ár) og 1981/82', *Jökull* 33: 141–5.
Rist, S. (1984) 'Jöklabreytingar 1964/65–1973/74 (10 ár), 1974/75–1981/82 (8 ár) og 1982/83', *Jökull* 34: 173–9.
Robin, G. de Q. (1976) 'Reconciliation of temperature–depth profiles in polar ice sheets with past surface temperatures deduced from oxygen-isotope profiles', *Journal of Glaciology* 16: 9–22.
Rodewald, M. (1973) *Der Trend der Meerestemperatur im Nordatlantik. Beilagen zur Berliner Wetterkarte SO 27/73 and SO 29/73*, Berlin, Institut für Meteorologie der Freien Universität.
Sigbjarnarson, G. (1970) 'On the recession of Vatnajökull', *Jökull* 20: 50–61.
Sigfúsdóttir, A. B. (1969) 'Temperature in Stykkishólmur 1846–1968', *Jökull* 19: 7–10.

Sigtryggsson, H. (1972) 'An outline of sea ice conditions in the vicinity of Iceland', *Jökull* 22: 1–11.
Skov, N.A. (1970) 'The ice cover of the Greenland Sea. An evaluation of oceanographic and meteorological causes to year-to-year variations', *Meddelelser om Grønland* 188, no. 2.
Strøm, H. (1762) *Physisk og oeconomisk Beskrivelse over Fogderiet Søndmør. Beliggende i Bergens Stift, i Norge*, 2 vols, Sorøe, 1762–6.
Teitsson, B. (1975) 'Bjarnfeldir í máldögum', in Teitsson, B., Thorsteinsson, B. and Tómasson, S. (eds) *Afmælisrit Björns Sigfússonar*, Reykjavík, Sögufélag, 23–46.
Thórarinsson, S. (1938) 'Über anomale Gletscherschwankungen mit besonderer Berücksichtigung des Vatnajökullgebietes', *Geologiska Föreningens i Stockholm Förhandlingar* 60: 490–506.
Thórarinsson, S. (1939) 'The ice dammed lakes of Iceland with particular reference to their values as indicators of glacial oscillation', *Geografiska Annaler* 21: 216–42.
Thórarinsson, S. (1943) 'Oscillations of the Iceland glaciers in the last 250 years', *Geografiska Annaler* 25: 1–54.
Thórarinsson, S. (1944) 'Tefrokronologiska studier på Island', *Geografiska Annaler* 26: 1–215.
Thórarinsson, S. (1956) *The Thousand Year Struggle Against Ice and Fire*, Reykjavík, Bókaúgáta Menningsarsjóds.
Thórarinsson, S. (1957) 'The jökulhlaup from the Katla area in 1955 compared with other jökulhlaups in Iceland', *Jökull* 7: 21–5.
Thórarinsson, S. (1958a) Flatarmál nokkurra íslenzkra jökla samkvaemt herforingjarádskortunum', *Jökull* 8: 25.
Thórarinsson, S. (1958b) 'The Oræfajökull eruption of 1362', *Acta Naturalia Islandica* 2, no. 2, Reykjavík.
Thórarinsson, S. (1959) 'Um möguleika á thví ad segja fyrir næsta Kötlugos', *Jökull* 9: 6–18.
Thórarinsson, S. (1960) 'Glaciological knowledge in Iceland before 1800. A historical outline', *Jökull* 10: 1–18.
Thórarinsson, S. (1964) 'Sudden advance of Vatnajökull outlet glaciers 1930–1964', *Jökull* 14: 76–89.
Thórarinsson, S. (1969) 'Glacier surges in Iceland, with special reference to the surges of Brúarjökull', *Canadian Journal of Earth Sciences* 6: 875–82.
Thórarinsson, S., Sæmundsson, K. and Williams, R. S. (1973) 'ERTS-1 image of Vatnajökull: analysis of glaciological, structural, and volcanic features', *Jökull* 23: 7–17.
Thoroddsen, T. (1884) 'Den grönländska drifisen vid Island', *Ymer* 4: 145–60.
Thoroddsen, T. (1892) 'Islands Jøkler i Fortid og Nutid', *Geografisk Tidsskrift* 11: 111–46.
Thoroddsen, T. (1895–6) 'Et to Hundrede Aar gammelt Skrift om islandske Jøkler', *Geografisk Tidsskrift* 13: 56–60.
Thoroddsen, T. (1905–6) *Island. Grundriss der Geographie und Geologie*, Petermanns Geographische Mitteilungen, Ergänzungsband 32, Heft 152/3.
Thoroddsen, T. (1914a) 'Islands Klima i Oldtiden', *Geografisk Tidsskrift* 22: 204–16.
Thoroddsen, T. (1914b) 'An account of the physical geography of Iceland', in Rosenvinge, L. K. and Warming, E. (eds) *The Botany of Iceland*, vol. 1, Copenhagen and London, 1912–18, 187–343.
Thoroddsen, T. (1915) 'Endnu nogle Bemærkninger om Islands Klima i Oldtiden', *Geografisk Tidsskrift* 23: 5–9.
Thoroddsen, T. (1916–17) *Árferdi á Íslandi í thúsund Ár*, 1916: pp. 1–192, 1917: pp. 193–432, Copenhagen, Hid Íslenska Fræðafjelag.
Thoroddsen, T. (1925) *Die Geschichte der isländischen Vulkane*, Det Kongelige Danske Videnskabernes Selskabs Skrifter, Naturvidenskabelig og Mathematisk Afdeling, Række 8, vol. 9.
Todtmann, E. M. (1960) Gletscherforschungen auf Island (Vatnajökull). Universität Hamburg. Abhandlungen aus dem Gebiet der Auslandskunde, Reihe C, 65 (19).
Vilmundarson, T. (1972) 'Evaluation of historical sources on sea ice near Iceland', in Karlsson, T. (ed.) *Sea Ice. Proceedings of an International Conference, Reykjavík, Iceland. May 10–13, 1971.*

3 Scandinavia

Aaland, J. (1932) *Nordfjord fra gamle dagar til no. Vol. 2: Dei einskilde bygder: Innvik–Stryn–Sandane*, Oslo, Ei Nemnd.

Aaland, J. (1973) *Nordfjord fra gamle dagar til no. Dei einskilde bygder: Innvik–Stryn–Sandane*, Oslo, Ei Nemnd.

Andersen, J. L. and Sollid, J. L. (1971) 'Glacial chronology and glacial geomorphology in the marginal zones of the glaciers, Midtdalsbreen and Nigardsbreen, South Norway', *Norsk Geografisk Tidsskrift* 25: 1–38.

Battagel, A. (1981) Supplement to *Privatarkiv* 273, Ritsen, Oslo, Riksarkivet.

Bell, W. T. and Ogilvie, A. E. J. (1978) 'Weather compilations as a source of data for the reconstruction of European climate during the medieval period', *Climatic Change* 1: 331–48.

Benedict, J. B. (1967) 'Recent glacial history of an alpine area in the Colorado Front range, USA. I. Establishing a lichen growth curve', *Journal of Glaciology* 6: 817–32.

Bing, K. R. (1899) 'Paa langs over Jostedalsbræen', *Den Norske Turistforening, Årbog*, pp. 101–9.

Bjørkvik, H. (1958) *Jordeige og jordleige i Ryfylke i eldre tid*, Stavanger, Rogaland Historie- og Ættesogelag.

Bjørkvik, H. and Holmsen, A. (1972) *Kven åtte jorda i den gamla leiglendingstida? Fordelinga av jordeigedomen i Noreg i 1661*, Trondheim, Tapir.

Blyth, J. R. (1982) 'Storofsen i Ottadalen', unpublished dissertation, Department of Geography, University of Cambridge.

Blytt, A. (1869) 'Botaniske Observationer fra Sogn', *Nytt Magazin for Naturvidenskaberne* 16: 81–226.

Bohr, C. (1820) 'Om Isbræerne i Justedalen og om Lodalskaabe', *Blandinger* 2, Christiania, 289–317. (English edition, 1827, *Edinburgh Philosophical Journal*, 225–61.)

Bray, J. R. (1982) 'Alpine glacial advance in relation to a proxy summer temperature index based mainly on wine harvest dates, AD 1453–1973', *Boreas* 11: 1–10.

Damon, P. E., Ferguson, C. W., Long, A. and Wallick, E. I. (1974) 'Dendrochronological calibration of the radiocarbon timescale', *American Antiquity* 39: 35–366.

Dass, P. (1739) *Nordlands Trompet*, Copenhagen.

Diplomatarium Norvegicum, vols 1–21, Christiania/Oslo 1849–1972.

Doughty, C. M. (1865) 'Memorandum on the summer motion of some glacier-streams in southern Norway, as observed by C. M. Doughty in 1864', *Proceedings of the Royal Geographical Society* 9: 109–11.

Doughty, C. M. (1866) 'The Jöstedal-Brae glaciers', *Geological Magazine* 3: 309–10.

Durocher, M. J. (1847) 'Études sur les glaciers du nord et du centre de l'Europe', *Annales des Mines*, Ser. 4, 12: 3–142.

Eide, T. O. (1955) 'Breden og bygda', *Norsk Tidsskrift for Folkelivsgransking* 5: 1–42.

Elven, R. (1978) 'Subglacial plant remains from the Omnsbreen glacier area, south Norway', *Boreas* 7: 83–9.

Fægri, K. (1933) 'Über die Längenvariationen einiger Gletscher des Jostedalsbre und die dadurch bedingten Pflanzensukzessionen', *Bergens Museums Åarbok,* Naturvidens-kapelig, rekke, 7.

Fægri, K. (1948) 'On the variations of western Norwegian glaciers during the last 200 years', *Procès-verbaux des séances de l'Assemblée Générale d'Oslo de l'Union Géodésique et Géophysique Internationale*, 293–303.

Forbes, J. D. (1853) *Norway and its Glaciers Visited in 1851*, Edinburgh, Adam & Charles Black.

Foss, M. (1750) In Berge, J. C. (ed.) (1802–3) *Justedalens kortelige Beskrivelse.*

Green, H. M. (1981) 'Lichens and lichenometry: an investigation of the lichens on the moraines of Svellnosbreen, and the use of lichenometry', unpublished BA dissertation, Department of Geography, University of Cambridge.

Griffey, N. J. and Matthews, J. A. (1978) 'Major Neoglacial glacier expansion episodes in southern Norway: evidences from moraine ridge stratigraphy with ^{14}C dates on buried palaeosols and moss layers', *Geografiska Annaler* 60A: 73–90.

Grove, J. M. (1966) 'The Little Ice Age in the Massif of Mont Blanc', *Transactions and Papers of the Institute of British Geographers* 40: 129–43.

Grove, J. M. (1972) 'The incidence of landslides, avalanches, and floods in western Norway during the Little Ice Age', *Arctic and Alpine Research* 4: 131–8.

Grove, J. M. (1985) 'The timing of the Little Ice Age in Scandinavia', in Tooley, M. J. and Sheail, C. M. (eds) *The Climatic Scene*, Hemel Hempsted, Allen & Unwin, 132–53.

Haakensen, N. (ed.) (1982) *Glasiologiske undersøkelser i Norge 1980*, Norges Vassdrags- og Elektrisitetsvesen, Hydrologisk Avdeling. Rapport, 1982, no. 1.

Haakensen, N. (ed.) (1984) *Glasiologiske undersøkelser i Norge 1981*, Norges Vassdrags- og Elektrisitetsvesen, Hydrologisk Avdeling. Rapport, 1984, no. 1.

Haakensen, N. and Wold, B. (eds) (1981) *Glasiologiske undersøkelser i Norge 1979*, Norges Vassdrags- og Elektrisitetsvesen, Hydrologisk Avdeling, Rapport. 1981, no. 3.

Haeberli, W. (ed.) (1985) *Fluctuations of Glaciers 1975–80*, vol. 4, a contribution to the International Hydrological Programme, Paris, IAHS and Unesco.

Hesselberg, T. and Birkeland, B. J. (1944) *Säkulare Schwankungen des Klimas von Norwegen*, Geofysiske Publikasjoner utgitt av Det Norske Videnskaps-Akademi i Oslo, XV.

Hesselberg, T. and Birkeland, B. J. (1956) *The Continuation of the Secular Variations of the Climate of Norway 1940–50*, Geofysiske Publikasjoner utgitt av Det Norske Videnskaps-Akademi i Oslo, XV.

Hoel, A. and Werenskiold, W. (1962) *Glaciers and Snowfields in Norway*, Norsk Polarinstitutt, Skrifter 114.

Hovstad, H. (1971) 'Pest og krise', *Den Norske Turistforening, Årbok*, pp. 33–9 and 41–9.

Ingram, M. J., Underhill, D. J. and Farmer, G. (1981) 'The use of documentary sources for the study of past climates', in Wigley, T. M. L., Ingram, M. J. and Farmer, G. (eds), *Climate and History: Studies in Past Climates and Their Impact on Man*, Cambridge University Press, 180–213.

Innes, J. L. (1982) 'Lichenometric use of an aggregated *Rhizocarpon* "species"', *Boreas* 11: 53–7.

Innes, J. L. (1983) 'Use of an aggregated Rhizocarpon "species" in lichenometry: an evaluation', *Boreas* 12: 183–90.

Innes, J. L. (1985) 'Lichenometric dating of moraine ridges in northern Norway: some problems of application', *Geografiska Annaler* 66A: 341–52.

International Study Group (1982) 'An inter-laboratory comparison of radiocarbon measurements in tree rings', *Nature* 298: 619–23.

'Jostedalsbreden kunde agtes som . . . Vand Magazin' (1758) *Danmarks og Norges oeconomiske Magazin* 2, 13: 287–90.

Karlén, W. (1973) 'Holocene glacier and climatic variations, Kebnekaise Mountains, Swedish Lapland', *Geografiska Annaler* 55A: 29–63.

Karlén, W. (1975) *Lichenometrisk datering i norra Skandinavien – metodens tillförlitlighet och regionala tillämpning*, Stockholms Universitet, Naturgeografiska Institutionen. Forskningsrapport 22.

Karlén, W. (1979) 'Glacier variations in the Svartisen area, northern Norway', *Geografiska Annaler* 61A: 11–28.

Karlén, W. (1981) 'A comment on John A. Matthews's article regarding ^{14}C dates of glacial variations', *Geografiska Annaler* 63A: 19–21.

Karlén, W. (1982) 'Holocene glacier fluctuations in Scandinavia', *Striae* 18: 26–34.

Karlén, W. and Denton, G. H. (1976) 'Holocene glacial variations in Sarek National Park, northern Sweden', *Boreas* 5: 25–56.

Kasser, P. (1967) *Fluctuations of Glaciers, 1959–1965*, International Commission of Snow and Ice of the International Association of Scientific Hydrology, Paris, IAHS and Unesco.
Kasser, P. (1973) *Fluctuations of Glaciers 1965–1970*, International Commission on Snow and Ice of the International Association of Hydrological Sciences, Paris, IAHS and Unesco.
King, C. A. M. (1959) 'Geomorphology in Austerdalen, Norway', *Geographical Journal* CXXV: 357–69.
Klein, J., Lerman, J. C., Damon, P. E. and Ralph, E. K. (1982) 'Calibration of radiocarbon dates', *Radiocarbon* 24: 103–50.
Kleiven, I. (1915) *Gamal Bondekultur i Gudbrandsdalen. Vol. I: Lom og Skjaak*, Kristiania.
Kraft, J. (1830) *Topographisk-statistisk Beskrivelse over Kongeriget Norge*, vol. 9, Kristiania.
Laberg, J. (1944) 'Jostedal', *Tidsskrift utgitt av Historielaget for Sogn.* 11: Leikanger, 5–85.
Laberg, J. (1948) 'Jostedal', *Tidsskrift utgitt av Historielaget for Sogn* 13: 200–24.
Ladurie, E. Le Roy (1971) *Times of Feast, Times of Famine: A History of Climate since the Year 1000*, New York, Doubleday.
Lamb, H. H. (1977) *Climate: Present, Past and Future*, Vol. 2: *Climatic History and the Future*, London, Methuen.
Larson, J. (1875) 'To Vestlandsruter', *Den Norske Turistforening, Aarbog*, 7–17.
Liestøl, O. (1963) 'Noen resultater av bremålinger i Norge 1962', *Norsk Polarinstitutt, Årbok 1962*: 187–90.
Liestøl, O. (1965) 'Noen resultater av bremålinger i Norge 1963', *Norsk Polarinstitutt, Årbok 1963*: 1985–92.
Liestøl, O. (1966a) 'Bremålinger i Norge 1964', *Norsk Polarinstitutt, Årbok 1964*: 155–64.
Liestøl, O. (1966b) 'Bremålinger i Norge 1965', *Norsk Polarinstitutt, Årbok 1965*: 135–42.
Liestøl, O. (1967) 'Storbreen Glacier in Jotunheimen, Norway', *Norsk Polarinstitutt, Skrifter* 141: 1–63.
Liestøl, O. (1968) 'Bremålinger i Norge i 1966', *Norsk Polarinstitutt, Årbok 1966*: 132–7.
Liestøl, O. (1969) 'Bremålinger i Norge i 1967', *Norsk Polarinstitutt, Årbok 1967*: 183–90.
Liestøl, O. (1970a) 'Glasiologiske undersøkelser i 1968', *Norsk Polarinstitutt, Årbok 1968*: 81–91.
Liestøl, O. (1970b) 'Glaciological work in 1969', *Norsk Polarinstitutt, Årbok 1969*: 116–28.
Liestøl, O. (1972) 'Glaciological work in 1970', *Norsk Polarinstitutt, Årbok 1970*: 240–51.
Liestøl, O. (1973) 'Glaciological work in 1971', *Norsk Polarinstitutt, Årbok 1971*: 67–75.
Liestøl, O. (1974) 'Glaciological work in 1972', *Norsk Polarinstitutt, Årbok 1972*: 125–35.
Liestøl, O. (1975) 'Glaciological work in 1973', *Norsk Polarinstitutt, Årbok 1973*: 181–92.
Liestøl, O. (1976) 'Glaciological work in 1974', *Norsk Polarinstitutt, Årbok 1974*: 183–94.
Liestøl, O. (1977a) 'Glaciological work in 1975', *Norsk Polarinstitutt, Årbok 1975*: 147–58.
Liestøl, O. (1977b) 'Glaciological work in 1976', *Norsk Polarinstitutt, Årbok 1976*: 297–304.
Liestøl, O. (1978) 'Glaciological work in Svalbard in 1977', *Norsk Polarinstitutt, Årbok 1977*: 271–7.
Liestøl, O. (1979a) 'Glaciological work in 1978', *Norsk Polarinstitutt, Årbok 1978*: 43–51.
Liestøl, O. (1979b) 'Svartisen. Fjell og Vidde', *Den Norske Turistforening, Årbok*: 137–43.
Liestøl, O. (1980) 'Glaciological work in 1979', *Norsk Polarinstitutt, Årbok 1979*: 43–51.
Liestøl, O. (1982a) 'Glaciological work in 1980', *Norsk Polarinstitutt, Årbok 1980*: 45–51.
Liestøl, O. (1982b) 'Glaciological work in 1981', *Norsk Polarinstitutt, Årbok 1981*: 45–52.
Liestøl, O. (1983) 'Glaciological work in 1982', *Norsk Polarinstitutt, Årbok 1982*: 37–43.
Lindblom, E. A. (1839) 'Vandring i Norrige, Sommaren År 1839', *Kungliga Svenska Vetenskaps-Academiens Handlingar*, 242–99.
Matthews, J. A. (1974) 'Families of lichenometric dating curves for the Storbreen gletschervorfeld, Jotunheimen, Norway', *Norsk Geografisk Tidsskrift* 28: 215–35.
Matthews, J. A. (1976) '"Little Ice Age" palaeotemperatures from high altitude tree growth in S. Norway', *Nature* 264: 243–5.
Matthews, J. A. (1980) 'Some problems and implications of ^{14}C dates from a podzol buried beneath an end moraine at Haugabreen, southern Norway', *Geografiska Annaler* 62A: 185–208.

Matthews, J. A. (1981) 'Natural ^{14}C age/depth gradient in a buried soil', *Naturwissenschaften* 68: 472–4.
Matthews, J. A. (1982) 'Soil dating and glacier variations: a reply to Wibjørn Karlén', *Geografiska Annaler* 64A: 15–20.
Matthews, J. A. (1984) 'Limitation of ^{14}C dates from buried soils in reconstructing glacier variations and Holocene climate', in Mörner, N.-A. and Karlén, W. (eds) *Proceedings of the Second Nordic Symposium on Climatic Changes and Related Problems, Stockholm*, Dordrecht, Reidel, 281–90.
Messerli, B., Messerli, P., Pfister, C. and Zumbühl, H. J. (1978) 'Fluctuations of climate and glaciers in the Bernese Oberland, Switzerland, and their geoecological significance, 1600 to 1975', *Arctic and Alpine Research* 10: 247–60.
Mottishead, D. N. and White, I. D. (1972) 'The lichenometric dating of glacier recession, Tunsbergdal, southern Norway', *Geografiska Annaler* 54A: 47–52.
Müller, F. (1977) *Fluctuations of Glaciers, 1970–1975*, International commission on Snow and Ice of the International Association of Hydrological Sciences, Paris, IAHS and Unesco.
Naumann, C. F. (1824) *Beyträge zur Kentniss Norwegens gesammelt auf Wanderungen der Sommermonate der Jahre 1821 und 1822*, vol. 2, Leipzig.
Neumann, J. (1923) 'Bemærkninger paa en Reise i Sogn og Søndfjord 1823', *Budstikken* 3, Række 5, 47–53, 67–73, 369–424 and 529–584.
Norsk historisk Leksikon (1974) edited R. Fladby, S. Imsen and H. Winge, Oslo, Cappelens Forlag.
Norske Rigs-Registranter 1523–1660 (1861–91), vol. VIII: 1647–48, Christiania, 1884; vol. IX: 1648–49, Christiania, 1887; vol. X: 1650–53, Christiania, 1887, 12 vols, Det norske historiske Kildeskriftfonds Skrifter, 5.
Norvik, J. (1962) 'Jostedalsbreen', in Hoel, A. and Werenskiold, W. (eds), *Glaciers and Snowfields in Norway*, Norsk Polarinstitutt, Skrifter 114, 86–91.
Østrem, G., Liestøl, D. and Wold, B. (1976) 'Glaciological investigations at Nigardsbreen, Norway', *Norsk Geografisk Tidsskrift* 30: 187–209.
Østrem, G. and Stanley, A. (1969) *Glacier Mass Balance Measurements*, A Guide prepared jointly by the Canadian Department of Energy, Mines and Resources and the Norwegian Resources and Electricity Board.
Øvrebø, E. (1970) 'Avtaksforretninger', Riksarkivet's Privat Arkiv. no. 273 Riksen, *Heimen* 15: 83–4.
Øyen, P. A. (1894) 'Bidrag til vore Bræegnes Geografi', *Nyt Magazin for Naturvidenskaberne* 35: 73–229.
Øyen, P. A. (1899) 'Bidrag til vore Bræegnes Geografi', *Nyt Magazin for Naturvidenskaberne* 37: 156–218.
Øyen, P. A. (1900a) 'Bidrag til vore Bræegnes Geografi', *Nyt Magazin for Naturvidenskaberne* 35: 73–229.
Øyen, P. A. (1900b) 'Om periodiske forandringer hos norske bræer', *Norges Geologiske Undersøgelse 28, Aarbog 1896–99.*
Øyen, P. A. (1907) 'Klima- und Gletscherschwankungen in Norwegen', *Zeitschrift für Gletscherkunde* 1: 46–61.
Øyen, P. A. (1908) 'Bidrag til vore bræegnes glacialgeologi', *Nyt Magazin for Naturvidenskaberne* 46: 301–59.
Pontoppidan, E. (1755) *The Natural History of Norway* (translated from the Danish original of 1752), 2 vols, London.
Pytte, R. (ed.) (1969) *Glasiologiske undersøkelser i Norge 1968*, Norges Vassdrags- og Elektrisitetsvesen, Hydrologisk Avdeling, Rapport, 1969, no. 3.
Ralph, E. K., Michael, H. N. and Han, M. C. (1973) 'Radiocarbon dates and reality', *Masca Newsletter* 9: 1–20.

Rekstad, J. (1891–2) 'Om Svartisen og dens gletschere', *Det Norske Geografiske Selskabs Aarbog 1891–2*: 71–86.
Rekstad, J. (1893) 'Beretning om en undersøkelse af Svartisen foretagen i somrene 1890–1', *Archiv for Mathematik og Naturvidenskab* 16: 266–321.
Rekstad, J. (1900) 'Om periodiske forandringer hos norske bræer', *Norges Geologiske Undersøgelse*, 28, *Aarbog 1896–9*.
Rekstad, J. (1901) 'Iagttagelse fra Bræer i Sogn og Nordfjord', *Norges Geologiske Undersøgelse* 34, *Aarbog no. 3*. Reprinted in 1902.
Rekstad, J. (1902) 'Iagttagelse fra Bræer i Sogn og Nordfjord', *Norges Geologiske Undersøgelse, Aarbog for 1902, no.3*: 9–45.
Rekstad, J. (1904) 'Fra Jostedalsbreen', *Bergens Museums Årbok 1904, no. 1*: 1–95.
Rogstad, O. (1941) 'Jostedalsbreens tilbakegang. Forsøk på beregning av bremassens minkning fra 1900 til 1940', *Norsk Geografisk Tidsskrift* 8: 273–93.
Roland, E. and Haakensen, N. (1985) *Glasiologiske undersøkelser i Norge 1982*, Norges Vassdrags- og Elektrisitetsvesen, Hydrologisk Avdeling, Rapport no. 1.
Samlinger til det norske Folks Sprog og Historie (1838), vol. V, Christiania.
Sandnes, J. (1971) *Ødetid og gjenreisning. Trøndsk busetningshistorie c.*1200–1660, Oslo–Bergen–Tromsø, Universitetsforlaget.
Schøning, G. (1761) 'Kort Beretning om endeel Uaar og Misvær', *Det Trondheimske Selskabs Skrifter*, Første Deel: 129–62, Copenhagen.
Schweingruber, F. H., Fritts, H. C., Broker, O. G., Drew, L. G. and Shar, E. (1978) 'The X-ray technique as applied to dendroclimatology', *Tree Ring Bulletin* 38: 61–91.
Seue, C. de (1870) *Le Névé de Jostedal et ses glaciers*, S. A. Sexe, Christiania.
Sexe, S. A. (1869) *Boiumbræen i Juli 1868*, Universitets-Program for Förste Semester, Christiania.
Shairp, J. C., Tait, P. G. and Adams-Reilly, A. (1873) *Life and Letters of James David Forbes, F.R.S.*, London, Macmillan.
Slåstad, T. (1957) *'Arringundersøkelser i Gudbrantsdalen', Meddelelser fra Det Norske Skogforsøksvesen* 48: 571–620.
Slingsby, W. C. (1904) *Norway, the Northern Playground. Sketches of Climbing and Mountain Exploration in Norway between 1872 and 1903*, Edinburgh.
Smith, C. (1817) 'Nogle Iagttagelser, især over Iisfjeldene, paa en Fjeldreise i Norge 1812', Topographiske-statistiske Samlinger, utgivne af Det Kongelige Selskap til Norges Vel 2: 1–62.
Sommerfeldt, W. (1972) *Ofsen i 1789 og virkninger av den i Fron*, Avhandling til embedseksamen, geografi hovedfag, våren 1943. Fron Historielag, Otta.
Stuiver, M. (1978) 'Radiocarbon timescale tested against magnetic and other dating methods', *Nature* 273: 271–4.
Theakstone, W. H. (1965) 'Recent changes in the glaciers of Svartisen', *Journal of Glaciology* 5: 411–31.
Thórarinsson, S. (1943) 'Oscillations of the Iceland glaciers in the last 250 years', *Geografiska Annaler* 25: 1–54.
Tvede, A. M. and Liestøl, O. (1977) 'Blomsterskardbreen, Folgefonni, mass balance and recent fluctuations', *Norsk Polarinstitutt, Årbok 1976*: 225–33.
Wiingaard, H. (1762) 'Om Justedalens Sneebræ eller Iisbræ', in H. Strøm (ed.), *Søndmørs Beskrivelse*, Copenhagen, 1: 46–7.
Wilson, J. M. (1872) 'On the forms of valleys and lake-basins in Norway', *Geological Magazine* 9, 481–4.
Wold, B. (1982) 'Breer i Norge. Fjell og vidde', *Den Norske Turistforening, Årbok*: 17–21.

ARCHIVAL REFERENCES

At Riksarkivet, Oslo

Matrikkel over Helgeland Fogderi, 1723 vol. 170. Unpaginated.
Matrikkel over Nordfjord Fogderi, 1667 vol. 44, fos. 35–48.
Matrikkel over Nordfjord Fogderi, 1723 vol. 146, fos. 126–39.
Rentekammeret, Affældnings Forretninger, Hardanger og Sunnhords Fogderier, 1702–84.
Rentekammeret, Affældnings Forretninger, Nordfjord Fogderi, Pakke 5, 1734–61.
Rentekammeret, Fogderegnskaper, Sunnfjord–Nordfjord, 1702.
Rentekammeret, Ordningsavdeling, Affældnings Forretninger, Bergens Stiftamt, Pakke 3, 1702–84.
Kongelige Resolusjoner, Bergens Stift, 1740–3.
Kongelige Resolusjoner, Bergens Stiftkontor, 1740–5.

At Statsarkivet, Bergen

Tingbok for Hardanger Fogderi, 1677, fos. 14b–15.
Tingbok for Indre Sogns Fogderi, no. 14, 1684, fos. 38–38b.
Tingbok for Nordfjord Sorenskriveri, no. 6, 1721–3, fos. 154–5.
Tingbok for Nordfjord Sorenskriveri, no. 8, 1726–8, fo. 66.

4 The massif of Mont Blanc

Adams-Reilly, A. (1864) 'A rough survey of the chain of Mont Blanc', *Alpine Journal* 1: 257–74.

Arnod, P. A. (1691) 'Relation des passages de tout le circuit du Duché d'Aoste venant des Provinces circonvoisines, avec une sommaire description des montagnes' (manuscript kept at the Archivio di Stato di Torino, dated April 1691 with notes added in 1694), *Archivum Augustinium 1968* 1: 11–72.

Baretti, M. (1880) 'Il ghiacciaio del Miage (versante italiano del Monte Bianco)', *Memorie della Reale Accademia delle Scienze di Torino*, ser. 2, 32: 269–303.

Blanchard, M. R. (1913) 'La Crue glaciaire dans les Alpes de Savoie au XVIIIe siècle', *Recueil de l'Institut de Géographie Alpine* 1: 443–54.

Bless, R. (1984) *Beiträge zur spät- und post-glazialen Geschichte der Gletscher in nordöstlichen Mont Blanc Gebiet*, Geographisches Institut der Universität Zürich.

Blümcke, A. and Hess, H. (1895) 'Der Hochjochferner im Jahre 1893', *Zeitschrift des Deutschen und Österreichischen Alpenvereins* 26: 16–24.

Bordier, M. B. (1773) *Voyage pitoresque aux glaciers de Savoye, fait en 1772*, Geneva, L.-A. Caille.

Bourrit, M.-T. (1773) *Descriptions des glaciers de Savoie*, Geneva.

Bourrit, M.-T. (1776a) *Relation of a Journey to the Glaciers in the Duchy of Savoy*, translated from the French by C. and F. Davy, 3rd edn, Dublin, Cross, Chamberlain & Hoey.

Bourrit, M.-T. (1776b) *Descriptions des aspects du Mont-Blanc*, Lausanne, Société Thypographique.

Bourrit, M.-T. (1787) *Nouvelles descriptions des glaciers, vallées du glace et glaciers qui forment la grande chaîne des Alpes de Savoye, de Suisse et d'Italie*, 3 vols, Geneva.

Bouverot, M. (1958) 'Notice sur les variations des glaciers du Mont-Blanc', *Union Géodésique et Géophysique, Association Internationale d'Hydrologie Scientifique, Assemblée Générale de Toronto* 4: 331–43. IAHS publ. no. 46.

Capello, C. F. (1936) 'La glaciazone attuale nel massiccio del Monte Bianco. Caratteri morfologici e morfometrici dei ghiacciai sul versante italiano', *Bollettino del Comitato Glaciologico Italiano*, ser. 1, 16: 153–230.

Capello, C. F. (1941) 'Studio sul ghiacciaio della Brenva (Monte Bianco) 1920–1940', *Bollettino del Comitato Glaciologico Italiano*, ser. 1, 21: 129–53.

Capello, C. F. (1958) 'Franc-Valanghe di ghiaccio nel gruppo del Monte Bianco', *Bollettino del Comitato Glaciologico Italiano*, ser. 2, 8: 125–38.
Capello, C. F. (1966) 'Relazioni della campagna glaciologica 1963. Gruppo Monte Bianco, Valle Veni', *Bollettino del Comitato Glaciologico Italiano*, ser. 2, 13: 35–6.
Capello, C. F. (1971) 'Il rilievo stereofotogrammetrico del ghiacciaio della Brenva', *Bollettino del Comitato Glaciologico Italiano*, ser. 2, 19: 17–30.
Cerruti, A. V. (1971) 'Osservationi sul progresso dei ghiacciai del Monte Bianco nell ultimo decennio', *Bollettino de Comitato Glaciologico Italiano*, ser. 2, 19: 251–72.
Cerruti, A. V. (1977) 'Variazioni climatiche, alimentazione ed oscillazioni glaciali sul Massiccio del Monte Bianco', *Bollettino del Comitato Glaciologico Italiano*, ser. 2, 25: 53–88.
Coolidge, W. A. B. (1908) *The Alps in Nature and History*, London, Methuen.
Corbel, J. (1963) 'Glaciers et climates dans le Massif du Mont-Blanc', *Revue de géographie alpine* 51: 321–60.
Corbel, J. and Ladurie, E. Le Roy (1963) 'Datation au ^{14}C d'une moraine du Mont Blanc', *Revue de géographie alpine* 51: 173–5.
D'Aubuisson, M. (1811) 'Statistique minéralogique du Département de la Doire', *Journal des Mines* (Paris) 29: 241–64.
De Saussure, H.-B. (1779–96) *Voyages dans les Alpes, précédés d'un essai sur l'histoire naturelle des environs de Genève*, 4 vols. Neufchâtel.
De Tillier, J.-B. (1968) *Historique de la vallée d'Aoste*, Aosta, ITLA.
Dollfus-Ausset (1867) *Matérieux pour l'étude des glaciers*, vol. 10, Paris, Savy.
Drygalski, E. von and Machatschek, F. (1942) *Gletscherkunde*, Vienna, Deuticke.
Favre, A. (1867) *Recherches géologiques dans les parties de la Savoie, du Piedmont et de la Suisse voisines de Mont-Blanc*, vol. 3, Paris.
Finsterwalder, R. (1959) 'Chamonix glaciers', letter in *Journal of Glaciology* 3: 547–8.
Forbes, J. D. (1843) *Travels through the Alps*. New edition 1900, revised and annotated by W. A. B. Coolidge, London, Adam & Charles Black.
Forbes, J. D. (1847) 'Twelfth letter on glaciers', *Edinburgh New Philosophical Journal* 42: 94–106.
Forel, F.-A. (1889) 'Les variations périodiques des glaciers des Alpes', *16. Rapport, Jahrbuch des Schweizer Alpenclub* 24: 462–4.
Forel, F.-A. (1901) 'Les glaciers du Mont-Blanc en 1780', *Annuaire du Club Alpin Français* 28: 425–35.
Glaister, R. M. (1951) 'The ice slide on the Glacier du Tour', *Journal of Glaciology* 1: 508–9.
Grove, J. M. (1966) 'The Little Ice Age in the massif of Mont Blanc', *Transactions and Papers of the Institute of British Geographers* 40: 129–43.
Guex, J. (1929) 'Au Glacier du Trient, 1878–1928', *Die Alpen* 5: 34–9.
Guichonnet, P. (1955) 'Le Cadastre Savoyard de 1738 et son utilisation pour les recherches d'histoire et de géographie sociales', *Revue de géographie alpine* 43: 255–98.
Janin, B. (1970) *Le Col du Grand-Saint-Bernard. Climat et variations climatiques*, Aosta, Marguerettaz-Musiemui.
King, S. W. (1858) *Italian Valleys of the Pennine Alps*, London, Murray.
Ladurie, E. Le Roy (1971) *Times of Feast, Times of Famine: A History of Climate since the Year 1000*, New York, Doubleday.
Le Masson, I. (1697) *La Vie de Messire Jean d'Arenthon d'Alex, Evêque et Prince de Genève*, Lyons, pp. 147–8.
Lesca, C. (1972a) 'Determinazione delle variazioni dal 1965 al 1970 della lingua terminale del Ghiacciaio del Brouillard mediante rilievi aerofotogrammetrici', *Bollettino del Comitato Glaciologico Italiano*, ser. 2, 20: 87–92.
Lesca, C. (1972b) 'L'espansione della lingua terminale del Ghiacciaio della Brenva in base ai rilievi fotogrammetrici del 1959, 1970 e 1971', *Bollettino del Comitato Glaciologico Italiano*, ser. 2, 20: 93–100.

Lesca, C. and Armando, E. (1972) 'Determinazione delle variazioni superficiali e volumetriche dal 1965 al 1970 e controllo della velocità di propagazione delle onde sismiche sul Ghiacciaio de La Lex Blanche', *Bollettino del Comitato Glaciologico Italiano*, ser. 2, 20: 65–86.
Letonnelier, G. (1913) 'Documents relatifs aux variations des glaciers dans les Alpes Français', *Ministère de l'Instruction Publique. Bulletin de la Section de Géographie* 1–2: 283–95.
Lliboutry, L. (1958) 'La dynamique de la Mer de Glace et la vague de 1891–95 d'après les mesures de Joseph Vallot', *Union Géodésique et Géophysique Internationale, IAHS Symposium de Chamonix*, IAHS Publ. no. 47: 125–38.
Lliboutry, L. (1965) *Traité de Glaciologie*, vol. 2, Paris, Masson et Cie.
Lliboutry, L. and Reynaud, L. (1981) '"Global dynamics" of a temperate valley glacier, Mer de Glace, and past velocities deduced from Forbes' bands', *Journal of Glaciology* 27: 207–26.
Marengo, G. G. (1881) 'Monographia del ghiacciaio della Brenva', *Bollettino del Club Alpino Italiano* 15: 3–9.
Martel, P. (1743) *Voyage aux glacières de Faucigny.*
Martel, P. (1744) *An Account of the Glaciers or Ice Alps in Savoy. In Two Letters. As laid before the Royal Society, London*, London, Peter Martel.
Matthes, F. E. (1942) 'Glaciers', in Meinzer, O. E. (ed.) *Hydrology*, New York, Dover Publications, pp. 149–219.
Mayr, F. (1969) 'Die postglazialen Gletscherschwankungen des Mont Blanc-Gebietes', *Zeitschrift für Geomorphologie*, Supplementband 8: 31–57.
Messines du Sourbier, M. (1950) 'Note sur l'éboulement du Glacier du Tour (Haute-Savoie) 14 août 1949', *La Houille Blanche*, No. spécial: 242–5.
Montagnier, H. F. (1920–1 and 1921–2) 'The early history of the Col du Géant and the legend of the Col Major', *Alpine Journal* 33: 323–40 and 34: 348–79.
Mougin, P. (1908–9) 'Les variations de longueur du Glacier des Bossons', *Zeitschrift für Gletscherkunde* 3: 144–8.
Mougin, M. (1910) *Études glaciologiques en Savoie 2: Direction de l'hydrologues et des améliorations agricoles.*
Mougin, P. (1925) *Études glaciologiques en Savoie 5.*
Mougin, P. and Bernard, C. (1922) *Glacier de Tête-Rousse, Études glaciologiques en Savoie* 4.
Orombelli, G. and Porter, S. C. (1982) 'Late Holocene fluctuations of Brenva Glacier', *Geografia Fisica e Dinamica Quaternaria* 5: 14–37.
Orombelli, G. and Porter, S. C. (1983) 'Lichen growth curves for the southern flank of the Mont Blanc massif, western Italian Alps', *Arctic and Alpine Research* 15: 193–200.
Porro, F. (1898) 'Notizie fui lavori della Commissione detta dal C. A. I. per lo Studio dei Ghiacciai Italiani', *Atti 3: Congresso Geogr. Firenze*, 130–3.
Porro, F. (1902) 'Richerche preliminari supra i ghiacciaio del Monte Bianco', *Bollettino del Comitato Glaciologico Italiano* 39: 863–78, 913–17.
Porro, F. (1914) 'Primi studi topografici sul ghiacciaio del Miage', *Bollettino del Comitato Glaciologico Italiano* 1: 31–44.
Porter, S. C. and Orombelli, G. (1980) 'Catastrophic rockfall of September 12, 1717 on the Italian flank of the Mont Blanc massif', *Zeitschrift für Geomorphologie* N.F. 24: 200–18.
Porter, S. C. and Orombelli, G. (1981) 'Alpine rockfall hazards', *American Scientist* 69: 67–75.
Rabot, C. (1902) 'Essai de chronologie des variations glaciaires', *Bulletin de géographie historique et descriptive*, 285–327.
Rabot, C. (1915) 'Récents travaux glaciaires dans les Alpes françaises', *La Géographie* 30: 257–68.
Rabot, C. (1920) 'Les catastrophes glaciaires dans la vallée de Chamonix au XVIIe siècle', *La Nature*, 28 August.
Revelli, P. (1911) 'Le fronti di 7 ghiacciai del versante italiano del Monte Bianco nel 1910', *Rivista del Club Alpino Italiano* 30: 254–8.
Revelli, P. (1912) 'Le fronti di sette ghiacciai nel versante italiano del Monte Bianco', *Rivista del Club Alpino Italiano* 31: 237–9.

Reynaud, L. (1977) 'Glacier fluctuations in the Mont Blanc area (French Alps)', *Zeitschrift für Gletscherkunde und Glazialgeologie* 13: 155–66.

Reynaud, L. (1979) 'Reconstruction of past velocities of Mer de Glace using Forbes Bands', *Zeitschrift für Gletscherkunde und Glazialgeologie* 15: 149–63.

Reynaud, L. (1982) *La Mer de Glace et les glaciers du Mont Blanc*, Le Comité Scientifique des Réserves Naturelles du Département de Haute-Savoie.

Reynaud, L. (1984) 'Mesures des fluctuations glaciaires dans les Alpes françaises. Collecte des données et résultats', *La Houille Blanche* 39: 519–25.

Reynaud, L., Vallon, M., Martin, S. and Letreguilly, A. (1984) 'Spatio temporal distribution of the glacial mass balance in the Alpine, Scandinavian and Tien Shan areas', *Geografiska Annaler* 66A: 239–47.

Röthlisberger, F. (1974) 'Étude des variations climatiques d'après l'histoire des cols glaciaires. Le Col d'Herens (Valais, Suisse)', *Bollettino del Comitato Glaciologico Italiano*, ser. 2, 22: 9–34.

Sacco, F. (1918) 'I ghiacciai italiani del Gruppo del Monto Bianco', *Bollettino del Comitato Glaciologico Italiano*, ser. 1, 3: 21–102.

Sessiano, J. (1982) 'Le Glacier des Bossons; la forte crue de 1981–2 et une estimation de vitesse sur 30 ans', *Revue de géographie alpine* 70: 431–8.

Stuiver, M. (1982) 'A high-precision timescale tested against magnetic and other dating methods', *Radiocarbon* 24: 1–26.

Vaccarone, L. (1881) 'I Valichi nel Ducato d'Aosta a nel secolo XVII', *Bollettino del Club Alpino Italiano*, 181–93.

Vaccarone, L. (1884) *La vie delle Alpi Occidentali negli antichi tempi. Richerche e studi publicati su documenti inediti Candeletti*, Turino.

Valbusa, U. (1921) 'La catàstrofe del Monte Bianco e del Ghiacciaio della Brenva del 14 e 19 Novembre 1920', *Bollettino della Reale Società Geografica Italiana*, ser. V, 10: 95–144 and 151–62.

Valbusa, U. (1931) 'La prima frana-valanga del Monte Bianco sul Ghiacciaio della Brenva (14 Novembre 1920)', *Bollettino della Reale Società Geografica Italiana*, ser. VI, 8: 118–25.

Valla, F. (1977) 'Recent work–France–glaciers', *Ice* 54: 4.

Vallot, J. (1900) 'Expériences sur la Marche et les variations de la Mer de Glace', *Annales de l'Observatoire Météorologique Physique et Glaciaire du Mont Blanc* 4: 35–137.

Vallot, J. (1908) 'Variations de la Mer de Glace de Chamonix depuis cent ans comparées à celles de la période glaciaire', *IX Congrès Internationale de Géographie, Genève 1908*: 343–7.

Vallot, J. (1922) *Évolution de la cartographie de la Savoie et du Mont Blanc*, Paris, Barrère.

Vanni, M. (1942) 'Le variazioni dei ghiacciai italiani nel 1941', *Bollettino del Comitato Glaciologico Italiano*, ser. 1, 22: 29–34.

Vanni, M. (1945) 'Le variazioni dei ghiacciai italiani nel 1942', *Bollettino del Comitato Glaciologico Italiano*, ser. 1, 23: 87–92.

Vanni, M. (1950) 'Le variazioni dei ghiacciai italiani negli anni 1947–1948–1949', *Bollettino del Comitato Glaciologico Italiano*, ser. 2, 1: 103–13.

Vanni, M. (1952) 'Le variazioni dei ghiacciai italiani nel 1951', *Bollettino del Comitato Glaciologico Italiano*, ser. 2, 3: 91–9.

Vanni, M. (1954) 'Le variazioni dei ghiacciai italiani nel 1953', *Bollettino del Comitato Glaciologico Italiano*, ser. 2, 5: 105–18.

Vanni, M. (1956) 'Le variazioni dei ghiacciai italiani nel 1955', *Bollettino del Comitato Glaciologico Italiano*, ser. 2, 7: 91–103.

Vanni, M. (1958) 'L'Activité du Comité Glaciologique et les variations des glaciers italiens en 1956', *Union Géodesique et Géophysique, Association Internationale d'Hydrologie Scientifique, Assemblée Générale de Toronto 1957*, 4: 315–19. (IAHS Publ. no. 46).

Vanni, M. (1961) 'Le variazioni dei ghiacciai italiani negli anni 1958 e 1959', *Bollettino del Comitato Glaciologico Italiano*, ser. 2, 9: 159–64.

Vanni, M. (1966) 'Le variazioni dei ghiacciai italiani nel 1962', *Bollettino del Comitato Glaciologico Italiano*, ser. 2, 12: 3–9.
Vanni, M. (1967) 'Le variazioni dei ghiacciai italiani nel 1964', *Bollettino del Comitato Glaciologico Italiano*, ser. 2, 14: 3–12.
Vanni, M. (1969) 'Le variazioni dei ghiacciai italiani nel 1965', *Bollettino del Comitato Glaciologico Italiano*, ser. 2, 15: 3–8.
Vanni, M. (1970) 'Le variazioni dei ghiacciai italiani negli anni 1966–1967–1968', *Bollettino del Comitato Glaciologico Italiano*, ser. 2, 16: 5–21.
Vanni, M. (1971) 'Le variazioni dei ghiacciai italiani nel 1971 (1970–71)', *Bollettino del Comitato Glaciologico Italiano*, ser. 2, 19: 9–16.
Venetz, J. (1833) 'Mémoires sur les variations de la température dans les Alpes de la Suisse', rédigé en 1821, Zürich, Orelli Füssli.
Veyret, P. (1971) 'Processus de l'érosion et de l'accumulation glaciaires en action. Observations sur certains glaciers en crue du Massif du Mont Blanc', *Revue de géographie alpine* 59: 155–70.
Veyret, P. (1974) 'Les Glaciers du Massif du Mont Blanc (versant nord) durant les étés 1971, 1972, 1973', *Revue de géographie alpine* 62: 137–51.
Veyret, P. (1981) 'Les Glaciers du Massif du Mont Blanc (versant nord) durant les étés 1978, 1979, 1980', *Revue de géographie alpine* 69: 385–91.
Viollet-le-Duc, E. M. (1876) *Le Massif du Mont Blanc: étude sur sa constitution géodésique et géologique, sur ses transformations, et sur l'état ancien et moderne de ses glaciers*, Paris, Librairie Polytechnique Baudry.
Virgilio, F. (1883) 'Sui recenti studi circa la variazione periodiche del ghiacciai', *Bollettino del Club Alpino Italiano* 16: 275–90.
Vivian, R. (1971) 'Les Variations récentes des glaciers dans les Alpes françaises (1900–1970). Possibilités de prévision', *Revue de géographie alpine* 59: 229–42.
Vivian, R. (1975) *Les Glaciers des Alpes Occidentales*, Grenoble, Allier.
Windham, Mr (attributed) (1744) 'A letter from an English Gentleman to Mr Arland, a celebrated painter at Geneva, giving an Account of a Journey to the Glaciers, or Ice Alps in Savoy, written in the year 1741. Translated from the French', in *An Account of the Glaciers or Ice Alps in Savoy, In Two Letters. As laid before the Royal Society, London*, London, P. Martel.

5 The Little Ice Age in the Ötztal, eastern Alps

Blümcke, A. and Hess, H. (1895) 'Der Hochjochferner im Jahre 1893', *Zeitschrift des Deutschen und Österreichischen Alpenvereins* 26: 16–24.
Blümcke, A. and Hess, H. (1897) 'Nachmessungen am Vernagt- und Guslarferner', *Wissenschaftliche Ergänzungshefte zur Zeitschrift des Deutschen und Österreichischen Alpenvereins* 1: 99–112.
Blümcke, A. and Hess, H. (1899) 'Untersuchungem an Hintereisferner', *Wissenschaftliche Ergänzungshefte zur Zeitschrift des Deutschen und Österreichischen Alpenvereins* 1.
Brunner, K. and Rentsch, H. (1972) 'Die Änderungen von Fläche, Höhe und Volumen am Vernagt- und Guslarferner von 1889–1912–1938–1969', *Zeitschrift für Gletscherkunde und Glazialgeologie* 8: 11–25.
De Tillier, J.-B. (1968) *Historíque de la vallée d'Aoste*, Aosta, ITLA.
Finsterwalder, R. (1953) 'Die zahlenmässige Erfassung des Gletscherrückgangs an Ostalpengletschern', *Zeitschrift für Gletscherkunde und Glazialgeologie* 2: 189–239.
Finsterwalder, R. (1972) 'Begleitwort zur Karte des Vernagtferners 1:10,000 vom Jahre 1969', *Zeitschrift für Gletscherkunde und Glazialgeologie* 8: 5–10.
Finsterwalder, R. (1978) 'Beiträge zur Gepatschfernervermessung', *Zeitschrift für Gletscherkunde und Glazialgeologie* 14: 153–9.
Finsterwalder, R. and Rentsch, H. (1976) 'Die Erfassung der Höhenänderung von Ostalpengletschern in den Zeiträumen 1950–1959–1969', *Zeitschrift für Gletscherkunde und Glazialgeologie*, 12: 29–35.

Finsterwalder, R. and Rentsch, H. (1981) 'Zur Höhenänderung von Ostalpengletschern im Zeitraum 1969–1979', *Zeitschrift für Gletscherkunde und Glazialgeologie* 16 (1980): 111–15.

Finsterwalder, S. (1888) 'Gletscherkunde: 1) Der Gepatschferner', *Mitteilungen des Deutschen und Österreichischen Alpenvereins* 14: 49–50.

Finsterwalder, S. (1889) 'Aus den Tagebüchern eines Gletschervermessers', *Zeitschrift des Deutschen Alpenvereins* 20: 259–83.

Finsterwalder, S. (1891) 'Die Pegelstation Vent', *Mitteilungen des Deutschen und Österreichischen Alpenvereins* 7: 65–6.

Finsterwalder, S. (1897a) 'Der Vernagtferner, seine Geschichte und seine Vermessung in den Jahren 1888 und 1889 (mit einer Karte des Ferners 1:10,000', *Wissenschaftliche Ergänzungshefte zur Zeitschrift des Deutschen und Österreichischen Alpenvereins* 1: 1–96.

Finsterwalder, S. (1897b) 'Vom Gepatsch, Weisssee und Lantauferer Ferner', *Mitteilungen des Deutschen und Österreichischen Alpenvereins* 23: 94–5.

Finsterwalder, S. (1928) 'Ergebnisse der Vermessung Grösseder Gletscher und ihren zeitlichen Änderungen', *Zeitschrift für Gletscherkunde und Glazialgeologie* 16: 20–41.

Finsterwalder, S. and Schunck, H. (1888) 'Der Gepatschferner', *Zeitschrift des Deutschen und Österreichischen Alpenvereins* 19: 50–7.

Frignet, E. (1846) *Essai sur le phénomène erratique en Tyrol suivi d'une relation historique de l'écoulement du lac de Rofen-Eis dans L'Oetzthal*, Strasbourg and Paris.

Haeberli, W. (ed.) (1985) *Fluctuations of Glaciers 1975–1980*, International Commission on Snow and Ice of the International Association of Hydrological Sciences, Paris, IAHS and Unesco.

Hess, H. (1918) 'Der Stausee des Vernagtferners im Jahre 1848', *Zeitschrift für Gletscherkunde* 12: 135–8.

Hess, H. (1923) 'Der Hintereisferner 1893 bis 1922. Ein Beitrag zur Lösung des Problems der Gletscherbewegung', *Zeitschrift für Gletscherkunde* 13: 145–203.

Hess, H. (1930) 'Beobachtungen am Hintereis- und Vernagtferner 1919–1929', *Zeitschrift für Gletscherkunde* 18: 220–6.

Hoinkes, H. C. (1955) 'Measurements of ablation and heat balance on Alpine glaciers, with some remarks on the cause of glacier recession in the Alps', *Journal of Glaciology* 2: 497–501.

Hoinkes, H. C. (1969) 'Surges of the Vernagtferner in the Ötztal Alps since 1559', *Canadian Journal of Earth Sciences* 6: 853–61.

Hoinkes, H. C. (1970) 'Methoden und Möglichkeiten von Massenhaushaltsstudien auf Gletschern. Ergebnisse der Messreihe Hintereisferner (Ötztaler Alpen) 1953–1968', *Zeitschrift für Gletscherkunde und Glazialgeologie* 6: 37–90.

Hoinkes, H. C. and Rudolph, R. (1962a) 'Mass balance studies on the Hintereisferner, Ötztal Alps 1952–1961', *Journal of Glaciology* 4: 266–80.

Hoinkes, H. C. and Rudolph, R. (1962b) 'Variations in the mass balance of Hintereisferner (Oetztal Alps), 1952–1961, and their relation to variations of climatic elements', *Union Géodesique et Géophysique Internationale. Association Internationale d'Hydrologie Scientifique, Colloque d'Obergurgl*, IAHS Publ. no. 58. 16–28.

Kasser, P. (ed.) (1967) *Fluctuations of Glaciers 1959–1965*, International Commission on Snow and Ice of the International Association of Hydrological Sciences, Paris, IAHS and Unesco.

Kasser, P. (ed.) (1973) *Fluctuations of Glaciers 1965–1970*, International Commission on Snow and Ice of the International Association of Hydrological Sciences, Paris, IAHS and Unesco.

Kerschensteiner, G. and Hess, H. (1892) 'Die Vermessung des Hochjochferners', *Zeitschrift des Deutschen und Österreichischen Alpenvereins* 23: 17–28.

Kinzl, H. (1978) 'Die Gletscher der Österreichischen Alpen 1976/77', *Zeitschrift für Gletscherkunde und Glazialgeologie* 14: 231–8.

Kinzl, H. (1979) 'Die Gletscher der Österreichischen Alpen 1977/8', *Zeitschrift für Gletscherkunde und Glazialgeologie* 15: 105–13.

Klebelsberg, R. (1920) 'Übersicht uber die Ergebnisse der Messungen an den Ötztaler Gletschern im den 11 Jahren 1909–19', *Zeitschrift für Gletscherkunde* 11: 192–5.

Klebelsberg, R. (1926) 'Die Ostalpengletscher 1920–25', *Mitteilungen des Deutschen und Österreichischen Alpenvereins* 52: 61–4.

Klebelsberg, R. (1943) 'Die Alpengletscher in den letzten dreissig Jahren (1911–1941)', *Petermanns Geographische Mitteilungen* 89: 23–32.

Klebelsberg, R. (1949) 'Von den Gletschern auf Blatt Gurgl und den Gletschermessungen des Alpenvereins', *Jahrbuch des Österreichischen Alpenvereins*, 74: 30–6.

Klien, H. (1968) *Ötztaler Alpen*, 5th edn, Munich, Burgverlag Rudolf Rother.

Kruss, P. D. and Smith, I. N. (1982) 'Numerical modelling of the Vernagtferner and its fluctuations', *Zeitschrift für Gletscherkunde und Glazialgeologie* 18: 93–106.

Kuhn, M. (1980) 'Begleitworte zur Karte des Hintereisferners 1979, 1:10,000', *Zeitschrift für Gletscherkunde und Glazialgeologie* 16: 117–24.

Kuhn, M., Markl, G., Kaser, G., Nickus, U., Obleitner, F. and Schneider, H. (1985) 'Fluctuations of climate and mass balance: different responses of two adjacent glaciers', *Zeitschrift für Gletscherkunde und Glazialgeologie* 21: 409–16.

Mariétan, I. (1959) 'La vie et l'œuvre de l'Ingénieur Ignace Venetz, 1788–1859', *Bulletin de la Murithienne, Société Valaisanne des Sciences Naturelles* 76: 3–51.

Meier, M. F. (1962)'Proposed definitions for glacier mass budget terms', *Journal of Glaciology* 4: 252–63.

Miller, H. (1972) 'Ergebnisse von Messungen mit der Methode der Refraktions-Seismik auf dem Vernagt- und Guslarferner', *Zeitschrift für Gletscherkunde und Glazialgeologie* 8: 27–41.

Müller, F. (ed.) (1977) *Fluctuations of Glaciers 1970–1975*, International Commission on Snow and Ice of the International Association of Hydrological Sciences, Paris, IAHS and Unesco.

Patzelt, G. (1970) 'Die Längenmessungen an den Gletschern der Österreichischen Ostalpen 1890–1969', *Zeitschrift für Gletscherkunde und Glazialgeologie* 6: 151–9.

Patzelt, G. (1973) 'Die neuzeitlichen Gletscherschwankungen in der Venedigergruppe (Hohe Tauern, Ostalpen)', *Zeitschrift für Gletscherkunde und Glazialgeologie* 9: 5–57.

Patzelt, G. (1979) 'Die Gletscher der Österreichischen Alpen 1978/79', *Zeitschrift für Gletscherkunde und Glazialgeologie* 15: 235–46.

Patzelt, G. (1980) 'Die Gletscher der Österreichischen Alpen 1979/80', *Zeitschrift für Gletscherkunde und Glazialgeologie* 16: 267–80.

Patzelt, G. (1981) 'Die Gletscher der Österreichischen Alpen 1980/81', *Zeitschrift für Gletscherkunde und Glazialgeologie* 17: 227–40.

Patzelt, G. (1983) 'Die Gletscher der Österreichischen Alpen 1982/83', *Zeitschrift für Gletscherkunde und Glazialgeologie* 19: 173–88.

Patzelt, G. (1984) 'Die Gletscher der Österreichischen Alpen 1983/84', *Zeitschrift für Gletscherkunde und Glazialgeologie* 20: 207–21.

Posamentier, H. W. (1977) 'A new climatic model for glacier behavior of the Austrian Alps', *Journal of Glaciology* 18: 57–65.

Posamentier, H. W. (1979) 'Climate versus glacier termini behavior in the Austrian Alps: reply to comments by M. Kuhn', *Journal of Glaciology* 22: 412–13.

Rentsch, H. (1982) 'Die Orthophotokarte Vernagtferner 1979', *Zeitschrift für Gletscherkunde und Glazialgeologie* 18: 85–91.

Richter, E. (1885) 'Beobachtungen an den Gletschern der Ostalpen: 2. Die Gletscher der Ötztaler Gruppe im Jahre 1883', *Zeitschrift des Deutschen und Österreichischen Alpenvereins* 16: 54–65.

Richter, E. (1888) *Die Gletscher der Ostalpen*, Stuttgart, Engelhorn.

Richter, E. (1891) 'Geschichte der Schwankungen der Alpengletscher', *Zeitschrift des Deutschen und Österreichischen Alpenvereins* 22: 1–74.

Richter, E. (1892) 'Urkunden über die Ausbrüche des Vernagt- und Gurglergletscher im 17 und 18 Jahrhundert, aus den Innsbrucker Archiven herausgegeben', *Forschungen zur deutschen Landes- und Volkskunde* 6: 349–409.

Richter, E. (1893–4)) *Die Erschliessung der Ostalpen*, 3 vols, Berlin.

Rohrhofer, F. von (1953–4) 'Untersuchungen an Ötztaler Gletschern über den Rückgang 1850–1950 (Niederjoch-, Marzel-, Mutmal- und Schalfferner)', *Geographischer Jahresbericht aus Österreich* 25: 57–84.

Rudolph, R. (1963) 'Glaciological bibliography of the Central Oetztal Alps', *Bulletin of the International Association for Scientific Hydrology*, 8th year, 2: 132–9.

Schimpp, O. (1958) 'Der Eishaushalt am Hintereisferner in den Jahren 1952/3 und 1953/54', *Union Géodésique et Géophysique, Association Internationale d'Hydrologie Scientifique, Assemblée Générale de Toronto* 4: 301–14 (IAHS Publ. no. 46).

Schimpp, O. (1960) 'Der Haushalt des Hintereisferner (Ötztal). Akkumulation, Ablation, und Gletscherbewegung in den Jahren 1952/3, 1953/4', *Veröffentlichungen des Museum Ferdinaneum in Innsbruck* 39 (1959): 66–138.

Schlagintweit, H. and Schlagintweit, S. (1850) *Untersuchungen über die physicalische Geographie der Alpen*, Leipzig.

Sonklar, K. (1860) *Die Oetzthaler Gebirgsgruppe, mit besonderer Rücksicht auf Orographie und Gletscherkunde*, Gotha, Justus Perthes.

Srbik, R. von (1935) 'Der Stausee des Schalfferners (Ötztaler Alpen)', *Zeitschrift für Gletscherkunde* 22: 214–17.

Srbik, R. von (1936) 'Der Stauseebecken des Schälfferners (Ötztaler Alpen)', *Zeitschrift für Gletscherkunde* 24: 191–4.

Srbik, R. von (1937) 'Vorfeldeinbrüche bei einigen Ötztaler Gletschern', *Zeitschrift für Gletscherkunde* 25: 224–7.

Srbik, R. von (1941) 'Grössere Eiseinbrüche bei einigen Ötztaler Gletschern', *Zeitschrift für Gletscherkunde* 27: 166–8.

Srbik, R. von (1942a) 'Übersicht der Messungen an den Ötztaler Gletschern bei Gurgl und Vent im letzten Jahrzehnt', *Zeitschrift für Gletscherkunde* 28: 145–50.

Srbik, R. von (1942b) 'Rückzug von Gletscherzungen in Felsschluchten. Zwei Beispiele aus den Ötztaler Alpen', *Zeitschrift für Gletscherkunde* 28: 150–5.

Steinhauser, F. (1957) 'Die säkularen Änderungen der Sonnenscheindauer in den Ostalpen', *Jahresbericht des Sonnblick-Vereins* 51–3: 3–27.

Stolz, O. (1928) 'Anschauung und Kenntnis des Hochgebirges vor Erwachen des Alpinismus', *Zeitschrift des Deutschen und Österreichischen Alpenvereins* 59: 14–24.

Stotter, M. (1846) *Die Gletscher des Vernagtthales in Tirol und ihre Geschichte*, Innsbruck, Wagner.

Vivian, R. (1975) *Les Glaciers des Alpes Occidentales. Étude géographique: L'emprise de la glaciation actuelle et ses fluctuations récentes; le rôle des eaux; l'aménagement du paysage montagnard par les glaciers*, Grenoble, Allier.

Walcher, J. (1773) *Nachrichten von den Eisbergen in Tirol*, Vienna, Kurzböch.

6 Swiss glacier fluctuations and Little Ice Age weather and climate

Amann, H. (1982) 'Hans Conrad Escher de la Linth, explorateur des Alpes', *Les Alpes* 58: 64–7.

Atlantis (1974) *Vues et panoramas de la Suisse dessinés par Hans Conrad Escher de la Linth*, Zürich, Atlantis.

Aubert, D. (1980) 'Les stades de retrait des glaciers du Haut-Valais', *Bulletin de la Murithienne, Société Valaisanne des Sciences Naturelles* 97: 101–69.

Becker, B. (1969) 'Phänologische Beobachtungen an Reben und ihre praktische Anwendung zur Gütekartierung von Weinbergslagen', *Weinwissenschaft* 24: 142.

Bell, B. (1980) 'Analysis of viticultural data by cumulative deviations', *Journal of Interdisciplinary History* 10: 851–8.

Bell, W. T. and Ogilvie, A. E. J. (1978) 'Weather compilations as a source of data for the reconstruction of European climate during the medieval period', *Climatic Change* 1: 331–48.

Bray, J. R. (1982) 'Alpine glacial advance in relation to a proxy summer temperature index based mainly on wine harvest dates, A.D. 1453–1973', *Boreas* 11: 1–10.

Bridel, P. S. (1818a) 'Course à l'éboulement du glacier de Giétroz et au lac de Mauvoisiau fond de la vallée de Bagnes, 16 Mai 1819.'

Bridel, P. S. (1818b) 'Fragments relatifs à la débâcle de 1819 qui a ravagé la vallée de Bagnes, dans le canton du Valais.'

Britton, C. E. (1937) *A Meteorological Chronicle to AD 1450*, Meteorological Office Geophysical Memoir 70, London, HMSO.

Delibrias, G., Ladurie, M. Le Roy and Ladurie, E. Le Roy (1975) 'La Forêt fossile de Grindelwald: nouvelles datations', *Inter-Sciences*, 137–47.

De Vries, J. (1977) 'Histoire du climat et économie: des faits nouveaux, une interprétation différente', *Annales E.S.C.* 32: 198–226.

Dufour, M. L. (1870) 'Notes sur le problème de la variation du climat', *Bulletin de la Société Vaudoise des Sciences Naturelles* 10: 359–436.

Flohn, H. (1949) 'Klima und Witterungsablauf in Zürich im 16. Jahrhundert', *Vierteljahrsschrift der Naturforschenden-Gesellschaft in Zürich* 94: 28–41.

Furrer, G., Gamper-Schollenberger, B. and Suter, J. (1980) 'Zur Geschichte unserer Gletscher in der Nacheiszeit', in *Das Klima. Auf dem Wege zum Verständnis der Klima-Mechanismen und ihrer Beeinflussung durch die Menschen*, New York, Springer-Verlag, 91–107.

Gamper, M. and Suter, J. (1978) 'Der Einfluss von Temperaturänderungen auf die Länge von Gletscherzungen', *Geographica Helvetica* 4: 183–9.

Grémaud, J. (1878) *Documents relatifs à l'histoire du Valais III*, Mémoires et documents publiés par la Société de la Suisse Romande 31.

Gruner, G. S. (1760) *Histoire naturelle des glacières de Suisse.*

Hennig, R. (1904) 'Katalog bemerkenswerter Witterungsereignisse von den ältesten Zeiten bis zum Jahre 1800', *Abhandlungen des Königlich-Preussischen Meteorologischen Instituts* 2.

Hess, P. and Brezowsky, H. (1952) 'Katalog der Grosswetterlagen Europas', *Berichte des Deutschen Wetterdienstes in der US-Zone* 5: 33.

Hoinkes, H. (1955) 'Measurements of ablation and heat balance on alpine glaciers', *Journal of Glaciology* 2: 497–501.

Hoinkes, H. C. (1968) 'Glacier variation and weather', *Journal of Glaciology* 7: 3–19.

Holzhauser, H. (1978) *Zur Geschichte des Fieschergletschers*, Diplomarbeit, Geographisches Institut der Universität Zürich.

Holzhauser, H. (1980) 'Beitrag zur Geschichte des Grösser Aletschgletschers', *Geographica Helvetica* 35: 17–24.

Holzhauser, H. (1982) 'Neuzeitliche Gletscherschwankungen', *Geographica Helvetica* 37: 115–26.

Holzhauser, H. (1984) 'Rekonstruktion von Gletscherschwankungen mit Hilfe fossiler Hölzer', *Geographica Helvetica* 39: 3–15.

Hopkins, G. M. (1953) *Poems and Prose of Gerard Manley Hopkins*, London, Penguin.

Hopp, R. J. (1974) 'Plant phenology observations', in Leith, H. (ed.) *Phenology and Seasonality Modelling*, New York, Springer-Verlag, 25–43.

Ingram, M. J., Underhill, D. J. and Farmer, G. (1981) 'The use of documentary sources for the study of past climates', in Wigley, T. M. L., Ingram, M. J. and Farmer, G. (eds) *Climate and History: Studies in Past Climates and Their Impact on Man*, Cambridge University Press, 180–213.

Kasser, P. (1959) 'Der Einfluss von Gletscherrückgang und Gletschervorstoss auf den Wasserhaushalt', *Wasser und Energiewirtschaft* 6: 2–16.

Kinzl, H. (1932) 'Die grössten nacheiszeitlichen Gletschervorstösse in den Schweizer Alpen und in der Mont Blanc-Gruppe', *Zeitschrift für Gletscherkunde* 20: 269–397.

Ladurie, E. Le Roy (1971) *Times of Feast, Times of Famine: A History of Climate since the Year 1000*, New York, Doubleday.

Ladurie, E. Le Roy and Baulant, M. (1980) 'Grape harvests from the fifteenth through the nineteenth centuries', *Journal of Interdisciplinary History* 10: 839–49.

Lamb, H. H. (1963) 'On the nature of certain climatic epochs which differed from the modern (1900–39) normal', in *Changes of Climate: Proceedings of the UNESCO/WMO Rome Symposium 1961*, Arid Zone Research Series 10, 125–50.

Lamb, H. H. (1965) 'The early medieval warm epoch and its sequel', *Palaeogeography, Palaeoclimatology, Palaeoecology* 1: 13–37.

Lindgren, S. and Neumann, J. (1981) 'The cold wet year 1695. A contemporary German account', *Climatic Change* 3: 173–87.

Lliboutry, L. (1974) 'Multivariate statistical analysis of glacier annual balances', *Journal of Glaciology* 13: 371–92.

Lütschg, O. (1925) *Über Niederschlag und Abfluss im Hochgebirge*, Secretariat de l'Association Hydraulique Suisse.

Lütschg, O. (1926) *Über Niederschlag und Abfluss im Hochgebirge, Sonderstellung des Mattmarkgebietes*, Schweizerische Wasserwirtschaftverband, Verbandschrift C no. 14. Veröffentlichung der Schweizerischen meteorologischen Zentralanstalt in Zürich, Zürich Sekretariat des Schweizerischen Wasserwirtschaftverbandes.

Manley, G. (1961) 'Meteorological factors in the great glacial advance (1690–1720)', *International Association for Scientific Hydrology* 54: 388–91. Union Géodésique et Geophysique Internationale. IAHS Assemblée Générale de Helsinki, IAHS Pub. no. 54.

Manley, G. (1969) 'Snowfall in Britain over the past 300 years', *Weather* 24: 428–37.

Manley, G. (1974) 'Central England temperatures: monthly means 1659–1973', *Quarterly Journal of the Royal Meteorological Society* 100: 389–405.

Mariétan, I. (1959) 'La Vie et l'œuvre de l'Ingénieur Ignace Venetz 1788–1859', *Bulletin de la Murithienne, Société Valaisanne des Sciences Naturelles* 76: 1–51.

Mariétan, I. (1965) 'Mattmark et le glacier d'Allalin', *Bulletin de la Murithienne, Société Valaisanne des Sciences Naturelles* 82: 129–48.

Mercanton, P.-L. (1916) 'Mensurations au Glacier du Rhône 1874–1915', *Nouveaux Mémoires de la Société Helvétique des Sciences Naturelles* 52: 1–190.

Messerli, B., Zumbühl, H. J., Ammann, K., Kienholz, H., Oeschger, H., Pfister, C. and Zurbuchen, M. (1975) 'Die Schwankungen des Unteren Grindelwaldgletschers seit dem Mittelalter. Ein interdisziplinärer Beitrag zur Klimageschichte', *Zeitschrift für Gletscherkunde und Glazialgeologie* 11: 3–110.

Messerli, B., Messerli, P., Pfister, C. and Zumbühl, H. J. (1978) 'Fluctuations of climate and glaciers in the Bernese Oberland, Switzerland, and their geoecological significance, 1600 to 1975', *Arctic and Alpine Research* 10: 247–60.

Müller, F., Ohmura, A., Schroff, K., Funk, M., Phirter, K., Bernath, A. and Steffen, K. (1980) 'Combined ice, water and energy balances of the Swiss Alps – the Rhônegletscher project', in Müller, F., Bridel, L. and Schwabe, E. (eds) 'Geography in Switzerland', *Geographica Helvetica* 35, no. 5 (special issue): 57–69.

Oeschger, H. and Röthlisberger, H. (1961) 'Datierung eines ehemaligen Standes des Aletschgletschers durch Radioaktivitätsmessung an Holzproben und Bemerkungen zu Holzfunden an weiteren Gletschern', *Zeitschrift für Gletscherkunde und Glazialgeologie* 4: 191–205.

Parry, M. L. (1978) *Climatic Change, Agriculture and Settlement*. Folkestone, Dawson.

Patzelt, G. and Bortenschlager, S. (1973) 'Die postglazialen Gletscher- und Klimaschwankungen in der Venedigergruppe (Hohe Tauern, Ostalpen), mit sechs Pollendiagrammen von S. Bortenschlager', *Zeitschrift für Geomorphologie* 16: 25–72.

Pfister, C. (1975) *Agrarkonjunktur und Witterungsverlaut im westlichen schweizer Mittelland*, Geographica Bernensia, 2, Bern.

Pfister, C. (1977) 'Zum Klima des Raumes Zürich im späten 17. und frühen 18. Jahrhundert', *Vierteljahrsschrift der Naturforschenden Gesellschaft in Zürich* 122: 447–71.

Pfister, C. (1978a) 'Fluctuations in the duration of snow-cover in Switzerland since the late seventeenth century', in Frydendahl, K. (ed.) *Proceedings of the Nordic Symposium on Climatic Changes and Related Problems, Copenhagen*, Danish Meteorological Institute Climatological Paper 4: 1–6.

Pfister, C. (1978b) 'Die älteste Niederschlagsreihe Mitteleuropas: Zürich 1708–1754', *Meteorologische Rundschau* 31: 56–62.

Pfister, C. (1978c) 'Climate and economy in eighteenth century Switzerland', *Journal of Interdisciplinary History* 9: 223–43.

Pfister, C. (1979a) 'The reconstruction of past climate: the example of the Swiss historical weather documentation project (16th to early 19th century)', International Conference on Climate and History, Climatic Research Unit, University of East Anglia, Norwich, *Review Papers*, pp. 128–47.

Pfister, C. (1979b) 'Getreide-Erntebeginn und Frühsommertemperaturen im schweizerischen Mittelland seit dem 17. Jahrhundert', *Geographica Helvetica* 34: 23–35.

Pfister, C. (1980a) 'The Little Ice Age thermal and wetness indices for central Europe', *Journal of Interdisciplinary History* 10: 665–96.

Pfister, C. (1980b) 'The climate of Switzerland in the last 450 years', in Müller, F., Bridel, L. and Schwabe, E. (eds) 'Geography in Switzerland', *Geographica Helvetica* 35, no. 5 (special issue): 15–20.

Pfister, C. (1981) 'An analysis of the Little Ice Age climate in Switzerland and its consequences for agricultural production', in Wigley, T. M. L., Ingram, M. J. and Farmer, G. (eds) *Climate and History: Studies in Past Climates and Their Impact on Man*, Cambridge University Press, 214–48.

Portmann, J. P. (1975) 'Notices glaciologiques: Aperçu historique I (1880–1900)', *Les Alpes* 51: 182–8.

Portmann, J. P. (1976) 'Notices glaciologiques: Aperçu historique II (1901–1920)', *Les Alpes* 52: 157–64.

Portmann, J. P. (1978) 'Notices glaciologiques: Aperçu historique III (1921–40)', *Les Alpes* 54: 123–30.

Portmann, J. P. (1980a) 'Notices glaciologiques: Aperçu historique IV (1941–60)', *Les Alpes* 56: 36–42.

Portmann, J. P. (1980b) 'Notices glaciologiques: Aperçu historique V (1961–80)', *Les Alpes* 56: 66–72.

Posamentier, H. W. (1977) 'A new climatic model for glacier behavior of the Austrian Alps', *Journal of Glaciology* 18: 57–67.

Primault, B. (1969) 'Le Climat et la viticulture', *International Journal of Biometeorology* 13: 7–24.

Reynaud, L. (1980) 'Can the linear balance model be extended to the whole Alps?', in Müller, F. (ed.) *World Glacier Inventory. Proceedings of the Riederalp Workshop, 1978*, International Association for Hydrological Sciences, Publ. no. 126, 273–84.

Reynaud, L. (1983) 'Recent fluctuations of Alpine Glaciers and their meteorological causes: 1880–1980', in Street-Perrot, A., Beran, M. and Ratcliff, R. (eds) *Variations in the Global Water Budget*, Dordrecht, Reidel, 195–202.

Richter, E. (1891) 'Geschichte der Schwankungen der Alpengletscher', *Zeitschrift des Deutschen und Österreichischen Alpenvereins* 22: 1–74.

Röthlisberger, F. (1974) 'Étude des variations climatiques d'après l'histoire des cols glaciaires. Le Col d'Hérens (Valais, Suisse)', *Bollettino del Comitato Glaciologico Italiano*, ser. 2, 22: 9–34.

Röthlisberger, F., Haas, P., Holzhauser, H., Keller, W., Bircher, W. and Renner, F. (1980) 'Holocene climatic fluctuations – radiocarbon dating of fossil soils (fAh) and woods from moraines and glaciers in the Alps', in 'Geography in Switzerland', *Geographica Helvetica* 35: 21–52.

Rudloff, H. von (1967) *Die Schwankungen und Pendelungen des Klimas in Europa seit dem Beginn der regelmässigen Instrumenten-Beobachtungen (1670)*, Braunschweig, Vieweg.

Ruppen, P. J., Imseng, G. and Imseng, W. (1979) 'Saaser Chronik 1200–1979', Saas-Fee, Verkehrsverein.
Schneebeli, W. (1976) 'Untersuchungen von Gletscherschwankungen im Val de Bagnes', *Die Alpen* 52: 5–58.
Schweingruber, F. H. and Röthlisberger, F. (1978) 'Long chronologies by X-ray densitometry for sub-alpine conifers and their value for indicating Holocene temperatures', in Fletcher, J. (ed.) *British Archaeological Reports, International Series* 51: 115–16.
Schweingruber, F. H., Bräker, O. U. and Schar, E. (1978a) 'Dendroclimatic studies in Great Britain and in the Alps', in *Evolution of Planetary Atmospheres and Climatology of the Earth*, Toulouse, Centre National d'Études Spatiales.
Schweingruber, F. H., Bräker, O. U. and Schär, E. (1978b) 'Dendroclimatic studies on conifers from central Europe and Great Britain', *Boreas* 8: 427–52.
Schweingruber, F. H., Fritts, H. C., Bräker, O. U., Drew, L. G. and Schar, E. (1978) 'The X-ray technique as applied to dendrochronology', *Tree Ring Bulletin* 38: 61–91.
Sion (Valais, Switzerland) archives, Département des Travaux Publics, 'Débâcle du Lac de Giétroz, 16 Juin 1818', Fasc. 1. R21. 28.
Swan, S. L. (1981) 'Mexico in the Little Ice Age', *Journal of Interdisciplinary History* 11: 633–48.
Tschudi, Aegidius (1538) 'De prisca ac vera Alpina Rhaetia, cum caetero Alpinarum gentium tracta descriptio', Basle, Basilae.
Tufnell, L. (1984) *Glacier Hazards*, London, Longman.
Venetz, I. (1833) *Mémoire sur les variations de la température dans les Alpes de la Suisse. Rédigé en 1821*, Zürich, Orelli Füssli.
Vivian, R. (1966) 'La catastrophe du glacier Allalin', *Revue de géographie alpine* 54: 97–112.
Wills, A. (1856) *Wandering in the High Alps*, London.
Zumbühl, H. J. (1980) *Die Schwankungen der Grindelwaldgletscher in den historischen Bild- und Schriftquellen des 12 bis 19 Jahrhunderts*, Denkschriften der Schweizerischen Naturforschenden-Gesellschaft in Zürich, 92.

7 The Little Ice Age in Asia

THE LITTLE ICE AGE – A GLOBAL EVENT?

Haeberli, W. (ed.) (1985) *Fluctuations of Glaciers 1975–80*, International Commission on Snow and Ice of the International Association of Hydrological Sciences, Paris, IAHS and Unesco.
Kasser, P. (ed.) (1967) *Fluctuations of Glaciers 1959–65*, International Commission on Snow and Ice of the International Association of Hydrological Sciences, Paris, IAHS and Unesco.
Kasser, P. (ed.) (1973) *Fluctuations of Glaciers 1965–70*, International Commission on Snow and Ice of the International Association of Hydrological Sciences, Paris, IAHS and Unesco.
Müller, F. (ed.) (1977) *Fluctuations of Glaciers 1970–75*, International Commission on Snow and Ice of the International Association of Hydrological Sciences, Paris, IAHS and Unesco.

THE USSR

Freshfield, D. W. (1869) *Travels in the Central Caucasus and Bushan*, London, Longman.
Freshfield, D. W. and Sella, V. (1896) *The Exploration of the Caucasus*, 2 vols, London, Edward Arnold.
Grosswal'd, M. G. and Kotlyakov, V. M. (1969) 'Present-day glaciers in the U.S.S.R. and some data on their mass balance', *Journal of Glaciology* 8: 9–22.
Horvath, E. (1975) 'Glaciers of Kol'skiy Poluostrov (Kola Peninsula), Ural'skiy Khrebet (Ural Mountains), and Poluostrov Taymyr (Tamyr Peninsula)'; 'Glaciers of the Kavkaz (Caucasus)';

'Glaciers of Pamir-Alay'; 'Glaciers of the Tyan'-Shan''; 'Glaciers of Altaysko-Sayanskaya Gornaya Strana', in Field, W. O. (ed.), *Mountain Glaciers of the Northern Hemisphere* CRREL, Hanover, New Hampshire, vol. 1, 185–310.

Kotlyakov, V. M. (1980a) 'Problems and results of studies of mountain glaciers in the Soviet Union', in Müller, F. (ed.) *World Glacier Inventory. Proceedings of the Workshop at Riederalp, 1978*, International Association of Hydrological Sciences, Publ. no. 126, 129–37.

Kotlyakov, V. M. (1980b) 'Recent work: USSR', *Ice*, 64: 12–17.

Kotlyakov, V. M. and Krenke, A. N. (1979) 'The régime of the present-day glaciation of the Caucasus', *Zeitschrift für Gletscherkunde und Glazialgeologie* 15: 7–21.

Kotlyakov, V. M. and Krenke, A. N. (1981) 'Glaciation actuelle et climat du Caucase', *Revue de géographie alpine* 69: 241–64.

Makarevich, K. G. (1962) 'The regime of the glaciers in the Zailiisky Alatau in recent decades', in *Variations of the Regime of Existing Glaciers*, Symposium of Obergurgl, International Association of Scientific Hydrology, Publ. no. 581, 249–61.

Makarevich, K. G. (1980) 'Studies of glacier fluctuations in the USSR', in *World Glacier Inventory. Proceedings of the Workshop at Riederalp, 1978*, International Association of Hydrological Sciences, Publ. no. 126, 143–8.

Revyakin, V. S. and Revyakina, N. V. (1976) 'Some peculiarities of glacier regression of intercontinental mountain area', in Lvovitch, M. I., Kotlyakov, V. M. and Rauner, Yu. L., *International Geography '76. Climatology, Hydrology, Glaciology*, Section 2, 23rd International Geographical Congress, 279–83, Moscow, Vneshtorgizdat.

Röthlisberger, F. (1974) 'Études des variations climatiques d'après l'histoire des cols glacières. Le Col d'Hérens (Valais, Suisse)', *Bollettino de Comitato Glaciologico Italiano* 22: 9–34.

Zabirov, R. D. (1955) *Glaciation of the Pamirs*, Moscow, State Publishing House of Geographical Literature. (In Russian.)

THE HIMALAYA

Ageta, Y., Ohata, T., Tanaka, Y., Ikegami, K. and Higuchi, K. (1980) 'Mass balance of glacier AX010 in Shorong Himal, East Nepal, during the summer monsoon', *Seppyo, Journal of Japanese Society of Snow and Ice* 4: 34–41.

Auden, J. B. (1935) 'The snout of the Biafo Glacier in Baltistan', *Records of the Geological Survey of India* 68: 400–13.

Batura Glacier Investigation Group (1979) 'The Batura Glacier in the Karakoram Mountains and its variations', *Scientia Sinica* 22: 958–74.

Best, F., Gruber, G. and Kick, W. (1981) 'Das Ende des Chogo-Lungma-Gletschers 1979, Beobachtungen zu Veränderungen an einem der grossen Eisströme des Karakorums mittels Photogrammetrie (1902–1954–1970–1979)', *Zeitschrift für Gletscherkunde und Glazialgeologie* 17: 177–89.

Bose, R. N., Dutta, N. P. and Lahiri, S. M. (1971) 'Refraction seismic investigation at Zemu Glacier, Sikkim', *Journal of Glaciology* 10: 113–19.

Conway, W. M. (1894) *Climbing in the Himalayas*, 3 vols, London, Fisher Unwin.

Cotter, G. de P. and Coggin-Brown, J. (1907) 'Notes on certain glaciers in Kumaon', *Records of the Geological Survey of India* 35: 148–57.

De Filippi, F. (1912) *Karakoram and the Western Himalaya. An Account of the Expedition of H.R.H. the Duke of Abruzzi*, London, Constable.

De Filippi, F. (1932) *The Italian Expedition to the Himalaya, Karakoram and Eastern Turkestan (1913–14)*, London, Edward Arnold.

Desio, A. (1930) 'Geological work of the Italian expedition to the Karakoram', *Geographical Journal* 75: 402–11.

De Terra, H. and Hutchinson, G. E. (1934) 'Evidence of recent climatic change shown by Tibetan highland lakes', *Geographical Journal* 84: 311–20.
Drew, F. (1875) *The Junmoo and Kashmir Territories*, London, Stanford.
Dutt, G. N. (1961) 'The Bara Shigri Glacier, Kangra District, East Punjab, India', *Journal of Glaciology* 3, 1007–15.
Egerton, P. H. (1864) *Journal of a Tour through Spiti*, London, Cundall Downes.
Featherstone, B. K. (1926) 'The Biafo Glacier', *Geographical Journal* 67: 351–4.
Finsterwalder, R. (1937) 'Die Gletscher des Nanga Parbat', *Zeitschrift für Gletscherkunde* 25: 57–108.
Finsterwalder, R. (1960) 'German glaciological and geological expeditions to the Batura Mustagh and Rakaposhi Range', *Journal of Glaciology* 3: 787–8.
Freshfield, D. W. (1902) 'The glaciers of Kanchenjunga', *Geographical Journal* 19: 453–75.
Freshfield, D. W. (1903) *Round Kanchenjunga*, London, Edward Arnold.
Fushimi, H. (1977) 'Glaciations in the Khumbu Himal (1)', *Seppyo, Journal of Japanese Society of Snow and Ice* 39: 60–7.
Fushimi, H. (1978) 'Glaciations in the Khumbu Himal (2)', *Seppyo, Journal of Japanese Society of Snow and Ice* 40: 71–7.
Fushimi, H. and Ohata, T. (1980) 'Fluctuations of glaciers from 1970 to 1978 in the Kuhumbu Himal, east Nepal', *Seppyo, Journal of Japanese Society of Snow and Ice* 41: 77–81.
Fushimi, H., Ohata, T. and Higuchi, K. (1981) 'Recent fluctuations of glaciers in the eastern part of Nepal Himalaya', in Allison, I. (ed.), *Sea Level, Ice, and Climatic Change. Proceedings of the Canberra Symposium*, International Association of Hydrological Sciences, Publ. no. 131, 21–9.
Gardner, J. S. (1986) 'Recent fluctuations of Rakhiot glacier, Nanga Parbat, Punjab Himalaya, Pakistan', *Journal of Glaciology* 32: 527–9.
Gilbert, L. B. and Auden, J. B. (1932–3) 'Note on a glacier in the Arwa Valley, British Garhwal', *Records of the Geological Survey of India* 66: 388–404.
Godwin-Austen, H. H. (1864) 'On the glaciers of the Mustagh Range', *Journal of the Royal Geographical Society* 34: 19–56.
Goudie, A. S., Jones, D. K. C. and Brunsden, D. (1984) 'Recent fluctuations in some glaciers of the Western Karakoram mountains, Hunza, Pakistan', in Miller, K. J. (ed.), International Karakoram Project, Cambridge University Press, 411–55.
Gunn, J. F. (1930) 'The Shyok flood 1929', *The Himalayan Journal* 2: 35–47.
Haeberli, W. (ed.) (1985) *Fluctuations of Glaciers 1975–80*, International Commission on Snow and Ice of the International Association of Hydrological Sciences, Paris, IAHS and Unesco.
Hayden, H. H. (1907) 'Notes on certain glaciers in north-west Kashmir', *Records of the Geological Survey of India* 35: 127–37.
Hewitt, K. (1967) 'Ice-front deposition and the seasonal effect: a Himalayan example', *Transactions of the Institute of British Geographers* 42: 93–106.
Hewitt, K. (1969) 'Glacier surges in the Karakoram Himalaya (central Asia)', *Canadian Journal of Earth Sciences* 6: 1009–18.
Hewitt, K. (1982) 'Natural dams and outburst floods of the Karakoram Himalaya', in Glen, J. W. (ed.) *Hydrological Aspects of Alpine and High-Mountain Areas. Proceedings of the Exeter Symposium*, International Association of Hydrological Sciences, Publ. no. 138, 259–69.
Higuchi, K., Fushimi, H., Ohata, T., Takenaka, S., Iwata, S., Yokoyama, K., Higuchi, H., Nagoshi, A. and Lozawa, T. (1980) 'Glacier inventory in the Dudh Kosi region, East Nepal', in *World Glacier Inventory. Proceedings of the Workshop at Riederalp, 1978*, International Association of Hydrological Sciences, Publ. no. 126, 95–100.
Holland, T. H. (1907) 'A preliminary survey of certain glaciers in the north-west Himalaya. Introduction', *Records of the Geological Survey of India* 35: 123–6.
Hughes, J. M. R. (1982) 'Vegetation of the Bara Shigri and Sonapani Glacier forefield; a study of successional plant assemblages', unpublished BA dissertation, Department of Geography, University of Cambridge.

Ikegami, K. and Inoue, J. (1978) 'Mass balance studies on Kongma Glacier, Khumbu Himal', *Seppyo, Journal of Japanese Society of Snow and Ice* 40: 12–16.
Jangpangi, B. S. and Vohra, E. P. (1962) 'The retreat of the Shunkalpa (Ralam) Glacier in Central Himalaya, Pithoragarh District, Uttar Pradesh, India', in *Variations in the Regime of Existing Glaciers*, Symposium of Obergurgl, International Association of Scientific Hydrology, Publ. no. 58, 234–8.
Kasser, P. (ed.) (1967) *Fluctuations of Glaciers 1959–65*, International Commission on Snow and Ice of the International Association of Hydrological Sciences, Paris, IAHS and Unesco.
Kasser, P. (ed.) (1973) *Fluctuations of Glaciers 1965–70*, International Commission on Snow and Ice of the International Association of Hydrological Sciences, Paris, IAHS and Unesco.
Kick, W. (1960) 'The first glaciologists in Central Asia, according to new studies in the Department of Manuscripts at the Bavarian State Library', *Journal of Glaciology* 3: 687–92.
Kick, W. (1962) 'Variations of some Central Asiatic glaciers', in *Variations of the Regime of Existing Glaciers*, Symposium of Obergurgl, International Association of Scientific Hydrology, Publ. no. 58, 223–9.
Kick, W. (1967) *Schlagintweits Vermessungsarbeiten am Nanga Parbat 1856*, Deutsche Geodätische Kommission bei der Bayerischen Akademie der Wissenschaften, ser. C, no. 97.
Kick, W. (1969) 'Alexander von Humboldts Wirken für die Hochgebirgsforschung in Asien, besonders über die Brüder Schlagintweit', *Petermanns Geographische Mitteilungen* 113: 89–99.
Kick, W. (1972) 'Auswertung photographischer Bilder für die Untersuchung und Messung von Gletscheränderungen, mit Beispielen aus dem Kaukasus und dem Karakorum', *Zeitschrift für Gletscherkunde und Glazialgeologie* 8: 147–67.
Kick, W. (1975) 'Application of geodesy, photogrammetry, history and geography to the study of long-term mass balances of Central Asiatic glaciers', in *Snow and Ice. Proceedings of the Moscow Symposium, August 1971*, International Association of Hydrological Sciences, Publ. no. 104, 150–60.
Kick, W. (1980) 'Material for a glacier inventory of the Indus drainage basin – the Nanga Parbat massif', in Müller, F. (ed.) *World Glacier Inventory. Proceedings of the Workshop at Riederalp, 1978*. International Association of Hydrological Sciences, Publ. no. 126, 105–9.
Krenek, L. and Bhawan, V. (1945) 'Recent and past glaciation of Lahaul', *Indian Geographical Journal* 26: 93–102.
Kumar, G., Mehdi, S. H. and Prakash, G. (1975) 'Observations on some glaciers of Kumaon, Himalaya, U.P.', *Records of the Geological Survey of India* 106: 231–9.
Kurien, T. K. and Munshi, M. M. (1962) 'Sonapani Glacier of Lahaul, Kangra District, Punjab, India', in *Variations of the Regime of Existing Glacier*, Symposium of Obergurgl, International Association of Scientific Hydrology, Publ. no. 58, 239–44.
La Touche, T. H. D. (1910) 'Notes on certain glaciers in Sikkim', *Records of the Geological Survey of India* 60: 52–62.
Loewe, F. (1961) 'Glaciers of Nangaparbat', *Pakistan Geographical Review*, 16: 19–24.
Longstaff, T. G. (1910) 'Glacier exploration in the eastern Karakoram', *Geographical Journal* 35: 622–58.
Ludlow, F. (1929–30) 'The Shyok Dam in 1928', *Himalayan Journal* 1: 4–10.
Lynam, J. P. O'F. (1960) 'The Kulu–Lahul–Spiti watershed', *Geographical Journal* 126: 481–2.
Mackley, I. and McIntyre, N. F. (1980) 'The Cambridge Bara Shigri Expedition', unpublished report, Cambridge, Scott Polar Research Institute.
Mason, K. (1923) 'Kishen Singh and the Indian explorers', *Geographical Journal* 62: 429–40.
Mason, K. (1930) 'Glaciers of the Karakoram and neighbourhood', *Records of the Geological Survey of India* 63: 214–78.
Mayewski, P. A. and Jeschke, P. A. (1979) 'Himalayan and Trans-Himalayan glacier fluctuations since AD 1812', *Arctic and Alpine Research* 11: 267–87.
Mayewski, P. A., Pregent, G. P., Jeschke, P. A. and Ahmad, N. (1980) 'Himalayan and Trans-

Himalayan glacier fluctuations and the south Asian monsoon record', *Arctic and Alpine Research* 12: 171–82.

Miller, M. M (1970) *Glaciology of the Khumbu Glacier and Mount Everest. National Geographic Society Projects, Research Reports, 1961–1962*, Washington, DC, National Geographic Society.

Müller, F. (1970) 'Inventory of glaciers in the Mount Everest region', in *Perennial Ice and Snow Masses*, Unesco/IAHS Technical Papers in Hydrology, no. 1, 47–59.

Müller, F. (ed.) (1977) *Fluctuations of Glaciers 1970–1975*, International Commission on Snow and Ice of the International Association of Hydrological Sciences, Paris, IAHS and Unesco.

Müller, F. (1980) 'Present and late Pleistocene equilibrium line altitudes in the Mt Everest region – an application of the glacier inventory', in Müller, F. (ed.) *World Glacier Inventory. Proceedings of the Riederalp Workshop, September 1978*, International Association of Hydrological Sciences, Publ. no. 126, 75–94.

Neve, E. F. (1907) 'Mt. Kolahoi and its northern glacier', *Alpine Journal* 25: 39–42.

Odell, N. E. (1963) 'The Kolahoi northern glacier, Kashmir', *Journal of Glaciology* 4: 633–5.

Ono, Y. (1984) 'Annual moraine ridges and recent fluctuation of Yala (Dakpatsen) Glacier, Langtang Himal', in Higuchi, K. (ed.) *Glacial Studies in the Langtang Valley*, Data Center for Glacier Research, Japanese Society of Snow and Ice, Publ. no. 2, 73–83.

Ono, Y. (1985) 'Recent fluctuations of the Yala (Dakpatsen) Glacier, Langtang Himal, reconstructed from annual moraine ridges', *Zeitschrift für Gletscherkunde und Glazialgeologie* 21: 251–8.

Raina, V. K., Bhattachorya, V. and Pattnaik, S. (1973) 'Zemu Glacier', *Records of the Geological Survey of India* 105: 95–106.

Schlagintweit, H. A. and Schlagintweit, R. (1860–6) *Results of a Scientific Mission to India and High Asia*, 4 vols, Leipzig and London.

Shi Ya-Feng (1979) 'The Batura Glacier in the Karakoram Mountains and its variations', *Scientia Sinica* 22: 558–74.

Shi Ya-Feng and Wang Jintai (1979) 'The fluctuations of climate, glaciers and sea-level since the late Pleistocene in China', *Scientia Sinica* 22: 1–21.

Shi Ya-Feng and Wang Wenying (1980) 'Research on snow cover in China and the avalanche phenomena of Batura Glacier in Pakistan', *Journal of Glaciology* 26: 25–30.

Shipton, E. (1935) 'Nanda Devi and the Ganges watershed', *Geographical Journal* 85: 305–22.

Skrikantia, S. V. and Padhu, R. N. (1963) 'Recession of the Bara Shigri Glacier', in *Symposium on Himalayan Geology*, Geological Survey of India, 97–100.

Smythe, F. S. (1932) 'Exploration in Garhwal around Kamet', *Geographical Journal* 79: 1–16.

Strachey, R. (1847) 'A description of the glaciers of the Pindari and Kuphinee rivers in the Kumaon Himalaya', *Journal of the Asiatic Society of Bengal* 16: 203.

Strachey, R. (1900) 'Narrative of a journey to the lakes Rakas-Tal and Manasarowar in western Tibet undertaken in September 1848', *Geographical Journal* 15: 150–70.

Tewari, A. P. (1971) 'A short report on glacier studies in the Himalaya mountains by the Geological Survey of India', *Studia Geomorphologica Carpatho-Balcanica* 5: 173–81.

Tewari, A. P. (1973) 'Recent changes in the position of the snout of the Pindari Glacier (Kumaon Himalaya, Almora District, Uttar Pradesh, India)', in *The Role of Snow and Ice in Hydrology*, Paris, Unesco-WMO-IAHS, 1144–49. (Proceedings of the Banff Symposia, September 1972, vol. 2, IHAS Publ. 107).

Tyacke, R. H. (1893) *A Sportsman's Manual*, Calcutta.

Vigne, G. T. (1842) *Travels in Kashmir, Ladak, Iskardo*, vol. 2 London: 285–7.

Visser, P. C. (1934) 'The Karakoram and Turkestan Expedition of 1929–30', *Geographical Journal* 84: 281–95.

Visser, P. C. and Visser-Hooft, J. (1938) *Karakoram; Wissenschaftliche Ergebnisse der Niederlandischen Expeditionen in den Karakoram und die aufgrenzenden Gebiete in die Jahren 1922, 1925, 1929/30 und 1935*, vol. 2, Leiden, Brill.

Vohra, C. P. (1980) 'Some problems of glacier inventory in the Himalayas', in *World Glacier Inventory*. Proceedings of the Workshop at Riederalp, 1978, International Association of Hydrological Sciences, Publ. no. 176: 67–74.

Walker, H. and Pascoe, E. H. (1907) 'Notes on certain glaciers in Lahaul', *Records of the Geological Survey of India* 35: 139–57.

Ward, F. K. (1934) 'The Himalaya east of the Tsangpo', *Geographical Journal* 84: 369–97.

Wissmann, H. von (1959) 'Die heutige Vergletscherung und Schneegrenze in Hochasien, mit Hinweisen auf die Vergletscherung der letzen Eiszeit', Akademie der Wissenschaften und der Literatur in Mainz, *Abhandlungen der Mathematisch-Naturwissenschaftlichen Klasse* 14: 1105–407.

Workman, W. H. (1905) 'From Srinagar to the sources of the Chogo Lungma Glacier', *Geographical Journal* 25: 245–68.

Workman, F. B. and Workman, W. H. (1908) *Ice-bound Heights of the Mustagh*, London.

Workman, F. B. and Workman, W. H. (1910) *The Call of the Snowy Hispar*, London, Constable.

Zhang Xiangsong and Shih Ya-feng (1980) 'Changes in the Batura Glacier in Quaternary and recent times', Academia Sinica, Lanzhou Institute of Glaciology and Cryopedology and Desert Research, 173–90.

CHINA

Ersi, K. (1985) 'A preliminary glacio-hydrological comparison between some glaciers in the Swiss Alps and the Chinese Tianshan', *Arbeitsheft* 7, Zürich, Versuchsanstalt für Wasserban, Hydrologie und Glaziologie an der Eidgenössichen Technischen Hochschule.

Hsieh Tze-chu and Fei Ching-shen (1980) 'Recent research on the distribution and fluctuations of the glaciers in Chilien Shan', in *World Glacier Inventory. Proceedings of the Workshop at Riederalp, 1978*, International Association of Hydrological Sciences, Publ. no. 126, 117–20.

Li Chi-chun and Cheng Pen-hsing (1980) 'Recent research on glaciers on the Chinghai-Tibet Plateau', in Müller, F. (ed.) *World Glacier Inventory. Proceedings of the Workshop at Riederalp, 1978*, International Association of Hydrological Sciences, Publ. no. 126, 121–7

Shih Ya-feng, Hsieh Tze-chu, Cheng Pen-hsing and Li Chi-chun (1980) 'Distribution, features and variations of glaciers in China', in *World Glacier Inventory. Proceedings of the Workshop at Riederalp, 1978*, International Association of Hydrological Sciences, Publ. no. 126, 111–16.

Shih Ya-feng and Wang Jingtai (1979) *The Fluctuations of Climate, Glaciers and Sea-level since the Late Pleistocene in China*, Langzhou Institute of Glaciology and Cryopedology and Desert Research.

Shih Ya-feng, Wang Zongtai and Liu Chaohai (1981) 'Progress and problems of glacier inventory in China', *Zeitschrift für Gletscherkunde und Glazialgeologie* 17: 191–8.

Shih Ya-feng and Zhang Xiangsong (1984) *Guide to the Tienshan Glaciological Station of China. Glaciers in the Urumqi Valley and Related Phenomena*, Lanzhou Institute of Glaciology and Cryopedology and Desert Research, Academia Sinica.

Wang Wenying (1983) 'Glaciers in the north-eastern part of the Ch'ing-hai–Hsi-tsang (Qinghai–Xizang) Plateau (Tibet) and their *Journal of Glaciology* 29: 383–91.

Zhang Xiangsong, Zheng Benxiang and Xie Zichu (1980) 'Recent variations of existing glaciers on the Qinghai-Xizang Plateau', *Proceedings of the Symposium on Qinghai–Xizang (Tibet) Plateau. Abstracts, Academia Sinica*, 227–8.

Zhang Xiangsong, Zheng Benxiang and Xie Zichu (1981) 'Recent variations of existing glaciers on the Qinghai–Xizang (Tibet) Plateau', Geological and Ecological Studies of Qinghai–Xizang Plateau, vol. 2, Beijing, Science Press 1625–9.

8 The Little Ice Age in North America and Greenland

NORTH AMERICA

Alford, D. L. (1974) 'Cirque glaciers of the Colorado forest range: mesoscale aspects of a glacier environment', *Dissertation Abstracts International* B 34: 3341 B.

Alt, B. T. (1978) 'Synoptic climatic controls of mass-balance variations on Devon Island ice cap', *Arctic and Alpine Research* 10: 61–80.

Andrews, J. T. and Barnett, D. M. (1979) 'Holocene (Neoglacial) moraine and proglacial lake chronology, Barnes Ice Cap, Canada', *Boreas* 8: 341–58.

Andrews, J. T., Davis, P. T. and Wright, C. (1976) 'Little Ice Age permanent snowcover in the eastern Canadian Arctic: extent mapped from Landsat-1 imagery', *Geografiska Annaler* 58A: 71–81.

Angell, J. K. and Korshover, J. (1983) 'Global temperature variation in the troposphere and stratosphere 1958–1982', *Monthly Weather Review* 111: 901–21.

Benedict, J. B. (1968) 'Recent glacial history of an alpine area in the Colorado Front Range, U.S.A. II. Dating the glacial deposits', *Journal of Glaciology* 7: 77–87.

Benedict, J. B. (1973) 'Chronology of cirque glaciation, Colorado Front Range', *Quaternary Research* 3: 584–99.

Bengtson, K. B. (1962) 'Recent history of the Brady Glacier, Glacier Bay National Monument, Alaska, U.S.A.', in *Variations in the Regime of Existing Glaciers*, Symposium of Obergurgl, International Association of Scientific Hydrology, Publ. no. 58, 78–87.

Bradley, R. S. (1973) 'Recent freezing level changes and climatic deterioration in the Canadian Arctic archipelago', *Nature* 243: 398–9.

Bradley, R. S. and England, J. (1978) 'Recent climatic fluctuations of the Canadian High Arctic and their significance for glaciology', *Arctic and Alpine Research* 10: 715–31.

Bradley, R. S. and Miller, G. H. (1972) 'Recent climatic change and increased glacierization in the eastern Canadian Arctic', *Nature* 237: 385–7.

Bray, J. R. (1964) 'Chronology of a small glacier in eastern British Columbia', *Science* 144: 287–8.

Bray, J. R. and Struik, G. J. (1963) 'Forest growth and glacial chronology in Eastern British Columbia, and their relation to recent climatic trends', *Canadian Journal of Botany* 41: 1245–71.

Brunger, A. G., Nelson, J. G. and Ashwell, I. Y. (1967) 'Recession of the Hector and Peyto Glaciers: further studies in the Drummond Glacier, Red Deer valley area, Alberta', *Canadian Geographer* 11: 35–48.

Burbank, D. W. (1981) 'A chronology of late Holocene glacier fluctuations on Mount Rainier, Washington', *Arctic and Alpine Research* 13: 369–86.

Burbank, D. W. (1982) 'Correlations of climate, mass balances, and glacial fluctuations at Mount Rainier, Washington, U.S.A., since 1850', *Arctic and Alpine Research* 14: 137–48.

Burns, P. E., Haworth, L. A., Calkin, P. E. and Ellis, J. M. (1983) 'Glaciology of Grizzly Glacier, Brooks Range, Alaska', *Abstracts of the 12th Arctic Workshop. University of Massachusetts at Amherst*, 19.

Calkin, P. E. and Ellis, J. M. (1980) 'A lichenometric dating curve and its application to Holocene glacier studies in the central Brooks Range, Alaska', *Arctic and Alpine Research* 12: 245–64.

Carrara, P. E. and McGimsey, R. G. (1981) 'The late-Neoglacial histories of the Agassiz and Jackson glaciers, Glacier National Park, Montana', *Arctic and Alpine Research* 13: 183–96.

Cavell, E. (1983) *Legacy in Ice. The Vaux Family and the Canadian Alps*, Banff, The Whyte Foundation.

Cooper, W. S. (1937) 'The problem of Glacier Bay, Alaska; a study of glacier variations', *Geographical Review* 27: 37–62.

Cooper, W. S. (1942) 'Vegetation of the Prince William Sound Region Alaska; with a brief excursion into Post-Pleistocene climatic history', *Ecological Monographs* 12: 1–22.

Cropper, J. P. (1982) 'Climatic reconstructions (1801 to 1938) inferred from tree-ring width chronologies of the North American Arctic', *Arctic and Alpine Research* 14: 223–41.

Davies, P. T. (1980) 'Late Holocene glacial, vegetational and climatic history of Pangnirtung and Kingnait Fiord area, Baffin Island, Canada', unpublished PhD thesis, University of Colorado.

Denton, G. H. (1975a) 'Glaciers of the Interior Ranges of British Columbia', in Field, W. O. (ed.) *Mountain Glaciers of the Northern Hemisphere*, CRREL, Hanover, New Hampshire, vol. 1, 655–70.

Denton, G. H. (1975b) 'Glaciers of the Cascade and Olympic Mountains', in Field, W. O. (ed.) *Mountain Glaciers of the Northern Hemisphere*, CRREL, Hanover, New Hampshire, vol. 1, 561–600.

Denton, G. H. and Karlén, W. (1977) 'Holocene glacial and tree-line variations in the White River valley and Skolai Pass, Alaska and Yukon Territory', *Quaternary Research* 7: 63–111.

Derksen, S. J. (1976) *Glacial Geology of the Brady Glacier Region, Alaska*, Report 60, Institute of Polar Studies (Ohio State University, Columbus, Ohio).

Dethier, D. P. and Frederick, J. E. (1981) 'Mass balance of the "Vesper" glacier, Washington, U.S.A.', *Journal of Glaciology* 27: 271–82.

Dowdeswell, J. A. (1984) 'Late Quaternary chronology for the Watts Bay area, Frobisher Bay, southern Baffin Island, N.W.T., Canada', *Arctic and Alpine Research* 16: 311–20.

Dowdeswell, J. A. and Morris, S. E. (1983) 'Multivariate statistical approaches to the analysis of late Quaternary relative age data', *Progress in Physical Geography* 7: 157–76.

Dyson, J. L. and Gibson, G. R. (1939) 'Grinell Glacier, Glacier National Park, Montana', *Bulletin of the Geological Society of America* 50: 681–96.

Field, W. O. (1932) 'The glaciers of the northern part of Prince William Sound, Alaska', *Geographical Review* 22: 361–8.

Field, W. O. (ed.) (1975) *Mountain Glaciers of the Northern Hemisphere*, 2 vols and atlas, CRREL, Hanover, New Hampshire.

Field, W. O. (1979) 'Observations of glacier variations in Glacier Bay National Monument', in Linn, R. M. (ed.) *Proceedings of the First Conference on Scientific Research in the National Parks, New Orleans, November 9–12 1976*, US Department of the Interior, National Parks Transactions and Proceedings Services 5: 803–8.

Gardner, J. (1978) 'Wenkchemna Glacier: ablation complex and rock glacier in the Canadian Rocky Mountains', *Canadian Journal of Earth Sciences* 15: 1200–4.

Haeberli, W. (ed.) (1985) *Fluctuations of Glaciers 1978–80*, International Commission on Snow and Ice of the International Association of Hydrological Sciences, Paris, IAHS and Unesco.

Hamilton, T. D. (1965) 'Alaskan temperature fluctuations and trends; an analysis of recorded data', *Arctic* 18: 105–17.

Harrison, A. E. (1956a) 'Fluctuations of the Nisqually Glacier, Mt. Rainier, Washington, during the last two centuries', *International Union of Geodesy and Geophysics, International Association of Scientific Hydrology, General Assembly of Rome, 1954*, 4, Pub. no. 39: 506–10.

Harrison, A. E. (1956b) 'Glacial activity in the western United States', *Journal of Glaciology* 2: 666–8.

Harrison, A. E. (1956c) 'Fluctuations of the Nisqually Glacier, Mt. Rainier, Washington, since 1750', *Journal of Glaciology* 2: 675–83.

Hattersley-Smith, G. and Serson, H. (1973) 'Reconnaissance of a small ice cap near St Patrick Bay, Robeson Channel, northern Ellesmere Island, Canada', *Journal of Glaciology* 12: 417–21.

Haworth, L. A., Calkin, P. E., Lamb, B. and Ellis, J. M. (1983) 'Holocene glacier variations across the Central Brooks Range, Alaska', *Abstracts of the 12th Arctic Workshop, University of Massachusetts, Amherst*, 36–7.

Heusser, C. J. (1956) 'Postglacial environments in the Canadian Rocky Mountains', *Ecological Monographs* 26: 263–302.

Heusser, C. J. (1957) 'Variations of Blue, Hoh, and White glaciers during recent centuries', *Arctic* 10: 139–50.

Heusser, C. J. and Marcus, M. G. (1964) 'Historical variations of Lemon Creek Glacier, Alaska, and their relationship to the climatic record', *Journal of Glaciology* 5: 77–86.

Hubley, R. C. (1956) 'Glaciers of the Washington Cascade and Olympic Mountains; their present activity and its relation to local climatic trends', *Journal of Glaciology* 2: 669–74.

Ives, J. D. (1962) 'Indications of recent extensive glacierization in north-central Baffin Island, N.W.T.', *Journal of Glaciology* 4: 197–205.

Johnson, A. (1960) 'Variation in surface elevation of the Nisqually Glacier, Mt. Rainier, Washington', *Bulletin of the International Association of Scientific Hydrology*, no. 19: 54–60.

Johnson, A. (1980) *Grinnell and Sperry Glaciers, Glacier National Park, Montana – a Record of Vanishing Ice*, US Geological Survey Professional Paper 1180.

Kasser, P. (ed.) (1967) *Fluctuations of Glaciers 1959–65*, International Commission on Snow and Ice of the International Association of Hydrological Sciences, Paris, IAHS and Unesco.

Kasser, P. (ed.) (1973) *Fluctuations of Glaciers 1965–70*, International Commission on Snow and Ice of the International Association of Hydrological Sciences, Paris, IAHS and Unesco.

Kearney, M. S. and Luckman, B. H. (1981) 'Evidence for late Wisconsin–early Holocene climatic/vegetational change in Jasper National Park, Alberta', in Mahaney, W. C. (ed.) *Quaternary Paleoclimate*, Norwich, Geo Abstracts, 85–105.

King, L. (1983) 'Contribution to the glacial history of the Borup Fiord area, northern Ellesmere Island, N.W.T., Canada', in Schroeder-Lanz, H. (ed.) *Late- and Postglacial Oscillations of Glaciers: Glacial and Periglacial Forms*, Rotterdam, Balkema, 305–23.

Kite, G. W. and Reid, I. A. (1977) 'Volumetric change of the Athabasca Glacier over the last 100 years', *Journal of Hydrology* 32: 279–94.

Klotz, O. J. (1899) 'Notes on glaciers of southernmost Alaska and adjoining territories', *Geographical Journal* 14: 523–34.

Koerner, J. M. (1980) 'The problem of lichen-free zones in Arctic Canada', *Arctic and Alpine Research* 12: 87–94.

Kukla, G. and Gavin, J. (1980) 'Recent secular variations of snow and sea ice cover', in Müller, F. (ed.) *World Glacier Inventory. Proceedings of Riederalp Workshop, 1978*, International Association of Hydrological Sciences, Publ. no. 126, 249–58.

La Chapelle, E. (1962) 'Assessing glacier mass budgets by reconnaissance aerial photography', *Journal of Glaciology* 4: 290–7.

La Pérouse, J. F. de G. (1979) *Voyage Round the World, performed in the years 1785, 1786, 1787, and 1788 by the Boussole and Astrolabe*, vol. 2 with atlas. Translated from French, London, Hamilton.

Lawrence, D. B. (1950) 'Glacier fluctuation for six centuries in southeastern Alaska and its relation to solar activity', *Geographical Review* 40: 191–223.

Lawrence, D. B. (1951) 'Glacier fluctuation in northwestern North America within the past six centuries', *International Union of Geodesy and Geophysics, International Association of Scientific Hydrology, General Assembly of Brussels*, International Association of Scientific Hydrology, Publ. no. 32: 161–6.

Lawrence, D. B. (1958) 'Glaciers and vegetation in southeastern Alaska', *American Scientist* 46: 89–122.

Luckman, B. H. (1977) 'Lichenometric dating of Holocene moraines at Mount Edith Cavell, Jasper, Alberta', *Canadian Journal of Earth Sciences* 14: 1809–22.

Luckman, B. H. and Osborn, G. D. (1979) 'Holocene glacier fluctuations in the middle Canadian Rocky Mountains', *Quaternary Research* 11: 52–77.

Meier, M. F. (1963) 'History of Mount Rainier glaciers', in Meier, M. F. (ed.) *The Glaciers of Mount Rainier; International Union of Geodesy and Geophysics Study Tour, Sept. 2–5, 1963*, 12–14.

Meier, M. F. and Post, A. S. (1962) 'Recent variations in mass net budgets of glaciers in western North America', *Variations of the Regimen of Existing Glaciers*, Symposium of Obergurgl, International Association of Scientific Hydrology, Publ. no. 58, 63–77.

Meier, M. F., Post, A., Rasmussen, L. A., Sikonia, W. G. and Mayo, L. R. (1979) *Retreat of the*

Columbia Glacier, Alaska. A Preliminary Prediction, US Geological Survey Open-file Report 80–10, Tacoma, Wash.

Meier, M. F., Rasmussen, L. A., Post, A., Brown, C. S., Sikonia, W. G., Mayo, L. R. and Trabant, D. C. (1980) *Predicted Timing of the Disintegration of the Lower Reaches of Columbia Glacier, Alaska*, US Geological Survey Open-file Report 80–582, Tacoma, Wash.

Meier, M. F. and Tangborn, W. V. (1965) 'Net budget and flow of South Cascade Glacier, Washington', *Journal of Glaciology* 5: 547–66.

Mercer, J. H. (1961) 'The response of fjord glaciers to changes in the firn limit', *Journal of Glaciology* 3: 850–8.

Miller, C. D. (1969) 'Chronology of Neoglacial moraines in the Dome Peak area, North Cascade Range, Washington', *Arctic and Alpine Research* 1: 49–65.

Miller, G. H. (1973) 'Late Quaternary glacial and climatic history of northern Cumberland Peninsula, Baffin Island, N.W.T., Canada', *Quaternary Research* 3: 561–83.

Miller, M. M (1958) 'The role of diastrophism in the regimen of glaciers in the St. Elias District, Alaska', *Journal of Glaciology* 3: 292–7.

Miller, M. M. (1964) 'Inventory of terminal position changes in Alaskan coastal glaciers since the 1750's', *Proceedings of the American Philosophical Society* 108: 257–73.

Miller, M. M. (1965) *Progress Report on National Geographic Society Alaskan Glacier Commemorative Project, Summer Phase – 1964*, Committee for Research and Exploration, National Geographic Society, Washington, DC.

Miller, M. M. (1970) '1946–62 survey of the regional pattern of Alaskan glacier variations', *National Geographic Society Research Reports, 1961–1962 Projects*, Washington, DC, National Geographic Society, 167–89.

Muller, D. S. (1980) 'Glacial geology and Quaternary history of southeast Meta Incognita Peninsula, Baffin Island, Canada', unpublished MSc. thesis, University of Colorado.

Müller, F. (1962) 'Glacier mass-budget studies on Axel Heiberg Island, Canadian Arctic Archipelago', in *Variations of the Regime of Existing Glaciers*, Obergurgl Symposium, International Association of Scientific Hydrology, Publ. no. 58, 131–42.

Müller, F. (ed.) (1977) *Fluctuations of Glaciers 1970–75*, International Commission on Snow and Ice of the International Association of Hydrological Sciences, Paris, IAHS and Unesco.

Nelson, J. G., Ashwell, I. Y. and Brunger, A. G. (1966) 'Recession of the Drummond Glacier, Alberta', *Canadian Geographer* 10: 71–81.

Ommanney, C. S. L. (1980) 'The inventory of Canadian glaciers: procedures, techniques, progress and applications', in Müller, F. (ed.) *World Glacier Inventory. Proceedings of Riederalp Workshop, 1978*, International Association of Hydrological Sciences, Publ. no. 126, 35–44.

Osborn, G. and Taylor, J. (1975) 'Lichenometry on calcareous substrates in the Canadian Rockies', *Quaternary Research* 5: 111–20.

Østrem, G. (1966) 'Mass balance studies on glaciers in western Canada, 1965', *Geographical Bulletin* 8: 81–107.

Péwé, T. L. (1975) *Quaternary Geology of Alaska*, US Geological Survey Professional Paper 835.

Porter, S. C. (1981) 'Lichenometric studies in the Cascade Range of Washington: establishment of *Rhizocarpon geographicum* growth curves at Mount Rainier', *Arctic and Alpine Research* 13: 11–23.

Post, A. S. (1965) 'Alaskan glaciers: recent observations in respect to the earthquake-advance theory', *Science* 148: 366–8.

Post, A. S. (1969) 'Distribution of surging glaciers in western North America', *Journal of Glaciology* 8: 229–40.

Raub, W. B., Post, A., Brown, C. S. and Meier, M. F. (1980) 'Perennial ice masses of the Sierra Nevada, California', in Müller, F. (ed.) *World Glacier Inventory. Proceedings of Riederalp Workshop, 1978*, International Association of Hydrological Sciences, Publ. no. 126, 33–4.

Reger, R. D. and Péwé, T. L. (1969) 'Lichenometric dating in the central Alaska Range', in Péwé,

T. L. (ed.) *The Periglacial Environment Past and Present*, Montreal, McGill–Queen's University Press, 223–47.
Reid, H. F. (1896) 'Glacier Bay and its glaciers', *US Geological Survey, 16th Annual Report*, Part 1, 415–61.
Reid, H. F. (1897–1916) 'Variations of glaciers 1–20', *Journal of Geology*, Series of Occasional Reports.
Russell, I. C. (1897) *Glaciers of North America*, Boston and London, Ginn.
Russell, I. C. (1898) 'Glaciers of Mount Rainier', *US Geological Survey, 18th Annual Report*, Part 2, 355–409.
Sherzer, W. H. (1905) 'Glacial studies in the Canadian Rockies and Selkirks', *Smithsonian Miscellaneous Collections* 47: 453–96.
Sigafoos, R. S. and Hendricks, E. L. (1961) *Botanical Evidence of the Modern History of Nisqually Glacier, Washington*, US Geological Survey Professional Paper 387-A.
Sigafoos, R. S. and Hendricks, E. L. (1972) *Recent Activity of Glaciers of Mount Rainier, Washington*, US Geological Survey Professional Paper 387-B.
Tangborn, W. (1980) 'Two models for estimating climate–glacier relationships in the North Cascades, Washington, U.S.A', *Journal of Glaciology* 25: 3–21.
Tarr, R. S. and Martin, L. (1914) *Alaskan Glacier Studies of the National Geographic Society in the Yakutat Bay, Prince William Sound and Lower Copper River Regions*, Washington, DC, National Geographic Society.
Thorington, J. M. (1927) 'The Lyell and Freshfield Glaciers', *Smithsonian Miscellaneous Collections* 78: 1–8.
Vancouver, G. (1798) *A Voyage of Discovery to the North Pacific Ocean, and Round the World in which the Coast of North-West America has been Carefully Examined and Accurately Surveyed . . . Performed in the Years 1790, 1791, 1792, 1793, 1794 and 1795, etc.*, 3 vols, London, G. G. & J. Robinson & J. Edwards.
Vaux, W. S. (1907) 'Modern glaciers; their movements and methods of observing them', *Proceedings of the Engineers Club of Philadelphia* 24.
Veatch, F. M. (1969) *Analysis of a 24-Year Photographic Record of Nisqually Glacier, Mount Rainier National Park, Washington*, US Geological Survey Professional Paper 631.
Viereck, L. A. (1968) 'Botanical dating of recent glacial activity in western North America', in Osburn, W. H. and Wright, H. E. (eds) *Arctic and Alpine Environment*, Bloomington, Ind., Indiana University Press, 189–204. Volume 10. Proceedings 7th Congress International Association for Quaternary Research.
Weller, G. (1980) 'Recent work: U.S.A.: Columbia Glacier', *Ice* 64: 3.
Wheeler, A. O. (1910) 'Motion of the Yoho Glacier', *Canadian Alpine Journal* 2, part 2: 121–5.
Wheeler, A. O. (1931) 'Glacial change in the Canadian Cordillera. The 1931 expedition', *Canadian Alpine Journal* 20: 120–42.
Wheeler, A. O. (1933) 'Records of glacial observations in the Canadian Cordillera, 1933 and 1934', *Canadian Alpine Journal* 22: 172–87.
Wilcox, R. E. (1965) 'Volcanic ash chronology', in Wright, H. E. and Frey, D. G. (eds) *The Quaternary of the United States*, Princeton University Press, 807–16.
Williams, L. D. (1979) 'An energy balance model of potential glacierization of northern Canada', *Arctic and Alpine Research* 11: 443–56.

GREENLAND

Beschel, R. E. (1961) 'Dating rock surfaces by lichen growth and its application to glaciology and physiography (lichenometry)', in G. O. Rausch (ed.) *Geology of the Arctic II*, Toronto University Press, vol. 2, 1044–62.

Beschel, R. E. and Weidick, A. (1973) 'Geobotanical and geomorphological reconnaissance in West Greenland, 1961', *Arctic and Alpine Research* 5: 311–19.
Davies, W. E. and Krinsley, D. B. (1962) 'The recent regimen of the ice cap margin in North Greenland', in *Variations of the Regime of Existing Glaciers*, Obergurgl Symposium, International Association of Scientific Hydrology, Publ. no. 58, 119–30.
Fitch, F. J., Kinsman, D. J. J., Sheard, J. W. and Thomas, D. (1962) 'Glacier re-advance on Jan Mayen', in *Variations of the Regime of Existing Glaciers*, Obergurgl Symposium, International Association of Scientific Hydrology, Publ. no. 58, 201–11.
Gad, F. (1970) *The History of Greenland. I, Earliest Times to 1700*, London, Hurst.
Gad, F. (1973) *The History of Greenland. II, 1700–1782*, London, Hurst.
Gordon, J. E. (1980) 'Recent climatic trends and local glacier margin fluctuations in West Greenland', *Nature* 284: 157–9.
Gordon, J.E. (1981) 'Glacier margin fluctuations during the 19th and 20th centuries in the Íkamiut kangerdluarssuat area, West Greenland', *Arctic and Alpine Research* 13: 47–62.
Gribbon, P. W. F. (1979) 'Cryoconite holes on Sermikavsak, West Greenland', *Journal of Glaciology* 22: 177–81.
Hovgaard, W. (1925) 'The Norsemen in Greenland. Recent discoveries at Herjolfsnes', *Geographical Review* 15: 605–16.
Nörlund, P. (1924) 'Buried Norseman at Herjolfsnes. An Archaeological and Historical Study', *Meddelelser om Grønland* 67, no. 1: 1–270.
Paars, C. E. (1729), in Bobé, L. (ed.) (1936) 'Diplomatarium Groenlandicum: 1492–1814. Aktstykker og Breve til Oplysning om Grønlands Besejling, Kolonisation og Missionering', *Meddelelser om Grønland* 55, no. 3: 186–9.
Rabot, C. (1897) 'Les Variations de longueur des glaciers dans les régions arctiques et boréales', *Archives des Sciences Physiques et Naturelles.*
Rink, H. (1852–7) *Grønland geographisk og statistisk beskrevet*, 2 vols, Copenhagen, Høst.
Sharp, R. P. (1956) 'Glaciers in the Arctic', *Arctic* 9: 78–117.
Weidick, A. (1959) 'Glacial variations in West Greenland in historical time, Part 1, Southwest Greenland', *Meddelelser om Grønland* 158, no. 4: 1–196.
Weidick, A. (1963) 'Ice margin features in the Julianehåb district, South Greenland', *Meddelelser om Grønland* 165, no. 3: 1–133.
Weidick, A. (1968) 'Observations on some Holocene glacier fluctuations in West Greenland', *Meddelelser om Grønland* 165, no. 6: 1–202.
Weidick, A. (1982) 'Klima og gletscherændringer i det sydlige Vestgrønland i de sidste 1000 år', *Meddelelser om Grønland* 30: 235–51.
Wright, J. W. (1939) Contributions to the glaciology of North-West Greenland', *Meddelelser om Grønland* 125, no. 3: 1–43.

9 Glaciers in low latitudes and the southern hemisphere

EAST AFRICA

Bhatt, N., Hastenrath, S. and Kruss, P. (1980) 'Ice thickness determination at Lewis Glacier, Mount Kenya: seismology, gravimetry, dynamics', *Zeitschrift für Gletscherkunde und Glazialgeologie* 16: 213–28.
Charnley, F. E. (1959) 'Some observations on the glaciers of Mt. Kenya', *Journal of Glaciology* 3: 483–92.
Davies, T. D., Brimblecombe, P. and Vincent, C. E. (1977) 'The first ice core from East Africa', *Weather* 32: 386–90.
Hastenrath, S. (1975) 'Glacier recession in East Africa', *Proceedings of the WMO/IAMAP Symposium on Long-Term Climatic Fluctuations, Norwich, 18–23rd August 1975*, WMO no. 421: 135–42.

Hastenrath, S. (1980) 'Tropical glacier systems and climate', Internationales Alfred-Wegener-Symposium, *Berliner Geowissenschaftliche Abhandlungen* A19, 72–6.

Hastenrath, S. (1984) *The Glaciers of Equatorial East Africa*, Dordrecht, Reidel.

Hastenrath, S. (1985) 'A review of Pleistocene to Holocene glacier variations in the tropics', *Zeitschrift für Gletscherkunde und Glazialgeologie* 21: 183–94.

Hastenrath, S. and Caukwell, R. A. (1979) 'Variations of Lewis Glacier, Mount Kenya, 1974–78', *Erdkunde* 33: 292–7.

Hastenrath, S. and Patnaik, J. K. (1980) 'Radiation measurements at Lewis Glacier, Mount Kenya, Kenya', *Journal of Glaciology* 25: 439–44.

Kruse, P. (1984) 'Terminus response of Lewis Glacier, Mount Kenya, Kenya, to sinusoidal net-balance forcing', *Journal of Glaciology* 30: 212–17.

Sampson, D. M. (1971) 'The geology, vulcanology and glaciology of Kilimanjaro', in Mitchell, J. (ed.) *Guide Book to Mount Kenya*, Mountain Club of Kenya, Nairobi, Kenya, 155–69.

Sansom, H. A. (1952) *The Trend of Rainfall in East Africa*, East African Meteorological Department, Technical Memorandum 1.

Temple, P. H. (1968) 'Further observations on the glaciers of Ruwenzori', *Geografiska Annaler* 50A: 136–50.

Thompson, L. G. (1981) 'Ice core studies from Mt Kenya, Africa, and their relationship to other tropical ice core studies', in Allison, I. (ed.) *Sea-level, Ice, and Climatic Change. Proceedings of the Canberra Symposium, December 1979*, International Association of Scientific Hydrology, Publ. no. 131, 55–62.

Thompson, L. G. and Hastenrath, S. L. (1981) 'Climatic ice core studies at Lewis Glacier, Mount Kenya', *Zeitschrift für Gletscherkunde und Glazialgeologie* 17: 115–23.

Vincent, C. E., Davies, T. D. and Beresford, A. K. C. (1979) 'Recent changes in the level of Lake Naivasha, Kenya, an indicator of Equatorial Westerlies over East Africa', *Climatic Change* 2: 175–89.

Vincent, C. E., Davies, T. D. and Brimblecombe, P. (1979) 'The Lewis Glacier (Mt Kenya) and possible links with tropical climate', in Allison, I. (ed.) *Sea-level, Ice, and Climatic Change. Proceedings of the Canberra Symposium, December 1979*, International Association of Scientific Hydrology, Publ. no. 131, 63–78.

Whittow, J. B., Shepherd, A., Goldthorpe, J.E. and Temple, P. H. (1963) 'Observations on the glaciers of the Ruwenzori', *Journal of Glaciology* 4: 581–616.

NEW GUINEA

Allison, I. and Bennett, J. (1976) 'Climate and microclimate', in Hope, G. S., Peterson, J. A., Radok, H. and Allison, I. (eds) *The Equatorial Glaciers of New Guinea*, Rotterdam, Balkema, 61–80.

Allison, I. and Kruss, P. (1977) 'Estimation of recent climate change in Irian Jaya by numerical modeling of its tropical glaciers', *Arctic and Alpine Research* 9: 49–60.

Allison, I. and Peterson, J. A. (1976) 'Ice areas on Mt. Jaya: their extent and recent history', in Hope, G. S. Peterson, J. A., Radok, U. and Allison, I. (eds) *The Equatorial Glaciers of New Guinea*, Rotterdam, Balkema, 27–38.

Hope, G. S., Peterson, J. A., Radok, U. and Allison, I. (eds) (1976) *The Equatorial Glaciers of New Guinea*, Rotterdam, Balkema. Results of the 1971–1973 Australian Universities' Expeditions to Irian Jaya: survey, glaciology, meteorology, biology and palaeoenvironments.

Kol, E. and Peterson, J. A. (1976) 'Cryobiology', in Hope, G. S., Peterson, J. A., Radok, U. and Allison, I. (eds) *The Equatorial Glaciers of New Guinea*, Rotterdam, Balkema, 81–91.

Mitchell, J. M. (1961) 'Recent secular change of global temperature', *Annals of the New York Academy of Science* 95: 235–50.

Peterson, J. A., Hope, G. S. and Mitton, R. (1973) 'Recession of snow and ice fields of Irian Jaya, Republic of Indonesia', *Zeitschrift für Gletscherkunde und Glazialgeologie* 9: 73–87.

SOUTH AMERICA

Broggi, J. A. (1943) 'La deglacion actual de los Andes del Peru', *Boletín de la Sociedad Geológica del Perú* 14–15: 59–90.
Cabot, T. D. (1939) 'The Cabot expedition to the Sierra Nevada de Santa Marta of Colombia', *Geographical Review* 29: 587–621.
Clapperton, C. M. (1972) 'The Pleistocene moraine stages of west-central Peru', *Journal of Glaciology* 11: 255–63.
Clapperton, C. M. (1983) 'The glaciation of the Andes', *Quaternary Science Review*, 2: 83–155.
Groveman, B. S. and Landsberg, H. E. (1979) 'Simulated Northern Hemisphere temperature departures 1579–1880', *Geographical Research Letters* 6: 767–9.
Hastenrath, S. (1981) *The Glaciation of the Equadorian Andes*, Rotterdam, Balkema.
Hastenrath, S. (1985) *Climate and Circulation of the Tropics*, Dordrecht, Reidel.
Hauthol, R. (1911) *Reisen in Bolivien und Peru*, Wissenschaftliche Veröffentlichungen der Gesellschaft für Erdkunde zu Leipzig 7.
Jordan, E. (1985) 'Recent glacier distribution and present climate in the Central Andes of South America', *Zeitschrift für Gletscherkunde und Glazialgeologie* 21: 213–24.
Kinzl, H. (1949) 'Die Vergletscherung in der Südhälfte der Cordillera Blanca (Peru). Begleitworte zu einer stereophotogrammetrischen Karte 1:100,000', *Zeitschrift für Gletscherkunde und Glazialgeologie* 1: 1–28.
Kinzl, H. (1955) 'Neues von der Huayhuash-Kordillere (Peru)', *Jahrbuch des Österreichischen Alpenvereins* 80: 123–31.
Kinzl, H. (1970) 'Gründung eines glaziologischen Institutes in Peru', *Zeitschrift für Gletscherkunde und Glazialgeologie* 6: 245–6.
Lliboutry, L., Arnao, B. M., Pautre, A. and Schneider, B. (1977) 'Glaciological problems set by the control of dangerous lakes in Cordillera Blanca, Peru. I. Historical failures of morainic dams, their causes and prevention', *Journal of Glaciology* 18: 239–54.
Melbourne, W. H. (1960) 'Exploration and survey on the Bolivia–Peru border', *Geographical Journal* 126: 455–8.
Mercer, J. H. (1967) *Southern Hemisphere Glacier Atlas*, New York, Department of Exploration and Field Research, American Geographical Society.
Meyer, H. (1907) *In den Hoch-Anden von Ecuador: Chimborazzo, Cotopaxi, etc.*, Berlin.
Meyer, H. (1908) *In the High Andes of Ecuador: Chimborazo, Cotopaxi, etc.*, London, William & Norgate.
Reiss, W. and Stubel, A. (1886) *Reisen in Süd-Amerika, 1. Skizzen aus Ecuador*, Berlin, Asher.
Schubert, C. (1972) 'Geomorphology and glacier retreat in the Pico Bolívar area, Sierra Nevada de Mérida, Venezuela', *Zeitschrift für Gletscherkunde und Glazialgeologie* 8: 189–202.
Sievers, W. (1908) 'Vergletscherung der Cordilleren des tropischen Südamerika', *Zeitschrift für Gletscherkunde* 2: 271–84.
Sievers, W. (1914) *Reise in Peru und Ecuador, ausgeführt 1909*, Wissenschaftliche Veröffentlichungen der Gesellschaft für Erdkunde zu Leipzig, 8, Munich and Leipzig, Dunker & Humlot.
Smith, C. T. (1957) 'Interim report on certain aspects of glaciation in the Cordillera Blanca, by reference to the Glaciers of Cordormina, Padiash, Pucaranva and Atlante', typescript on file at World Data Centre A, Glaciology, American Geographical Society, New York.
Spann, H. J. (1948) 'Report on glacier recession in Peru', *International Union of Geodesy and Geophysics, International Association of Scientific Hydrology, General Assembly of Oslo*, International Association of Scientific Hydrology, Publ. no. 30, 283–4.

Thompson, L. G. (1980) 'Glaciological investigations of the tropical Quelccaya Ice Cap, Peru', *Glaciological Journal* 25: 69–84.
Thompson, L. G., Mosley-Thompson, E., Dansgaard, W. and Grootes, P. M. (1986) 'The Little Ice Age as recorded in the stratigraphy of the tropical Quelccaya ice cap', *Science* 234: 361–4.
Thompson, L. G., Mosley-Thompson, E., Grootes, P. M., Pourchet, M. and Hastenrath, S. (1984) 'Tropical glaciers: potential for ice core paleoclimatic reconstruction', *Journal of Geophysical Research* 89: 4638–46.
Whymper, E. (1892) *Travels amongst the Great Andes of the Equator*, New York, Scribner's.
Wood, W. A. (1940) 'The Sierra Nevada de Santa Marta, Colombia', *Arctic and Alpine Research* 4: 21–8.
Wood, W. A. (1970) 'Recent glacier fluctuations in the Sierra Nevada de Santa Marta, Colombia', *Geographical Review* 60: 374–92.

SOUTHERN ANDES

Agostini, A. (1941) *Andes Patagónicos*, Buenos Aires.
Aguado, C. (1984) 'Inventario de glaciares del Rio San Juan', unpublished report, Mendoza, IANIGLA.
Bader, H. (1973) 'Request for photographs of Cordilleran glaciers', *Journal of Glaciology* 12: 526.
Bertone, M. (1960) *Inventario de los glaciares existentes en la vertiente Argentina entre nos paralelos 47°31′ y 51°S*, Buenos Aires, Instituto Nacional del Hielo, Publ. no. 3, Continental patagonica.
Cobos, D. R. (1981) *Glacier Inventory of the Atuel River Basin*, IANIGLA-CONISET, Mendoza, Rotterdam, Balkema.
Cobos, D. R. and Boninsegna, J. A. (1983) 'Fluctuations of some glaciers in the upper Atuel River basin, Mendoza, Argentina', in Rabassa, J. (ed.) *Quaternary of South America and Antarctic Peninsula*, Rotterdam, Balkema, vol. 1, 61–82.
Cobos, D. R. and Boninsegna, J. A. (1984) 'Paleoclima Cordillerano de las ultimas centurias en la zona suroeste de Mendoza, República Argentina', *Segundo Reunión Grupo Periglacial Argentina, Anales* 6: 51–65.
Corte, A. E. (1980) 'Glaciers and glaciolithic systems of the Central Andes', in Müller, F. (ed.) *World Glacier Inventory. Proceedings of the Workshop at Riederalp, 1978*, International Association of Hydrological Sciences, Publ. no. 126, 11–24.
Corte, A. E. and Espizua, L. (1981) *Inventario di Glaciares de la Cuenca del Rio*, IANIGLA-CONISET, Mendoza.
Groeber, P. (1947) 'Observaciones geológicas a lo largo del meridiano 70°', *Revista Asociación Geológica Argentina* 2: 164–74.
Groeber, P. (1954) 'Bosquejo paleogeográfico de los glaciares de Diamante y Atuel', *Revista Asociación Geológica Argentina* 9: 89–108.
Hatcher, J. B. (1903) *Reports of the Princeton University Expeditions to Patagonia 1896–99. Vol. 1. Narrative and Geography*, Princeton University Press.
Hauthol, R. (1895) 'Observaciones generales sobra algunos ventisqueros de la Cordillera de los Andes', *Revista Museo de la Plata* 6: 109–16.
Heusser, C. J. (1960) 'Late-Pleistocene environments of the Laguna de San Rafael area, Chile', *Geographical Review* 50: 555–77.
Heusser, C. J. (1961) *Final report of the American Geographical Society South Chile Expedition, 1959*, mimeo., New York, American Geographical Society.
Klein, J. K., Lerman, J. C., Damon, P. E. and Ralph, E. K. (1983) 'Calibration of radiocarbon dates', *Radiocarbon* 24: 103–50.
Lawrence, D. B. and Lawrence, E. G. (1959) *Recent Glacier Variations in Southern South America*, American Geographical Society Southern Chile Expedition, Tehnical Report 1, New York, American Geographical Society.

Liss, C.-C. (1970) 'Der Morenogletscher in der patagonischen Kordillere', *Zeitschrift für Gletscherkunde und Glazialgeologie* 6: 161–80.
Lliboutry, L. (1953) 'More about advancing and retreating glaciers in Patagonia', *Journal of Glaciology* 2: 168–72.
Lliboutry, L. (1956) *Nieves y Glaciares de Chile: Fundamentos de Glaciología*, Santiago, Universidad de Chile.
Lliboutry, L., Gonzales, O. and Simken, J. (1958) 'Les glaciers du désert chilien', International Association of Scientific Hydrology, General Assembly of Toronto, 1957, IV, 'Snow and Ice', Publ. no. 46: 291–300.
Magnani, M. (1961) 'Sulle variazioni dei ghiacciai Patagonici', *Atti del 18th Congresso Geografico Italiano, Trieste*, 1–11.
Martinic, B. M. (1982) *Hielo Patagónico Sur*, Instituto de la Patagonia, Serie Monografias 12.
Mercer, J. H. (1962) 'Glacier variations in the Andes', *Glaciological Notes* 12: 9–31.
Mercer, J. H. (1965) 'Glacier variations in southern Patagonia', *Geographical Review* 55: 390–413.
Mercer, J. H. (1967) *Southern Hemisphere Glacier Atlas*, Technical Report 67-76-ES, Earth Sciences Laboratory ES-33, US Army Natick Laboratories.
Mercer, J. H. (1968) 'Variations in some Patagonian glaciers since the Late-Glacial, I', *American Journal of Science* 266: 91–109.
Mercer, J. H. (1970) 'Variations of some Patagonian glaciers since the Late Glacial, II', *American Journal of Science* 269: 1–25.
Mercer, J. H. (1976) 'Glacial history of southernmost South America', *Quaternary Research* 6: 125–66.
Muller, E. H. (1959) *Glacial Geology of the Laguna San Rafael Area*, American Geographical Society Southern Chile Expedition 1959, Technical Report, mimeo., New York, American Geographical Society.
Nichols, R. L. and Miller, M. M. (1951) 'Glacial geology of Ameghino, Valley, Lago Argentino, Patagonia', *Geographical Review* 41: 274–94.
Nichols, R. L. and Miller, M. M. (1952) 'The Moreno Glacier, Lago Argentino, Patagonia. Advancing glaciers and nearby simultaneously retreating glaciers', *Journal of Glaciology* 2: 41–50.
Rabassa, J., Rubulis, S. and Suarez, J. (1978) 'Los glaciares del Monte Tronador Parque Nacional Nahuel Huapi (Rio Negro, Argentina)', *Anales de Parques Nacionales* 14: 259–318.
Shipton, E. (1962) 'Across the Patagonian Icecap', *American Alpine Journal* 13: 119–28.

NEW ZEALAND

Birkeland, P. W. (1981) 'Soil data and the shape of the lichen growth-rate curve for the Mt Cook area (Note)', *New Zealand Journal of Geology and Geophysics* 24: 443–5.
Burrows, C. J. (1973) 'Studies of some glacial moraines in New Zealand – 2. Ages of moraines of the Mueller, Hooker and Tasman glacier (S79)', *New Zealand Journal of Geology and Geophysics* 16: 831–55.
Burrows, C. J. (1975) 'Late Pleistocene and Holocene moraines of the Cameron Valley, Arrowsmith Range, Canterbury, New Zealand', *Arctic and Alpine Research* 7: 125–40.
Burrows, C. J. (1977) 'Late Pleistocene and Holocene glacial episodes in the South Island, New Zealand and some climatic implications', *New Zealand Geographer* 33: 34–8.
Burrows, C. J. (1980) 'Radiocarbon dates for post-Otiran glacial activity in New Zealand', *New Zealand Journal of Geology and Geophysics* 23: 239–48.
Burrows, C. J. and Greenland, D. E. (1979) 'An analysis of the evidence for climatic change in New Zealand in the last thousand years: evidence from diverse natural phenomena and from instrumental records', *Journal of the Royal Society of New Zealand* 9: 321–73.

Burrows, C. J. and Gellatly, A. F. (1982) 'Holocene glacial activity in New Zealand', *Striae* 18: 41–7.
Burrows, C. J. and Lucas, J. (1967) 'Variations in two New Zealand glaciers during the past 800 years', *Nature* 216: 467–8.
Burrows, C. J. and Orwin, J. (1971) 'Studies on some glacial moraines in New Zealand – 1. The establishment of lichen-growth curves in the Mount Cook area', *New Zealand Journal of Science* 14: 327–35.
Chinn, T. J. H. (1981) 'Use of rock weathering-rind thickness for Holocene absolute age-dating in New Zealand', *Arctic and Alpine Research* 13: 33–45.
Chinn, T. J. H. and Whitehouse, I. E. (1980) 'Glacier snow line variations in the Southern Alps, New Zealand', in Müller, F. (ed.) *World Glacier Inventory. Proceedings of Riederalp Workshop, 1978*, International Association of Hydrological Sciences, Publ. no. 126, 219–28.
Douglas, C. E. and Harper, A. P. (1985) 'The Westland Alps, New Zealand', *Geographical Journal* 5: 61–8.
Dowdeswell, J. A. and Morris, S. E. (1983) 'Multivariate statistical approaches to the analysis of late Quaternary relative age data', *Progress in Physical Geography* 7: 157–76.
Gellatly, A. F. (1982) 'The use of lichenometry as a relative-age dating method with specific reference to Mount Cook National Park, New Zealand', *New Zealand Journal of Botany* 20: 343–53.
Gellatly, A. F. (1983) 'Revised dates for 2 recent moraines of the Mueller Glacier, Mt Cook National Park (Note)', *New Zealand Journal of Geology and Geophysics* 26: 311–15.
Gellatly, A. F. (1984) 'The use of rock weathering-rind thickness to redate moraines in Mount Cook National Park, New Zealand', *Arctic and Alpine Research* 16: 225–32.
Gellatly, A. F. (1985a) 'Historical records of glacier fluctuations in Mt Cook National Park, New Zealand: a century of change', *Geographical Journal* 151: 86–99.
Gellatly, A. F. (1985b) 'Glacier fluctuations in the central Southern Alps, New Zealand: documentation and implications for environmental change during the last 1000 years', *Zeitschrift für Gletscherkunde und Glazialgeologie* 21: 259–64.
Gellatly, A. F. and Norton, D. A. (1984) 'Possible warming and glacier recession in the South Island, New Zealand', *New Zealand Journal of Science* 27: 381–8.
Harper, A. P. (1893) 'Exploration and character of the principal New Zealand glaciers', *Geographical Journal* 1: 32–42.
Harper, A. P. (1896) *Pioneer Work in the Alps of New Zealand*, London, Unwin.
Hessell, J. W. D. (1983) 'Climatic effects on the recession of the Franz Josef Glacier', *New Zealand Journal of Science* 26: 315–20.
Kasser, P. (ed.) (1973) *Fluctuations of Glaciers 1965–1970*, International Commission on Snow and Ice of the International Association of Hydrological Sciences, Paris, IAHS and Unesco.
Klein, J., Lerman, J. C., Damon, P. E. and Ralph, E. K. (1982) 'Calibration of radiocarbon dates', *Radiocarbon* 24: 103–50.
Paul, J. (1974) 'Twelve watercolours of glaciers in the Province of Canterbury. Julius Haast and John Gully collaborators', *Turnbull Library Record* 7.
Salinger, M. J. (1979) 'New Zealand climate: the temperature record, historical data and some agricultural implications', *Climatic Change* 2: 109–26.
Salinger, M. J. (1982) 'On the suggestion of post-1950 warming over New Zealand', *New Zealand Journal of Science* 25: 77–86.
Salinger, M. J. and Gunn, J. M. (1975) 'Recent climatic warming around New Zealand', *Nature* 256: 396–8.
Salinger, M. J., Heine, M. J. and Burrows, C. J. (1983) 'Variations of the Stocking (Te Wae Wae) Glacier, Mount Cook, and climatic relationships', *New Zealand Journal of Science* 26: 321–38.
Sara, W. A. (1968) 'Franz Josef and Fox Glaciers, 1951–1967', *New Zealand Journal of Geology and Geophysics* 11: 768–80.

Sara, W. A. (1970) *Glaciers of Westland National Park. A New Zealand Geological Survey Handbook*, Information Series 75, New Zealand Department of Scientific and Industrial Research.

Suggate, R. P. (1950) 'Franz Josef and other glaciers of the Southern Alps, New Zealand', *Journal of Glaciology* 1: 422–9.

Thompson, R. D. and Kells, B. R. (1973) 'Mass balance studies on the Whakapapanui glacier, New Zealand', in *The Role of Snow and Ice in Hydrology. Proceedings of the Banff Symposia, September 1972*. Unesco-WMO-AIHS, vol. 1, 383–93 (IAHS Publ. no. 107).

Trenberth, K. E. (1976) 'Fluctuations and trends in indices of Southern Hemisphere circulation', *Quarterly Journal of the Royal Meteorological Society* 102: 65–75.

Von Haast, J. (1864) 'Notes on the mountains and glaciers of the Canterbury Province, New Zealand', *Journal of the Royal Geographical Society* 40: 433–41.

Von Haast, J. (1879) *A Geology of the Provinces of Canterbury and Westland, New Zealand. A Report Comprising the Results of Official Exploration*, Christchurch.

Wardle, P. (1973) 'Variations of the glaciers of Westland National Park and the Hooker Range, New Zealand', *New Zealand Journal of Botany* 11: 349–88.

SUB-ANTARCTIC ISLANDS

Allison, I. F. (1980) *A Preliminary Investigation of the Physical Characteristics of the Vahsel Glacier, Heard Island*, Australian National University Research Expeditions Scientific Report, Series A4, Publ. no. 128.

Allison, I. F. and Keage, P. L. (1986) 'Recent changes in the glaciers of Heard Island', *Polar Record* 23 (144): 255–71.

Budd, G. M. (1970) 'Heard Island reconnaissance, 1969', *Polar Record* 15 (96): 335–6.

Budd, G. M. and Stephenson, P. J. (1970) 'Recent glacier retreat on Heard Island', in *International Symposium on Antarctic Glaciological Exploration (ISAGE), Hanover, N. H., 1968*, International Association for Scientific Hydrology, Publ. no. 86, 449–58.

Chinn, T. J. H. (1981) *Hydrology and Climate in the Ross Sea Area*, Report 418, Water and Soil Science Centre, Christchurch, Ministry of Works and Development.

Clapperton, C. M., Sugden, D. E., Birnie, R. V., Hanson, J. D. and Thom, G. (1978) 'Glacier fluctuations in South Georgia and comparison with other island groups in the Scotia Sea', in Van Zinderen Bakker, E. M. (ed.) *Antarctic Glacial History and World Palaeoenvironments*. Proceedings of the 10th INQUA Congress, Birmingham 1977, 95–104, Rotterdam, Balkema.

Kasser, P. (ed.) (1973) *Fluctuations of Glaciers 1965–70*, International Commission on Snow and Ice of the International Association of Hydrological Sciences, Paris, IAHS and Unesco.

Klein, J. J. C., Damon, P. E. and Ralph, E. K. (1982) 'Calibration of radiocarbon dates', *Radiocarbon* 24: 103–50.

Liestøl, O. (1967) *Storbreen glacier in Jotunheimen, Norway*, Norsk Polarinstitutt, Skrifter 141.

Mercer, J. H. (1962) 'Glacier variations in the Antarctic', *Glaciological Notes* 11: 5–29.

Morgan, B. I. (1985) 'An oxygen isotope-climate record from the Law Dome, Antarctica', *Climatic Change* 7: 415–26.

Orheim, O. (1972) *A 200-year Record of Glacier Mass Balance at Deception Island, Southwest Atlantic Ocean, and its Bearing on Models of Global Climatic Change*, Report 42, Ohio State University Research Foundation, Institute of Polar Studies.

Orheim, O. (1977) 'Global glacier mass balance variations during the past 300 years', in Dunbar, M. J. (ed.) *Polar Oceans. Proceedings of the Polar Oceans Conference, Montreal 1974*, Calgary, Arctic Institute of North America, 667–81.

Parkinson, C. L. and Cavalieri, D. J. (1982) 'Interannual sea-ice variations and sea-ice/atmosphere interactions in the Southern Ocean, 1973–1975', *Annals of Glaciology* 3: 249–54.

Radok, U. and Watts, D. (1975) 'A synoptic background to glacier variations of Heard Island',

Snow and Ice. Proceedings of the Moscow Symposium, August 1971, International Association of Hydrological Sciences, Publ. no. 104, 42–56.

10 The glacial history of the Holocene

Aario, L. (1945) 'Ein nachwärmezeitlicher Gletschervorstoss in Oberfernau in den Stubaier Alpen', *Acta Geographica* 9: 1–31.
Ackert, R. P. (1984) 'Ice-cored lateral moraines in Tarfala Valley, Swedish Lappland', *Geografiska Annaler* 66A: 79–88.
Alexander, M. J. and Worsley, P. (1973) 'Stratigraphy of a Neoglacial end moraine in Norway', *Boreas* 2: 117–42.
Alley, N. F. (1976a) 'Post-Pleistocene glaciation in the interior of British Columbia', *Programme and Abstracts, Geological Association of Canada (Cordilleran Section)*, p. 6.
Alley, N. F. (1976b) 'The palynology and paleoclimatic significance of a dated core of Holocene peat, Okanagan Valley, southern British Columbia', *Canadian Journal of Earth Sciences* 13: 1131–44.
Alley, N. F. (1980) 'Holocene and latest Pleistocene cirque glaciations in the Shuswap Highland, British Columbia: discussion', *Canadian Journal of Earth Sciences* 17: 797–8.
Andersen, B. G. (1980) 'The deglaciation of Norway after 10,000 BP', *Boreas* 9: 211–16.
Andrews, J. T. (1982a) 'Holocene glacier variations in the eastern Canadian Arctic: a review', *Striae* 16: 9–14.
Andrews, J. T. (1982b) 'Chronostratigraphic division of the Holocene, Arctic Canada', *Striae* 16: 56–64.
Andrews, J. T. and Barnett, D. M. (1979) 'Holocene (Neoglacial) moraine and proglacial lake chronology, Barnes Ice Cap, Canada', *Boreas* 8: 341–58.
Andrews, J. T. and Ives, J. D. (1972) 'Late- and Postglacial events (< 10,000 BP) in the eastern Canadian Arctic with particular reference to the Cockburn moraines and break-up of the Laurentide ice sheet', in Vasari, Y., Hyvärinen, H. and Hicks, S. (eds) *Climatic Changes in Arctic Areas During the Last 10,000 Years. A Symposium held at Oulanka and Kevo, 4–10 October 1971*, Acta Universitatis Ouluensis, Series A, Scientiae Rerum Naturalium 3, Geologica 1, Oulu, Finland, University of Oulu, 149–74.
Andrews, J. T. and Miller, G. H. (1972) 'Quaternary history of northern Cumberland Peninsula, Baffin Island, NWT, Canada: Part IV. Maps of the present glaciation limits and lowest equilibrium line altitude for north and south Baffin Island', *Arctic and Alpine Research* 4: 45–59.
Andrews, J. T., Carrara, P. E., King, F. B. and Stuckenrath, R. (1975) 'Holocene environmental changes in the alpine zone, northern San Juan Mountains, Colorado: evidence from bog stratigraphy and palynology', *Quaternary Research* 5: 173–97.
Andrews, J. T., Davis, P. T., Mode, W. N., Nichols, H. and Short, S. K. (1981) 'Relative departures in July temperatures in northern Canada for the past 6,000 years', *Nature* 289: 264–7.
Andrews, J. T., Funder, S., Hjort, C. and Imrie, J. (1974) 'Comparison of the glacial chronology of eastern Baffin Island, East Greenland, and the Camp Century accumulation record', *Geology* 2: 355–8.
Bachmann, F. and Furrer, G. (1971) 'Solifluktionsdecken im schweizerischen Nationalpark und ihre Beziehung zur postglazialen Landschaftsentwicklung', *Geographia Helvetica* 26: 122–8.
Baker, B. H. (1967) *Geology of the Mount Kenya Area*, Geological Survey of Kenya Report 79.
Barnosky, C. W. (1984) 'Late Pleistocene and early Holocene environmental history of southwestern Washington State, USA', *Canadian Journal of Earth Sciences* 21: 619–29.
Barsch, D. (1971) 'Rock-glaciers and ice-cored moraines', *Geografiska Annaler* 53A: 203–6.
Beget, J. E. (1984) 'Tephrochronology of late Wisconsin deglaciation and Holocene glacier

fluctuations near Glacier Peak, North Cascade Range, Washington', *Quaternary Research* 21: 304–16.
Benedict, J. B. (1967) 'Recent glacial history of an alpine area in the Colorado Front Range, USA. I, Establishing a lichen-growth curve', *Journal of Glaciology* 6: 817–32.
Benedict, J. B. (1968) 'Recent glacial history of an alpine area in the Colorado Front Range, USA. II, Dating the glacial deposits', *Journal of Glaciology* 7: 77–87.
Benedict, J. B. (1973) 'Chronology of cirque glaciation, Colorado Front Range', *Quaternary Research* 3: 584–99.
Benedict, J. B. (1976) 'Khumbu Glacier series, Nepal', *Radiocarbon* 18: 177–8.
Benedict, J. B. (1981) 'The Fourth of July Valley: glacial geology and archeology of the timberline ecotone', Ward CO: Centre for Mountain Archeology Research.
Bezinge, A. (1974) *Vieux Troncs morainiques et climat post-glaciaire sur les Alpes*, Étude presentée, Société Hydrotechnique de France.
Bezinge, A. and Vivian, R. (1976a) *Troncs fossiles morainiques et climat de la période Holocène en Europe.* Étude presentée, Société Hydrotechnique de France.
Bezinge, A. and Vivian, R. (1976b) 'Sites sous-glaciaires et climat de la période Holocène en Europe', *La Houille Blanche* 226: 441–59.
Birkeland, P. W. (1981) 'Soil data and the shape of the lichen growth rate curve for the Mt. Cook area (Note)', *New Zealand Journal of Geology and Geophysics* 24: 443–50.
Birkeland, P. W. (1982a) 'Subdivision of Holocene glacial deposits, Ben Ohau Range, New Zealand, using relative dating methods', *Geological Society of America Bulletin* 93: 433–49.
Birkeland, P. W. (1982b) 'Multiparameter relative dating methods to differentiate Holocene glacial deposits, Ben Ohau Range, NZ', *United States Geological Survey Bulletin.*
Birkeland, P. W. and Miller, I. D. (1973) 'Re-interpretation of the type Temple Lake moraine and other Neoglacial deposits, southern Wind River Mountains, Wyoming', *Geological Society of America, Abstracts with Programs*, Rocky Mountain Section 5: 465–6.
Birkeland, P. W., Crandell, D. R. and Richmond, G. M. (1971) 'Status of correlation of Quaternary stratigraphic units in the western conterminous United States', *Quaternary Research* 1: 208–27.
Björnsson, H. (1979) 'Glaciers in Iceland', *Jökull* 29: 74–80.
Bless, R. (1982) 'Postglaziale Schwankungen des Glaciers d'Argentière', *Physische Geographie* 1: 187–94.
Bless, R. (1984) *Beiträge zur Spät- und Post-Glazialen Geschichte der Gletscher im Nordöstlichen Mont Blanc Gebiet*, Geographisches Institut der Universität Zürich.
Borns, H. W. and Goldthwait, R. P. (1966) 'Late Pleistocene fluctuations of Kaskawulsh glacier, southwestern Yukon Territory, Canada', *American Journal of Science* 264: 600–19.
Bortenschlager, S. (1970) 'Waldgrenz- und Klimaschwankungen im pollenanalytischen Bild des Gurgler Rotmooses. Mitteilungen Ostalpen-Dinar', *Gesellschaft für Vegetationskunde* 11: 19–26.
Bortenschlager, S. and Patzelt, G. (1960) 'Warmezeitliche Klima- und Gletscherschwankungen im Pollenprofil eines hochgelegenen Moores (2270 m) der Venedigergruppe', *Eiszeitalter und Gegenwart* 20: 116–22.
Bryson, R. A., Wendland, D. A. B. and Wendland, W. M. (1970) 'The character of late-glacial and post-glacial climatic changes', in *Pleistocene and Recent Environments of the Great Plains*, University of Kansas, Department of Geology Special Publication no. 3, 53–74.
Burrows, C. J. (1973) 'Studies of some glacial moraines in New Zealand–2: Ages of moraines of the Mueller, Hooker and Tasman glaciers', *New Zealand Journal of Geology and Geophysics* 16: 831–56.
Burrows, C. J. (1975) 'Late Pleistocene and Holocene moraines of the Cameron Valley, Arrowsmith Range, Canterbury, New Zealand', *Arctic and Alpine Research* 7: 125–40.
Burrows, C. J. (1977) 'Late Pleistocene and Holocene glacial episodes in South Island, New Zealand and some climatic implications', *New Zealand Geographer* 33: 34–9.

Burrows, C. J. (1979) 'A chronology for cool-climate episodes in the Southern Hemisphere 12,000–1,000 yr BP', *Palaeogeography, Palaeoclimatology, Palaeoecology* 27: 287–347.

Burrows, C. J. (1980) 'Radiocarbon dates for post-Otiran glacial activity in the Mount Cook region, New Zealand', *New Zealand Journal of Geology and Geophysics* 23: 239–48.

Burrows, C. J. (1983) 'Radiocarbon dates from late quaternary deposits in the Cass district, Canterbury, New Zealand', *New Zealand Journal of Botany* 21: 443–54.

Burrows, C. J. and Gellatly, A. F. (1982) 'Holocene glacial activity in New Zealand', *Striae* 18: 41–7.

Burrows, C. J. and Greenland, D. E. (1979) 'An analysis of evidence for climatic change in New Zealand for the last thousand years. Evidence from diverse natural phenomena and from instrumental records', *Journal of the Royal Society of New Zealand* 9: 321–73.

Burrows, C. J. and Russell, J. R. (1975) 'Moraines of the upper Rakaia valley', *Journal of the Royal Society of New Zealand* 5: 463–77.

Burrows, C. J., Chinn, T. and Kelly, M. (1976) 'Glacial activity in New Zealand near the Pleistocene–Holocene boundary in the light of new radiocarbon dates', *Boreas* 5: 57–60.

Calkin, P. E. and Ellis, J. M. (1980) 'A lichenometric dating curve and its application to Holocene glacier studies in the central Brooks Range, Alaska', *Arctic and Alpine Research* 12: 245–64.

Calkin, P. E. and Ellis, J. M. (1981) 'A cirque glacier chronology based on emergent lichens and mosses', *Journal of Glaciology* 27: 511–15.

Calkin, P. E. and Ellis, J. M. (1982) 'Holocene glacial chronology of the Brooks Range, northern Alaska', *Striae* 18: 3–8.

Calkin, P. E. and Haworth, L.A. (in press) 'Comparison of some lichenometrically supported Holocene glacial chronologies, Alaska', INQUA.

Calkin, P. E., Ellis, J. M., Haworth, L. A. and Burns, P. E. (1985) 'Cirque glacier regime and Neoglaciation, Brooks Range, Alaska', *Zeitschrift für Gletscherkunde und Glazialgeologie* 21: 371–8.

Carrara, P. and Andrews, J. T. (1972) 'The Quaternary history of northern Cumberland Peninsula, Baffin Island, NWT: Part I, the late- and neo-glacial deposits of the Akudlermuit and Boas glaciers', *Canadian Journal of Earth Sciences* 9: 403–14.

Chinn, T. J. H. (1981) 'Use of rock weathering-rind thickness for Holocene absolute age-dating in New Zealand', *Arctic and Alpine Research* 13: 33–45.

Clapperton, C. M. (1971a) 'Antarctic link with the Andes', *Geographical Magazine* 44: 125–30.

Clapperton, C. M. (1971b) *Geomorphology of the Stromness Bay–Cumberland Bay Area, South Georgia*, British Antarctic Survey, Scientific Reports 70.

Clapperton, C. M. (1976) 'Glacial fluctuations in South Georgia and Peru', *Quaternary Newsletter* 20: 10.

Coetzee, J. A. (1967) 'Pollen analytical studies in east and southern Africa', *Palaeoecology of Africa* 3.

Corner, G. D. (1980) 'Preboreal deglaciation chronology and marine limits of the Lyngen–Storfjord Area, Troms', *Boreas* 9: 239–49.

Craig, B. G. and Fyles, J. G. (1960) *Pleistocene Geology of Arctic Canada*, Geological Survey of Canada Paper 60–10.

Currey, D. R. (1969) 'Holocene climatic and glacial history of the central Sierra Nevada, California', *Geological Society of America Special Paper 123 (INQUA volume)*: 1–47.

Currey, D. R. (1974) 'Probable pre-Neoglacial age of type Temple lake moraine, Wyoming', *Arctic and Alpine Research* 6: 293–300.

Davis, P. T. (1985) 'Neoglacial moraines on Baffin Island', in Andrews, J. T. (ed.) *Quaternary Environments: Eastern Canadian Arctic, Baffin Bay and Western Greenland*, Boston, Allen & Unwin, 682–718B.

Denton, G. H. (1970) 'Late Wisconsin glaciation in northwestern North America: ice recession and origin of Paleo-Clovis complex', *American Quaternary Association, Abstracts, First Meeting, Boseman, Montana, 28th August–1st September*, 34–5.

Denton, G. H. (1974) 'Quaternary glaciations of the White River Valley, Alaska, with a regional

synthesis for the northern St Elias Mountains, Alaska and Yukon territory', *Geological Society of America Bulletin* 85: 871–92.
Denton, G. H. and Karlén, W. (1973a) 'Lichenometry: its application to Holocene moraine studies in southern Alaska and Swedish Lappland', *Arctic and Alpine Research* 5: 347–72.
Denton, G. H. and Karlén, W. (1973b) 'Holocene climatic variations – their pattern and possible causes', *Quaternary Research* 3: 155–205.
Denton, G. H. and Karlén, W. (1977) 'Holocene glacial and treeline variations in the White River valley and Skolai Pass, Alaska and Yukon territory', *Quaternary Research* 7: 63–111.
Denton, G. H. and Stuiver, M. (1966) 'Neoglacial chronology, northeastern St Elias Mountains, Canada', *American Journal of Science* 264: 577–99.
Donner, J. J. (1969) 'A profile across Fennoscandia of late Weichselian and Flandrian shorelines', *Commentationes Physico-Mathematicae* 36: 2–23.
Duford, J. M. and Osborn, G. D. (1978) 'Holocene and latest Pleistocene cirque glaciation in the Shuswap Highland, BC', *Canadian Journal of Earth Sciences* 15: 865–73.
Duford, J. M. and Osborn, G. D. (1980) 'Holocene and latest Pleistocene cirque glaciations in the Shuswap Highland, British Columbia: Reply', *Canadian Journal of Earth Sciences* 17: 799–800.
Dugmore, A. and Maizels, J. (1985) 'A date with tephra', *Geographical Magazine*, October: 532–8.
Easterbrook, D. J. (1974) 'Comparison of the Late Pleistocene glacial fluctuation', in Sibrava, V. (ed.) *Quaternary Glaciation in the Northern Hemisphere* (Project 73/1/24, formerly PA 7145, Correlation of European and American Glaciation), Report 1 of the session in Cologne, September 1973, IUG-Unesco International Geological Correlation Programme, Geological Survey, Prague, 96–109.
Einarsson, T., Hopkins, D. M. and Doell, R. R. (1967) *The Stratigraphy of Tjörnes, Northern Iceland, and the History of the Bering Land Bridge*, Stanford, Calif., Stanford University Press, 312–25.
Ellis, J. M. and Calkin, P. E. (1979) 'Nature and distribution of glaciers, Neoglacial moraines and rock glaciers, east-central Brooks Range, Alaska', *Arctic and Alpine Research* 11: 403–20.
Ellis, J. M. and Calkin, P. E. (1984) 'Chronology of Holocene glaciation, central Brooks Range, Alaska', *Geological Society of America Bulletin* 95: 897–912.
Falconer, G., Ives, J. D., Løken, O. H. and Andrews, J. T. (1965) 'Major endmoraines in eastern and central Arctic Canada', *Geographical Bulletin* 7: 137–53.
Ferguson, A. J. (1978) 'Late Quaternary geology of the Upper Elk Valley, British Columbia', unpublished MSc. thesis, University of Calgary, Canada.
Fulton, R. J. (1971) *Radiocarbon Geochronology of Southern British Columbia*, Geological Survey of Canada Paper 71–137.
Fulton, R. J. (1975) *Quaternary Geology and Geomorphology, Nicola-Vernon Area, British Columbia*, Geological Survey of Canada Memoir 308.
Furrer, G. and Holzhauser, H. (1984) 'Gletscher- und klimageschichtliche Auswertung fossiler Hölzer', *Zeitschrift für Geomorphologie* 50: 117–36.
Furrer, G., Gamper-Schollenberger, B. and Suter, J. (1980/1) 'Zur Geschichte unserer Gletscher in der Nacheiszeit. Methoden und Ergebnisse', in Oeschger, H., Messerli, B. and Suter, J. (eds) *Das Klima*, Berlin, Springer-Verlag, 91–107.
Fushimi, H. (1977) 'Glaciation in the Khumbu Himal (1)', *Seppyo, Journal of Japanese Society of Snow and Ice* 39: 60–7.
Fushimi, H. (1978) 'Glaciations in the Khumbu Himal (2)', *Seppyo, Journal of Japanese Society of Snow and Ice* 40: 71–7.
Gamper, M. and Suter, J. (1982) 'Postglaziale Klimageschichte der Schweizer Alpen', *Geographica Helvetica* 37: 105–114.
Gellatly, A. F. (1982) 'The use of lichenometry as a relative-age dating method with specific reference to Mount Cook National Park, New Zealand', *New Zealand Journal of Botany* 20: 343–53.

Gellatly, A. F. (1984) 'The use of rock weathering-rind thickness to redate moraines in Mount Cook National Park, New Zealand', *Arctic and Alpine Research* 16: 225–32.

Gellatly, A. F. (1985) 'Glacial fluctuations in the central southern Alps, New Zealand: documentation and implications for environmental change during the last 1000 years', *Zeitschrift für Gletscherkunde und Glazialgeologie* 21: 259–64.

Gellatly, A. F., Röthlisberger, F. and Geyh, M. A. (1985) 'Holocene glacier variations in New Zealand (South Island)', *Zeitschrift für Gletscherkunde und Glazialgeologie* 21: 265–73.

Geyh, M. A., Röthlisberger, F. and Gellatly, A. (1985) 'Reliability tests and interpretation of ^{14}C dates from palaeosols in glacier environments', *Zeitschrift für Gletscherkunde und Glazialgeologie* 21: 275–81.

Godwin, H. (1956) 'The climatic optimum', *The History of the British Flora: a factual basis for phytogeography*, Cambridge, Cambridge University Press, 45–63.

Graf, W. I. (1971) 'Quantitative analysis of Pinedale landforms, Beartooth Mountains, Montana and Wyoming', *Arctic and Alpine Research* 3: 253–61.

Griffey, N. J. (1975) 'Investigation of the Neoglacial deposits of the Okstindan glaciers 1–7', in Parry, R. B. and Worsley, P. (eds) *Okstindan Research Project Preliminary Report 1973*, Reading University, 1–7.

Griffey, N. J. (1976) 'Stratigraphical evidence for an early Neoglacial glacier maximum at Steikvassbreen, Okstindan, north Norway', *Norsk Geologisk Tidskrift* 56: 187–94.

Hack, J. T. (1943) 'Antiquity of the Finley site', *American Antiquity* 8: 235–41.

Hamilton, T. D. (1969) 'Glacial geology of the lower Alatna valley, Brooks Range, Alaska', *Geological Society of America Special Paper* 123: 181–223.

Hamilton, T. D. and Porter, S. C. (1975) 'Itkillic glaciation in the Brooks Range, northern Alaska', *Quaternary Research* 5: 471–97.

Haworth, L. A. and Calkin, P. E. (1986) 'Periodic formation of moraines during late Holocene time, Brooks Range, Alaska', *AMQUA Abstracts*.

Heine, K. (1975) *Studien zur jungquartären Glazialmorphologie mexikanischer Vulkane, mit einem Ausblick auf die Klimaentwicklung*, Das Mexiko-Projekt der DPG 7, Wiesbaden.

Heine, K. (1976) 'Blockgletscher- und Blockzungen-Generationen am Nevado de Toluca, Mexiko', *Zeitschrift der Gesellschaft für Erdkunde zu Berlin* 107: 330–52.

Heine, K. (1983) 'Spät- und postglaziale Gletscherschwankungen in Mexico. Befunde und paläoklimatische Deutung', in H. Schroeder-Lanz (ed.) *Late and Postglacial Oscillation of Glaciers: Glacial and Periglacial*, Rotterdam, Balkema, 291–304.

Heuberger, H. (1966) *Gletschergeschichtliche Untersuchungen in den Zentralalpen zweischen Sellrain und Ötztal*, Wissenschaftliche Alpenvereinshefte, 20.

Heuberger, H. (1974) 'Alpine Quaternary glaciation', in Ives, J. D. and Barry, R. G. (eds) *Arctic and Alpine Environments*, London, Methuen, 318–38.

Heuberger, H. and Beschel, R. (1958) 'Beiträge zur Datierung alter Gletscherstände in Hochstubai (Tirol)', *Schlern Schriften* 190, 73–100.

Heusser, C. J. (1974) 'Vegetation and climate of the southern Chilean lake district during and since the last interglaciation', *Quaternary Research* 4: 290–315.

Heusser, C. J. (1978) 'Palynology of Quaternary deposits of the lower Bogachiel river area, Olympic Peninsula, Washington', *Canadian Journal of Earth Sciences* 15: 1568–78.

Heusser, C. J. (1983) 'Holocene vegetation history of the Prince William Sound Region South-Central Alaska', *Quaternary Research* 19: 337–55.

Heusser, C. J. and Streeter, S. S. (1980) 'A temperature and precipitation record of the past 16,000 years in southern Chile', *Science* 210: 1345–6.

Holzhauser, H. (1982) 'Neuzeitliche Gletscherschwankungen', *Geographica Helvetica* 37: 115–26.

Holzhauser, H. (1984) 'Rekonstruktion von Gletscherschwankungen mit Hilfe fossiler Hölzer', *Geographica Helvetica* 39: 3–15.

Hope, G. S. and Peterson, J. A. (1975) 'Glaciation and vegetation in the high New Guinea mountains', *Bulletin of the Royal Society of New Zealand* 82: 329–410.

Hope, G. S., Peterson, J. A., Radok, U. and Allison, I. (1976) *The Equatorial Glaciers of New Guinea*, Rotterdam, Balkema.

Hopkins, D. M. (1967) 'The Cenozoic history of Beringia – a synthesis', in Hopkins, D. M. (ed.) *The Bering Land Bridge*, Stanford, Calif., Stanford University Press, 451–84.

Innes, J. L. (1984) 'Lichenometric dating of moraine ridges in northern Norway: some problems of application', *Geografiska Annaler* 66A: 341–52.

Iwata, S. (1976) 'Late Pleistocene and Holocene moraines in the Sugarmatta (Everest) region, Khumbu Himal', *Seppyo, Journal of Japanese Society of Snow and Ice* 38: 109–14.

Johansson, L. and Holmgren, K. (1985) 'Dating of a moraine on Mount Kenya', *Geografiska Annaler* 67A: 123–8.

Karlén, W. (1973) 'Holocene glacier and climate variations, Kebnekaise Mountains, Swedish Lappland', *Geografiska Annaler* 55A: 29–63.

Karlén, W. (1976) 'Lacustrine sediments and tree-limit variations as indicators of Holocene climatic fluctuations in Lappland, northern Sweden', *Geografiska Annaler* 58A: 1–34.

Karlén, W. (1979) 'Deglaciation dates from northern Swedish Lappland', *Geografiska Annaler* 61A: 203–10.

Karlén, W. (1985) 'Glacier and climate fluctuations on Mount Kenya, East Africa', *Zeitschrift für Gletscherkunde und Glazialgeologie* 21: 195–201.

Kearney, M. S. (1981) 'Late Quaternary vegetation and environments of Jasper National Park, Alberta', unpublished PhD thesis, University of Western Ontario, Canada.

Kearney, M. S. and Luckman, B. H. (1981) 'Evidence of late Wisconsin–early Holocene climatic/vegetational change in Jasper National Park, Alberta', in Mahaney, W. C. (ed.) *Quaternary Paleoclime*, Norwich, Geoabstracts, 85–105.

Kind, N. V. (1967) 'Radiocarbon chronology in Siberia', in Hopkins, D. M. (ed.) *The Bering Land Bridge*, Stanford, Calif., Stanford University Press, 172–92.

King, L. (1974) *Studien zur postglazialen Gletscher- und Vegetationsgeschichte des Sustenpassgebietes*, Basler Beiträge zur Geographie 18.

King, L. (1983) 'Contribution to the glacial history of the Borup fiord area, northern Ellesmere Is., NWT, Canada', in Schroeder-Lanz, H. (ed.) *Late- and Postglacial Oscillations of Glaciers: Glacial and Periglacial Forms*, Rotterdam, Balkema, 305–23.

Kinzl, H. (1932) 'Die grössten nacheiszeitlichen Gletschervorstösse in den Schweizer Alpen und in der Mont-Blanc-Gruppe', *Zeitschrift für Gletscherkunde* 20: 269–397.

Kiver, E. P. (1972) 'Two late Pinedale advances in the south Medicine Bow Mountains, Colorado', *Contributions to Geology at the University of Wyoming* 11: 1–8.

Livingstone, D. A. (1967) 'Postglacial vegetation of the Ruwenzori Mountains in equatorial Africa', *Ecological Monographs* 37: 25–52.

Locke, W. W. (1980) 'The Quaternary geology of the Cape Dyer area, southeasternmost Baffin Island, Canada', PhD thesis, University of Colorado.

Luckman, B. H. and Osborn, G. D. (1979) 'Holocene glacier fluctuations in the middle Canadian Rocky Mountains', *Quaternary Research* 11: 52–77.

Lundqvist, J. (1969) *Beskrivning till jordartskarta över Jämtlands län*, Sveriges Geologiska Undersökning, series Ca, 45.

McCoy, W. D. (1983) 'Holocene glacier fluctuations in the Torngat Mts., northern Labrador', *Géographie physique et quaternaire* 37: 211–16.

Mack, R. N., Okazaki, R. and Valastro, S. (1979) 'Bracketing dates for two ash falls from Mount Mazama', *Nature* 279: 28–9.

Mahaney, W. C. (1972) 'Audubon: new name for Colorado Front Range Neoglacial deposits formerly called "Arikaree"', *Arctic and Alpine Research* 4: 355–7.

Mahaney, W. C. (1973) 'Neoglacial chronology of the Fourth of July Cirque, central Colorado Front Range', *Geological Society of America Bulletin* 84: 161–70.
Mahaney, W. C. (1985) 'Late glacial and Holocene paleoclimate of Mount Kenya, East Africa', *Zeitschrift für Gletscherkunde und Glazialgeologie* 21: 203–11.
Maizels, J. K. and Dugmore, A. J. (1985) 'Lichenometric dating and tephrochronology of sandur deposits, Sólheimajökull area, southern Iceland', *Jökull* 35: 69–77.
Mangerud, J. (1970) 'Late Weichselian vegetation and ice-front oscillations in the Bergen district, western Norway', *Norsk Geografisk Tidsskrift* 24: 121–48.
Matthes, F. E. (1941) 'Rebirth of the glaciers of Sierra Nevada during late Post-Pleistocene time' (abstract), *Geological Society of America Bulletin* 52: 2030.
Matthews, J. A. and Dresser, P. Q. (1983) 'Intensive ^{14}C dating of a buried paleosol horizon', *Geologiska Föreningens i Stockholm Förhandlingar* 105: 59–63.
Matthews, J. A. and Shakesby, R. (1984) 'The status of the "Little Ice Age" in southern Norway. Relative age dating of Neoglacial moraines with Schmidt hammer and lichenometry', *Boreas* 13: 333–46.
Matthews, W. H. (1951) 'Historic and prehistoric fluctuations of alpine glaciers in the Mount Garibaldi map-area, southwestern British Columbia', *Journal of Geology* 59: 357–80.
Mayr, F. (1964) 'Untersuchungen über Ausmass und Folgen der Klima- und Gletscherschwankungen seit dem Beginn der postglazialen Wärmezeit', *Zeitschrift für Geomorphologie* 8: 257–85.
Mayr, F. (1968) 'Postglacial glacial fluctuations and correlative phenomena in the Stubai Mountains, eastern Alps, Tyrol', in Richmond, G. M. (ed.), *INQUA 1965, Glaciation of the Alps*, University of Colorado Studies, Series in Earth Science no. 7, 143–65.
Mayr, F. (1969) 'Die postglazialen Gletscherschwankungen des Mont Blanc-Gebietes', *Zeitschrift für Geomorphologie*, Supplementband 8: 31–57.
Mercer, J. H. (1956) 'Geomorphology and glacial history of southernmost Baffin Island', *Geological Society of America Bulletin* 67: 553–70.
Mercer, J. H. (1965) 'Glacier variations in southern Patagonia', *Geographical Review* 55: 390–413.
Mercer, J. H. (1968) 'Variations of some Patagonian glaciers since the late-glacial', *American Journal of Science* 266: 91–109.
Mercer, J. H. (1970) 'Variations of some Patagonian glaciers since the late-glacial, II', *American Journal of Science* 269: 1–25.
Mercer, J. H. (1972) 'The lower boundary of the Holocene', *Quaternary Research* 2: 15–24.
Mercer, J. H. (1976) 'Glacial history of southernmost South America', *Quaternary Research* 6: 125–66.
Mercer, J. H. and Ager, M. A. (1983) 'Glacial and floral changes in southern Argentina since 14,000 years ago', *National Geographic Society Reports* 13: 457–77.
Mercer, J. H. and Palacios, O. (1977) 'Radiocarbon dating of the last glaciation in Peru', *Geology* 5: 600–4.
Miller, C. D. (1969) 'Chronology of Neoglacial moraines in the Dome Peak area, North Cascade Range, Washington', *Arctic and Alpine Research* 1: 49–65.
Miller, C. D. (1973) 'Chronology of Neoglacial deposits in the northern Sawatch range, Colorado', *Arctic and Alpine Research* 5: 385–400.
Miller, G. H. (1973) 'Late Quaternary glacial and climatic history of northern Cumberland Peninsula, Baffin Island, NWT, Canada', *Quaternary Research* 3: 561–83.
Miller, G. H. and Andrews, J. T. (1972) 'Quaternary history of northern Cumberland Peninsula, east Baffin Island, NWT, Canada: Part VI. Preliminary lichen growth curve for *Rhizocarpon geographicum*', *Geological Society of America Bulletin* 83: 1133–8.
Mörner, N.-A. (1980) 'A 10,700 year paleotemperature record from Gotland and Pleistocene/Holocene boundary events in Sweden', *Boreas* 9: 283–7.

Moss, J. H. (1951) 'Late glacial advances in the southern Wind River Mountains, Wyoming', *American Journal of Science* 12: 865–83.
Müller, F. (1958) 'Eight months of glacier and soil research in the Mount Everest region', in *The Mountain World 1958/9*, New York, Harper, 191–208.
Mullineaux, D. R. (1974) *Pumice and Other Pyroclastic Deposits in Mount Rainier National Park, Washington*, US Geological Survey Bulletin 1236.
Nichols, H. (1974) 'Arctic North American palaeoecology: the recent history of vegetation and climate deduced from pollen analysis', in Ives, J. D. and Barry, R. J. (eds) *Arctic and Alpine Environments*, London, Methuen, 637–67.
Nichols, H. (1975) *Palynological and paleoclimatic study of the Late Quaternary displacement of the forest–tundra ecotone in Keewatin and Mackenzie, N.W.T., Canada*, Occasional Paper 15, Institute of Arctic and Alpine Research, University of Colorado.
Olyphant, G. A. (1985) 'Topoclimate and distribution of neoglacial facies in the Indian Peaks section of the Front Range, Colorado, USA', *Arctic and Alpine Research* 17: 69–78.
Ono, Y. (1985) 'Recent fluctuations of the Yala (Dakpatsen) Glacier, Langtang Himal, reconstructed from annual moraine ridges', *Zeitschrift für Gletscherkunde und Glazialgeologie* 21: 251–8.
Osborn, G. D. (1982) 'Holocene glacier and climate fluctuations in the southern Canadian Rocky Mountains: a review', *Striae* 18: 15–25.
Osborn, G. D. (1984) '2000-year history of the Bugaboo Glacier, Purcell Mountains, British Columbia', *American Quaternary Association, Eighth Biennial Meeting, University of Colorado, Boulder, Program and Abstracts*, 97.
Østrem, G. (1964) 'Ice-cored moraines in Scandinavia', *Geografiska Annaler* 46: 282–337.
Outcalt, S. I. (1964) 'Two Ural-type glaciers in Rocky Mountain National Park, Colorado', unpublished MA thesis, Department of Geography, University of Colorado.
Page, N. R. (1968) 'Atlantic/early sub-Boreal glaciation in Norway', *Nature* 219: 694–7.
Patzelt, G. (1974) 'Holocene variations of glaciers in the Alps', Colloques Internationaux du Centre National de la Recherche Scientifique no. 219: *Les méthodes quantitative d'étude des variations du climat au cours du Pléistocene*, 51–9.
Patzelt, G. (1977) 'Der zeitliche Ablauf und das Ausmass postglazialer Klimaschwankungen in den Alpen', in Frenzel, B. (ed.) *Dendrochronologie und postglaziale Klimaschwankungen in Europa*, Erdwissenschasiche Forschung no. 13, 248–59.
Perrott, R. A. (1982) 'A high altitude pollen diagram from Mt Kenya: implications for the history of glaciation', *Paleoecology of Africa* 14: 77–83.
Peterson, J. A., Hope, G. S. and Mitton, R. (1973) 'Recession of snow and ice fields of Irian Jaya, Republic of Indonesia', *Zeitschrift für Gletscherkunde und Glazialgeologie* 9: 73–87.
Porter, S. C. (1975) 'Equilibrium-line altitudes of late Quaternary glaciers in the Southern alps, NZ', *Quaternary Research* 5: 27–47.
Porter, S. C. (1976) 'Pleistocene glaciation in the southern part of the North Cascade Range, Washington', *Geological Society of America Bulletin* 87: 61–75.
Porter, S. C. (1978) 'Glacier peak tephra in the North Cascade Range, Washington: stratigraphy, distribution and relationship to late-glacial events', *Quaternary Research* 10: 30–41.
Porter, S. C. and Denton, G. H. (1967) 'Chronology of Neoglaciation in the North American Cordillera', *American Journal of Science* 265: 177–210.
Porter, S. C. and Orombelli, G. (1986) 'Glacier contraction during the middle Holocene in the western Italian Alps: evidence and implications', *Geology* 13: 296–8.
Rampton, V. (1970) 'Neoglacial fluctuations of the Natazhat and Klutlan glaciers, Yukon Territory, Canada', *Canadian Journal of Earth Sciences* 7: 1236–63.
Renner, F. (1982) 'Beiträge zur Gletschergeschichte des Gotthardgebietes und dendroklimatologische Analysen an fossilen Hölzern', *Physische Geographie* 8.

Richmond, G. M. (1960a) 'Glaciation of the east slope of Rocky Mountain National Park, Colorado', *Geological Society of America Bulletin* 71: 1371–81.
Richmond, G. M. (1960b) *Correlation of Alpine and Continental Glacial Deposits at Glacier National Park, Montana*, US Geological Survey Professional Paper 400-B, B223–4.
Richmond, G. M. (1962) *Quaternary Stratigraphy of La Sal Mountains, Utah*, US Geological Survey Professional Paper 324.
Richmond, G. M. (1964) *Glaciation of Little Cottonwood and Bells Canyons, Wasatch Mountains, Utah*, US Geological Survey Professional Paper 454-D.
Richmond, G. M. (1965) 'Glaciation of the Rocky Mountains', in Wright, H. E. and Frey, D. G. (eds) *The Quaternary of the United States*, Princeton University Press, 217–30.
Röthlisberger, F. (1976) 'Gletscher- und Klimaschwankungen im Raum Zermatt, Ferpècle und Arolla', *Die Alpen* 52: 59–152.
Röthlisberger, F. (1986) *10,000 Jahre Gletschergeschichte der Erde*, Aarau, Verlag Sauerländer.
Röthlisberger, F. and Geyh, M.A. (1985) 'Glacier variations in Himalayas and Karakorum', *Zeitschrift für Gletscherkunde und Glazialgeologie* 21: 237–49.
Röthlisberger, F. and Geyh, M. A. (in preparation) *Gletscherschwankungen der letzten 10,000 Jahre – Ein Vergleich zwischen Nord- und Südhemisphäre (Alpen, Himalaya, Alaska, Südamerika, Neuseeland)*, Aarau, Verlag Sauerländer.
Röthlisberger, F., Haas, P., Holzhauser, H., Keller, W., Bircher, W. and Renner, F. (1980) 'Holocene climatic fluctuations – radiocarbon dating of fossil soils (fAh) and wood from moraines and glaciers in the Alps', in Müller, F., Bridel, L. and Schwabe, E. (eds) 'Geography in Switzerland', *Geographica Helvetica* 35, no. 5 (special issue): 21–52.
Röthlisberger, H. and Oeschger, H. (1961) 'Datierung eines ehemaligen Standes des Aletschgletschers durch Radioaktivitätsmessung an Holzproben und Bemerkungen zu Holzfunden an weiteren Gletschern', *Zeitschrift für Gletscherkunde und Glazialgeologie* 4: 191–205.
Ryder, J. M. and Thomson, B. (1986) 'Neoglaciation in the southern Coast Mountains of British Columbia; chronology prior to the late Neoglacial maximum', *Canadian Journal of Earth Sciences* 23: 273–87.
Schneebeli, W. (1976) 'Untersuchungen von Gletscherschwankungen im Val des Bagnes', *Die Alpen* 52: 5–58.
Shackleton, N. J. and Opdyke, N. D. (1973) 'Oxygen isotope and palaeomagnetic stratigraphy of Equatorial Pacific core V28-238', *Quaternary Research* 3: 39–55.
Sharp, M. and Dugmore, A. (1985) 'Holocene glacier fluctuations in eastern Iceland', *Zeitschrift für Gletscherkunde und Glazialgeologie* 21: 341–9.
Sharp, R. P. (1951) 'Glacial history of Wolf Creek, St Elias Range, Canada', *Journal of Geology* 59: 97–115.
Shi Yafeng and Wang Jingtai (1979) 'The fluctuations of climate, glaciers and sea-level since the late Pleistocene in China', unpublished manuscript.
Shi Yafeng and Zhang Xiangsong (1984) *Guide to the Tianshan Glaciological Station of China; Glaciers in the Urumqi Valley and Related Phenomena*. Lanzhou Institute of Glaciology and Cryopedology and Desert Research, Academia Sinica.
Smith, D. R. and Leeman, W. P. (1982) 'Mineralogy and phase chemistry of Mount St Helens tephra sets W and Y as keys to their identification', *Quaternary Research* 17: 211–27.
Sollid, J. L. and Sørbel, L. (1975) 'Younger Dryas ice-marginal deposits in Trøndelag, central Norway', *Norsk Geografisk Tidsskrift* 29: 1–9.
Stingl, H. and Garleff, K. (1985) 'Glacier variations and climate of the Late Quaternary in the subtropical and mid-latitude Andes of Argentina', *Zeitschrift für Gletscherkunde und Glazialgeologie* 21: 225–8.
Suggate, R. P. and Moar, N. T. (1970) 'Revision of the chronology of the late Otira Glacial', *New Zealand Journal of Geology and Geophysics* 13: 742–6.

Tedrow, J. C. F. and Walton, G. F. (1964) 'Some Quaternary events of northern Alaska', *Arctic* 17: 268–71.

Ten Brink, N. W. and Weidick, A. (1974) 'Greenland ice sheet history since the last glaciation', *Quaternary Research* 4: 429–40.

Terasmae, J. (1967) 'Postglacial chronology and forest history in the northern Lake Huron and Lake Superior regions', in Cushing, E. J. and Wright, H. E. (eds) *Quaternary Paleoecology*, New Haven, Conn., Yale University Press, 45–50.

Thompson, L. G., Mosley-Thompson, E., Dansgaard, W. and Grootes, P. M. (1986) 'The Little Ice Age as recorded in the stratigraphy of the Tropical Quelccaya Icecap', *Science* 2, 3, 4: 361–4.

Thórarinsson, S. (1956) 'On the variations of Svínafellsjökull, Skaftafellsjökull and Kvíárjökull in Öræfi', *Jökull* 6: 1–15.

Thórarinsson, S. (1958) 'The Öræfajökull eruption of 1362', *Acta Naturalia Islandica* II, no. 2.

Thórarinsson, S. (1964) 'On the age of the terminal moraines of Brúarjökull and Hálsajökull', *Jökull* 14: 67–75.

Waitt, R. B., Yount, J. C. and Davis, P. T. (1982) 'Regional significance of an early Holocene moraine in Enchantment Lakes basin, North Cascade Range, Washington', *Quaternary Research* 17: 191–210.

Waldrop, H. A. (1964) 'Arapaho glacier: a sixty year record', University of Colorado Studies, Series in Geology.

Wardle, P. (1973) 'Variation of the glaciers of Westland National Park and the Hooker range, New Zealand', *New Zealand Journal of Botany* 11: 349–88.

Wardle, P. (1978) 'Further radiocarbon dates from Westland National Park and the Hooker Range, New Zealand', *New Zealand Journal of Botany* 16: 147–52.

Warner, B. G., Hebda, R. J. and Hann, B. J. (1984) 'Postglacial paleological history of a cedar swamp, Manitoulin Island, Ontario, Canada', *Palaeogeography, Palaeoclimatology, Palaeoecology* 45: 301–43.

Westgate, J. E. (1975) 'Holocene tephra layers in southwestern Canada: new units at revision of age estimates' (abstract), *Abstracts with programs, Geological Society of America, North Central Section, 9th Annual Meeting.*

Westgate, J. E. (1977) 'Compositional variability of Glacier Peak and Mazama tephra in western North America and its stratigraphic significance', *Abstracts X INQUA Conference, Birmingham, England.*

Williams, J. (1973) 'Neoglacial chronology of the Fourth of July Cirque, central Colorado Front Range: discussion', *Geological Society of America Bulletin* 84: 3761–6.

Williams, V. S. (1983) 'Present and former equilibrium line altitudes near Mount Everest, Nepal and Tibet', *Arctic and Alpine Research* 15: 201–11.

Worsley, P. (1974) 'On the significance of the age of a buried tree stump by Engabreen, Svartisen', *Norsk Polarinstitutt, Årbok 1972*: 111–17.

Worsley, P. and Alexander, M. J. (1976) 'Glacier and environmental changes – neoglacial data from the outermost moraine ridges at Engabreen, Northern Norway', *Geografiska Annaler* 58A: 55–69.

Zhen Ben-xing and Li Jijun (1982) 'Quaternary glaciation of the Qinghai-Xizang Plateau', *Geological and Ecological Studies of Qinghai-Xizang Plateau*, vol. II, Beijing, Science Press, 1631–40.

Zoller, H. (1960) 'Pollenanalytische Untersuchungen der Vegetationsgeschichte der insubrischen Schweiz', *Denkschriften der Schweizerischen Naturforschenden Gesellschaft* 83: 45–156.

Zoller, H. (1972) 'Zur Grenze Pleistozän/Holozän in den östlichen Schweizer Alpen', *Bericht der Deutschen botanischen Gesellschaft* 85: 59–67.

Zoller, H. (1977) 'Alter und Ausmass postglazialer Klimaschwankungen in den Schweizer Alpen',

in Frenzel, B. (ed.) *Dendrochronologie und postglaziale Klimaschwankungen in Europa*, Erdwissenschaftliche Forschung 13: 271–81.
Zoller, H. and Kleiber, H. (1971) 'Vegetationsgeschichtliche Untersuchungen in der montanen und subalpinen Stufe der Tessintäler', *Verhandlungen der Naturforschenden Gesellschaft in Basel* 81: 90–154.
Zoller, H., Schindler, C. and Röthlisberger, H. (1966) 'Postglaziale Gletscherstände und Klimaschwankungen in Gotthardmassiv und Vorderrheingebiet', *Verhandlungen der Naturforschenden Gesellschaft in Basel* 77: 97–164.

11 Little Ice Age and other Holocene phases of glacial advance: a consideration of their possible causes

Ahlmann, H. W. (1953) *Glacier Variations and Climatic Fluctuations*, Bowman Memorial Lecture, New York, American Geographical Society.
Angione, R. I., Medeiros, E. J. and Roussen, R. G. (1976) 'Stratosphere ozone as received from the Chappius Band', *Nature* 261: 289–90.
Beget, J. E. (1981) 'Early Holocene glacier advances in the North Cascade range, Washington', *Geology* 9: 409–13.
Beget, J. E. (1983) 'Radiocarbon-dated evidence of worldwide early Holocene climatic change', *Geology* 11: 387–93.
Bjerknes, J. (1968) 'Atmosphere–ocean interaction during the "Little Ice Age" (seventeenth to nineteenth centuries AD)', *Meteorological Monographs* 8: 37–62.
Bradley, R. S. (1985) *Quaternary Paleoclimatology. Methods of Paleoclimatic Reconstruction*, London, Allen & Unwin.
Bradley, R. S. and England, J. (1978) 'Volcanic dust influence on glacier mass balance at high latitudes', *Nature* 271: 736–8.
Bray, J. R. (1974a) 'Volcanism and glaciation during the past 40 millennia', *Nature* 252: 679–80.
Bray, J. R. (1974b) 'Glacial advance relative to volcanic activity since 1500 AD', *Nature* 248: 42–3.
Broecker, W. S., Peteet, D. M. and Rind, D. (1985) 'Does the ocean–atmosphere system have more than one stable mode of operation?', *Nature* 315: 21–6.
Bryson, R. A. and Goodman, B. M. (1980) 'Volcanic activity and climatic change', *Science* 207: 1041–4.
Bucha, V. (1970) 'Evidence for changes in the Earth's magnetic field intensity', *Philosophical Transactions of the Royal Society of London* A269: 47–55.
Butler, D. R. (1984) 'An early Holocene cool climatic episode in eastern Idaho', *Physical Geography* 4: 86–98.
Butler, D. R. (1983) 'Late quaternary glaciation and paleoenvironmental changes in adjacent valleys, east-central Lemhi Mountains, Idaho', *Dissertation Abstracts International* 43: 3899B.
Dansgaard, W. (1984) 'Past climates and their relevance to the future', in Flohn, H. and Fantechi, R. (eds) *The Climate of Europe: Past, Present and Future*, Dordrecht, Reidel, 207–47.
Davis, M. B. and Botkin, D. B. (1985) 'Sensitivity of cool temperate forests and their fossil pollen record to rapid temperature change', *Quaternary Research* 23: 327–40.
Delmas, R. and Boutron, C. (1980) 'Are the past variations of the stratospheric sulfate burden recorded in Central Antarctic snow and ice layers?', *Journal of Geophysical Research* 85, C10: 5645–9.
Eddy, J. A. (1976) 'The Maunder Minimum', *Science* 192: 1189.
Eddy, J. A. (1977) 'Historical evidence for the existence of the solar cycle', in White, O. R. (ed.) *The Solar Output and its Variation*, Boulder, Colo., Associated University Press, 22–5.
Eddy, J. A. (1980) 'Climate and the role of the sun', *Journal of Interdisciplinary History* 10: 725–47.

Eddy, J. A., Gilman, P. A. and Trotter, D. E. (1976) 'Solar rotation during the Maunder Minimum', *Solar Physics* 46: 3–14.

Gellatly, A. F. (1982) 'The use of lichenometry as a relative age-dating method with specific reference to Mt Cook National Park, New Zealand', *New Zealand Journal of Botany* 20: 343–53.

Gellatly, A. F. and Norton, D. A. (1984) 'Possible warming and glacier recession in the South Island, New Zealand', *New Zealand Journal of Science* 27: 381–8.

Gellatly, A. F., Röthlisberger, F. and Geyh, M. A. (1985) 'Holocene glacier variations in New Zealand (South Island)', *Zeitschrift für Gletscherkunde und Glazialgeologie* 21: 265–73.

Hammer, C. U. (1980) 'Acidity of polar ice cores in relation to absolute dating, past volcanism and radio-echoes', *Journal of Glaciology* 25: 359–72.

Hammer, C. H., Clausen, H. B. and Dansgaard, W. (1980) 'Greenland ice sheet evidence of post-glacial volcanism and its climatic impact', *Nature* 288: 230–5.

Hammer, C. H., Clausen, H. B. and Dansgaard, W. (1981) 'Past volcanism and climate revealed by Greenland ice cores', *Journal of Volcanology and Geothermal Research* 11: 3–10.

Harvey, L. D. D. (1980) 'Solar variability as a contributary factor to Holocene climatic change', *Progress in Physical Geography* 487–530.

Hays, J. D., Imbrie, J. and Shackleton, N. J. (1976) 'Variations in the Earth's orbit; pacemaker of the ice ages', *Science* 194: 1121–32.

Herr, R. B. (1978) 'Solar rotation determined from Thomas Harriot's sunspot observations of 1611–1613', *Science* 202: 1079–81.

Heusser, C. J., Heusser, L. E. and Peteet, D. M. (1985) 'Late Quaternary climatic change in the American North Pacific Coast', *Nature* 315: 485–7.

Hirschboeck, K. K. (1980) 'A new worldwide chronology of volcanic eruptions', *Palaeogeography, Palaeoclimatology, Palaeoecology* 29: 223–41.

Holzhauser, H. (1982) 'Neuzeitliche Gletscherschwankungen', *Geographica Helvetica* 35: 115–26.

Holzhauser, H. (1984) 'Rekonstruktion von Gletscherschwankungen mit Hilfe fossiler Hölzer', *Geographica Helvetica* 39: 3–15.

Imbrie, J. and Imbrie, J. Z. (1980) 'Modeling the climatic response to orbital variations', *Science* 207: 943–53.

King, J. W. (1974) 'Weather and the Earth's magnetic field', *Nature* 247: 131–4.

Kuhn, G. G. and Shepherd, F. P. (1983) 'Importance of Phreatic Volcanism in producing abnormal weather conditions', *Shore and Beach* 51: 19–29.

Kukla, G. J. and Kukla, H. J. (1974) 'Increased surface albedo in the Northern Hemisphere', *Science* 183: 709–14.

Kutzbach, J. E. (1983) 'Monsoon climates of the early Holocene: climate experiments with the Earth's orbital parameters for 9000 years ago', *Science* 214: 59–64.

Lal, D. and Revelle, R. (1984) 'Atmospheric PCO_2 changes recorded in lake sediments', *Nature* 308: 344–6.

Lal, D. and Venkatavaradan, V. S. (1970) 'Analysis of the causes of ^{14}C variations in the atmosphere', in Olsson, I. U. (ed.) *Radiocarbon Variations and Absolute Chronology*, Nobel Symposium 12, Chichester and New York, Wiley, 548–69.

LaMarche, V. C. and Hirschboeck, K. K. (1984) 'Frost rings in trees as records of major volcanic eruptions', *Nature* 307: 121–6.

Lamb, H. H. (1970) 'Volcanic dust in the atmosphere, with a chronology and assessment of its meteorological significance', *Philosophical Transactions of the Royal Society of London* A266: 425–533.

Lamb, H. H. (1972) *Climate: Present, Past and Future. Volume 1, Fundamentals and Climate Now*, London, Methuen.

Lamb, H. H. (1979) 'Climatic variation and changes in wind and ocean circulation: the Little Ice Age in the Northeast Atlantic', *Quaternary Research* 11: 1–20.

Lamb, H. H. and Johnson, A. I. (1959) 'Climatic variation and observed changes in the general circulation. Parts I and II', *Geografiska Annaler* 41: 94–134.
Lamb, H. H. and Johnson, A. I. (1961) 'Climatic variation and observed changes in the general circulation. Part III', *Geografiska Annaler* 43: 363–400.
Landsberg, H. E. and Albert, J. M. (1974) 'The summer of 1816 and volcanism', *Weatherwise* 27: 63–6.
Lorenz, E. N. (1968) 'Climatic determinism', *Meteorological Monographs* 8: 1–3.
Lorenz, E. N. (1976) 'Nondeterministic theories of climatic change', *Quaternary Research* 6: 495–506.
McIntyre, A., Kipp, N. G., with Be, A. W. H., Crowley, T., Kellogg, T., Gardner, J. V., Prell, W. and Ruddiman, W. F. (1976) 'The glacial North Atlantic 18,000 years ago: a CLIMAP reconstruction', *Geological Society of America Memoir* 145: 43–76.
Mass, C. and Schneider, S. H. (1977) 'Statistical evidence on the influence of sunspots and volcanic dust on long-term temperature records', *Journal of Atmospheric Science* 34: 1995–2008.
Matson, M. and Wiesnet, D. R. (1981) 'New data base for climate studies', *Nature* 289: 451–5.
Mitchell, J. M. (1971) 'The effect of atmospheric aerosols on climate with special reference to temperature near the earth's surface', *Journal of Applied Meteorology* 10: 703–14.
Mitchell, M. J. (1976) 'An overview of climatic variability and its causal mechanics', *Quaternary Research* 6: 481–93.
Murrow, P., Rose, W. I. and Self, S. (1980) 'Determination of the total grain size distribution in a volcanic eruption column and its implication in stratospheric aerosol perturbation', *Geophysical Research Letters* 7: 893–6.
Newell, R. E. (1981) 'Further studies of the atmospheric temperature change produced by Mt Agung volcanic eruption in 1963', *Journal of Volcanology and Geothermal Research* 11: 61–6.
Newhall, C. G. and Self, S. (1982) 'The volcanic explosivity index (VEI): an estimate of explosive magnitude for historical volcanism', *Journal of Geophysical Research* 87, no. C2: 1231–8.
Oeschger, H., Houtermans, J., Loosk, H. and Wahle, M. (1970) 'The constancy of cosmic radiation from isotope studies in meteorites and on earth', in Olsson, I. U. (ed.) *Radiocarbon Variations and Absolute Chronology*, Nobel Symposium 12, Chichester and New York, Wiley, 471–98.
Patzelt, G. (1985) 'The period of glacier advances in the Alps, 1965 to 1980', *Zeitschrift für Gletscherkunde und Glazialgeologie* 21: 403–7.
Pollack, J. B., Toon, O. B., Sagan, C., Summers, A., Baldwin, B. and Van Camp, W. (1976) 'Volcanic explosions and climatic change: a theoretical assessment', *Journal of Geophysical Research* 81, no. 6: 1071–83.
Porter, S. C. (1981) 'Recent glacier variations and volcanic eruptions', *Nature* 291: 139–42.
Potter, G. L., Elsaesser, H. W., MacCracken, M. C. and Ellis, J. S. (1981) 'Albedo change by man: test of climatic effect', *Nature* 291: 47–9.
Rampino, M. R. and Self, S. (1982) 'Historic eruptions of Tambora (1815), Krakatau (1883) and Agung (1963)', *Quaternary Research* 18: 127–43.
Rampino, M. R., Self, S. and Fairbridge, R. W. (1979) 'Can rapid climatic change cause volcanic eruptions?', *Science* 206: 826–9.
Reynaud, L., Vallon, M., Martin, S. and Letreguilly, A. (1984) 'Spatiotemporal distribution of the glacial mass balance in the Alpine, Scandinavian and Tien Shan areas', *Geografiska Annaler* 66A: 239–47.
Rose, W. I., Wanderman, R. L., Hoffman, M. F. and Gale, L. (1983) 'A volcanologist's review of atmospheric hazards of volcanic activity: Fuego and Mount St Helens', *Journal of Volcanology and Geothermal Research* 17: 133–57.
Ruddiman, W. F. and McIntyre, A. (1981) 'Oceanic mechanisms for amplification of the 23,000-year ice-volume cycle', *Science* 212: 617–27.
Sagan, C., Toon, O. B. and Pollack, J. B. (1979) 'Anthropogenic albedo changes and the Earth's climate', *Science* 206: 1363–8.

Sakurai, K. (1977) 'Equatorial solar rotation and its relation to climatic changes', *Nature* 269: 401–2.
Salinger, M. J. (1976) 'New Zealand temperatures since 300 AD', *Nature* 260: 310–11.
Self, S., Rampino, M. A. and Borbera, J. J. (1981) 'The possible effects of large 19th and 20th century volcanic eruptions on zonal and hemispherical surface temperatures', *Journal of Volcanology and Geothermal Research* 11: 41–60.
Simkin, T., Siebert, L., McClelland, L., Bridge, D., Newhall, C. and Lettau, J. H. (1981) *Volcanoes of the World – A Regional Directory, Gazetteer and Chronology of Volcanism during the last 10,000 Years*, Stroudsberg, Pa, Hutchinson Ross.
Stuiver, M. (1978) 'Radiocarbon timescale tested against magnetic and other dating methods', *Nature* 273: 271–4.
Stuiver, M. and Quay, P. D. (1980) 'Changes in atmosphere Carbon 14 attributed to a variable sun', *Science* 207: 11–19.
Suess, H. E. (1968) 'Climatic changes, solar activity and the cosmic-ray production rate of natural radiocarbon', *Meteorological Monographs* 8: 146–50.
Suess, H. E. (1970) 'The three causes of the secular C14 fluctuations, their amplitudes and time constants', in Olsson, I. U. (ed.) *Radiocarbon Variations and Absolute Chronology*, Nobel Symposium 12, Chichester and New York, Wiley, 595–605.
Taylor, B. L., Tsvi Gal-Chen and Schneider, S. H. (1980) 'Volcanic eruptions and long-term temperature records: an empirical search for cause and effect', *Quarterly Journal of the Royal Meteorological Society* 106: 175–99.
Wallin, G., Ericson, D. B. and Ryan, W. B. (1971) 'Variation in magnetic intensity and climatic change', *Nature* 232: 549–50.
Watts, R. G. (1985) 'Global climate variation due to fluctuations in the rate of deep water formation', *Journal of Geophysical Research* 90, no. D5: 8067–70.
Weertman, J. (1976) 'Milankovitch solar radiation variations and the Ice Age ice sheet sizes', *Nature* 261: 17–20.
Weyl, P. K. (1968) 'The role of the oceans in climatic change. A theory of the ice ages', *Meteorological Monographs* 8, no. 30: 37–62.
Wigley, T. M. L. and Jones, P. D. (1981) 'Detecting Co_2-induced climatic change', *Nature* 292: 205–8.

12 Consequences of the Little Ice Age climatic fluctuation

PHYSICAL CONSEQUENCES

Aaland, J. (1932) *Nordfjord, Innvik–Stryn–Sandane*, 2 vols, Oslo, Ei Nemnd.
Ackert, R. P. (1984) 'Ice-cored lateral moraines in Tarfala valley, Swedish Lappland', *Geografiska Annaler* 66A: 79–88.
Butzer, K. W. (1980) 'Holocene alluvial sequences: problems of dating and correlation', in Cullingford, R. A., Davidson, D. A. and Lewin, J. (eds) *Timescales in Geomorphology*, New York, Wiley, 131–42.
Clark, J. A., Farrell, W. E. and Peltier, W. R. (1978) 'Global changes in postglacial sea level: a numerical calculation', *Quaternary Research* 9: 265–87.
Fairbridge, R. W. (1976) 'Effects of Holocene climatic change on some tropical geomorphic processes', *Quaternary Research* 6: 529–56.
Filion, L. (1984) 'A relationship between dunes, fire and climatic record in the Holocene deposits of Quebec', *Nature* 309: 543–6.
Gamper, M. and Suter, J. (1982) 'Postglaziale Klimageschichte der Schweizer Alpen', *Geographica Helvetica* 37: 105–14.

Gardner, J. (1969) 'Snowpatches: their influence on mountain wall temperatures and the geomorphic implications', *Geografiska Annaler* 51A: 114–20.

Gardner, J. (1970) 'Geomorphic significance of avalanches in the Lake Louise area, Alberta, Canada', *Arctic and Alpine Research* 2: 135–44.

Gornitz, V., Lebedeff, S. and Hansen, J. (1982) 'Global sealevel trend in the past century', *Science* 215: 1611–14.

Gregory, K. T. (ed.) (1983) *Background to Paleohydrology*, Chichester and New York, Wiley.

Grove, J. M. (1972) 'The incidence of landslides, avalanches and floods in western Norway during the Little Ice Age', *Arctic and Alpine Research* 4: 131–8.

Grove, J. M. (1985) 'The timing of the Little Ice Age in Scandinavia', in Tooley, M. J. and Sheail, G. M. (eds) *The Climatic Scene*, London, Allen & Unwin, 132–53.

Grove, J. M. and Battagel, A. (1983) 'Tax records from western Norway as an index of Little Ice Age environmental and economic deterioration', *Climatic Change* 5: 265–82.

Hastenrath, S. (1981) *The Glaciation of the Ecuadorian Andes*, Rotterdam, Balkema.

Inman, D. L. (1983) 'Applications of coastal dynamics to the reconstruction of paleocoastlines in the vicinity of La Jolla, California', in Masters, P. M. and Flemming, N. C. (eds) *Quaternary Coastlines and Marine Archeology: Towards the Prehistory of Land Bridges and Continental Shelves*, London and New York, Academic Press, 1–49.

Innes, J. L. (1985) 'Lichenometric dating of debris flow deposits on alpine colluvial fans in southwest Norway', *Earth Surface Processes and Landforms* 10: 519–24.

Jónsson, Ó. (1957) *Skriduföll og Snjóflód*, 2 vols, Akureyri, Nordri.

Laely, A. (1951) *Lawinenchronik der Landschaft Davos*, Davos, Heimatkunde.

Lamb, H. H. (1977) *Climate: Present, Past and Future. II, Climatic History and the Future*, London, Methuen.

Lamb, H. H. (1980) 'Climatic fluctuations in historical time and their connections with transgressions of the sea, storm floods and other coastal changes', in Verhalst, A. and Gottschalk, M. K. E. (eds) *Transgressies en Occupatiegeschiedenis in de Kustgebieden van Nederland en Belgie*, Ghent, Rijkuniversiteit, 251–84.

Lamb, H. H. (1984) 'Climate in the last thousand years: natural climatic fluctuations and change', in Flohn, H. and Fantechi, R. (eds) *The Climate of Europe: Past, Present and Future*, Dordrecht, Reidel, 25–64.

Lamb, H. H. (1985) 'The Little Ice Age period and the great storms within it', in Tooley, M. J. and Sheail, G. M. (eds) *The Climatic Scene*, London, Allen & Unwin, 104–31.

Leonard, E. M. (1985) 'Glaciological and climatic controls on lake sedimentation, Canadian Rocky Mountains', *Zeitschrift für Gletscherkunde und Glazialgeologie* 21: 35–42.

Maizels, J. K. (1979) 'Proglacial aggradation and changes in braided channel patterns during a period of glacier advance: an Alpine example', *Geografiska Annaler* 61A: 87–101.

Manley, G. (1949) 'The snowline in Britain', *Geografiska Annaler* 31: 179–93.

Manley, G. (1969) 'Snowfall in Britain over the past 300 years', *Weather* 24: 428–37.

Manley, G. (1971) 'The mountain snows of Britain', *Weather* 26: 192–200.

Meier, M. F. (1984) 'Contribution of small glaciers to global sea level', *Science* 226: 1418–21.

Ono, Y. (1984) 'Annual moraine ridges and recent fluctuations of Yala (Dakpatsen) Glacier, Langtang Himal', in *Glacier Studies in the Langtang Valley*, Data Centre for Glacier Research, Japanese Society of Snow and Ice, Publ. no. 2, 73–83.

Pennant, T. (1771) *A Tour of Scotland, 1769*, Chester.

Porter, S. C. and Orombelli, G. (1981) 'Alpine rockfall hazards', *American Scientist* 67: 69–75.

Rapp, A. (1959) 'Avalanche boulder tongues in Lappland', *Geografiska Annaler* 41: 34–48.

Rapp, A. (1960a) 'Recent development of mountain slopes in Kärkevagge and surroundings, northern Scandinavia', *Geografiska Annaler* 42: 65–200.

Rapp, A. (1960b) 'Talus slopes and mountain walls at Templefjorden, Spitsbergen', *Norsk Polarinstitutt, Skrifter* 119.

Reheis, M. J. (1975) 'Source, transportation and deposition of debris on Arapaho Glacier, Front Range, Colorado, USA', *Journal of Glaciology* 14, 407–20.
Starkel, L. (1983) 'The reflections of hydrological changes in the fluvial environment of the temperate zone during the last 15,000 years', in Gregory, K. J. (ed.) *Background to Paleohydrology*, Chichester and New York, Wiley, 213–35.
Starkel, L. (1985) 'Late glacial and postglacial history of river valleys in Europe as a reflection of climatic changes', *Zeitschrift für Gletscherkunde und Glazialgeologie* 21: 159–64.
Sugden, D. E. (1977) 'Did glaciers form in the Cairngorms in the seventeenth to nineteenth centuries?', *Cairngorm Club Journal* 18: 189–201.
Tangborn, W. (1980) 'Two models for estimating climate–glacier relationships in the North Cascades, Washington', *Journal of Glaciology* 25: 3–21.
Taylor, J. (1618) *The Penniless Pilgrimage.*
Tooley, M. J. (1978) 'Interpretation of Holocene sea level changes', *Geologiska Föreningens i Stockholm Förhandlingar* 100: 203–12.
Tooley, M. J. (1985) 'Climate, sea-level and coastal changes', in Tooley, M. J., Sheail, G. M. (eds) *The Climatic Scene*, London, Allen & Unwin, 206–34.
Waldrop, H. A. (1964) *Arapaho Glacier, a sixty year record*, University of Colorado Studies, Series in Geology.

BIOLOGICAL CONSEQUENCES

Bernabo, J. C. (1981) 'Quantitative estimates of temperature changes over the last 2700 years in Michigan based on pollen data', *Quaternary Research* 15: 143–59.
Beverton, R. J. H. and Lee, A. J. (1965) 'Hydrographic fluctuations in the North Atlantic Ocean and some biological consequences', in Johnson, C. G. and Smith, L. P. (eds) *The Biological Significance of Climatic Changes in Britain*, London, Academic Press, 79–107.
Birks, H. J. B. (1981) 'The use of pollen analysis in the reconstruction of past climates: a review', in Wigley, T. M. L., Ingram, M. J. and Farmer, G. (eds) *Climate and History. Studies in Past Climates and Their Impact on Man*, Cambridge, Cambridge University Press, 111–38.
Bradley, R. S. (1985) *Quaternary Paleoclimatology. Methods of Paleoclimatic Reconstruction*, London, Allen & Unwin.
Burton, J. (1981) 'Half north, half south: Britain's rich wildlife', *New Scientist*, 2 April, 16–18.
Corbet, G. B. (1974) 'The distribution of mammals in historic times', in Hawksworth, D. L. (ed.) *The Changing Flora and Fauna of Britain*, Systematics Association, Special volume 6, London, Academic Press, 179–202.
Crisp, D. J. (1965) 'Observations on the effects of climate and weather on marine communities', in Johnson, C. G. and Smith, L. P. (eds) *The Biological Significance of Climatic Change in Britain*, London, Academic Press, 63–77.
Davis, M. B. and Botkin, D. B. (1985) 'Sensitivity of cool-temperate forests and their fossil pollen record to rapid temperature change', *Quaternary Research* 23: 327–40.
Ford, M. J. (1983) *The Changing Climatic Response of the Natural Fauna and Flora*, London, Allen & Unwin.
Grimm, E. C. (1983) 'Chronology and dynamics of vegetation change in the prairie-woodland region of southern Minnesota, USA', *New Phytologist* 93: 311–50.
Jakobsson, J. (1969) 'On herring migrations in relation to changes in sea temperature', *Jökull* 19: 134–45.
Kalela, O. (1949) 'Changes in geographical ranges in the avifauna of northern and central Europe in relation to recent changes in climate', *Bird Banding* 20: 77–103.
Kristjánsson, L. (1969) 'The ice drifts back to Iceland', *New Scientist* 41: 508–9.
Lamb, H. H. (1977) *Climate: Present, Past and Future, Vol. 2, Climatic History and the Future*, London, Methuen.

Lamb, H. H. (1979) 'Climatic variation and changes in wind and ocean circulation: the Little Ice Age in the northeast Atlantic', *Quaternary Research* 11: 1–20.
Nichols, H. (1976) 'Historical aspects of the northern Canadian treeline', *Arctic* 29: 38–47.
Patzelt, G. (1974) 'Holocene variations of glaciers in the Alps. Colloques Internationaux du Centre de la Recherche Scientifique 219: *Les méthodes quantitatives d'étude des variations du climat au cours du Pléistocene*, 51–9.
Payette, S., Filion, L., Gautier, L. and Boutin, Y. (1985) 'Secular climatic change in old-growth treeline vegetation of northern Quebec', *Nature* 315: 135–8.
Sharrock, J. T. R. (1974) 'The changing status of breeding birds in Britain and Ireland', in Hawksworth, D. L. (ed.) *The Changing Flora and Fauna of Britain*, Systematics Association, Special volume 6, London, Academic Press, 203–20.
Swain, A. M. (1978) 'Environmental changes during the past 2000 years in north-central Wisconsin: analysis of pollen, charcoal and seeds from varved lake sediments', *Quaternary Research* 10: 55–68.
Webb, T. (1980) 'The reconstruction of climatic sequences from botanical data', *Journal of Interdisciplinary History* 10: 749–72.
Williamson, K. (1974) 'New bird species admitted to British and Irish lists since 1800', in Hawksworth, D. L. (ed.) *The Changing Flora and Fauna of Britain*, Systematics Association, Special volume 6, London, Academic Press.

HUMAN CONSEQUENCES

Bach, W., Pankroth, J. and Schneider, S. H. (1981) *Food–Climate Interactions*. Dordrecht, Reidel.
Baker, A. R. H. (1966) 'Evidence in the "Nonarium Inquisitiones" of contracting arable lands in England during the early fourteenth century', *Economic History Review* 1: 518–32.
Bell, W. T. and Ogilvie, A. E. J. (1978) 'Weather compilations as a source of data for the reconstruction of European climate during the medieval period', *Climatic Change* 1: 331–48.
Beresford, G. (1981) 'Climatic change and its effect upon the settlement and desertion of medieval villages in Britain', in Delano Smith, C. and Parry, M. (eds) *Consequences of Climatic Change*, Department of Geography, University of Nottingham, 30–9.
Beresford, M. and Hurst, J. G. (1971) *Deserted Medieval Villages*, Guildford, Lutterworth Press.
Bergthórsson, P. (1969) 'An estimate of drift ice and temperature in Iceland in 100 years', *Jökull* 19: 94–101.
Bergthórsson, P. (1985) 'Sensitivity of Icelandic agriculture to climatic variations', *Climatic Change* 7: 111–27.
Bindschadler, R. (1980) 'The predicted behavior of Griesgletscher, Wallis, Switzerland, and its possible threat to a nearby dam', *Zeitschrift für Gletscherkunde und Glazialgeologie* 16: 45–59.
Bourke, A. (1984) 'Impact of climatic fluctuations on agriculture', in Flohn, H. and Fantechi, R. (eds) *The Climate of Europe: Past, Present and Future*, Dordrecht, Reidel, 269–314.
Britton, C. E. (1937) *A Meteorological Chronology to AD 1450*. Meteorological Office, Geophysical Memoirs no. 70, London, HMSO.
Brooks, C. E. P. (1949) *Climate Through the Ages*, London, Benn.
Claxton, R. H. and Hecht, A. D. (1978) 'Climatic and human history in Europe and Latin America: an opportunity for comparative study', *Climatic Change* 1: 195–203.
Czelnai, R. (1980) 'Climate and society: the Great Plains of the Danube', in Ausubel, J. and Biswas, A. K. (eds) *Climatic Constraints and Human Activities*, Oxford, Pergamon Press, 149–80.
De Vries, J. (1980) 'Measuring the impact of climate on history: the search for appropriate methodologies', *Journal of Interdisciplinary History* 10: 599–630.
Drake, M. (1969) *Population and Society in Norway, 1735–1865*, Cambridge University Press.
Dyrvik, S., Mykland, K. and Oldervoll, J. (1976) *The Demographic Crises in Norway in the 17th and 18th Centuries*, Bergen, Oslo and Tromsø, Universitetsforlaget.

Easton, C. (1928) *Les Hivers dans l'Europe Occidentale*, Leiden, Brill.
Farmer, G. and Wigley, T. M. L. (1983) *The Reconstruction of European Climate on Decadal and Shorter Time Scales*, Commission of the European Communities.
Fischer, D. H. (1980) 'Climate and history: priorities for research', *Journal of Interdisciplinary History* 10: 821–30.
Fisher, D. A. and Jones, S. J. (1971) 'The possible future behaviour of the Berendon Glacier, Canada – a further study', *Journal of Glaciology* 10: 85–92.
Fridriksson, S. (1969) 'The effects of sea ice on flora, fauna and agriculture', *Jökull* 19: 146–57.
Gad, F. (1970) *A History of Greenland. Vol. I, Earliest Times to 1700*. London, Hurst.
Gissel, S., Jutikkala, E., Österberg, E., Sandnes, J. and Teitsson, B. (1981) *Desertion and Land Colonisation in the Nordic Countries c.1300–1600*. Stockholm, Almqvist & Wiksell.
Grove, J. M. (1972) 'The incidence of landslides, avalanches and floods in western Norway during the Little Ice Age', *Arctic and Alpine Research* 4: 131–8.
Grove, J. M. (1987) 'Glacier fluctuations and hazards', *Geographical Journal* 153: 351–69.
Grove, J. M. and Battagel, A. (1983) 'Tax records from western Norway as an index of Little Ice Age environmental and economic deterioration', *Climatic Change* 5: 265–82.
Huntington, E. (1907) *The Pulse of Asia*, Boston, Yale University Press.
Huntington, E. (1915) *Civilization and Climate*, New Haven, Conn., Yale University Press.
Jakobsson, J. (1969) 'On herring migration in relation to changes in sea temperature', *Jökull* 19: 134–45.
Jansen, H. M. (1972) 'A critical account of the written and archeological sources' evidence concerning the Norse settlements in Greenland', *Meddelelser om Grønland* 182, no. 4.
Jutikkala, E. (1955) 'The great Finnish famine of 1696–97', *Scandinavian Economic History Review* 3: 48–63.
Kershaw, I. (1973) 'The great famine and agrarian crisis in England, 1315–1322', *Past and Present* 59: 1–50.
Kraft, J. (1830) *Topographisk-statistisk Beskrivelse over Kongeriget Norge*, vol. 4.
Kristjánsson, L. (1969) 'The ice drifts back to Iceland', *New Scientist* 41: 508–9.
Ladurie, E. Le Roy (1971) *Times of Feast, Times of Famine. A History of Climate since the Year 1000*, New York, Doubleday.
Lamb, H. H. (1965) 'The early medieval warm epoch and its sequel', *Palaeogeography, Palaeoclimatology, Palaeoecology* 1: 13–37.
Lamb, H. H. (1977) *Climate: Present, Past and Future, Vol. II, Climatic History and the Future*, London, Methuen.
Lamb, H. H. (1982) *Climate, History and the Modern World*, London, Methuen.
Lamb, H. H. (1984) 'Climate in the last thousand years: natural climatic fluctuations and change', in Flohn, H. and Fantechi, R. (eds) *The Climate of Europe: Past, Present and Future*, Dordrecht, Reidel, 25–64.
Lindgren, S. and Neumann, J. (1981) 'The cold and wet year 1695 – a contemporary German account', *Climatic Change* 3: 173–85.
Linehan, C. D. (1966) 'Deserted sites and rabbit-warrens on Dartmoor', *Medieval Archaeology* 10: 113–44.
McGovern, T. H. (1981) 'The economics of extinction in Norse Greenland', in Wigley, T. M. L., Ingram, M. J. and Farmer, G. (eds) *Climate and History. Studies in Past Climates and Their Impact on Man*, Cambridge University Press, 404–33.
Manley, G. (1945) 'The effective rate of altitudinal change in temperate Atlantic climates', *Geographical Review* 35: 408–17.
Manley, G. (1951) 'The range of variation of the British climate', *Geographical Journal* 117: 43–68.
Manley, G. (1953) 'The mean temperature of central England, 1698–1952', *Quarterly Journal of the Royal Meteorological Society* 79: 242–61.
Manley, G. (1958) 'The great winter of 1740', *Weather* 13: 11–17.

Manley, G. (1974) 'Central England temperatures: monthly means 1659 to 1973', *Quarterly Journal of the Royal Meteorological Society* 100: 389–405.

Ogilvie, A. E. J. (1981) 'Climate and economy in eighteenth century Iceland', in Delano Smith, C. and Parry, M. (eds) *Consequences of Climatic Change*, Department of Geography, University of Nottingham, 54–69.

Parry, M. L. (1975) 'Secular climatic change and marginal land', *Transactions of the Institute of British Geographers* 64: 1–13.

Parry, M. L. (1976) 'The significance of the variability of summer weather in upland Britain', *Weather* 31: 212–17.

Parry, M. L. (1978) *Climatic Change, Agriculture and Settlement*, Folkestone, Dawson.

Parry, M. L. (1981) 'Climatic change and the agricultural frontier: a research strategy', in Wigley, T. M. L., Ingram M. J. and Farmer, G. (eds) *Climate and History. Studies in Past Climates and Their Impact on Man*, Cambridge University Press, 319–36.

Parry, M. L. and Carter, T. R. (1985) 'The effect of climatic variation on agricultural risk', *Climatic Change* 7: 95–110.

Pfister, C. (1975) *Agrokonjunktur und Witterungsverlauf im westlichen Schweizer Mittelland, 1755–1797*, Geographisches Institut der Universität Bern.

Pfister, C. (1978) 'Climate and economy in eighteenth century Switzerland', *Journal of Interdisciplinary History* 9: 223–43.

Pontoppidan, E. (1972) *The Natural History of Norway*, trans. into English 1755, London, 2 vols.

Post, J. D. (1977) *The Last Great Subsistence Crisis in the Western World*, Baltimore, Johns Hopkins University Press.

Post, J. D. (1980) 'The impact of climate on political, social and economic change: a comment', *Journal of Interdisciplinary History* 10: 719–23.

Postan, M. M. (1975) *The Medieval Economy and Society*, 2nd edn, Harmondsworth, Penguin.

Postan, M. M. and Hatcher, J. (1978) 'Populations and class relations in feudal society', *Past and Present* 78: 24–37.

Rabb, T. K. (1980) 'The historian and the climatologist', *Journal of Interdisciplinary History* 10: 831–7.

Sandnes, J. (1971) *Ødetid og gjenreisning. Trøndsk busetningshistorie c.1200–1600*, Oslo-Bergen-Tromsø, Skrifter utgitt av Norsk Agrarhistorisk Forskergruppe I.

Sinclair, J. (ed.) (1791–99) *The Statistical Account of Scotland 1791–99*, Edinburgh, 21 vols.

Swan, S. L. (1981) 'Mexico in the Little Ice Age', *Journal of Interdisciplinary History* 11: 633–48.

Takashashi, K. and Yoshino, M. M. (1978) *Climatic Change and Food Production*, University of Tokyo Press.

Thórarinsson, S. (1956) *The Thousand Years Struggle against Ice and Fire*, Reykjavík, Bókaútgáfa Menningarsjóds.

Utterström, G. (1955) 'Climatic fluctuations and population problems in early modern history', *The Scandinavian Economic History Review* 3: 3–47.

Van Bath, S. (1963) *The Agrarian History of Western Europe, AD 500–1800*, London, Edward Arnold.

Vibe, C. (1967) *Arctic Animals in Relation to Climatic Fluctuations*, Meddelelser om Grønland 170, no. 5.

Wigley, T. M. L. (1985) 'Impact of extreme events', *Nature* 316: 106–7.

Wigley, T. M. L., Ingram, M. J. and Farmer, G. (eds) (1981) *Climate and History. Studies in Past Climates and Their Impact on Man*, Cambridge University Press.

Yamamoto, T. (1971a) 'On the climatic change in the XV and XVI centuries in Japan', *Geophysical Magazine* 35: 187–206.

Yamamoto, T. (1971b) 'On the nature of the climatic change in Japan since the Little Ice Age around 1800 AD', *Journal of the Meteorological Society of Japan* 49: 798–812.

Geographical index

(F = Figure, P = Plate, T = Table)

Source index

(F = Figure, P = Plate, T = Table)

Subject index

(F = Figure, P = Plate, T = Table)